Sensors for Environmental Monitoring, Identification, and Assessment

Khursheed Ahmad Wani
Government Degree College Thindim Kreeri, India

IGI Global
PUBLISHER of TIMELY KNOWLEDGE

A volume in the Advances in Environmental Engineering and Green Technologies (AEEGT) Book Series

Published in the United States of America by
IGI Global
Engineering Science Reference (an imprint of IGI Global)
701 E. Chocolate Avenue
Hershey PA, USA 17033
Tel: 717-533-8845
Fax: 717-533-8661
E-mail: cust@igi-global.com
Web site: http://www.igi-global.com

Library of Congress Cataloging-in-Publication Data

CIP DATA PROCESSING

ISBN 9798369319307(hc) | ISBN 9798369347744(sc) | eISBN 9798369319314

This book is published in the IGI Global book series Advances in Environmental Engineering and Green Technologies (AEEGT) (ISSN: 2326-9162; eISSN: 2326-9170)

British Cataloguing in Publication Data
A Cataloguing in Publication record for this book is available from the British Library.

Advances in Environmental Engineering and Green Technologies (AEEGT) Book Series

Sang-Bing (Jason) Tsai
Zhongshan Institute, University of Electronic Science and Technology of China, China & Wuyi University, China
Ming-Lang Tseng
Lunghwa University of Science and Technology, Taiwan
Yuchi Wang
University of Electronic Science and Technology of China Zhongshan Institute, China

ISSN:2326-9162
EISSN:2326-9170

Mission

Growing awareness and an increased focus on environmental issues such as climate change, energy use, and loss of non-renewable resources have brought about a greater need for research that provides potential solutions to these problems. Research in environmental science and engineering continues to play a vital role in uncovering new opportunities for a "green" future.

The **Advances in Environmental Engineering and Green Technologies (AEEGT)** book series is a mouthpiece for research in all aspects of environmental science, earth science, and green initiatives. This series supports the ongoing research in this field through publishing books that discuss topics within environmental engineering or that deal with the interdisciplinary field of green technologies.

Coverage

- Waste Management
- Pollution Management
- Radioactive Waste Treatment
- Alternative Power Sources
- Cleantech
- Air Quality
- Renewable Energy
- Electric Vehicles
- Industrial Waste Management and Minimization
- Contaminated Site Remediation

IGI Global is currently accepting manuscripts for publication within this series. To submit a proposal for a volume in this series, please contact our Acquisition Editors at Acquisitions@igi-global.com or visit: http://www.igi-global.com/publish/.

The Advances in Environmental Engineering and Green Technologies (AEEGT) Book Series (ISSN 2326-9162) is published by IGI Global, 701 E. Chocolate Avenue, Hershey, PA 17033-1240, USA, www.igi-global.com. This series is composed of titles available for purchase individually; each title is edited to be contextually exclusive from any other title within the series. For pricing and ordering information please visit http://www.igi-global.com/book-series/advances-environmental-engineering-green-technologies/73679. Postmaster: Send all address changes to above address.

Titles in this Series

For a list of additional titles in this series, please visit: http://www.igi-global.com/book-series/advances-environmental-engineering-green-technologies/73679

Reshaping Environmental Science Through Machine Learning and IoT
Rajeev Kumar Gupta (Pandit Deendayal Energy University, India) Arti Jain (Jaypee Institute of Information Technology, India) John Wang (Montclair State University, USA) and Rajesh Kumar Pateriya (Maulana Azad National Institute of Technology, India)
Engineering Science Reference • copyright 2024 • 438pp • H/C (ISBN: 9798369323519) • US $295.00 (our price)

Advancements in Climate and Smart Environment Technology
Jamal Mabrouki (Faculty of Science, Mohammed V University in Rabat, Morocco) and Mourade Azrour (Faculty of Sciences and Techniques, Moulay Ismail University of Meknes, Errachidia, Morocco)
Engineering Science Reference • copyright 2024 • 292pp • H/C (ISBN: 9798369338070) • US $295.00 (our price)

Harnessing NanoOmics and Nanozymes for Sustainable Agriculture
Vishnu D. Rajput (Academy of Biology and Biotechnology, Southern Federal University, Rostov-on-Don, Russia) Abhishek Singh (Faculty of Biology, Yerevan State University, Armenia) Karen Ghazaryan (Faculty of Biology, Yerevan State University, Armenia) Athanasios Alexiou (Department of Science and Engineering, Novel Global Community Educational Foundation, Hebersham, Australia) and Abdel Rahman Mohammad Said Al-Tawaha (Department of Biological Sciences, Al Hussein bin Talal University, Jordan)
Engineering Science Reference • copyright 2024 • 531pp • H/C (ISBN: 9798369318904) • US $345.00 (our price)

Machine Learning and Computer Vision for Renewable Energy
Pinaki Pratim Acharjya (Haldia Institute of Technology, India) Santanu Koley (Haldia Institute of Technology, India) and Subhabrata Barman (Haldia Institute of Technology, India)
Engineering Science Reference • copyright 2024 • 333pp • H/C (ISBN: 9798369323557) • US $295.00 (our price)

Convergence Strategies for Green Computing and Sustainable Development
Vishal Jain (Sharda University, India) Murali Raman (Asia Pacific University of Technology, Malaysia) Akshat Agrawal (Amity University, Guragon, India) Meenu Hans (K.R. Mangalam University, India) and Swati Gupta (Center of Excellence, K.R. Mangalam University, India)
Engineering Science Reference • copyright 2024 • 332pp • H/C (ISBN: 9798369303382) • US $270.00 (our price)

Biosorption Processes for Heavy Metal Removal
Pinki Saini (University of Allahabad, India)
Engineering Science Reference • copyright 2024 • 371pp • H/C (ISBN: 9798369316184) • US $275.00 (our price)

IGI Global
PUBLISHER of TIMELY KNOWLEDGE

701 East Chocolate Avenue, Hershey, PA 17033, USA
Tel: 717-533-8845 x100 • Fax: 717-533-8661
E-Mail: cust@igi-global.com • www.igi-global.com

Dedicated to my parents and grandfather, whose courage and hard work made me a dedicated and hardworking human being.

Editorial Advisory Board

Table of Contents

Foreword xix

Preface xxi

Chapter 1
Technology of Sensors: Ways Ahead 1
Surendra Prakash Gupta, Shri Vaishnav Vidhyapeeth Viswavidhayalaya, India
Ankur Bhardwaj, Shri Vaishnav Vidhyapeeth Viswavidhayalaya, India

Chapter 2
Air Sensors and Their Capabilities 10
Prashantkumar Bharatbhai Sathvara, Nims Institute of Allied Medical Science and Technology, Jaipur, India
Anuradha Jayaraman, Nims Institute of Allied Medical Science and Technology, Jaipur, India
Sandeep Tripathi, Nims Institute of Allied Medical Science and Technology, Jaipur, India
Sanjeevi Ramakrishnan, Nims Institute of Allied Medical Science and Technology, Jaipur, India

Chapter 3
Pioneering Sensing Solutions for In-Situ Monitoring of Volatile Organic Compounds in Soil and Groundwater Using Chemiresistor Sensor Technology 25
Periasamy Palanisamy, Nehru Institute of Engineering and Technology, Anna University, India
Maheswaran M., Nehru Institute of Engineering and Technology, Anna University, India

Chapter 4
Optical and Electrochemical Chemosensors for Identification of Carbon Dioxide Gas 42
Kirandeep Kaur, Maharaja Ranjit Singh Punjab Technical University, Bathinda, India
Sudhanshu Pratap Singh, Maharaja Ranjit Singh Punjab Technical University, Bathinda, India
Navdeep Singh Gill, Australian Maritime College, University of Tasmania, Australia

Chapter 5
Water Pollutants, Sensor Types, and Their Advantages and Challenges 78
Sanjeevi Ramakrishnan, Nims Institute of Allied Medical Science and Technology, Nims University, Jaipur, India
Prashantkumar Sathvara, Nims Institute of Allied Medical Science and Technology, Nims University, Jaipur, India
Sandeep Tripathi, Nims Institute of Allied Medical Science and Technology, Nims University, Jaipur, India
Anuradha Jayaraman, Nims Institute of Allied Medical Science and Technology, Nims University, Jaipur, India

Chapter 6
Sensors for Monitoring Water Pollutants 102
Surjit Singha, Kristu Jayanti College (Autonomous), India

Chapter 7
Smart Sensors for Water Quality Monitoring Using IoT 111
Kumud, D.S. College, Aligarh, India

Chapter 8
Sensors in the Marine Environments and Pollutant Identifications and Controversies 124
Minakshi Harod, Devi Ahilya Vishwavidyalaya, India

Chapter 9
Removal of Heavy Metals From Waste Water Using Different Biosensors 132
Kamal Kishore, Department of Chemistry & Biochemistry, Eternal University, India
Yogesh Kumar Walia, Department of Chemistry, Career Point University, India

Chapter 10
Sensors in the Oceans, Pollutant Identifications, and Controversies: Radio-Isotopic Tracing of Pollutants in Marine Ecosystems and Controveries 151
Sanchari Biswas, Amity University, Kolkata, India

Chapter 11
Identification of Different Pollutants in Lotic and Lentic Ecosystems by Biosensors 167
Afaq Majid Wani, College of Forestry, India
Pritam Kumar Barman, College of Forestry, India
Satyendra Nath, College of Forestry, India

Chapter 12
Limnology and the Science of Biosensors 194
Hemendra Wala, Devi Ahilya Vishwavidyalaya, India

Chapter 13
Biosensors for Environmental Monitoring 204
Rajni Gautam, K.R. Mangalam University, India

Chapter 14
Advances in Sensor Technologies for Detecting Soil Pollution 218
Ranjit Singha, Christ University, India
Surjit Singha, Kristu Jayanti College (Autonomous), India
V. Muthu Ruben, Christ University, India
Alphonsa Diana Haokip, Christ King High School, India

Chapter 15
Sensors for Waste Management 231
Ankur Bhardwaj, Department of Life Sciences, Shri Vaishnav Institute of Science, India
Surendra Prakash Gupta, Department of Life Sciences, Shri Vaishnav Institute of Science, India

Chapter 16
Environmental Sensors: Safeguarding the Ecosystem by Monitoring Sanitary Pad Disposal 251
Deepa V. Jose, Christ University, India

Chapter 17
Bioindicators of Environmental Pollution With Emphasis to Wetlands of Kashmir 269
Kounsar Jan, Government Degree College Thindim Kreeri, India
Javid Majeed Wani, Government Degree College Dangiwacha, India

Chapter 18
Catharanthus roseus L. and Ocimum sanctum L. as Sensors for Air Pollution 284
Ab Qayoom Mir, Government Degree College Hajin, India
Javid Manzoor, Shri JJT University, India

Chapter 19
Harnessing Nature: Whole Cell Biosensors for Environmental Monitoring 317
Naseema Banu A., Ethiraj College for Women, India
Sangeetha Vani G., Ethiraj College for Women, India
Sarah Grace P., Ethiraj College for Women, India

Chapter 20
Impact of Urbanization on Environment and Health Role of Different Environmental Sensors 333
Madhumita Hussain, Sophia Girls' College (Autonomous), Ajmer, India

Chapter 21
Environmental Sensors in Extreme Environments: Scope and Validity 359
Aashish Verma, SAGE University, India
Sonam Verma, Rabindranath Tagore University, India

Detailed Table of Contents

Foreword xix

Preface xxi

Chapter 1
Technology of Sensors: Ways Ahead 1
Surendra Prakash Gupta, Shri Vaishnav Vidhyapeeth Viswavidhayalaya, India
Ankur Bhardwaj, Shri Vaishnav Vidhyapeeth Viswavidhayalaya, India

The technology of sensors is continuously evolving to meet the demands of various fields such as safety, pollution, medical engineering, and more. Sensors play a significant role in improving human working styles and routine tasks by converting real-world actions into electrical signals. They can be used to measure a wide range of environmental parameters, including air quality, water quality, soil quality, noise levels, and weather conditions. Sensors can be used to monitor environmental conditions in real time, or they can be used to collect data over time to identify trends and patterns. Sensor data can also be used to create models of environmental systems, which can be used to predict future conditions and assess the impact of human activities. This chapter is focused on the recent development in sensor technology that has the potential to revolutionize environmental monitoring, identification, and assessment. By providing real-time data on environmental conditions, sensors can help us to better understand our environment and make informed decisions about how to protect it.

Chapter 2
Air Sensors and Their Capabilities 10
Prashantkumar Bharatbhai Sathvara, Nims Institute of Allied Medical Science and Technology, Jaipur, India
Anuradha Jayaraman, Nims Institute of Allied Medical Science and Technology, Jaipur, India
Sandeep Tripathi, Nims Institute of Allied Medical Science and Technology, Jaipur, India
Sanjeevi Ramakrishnan, Nims Institute of Allied Medical Science and Technology, Jaipur, India

In this comprehensive chapter on air quality, the paramount significance and intricate challenges associated with this critical environmental and public health concern are explored. Focusing on the mission to improve air quality, the chapter underscores the profound impact of poor air quality on human health, particularly in densely populated urban areas. Emphasis is placed on safeguarding ecosystems and biodiversity through the reduction of air pollution, acknowledging its widespread ecological implications. The chapter delves into international and national regulations and initiatives, such as the Clean Air Act and the Paris Agreement, representing concerted global efforts to combat air pollution and address climate change. A detailed examination of concerns related to various air pollutants, including Particulate Matter, gases like NO2 and SO2, and airborne toxic substances, sheds light on the health risks and environmental consequences.

Chapter 3
Pioneering Sensing Solutions for In-Situ Monitoring of Volatile Organic Compounds in Soil and Groundwater Using Chemiresistor Sensor Technology 25
Periasamy Palanisamy, Nehru Institute of Engineering and Technology, Anna University, India
Maheswaran M., Nehru Institute of Engineering and Technology, Anna University, India

The monitoring and classification of volatile organic compounds (VOCs) in soil and groundwater is a critical task in guaranteeing environmental protection and remediation of contaminated sites. Traditional methods of sample collection and off-site analysis could be expensive, laborious, and may not accurately represent in-situ conditions. To address these encounters, state-of-the-art sensing solutions using chemiresistor sensor technology have been developed for instantaneous, uninterrupted, and long-term monitoring of VOCs in the subsurface. The chemiresistor comprises of a chemically-sensitive polymer dissolved in a solvent and mixed with conductive carbon particles. The resultant ink is deposited onto thin-film platinum traces on a solid substrate. When VOCs come into interaction with the polymers, they are absorbed, initiating the polymers to swell and change the resistance of the electrode. This variation in resistance can be measured and recorded, providing information about the concentration of the VOCs.

Chapter 4
Optical and Electrochemical Chemosensors for Identification of Carbon Dioxide Gas 42
Kirandeep Kaur, Maharaja Ranjit Singh Punjab Technical University, Bathinda, India
Sudhanshu Pratap Singh, Maharaja Ranjit Singh Punjab Technical University, Bathinda, India
Navdeep Singh Gill, Australian Maritime College, University of Tasmania, Australia

Intelligent sensing and vigilant monitoring of CO2 gas is immensely significant, as the concentration of CO2 gas in atmosphere is increasing day by day due to anthropogenic activities which enhances the natural greenhouse effect and makes the earth warmer. Besides global warming, CO2 is also a toxicant in enclosed environments causing asphyxiation by hypoxia, unconsciousness almost instantaneously, and respiratory arrest within one minute. Another area where sensing of CO2 is imperative are agriculture and food industry. For these conditions, sensors should be capable of working under extreme temperatures, pressure, and interference due to the inherent complex materials and microorganisms. This chapter focuses on the information regarding the different types of CO2 sensors available with special emphasis on electrochemical sensors and optical chemical sensors, which showed fluorescent and spectrophotometric variation on detection of CO2, a detailed analysis of their detection process and sensing mechanism.

Chapter 5
Water Pollutants, Sensor Types, and Their Advantages and Challenges 78
Sanjeevi Ramakrishnan, Nims Institute of Allied Medical Science and Technology, Nims University, Jaipur, India
Prashantkumar Sathvara, Nims Institute of Allied Medical Science and Technology, Nims University, Jaipur, India
Sandeep Tripathi, Nims Institute of Allied Medical Science and Technology, Nims University, Jaipur, India
Anuradha Jayaraman, Nims Institute of Allied Medical Science and Technology, Nims University, Jaipur, India

Water pollution is a global crisis impacting ecosystems, health, and economies. This chapter explores strategies to combat it, stressing advanced water quality sensors' vital role. It scrutinizes pollutants, emphasizing modern sensor tech's importance in ensuring water safety. Tackling pollution is crucial for biodiversity, human health, and clean water access. Pollutants include heavy metals, chemicals, pathogens, and sediments, requiring precise monitoring by sensors using various technologies. They offer real-time detection and response, covering chemical, biological, physical, remote sensing, and IoT-enabled sensors. Challenges like maintenance persist, requiring protocols and training. Collaboration and sensor tech are pivotal in ensuring cleaner water. This chapter highlights technology's role in managing water quality, emphasizing innovation for safeguarding this vital resource.

Chapter 6
Sensors for Monitoring Water Pollutants 102
Surjit Singha, Kristu Jayanti College (Autonomous), India

This chapter presents an overview of water pollutants and sensor technologies for monitoring them. The chapter emphasizes detection and quantification techniques while discussing chemical, physical, and biological contaminants in surface and groundwater. In addition to examining real-time monitoring advancements, this study delves into critical sensors, including spectroscopic, electrochemical, biosensor, and remote sensing technologies that are emerging, lab-on-a-chip, and nanomaterials. An analysis is conducted on the prospects of water pollutant sensors that progressively improve sensitivity, selectivity, and cost-effectiveness. This extensive evaluation enhances comprehension and resolution of water pollution issues while advocating for sustainable water management strategies that benefit ecosystems and human health.

Chapter 7
Smart Sensors for Water Quality Monitoring Using IoT 111
Kumud, D.S. College, Aligarh, India

Due to the increasing contamination and pollution of drinking water, water pollution has emerged as a major concern in recent years. Infectious illnesses spread by contaminated water have a domino effect on ecological life cycles. Early detection of water contamination allows for the implementation of appropriate solutions, therefore preventing potentially disastrous circumstances. It is important to monitor the water quality in real-time to ensure a steady supply of clean water. Improvements in sensor technology, connectivity, and the internet of things (IoT) have led to a rise in the importance of smart solutions for water pollution monitoring. This chapter presents a comprehensive overview of recent developments in the field of smart water pollution monitoring systems. An efficient and cost-effective smart water quality monitoring system that continuously checks quality indicators is proposed in this research. After running the model on three different water samples, the parameters are sent to the server in the cloud for further processing.

Chapter 8
Sensors in the Marine Environments and Pollutant Identifications and Controversies 124
Minakshi Harod, Devi Ahilya Vishwavidyalaya, India

The quality of marine environments is influenced by a range of anthropogenic and natural hazards, which may adversely affect human health, living resources, and the general ecosystem. The most common anthropogenic wastes found in marine environments are dredged spoils, sewage, and industrial and municipal discharges. These wastes generally contain a wide range of pollutants, notably heavy metals, petroleum hydrocarbons, polycyclic aromatic hydrocarbons, and others. Real-time measurements of pollutants, toxins, and pathogens across a range of spatial scales are required to adequately monitor these hazards, manage the consequences, and to understand the processes governing their magnitude and distribution. Significant technological advancements have been made in recent years for the detection and analysis of such marine hazards. This chapter aims to review the availability and application of sensor technology for the detection of marine hazards and for observing marine ecosystem status.

Chapter 9
Removal of Heavy Metals From Waste Water Using Different Biosensors 132
Kamal Kishore, Department of Chemistry & Biochemistry, Eternal University, India
Yogesh Kumar Walia, Department of Chemistry, Career Point University, India

Recently, much effort has been made to reach to an effective strategy for wastewater monitoring. Several pieces of evidence support the special role of biosensors in plans for the administration of water resources. Concerning this fact, there are some technical and practical limitations and complications, which should be overcome to develop more efficient and commercial applicable biosensors. To achieve this goal for the detection of a broad range of wastewater pollutants, it is necessary to design novel sensing systems with larger detection range and capability for the simultaneous detection of several compounds. Additionally, the limit of detection in the lower concentration range should be possible, and also biosensor should have long-storage stability. This chapter explores the various ways by which heavy metals can be removed from wastewater. Different biosensors are under investigation that can be used to remove different pollutants form different ecosystems. This will help to solve the problem of water pollution and will also help to reduce human health impact.

Chapter 10
Sensors in the Oceans, Pollutant Identifications, and Controversies: Radio-Isotopic Tracing of Pollutants in Marine Ecosystems and Controveries 151
Sanchari Biswas, Amity University, Kolkata, India

Oceans are the largest means of survival for millions of people and also the source of many life forms. Human activities have made the environmental conditions in marine habitats more dire for the last fifty years. The discharge of agricultural nutrients, heavy metals, and persistent organic pollutants (plastics, pesticides) threaten the coastal zones. Chemical compounds containing one or more radioisotope atoms are known as radiotracers, which are particularly useful for identifying and analysing pollutants as they can readily identify trace amounts of a particular radioisotope and short-lived isotope decays. It is thus important to identify such sources of contaminants by quantifying essential pollutants separately and gathering dependable information regarding their origin, movement, and ultimate destination. Nuclear and isotope techniques help in gathering such data. This book chapter gives an overview of the modern techniques available for probing the various contaminants across marine ecosystems and several drawbacks and controversies associated with the same.

Chapter 11
Identification of Different Pollutants in Lotic and Lentic Ecosystems by Biosensors........................ 167
Afaq Majid Wani, College of Forestry, India
Pritam Kumar Barman, College of Forestry, India
Satyendra Nath, College of Forestry, India

Environmental pollution is becoming a major global concern, especially about new pollutants, poisonous heavy metals, and other dangerous agents. Pollutants have a profound impact on ecosystems and present serious threats to the health of both the natural world and human communities. Water is one of the most important resources on the planet, since it is required for all species' survival and well-being. Surface water in an aquatic system is referred to as an inland water environment and is divided into lentic and lotic systems. In contrast to lotic water ecosystems, which share continuous habitats through the connection of many basins in unidirectional flow within the dendritic structure of river networks, lentic water ecosystems display discontinuous habitats as aquatic matrices inside the terrestrial system. The lentic water ecology is more similar to terrestrial waters than the lentic water ecosystem, which differs greatly. Aquatic ecosystems are diverse and, despite making up just a little of the planet's surface, are essential for several reasons.

Chapter 12
Limnology and the Science of Biosensors.. 194
Hemendra Wala, Devi Ahilya Vishwavidyalaya, India

Increasing concern about levels of pollution in the aquatic environment has led to the adoption of a number of preventive measures to assist in maintaining the quality of water bodies. The development of new user-friendly, portable, and low-cost bioanalytical methods is the focus of research, and biosensors are in the forefront of these research works. Biosensors have various prospective and existing applications in the detection of contaminants in the aquatic environment by transducing a signal. Biosensors are able to detect a wide range of analytes in complex matrices and have proven a great potential in environment monitoring, clinical diagnostics and food analysis Hence, the aim of this work is to provide a description of the state of the art about the development and application of biosensors to detect contaminants in freshwater ecosystems.

Chapter 13
Biosensors for Environmental Monitoring 204
Rajni Gautam, K.R. Mangalam University, India

Environmental monitoring is essential to safeguard our planet's ecosystems, public health, and natural resources. Biosensors have emerged as powerful tools for assessing environmental parameters due to their sensitivity, specificity, and versatility. From the detection of pollutants to the monitoring of water and air quality, biosensors offer a wide array of applications that contribute to comprehensive environmental assessment. In this chapter, the principles, applications, and significance of biosensors in the context of environmental monitoring are explored in detail. Future prospects and challenges are discussed as well.

Chapter 14
Advances in Sensor Technologies for Detecting Soil Pollution 218
Ranjit Singha, Christ University, India
Surjit Singha, Kristu Jayanti College (Autonomous), India
V. Muthu Ruben, Christ University, India
Alphonsa Diana Haokip, Christ King High School, India

The present chapter elucidates progressions in the surveillance of soil pollution, with a specific emphasis on integrated systems and sensor technologies. Future trends (e.g., enhanced selectivity, regulatory adoption), deployment platforms (field-deployable, wireless networks), and sensor types (electrochemical,

optical, and biosensors) are discussed. Increasing sensitivity and specificity, facilitating on-site, real-time analysis, and integrating sensing with remediation strategies are priorities. The discourse highlights the revolutionary capacity that soil pollution sensors possess to propel environmental monitoring and management forward. Collaboration among stakeholders is critical for successfully implementing sensor-based approaches and driving innovation.

The globe is becoming more and more urbanised and industrialised, making waste management an urgent worldwide challenge. Traditional waste management methods have proven to be insufficient in addressing the challenges posed by increasing waste volumes, environmental concerns, and resource scarcity. One potential answer to these problems is the use of sensor technologies in waste management (WM) systems. It explores the various types of sensors used in waste management applications, ranging from simple bin-level sensors to advanced technologies such as remote sensing and IoT-based systems. The chapter also discusses the key advantages of sensor-driven waste management, including improved efficiency, cost reduction, and enhanced environmental sustainability. As WM continues to evolve in response to the demands of the 21st century, this chapter underscores the pivotal role that sensor technologies play in revolutionizing the industry, a glimpse into a more sustainable and efficient future for WM practices worldwide.

This chapter focuses on the applications of environmental sensors in general and their role in identifying and addressing the issues related to the improper disposal of sanitary pads, which is a growing concern. It also gives an overview of the pollutants associated with it, and the role that environmental sensors can play in mitigating this problem. By harnessing the power of advanced sensing technologies, we can gain a better understanding of the environmental impact of sanitary pad disposal and work towards sustainable solutions. This chapter aims to provide valuable insights and guidance for researchers and practitioners working to create a cleaner and healthier environment and generate self-awareness for individuals in safeguarding ecosystem.

The use of bioindicators has grown in recent years, and they have provided a wealth of valuable data that has improved water resource management. One way to measure the quality of an environment is by looking at how well a species (or group of species) can adapt to different kinds of chemical, physical, and biological stresses. A further benefit of bioindicators is their capacity to detect the indirect biotic impacts of contaminants, a feat that is not accomplished by many physical or chemical tests. When used

emphasizing the pivotal role played by various environmental sensors in monitoring and mitigating these effects. This chapter delves into the types and functionalities of environmental sensors employed to monitor urbanization's impact. Air quality sensors, water quality sensors, noise monitors, and solid waste sensors contribute valuable data to assess pollution levels, track environmental changes, and evaluate the overall well-being of urban ecosystems. The integration of real-time data from these sensors facilitates the formulation of effective policies and interventions to curb environmental degradation and enhance public health.

Chapter 21
Environmental Sensors in Extreme Environments: Scope and Validity 359
Aashish Verma, SAGE University, India
Sonam Verma, Rabindranath Tagore University, India

Many potentially harmful chemicals, released by industries and human activities, can contaminate water, soil, or air, and further impact the environment and public health. Real-time and in situ monitoring of various contaminants such as heavy metals, pesticides, pathogens, toxins, particulate matters, radioisotopes, volatile organic compounds, crude oil, and agricultural chemicals at low levels is mandatory in the fields of industrial plants, automotive technologies, medicine and health, water and air quality control, natural soil/land/sea, and so forth. Consequently, the monitoring of environmental pollutants became a priority. For this aim, sensors have captivated the attention of many scientists in modern times by virtue of their eco-friendliness, cost-effectiveness, miniaturization ability, and rapidness. Environmental samples, however, are very complex and unexpectedly relative to other ecosystems. Thus far, environmental sensors have been developed with greater sensitivity, simpler and more efficient detection, better environmental adaptation and etc. for pollutant detection.

Chapter 22
Examining the Effects of Forest Fires: A Framework for Integrating System Dynamics and Remote Sensing Approaches 370
Müjgan Bilge Eriş, Yeditepe University, Turkey
Duygun Fatih Demirel, İstanbul Kültür University, Turkey
Eylül Damla Gönül Sezer, Yeditepe University, Turkey

Forest fires have been a major concern for many countries over an extended period of time due to natural and human induced factors. In recent years, detection of forest fires has progressively shifted toward advanced technologies where the remote sensing approaches are fully operational. To enhance fire management strategies, it is crucial to gain a comprehensive understanding of the fire dynamics and its consequences on the environment, operational sources, and economic sectors. Therefore, this chapter develops an integrated framework to predict and analyze the effects of forest fires by using system dynamics approach and remote sensing technology, ultimately leading to the establishment of a conceptual model and conclusive insights.

Compilation of References 397

About the Contributors 471

Index 479

Foreword

Environmental monitoring, identification, and assessment rely on various sensors capable of detecting and measuring different parameters to identify and assess various pollutants in different ecosystems. These sensors detect pollutants such as particulate matter (PM), volatile organic compounds (VOCs), carbon monoxide (CO), nitrogen dioxide (NO_2), sulfur dioxide (SO_2), and ozone (O_3) in the air. They help in assessing air pollution levels and ensuring compliance with air quality standards. Water quality sensors measure parameters such as pH, dissolved oxygen (DO), conductivity, turbidity, and various contaminants (e.g., heavy metals, pesticides, bacteria) in water bodies. They play a crucial role in monitoring the health of aquatic ecosystems, drinking water sources, and industrial discharge and even in extreme environments. Noise sensors, or sound level meters, measure the intensity of sound in the environment. They help assess noise pollution levels and identify sources of excessive noise that may impact human health and wildlife. Radiation sensors detect ionizing radiation levels in the environment, including gamma rays, X-rays, and alpha and beta particles. They are used in nuclear power plants, medical facilities, and areas affected by radioactive contamination to ensure public safety. Light sensors measure ambient light levels, including both natural and artificial light sources. They are used in applications such as monitoring light pollution, optimizing indoor lighting systems for energy efficiency, and studying plant growth. Weather sensors measure various meteorological parameters such as atmospheric pressure, wind speed and direction, rainfall, and solar radiation. They are critical for weather forecasting, climate research, and agriculture. Soil sensors measure soil moisture, temperature, pH, and nutrient levels. They are used in agriculture, environmental science, and geotechnical engineering to monitor soil health, optimize irrigation practices, and assess soil contamination. Gas sensors detect the presence and concentration of gases in the environment, including combustible gases, toxic gases, and greenhouse gases. They are used in industrial safety monitoring, indoor air quality assessment, and environmental pollution detection. These sensors are often deployed in networks or integrated into monitoring devices and systems to provide real-time data for environmental assessment and decision-making. Advances in sensor technology, including miniaturization, wireless connectivity, and data analytics, are continually improving their accuracy, reliability, and applicability in environmental monitoring efforts. The book presents amalgam of 22 chapters on different sensors that will revolutionize environmental research and provide a roadmap for tackling pollution head-on. This comprehensive guide is poised to make a significant impact on scholars, environmentalists, planners, researchers, industrialists, and academics globally. By delving into the diverse realms of environmental sensors, the book equips readers with the knowledge and tools necessary to identify pollutants in varied ecosystems and adopt sustainable approaches for clean-up. Its recommended topics cover critical areas such as air pollution, noise pollution, waste management, advancements in sensor technology, and the detection of pollutants in soil, water, air, and oceans that will solve different problems of environmental contamination that were previously unsolved.

Ebru Kafkas
University of Çukurova, Adana, Turkey

Preface

Sensors for Environmental Monitoring, Identification, and Assessment, curated by Khursheed Ahmad Wani, delves into the pressing issue of environmental contamination that continues to challenge our ecosystems worldwide. As editor, it is my privilege to present this comprehensive reference book that addresses the urgent need for effective environmental monitoring and assessment tools.

In our modern era, the proliferation of air, water, and soil contaminants has reached alarming levels, posing significant threats to both environmental equilibrium and human health. Despite numerous attempts by scholars to mitigate these issues, we find ourselves still in the nascent stages of understanding and combating environmental pollutants. Many researchers grapple with identifying the types of pollutants and assessing their impact on ecosystems, underscoring the crucial role of environmental sensors in this endeavor.

This book serves as a beacon of knowledge, illuminating the path for researchers, environmentalists, planners, industrialists, and academics from diverse fields. By exploring a myriad of sensor technologies, ranging from chemical sensors to hyperspectral imaging, it equips readers with the tools necessary to detect and address environmental contaminants across various ecosystems.

The intended audience encompasses individuals and institutions globally, spanning both developed and developing nations. From seasoned environmental experts to aspiring students in fields such as Environmental Science, Agriculture, Economics, and Nanotechnology, this book offers valuable insights and practical solutions.

Throughout its pages, readers will encounter in-depth discussions on topics including the scope and validity of environmental sensors in extreme conditions, the efficacy of indoor pollution sensors, and the evolving technology of sensor development. Additionally, the book explores cutting-edge research on topics such as chemical sensor arrays, real-time detection of volatile organic compounds, and the application of sensors in waste management and limnology.

By fostering collaboration and knowledge exchange, this book endeavors to inspire meaningful action towards a cleaner, healthier planet. I extend my gratitude to the contributors whose expertise and dedication have enriched this compilation. It is my hope that this book will serve as a catalyst for transformative research and contribute to the collective effort of safeguarding our environment for generations to come.

Organization of the Book

Chapter 1 focuses on recent developments in sensor technology and their potential to revolutionize environmental monitoring, identification, and assessment. The chapter examines the evolving role of sensors in various fields, including safety, pollution control, medical engineering, and more. It emphasizes the importance of real-time data collection and analysis for understanding environmental conditions and making informed decisions to protect the environment.

In Chapter 2, the paramount significance and intricate challenges associated with this critical environmental and public health concern are explored. Focusing on the mission to improve air quality, the chapter underscores the profound impact of poor air quality on human health, particularly in densely populated urban areas. Emphasis is placed on safeguarding ecosystems and biodiversity through the reduction of air pollution, acknowledging its widespread ecological implications. The chapter delves into international and national regulations and initiatives, such as the Clean Air Act and the Paris Agreement, representing concerted global efforts to combat air pollution and address climate change. A detailed examination of concerns related to various air pollutants, including Particulate Matter, gases like NO_2 and SO_2, and airborne toxic substances, sheds light on the health risks and environmental consequences.

Chapter 3 describes chemiresistor sensor technology that have been developed for instantaneous, uninterrupted, and long-term monitoring of VOCs in the subsurface. The chemiresistor comprises of a chemically-sensitive polymer dissolved in a solvent and mixed with conductive carbon particles. The resultant ink is deposited onto thin-film platinum traces on a solid substrate. When VOCs come into interaction with the polymers, they are absorbed, initiating the polymers to swell and change the resistance of the electrode. This variation in resistance can be measured and recorded, providing information about the concentration of the VOCs.

Chapter 4 focuses on optical and electrochemical chemosensors for the identification of carbon dioxide gas. The chapter emphasizes the importance of intelligent sensing and monitoring of CO_2 gas due to its increasing concentration in the atmosphere and its harmful effects on the environment and human health. Various types of CO_2 sensors, including electrochemical and optical sensors, are discussed, along with their detection processes and sensing mechanisms. The chapter aims to provide insights into the development of sensors capable of detecting CO_2 gas under extreme conditions, such as those found in agriculture, food industries, and enclosed environments.

Chapter 5 provides an overview of water pollutants and sensor technologies for monitoring them. The chapter examines detection and quantification techniques for chemical, physical, and biological contaminants in surface and groundwater. It explores advancements in real-time monitoring and critical sensor technologies, including spectroscopic, electrochemical, biosensor, and remote sensing technologies. The chapter emphasizes the importance of improving sensor sensitivity, selectivity, and cost-effectiveness to address water pollution issues and promote sustainable water management strategies.

Chapter 6 explores the increasing concern about levels of pollution in the Aquatic environment, which has led to the adoption of a number of preventive measures to assist in maintaining the quality of water bodies. The development of new user-friendly, portable and low-cost bioanalytical methods is in the focus of research and biosensors are in the forefront of these research works. Biosensors have various prospective and existing applications in the detection of contaminants in the aquatic environment by transducing a signal. Biosensors are able to detect a wide range of analytes in complex matrices and have proven a great potential in environment monitoring, clinical diagnostics and food analysis Hence, the aim of this work is to provide a description of the state of the art about the development and application of biosensors to detect contaminants in fresh water ecosystems.

Chapter 7 presents a comprehensive overview of recent developments in the field of smart water pollution monitoring systems. An efficient and cost-effective smart water quality monitoring system that continuously checks quality indicators is proposed in this research. After running the model on three different water samples, the parameters are sent to the server in the cloud for further processing. Due to the increasing contamination and pollution of drinking water, water pollution has emerged as a major concern in recent years. Infectious illnesses spread by contaminated water have a domino effect on ecological life cycles. Early detection of water contamination allows for the implementation of appropriate solutions, therefore preventing potentially disastrous circumstances. It is important to monitor the water quality in real-time to ensure a steady supply of clean water. Improvements in sensor technology, connectivity, and the Internet of Things (IoT) have led to a rise in the importance of smart solutions for water pollution monitoring.

Chapter 8 aims to review the availability and application of sensor technology for the detection of marine hazards and for observing marine ecosystem status. The quality of marine environments is influenced by a range of anthropogenic and natural hazards, which may adversely affect human health, living resources and the general ecosystem. The most common anthropogenic wastes found in marine environments are dredged spoils, sewage and industrial and municipal discharges. These wastes generally contain a wide range of pollutants notably heavy metals, petroleum hydrocarbons, polycyclic aromatic hydrocarbons and other. Real-time measurements of pollutants, toxins, and pathogens across a range of spatial scales are required to adequately monitor these hazards, manage the consequences, and to understand the processes governing their magnitude and distribution. Significant technological advancements have been made in recent years for the detection and analysis of such marine hazards.

In Chapter 9, the removal of heavy metals from wastewater using biosensors is examined. The chapter explores the role of biosensors in wastewater monitoring and highlights the technical and practical limitations that need to be addressed to develop more efficient sensing systems. Various biosensors under investigation for detecting different pollutants are discussed, with a focus on their potential to solve water pollution problems and reduce human health impacts.

Chapter 10 presents an overview of the modern techniques available for probing the various contaminants across marine ecosystems and several drawbacks and controversies associated with the same. It is thus important to identify such sources of contaminants by quantifying essential pollutants separately and gathering dependable information regarding their origin, movement, and ultimate destination. Now a day's different types of Nuclear sensors have been developed to identify pollutants in varied environments.

In Chapter 11, the focus shifts to the identification of different pollutants in lotic and lentic ecosystems using biosensors. The chapter examines the diverse pollutants threatening aquatic ecosystems and discusses the role of biosensors in detecting and monitoring these pollutants. By exploring various biosensor technologies, the chapter aims to provide insights into addressing water pollution and minimizing its impact on human health and the environment.

Chapter 12 addresses water pollution as a global crisis impacting ecosystems, health, and economies. The chapter explores strategies to combat water pollution, emphasizing the vital role of advanced water quality sensors. It examines various pollutants and discusses modern sensor technologies' importance in ensuring water safety and managing pollution. The chapter highlights collaboration and innovation as key factors in managing water quality and safeguarding this vital resource.

In Chapter 13, the principles, applications, and significance of biosensors in the context of environmental monitoring are explored in detail. Future prospects and challenges are discussed as well. Environmental monitoring is essential to safeguard our planet's ecosystems, public health, and natural resources. Biosensors have emerged as powerful tools for assessing environmental parameters due to their sensitivity, specificity, and versatility. From the detection of pollutants to the monitoring of water and air quality, biosensors offer a wide array of applications that contribute to comprehensive environmental assessment.

In Chapter 14, the focus is on the advancements in sensor technologies aimed at detecting soil pollution. The discussion delves into integrated systems and sensor technologies, exploring future trends like enhanced selectivity and regulatory adoption. The chapter emphasizes the importance of increasing sensitivity and specificity, facilitating on-site, real-time analysis, and integrating sensing with remediation strategies to effectively manage soil pollution. Collaboration among stakeholders is highlighted as crucial for successful implementation of sensor-based approaches and driving innovation in environmental monitoring and management.

Chapter 15 explores the use of sensor technologies in waste management applications. The chapter discusses various types of sensors used in waste management, from simple bin-level sensors to advanced technologies like remote sensing and IoT-based systems. It highlights the advantages of sensor-driven waste management, including improved efficiency, cost reduction, and enhanced environmental sustainability. The chapter underscores the pivotal role of sensor technologies in revolutionizing waste management practices and achieving a more sustainable future.

Chapter 16 delves into the application of environmental sensors in addressing the issue of improper disposal of sanitary pads. It discusses the pollutants associated with this problem and explores the role of environmental sensors in mitigating it. By harnessing advanced sensing technologies, the chapter aims to provide insights into the environmental impact of sanitary pad disposal and offers guidance for creating sustainable solutions to safeguard the ecosystem.

Chapter 17 centers on the use of bioindicators to assess environmental pollution, particularly in wetlands. It highlights how bioindicators offer valuable insights into the quality of an environment by measuring species' adaptability to various stresses. The chapter underscores the significance of bioindicators in detecting indirect biotic impacts of contaminants, providing a nuanced understanding of environmental problems. Various plankton species are discussed as indicators of water pollution, showcasing the potential of bioindicators to improve water resource management.

In Chapter 18, the focus shifts to the impact of air pollution on plant species, with specific attention given to *Catharanthus roseus* L. and *Ocimum sanctum* L. The chapter examines how urbanization and industrial activities contribute to air pollution, posing ecological challenges for various plant species. It explores the biochemical, physiological, morphological, and anatomical reactions of these plants to air pollutants, aiming to understand their response to different pollution levels. The research aims to shed light on how these plant species can serve as indicators of air pollution, contributing to environmental monitoring efforts.

Chapter 19 discusses the role of whole-cell biosensors in environmental monitoring, emphasizing their specificity, sensitivity, and portability. The chapter explores the use of genetically engineered microbial cells as biosensors to detect pollutants in air, water, and soil. It showcases the customizable nature of whole-cell biosensors and their applications in addressing environmental issues, offering a comprehensive understanding of their principles and functionalities.

In Chapter 20, the impact of urbanization on the environment and public health is explored, with a focus on the role of environmental sensors in monitoring and mitigating these effects. The chapter delves into various types of sensors used to assess pollution levels in urban areas, including air quality sensors, water quality sensors, noise monitors, and solid waste sensors. By integrating real-time data from these sensors, the chapter highlights how policymakers can formulate effective strategies to address environmental degradation and improve public health in urban settings.

Chapter 21 focuses on many potentially harmful chemicals, released by industries and human activities, which can contaminate water, soil, or air and further impact the extreme environments. For this aim, sensors have captivated the attention of many scientists in modern time by virtue of their eco-friendliness, cost-effectiveness, miniaturization ability, and rapidness. Environmental samples, however are very complex and unexpectedly relative to other ecosystems. Thus far, environmental sensors have been developed with greater sensitivity, simpler and more efficient detection, better environmental adaptation and etc. for pollutant detection in extreme environments.

In Chapter 22, the focus is on understanding and predicting the effects of forest fires using a comprehensive framework integrating system dynamics and remote sensing approaches. The chapter aims to develop a conceptual model to analyze the consequences of forest fires on the environment, operational sources, and economic sectors. By combining system dynamics and remote sensing technology, the chapter seeks to provide conclusive insights into forest fire management strategies.

IN CONCLUSION

As editor of this edited reference book on *Sensors for Environmental Monitoring, Identification, and Assessment,* I am honored to present a culmination of diverse perspectives, insights, and advancements in sensor technologies aimed at safeguarding our planet's ecosystems.

Throughout the chapters, esteemed contributors have delved into critical issues such as soil pollution, water quality, air pollution, waste management, and the impacts of urbanization and forest fires. They have highlighted the indispensable role of sensor technologies in detecting, monitoring, and mitigating environmental pollutants, paving the way for more effective environmental management strategies.

From the exploration of bioindicators in wetlands to the application of whole-cell biosensors for environmental monitoring, each chapter offers valuable contributions to our understanding of environmental challenges and solutions. The chapters on identifying pollutants in lotic and lentic ecosystems, removing heavy metals from wastewater, and monitoring water pollutants underscore the urgency of addressing water pollution, a global crisis impacting biodiversity and human health.

Moreover, the discussions on sensors for air pollution, sanitary pad disposal, and carbon dioxide gas detection shed light on the diverse applications of sensor technologies in tackling specific environmental issues. The chapters on waste management sensors and advancements in sensor technology highlight the transformative potential of sensor-driven approaches in revolutionizing waste management practices and enhancing environmental sustainability.

As I conclude this preface, I extend my heartfelt appreciation to all the contributors who have shared their expertise and insights, making this edited reference book a comprehensive resource for researchers, practitioners, policymakers, and students alike. I hope that the knowledge and innovations presented in this book will inspire collaboration, drive innovation, and contribute to the collective effort of safeguarding our environment for future generations.

Together, let us harness the power of sensor technologies to create a cleaner, healthier, and more sustainable planet for all.

Khursheed Ahmad Wani
University of Kashmir, India

Chapter 1
Technology of Sensors:
Ways Ahead

Surendra Prakash Gupta
https://orcid.org/0000-0002-4873-6346
Shri Vaishnav Vidhyapeeth Viswavidhayalaya, India

Ankur Bhardwaj
https://orcid.org/0000-0003-0687-8810
Shri Vaishnav Vidhyapeeth Viswavidhayalaya, India

ABSTRACT

The technology of sensors is continuously evolving to meet the demands of various fields such as safety, pollution, medical engineering, and more. Sensors play a significant role in improving human working styles and routine tasks by converting real-world actions into electrical signals. They can be used to measure a wide range of environmental parameters, including air quality, water quality, soil quality, noise levels, and weather conditions. Sensors can be used to monitor environmental conditions in real time, or they can be used to collect data over time to identify trends and patterns. Sensor data can also be used to create models of environmental systems, which can be used to predict future conditions and assess the impact of human activities. This chapter is focused on the recent development in sensor technology that has the potential to revolutionize environmental monitoring, identification, and assessment. By providing real-time data on environmental conditions, sensors can help us to better understand our environment and make informed decisions about how to protect it.

INTRODUCTION

In recent years, there has been significant progress in sensor technology, with a focus on wearable sensing devices. A key advancement is the integration of ambient light sensors into LCD displays using advanced fabrication techniques and sensor structures like amorphous silicon and low-temperature polycrystalline silicon. These sensors are valuable for data collection and processing. However, a challenge lies in optimizing sensor coverage while minimizing costs and energy usage. This involves designing sensors

DOI: 10.4018/979-8-3693-1930-7.ch001

effectively and ensuring efficient data transfer between them. Different strategies such as solving complex optimization problems and using weighted criteria convolution have been suggested. Mathematical models and IT systems have been developed to control sensor coverage radius for optimal results.

In the field of environmental monitoring and analysis, sensor-based technology is essential. It permits real-time data gathering from multiple sources, including wireless sensor networks (WSNs) (Dai et al., 2021) and the fusion of data from multiple sensors (Yin, 2022). The quality of the air can be monitored thanks to these technologies (Yuan et al., 2022), landscape patterns, and indoor environment quality (IEQ) (Liu et al., 2021). They additionally aid in the assessment of the general functioning of the building and the welfare of its occupants. This technology is used to optimize the computational efficiency of deep learning models on embedded platforms for mobile robots. It enables the autonomous positioning and navigation of autonomous guided vehicles (AGVs) in dynamic storage environments. In addition, this technology provides valuable insights into the environment, enabling effective monitoring, analysis, and decision-making processes.

SENSORS TOOLS FOR IDENTIFICATION, MONITORING, AND ASSESSMENT FOR ENVIRONMENTAL ASPECTS

Sensors and monitoring tools are essential for identifying, monitoring, and assessing various environmental aspects. In the field of underground mining, real-time monitoring and assessment of climatic conditions using sensors and Geographic Information System (GIS) tools can help identify potential hazards and create a safe working environment (Jha & Tukkaraja, 2020). In the context of pharmaceutical contamination, screen-printed voltammetric sensors offer a simple, reliable, and cost-effective method for monitoring painkiller residues in environmental water samples (Tyszczuk-Rotko et al., 2022). Green analytical chemistry emphasizes the use of bio-chemical sensors for sustainable qualitative and quantitative analysis, with examples including the use of green nanoparticles, colorimetric assays with smart phone cameras, and sensor arrays coupled with machine learning algorithms (del Valle, 2020). In coastal areas, high-resolution hydrodynamic models and remote sensing products can be used to monitor and describe the evolution of oil spills, providing valuable information for validation and operational applications. Environmental monitoring and control rely heavily on in-situ and real-time measurement techniques, which enable global earth monitoring systems, early warning systems, and automation for surface water and wastewater management.

WATER QUALITY ASSESSMENT

Numerous studies have looked into the use of sensors for measuring water quality. There are several sources of waste residue contamination of surface and subsurface waters, including the pharmaceutical industry, veterinary and agricultural practices, homes, and healthcare facilities. A portable, low-cost sensor-based system for accurate data collection and real-time water quality monitoring has drawn a lot of interest. In a prior study, wireless sensor networks were utilized to develop a real-time water quality monitoring system using turbidity, TDS, and pH sensors (Irawan et al., 2021); Mokua et al., 2021).

In addition to tiny particles and pollutants like heavy metals, microplastics, and detergents, sewage effluent is a complex mixture of water. These particles alter the surface tension of water according to the quantity of impurities present. It is helpful to have appropriate criteria for water monitoring since Sridhar and Reddy had established the connection between surface tension and water-borne contamination (Sridhar & Reddy, 1984).

AIR QUALITY

Air quality monitoring is important for assessing pollution levels and their impact on public health and wellbeing. Recent technical developments provide new solutions for smart sensors that can measure pollutants in real-time and at small scales and have attracted the interest of a broad range of environmental researchers as well as authorities and local communities (Nagendra et al., 2021) . Wireless Sensor Networks (WSNs) offer a solution by providing high-resolution data collection and analysis capabilities. These networks can detect, calculate, and gather information about pollutants, enabling assessment at high resolutions (Broday, 2017). However, sensors measurements can be affected by environmental factors and require frequent calibrations.

SENSOR-BIOLOGY INTERFACE

A sensor-biology interface refers to a device or system that connects sensors to biological entities, such as the human body or biological reaction containers, in order to collect and process data. These interfaces are designed to improve the precision and functionality of sensor data collection. They can support multiple types of sensors without the need for switches, reducing manufacturing and maintenance costs (Wang et al., 2023). The sensitivity and stretchability of sensor are the two most important parameters to assess the performance of the device. Recently, Graphenes nanoparticles (GNs) based sensor are exploited as the appealing conductive nanofillers. However, the weak interaction between GNs and polymer matrix affect the sensitivity of the strain sensor.

The sensor interface can be enhanced with the use of other composite modifications and the ultrasonic method. By making these changes, the sensor's electrical and mechanical qualities can be enhanced. A multifunctional biological electric sensor that can measure pressure, bioelectricity, and environmental parameters is one example. Additionally, a bioelectricity signal adjusting controller is included to increase the bioelectricity signal detection precision. According to McGrath et al., an additional illustration is a sensor interface device that attaches to a sensor's pin and has circuits for both detection and power supply to accommodate various sensor kinds (McGrath et al., 2014). A bioreactor sensing system additionally consists of a biological reaction container equipped with sensing apparatus for monitoring the contents of the container and monitoring interfaces. These sensor-biology interfaces are essential for transforming the electrical signals that sensors record into data that can be used for tracking and analysis.

IoT AND SENSOR IN ENVIRONMENT

Managing sensors spread in the environment can be a valuable process for a variety of applications in domains such as agriculture, smart homes, medical technology, and healthcare, where context-aware features and a smart interpretation of the surrounding environment is necessary (Kapitsaki et al., 2021).

These sensors, deployed in large-scale IoT networks, enable the collection and monitoring of data in real-world applications such as environment monitoring, transportation, urban security, smart energy management, agriculture, and health care.

Recent advances in sensor technology and wireless communications have opened up new possibilities for deploying extensive networks of sensors, significantly expanding their potential applications. These developments are particularly beneficial for environmental monitoring purposes as they enable more comprehensive and effective data collection and analysis. The Internet of Things (IoT) and sensor technology find significant applications in environmental monitoring and control. IoT systems, consisting of interconnected devices and sensors, can be used to manage various aspects of the environment such as temperature, lighting, air quality and water quality. In industrial environments, IoT sensors can monitor air and water quality and provide real-time data on pollutants and water parameters. This information can help industry take proactive measures to reduce pollution and improve air and water quality. First, IoT sensor networks coupled with edge computing (IoTEC) can significantly reduce data latency, reduce data transmission, and increase power duration, leading to cost reduction in environmental monitoring (Roostaei et al., 2023). Secondly, the concept of Green IoT (G-IoT) aims to replace current technologies, IoT and economies with greener alternatives, thereby contributing to sustainable development and a greener world (Prakash & Singh, 2023).

Thirdly, the successful monitoring and management of air quality, radiation pollution, and water pollution is made possible by the integration of IoT devices and wireless sensors with smart environment monitoring (SEM) systems. This promotes sustainable development and a healthier society. Additionally, IoT applications for smart environmental monitoring such as soil, water, and air monitoring are essential for enhancing environmental quality and tackling environmental and industrial challenges. Furthermore, using sensors, IoT systems can monitor and analyze various pollutants, like nitrogen dioxide (NO_2) and carbon monoxide (CO), to determine how they affect respiratory diseases and assist decision-making by authorities (Ramachandran, 2023).

TYPES OF SENSORS

Small sensors known as miniature sensors find use in a variety of fields, including medical, defense, and personal electronics. An A.I. powered sensors make use of artificial intelligence technology to improve their sensing performance and facilitate wise decision-making. Conversely, wireless sensors are those that don't require wired connections to function, giving them more flexibility and making deployment simpler. Numerous energy harvesting technologies, such as those that use solar, thermal, mechanical, or chemical energy, can power these sensors. In general, the creation of miniature, A.I. powered and wireless sensors have created new avenues for sensor applications across multiple domains, providing enhanced functionality and convenience.

Our comprehension and engagement with the surroundings are being transformed by the mutually beneficial relationship between the environment and miniature sensors. Because they use thermal inspection for in-situ analysis, these sensors are regarded as non-destructive testing tools. With the help of these sensors, we can get up-to-date information on a variety of environmental parameters, including humidity, temperature, and air quality. We can use this important information to protect the environment and enhance our general well-being by making wise decisions and acting appropriately. It is essential to comprehend the analysis of microplastics (MPs) and nanoplastics (NPs) found in the environment in order to fully utilize the potential of miniaturized sensors. Because of their potential to impact both living organisms and the ecosystem in which they reside, MPs and NPs have been identified as hazardous substances. Because they can be ingested by both marine and land animals and linger in the environment for extended periods of time, they pose an ecological risk. Consuming tainted seafood along with other foods and drinks allows them to also enter the human food chain. The most important environmental issues facing the world today are the removal and identification of MPs and NPs. They are dependable, sensitive, accurate, and precise. For environmental data integrity to be guaranteed and for efficient decision-making in a variety of domains, including agriculture, urban planning, industrial monitoring, and wildlife conservation, miniaturized sensors must possess accurate and dependable performance characteristics.

In addition to having the potential to drastically alter how humans interact with our surroundings, miniature sensors have already started to revolutionize a number of industries. With the use of these sensors, farmers will be able to optimize irrigation, increase crop yield, and preserve water resources with real-time data on soil moisture. Smarter and more sustainable cities can be developed by using these sensors to monitor noise levels and air quality by urban planners. Miniaturized sensors make industrial monitoring safer for workers and less likely to cause environmental accidents by identifying and reporting hazardous conditions. Smaller spectrometers, a new breed of portable diagnostic tools, can be connected to automated data processing systems, simplifying the interpretation of the findings for non-specialist technicians. Near infrared (NIR) spectrometers are designed to enable fast, non-invasive readings and are used in the development of miniature analytical instruments. The fields of agro-food, pharmaceutical, and forensic diagnostics have all made use of miniature NIR spectrometers (Catelli et al. by 2020). Micro spectrometers have also become more portable and smaller as a result of the incorporation of nanomaterials into their design.

Additionally, by incorporating miniature sensors into wildlife conservation initiatives additionally enables the observation of animal behavior and environmental conditions, contributing to the maintenance of vulnerable species and ecosystems. These sensors' potential applications in sustainability and environmental protection will only grow as technological developments continue to improve their capabilities. The performance and adaptability of miniaturized sensors can be further enhanced by continuous research and development, as this emphasizes.

With the use of environmental sensors and artificial intelligence, AIEn Sensor is a cutting-edge technology that collects and analyzes environmental data. Among the many environmental variables that this cutting-edge system can identify are temperature, humidity, noise levels, and air quality. One of the most amazing features of the AIEnsensor is its ability to send real-time data to a centralized dashboard; by continuously monitoring these parameters, users can watch and study the environment from any location. The AIEn sensor is an invaluable resource for businesses, academic institutions, and environmental organizations due to its high degree of accessibility and connectivity. Furthermore, the AIEnsensor's design promotes sustainability by utilizing low energy consumption and recyclable materials. It can be used for extended periods of time in a variety of environmental settings due to its sturdy construction, which ensures dependability and longevity. The AIEn sensor can be used for a wide range of tasks, from monitoring urban air quality to assessing indoor environments for safety and comfort. Because of the rapid advancement of sensor technology, which continuously enhances our comprehension and response to environmental challenges, the AIEn sensor is at the forefront of innovation.

SENSORS IN ENVIRONMENT MONITORING: SIGNIFICANCE AND APPLICATIONS

Maintaining the sustainability of our natural resources and comprehending how human activity affects the environment depend on environmental monitoring. Real-time data on pollution levels, weather patterns, air and water quality, and other topics is provided by sensors, which are essential in this field. Decisions to reduce environmental risks and safeguard ecosystems can be made by scientists and policymakers with the help of this data collection and analysis. The concentration of pollutants like particulate matter, nitrogen dioxide, and ozone is measured by sensors, which are widely used in air quality monitoring. In addition to assisting in determining the health risks connected to air pollution, this data is helpful in locating the sources of pollution and developing efficient response strategies.

In the realm of water quality monitoring, sensors are deployed to measure parameters like pH, dissolved oxygen, turbidity, and nutrient levels in bodies of water. This data is vital for managing water resources, safeguarding aquatic habitats, and ensuring safe drinking water for communities. Furthermore, sensors are integral to monitoring weather and climate conditions, aiding in early detection of extreme weather events, forecasting climate trends, and assessing the impact of climate change on ecosystems. With the advancements in sensor technology, the applications of environmental monitoring continue to expand, encompassing areas such as precision agriculture, wildlife conservation, and urban planning. These developments underscore the growing significance of sensors in environmental monitoring and the need for continued research and innovation in this field.

FUTURE SCOPE OF SENSORS IN ENVIRONMENTAL MONITORING

As a vital component of environmental monitoring, sensors provide data in real time on a range of environmental parameters, including temperature, humidity, pollution levels, and the quality of the air and water. Despite the immense potential of sensors, there is ample space in the sensing field to develop quick, inexpensive sensors that have negligible to no environmental impact. Future environmental monitoring applications have a lot of promise thanks to the development of sensor technology. The integration and downsizing of sensors into wearable technology is a crucial area of future development that will enable people to track their individual exposure to environmental contaminants. This could have significant implications for public health and awareness.

Furthermore, the use of advanced data analytics and machine learning algorithms can enable the interpretation of complex environmental data collected by sensors, leading to more accurate predictions and early warning systems for natural disasters and environmental hazards. In addition, the development of low-cost, energy-efficient sensors will make it possible to deploy large-scale sensor networks for monitoring environmental parameters in remote and inaccessible locations. This could provide valuable data for conservation efforts and ecosystem management.

Moreover, new opportunities for thorough and real-time environmental monitoring on a global scale will arise from the integration of sensors with emerging technologies like the Internet of Things and remote sensing satellites. As deforestation, climate patterns, and ecosystem changes occur, a growing number of remote sensors including satellites and drones will be used in large-scale environmental monitoring programs. In general, sensors have a bright future in environmental monitoring, helping to reduce the negative effects of human activity on the environment and deepen our understanding of the natural world. Sensors raise awareness of environmental challenges as technology advances by improving our understanding and ability to solve environmental problems.

REFERENCES

Broday, D. M. (2017). Wireless Distributed Environmental Sensor Networks for Air Pollution Measurement-The Promise and the Current Reality. *Sensors (Basel)*, *17*(10), 2263. Advance online publication. doi:10.3390/s17102263 PMID:28974042

Catelli, E., Sciutto, G., Prati, S., Chavez Lozano, M. V., Gatti, L., Lugli, F., Silvestrini, S., Benazzi, S., Genorini, E., & Mazzeo, R. (2020). A new miniaturised short-wave infrared (SWIR) spectrometer for on-site cultural heritage investigations. *Talanta*, *121112*, 121112. Advance online publication. doi:10.1016/j.talanta.2020.121112 PMID:32797874

Dai, B., Li, C., Lin, T., Wang, Y., Gong, D., Ji, X., & Zhu, B. (2021). Field robot environment sensing technology based on TensorRT. In *Intelligent Robotics and Applications* (pp. 370–377). Springer International Publishing. doi:10.1007/978-3-030-89095-7_36

del Valle, M. (2020). *Sensors as Green Tools*.

Irawan, Y., Febriani, A., Wahyuni, R., & Devis, Y. (2021). Water quality measurement and filtering tools using arduino Uno, PH sensor and TDS meter sensor. [JRC]. *Journal of Robotics and Control*, *2*(5). Advance online publication. doi:10.18196/jrc.25107

Jha, A., & Tukkaraja, P. (2020). Monitoring and assessment of underground climatic conditions using sensors and GIS tools. *International Journal of Mining Science and Technology*, *30*(4), 495–499. doi:10.1016/j.ijmst.2020.05.010

Kapitsaki, G. M., Achilleos, A. P., Aziz, P., & Paphitou, A. C. (2021). SensoMan: Social Management of Context Sensors and Actuators for IoT. *Journal of Sensor and Actuator Networks*, *4*(4), 68. doi:10.3390/jsan10040068

Liu, Y., Zhang, D., He, K., Gao, Q., & Qin, F. (2021). Research on Land Use Change and Ecological Environment Effect Based on Remote Sensing Sensor Technology. *Journal of Sensors*, *2021*, 1–11. Advance online publication. doi:10.1155/2021/4351733

McGrath, M. J., Scanaill, C. N., & Nafus, D. (2014). *Sensor Technologies: Healthcare, Wellness and Environmental Applications*. Apress.

Mokua, N., Maina, C., & Kiragu, H. (2021). A raw water quality monitoring system using wireless sensor networks. *International Journal of Computer Applications*, *174*(21), 35–42. doi:10.5120/ijca2021921113

Nagendra, S., Schlink, S. M., & Khare, M. (2021). Air quality measuring sensors. In Urban Air Quality Monitoring, Modelling and Human Exposure Assessment (pp. 89–104). Springer Singapore. doi:10.1007/978-981-15-5511-4_7

Prakash, R., & Singh, G. (2023). GREEN INTERNET OF THINGS (G-IoT) FOR SUSTAINABLE ENVIRONMENT. [IJMR]. *EPRA International Journal of Multidisciplinary Research*, *9*(5), 1–1. doi:10.36713/epra13324

Ramachandran, A. (2023). *Modeling of Internet of Things Enabled Sustainable Environment Air Pollution Monitoring System*. . doi:10.30955/gnj.004707

Roostaei, J., Wager, Y. Z., Shi, W., Dittrich, T., Miller, C., & Gopalakrishnan, K. (2023). IoT-based Edge Computing (IoTEC) for Improved Environmental Monitoring. *Sustainable Computing : Informatics and Systems*, *38*, 100870. Advance online publication. doi:10.1016/j.suscom.2023.100870 PMID:37234690

Sridhar, M. K. C., & Rami Reddy, C. (1984). Surface tension of polluted waters and treated wastewater. *Environmental Pollution. Series B. Chemical and Physical*, *7*(1), 49–69. doi:10.1016/0143-148X(84)90037-5

Tyszczuk-Rotko, K., Kozak, J., & Czech, B. (2022). Screen-Printed Voltammetric Sensors—Tools for Environmental Water Monitoring of Painkillers. *Sensors (Basel)*, *22*(7), 2437. doi:10.3390/s22072437 PMID:35408052

Wang, J., Liu, Z., Zhou, Y., Zhu, S., Gao, C., Yan, X., Wei, K., Gao, Q., Ding, C., Luo, T., & Yang, R. (2023). A multifunctional sensor for real-time monitoring and pro-healing of frostbite wounds. *Acta Biomaterialia*, *172*, 330–342. doi:10.1016/j.actbio.2023.10.003 PMID:37806374

Yin, F. (2022). Practice of air environment quality monitoring data visualization technology based on adaptive wireless sensor networks. *Proceedings of the ... International Wireless Communications & Mobile Computing Conference / Association for Computing Machinery. International Wireless Communications & Mobile Computing Conference*, *2022*, 1–12. 10.1155/2022/4160186

Yuan, C., Wang, Y., & Liu, J. (2022). Research on multi-sensor fusion-based AGV positioning and navigation technology in storage environment. *Journal of Physics: Conference Series*, *2378*(1), 012052. doi:10.1088/1742-6596/2378/1/012052

List of Abbreviations

IoT= Internet of Things
LCD= Liquid Crystal Display
AGV= Automated guided Vehicles
IEQ=Indoor environment quality
WSN= Wireless sensor network
GIS= Geographic Information System
GNs = Graphenes nanoparticles
MP = microplastics
NP=-nanoplastics
NIR=Near infrared
SEM= smart environment monitoring

Chapter 2
Air Sensors and Their Capabilities

Prashantkumar Bharatbhai Sathvara
https://orcid.org/0000-0002-3548-5584
Nims Institute of Allied Medical Science and Technology, Jaipur, India

Anuradha Jayaraman
Nims Institute of Allied Medical Science and Technology, Jaipur, India

Sandeep Tripathi
Nims Institute of Allied Medical Science and Technology, Jaipur, India

Sanjeevi Ramakrishnan
Nims Institute of Allied Medical Science and Technology, Jaipur, India

ABSTRACT

In this comprehensive chapter on air quality, the paramount significance and intricate challenges associated with this critical environmental and public health concern are explored. Focusing on the mission to improve air quality, the chapter underscores the profound impact of poor air quality on human health, particularly in densely populated urban areas. Emphasis is placed on safeguarding ecosystems and biodiversity through the reduction of air pollution, acknowledging its widespread ecological implications. The chapter delves into international and national regulations and initiatives, such as the Clean Air Act and the Paris Agreement, representing concerted global efforts to combat air pollution and address climate change. A detailed examination of concerns related to various air pollutants, including Particulate Matter, gases like NO2 and SO2, and airborne toxic substances, sheds light on the health risks and environmental consequences.

INTRODUCTION

Air pollution has emerged as a critical global concern, posing significant threats to public health, the environment, and climate change. The primary contributors to this issue are diverse, encompassing in-

DOI: 10.4018/979-8-3693-1930-7.ch002

dustrial emissions, vehicular exhaust, deforestation, and agricultural practices. These activities release a myriad of pollutants, including particulate matter, nitrogen oxides, sulfur dioxide, ozone, and volatile organic compounds, into the atmosphere (Zhang & Srinivasan, 2020). The World Health Organization estimates that millions of premature deaths occur annually worldwide due to outdoor air pollution, disproportionately impacting vulnerable populations. International collaboration has become imperative to address this global challenge. Agreements like the Paris Agreement and Sustainable Development Goals underscore the shared responsibility among nations to combat air pollution collectively. These frameworks aim to establish global targets and encourage coordinated efforts to reduce emissions and mitigate environmental impact.

In the Indian context, the nation grapples with severe air pollution challenges driven by rapid urbanization, industrial growth, and agricultural practices. Sources of pollution in India include vehicular emissions, industrial activities, agricultural burning, and household pollution. The government has responded to this crisis by initiating programs such as the National Clean Air Programme, intending to curb pollution levels in key cities through measures like stricter emission norms, promotion of public transportation, and increased green cover. However, despite these efforts, challenges persist on multiple fronts. Enforcement of regulations remains inadequate, monitoring infrastructure is insufficient, and behavioral changes necessary to reduce pollution are slow to materialize. Achieving meaningful progress in tackling air pollution demands comprehensive and sustained efforts (Dhall et al., 2021).

To address this multifaceted problem, a holistic approach is crucial. Coordinated efforts are needed, involving sustainable urban planning to reduce emissions from transportation and industrial activities. Adoption of cleaner technologies, such as renewable energy sources and cleaner industrial processes, is vital for a sustainable future. Equally important is increasing public awareness and fostering a societal shift towards environmentally conscious practices. Protecting the health and well-being of populations globally, as well as in India, necessitates a paradigm shift in how societies perceive and interact with their environment. It requires not only stringent regulations and technological advancements but also active participation and awareness at the individual and community levels (Saini et al., 2021). Air pollution is a complex global challenge that demands collaborative and concerted efforts. It requires nations to work together, share responsibilities, and implement sustainable solutions. The Indian government's initiatives are commendable, but ongoing commitment and innovation are imperative to overcome persistent challenges. Only through coordinated international and local actions can the detrimental effects of air pollution be mitigated, ensuring a healthier and sustainable future for all.

Air quality detection is a critical component in the ongoing efforts to safeguard public health and preserve the environment. Utilizing specialized sensors and monitoring technologies, this process focuses on assessing the concentration of pollutants in the air, including particulate matter, nitrogen dioxide, sulfur dioxide, ozone, and volatile organic compounds. The significance of air quality detection stems from the severe risks associated with air pollution, ranging from respiratory diseases to environmental degradation and climate change. Various monitoring technologies, such as air quality sensors, satellite-based observations, and ground-based stations, contribute to comprehensive assessments. Parameters monitored encompass particulate matter, gaseous pollutants, and meteorological conditions. Applications of air quality detection include health protection, environmental monitoring, regulatory compliance, and influencing urban planning decisions for sustainable development (Javaid, Haleem, Rab, et al., 2021). Despite advancements, challenges like standardized calibration methods and data accuracy persist. The future involves integrating emerging technologies like artificial intelligence, expanding monitoring networks, and promoting public awareness to create a healthier and more sustainable global environment.

Air Quality, Vulnerable Populations, and Asthma

Asthma, a persistent medical condition, frequently leads to heightened disease activity, some instances of which necessitate hospitalization. Air quality indicators, including PM2.5, NO2, O3, and dampness-related contaminants, exert a significant influence on both the exacerbation of asthma and its overall progression. Notably, asthmatic children allocate approximately 60% of their waking hours to school environments (Schweitzer & Zhou, 2010). A recent expansive study revealed a synergistic correlation between co-exposure to elevated endotoxin levels and PM2.5, leading to an increase in emergency room visits, particularly concerning asthma among children. Furthermore, exposure to elevated concentrations of endotoxin and NO2 demonstrated a synergistic association with an upsurge in asthma attacks. It is noteworthy that these associations persisted, even when geometric mean concentrations of PM2.5, O3, and NO2 were below the standards set by the EPA NAAQ (Sathvara et al., 2023; Spurr et al., 2014).

Indoor air quality (IAQ) plays a pivotal role in respiratory health, particularly for vulnerable populations such as children, the elderly, and individuals with pre-existing conditions like asthma. This concise research note delves into the intricate connection between IAQ and asthma, shedding light on the impact of indoor air pollutants on those most susceptible. Common indoor pollutants include particulate matter, volatile organic compounds (VOCs), mold, and allergens. These pollutants originate from various sources within homes, such as cooking, cleaning products, and inadequate ventilation. Particular attention must be paid to the impact of these pollutants on vulnerable populations, as exposure can exacerbate asthma symptoms and lead to respiratory distress (DR. SANJEEVI RAMAKRISHNAN et al., 2023; Sanjeevi et al., 2017).

Children, with their developing respiratory systems, are especially vulnerable to the effects of indoor pollutants. Early exposure can impede lung development and increase the risk of respiratory issues, making IAQ crucial for their overall well-being. The elderly are prone to respiratory issues, and poor IAQ further exacerbates these challenges. Age-related factors, coupled with prolonged exposure to indoor pollutants, contribute to respiratory complications, emphasizing the need for targeted interventions. For those with asthma, indoor air quality directly impacts their respiratory health. Triggers such as dust mites, pet dander, and mold can provoke asthma attacks. Understanding and mitigating these triggers are essential to managing asthma in affected individuals (Saini et al., 2021).

Adequate ventilation is a fundamental strategy for improving IAQ. Properly functioning ventilation systems help in reducing indoor pollutants by allowing fresh outdoor air to replace the contaminated indoor air. Air purifiers equipped with HEPA filters can effectively remove particulate matter and allergens, providing a supplementary measure to enhance IAQ. However, choosing the right type of purifier and ensuring regular maintenance are crucial for optimal effectiveness. Simple lifestyle adjustments, such as reducing tobacco smoke indoors, using environmentally friendly cleaning products, and proper waste management, contribute significantly to IAQ improvement. Educating individuals about these practices is vital for fostering healthier indoor environments (Ankitkumar B Rathod et al., 2023; Bearg, 2019).

In conclusion, recognizing the nexus between IAQ, vulnerable populations, and asthma is imperative for public health. The impact of indoor pollutants on children, the elderly, and individuals with asthma underscores the need for targeted interventions. Implementing ventilation improvements, utilizing air purifiers, and promoting awareness of IAQ-friendly behaviors collectively form a holistic approach to mitigate the adverse effects of indoor air pollution. As we navigate the challenges posed by indoor pollutants, prioritizing research, policy initiatives, and public awareness campaigns will be pivotal in safeguarding the respiratory health of vulnerable populations.

Common Air Pollutants That Affect IAQ

The most prevalent air pollutants impacting Indoor Air Quality (IAQ) encompass O3, CO, CO2, SO2, NO2, PM, and VOCs. Each of these pollutants possesses distinct pathophysiologic mechanisms: Resulting from a chemical reaction between NO2 and VOCs exposed to sunlight, O3 concentrations can intensify in both hot and cold environments. Emitted by chemical solvents, electric utilities, and gasoline vapors, O3 induces lung inflammation and airway constriction. Vulnerable populations, including individuals with underlying diseases, children, and the elderly, face elevated risks from O3 exposure. An odorless, colorless, and tasteless toxic gas, CO stems from various sources such as unvented fuel and gas space heaters, leaky chimneys, tobacco smoke, and combustion devices. Exposure to CO can lead to fatigue, chest pain, reduced brain function, impaired vision, and fetal death. Designated as an anthropogenic air pollutant by the EPA and IPCC, CO2 is colorless and odorless, primarily originating from occupant respiration. Elevated CO2 concentrations, as indicated by the US EPA BASE, are associated with increased prevalence of Sick Building Syndrome (SBS) symptoms (Mamun & Yuce, n.d.). A major precursor to ambient PM2.5, SO2 concentrations derive from the combustion of sulfur-containing coal, oil, and gas. Short-term exposure to SO2 can cause respiratory illnesses, airway inflammation, and toxic symptoms. Asthmatics, children, and older adults are particularly susceptible. A highly reactive gas related to ozone and PM2.5 development, NO2 primarily enters the air from fuel combustion. Similar to sulfur dioxide, NO2 causes respiratory symptoms and airway inflammation, posing higher risks for asthmatics, children, and older adults. A mixture of solid and liquid particles in the air, PM categorizes into PM10, PM2.5, and PM1.0 based on size. PM10 affects upper respiratory tracts, while PM2.5 and PM1.0 can penetrate deeper into the respiratory system and even internal organs. PM is associated with various health issues and is estimated to cause 3.3 million deaths annually worldwide. A diverse set of hazardous organic chemicals participating in atmospheric reactions, VOCs are major contributors to SBS (Ramakrishnan & Jayaraman, 2019; Singh et al., 2021). Indoor VOC concentrations, especially benzene, formaldehyde, and toluene, can surpass outdoor levels. Asthmatics, young children, and the elderly are more susceptible to the carcinogenic, irritant, and toxic effects of VOC exposure. In addition to these pollutants, indoor temperature and relative humidity significantly influence IAQ. Studies have demonstrated a linear correlation between acceptability and enthalpy of IAQ, indicating that IAQ declines with increased temperature and humidity. Temperature has a stronger linear effect on IAQ than humidity, with lower relative humidity levels considered more acceptable for IAQ performance. Understanding these mechanisms is crucial for formulating effective strategies to mitigate the impact of indoor air pollutants on human health (Abbasi et al., 2012; Parry & Hubbard, 2023).

Air Quality Sensors, Measurement Tolerances

In recent years, air quality sensor technology has emerged from several laboratories for practical application, as they can be used to support real-time, spatial, and temporal data resolution for the monitoring of air concentration levels. Additionally, more and more companies provide their own air quality sensor products. The principles of operation for the low-cost gas-phase sensors are typically based on five major components (Anwar Abdelrahman Aly et al., 2016; Chow, 1995). Studies have shown that modern air quality sensor provide useful qualitative information for scientific research, as well as for end-users. However, due to the embedded technical uncertainties and lack of cross-validation and verification, there are certain limitations when comparing them to the expensive conventional equipment.

Air quality sensors play a pivotal role in monitoring and assessing the levels of various pollutants in the atmosphere, contributing valuable data for environmental health studies and regulatory compliance. However, achieving accurate measurements demands an understanding of the inherent tolerances associated with these sensors. This discussion explores the importance of air quality sensors, their measurement tolerances, and strategies to enhance precision in environmental monitoring. Air quality sensors, commonly known as air quality monitors or detectors, are devices designed to measure the concentration of pollutants present in the air. These pollutants may include particulate matter (PM), nitrogen dioxide (NO2), ozone (O3), sulfur dioxide (SO2), carbon monoxide (CO), and volatile organic compounds (VOCs). These sensors are deployed in various settings, ranging from industrial facilities and urban areas to homes and personal devices (R. Sanjeevi et al., 2022; Ródenas García et al., 2022).

Measurement tolerances refer to the allowable range of deviation in sensor readings from the true value. Several factors contribute to the tolerances in air quality sensor measurements: Different sensor technologies, such as electrochemical, optical, and semiconductor-based sensors, have varying levels of precision. Understanding the technology used in a sensor is crucial for interpreting measurement tolerances. Regular calibration is essential to maintain accuracy. Sensor calibration involves adjusting the device to a standard reference to ensure reliable and consistent measurements. Deviations from calibrated values may contribute to measurement tolerances. Factors like temperature, humidity, and air pressure can influence sensor performance (Cheng & Lee, 2016). Manufacturers often specify operational conditions, and deviations from these conditions may affect measurement accuracy. Over time, sensors may experience wear and tear, impacting their accuracy. Regular maintenance and replacement of sensors, when necessary, help minimize measurement tolerances. Some sensors may be sensitive to multiple pollutants, leading to cross-sensitivity. Understanding and compensating for cross-sensitivity are vital for accurate measurements of specific pollutants.

To enhance precision in air quality measurements and minimize tolerances, the following strategies are recommended: Schedule routine calibration checks to ensure that sensors are aligned with reference standards, reducing measurement deviations. Source sensors from reputable manufacturers with a track record of producing accurate and reliable devices. Follow recommended maintenance procedures to extend sensor lifespan. Consider the environmental conditions where sensors are deployed. Implement corrective measures or adjustments when sensors operate outside specified conditions. Implement data validation techniques to identify and address outliers or anomalous readings. This helps ensure that reported measurements are consistent and reliable (Abbasi et al., 2013; Cheng & Lee, 2016). Maintain detailed calibration records, including calibration dates and any adjustments made. This documentation aids in identifying trends in sensor performance over time.

In conclusion, understanding the tolerances associated with air quality sensors is imperative for reliable environmental monitoring. By incorporating precision-enhancing strategies, such as regular calibration and quality assurance measures, the accuracy and reliability of air quality measurements can be significantly improved, contributing to more informed decision-making in environmental management and public health (Villa et al., 2016).

AIR SENSORS AND THEIR CAPABILITIES

Air sensors are vital devices employed to detect and measure the presence and concentration of gases or particles in the air. They serve a crucial role in various applications, encompassing environmental monitoring, industrial process control, and healthcare. These sensors enable us to safeguard human health and preserve the environment by providing valuable insights into air quality (Javaid, Haleem, Singh, et al., 2021).

Classification of Air Sensors: Air sensors can be broadly classified into two main types based on their functionality:

1. Gas Sensors:

Gas sensors are specialized devices crafted to identify and measure the levels of particular gases in the air. They play a crucial role in various applications, with common examples targeting gases like carbon monoxide (CO), nitrogen dioxide (NO2), and sulfur dioxide (SO2). These sensors are vital for detecting hazardous gas leaks, maintaining workplace and residential air safety, and monitoring air pollution levels. By providing real-time data on gas concentrations, gas sensors contribute significantly to ensuring environmental health and safety. Their applications extend from industrial settings to everyday environments, enhancing our ability to respond promptly to potential threats and create healthier living and working conditions (Nikolic et al., 2020).

2. Particle Sensors:

Particle sensors are devices designed to detect and quantify particulate matter (PM) concentrations in the air. These sensors play a critical role in monitoring air quality by measuring the presence of particles of various sizes, such as PM10, PM2.5, and PM1.0. Commonly used in environmental monitoring, industrial settings, and indoor air quality assessments, particle sensors provide valuable insights into the levels of fine and coarse particles that may pose health risks. They operate based on diverse technologies, including optical methods and laser scattering, allowing for real-time measurement of particle concentrations (Wang et al., 2015). By offering data on airborne particulates, particle sensors contribute to our understanding of pollution sources, potential health impacts, and the effectiveness of pollution control measures. This information is essential for making informed decisions to mitigate the adverse effects of particulate pollution on public health and the environment.

Air sensors Technologies employ various sensing technologies to detect and measure gases or particles. Here are some commonly used technologies:

1. Electrochemical Sensors:

Electrochemical sensors employ chemical reactions to produce electrical signals directly correlated with the concentration of the targeted gas. This technology is extensively utilized for detecting toxic and flammable gases, including carbon monoxide (CO), nitrogen dioxide (NO2), and hydrogen sulfide (H2S). The sensors typically consist of a working electrode, a reference electrode, and an electrolyte. When the target gas interacts with the working electrode, a chemical reaction occurs, leading to a flow of electrons and the creation of an electrical signal (Murthy et al., 2022). This signal is then measured and translated into a concentration value. Due to their high sensitivity, accuracy, and specificity, electrochemical sensors are crucial in various applications, such as industrial safety, environmental monitoring, and indoor air quality assessments.

2. Metal Oxide Semiconductor (MOS) Sensors:

Metal Oxide Semiconductor (MOS) sensors operate on the principle of measuring alterations in electrical resistance resulting from the interaction between a target gas and a metal oxide semiconductor material. These sensors are widely employed for their versatility in detecting various gases, encompassing carbon monoxide (CO), nitrogen dioxide (NO2), and volatile organic compounds (VOCs). When the target gas contacts the semiconductor surface, it induces a change in the conductivity of the material, leading to a measurable shift in electrical resistance (*Metal Oxide Semiconductor - an Overview | ScienceDirect Topics*, n.d.). This change is then translated into a quantifiable gas concentration. MOS sensors are known for their effectiveness, cost-efficiency, and applicability in applications such as indoor air quality monitoring and industrial safety, where detecting a diverse range of gases is essential for maintaining a safe and healthy environment.

3. Optical Sensors:

Figure 1. Electrochemical sensors
(Murthy et al., 2022)

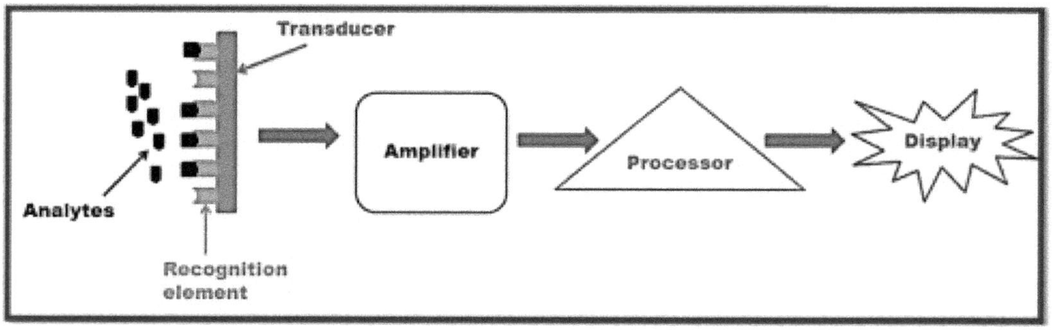

Figure 2. Metal Oxide Semiconductor (MOS) sensors
(Metal Oxide Semiconductor - an Overview | ScienceDirect Topics, n.d.)

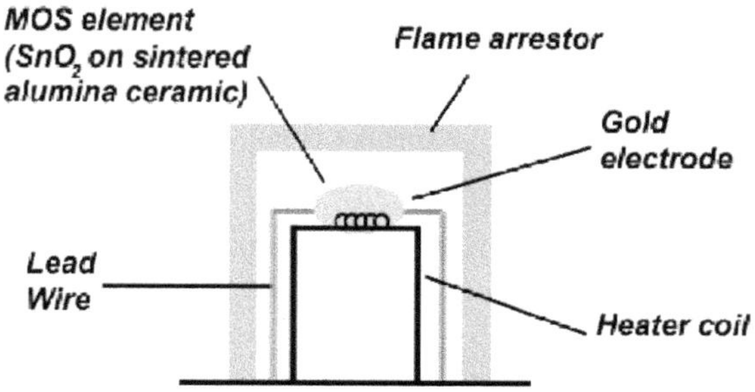

Optical sensors harness light to gauge the concentration of gases or particles in the air, employing methods like absorption, scattering, or fluorescence. In absorption-based optical sensors, specific wavelengths of light are absorbed by target substances, and the extent of absorption is correlated with their concentration. Scattering-based sensors measure changes in the direction of light as it encounters particles, providing insights into particle concentrations. Fluorescence-based sensors utilize the emission of light by certain substances when exposed to specific wavelengths, offering a distinctive signal for identification and quantification (Agarwal, 2017; Prashantkumar B. Sathvara et al., 2023). Optical sensors are versatile, allowing for the detection and quantification of various substances, including gases and particles. Their applications range from environmental monitoring and industrial safety to medical diagnostics, showcasing their significance in obtaining accurate and reliable data about air quality and composition.

4. Light Scattering Sensors:

Light scattering sensors utilize the principle of light interaction with particles in the air to assess their size and concentration. These sensors are particularly effective in detecting and monitoring particulate matter, including PM2.5 (fine particles with a diameter of 2.5 micrometers or smaller) and PM10 (particles with a diameter of 10 micrometers or smaller). When light encounters particles, the scattering pattern provides information about particle characteristics. Light scattering sensors can differentiate between various particle sizes, aiding in the identification of fine and coarse particulate matter. These sensors play a crucial role in environmental monitoring, providing real-time data on air quality and helping to assess the potential health impacts associated with airborne particles (Chicea et al., 2021). Their application extends to indoor air quality assessments, industrial settings, and regulatory compliance for maintaining a safe and healthy atmosphere.

5. Photoionization Sensors:

Photoionization sensors employ ultraviolet (UV) light to ionize particles in the air, measuring the resulting electrical current to determine particle concentration. Particularly effective in detecting and quantifying volatile organic compounds (VOCs) and other organic vapors, these sensors operate based on the principle that UV light ionizes molecules, creating positively charged ions. The generated ions

Figure 3. Optical sensors
(Agarwal, 2017)

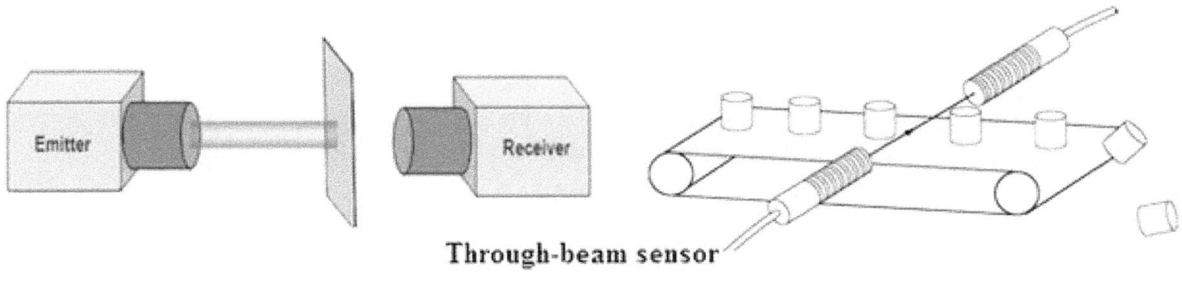

Figure 4. Light scattering sensors
(Chicea et al., 2021)

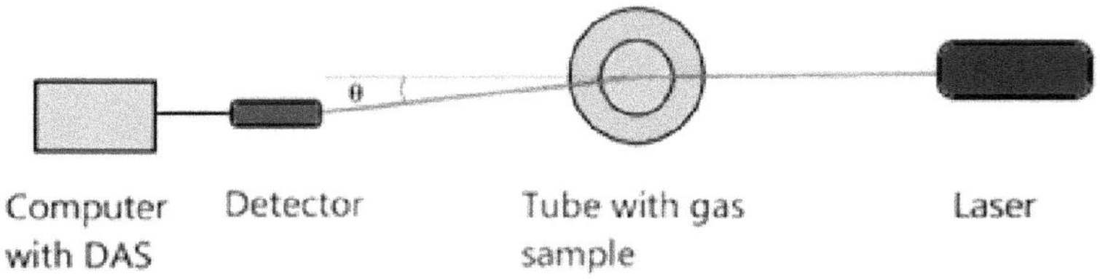

contribute to an electrical current, the magnitude of which is proportional to the concentration of the targeted particles. Photoionization sensors are widely used in applications where the detection of VOCs is crucial, such as environmental monitoring, industrial safety, and indoor air quality assessments (Fumian et al., 2020). Their high sensitivity and specificity make them valuable tools for identifying and mitigating potential health and safety risks associated with exposure to specific organic compounds.

6. Beta Attenuation Sensors:

Beta attenuation sensors utilize a radioactive source to measure the mass of particles present in the air. These sensors are distinguished by their high sensitivity and accuracy in detecting particulate matter, making them valuable tools for various applications, especially in continuous emissions monitoring systems (CEMS). The radioactive source emits beta particles, and as these particles pass through the air, their attenuation is measured. The extent of attenuation is proportional to the mass concentration of particles. Beta attenuation sensors are particularly effective in industrial settings where precise monitoring of particulate emissions is crucial for regulatory compliance and environmental impact assessment (Shukla & Aggarwal, 2022). Their continuous and real-time monitoring capabilities contribute to maintaining air quality standards, ensuring workplace safety, and minimizing environmental pollution.

Figure 5. Photoionization sensors
(Fumian et al., 2020)

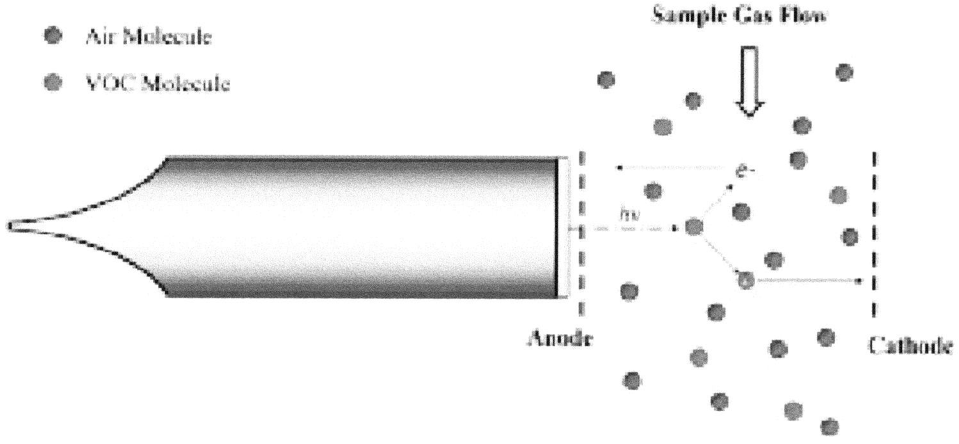

Figure 6. Beta attenuation sensors
(Shukla & Aggarwal, 2022)

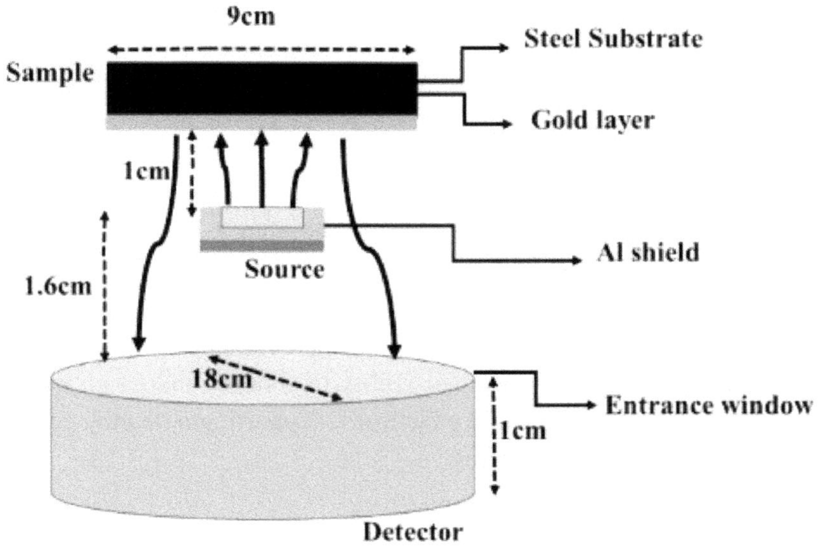

Applications of Air Sensors

Air sensors are extensively employed in a diverse range of applications, including:

1. Environmental Monitoring:

Air sensors are essential instruments for monitoring and evaluating air quality in diverse environments, including both indoor and outdoor settings. These sensors play a crucial role in identifying pollution sources, assessing air quality trends, and enforcing regulations to safeguard public health and the environment. In outdoor spaces, air sensors contribute to understanding the concentration levels of pollutants such as ozone, particulate matter, and nitrogen dioxide, aiding in the formulation of effective pollution control measures. Indoor air sensors are vital for detecting volatile organic compounds, carbon monoxide, and other indoor pollutants, ensuring a healthy indoor environment. By providing real-time data and insights, air sensors empower regulatory bodies, researchers, and the general public to make informed decisions for mitigating the impact of pollution and creating healthier living and working conditions.

2. Industrial Process Control:

In industrial settings, air sensors play a pivotal role in monitoring and controlling air quality throughout various processes. These sensors are instrumental in ensuring worker safety by detecting and quantifying the concentration of potentially harmful airborne substances. By providing real-time data on pollutant levels, air sensors contribute to preventing emissions that could adversely affect both the environment and human health. Additionally, in industries subject to regulatory standards, air sensors assist in maintaining compliance by continuously monitoring and reporting air quality parameters. Whether it's the detection of gases, particulate matter, or volatile compounds, these sensors enhance the overall safety and environmental performance of industrial operations, creating healthier working conditions and minimizing the impact of industrial activities on the surrounding air quality.

3. Healthcare:

In healthcare facilities, air sensors play a crucial role in maintaining a clean and safe environment for patients and staff. These sensors contribute significantly to infection control by preventing the spread of airborne infections. By continuously monitoring air quality, particularly in critical areas such as operating rooms and patient care spaces, air sensors aid in identifying potential contaminants and ensuring optimal conditions for patient recovery. The real-time data provided by these sensors enables swift responses to changes in air quality, contributing to improved patient outcomes and reducing the risk of healthcare-associated infections. With their ability to enhance indoor air quality management, air sensors are indispensable tools for healthcare facilities, supporting a healthy and safe environment conducive to patient healing and staff well-being.

4. Smart Cities and Buildings:

Air sensors play a vital role in the development of smart cities and buildings by providing real-time data on air quality. Integrated into these environments, these sensors continuously monitor pollutants and particulate matter levels, offering crucial insights for residents, workers, and policymakers. The accessibility of real-time air quality information empowers individuals to make informed decisions concerning their health and surroundings. Residents can adjust daily activities based on current air quality conditions, promoting a healthier lifestyle. Additionally, city planners and authorities can utilize this data to implement targeted interventions, improving overall environmental quality and urban sustainability. By fostering awareness and enabling proactive measures, air sensors contribute to the creation of healthier and more environmentally conscious communities within the framework of smart cities and buildings.

5. Research and Development:

Air sensors play a pivotal role in advancing scientific research and development by supplying valuable data on various aspects of the atmosphere. These sensors contribute crucial information on air pollution, aiding researchers in understanding the sources, levels, and distribution of pollutants. Additionally, in the context of climate change, air sensors assist in monitoring greenhouse gas concentrations and studying climate-related variables. The data collected by these sensors also enhances our understanding of atmospheric chemistry, helping scientists unravel complex interactions between different chemical compounds in the air. The insights gained from air sensor data contribute to the formulation of effective environmental policies and strategies for mitigating the impacts of air pollution and climate change. This synergy between air sensors and scientific research is essential for creating a sustainable and healthier environment.

CONCLUSION

Air sensors stand as indispensable tools in safeguarding human health and preserving the environment. Their diverse applications, ranging from environmental monitoring to industrial process control and healthcare, underscore their significance in maintaining clean air and protecting public well-being. Through continuous advancements in sensing technologies and data analytics, air sensors will continue to play a pivotal role in shaping a healthier and sustainable future for our planet.

Indoor air pollutants significantly impact human health, leading international agencies to continually develop guidelines for effective indoor air quality management. This study aims to enhance comprehension of major standards and guidelines concerning indoor air pollutants and their health implications. The paper reviews diverse pollutants, encompassing their specified limits, enforcement levels, applicable demographics, and operational protocols. Emphasizing the necessity of real-time, spatial, and temporal monitoring for comprehensive air quality management, the research underscores the importance of technology in achieving this goal. Additionally, the paper delves into the specifications of existing air quality sensor technologies, covering aspects like detection range, measurement tolerance, data resolution, response time, current consumption, and market pricing. While acknowledging the transformative potential of air quality sensors, the study highlights current limitations, emphasizing the need for further advancements before widespread regulatory implementation, particularly regarding robustness, repeatability, and standardized testing protocols. Gaseous sensors, compared to fine particulate matter sensors, exhibit additional uncertainties and data variations.

REFERENCE

Abbasi, T., Sanjeevi, R., Anuradha, J., & Abbasi, S. A. (2013). Impact of Al 3+ on sludge granulation in UASB reactor. *Indian Journal of Biotechnology*, *12*(2), 254–259.

Abbasi, T., Sanjeevi, R., Makhija, M., & Abbasi, S. A. (2012). Role of Vitamins B-3 and C in the Fashioning of Granules in UASB Reactor Sludge. *Applied Biochemistry and Biotechnology*, *167*(2), 348–357. doi:10.1007/s12010-012-9691-y PMID:22549583

Agarwal, T. (2017, March 15). *Different Types of Optical Sensors and Applications. ElProCus - Electronic Projects for Engineering Students*. Elprocuss. https://www.elprocus.com/optical-sensors-types-basics-and-applications/

Aly, A. A., Al-Omran, A. M., Sallam, A. S., Al-Wabel, M. I., & Al-Shayaa, M. S. (2016). Vegetation cover change detection and assessment in arid environment using multi-temporal remote sensing images and ecosystem management approach. *Solid Earth*, *7*(2), 713–725. doi:10.5194/se-7-713-2016

Ankitkumar, B Rathod, Anuradha, J., Prashantkumar B Sathvara, Tripathi, S., & Sanjeevi, R. (2023). Vegetational Change Detection Using Machine Learning in GIS Technique: A Case Study from Jamnagar (Gujarat). *Journal of Data Acquisition and Processing*, *38*(1), 1046–1061. doi:10.5281/zenodo.7700655

Bearg, D. W. (2019). *Indoor air quality and HVAC systems*. Routledge. https://www.taylorfrancis.com/books/mono/10.1201/9780203751152/indoor-air-quality-hvac-systems-david-bearg

Cheng, C.-C., & Lee, D. (2016). Enabling Smart Air Conditioning by Sensor Development: A Review. *Sensors (Basel)*, *16*(12), 2028. doi:10.3390/s16122028 PMID:27916906

Chicea, D., Leca, C., Olaru, S., & Chicea, L. M. (2021). An Advanced Sensor for Particles in Gases Using Dynamic Light Scattering in Air as Solvent. *Sensors (Basel)*, *21*(15), 15. Advance online publication. doi:10.3390/s21155115 PMID:34372352

Chow, J. C. (1995). Measurement methods to determine compliance with ambient air quality standards for suspended particles. *Journal of the Air & Waste Management Association*, *45*(5), 320–382. doi:10.1080/10473289.1995.10467369 PMID:7773805

Dhall, S., Mehta, B. R., Tyagi, A. K., & Sood, K. (2021). A review on environmental gas sensors: Materials and technologies. *Sensors International*, *2*, 100116. doi:10.1016/j.sintl.2021.100116

Fumian, F., Di Giovanni, D., Martellucci, L., Rossi, R., & Gaudio, P. (2020). Application of Miniaturized Sensors to Unmanned Aerial Systems, A New Pathway for the Survey of Polluted Areas: Preliminary Results. *Atmosphere (Basel)*, *11*(5), 471. doi:10.3390/atmos11050471

Javaid, M., Haleem, A., Rab, S., Pratap Singh, R., & Suman, R. (2021). Sensors for daily life: A review. *Sensors International*, *2*, 100121. doi:10.1016/j.sintl.2021.100121

Javaid, M., Haleem, A., Singh, R. P., Rab, S., & Suman, R. (2021). Significance of sensors for industry 4.0: Roles, capabilities, and applications. *Sensors International*, *2*, 100110. doi:10.1016/j.sintl.2021.100110

Mamun, A. A., & Yuce, M. R. (n.d.). Sensors and Systems for Wearable Environmental Monitoring towards IOT-enabled Applications. *RE:view*.

Murthy, H. A., Wagassa, A. N., Ravikumar, C., & Nagaswarupa, H. (2022). Functionalized metal and metal oxide nanomaterial-based electrochemical sensors. In *Functionalized Nanomaterial-Based Electrochemical Sensors* (pp. 369–392). Elsevier. doi:10.1016/B978-0-12-823788-5.00001-6

Nikolic, M. V., Milovanovic, V., Vasiljevic, Z. Z., & Stamenkovic, Z. (2020). Semiconductor gas sensors: Materials, technology, design, and application. *Sensors (Basel)*, *20*(22), 6694. doi:10.3390/s20226694 PMID:33238459

Parry, J., & Hubbard, S. (2023). Review of Sensor Technology to Support Automated Air-to-Air Refueling of a Probe Configured Uncrewed Aircraft. *Sensors (Basel)*, *23*(2), 995. doi:10.3390/s23020995 PMID:36679790

Prashantkumar, B. Sathvara, J. Anuradha, Sandeep Tripathi, & R. Sanjeevi. (2023). Impact of climate change and its importance on human performance. In Insights on Impact of Climate Change and Adaptation of Biodiversity (1st ed., pp. 1–9). KD Publication.

Ramakrishnan, S., & Jayaraman, A. (2019). Global Warming and Pesticides in Water Bodies. In *Handbook of Research on the Adverse Effects of Pesticide Pollution in Aquatic Ecosystems* (pp. 421–436). IGI Global. https://www.igi-global.com/chapter/global-warming-and-pesticides-in-water-bodies/213519

Ródenas García, M., Spinazzé, A., Branco, P. T. B. S., Borghi, F., Villena, G., Cattaneo, A., Di Gilio, A., Mihucz, V. G., Gómez Álvarez, E., Lopes, S. I., Bergmans, B., Orłowski, C., Karatzas, K., Marques, G., Saffell, J., & Sousa, S. I. V. (2022). Review of low-cost sensors for indoor air quality: Features and applications. *Applied Spectroscopy Reviews*, *57*(9–10), 747–779. doi:10.1080/05704928.2022.2085734

Saini, J., Dutta, M., & Marques, G. (2021). Sensors for indoor air quality monitoring and assessment through Internet of Things: A systematic review. *Environmental Monitoring and Assessment*, *193*(2), 66. doi:10.1007/s10661-020-08781-6 PMID:33452599

Sanjeevi, R., & Ankitkumar, B., Rathod, Prashantkumar B. Sathvara, Aviral Tripathi, J. Anuradha, & Sandeep Tripathi. (2022). Vegetational Cartography Analysis Utilizing Multi-Temporal Ndvi Data Series: A Case Study from Rajkot District (Gujarat), India. *Tianjin Daxue Xuebao (Ziran Kexue Yu Gongcheng Jishu Ban)/ Journal of Tianjin University Science and Technology, 55*(4), 490–497. https://doi.org/ doi:10.17605/OSF.IO/UGJYM

Sanjeevi, R., Haruna, M., Tripathi, S., Singh, B., & Jayaraman, A. (2017). Impacts of Global Carbon Foot Print on the Marine Environment. *International Journal of Engineering Research & Technology (Ahmedabad)*, *5*, 51–54.

Sathvara, P. B., Anuradha, J., Sanjeevi, R., Tripathi, S., & Rathod, A. B. (2023). Spatial Analysis of Carbon Sequestration Mapping Using Remote Sensing and Satellite Image Processing. In Multimodal Biometric and Machine Learning Technologies (pp. 71–83). doi:10.1002/9781119785491.ch4

Schweitzer, L., & Zhou, J. (2010). Neighborhood air quality, respiratory health, and vulnerable populations in compact and sprawled regions. *Journal of the American Planning Association*, *76*(3), 363–371. doi:10.1080/01944363.2010.486623

Shukla, K., & Aggarwal, S. G. (2022). A Technical Overview on Beta-Attenuation Method for the Monitoring of Particulate Matter in Ambient Air. *Aerosol and Air Quality Research*, *22*(12), 220195. doi:10.4209/aaqr.220195

Singh, D., Dahiya, M., Kumar, R., & Nanda, C. (2021). Sensors and systems for air quality assessment monitoring and management: A review. *Journal of Environmental Management*, *289*, 112510. doi:10.1016/j.jenvman.2021.112510 PMID:33827002

Spurr, K., Pendergast, N., & MacDonald, S. (2014). assessing the use of the air Quality health index by vulnerable populations in a 'low-risk'region: A pilot study. Canadian Journal of Respiratory Therapy: CJRT=. *Canadian Journal of Respiratory Therapy : CJRT*, *50*(2), 45. PMID:26078611

Villa, T., Gonzalez, F., Miljievic, B., Ristovski, Z., & Morawska, L. (2016). An Overview of Small Unmanned Aerial Vehicles for Air Quality Measurements: Present Applications and Future Prospectives. *Sensors (Basel)*, *16*(7), 1072. doi:10.3390/s16071072 PMID:27420065

Wang, Y., Li, J., Jing, H., Zhang, Q., Jiang, J., & Biswas, P. (2015). Laboratory evaluation and calibration of three low-cost particle sensors for particulate matter measurement. *Aerosol Science and Technology*, *49*(11), 1063–1077. doi:10.1080/02786826.2015.1100710

Zhang, H., & Srinivasan, R. (2020). A Systematic Review of Air Quality Sensors, Guidelines, and Measurement Studies for Indoor Air Quality Management. *Sustainability (Basel)*, *12*(21), 9045. doi:10.3390/su12219045

Chapter 3
Pioneering Sensing Solutions for In-Situ Monitoring of Volatile Organic Compounds in Soil and Groundwater Using Chemiresistor Sensor Technology

Periasamy Palanisamy
https://orcid.org/0000-0003-1069-5822
Nehru Institute of Engineering and Technology, Anna University, India

Maheswaran M.
Nehru Institute of Engineering and Technology, Anna University, India

ABSTRACT

The monitoring and classification of volatile organic compounds (VOCs) in soil and groundwater is a critical task in guaranteeing environmental protection and remediation of contaminated sites. Traditional methods of sample collection and off-site analysis could be expensive, laborious, and may not accurately represent in-situ conditions. To address these encounters, state-of-the-art sensing solutions using chemiresistor sensor technology have been developed for instantaneous, uninterrupted, and long-term monitoring of VOCs in the subsurface. The chemiresistor comprises of a chemically-sensitive polymer dissolved in a solvent and mixed with conductive carbon particles. The resultant ink is deposited onto thin-film platinum traces on a solid substrate. When VOCs come into interaction with the polymers, they are absorbed, initiating the polymers to swell and change the resistance of the electrode. This variation in resistance can be measured and recorded, providing information about the concentration of the VOCs.

DOI: 10.4018/979-8-3693-1930-7.ch003

INTRODUCTION

In the recent decades, organic thin film devices have reaped a lot of interest for their broad range of applications, which include printing inks, colorants, photoconductors, solar cells, electrochromic sensors, and gas sensors. The microfabrication of gas sensors that are established with the nanostructured materials and have a very low power consumption are technologies which are very necessary for today's sophistication(Wu et al., 2020). Such sensors are employed in a variety of industries, including the excavating industry for the purpose of detecting mine methane, the automotive industry for the tenacity of sensing combustion gases from vehicles, medical applications for bioelectronic noses which mimic human olfaction, and air quality monitoring for the purpose of estimating greenhouse gas emissions(Chen et al., 2021). Other applications of chemical sensors include the detection of gases, the observing of water and soil pollutants, the monitoring of temperature, speed, magnetic fields, and the regulation of emissions. such kind of electronic circuit element is known as a chemical sensor. This category of sensor endures a change in both its physical and chemical structure, when it captivates a chemical stimulant in the surface layer. Owing of this, the electrical characteristics of the sensors are changed, and such changes are translated into values that can be measured(Chen et al., 2021),(Hooshmand et al., 2023),(Hsieh & Yao, 2018).

Carbon dioxide (CO_2), nitrogen oxides (NO_x), sulphur oxides (SO_x), hydrocarbons (HC), carbon monoxide (CO), lead (Pb), and volatile organic compounds (VOCs) are some of the crucial contaminants that have the potential to have a detrimental effect on both the environment and human health, ensuing in respiratory illnesses(David & Niculescu, 2021). Combustion activities in industrial settings and at high temperatures, as well as the use of chemical poisons in agriculture without discernment associated to the ongoing innovation in technology, all have an impact on the features of the atmosphere, which ultimately results in harmful levels of air pollution. Air pollution in urban areas that is caused by nitrogen dioxide, sometimes known as NO_2, is a global problem. Long-term exposure to nitrogen dioxide (NO_2) has been shown to have negative effects on human health and is responsible for a wide range of disorders that affect the cardiovascular and respiratory systems. Personal sensitivity and vulnerability are determined by a combination of biological variables and genetic inclination as individuals. The factors of age, gender, education level, socioeconomic class, area of residence, and employment all have a role in determining an individual's susceptibility to the dangers of pollution. There is a significant effect that air pollution has on the health and well-being of individuals, especially on the most vulnerable groups, such as children, women, and those living in underdeveloped nations. The combustion processes used in high-temperature manufacturing environments and the unrestricted use of chemical toxins in agriculture, which are linked to ongoing technical advancements, have an impact on atmospheric characteristics and cause air pollution (David & Niculescu, 2021),(Hooshmand et al., 2023).

NO_2 pollution in urban areas is a global issue of great concern. Prolonged exposure to NO_2 leads to detrimental effects on human health and is responsible for many respiratory and cardiovascular disorders. Individual sensitivity and susceptibility are determined by biological variables and genetic predisposition. Socioeconomic level, place of residence, and employment are factors that contribute to individual susceptibility to pollution dangers. The presence of air pollution has detrimental effects on the health and overall state of well-being of individuals, especially those who are most susceptible such as children, women, and those residing in underdeveloped nations. Air pollution incurs significant economic costs that hinder sustainable development (David & Niculescu, 2021).

Engaging in economic development that tolerates air pollution and disregards the consequences on public health and the environment is morally wrong and cannot be maintained in the long term. Fur-

Figure 1. Gas sensor (Metal oxides to 2D materials)
(reproduced from (Kumar et al., 2020) under CC BY 4.0)

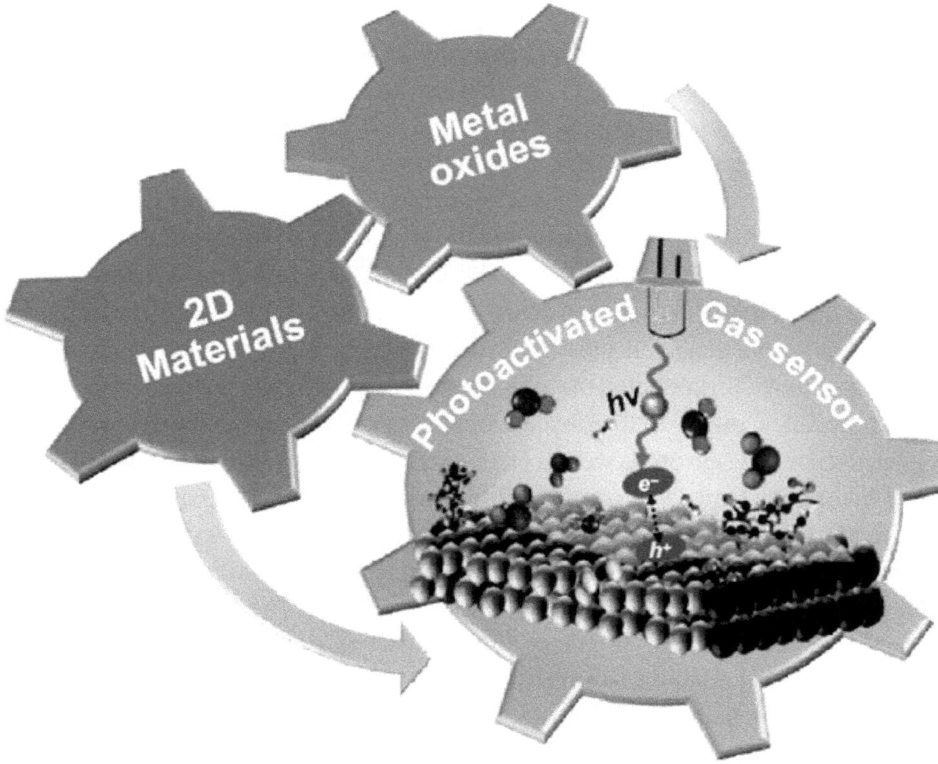

thermore, pollution-related disorders contribute approximately 7% of the total national health budget in terms of healthcare expenses(Caballero et al., 2012). The emission of hazardous chemicals and vapors poses a significant and concerning risk to the ecosystems inside the biosphere(Remoundou & Koundouri, 2009). Pristine air and water are crucial components for sustaining life on Earth, and the presence of air pollution poses a significant danger to the long-term viability of Earth's environment. Early identification and monitoring are necessary to address the concerning increase in pollutant concentrations in urban environments. The use of passive diffusion tube sampling methods has been implemented to examine the geographical and temporal fluctuations in NO_2 concentrations. The validity of measurements is crucially dependent on the automated techniques used for pipe extraction, preparation, and co-location(X. Zhang et al., 2022),(Caballero et al., 2012).

Urban areas with high levels of NO_2 pollution serve as concentrated locations for sample sites, offering specific geographic information and further data on the distribution, dispersion, and human exposure to pollutants in urban contexts. Utilizing a solitary NOx diffusion tube method simplifies the process of pinpointing areas with high concentrations of NO_2 by providing accurate geographical information on the locations where air quality is degraded(Caballero et al., 2012). This is achieved via sampling over a span of 12 months, including all variations throughout the seasons. Several literary techniques that compare metropolitan areas provide a preliminary tool for screening air quality and air pollution. This tool is designed to assist in conducting research on air quality evaluation. Utilizing these empirical models to calculate the conversion rate from NO_x to NO_2 is crucial for enhancing air quality and mitigating the possible harm to human health and ecosystems via air pollution mitigation strategies. Precise monitoring systems and gas sensors play a vital role in detecting gas in industrial environments as well as in buildings that demand high air quality (Norris & Larson, 1999), (Niepsch et al., 2022), (Vîrghileanu et al., 2020).

Typically, these devices are outfitted with a chemically selective layer that interacts with its environment in the capacity of a sensor layer (Fig.1). The contact causes a modification in the characteristics of the sensing layer, such as changes in conductivities, which occur due to the adsorption of the analyte molecule on the surface of the layer. An organic semiconductor gas sensor, using sophisticated nano- and microfabrication technology, offers many benefits over existing semiconductor gas sensors in terms of manufacture, cost, and operation at room temperature(Bai et al., 2015). The contribution of nanotechnologies is in the development of novel materials with a significantly increased surface area, which enhances the absorption of gaseous molecules and improves sensing performance. Graphene, a kind of carbon-based nanomaterial, has been extensively used as a sensor for detecting NO_2 at normal room temperature. Multiple methodologies have substantiated the effectiveness of electrically conductive polymeric materials and graphene composites, owing to their exceptional conductivity and extensive surface area. These materials have been used in gas sensors as well as electrochemical biosensors(Krishnakumar et al., 2009).

A recent set of laboratory experiments involved the use of polypyrrole/graphene oxide reduced and polypyrrole/graphene oxide covalently bound nanocomposites, combined with 4-carboxybenzene diazonium aryl salt nanocomposites. These nanocomposites were tested to assess their chemical properties in detecting NO_2 gases at room temperature(Guettiche et al., 2021). The nanocomposite ppy-graphene has been synthesized by incorporating graphene into the polymeric solution and during the polymerization process. Graphene and other carbon-based nanofillers have been used to enhance gas detection capabilities and increase both sensitivity and specificity. However, the clumping together of graphene particles negatively impacts the polymer matrix. Furthermore, other chemicals have undergone testing for the same purpose. Combinations of carbon nanotubes and ZnO (CNTs&/ZnO) have shown enhanced electrical performance when exposed to CO, with a 60% increase in reaction time compared to ZnO alone(Sara et al., 2022). Zinc oxide (ZnO) nanostructures has multifunctional features, including a large surface area, high crystallinity, and an ordered molecular structure, making them well-suited for various sensing and selective applications. ZnO nanocomposites have recently been used for the production of NO_2 gas sensors because of its temperature resistance, stability, and flexibility. These characteristics prompted several endeavors to enhance the semiconductor's performance for gas sensing applications via the development of novel morphologies and crystallographic forms of ZnO (Krishnan et al., 2019).

The volatile organic compound (VOC) sensor is a vital tool for indoor air quality (IAQ) monitoring, as it senses the contaminants that exhibit a threat to human health (Krishnan et al., 2019). VOCs are chemical ingredients with high vapor pressure at room temperature and are released into the environment, generally found in consumer products, gasolines, and industrial processes. Examples of VOCs comprises benzene, toluene, ethylene, formaldehyde, and xylene. VOCs can have adverse impacts on human wellness and the environment, including respiratory irritation, headaches, eye and nose irritation, and contribute to the creation of ground-level ozone, a major component of smog. VOC sensors work by means of a sensitive chemical (David & Niculescu, 2021) element or electronic components to identify the presence of VOCs (Fig.2).

When VOCs are contemporary in the air, they react with the chemicals present in the sensor and produce an electrical charge, which is then calibrated by the sensor and the concentration of VOCs can be determined. VOC sensors are used in several applications such as automotive emissions monitoring, indoor air quality monitoring, industrial process monitoring, and environmental monitoring. VOC sensors classically use one of two different types of method to measure the presence of VOCs: photoionization detectors (PIDs) or metal oxide semiconductor (MOS) sensors. Photoionization detectors detect the concentration of VOCs by ionizing the molecules of the VOCs with a UV light, which are then passed through a chamber anywhere they are sensed and measured by an electrostatic field. The measured ions are rehabilitated into a measureable electrical current and passed to a read-out device for further analysis(Pathak et al., 2023). There are 3 main types of common VOC gas sensors centred on their working principles: electrochemical VOC sensors, optical VOC sensors, and mass VOC sensors. Electrochemical VOC sensors encompass the adsorption or reaction (physical or chemical) of VOC gases on the surface of a gas-sensitive material, which results in fluctuations in its electrical properties. Amid the most broadly used types of VOC sensor based on semiconductor metal oxides is the conductive type, which plays a significant role in the current gas sensing field(Epping & Koch, 2023).

They can be divided into common two-electrode conductive detection systems in addition to three-electrode field-effect transistor detection systems. Based on the VOC gas-sensitive materials, they can be characterized as semiconductor metal oxides, conductive polymers, nanomaterials, and porous materials. The volatile organic compound sensor is crucial for confirming a healthy indoor environment and preventing harmful contaminants from entering the air. Semiconductor metal oxide conductivity sensors are amongst the primary and most mature gas sensors, developed in 1936. These sensors sense the gases by exploiting the property that the resistance or work function of a semiconductor changes when it comes into encounter with a gas. Though, they have boundaries such as working at high temperatures, poor gas selectivity, and prone to poisoning(P. Zhang et al., 2021),(Verma et al., 2024).

Zero-dimensional nanomaterial conductivity sensors are a hopeful material due to their exclusive physical and chemical properties. Gold nanoclusters demonstrates quantum dot behavior and surface interactions with ligands, providing a selective adsorption interface for VOCs. The electrical response characteristics of monolayer gold nanoclusters to VOCs are linked to variations in electronic conductivity between the gold cores triggered by adsorption of VOCs and activation energy. Organic thiol kinds and structures are premeditated and selected based on the interaction forces between unlike functionalized gold nanoclusters and VOCs. A VOC sensing array is fabricated based on the cross-selective response characteristics of diverse gold nanoclusters to VOCs. Conductivity gas sensors based on nanoporous materials, precisely nano-porous silicon photonic crystals, are also employed due to their high surface area and gas adsorption capabilities. Conductive polymer materials are generally used in gas sensors due to their specular electrical and optical properties, mechanical flexibility, and electrochemical redox characteristics. conjugated polymer materials such as phthalocyanine polymers, polypyrrole, polyaniline, and porphyrins and metalloporphyrin complexes are employed as gas sensor materials. Optical VOC sensors have the benefits of strong anti-electromagnetic interference, fast response, and easy implementation for online monitoring of organic gases. There are several types of optical sensors based on their working principles, together with reflective interference method, ultraviolet-visible absorption photometry, colorimetric method, fluorescence method, surface plasmon resonance method, and fiber optic sensing technology. Spectroscopic absorption gas sensors identify VOCs based on the intensity or displacement change of the absorption spectrum of gas-sensitive materials after adsorbing VOCs.

Color-based visual VOC sensors signify the characteristic information of smells in the form of images, also recognized as visual olfaction. The colorimetric output signal mode is the utmost straightforward sensing platform for evolving naked-eye detection technology, minimizing the need for signal conversion equipment modules. Presently reported sensing materials for VOC picturing include polydiacetylene paper chips, methylene yellow 6 nanofibers, Fabry-Perot interference micro-porous polymers, and supramolecu-

Figure 2. Schematic diagram of sensors response
(reproduced from (Isaac et al., 2022) under CC BY 4.0)

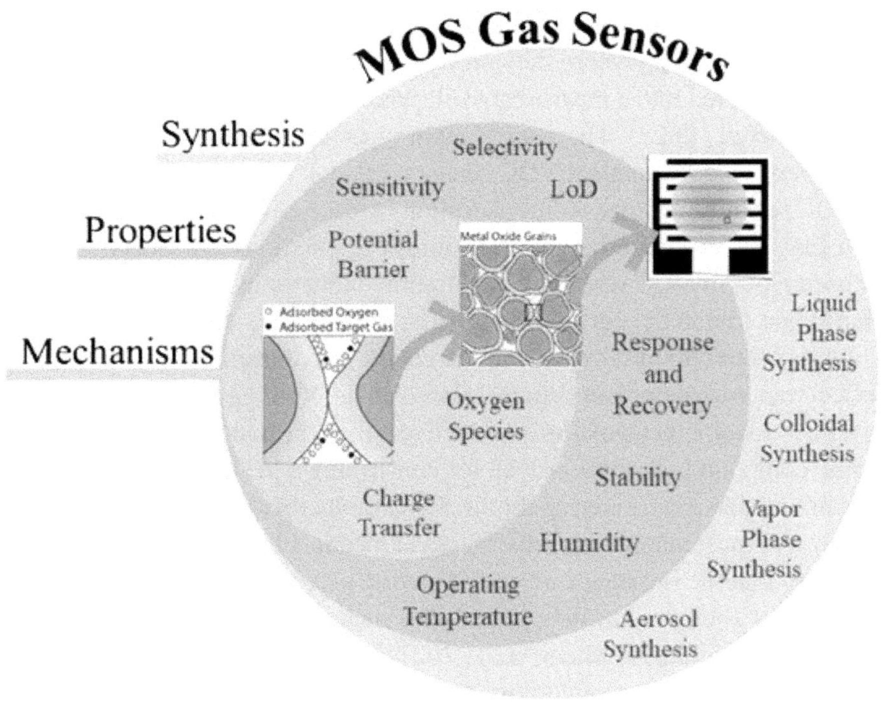

lar host-guest complexes. VOC sensors grounded on the optical interference principle involve photonic crystals (PCs) with periodic variations in refractive index in space. PCs have the function of filtering, selectively allowing certain bands of light to pass through and hindering other wavelengths of light.

Fluorescence VOC sensors are substantial advancements in analytical chemistry, contribute high sensitivity, good selectivity, and strong resistance to electromagnetic interference (Fig.3). Though, they often face tasks such as difficult cataloguing and poor repeatability due to peripheral factors such as humidity, polarity, and pH. Surface plasmon resonance (SPR) is a physical optical phenomenon of the momentary field that arises when light undergoes total internal reflection at the boundary between glass and a metal film, producing a momentary wave that can induce surface plasmon waves on the metal surface by generating free electrons. Underneath certain circumstances of incident angle or wavelength, the frequency and wave number of the surface plasmon wave and the temporary wave resonate, and the incident light is absorbed, subsequent in a resonance peak in the reflection spectrum. SPR technology is a novel gas detection method with the compensations of simple structure, high sensitivity, and wide detection range.

A recent published studies reports the development of a highly sensitive VOC sensor based on olfactory receptors reconstituted into a lipid bilayer and used in a precisely designed gas flow system for rapid parts per billion (ppb)–level detection. The study validates the potential for using biological odorant sensing in breath analysis systems and environmental monitoring. Olfactory receptors in living organisms can identify various VOCs with a level of detection corresponding to a single molecule, making them far greater in selectivity and sensitivity compared with current VOC sensors using artificial materials. VOC sensors work by sensing the changes in electrical conductivity when VOC molecules bind to the sensor's

Figure 3. Metal oxide nanomaterials for chemiresistive gas sensors
(reproduced from (L. Y. Zhu et al., 2023) under CC By 4.0)

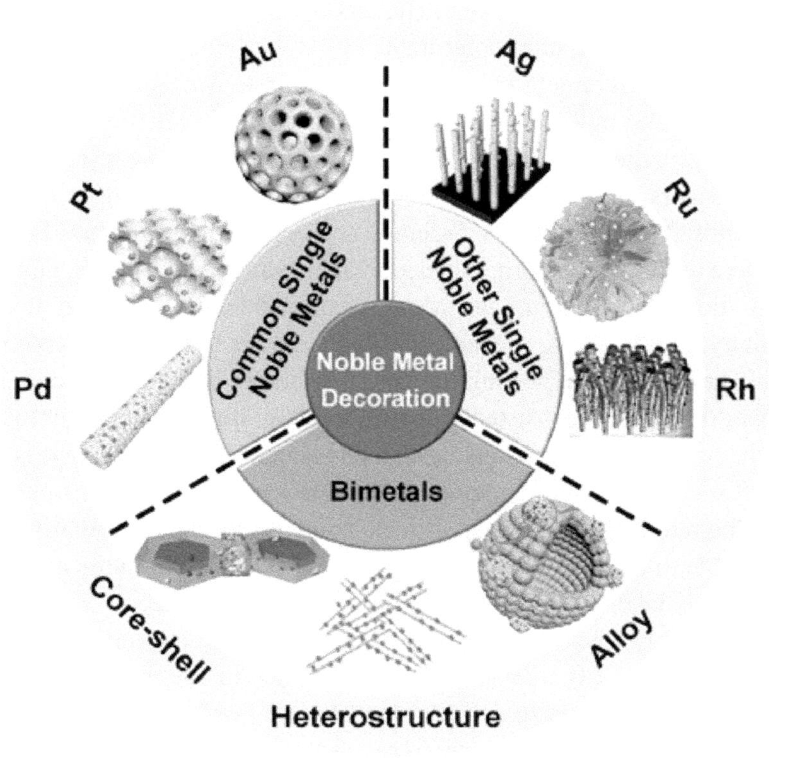

surface. The sensors are fabricated of a thin film of a conducting polymer, coated with a layer of a material that absorbs specific VOCs. When VOCs predicament to the absorber layer, they alter the electrical conductivity of the conducting polymer film, which can be measured and used to notice the presence and concentration of the VOC. Current advances in VOC sensor technology include the use of machine learning algorithms to advance sensor accurateness and the development of flexible, wearable sensors.

SENSOR MEASUREMENT SETUP

In order to carry out the gas-sensing measurements, the chemiresistor gas sensors were positioned within the gas chamber. Materials that are resistant to chemical reactions, such as polytetrafluoroethylene (PTFE) and stainless steel, have been used in the construction of the measuring gas chamber, which has intake and exit pipes with a diameter of 4 millimeters. Within the measuring chamber, quartz windows were used, and the constructions were lighted using ultraviolet light emitting diodes (UV LEDs) with a wavelength of 390 nanometers. A base that was fitted with a Pt100 sensor was put on top of four chemically resistive constructions of gas sensors in order to monitor the changes in temperature (Figure 4).

The use of a test gas bench was necessary in order to carry out the optical characterization of the layers that were susceptible to gas combinations. The test bench is comprised of a digital mass flow controller interface, a gas test chamber equipped with the required sensors, pressurized gas cylinders, a digital interface, a personal computer, and a Keithley measuring equipment. In order to establish stability of the sensor active surface and a stable baseline resistance, the gas chamber was illuminated by ultraviolet light for one or two hours while synthetic airflow was present. This was done in preparation for gas-sensing measurements using ultraviolet light.

For the duration of the measurement procedure, the flow rate over the sensing chamber was maintained at a constant rate of 200 milliliters per hour for the synthetic air flow meter counter 1 and 200 milliliters per hour for the synthetic air flow meter counter 2. After two hours (about 0.5 9 104 seconds), the chemiresistor gas sensors were subjected to low concentrations of nitrogen dioxide (500 parts per billion) while being exposed to continual UV radiation and dry synthetic air flow. Furthermore, the flow cycle was repeated three times, which corresponded to approximately ten hours. During the times when the flow of NOx was allowed to continue, the synthetic air counter 2 was closed. It was always open at the synthetic air counter number 1. To prevent interference from humidity and non-uniform pressure, the target gases were first mixed in a gas mixing chamber before being introduced into the gas sensor chamber. This chamber maintains a consistent pressure and humidity level in order to prevent any potential interference. In the presence of ultraviolet light, a physicochemical interaction is produced between oxides of nitrogen (NOx) and sensitive compounds that are located on the substrate of chemiresistors. As a consequence of the molecule of nitrogen dioxide being adsorbent on the surface of the ZnO/graft comb copolymer, it is able to ensnare the electrical charge that is present in the nanomaterial's conduction band, which ultimately leads to an increase in the sensor's resistance.

During the process of data gathering, a sensor was equipped with the capability to detect and record the operating temperature of the gas analyte that was contained inside the gas mixing chamber. It was possible to keep the temperature conditions consistent despite the fact that the temperature fluctuated between 23.5 and 24.6 degrees Celsius, with an average value of 23.7 degrees Celsius. The temperature

Figure 4. Sensor measurement setup
(reproduced from (Roy et al., 2010) under CC BY-NC-ND)

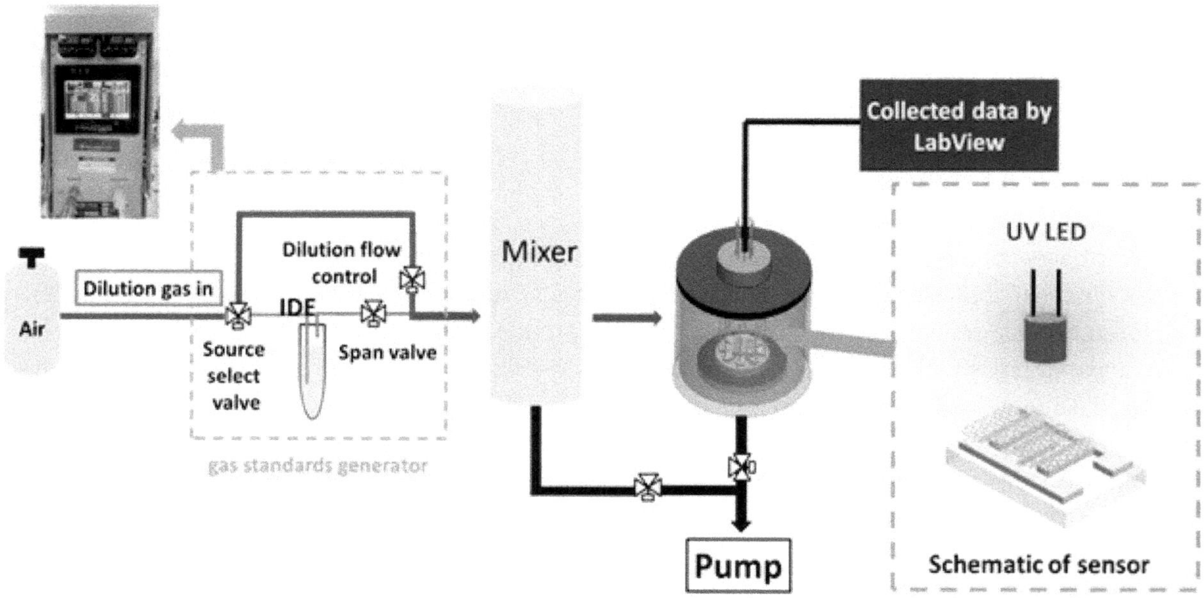

of the chamber decreased as the flow of nitrogen oxide gas was introduced, despite the fact that the synthetic air gas injection caused a modest rise in temperature during the first stage. It is quite likely that the variations in gas pressure were the source of the problem. Throughout the course of the measurements, the sensing structures were continuously irradiated by the ultraviolet light. The humidity was measured inside the chamber that was used for mixing the gas. The value of humidity that was measured fluctuated approximately 3.5% during the whole process of dosing the gaseous analytes and the carrier gas, and it remained steady throughout the time that the measurements were being taken.

ADVANTAGES OF IN-SITU MONITORING OF VOLATILE ORGANIC COMPOUND GASES

In-situ monitoring of VOCs using chemiresistor sensors offers several advantages over traditional methods. Firstly, it eliminates the need for manual sample collection and expensive off-site analysis. Instead, real-time monitoring provides continuous data, allowing for immediate detection of changes in VOC concentrations. This enables prompt response and intervention when necessary. Additionally, in-situ monitoring reduces the risk of evaporation and loss of VOCs during sample handling and storage, providing more accurate measurements of in-situ concentrations. Another advantage of in-situ monitoring is the ability to gather data in challenging environments, such as flammable or explosive sites (Wiśniewska & Szyłak-Szydłowski, 2022). Chemiresistor sensors are immune to electrical noises and can operate in these hazardous conditions safely. Moreover, the miniaturization and geometrical flexibility of the sensors allow for measurements in small sample volumes and remote sensing in inaccessible or harsh environments(Sara et al., 2022).

It has been shown (Davis et al., 2005) that the incorporation of a micro fabricated preconcentrator into a chemiresistor chemical sensor results in an increase in sensitivity and a reduction in the detection limits for m-xylene. In the absence of the preconcentrator, our PEVA chemiresistor demonstrated the capability to detect concentrations up to 0.12% of the saturated vapor pressure of m-xylene (at a temperature of 3 degrees Celsius). The detection limit was reduced to roughly five one-millionths (0.0005%) of the saturation vapor pressure of m-xylene (at 3σ) when a preconcentrator was added to the mixture. This improvement of more than two orders of magnitude may make it possible for the chemiresistor to be used in applications that demand lower detection limits. One example of such an application is the detection of controlled pollutants in geologic medium. It is important to note that there is a contrast between the detection limit, which is constrained by the noise values stated earlier, and the accuracy or uncertainty in the concentrations that are measured. The preconcentrator also has the benefit of knowing the exact timing of the heat pulse, which enables the baseline drift of the chemiresistor to be adjusted without the need for pumps, valves, and purge gases to be used in the field. Pumps and valves are used in the collection of data that is shown in Figure 4. If the low concentrations were to fluctuate slowly and in an unpredictable manner, it would be more challenging to differentiate between them. The preconcentrator timing enables a more definite detection of low concentrations, which is a significant advantage.

The calibration and prediction of chemiresistor arrays across various environmental conditions presents significant challenges(Rivera et al., 2003). Both intended sequences and non-designed sequences have statistically equal calibration results, with R2 values of 0:94 and 0.96, respectively, and CVSEPs of 750 and 650 ppm, respectively. In short-term prediction, the model derived from designed sequences has an R2 of ¼ 0:93 and a SEP of ¼ 900 ppm, while the model derived from non-designed sequences also predicts designed data with an R2 of ¼ 0:87 and a SEP of ¼ 1100 ppm. In long-term prediction, the model formed from the intended sequence produces an R2 of ¼ 0:91 with a SEP of ¼ 920 ppm, while the model built from non-designed data produces an R2 of ¼ 0:81 with a SEP of ¼ 1340 ppm. This difference in prediction statistics suggests that models constructed using the planned data set provide more accurate predictions for the given circumstances. A strategically constructed experiment involving random sampling sequences with repeat samples was used to calibrate microsensor data to TCE at various temperatures and humidity. This experiment was more reliable in forecasting TCE concentrations for both short and long periods following calibration. The study also demonstrated the effects of temperature and humidity on the chemiresistor response and PLS calibration. The calibration model's capacity to make accurate predictions cannot be evaluated solely based on leave-one-out cross-validation outcomes during calibration. To determine the model's capability, a genuine prediction is necessary.

A resistive sensor capable of detecting the presence of volatile organic compounds (VOCs) in water at a concentration of μg/L was prepared by (Rivadeneyra et al., 2016). The device was manufactured using a flexible substrate and printing procedures, specifically using silver electrodes and a resistive composite made of graphite and polystyrene. The study investigated various manufacturing parameters and their reaction to various volatile organic compounds. Higher resistance values were achieved with composites with reduced graphite concentration, but they also had worse repeatability. Composites with the maximum amount of graphite had higher repeatability. Lower mesh density resulted in lesser pattern definition and lower resistance values but also provided greater repeatability in a wider range of graphite content. Serpentine electrodes provided around 12% higher resistance than interdigitated electrodes. The dynamic reaction of these resistors showed a more rapid and intense response to toluene than pure water. The composite with a graphite content of fifty percent had the fastest reaction when printed with a 120 T/cm screen. The maximum sensitivity was found to be benzene, followed by o-xylene, p-xylene, m-xylene, and toluene.

A microfluidic water monitoring system with an integrated microhotplate gas sensor was fabricated to evaluate the performance of the water monitoring sensor platform [REMOVED HYPERLINK FIELD] (L. Zhu et al., 2007). The liquid/gas channel overlap was chosen based on observation of data, suggesting that larger overlaps do not provide significant improvements in gasphase analyte concentration. A syringe pump was used to deliver precise flow of water/solvent mixture in the liquid channel, and a mass flow controller was used to deliver dry air flow in the gas channel. A laptop computer provided temperature control and monitoring of conductometric response for the integrated gas sensor elements. The flow rate of dry air was set to the minimum value supported by the mass flow controller of 500 L/min, and the flow rate of liquid was set to a relatively high value of 20 L/min. The evaluation of toluene and 1,2-dichloroethane were of particular interest since these solvents are chemical contaminants listed by the US EPA. The Guterman–Boger set of algorithms was employed to train the large-scale Artificial Neural Networks (ANNs) for the methanol, toluene, and 1,2-dichloroethane TPS data measured using the integrated microfluidic platform. The resulting ANN model was applied to both the training and validation data sets.

Based on the measured data, approximate detection limits for the fabricated system are 1 ppm for methanol, 10 ppm for toluene, and 100 ppm for 1,2-dichloroethane. The maximum contamination levels (MCLs) in drinking water have been set by the EPA-enforced Safe Drinking Water Act at 0.2 ppm for toluene, and 0.9 ppb for 1,2-dichloroethane. There are several reasons for the relatively low sensitivity of the water monitoring system to both toluene and 1,2-dichloroethane. First, the overall gas phase mass transfer coefficients for toluene and 1,2-dichloroethane are substantially smaller than for methanol. Second, toluene and 1,2-dichloroethane are readily absorbed by the polycarbonate substrate, lowering their concentrations in both the liquid and vapor phases.

A water monitoring platform combines a silicon-based microhotplate sensor chip for conductometric measurement of organic solvents in the gas phase with a microfluidic two-phase flow network for solvent extraction. The system's sensitivity to a suite of volatile organic compounds (VOCs) diluted in water was evaluated, with detection limits of 1 ppm for methanol, 10 ppm for toluene, and 100 ppm for 1,2-dichloroethane. Although the sensitivity is lower than for drinking water supplies, the system could be useful for monitoring point source contaminant emissions. Further study is needed to optimize the selectivity of the system for specific analytes, especially in the presence of water vapor. The siliconin-plastic fabrication process could be used to integrate additional functionality into the sensor platform, enhancing its potential for unattended and remote operation (L. Zhu et al., 2007).

Microbial communities play a crucial role in developing sustainable bioremediation strategies, especially for sites contaminated with mixed chlorinated volatile organic compounds (CVOCs) (Hussain et al., 2023). The study found that sensitive taxa like Proteobacteria and Acidobacteriota are lost, while CVOCs-resistant taxa like Campilobacterota are enriched in contaminated sites. The abundance of crucial enzymes involved in CVOC sequential biodegradation also varies depending on the contamination level. Genera like Sulfurospirillum, Azospira, Trichlorobacter, Acidiphilium, and Magnetospririllum can survive higher levels of CVOC contamination, while pH, ORP, and temperature negatively affect their abundance and distribution. Dechloromonas, Thiobacillus, Pseudarcicella, Hydrogenophaga, and Sulfuritalea show a negative relationship with CVOC contamination, highlighting their sensitivity. These findings offer insights into the ecological responses, groundwater bacterial community, and functionality in response to mixed CVOC contamination.

The environmental impacts and costs of three options for remediating groundwater contaminated by volatile organic compounds (VOCs) at a closed pesticide manufacturing plant site was studied by (Ding et al., 2022). The options include a combination of MNA and BC (MNA + BC), BC, and pump and treat (PT). The environmental impacts were assessed using a Life Cycle Assessment (LCA) using the ReCiPe 2016 method, and the costs were evaluated using a Life Cycle Cost (LCC) method created in SimaPro. The LCA results show that the overall environmental impacts follow a sequence of PT > BC > MNA + BC, with MNA + BC showing evident primary impacts. The environmental hotspots in PT include cement, electricity, steel, and operation energy. In MNA + BC and BC, electricity for feedstock pyrolysis is the environmental hotspot, while the use of BC by-products has positive environmental credit. The LCC results show that PT yields the highest cost, with cement and electricity being the two most expensive items. The study provides scientific support for developing and optimizing green remediation solutions for VOCs contaminated groundwater.

Thermal conductive heating (TCH) is a popular method for removing organic contaminants from subsurface, but it is often considered unsustainable due to high carbon emissions, energy consumption, and high costs(Yang et al., 2023). TCH-ISCO, a combination of TCH and in situ chemical oxidation, has gained attention for its ability to achieve groundwater remediation goals at lower temperatures. However, there is a lack of quantitative assessment on the sustainability of TCH-ISCO based on field data. A quantitative life cycle assessment approach was developed to assess the sustainability of TCH-ISCO for chlorinated aliphatic hydrocarbons (CAHs) contaminated soil and groundwater remediation. The results showed that TCH-ISCO and TCH contributed to 68% and 93% of adverse environmental impacts, respectively. TCH-ISCO had significantly lower carbon emissions and direct costs. The overall sustainability scores of TCH-ISCO and TCH were 89.6 and 61.9, respectively, indicating its superior performance.

Light-Non-Aqueous phase liquids (LNAPLs) are significant soil contamination sources, and groundwater fluctuations can significantly impact their migration and release (Cavelan et al., 2024). Risk assessment is complex due to the continuous three-phase fluid redistribution caused by water table level variations. To improve monitoring methods, a lysimetric contaminated soil column was developed, combining in-situ monitoring, direct water and gas sampling, and analyses. The experiment assessed the effects of controlled rainfalls and water table fluctuation patterns on LNAPL vertical soil saturation distribution and release. The results showed that 7.5% of the contamination was remobilized towards the dissolved and gaseous phase after 120 days. Groundwater level variations were responsible for free LNAPL soil spreading and trapping, modifying dissolved LNAPL concentrations. However, part of the dissolved contamination was rapidly biodegraded, leaving only the most bio-resistant components in water. This highlights the need for new experimental devices to assess the effect of climate-related parameters on LNAPL fate at contaminated sites.

Ozone-permeable membranes were used as a novel in-situ treatment method for groundwater contamination caused by organic compounds(Bein et al., 2024). The researchers found that ozone depletion was faster in the presence of sub-stoichiometric benzoic acid (BA) than in non-aromatic 1,4-dioxane (DIOX), with lower removal of 5 mg $L{-}1$ BA (52.7%) compared to DIOX (60.6%). The study also found that reactive porous media did not significantly change in-situ DIOX oxidation, although a stronger impact was hypothesized. The researchers determined experimental ozone mass transfer coefficients and compared them to modeled values for different membrane types. A mathematical model was developed to support upscaling efforts. The study concluded that contaminant properties are crucial for the feasibility assessment of in-situ ozone membrane treatment technology.

A system for monitoring groundwater contamination by aromatic VOCs has been developed using a gas-water separation unit and APPI-FAIMS(Joksimoski et al., 2022). The system successfully reduced humidity in the sample flow to $\leq$1.6 ppmv before analyte ionization. Toluene was chosen as a model aromatic VOC to test the system's feasibility and the impact of humidity on the signal produced by APPI-FAIMS. With increased humidity, the toluene signal increased by about 30%, allowing for the formation and detection of water clusters. Similar effects were observed for benzene. However, single contaminants like indane and trimethylbenzenes were not detected even at high humidity. On-site, continuous groundwater monitoring of aromatic VOC contamination was successfully carried out with the gas-water separation APPI-FAIMS at low humidity, simplifying the monitoring of a specific, total aromatic VOC signal in groundwater.

A novel electrolyzed catalytic system (ECS) was designed to produce nanobubble-contained electrolyzed catalytic (NEC) water for the remediation of petroleum-hydrocarbon-contaminated soils and groundwater(Ho et al., 2023). The ECS uses high voltage (220 V) with direct current, titanium electrodes coated with iridium dioxide, and iron-copper hybrid oxide catalysts to enhance the hydroxyl radical production rate. The system uses electron paramagnetic resonance (EPR) and Rhodamine B (RhB) methods for the generated radical species and concentration determination. During operation, high concentrations of nanobubbles are produced due to the cavitation mechanism, which prevents bubble aggregations and extends their lifetime in NEC water. The radicals produced after the bursting of nanobubbles increase the radical concentration and subsequent petroleum hydrocarbon oxidation. Highly oxidized NEC water can be produced with a radical concentration of 9.5×10^{-9} M. In a pilot-scale study, the prototype system was applied to clean up petroleum-hydrocarbon polluted soils at a diesel-oil spill site via an on-site slurry-phase soil washing process. Results showed that up to 74.4% of TPH could be removed from soils after four rounds of NEC water treatment.

Traditional methods have limitations such as safety concerns, rebound of contaminants, and difficulty in reaching all contamination areas. To address these, Controlled-Release Biodegradable Polymer (CRBP) pellets containing $KMnO_4$ were designed and tested (Lamssali et al., 2023). The CRBP pellets were encapsulated in Polyvinyl Acetate (CRBP-PVAc) and Polyethylene Oxide (CRBP-PEO) at different weight percentages, baking temperatures, and time. The highest release percentage and rate were found in CRBP-PVAc pellets with 60% $KMnO_4$ and baked at 120°C for 2 minutes. Natural organic matter was also considered for in-field applications due to its potential reducing effect. CRBP pellets offer an innovative and sustainable solution for remediating contaminated groundwater systems, reducing the need for multiple injections and minimizing safety and handling concerns.

Chlorinated volatile organic compounds (CVOCs) are often combined with 1,4-dioxane, a solvent stabilizer, for biodegradation. However, anaerobic conditions are needed for CVOC biodegradation(Abaie et al., 2024). Conventional adsorbents like activated carbon and carbonaceous resins have high adsorption capacities for these compounds but lack selectivity, limiting their use for separation. This study examined two macrocyclic adsorbents, β-CD-TFN and Res-TFN, for selective adsorption of chlorinated ethenes in the presence of 1,4-dioxane. Both adsorbents showed rapid adsorption of CVOCs and minimal adsorption of 1,4-dioxane. Res-TFN had a higher adsorption capacity for CVOCs and was highly selective for CVOCs. Its greater adsorption and selectivity suggest it can be used as a selective adsorbent for CVOC separation from 1,4-dioxane, allowing separate biodegradation.

A microfabricated gas chromatograph was used to analyze sub-parts-per-billion concentrations of trichloroethylene (TCE) in mixtures, addressing the issue of TCE vapor intrusion in homes and offices(Chang et al., 2010). The system uses a MEMS focuser, dual MEMS separation columns, and MEMS interconnects, along with a microsensor array. It is interfaced to a non-MEMS front-end pretrap and high-volume sampler module to reduce analysis time. The response patterns generated from the sensor array are combined with chromatographic retention time to identify and differentiate VOC mixture components. A chemometric method based on multivariate curve resolution is also developed for partially resolved mixture components. Results show TCE concentration at 0.185 ppb, with a projected detection limit of 0.030 ppb.

CONCLUSION

The development of chemiresistor sensor technology for in-situ monitoring of volatile organic compounds in soil and groundwater is a substantial progress in environmental monitoring and remediation. This state-of-the-art sensing solution provides real-time, continuous, and long-term monitoring capabilities, reducing the reliance on manual sample collection and off-site analysis. The integration of chemiresistor sensors with a waterproof housing and the development of characterization methods contribute to the precise identification and characterization of VOCs in the subsurface. Field testing and collaborations with commercial and academic institutions further validate and enhance the performance of the sensor system. The continuous improvement and optimization of chemiresistor sensor technology hold great promise for the efficient and effective management of contaminated sites and the protection of environmental resources.

REFERENCES

Abaie, E., Kumar, M., Garza-Rubalcava, U., Rao, B., Sun, Y., Shen, Y., & Reible, D. (2024). Chlorinated volatile organic compounds (CVOCs) and 1,4-dioxane kinetics and equilibrium adsorption studies on selective macrocyclic adsorbents. *Environmental Advances*, *16*, 100520. doi:10.1016/j.envadv.2024.100520

Bai, S., Liu, H., Sun, J., Tian, Y., Luo, R., Li, D., & Chen, A. (2015). Mechanism of enhancing the formaldehyde sensing properties of Co3O4via Ag modification. *RSC Advances*, *5*(60), 48619–48625. doi:10.1039/C5RA05772H

Bein, E., Pasquazzo, G., Dawas, A., Yecheskel, Y., Zucker, I., Drewes, J. E., & Hübner, U. (2024). Groundwater remediation by in-situ membrane ozonation: Removal of aliphatic 1,4-dioxane and monocyclic aromatic hydrocarbons. *Journal of Environmental Chemical Engineering*, *12*(2), 111945. doi:10.1016/j.jece.2024.111945

Caballero, S., Esclapez, R., Galindo, N., Mantilla, E., & Crespo, J. (2012). Use of a passive sampling network for the determination of urban NO 2 spatiotemporal variations. *Atmospheric Environment*, *63*, 148–155. doi:10.1016/j.atmosenv.2012.08.071

Cavelan, A., Faure, P., Lorgeoux, C., Colombano, S., Deparis, J., Davarzani, D., Enjelvin, N., Oltean, C., Tinet, A. J., Domptail, F., & Golfier, F. (2024). An experimental multi-method approach to better characterize the LNAPL fate in soil under fluctuating groundwater levels. *Journal of Contaminant Hydrology*, *262*, 104319. doi:10.1016/j.jconhyd.2024.104319 PMID:38359773

Chang, H., Kim, S. K., Sukaew, T., Bohrer, F., & Zellers, E. T. (2010). Microfabricated gas chromatograph for Sub-ppb determinations of TCE in vapor intrusion investigations. *Procedia Engineering*, *5*, 973–976. doi:10.1016/j.proeng.2010.09.271

Chen, X., Leishman, M., Bagnall, D., & Nasiri, N. (2021). Nanostructured Gas Sensors: From Air Quality and Environmental Monitoring to Healthcare and Medical Applications. *Nanomaterials (Basel, Switzerland)*, *11*(8), 1927. Advance online publication. doi:10.3390/nano11081927 PMID:34443755

David, E., & Niculescu, V. C. (2021). Volatile Organic Compounds (VOCs) as Environmental Pollutants: Occurrence and Mitigation Using Nanomaterials. *International Journal of Environmental Research and Public Health*, *18*(24), 13147. doi:10.3390/ijerph182413147 PMID:34948756

Davis, C. E., Ho, C. K., Hughes, R. C., & Thomas, M. L. (2005). Enhanced detection of m-xylene using a preconcentrator with a chemiresistor sensor. *Sensors and Actuators. B, Chemical*, *104*(2), 207–216. doi:10.1016/j.snb.2004.04.120

Ding, D., Jiang, D., Zhou, Y., Xia, F., Chen, Y., Kong, L., Wei, J., Zhang, S., & Deng, S. (2022). Assessing the environmental impacts and costs of biochar and monitored natural attenuation for groundwater heavily contaminated with volatile organic compounds. *The Science of the Total Environment*, *846*, 157316. doi:10.1016/j.scitotenv.2022.157316 PMID:35842168

Epping, R., & Koch, M. (2023). On-Site Detection of Volatile Organic Compounds (VOCs). *Molecules*, 28(4), 1598. doi:10.3390/molecules28041598

Guettiche, D., Mekki, A., Lilia, B., Fatma-Zohra, T., & Boudjellal, A. (2021). Flexible chemiresistive nitrogen oxide sensors based on a nanocomposite of polypyrrole-reduced graphene oxide-functionalized carboxybenzene diazonium salts. *Journal of Materials Science Materials in Electronics*, *32*(8), 10662–10677. doi:10.1007/s10854-021-05721-z

Ho, W. S., Lin, W. H., Verpoort, F., Hong, K. L., Ou, J. H., & Kao, C. M. (2023). Application of novel nanobubble-contained electrolyzed catalytic water to cleanup petroleum-hydrocarbon contaminated soils and groundwater: A pilot-scale and performance evaluation study. *Journal of Environmental Management*, *347*, 119058. doi:10.1016/j.jenvman.2023.119058 PMID:37757689

Hooshmand, S., Kassanos, P., Keshavarz, M., Duru, P., Kayalan, C. I., Kale, İ., & Bayazit, M. K. (2023). Wearable Nano-Based Gas Sensors for Environmental Monitoring and Encountered Challenges in Optimization. *Sensors, 23*(20), 8648. doi:10.3390/s23208648

Hsieh, Y. C., & Yao, D. J. (2018). Intelligent gas-sensing systems and their applications. *Journal of Micromechanics and Microengineering*, *28*(9), 093001. doi:10.1088/1361-6439/aac849

Hussain, B., Chen, J. S., Huang, S. W., Sen Tsai, I., Rathod, J., & Hsu, B. M. (2023). Underpinning the ecological response of mixed chlorinated volatile organic compounds (CVOCs) associated with contaminated and bioremediated groundwaters: A potential nexus of microbial community structure and function for strategizing efficient bioremediation. *Environmental Pollution*, *334*, 122215. doi:10.1016/j.envpol.2023.122215 PMID:37473850

Isaac, N. A., Pikaar, I., & Biskos, G. (2022). Metal oxide semiconducting nanomaterials for air quality gas sensors: operating principles, performance, and synthesis techniques. *Microchimica Acta, 189*(5), 1–22. doi:10.1007/s00604-022-05254-0

Joksimoski, S., Kerpen, K., & Telgheder, U. (2022). Atmospheric pressure photoionization – High-field asymmetric ion mobility spectrometry (APPI-FAIMS) studies for on-site monitoring of aromatic volatile organic compounds (VOCs) in groundwater. *Talanta*, *247*, 123555. doi:10.1016/j.talanta.2022.123555 PMID:35613524

Krishnakumar, T., Jayaprakash, R., Pinna, N., Donato, N., Bonavita, A., Micali, G., & Neri, G. (2009). CO gas sensing of ZnO nanostructures synthesized by an assisted microwave wet chemical route. *Sensors and Actuators. B, Chemical*, *143*(1), 198–204. doi:10.1016/j.snb.2009.09.039

Krishnan, S., Tadiboyina, R., Chavali, M., Nikolova, M. P., Wu, R.-J., Bian, D., Jeng, Y.-R., Rao, P. T. S. R. K. P., Palanisamy, P., & Pamanji, S. R. (2019). *Graphene-Based Polymer Nanocomposites for Sensor Applications. Hybrid Nanocomposites*, (pp. 1–62). Springer. doi:10.1201/9780429000966-1

Kumar, R., Liu, X., Zhang, J., & Kumar, M. (2020). Room-Temperature Gas Sensors Under Photoactivation: From Metal Oxides to 2D Materials. Nano-Micro Letters, *12*(1), 1–37. doi:10.1007/s40820-020-00503-4

Lamssali, M., Luster-Teasley, S., Deng, D., Sirelkhatim, N., Doan, Y., Kabir, M. S., & Zeng, Q. (2023). Release efficiencies of potassium permanganate controlled-release biodegradable polymer (CRBP) pellets embedded in polyvinyl acetate (CRBP-PVAc) and polyethylene oxide (CRBP- PEO) for groundwater treatment. *Heliyon*, *9*(10), e20858. doi:10.1016/j.heliyon.2023.e20858 PMID:37867834

Niepsch, D., Clarke, L. J., Tzoulas, K., & Cavan, G. (2022). Spatiotemporal variability of nitrogen dioxide (NO2) pollution in Manchester (UK) city centre (2017–2018) using a fine spatial scale single-NOx diffusion tube network. *Environmental Geochemistry and Health, 44*(11), 3907–3927. doi:10.1007/s10653-021-01149-w PMID:34739651

Norris, G., & Larson, T. (1999). Spatial and temporal measurements of NO2 in an urban area using continuous mobile monitoring and passive samplers. *Journal of Exposure Analysis and Environmental Epidemiology, 9*(6), 586–593. doi:10.1038/sj.jea.7500063 PMID:10638844

Pathak, A. K., Swargiary, K., Kongsawang, N., Jitpratak, P., Ajchareeyasoontorn, N., Udomkittivorakul, J., & Viphavakit, C. (2023). Recent Advances in Sensing Materials Targeting Clinical Volatile Organic Compound (VOC) Biomarkers: A Review. *Biosensors, 13*(1), 114. doi:10.3390/bios13010114

Remoundou, K., & Koundouri, P. (2009). Environmental Effects on Public Health: An Economic Perspective. *International Journal of Environmental Research and Public Health, 6*(8), 2160–2178. doi:10.3390/ijerph6082160 PMID:19742153

Rivadeneyra, A., Fernández-Salmerón, J., Salinas-Castillo, A., Palma, A. J., & Capitán-Vallvey, L. F. (2016). Development of a printed sensor for volatile organic compound detection at µg/L-level. *Sensors and Actuators. B, Chemical, 230*, 115–122. doi:10.1016/j.snb.2016.02.047

Rivera, D., Alam, M. K., Davis, C. E., & Ho, C. K. (2003). Characterization of the ability of polymeric chemiresistor arrays to quantitate trichloroethylene using partial least squares (PLS): Effects of experimental design, humidity, and temperature. *Sensors and Actuators. B, Chemical, 92*(1–2), 110–120. doi:10.1016/S0925-4005(03)00122-9

Roy, S., Saha, H., & Sarkar, C. K. (2010). High sensitivity methane sensor by chemically deposited nanocrystalline ZnO thin film. *International Journal on Smart Sensing and Intelligent Systems, 3*(4), 605–620. doi:10.21307/ijssis-2017-411

Sara, S. M., Al-Dhahebi, A. M., & Mohamed Saheed, M. S. (2022). Recent Advances in Graphene-Based Nanocomposites for Ammonia Detection. *Polymers, 14*(23). doi:10.3390/polym14235125 PMID:36501520

Verma, A., Kumar, P., & Yadav, B. C. (2024). Fundamentals of electrical gas sensors. *Complex and Composite Metal Oxides for Gas VOC and Humidity Sensors, 1*, 27–50. doi:10.1016/B978-0-323-95385-6.00004-0

Vîrghileanu, M., Săvulescu, I., Mihai, B. A., Nistor, C., & Dobre, R. (2020). *Nitrogen Dioxide (NO2) Pollution Monitoring with Sentinel-5P Satellite Imagery over Europe during the Coronavirus Pandemic Outbreak*. MDPI. doi:10.3390/rs12213575

Wiśniewska, M., & Szyłak-Szydłowski, M. (2022). The Application of In Situ Methods to Monitor VOC Concentrations in Urban Areas—A Bibliometric Analysis and Measuring Solution Review. *Sustainability, 14*(14), 8815. doi:10.3390/su14148815

Wu, M., Hou, S., Yu, X., & Yu, J. (2020). Recent progress in chemical gas sensors based on organic thin film transistors. *Journal of Materials Chemistry. C, Materials for Optical and Electronic Devices, 8*(39), 13482–13500. doi:10.1039/D0TC03132A

Yang, Z., Wei, C., Song, X., Liu, X., Tang, Z., Liu, P., & Wei, Y. (2023). Thermal conductive heating coupled with in situ chemical oxidation for soil and groundwater remediation: A quantitative assessment for sustainability. *Journal of Cleaner Production*, *423*, 138732. doi:10.1016/j.jclepro.2023.138732

Zhang, P., Xiao, Y., Zhang, J., Liu, B., Ma, X., & Wang, Y. (2021). Highly sensitive gas sensing platforms based on field effect Transistor-A review. *Analytica Chimica Acta*, *1172*, 338575. doi:10.1016/j.aca.2021.338575 PMID:34119019

Zhang, X., Xia, Q., Lai, Y., Wu, B., Tian, W., Miao, W., Feng, X., Xin, L., Miao, J., Wang, N., Wu, Q., Jiao, M., Shan, L., Du, J., Li, Y., & Shi, B. (2022). Spatial effects of air pollution on the economic burden of disease: Implications of health and environment crisis in a post-COVID-19 world. *International Journal for Equity in Health*, *21*(1), 1–16. doi:10.1186/s12939-022-01774-6 PMID:36380331

Zhu, L., Meier, D., Boger, Z., Montgomery, C., Semancik, S., & DeVoe, D. L. (2007). Integrated microfluidic gas sensor for detection of volatile organic compounds in water. *Sensors and Actuators. B, Chemical*, *121*(2), 679–688. doi:10.1016/j.snb.2006.03.023

Zhu, L. Y., Ou, L. X., Mao, L. W., Wu, X. Y., Liu, Y. P., & Lu, H. L. (2023). Advances in Noble Metal-Decorated Metal Oxide Nanomaterials for Chemiresistive Gas Sensors: Overview. *Nano-Micro Letters, 15*(1), 1–75. doi:10.1007/s40820-023-01047-z

Chapter 4
Optical and Electrochemical Chemosensors for Identification of Carbon Dioxide Gas

Kirandeep Kaur
Maharaja Ranjit Singh Punjab Technical University, Bathinda, India

Sudhanshu Pratap Singh
Maharaja Ranjit Singh Punjab Technical University, Bathinda, India

Navdeep Singh Gill
https://orcid.org/0009-0002-9995-1756
Australian Maritime College, University of Tasmania, Australia

ABSTRACT

Intelligent sensing and vigilant monitoring of C_02 gas is immensely significant, as the concentration of C_02 gas in atmosphere is increasing day by day due to anthropogenic activities which enhances the natural greenhouse effect and makes the earth warmer. Besides global warming, C_02 is also a toxicant in enclosed environments causing asphyxiation by hypoxia, unconsciousness almost instantaneously, and respiratory arrest within one minute. Another area where sensing of C_02 is imperative are agriculture and food industry. For these conditions, sensors should be capable of working under extreme temperatures, pressure, and interference due to the inherent complex materials and microorganisms. This chapter focuses on the information regarding the different types of C_02 sensors available with special emphasis on electrochemical sensors and optical chemical sensors, which showed fluorescent and spectrophotometric variation on detection of C_02, a detailed analysis of their detection process and sensing mechanism.

INTRODUCTION

There exist only few analytically investigable targeted entities as crucial as carbon dioxide, as it is the primary chemical feed stock for life. Photosynthesis process convert carbon dioxide into fuel and feed,

DOI: 10.4018/979-8-3693-1930-7.ch004

primarily required for continual survival of all life forms.

$$CO_2 + H_2O \xrightarrow{\text{Sunlight}} C(H_2O) + O_2 \text{ (I)}$$

The backward chemical reaction of photosynthesis is the metabolic chemical reaction for most of the cells, liberating energy for biotic communities. Carbon dioxide is crucial for producing necessary chemicals for life and is commonly utilized as an indicator of life's presence and health measurement. This makes carbon dioxide an imperative analytical entity, as a measure of health. For instance, in medical domain blood analysis includes the analysis of dissolved oxygen, pH and CO_2 (Mills, 2009). In capnography, (the branch devoted to measurement of CO_2 levels in breath) the levels and temporary variation in CO_2 are monitored for diagnostic data (Gravenstein & Jaffe, 2021). Similarly, in anesthesiology monitoring of CO_2 have a crucial role to play. Contemporary studies show that the carbon dioxide levels in cellular environment is directly related to many biophysiological processes which include respiration, pH levels in biosystems, carcinogenesis, genetic information transformation, synthesis of nucleic acid and cell proliferations (Mills, 2009).

Although carbon dioxide is a primary requirement for existence of life but its continuous increasing levels in environment has significant impacts on global climate and human health. In recent years, because of its greenhouse effect, global warming is growing at a concerning pace and adversely affecting the season pattern all over the world. Oceans are assimilating CO_2 from the environment, which consequently causes ocean acidification, climate deterioration and environmental pollution. In addition excessive CO_2 inhalation culminate towards diseases such as asthma and bronchitis. Burning of fuels in transportation means, industries, power generation, domestic burning, degradative oxidation of biomass and fermentation are the major causes of carbon dioxide level increment in biosphere (Chen et al., 2016; Wang & Yang, 2014; Ziebart et al., 2014; Finn et al., 2012)

The gas under discussion is also one of the important indicators for determining the constitution along with safety standards of packaged food. After packaging of the food material, its deterioration starts immediately because of multiplication of microbes accompanying evolution of carbon dioxide gas as catabolic product. In modified atmospheric packaging (MAP) the food package is flushed with carbon dioxide and sealed. Low levels of oxygen thus prevents microbial spoilage allows the food item to stay fresh longer. Carbon dioxide act as active packaging gas in packaging mechanism as it can prevent the spoilage even in presence of oxygen. As a result the recognition and monitoring of carbon dioxide levels is highly imperative and this requires development of highly sensitive and selective sensing devices (Puligundla et al., 2012).

An efficient sensing device is defined by three basic characteristic feature selectivity, sensitivity and time of response (Kaur & Baral, 2014). Sensitivity may be defined as the capability of the sensing system to accurately quantify the analyte under the imposed conditions. The sensitivity of the sensor is controlled by the intense physico- chemical nature of the solvent system employed for the matrix preparations. Selectivity of the sensing material may be defined as its capability to recognize a targeted analyte free from interferences. Time of response is the count of quickness of attainment of maximum variation in signal upon varying the concentration of analyte. Additionally, reversibility, prolonged stability, dimensions and illumination source and some other factors that can manipulate the efficiency of the sensor. Thus, sensor can be stated precisely as a mechanism, which includes controlling and machine running electronic system, software coupled with networking module that detects and reacts to a variation in observable physico-chemical quantity, resulting an output which can be measured and directly proportional to the quantity (Wencel et al., 2014). The optical sensor must address some of the concerns:

1. They should be dissolved to form transparent solution, making them suitable for quantitative sensing spectrometrically.
2. They should be non-reactive chemically, photostable and thermally stable
3. Involve less complicated matrix preparation process
4. The matrix material could be moldable to various shapes.

Optical sensor have four basic modules:

1. Sample cross section unit, the area for the surficial chemical reaction to occur
2. The transduction unit
3. Digital signal analysis electronics unit
4. Signal output system

Sensor for gaseous analyte has two significant operations, first is to identify the analyte followed by the conversion of identification event into a functional signal. Finally the transduction unit convert functional signal to readable form. The chemical activity occurring at the sample cross section brings out the variation in properties of the chemical system like pH, resistivity, conductance, resulting from chemical reaction of the analyte and sensing motif (Neethirajan et al., 2009).

The qualitative and quantitative analytical investigations of CO_2 gas is often carried out using infrared spectroscopy, electrochemical method, gas chromatographic technique, mass spectrophotometry, and field effect transistors etc. When CO_2 is present in gas form, the Infrared spectroscopic method is usually employed. However, the method inherent the disadvantage of being unfriendly to pocket and sensitive to water vapors (Zhua et al., 2018). Severinghaus sensing technique for CO_2 sensing includes an aqueous bicarbonate solution containing glass electrodes enclosed in a gas permeable and water & electrolyte impermeable material. Carbon dioxide results in carbonic acid formation in aqueous medium followed by disintegration of carbonic acid into bicarbonate anion and a proton and thereby varying the pH of the electrolytic solution which is sensed by probe. In this process the carbon dioxide is recognized indirectly in its ionic form. Volatile acidic or basic gaseous material may interfere in pH measurements. Gas chromatographic technique absorb the carbon dioxide gas selectivity and elute through the column. While in mass spectrometry the analyte gas is bombarded with highly energetic electrons, converting gas in fragmentations which is further get separated based on charge and mass under the influence of magnetic field. Mass spectrometry is able to measure in near real time but it is not cost efficient, while gas chromatography is inexpensive as compared to mass spectrometry but it has long response time. (Qazi et al., 2012). Subsequent research suggested use of optical chemical sensors to get around the drawbacks of other approaches. Over the last ten or so years, numerous optical CO_2 detectors have been proposed working on the principle of origination of pH variation on addition of CO_2. These detectors often used in the form of planar or fiber optic machines and employing a range of sensing techniques which include fluorescence intensity detection, fluorescence resonance energy transfer, dual luminophore, DLR and many more.

Depending upon the mechanistic pathway of sensing, the sensing methods for carbon dioxide for carbon dioxide can be widely classified into two major types:

a. Optical sensor b. Electrochemical sensor

OPTICAL SENSORS

An optical sensor for carbon dioxide has following main components:

a. A dye molecule responsive towards pH (Deprotonated dye: A^{-1}, protonated dye: HA) these two forms either have dissimilar color/ optical profiles.
b. An aqueous/non aqueous solution as medium along with base/phase transfer reagent to solubilize the dye
c. A integument which is penetrable to gas and non-penetrable for ions

The optodes for carbon dioxide operate on similar principles as that of the Severinghaus type sensor. In a standard optode for CO_2, the diffusion of the gas occurs towards the solvent (liquid or gas phase) through the gas-penetrable layer and attains equilibration with the enclosed water layer that contains the pH-sensitive color inducing pigment (dye molecule). Enclosed aqueous medium containing dye molecule, on exposure to carbon dioxide establishes the equilibrium as per the following equations:

$$CO_2\,(g) \xrightarrow{K_a} CO_2\,(l) \quad K_a = 3.3 \times 10^{-2}\ mol/dm^3.atm \quad (II)$$

$$CO_2(aq) + H_2O \xrightarrow{K_b} H_2CO_3 \quad K_b = 2.6 \text{ X } 10^{-3}\ mol/dm^3.atm \quad (III)$$

$$H_2CO_3 \xrightarrow{K_c} H^+ + HCO_3^- \quad K_c = 1.72 \text{ X } 10^{-4}\ mol/dm^3.atm \quad (IV)$$

$$HCO_3^- \xrightarrow{K_d} H^+ + CO_3^{-2} \quad K_d = 5.59 \text{ X } 10^{-11}\ mol/dm^3.atm \quad (V)$$

When the base concentration is notably high, approximately 10^2 mol/dm^3, the above equilibrium equation results in a relation between P_{CO2} (ambient partial pressure of carbon dioxide) and concentration of H^+ ions present in enclosed water layer.

$$\alpha.\ P_{CO2} = [H_2CO_3] = [H^+][Na^+]/K_3 \quad \alpha = \text{constant} \quad (VI)$$

pH responsive dye molecule exists in protonated and deprotonated species results in following equilibrium:

$$HA \xleftrightarrow{K_a} H^+ + A^- \quad K_a\text{=acid dissociation constant of the dye (VII)}$$

Equations VI and VII gives the relation:

$$P_{CO2} = K'\,[AH]/[A^-] \quad K' = \text{constant}$$

The protonated and deprotonated species of dye molecules usually absorbs in quite different regions of electronic spectra and the molar absorptivity of protonated form is usually very small as compared deprotonated form at wavelength of maximum absorbance of deprotonated form. This yields a direct relation between absorbance of deprotonated form (A_d) and its concentration ([A]). On exposure to carbon dioxide the dye molecules usually remain in deprotonated form, then absorbance A_{do} of this form varies linearly with the entire dye concentration. Thus, the ambient partial pressure of the gas is related to the absorbance of deprotonated and protonated forms as follows:

$$P_{CO2} = K' (A_{do} - A_d)/A_d$$

Thus, measuring the absorbance of protonated and deprotonated form directly yields $P_{CO2.}$

Another type of optodes (sensing materials) for CO_2 are those which employs phase transfer agents like tetraoctyl ammonium hydroxide, ($Q^+ OH^-$), to solubilize deprotonated form of pH responsive dye in a hydrophobic solvent containing water insoluble polymeric matrix. The phase transfer agent ($Q^+ OH^-$) when added to dye (A^-), it form an ion pair Q^+A^-. This ion pair can be easily solubilized in non-aqueous solvents. A salient characteristic of this ion couple in not only to get solubilize in hydro-repellent organic solvents but in addition it associates some water molecules with it (it can be written as $Q^+A^-. xH_2O$) and set up the ensuing equilibrium:

$$Q^+A^-. xH_2O + CO_2 \xrightarrow{K''} Q^+HCO_3^-.(x\text{-}1)\ H_2O.HA \quad K'' = \text{equilibrium constant (VIII)}$$

To facilitate the penetration of the gas across the polymeric matrix usually a plasticizer like tributyl phosphate is used as a component in the film ensemble. This ensemble of dye, phase transfer reagent, plasticizer, hydrophobic polymer and organic solvent offers sufficient potential to develop dry carbon dioxide sensors for routine use. If the above equilibrium is assumed to be decisive equilibrium event, then a new measuring factor R can be defined taking into account the practically measurable optical event as given below:

$$R = (A_{A0} - A_A)/A_A = K''.P_{CO2}$$

A_{A0} = Absorbance of Q+ A- .xH_2O on non-existence of CO_2

A_A = Absorbance of Q+ A- .xH_2O in presence of CO_2.

Both the methods (first one measuring absorbance in aqueous medium (wet method) and other one in presence of organic solvent and phase transfer agents (dry method)) follows similar equilibrium equations as it has been assumed that at wavelength of maximum absorption only Q+ A- .xH_2O absorbs and the dye exists as deprotonated negatively charged species on non-exposure to carbon dioxide (Mills & Hodgen, 2005)

The most often employed and readily accessible fluorescent active indicator in sensing of carbon dioxide optically is 8-hydroxypyrene-1,3,6-trisulfonate(HPTS) (Dansby-Sparks et al., 2010). Unfortunately, there are situations where green emission and blue-spectrum excitation are not the best choices, especially when working with medium that have significant levels of scattering and autofluorescence.

In addition, its pKa of ~ 7.3 makes it more difficult to design sensors to resolve environmental carbon dioxide levels. To get beyond the aforementioned constraints other categories of dye molecules were tried which include, diketopyrrolopyrroles (Schutting et al., 2013), perylene bisimides (Pfeifer et al., 2018), azaphthalocyanines (Lochman et al., 2017) etc. but the modulation of spectroscopic and sensing abilities still remain a major challenge. Among the most often used fluorophoric molecules are 4,4-difluoro-4-bora-3a,4a-diaza-*s*-indacene (BODIPY)dyes, which have high fluorogenic quantum yields and sufficient molar absorptivity coefficients (Loudet et al., 2007). The flourophore become even more appealing as flourogenic sensor for sensing and (bio) imaging, when the absorptivity and emissivity spectral features are elongated into the near-infrared region. Taking into consideration these characteristics, a series of BODIPY pH responsive sensors DI,D2,D3,D4,D5 (Figure 1)were synthesized for sensing CO_2 optically. The core BODIPY chromophoric molecule which absorbs at 505 nm was derivatized to extend the π-conjugation and the resulted derivatives demonstrated absorption at longer wavelength i.e 635 and 665 nm along with large molar absorptivity, fluorescent quantum efficiency and unmatched photostability. The phenol functionalized BODIPY chromophore becomes more sensitive towards pH variations due to the initiation of fluorescent intensity quenching process, in its anionic form. Theses dyes were loaded on ethyl cellulose polymeric matrix coupled with tetraoctylammonium hydrogencarbonate. The functionalized receptor bearing carboxylic group at ortho location to hydroxyl functional group shown considerably sensitive behavior with detection limit of 0.009hPa while the other sensors based on other dye molecules showed LOD from 0.2 to 60 hPa and from 20 to 400 hPa.

di-OH-aza-BODIPYs (D6) (Figure 2) another phenolic functionalized derivative of BODIPY, was developed as colorimetric sensor by Schutting et al. (2015) and it displayed maximum absorption in the NIR region (670–700 nm in neutral state, 725–760 nm in mono-basic form, 785–830 nm in di-basic form), sufficiently large molar absorption coefficients estimated as 77 000 M^{-1} cm^{-1} and stable photonically. The protonable behavior can be tuned by substituting electron-withdrawing or electron-donating functional groups (pKa values 8.7–10.7) hence these sensors offers diverse span enabling their applicability in diversified fields. Another 4-anilineboron-dipyrromethene (BODIPY) based fluorophore (D7) (Figure 2) was introduced by Pan et al. (2014) as sensitive towards acidic pH sensing via fluorescence phenomenon which can be excited in the visible region following PET mechanistic pathway. The pH metric titration demonstrates 500 fold increment in fluorescent intensity in pH span 4.12–1.42 having pKa 3.24 in aqueous methanolic mixture (1:1, v/v) medium, that is necessary to carry out studies in acidic environment. The probe is also found to respond colorimetrically towards dissolved carbon dioxide (CO_2) gas.

Aggregation Induced Emission Based Small Florogenic Sensors for CO_2

Luminophoric materials demonstrating aggregation-induced emission (AIE) phenomenon show negligible electromagnetic radiative outflow in less concentrated solutions but exhibit measurable fluorescence when the ‘restriction of intramolecular rotation’ (RIR) is brought about by agglomeration or elevated viscous behavior. Aggregation-induced emission (AIE) is an unusual photophysical phenomenon, which occurs when nonemissive luminophores are efficiently driven to emit during the aggregate development. AIE was introduced by Luo et al. (2001), overcoming the limitation of previous aggregation-caused quenching materials. The functioning mechanistic process of AIE has been established as the restriction of intramolecular movements (RIM) .The unique solubility between an AIE/AEE (carrying amine as one of the functional group) compound’s neutral/salt state will cause a phase change to disaggregated form from an aggregated form when used for CO_2 sensing. It will also cause a noticeable increase in

Figure 1. Molecular structures of D1, D2, D3, D4 and D5

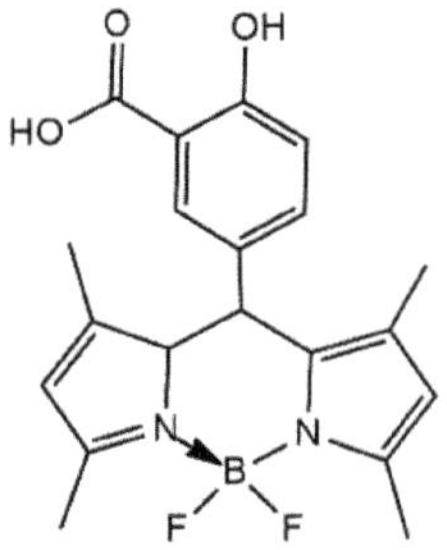

D1

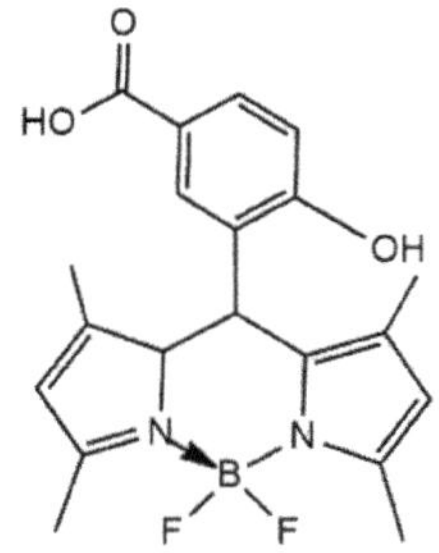

D2

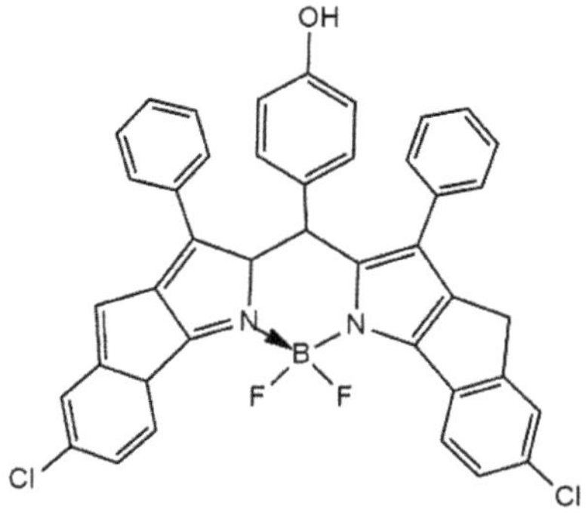

D3

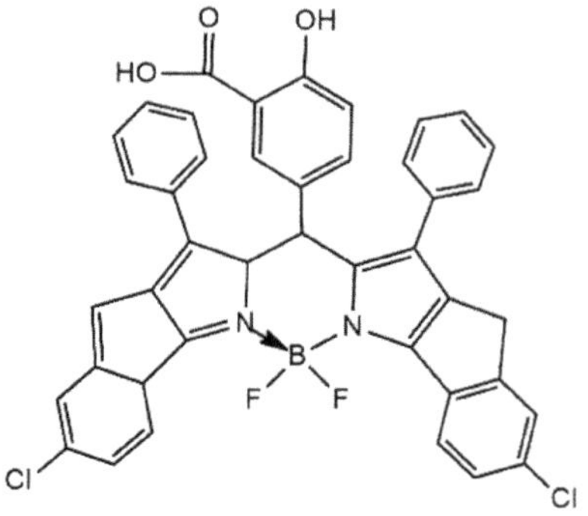

D4

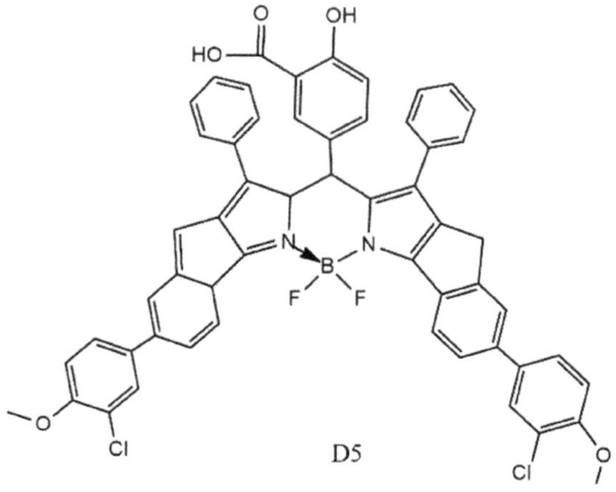

Figure 2. Molecular structures of D6 and D7

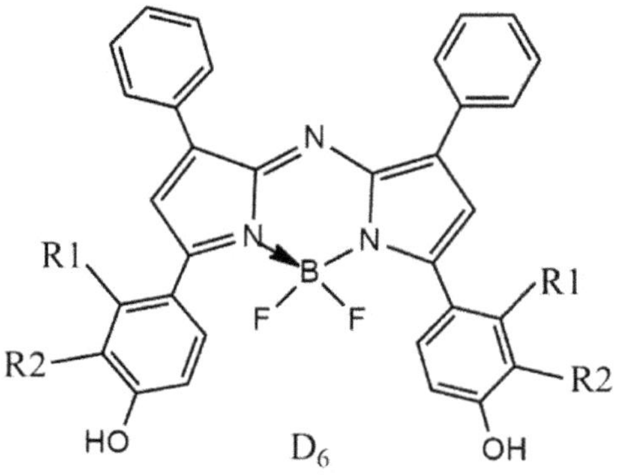

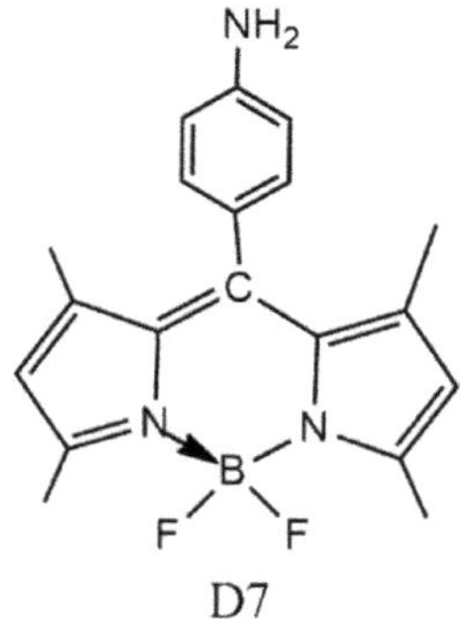

D7

R1=R2=H, di-OH-Complex

R1=H, R2=F, di-F-di-OH Complex

R2=H, R1=Cl, di-Cl-di-OH Complex

R1= CH_3, R2=OH, di-CH_3-di-OH Complex

viscosity due to the formation of electrolytes. These physical changes lead to a pronounced fluorescence integration and hence facilitate the detection of carbon dioxide.

One such AIE/AEE based fluorescent sensor employing dipropylamine solution of hexaphenylsilole as aggregation induced fluorescent system that can recognize carbon dioxide via variation in emission intensity as result of emergence of carbamate ionic liquid of high viscosity. Sun et al. (2013) prepared optically active probe utilizing tetraphenylethene (TPE) as an AIE molecule and 1,8-diazabi-cyclo-[5,4,0]-undec-7-ene (DBU) as an amidine containing molecule. The probe offers a simple and visible method of sensing and detecting CO_2 because they are sensitive to the rise in viscosity caused due to emergence of ionic liquid via specific interaction between amidine moiety and CO_2 (Figure 3). However, these techniques are inappropriate when dealing with biological systems.

With the intention of enhancing the analysis's sensitiveness, lower the LOD, and improving applicability aqueous medium, the phase transition and AIE/AEE features are used in combination to enhance the measurable signal changes in order to detect low concentration of carbon dioxide.

Wang et al. (2015), prepared the molecules based on 1,2,5-triphenyl-pyrrole (TPP) core containing a different number of tertiary amine moieties, 2-(dimethylamino)ethyl 4-(2,5-diphenyl-1H-pyrrol-1-yl) benzoate(TPP-DMAE), bis(2-(dimethylamino)ethyl) 4,4'-(1-phenyl-1H-pyrrole-2,5-diyl)dibenzoate(TPP-BDMAE) and tris(2-(dimethylamino)ethyl) 4,4',4''-(1H-pyrrole-1,2,5-triyl)tri-benzoate(TPP-TDMAE). The inherited AEE property and amine functional moiety by the TPP core molecule, TPP-TDMAE among various sensors showed the notable frequency variations on titrating even with a lesser concentration (0.031 to 5%) of CO_2.

Mishra et al. (2019) synthesized cyano derivitized p-phenylenevinylene molecule possessing tertiary amine fragment at terminal position, which aggregates and exhibits fluorogenic and Raman spectral response for C_02. The aggregation induced emission along with aggregation enhanced Raman scattering of the derivative in aqueous medium diminished on exposure to even a minute quantity of carbon dioxide resulting in its facile and quick recognition in various samples of neutral gases. Furthermore, this AIE based CO_2 assay found to be effective in identifying CO_2 in biological specimens.

Mai et al. (2020) prepared two carbon dioxide fluorogenic chemical signalers, TPE-amidine-L(liquid) and TPE-amidine-S(solid) based on (N,N-dimethyl-N'-(4-(1,2,2-triphenylvinyl)phenyl)acetimidamide. These sensing ensembles demonstrated substantial fluorescent variations initiated by CO_2 and both could be regenerated on exposure to triethylamine. The liquid probe showed the lowest CO_2 detection limit of

Figure 3. Representation of formation of ionic liquid on exposure to carbon dioxide through amidine containing molecule (DBU)

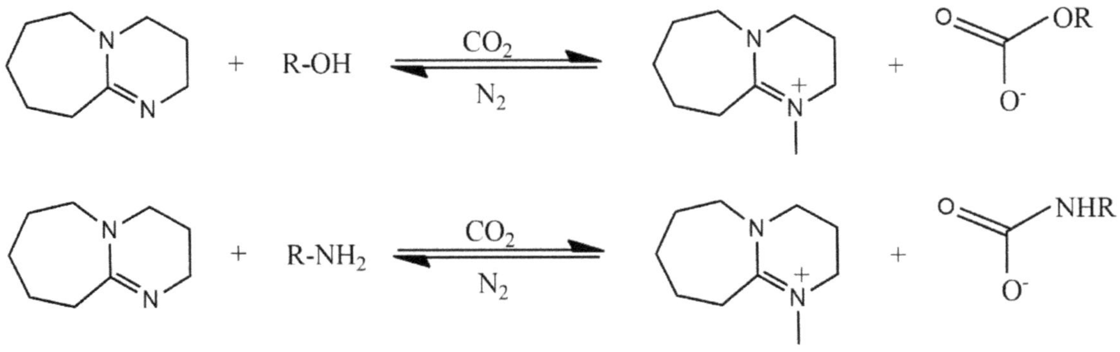

24.6 ppm, while the solid probe made from a polyacrylamide hydrogel framework is capable of serving as a convenient CO_2 fluorescent indicator for in situation monitoring of CO_2. Ma et al. (2015) developed an another fluorescent chemosensor for carbon dioxide gas exhibiting AIE phenomenon, which consists of sodium phenolic salt (ONa^+), as hydro loving moiety, incorporated onto tetraphenylethene (TPE) to obtain TPE-ONa, which is found to be soluble in aqueous medium and displays no emission. On passing CO_2 through the reaction medium, a notable color variation and fluorescence intensity increment is observed. The detection of gas by TPE-ONa in aqueous medium is fast and the detection limit is estimated to be 2.4×10^{-6} M. To realize the operationalization of TPE-ONa, it is conglomerated with sodium carboxymethyl cellulose in aqueous solution to fabricate a porous film and observed sufficient sensing ability. Additionally, TPE-ONa demonstrates low cytotoxicity for live cells and displays the potential to observe external CO_2 concentration variations in living cells.

One more TPE-derived sensing ensemble, sodium-4-(4-(1,2,2-triphenylvinyl)phenoxy)butane-1-sulfonate(TPE-SO_3^{-1}) in combination with a easily procurable polymer(chitosan) utilizes AIE phenomenon of TPE core and act as sensing device for carbon dioxide (Khandare et al., 2015). Dissolved carbon dioxide disintegrate to release H^+ ions resulting in protonation of chitosan which act as befitting counter ion for the (TPE-SO_3^{-1}). (TPE-SO_3^{-1}) molecules assemble with one another in a particular pattern on a polymeric branches of chitosan on account of electrostatic attractive forces between them, resulting in AIE leading to increased fluorescence intensity. Fluorescence emission of the sensor varies linearly with the span of aggregation and depended on the charge density of the polymeric entity, thus amount of dissolved carbon dioxide could be measured by it. This method was applied to live samples and the results displayed were satisfactory. The sensor demonstrated high sensitivity leading to LOD of 0.00127 hPa.

An additional, fluorogenic sensor, tris(2-(dimethylamino)ethyl)-4,4',4''-(1H-pyrrole-1,2,5-triyl) tribenzoate(TPP-TMAE), possessing AEE property synthesized by Chen et al. (2016) demonstrates extreme selectivity, specificity, and respond instantly even to minute quantity of carbon dioxide (CO_2). A notable decrement in TPP-TMAE fluorescent integration by almost 20 times in just 12 s may be attributed to disassembling mechanism coupled by neutralization of amine due to presence of carbon dioxide in aqueous medium. This fluorogenic sensor can quantitatively measure carbon dioxide concentration under the limits 0.031%-5%, which almost enclose the whole range of carbon dioxide gas concentration produced from cell metabolic reactions. This typical feature allows TPP-TMAE to be employed as a biomarker for in-situ measurement of the rate of the target gas production in the course of metabolic reactions a single living cell which enables distinguishability between cancer cells and normal cells.

α-Cyanostilbenes are another class of fluorogenic molecules demonstrating aggregation-based emission as a result of formation of aggregates due to π-π conjugated stacking effect. These molecular frameworks display intense fluorescent emission in combination with adequate photostability in agglomerated conditions in comparison to less concentrated solution. In addition, photoluminescence attributes of these molecules can be modulated by substituting functional arrangement of terminal aryls. Besides, these molecules can be tailored as a turn-on fluorophores, capable of using AIE feature, by substituting various substituents to assist self-assembling.

Jang et al. (2019) prepared another probe from α-cyanostilbene[(Z)-3-(4-(3-Aminopropoxy)phenyl)-2-(4-16 nitrophenyl)acrylonitrile], which consists of a primary amino group, for analyzing carbon dioxide quantitatively. It reacts with the target gas and facilitate turn-on fluorescence caused by aggregation-induced emission. More essentially, it does not require any additives like an amine medium for sensing of carbon dioxide gas, and exhibit LOD as ~26 ppm. The current sensing ensemble is further utilized in screening of adsorbents for target gas and it revealed that this system can be employed to create effective throughput screening technique to assess the effectiveness of CO_2 absorbing molecules. The amino group of the sensor reacts with the target gas (CO_2) resulting in formation of carbamic acid, which further reacts with other amine molecules forming salt bridges. Thus, the resulting electronic attractive forces between carbamate and ammonium salt produces self-congregation leads to increased fluorescence. This process can be reversed by bubbling nitrogen gas. As a result, this ensemble exhibiting reversible 'turn-on' fluorescence activity can be used repeatedly and enables the determination of CO_2 quantitatively (Figure 4).

Anion Activated Fluorescent Molecules for Carbon Dioxide Sensing

In addition to aggregation induced emission (AIE) mechanism for fluorescent molecule for recognizing the carbon dioxide, an ancillary mechanism is to activate the fluorescent molecule through anions. N-heterocyclic carbenes (NHCs) are typically formed through deprotonation of appropriate salts of imidazole fragment have the ability to react with carbon dioxide to produce imidazolium carboxylates (Ausdall et al., 2009). The molecules carrying imidazolium fragments react with anionic species present in the vicinity to form a firm $(C–H)^{+} \cdots X^{-}$ interaction (Xu et al., 2010). This particular bonds in combination with electrostatic interactions are believed to act as the foundation for the properties of molecular recognition. However, it was speculated that anion is responsible for the activation of imidazolium functional moiety and make them able to interact with CO_2 through the development of NHC (in absence of any additional base). This mechanism proceed in acidic imidazolium arrangement and demonstrated deprotonation-based absorptivity or emissivity within the visible region, it might serve as the foundation for a novel category of optical CO_2 sensing chemical devices. One such recognition mechanism is realized using tetrapropyl benzobisimidazolium salts (TBBI). Guo et al. (2012) submitted a benzobisimidazolium (BBI) salts that work as carbene containing ligating molecule towards metallic entities and found to be fluorescent many time. One of the reports established that the BBI derivative can behave as redox analog in some anion-catalyzed electron transfer mechanism consisting of calix[4]- pyrrole. The referred investigations show that structural alteration can be used to tune the inherent electronic characteristics of BBI cations. A vital property of these fluorogens is the possibility of functional modification at carbon number one (C1) of imidazole fragment that enable tuning of photophysical features demonstrated by these salts, as evidenced by the corresponding variations observed in photophysical properties. Taking these observations into consideration, it has been thought that functional substitution at C1 of derivatized BBI would enable the preparation of efficient sensor system for carbon dioxide gas. To accomplish the target the research group particularly added F^{-} ion, which is not a strong base, might result in a species with some NHC characteristics thus resulting in interaction with carbon dioxide gas. As a result of these associative type interactions optical variations occur which enables the detection. The researchers synthesized tetrapropylbenzobisimidazolium salt (TBBI) an organic-soluble NHC precursor. Spectrometric and theoretical investigation revealed that on addition of fluoride ions, N-heterocyclic carbene intermediate is formed that interact with carbon dioxide to result in imidazolium carboxylate. The method demonstrates a notable selective response towards fluoride ion with a low LOD of 30 ppm and quick response both fluorometrically and colorimetrically (Figure 5).

Kwon et al. (2009) prepared two naphthoimidazolium derivatives (D8, D9) (Figure 6) activated through fluoride and cyanide ions and employed to sense CO_2. In the pool of different ions only F^{-} and CN^{-1} ions are able to bring out the optical changes in acetonitrile medium. These variations can be assigned to [$(C–H)^{+}$–negatively charge specie] hydrogen bond type attractive forces associated with imidazolium

Figure 4. Carbon dioxide sensing through aggregation induced emission phenomenon by α-cyanostilbene based fluorescent chemosensor

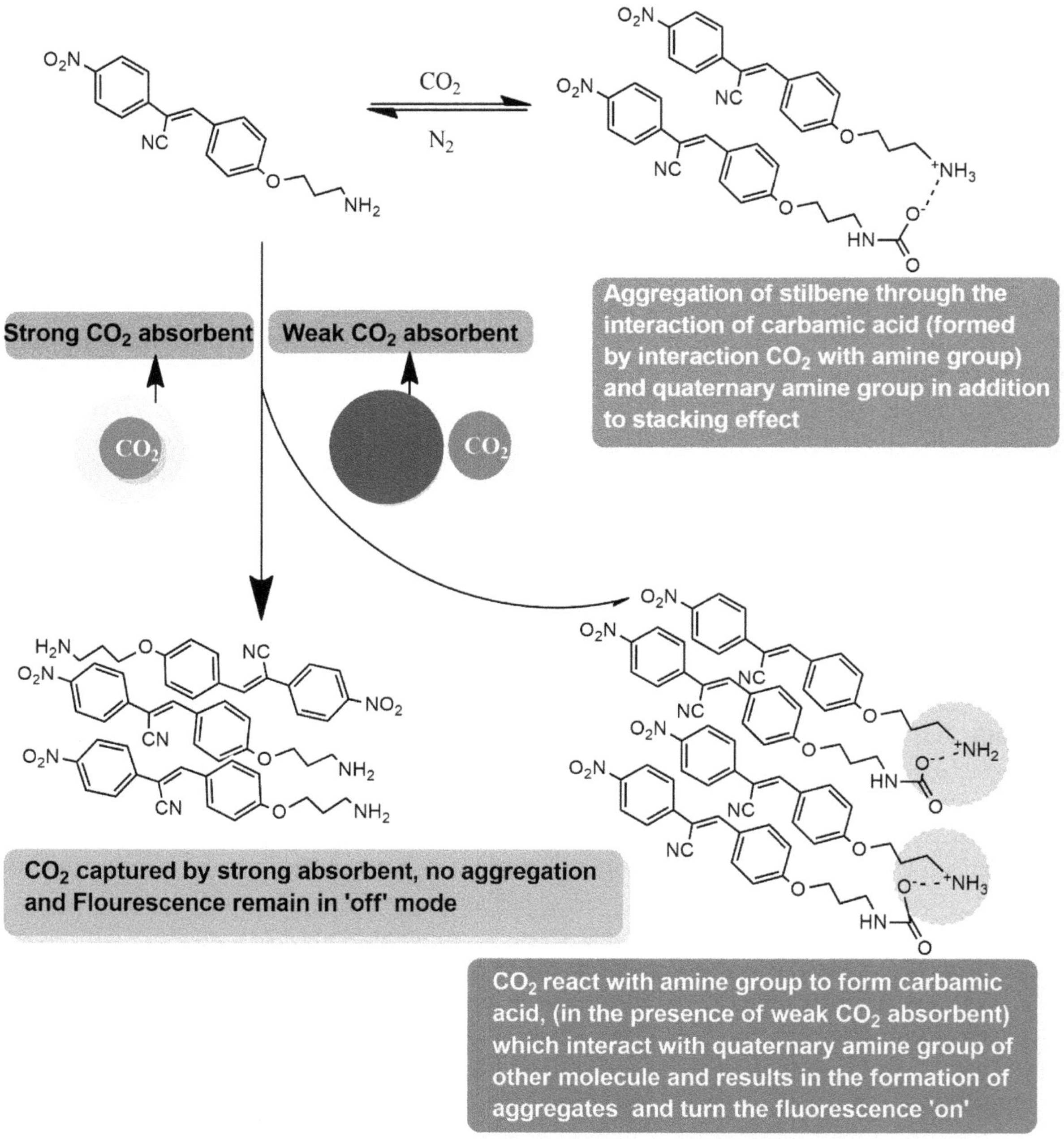

salts and anions. A notable observation is that the wavelength of maximum emission (465 nm) shifted hysochromically (375 nm) on exposure to F^- and CN^- ions. Additionally, on exposing CN^- activated fluorogen to CO_2 allows shifting towards red region for the emission maximum to 465 nm. The detection limit of the one of the derivatives towards carbon dioxide with F^- activated sensor was estimated

Figure 5. Plausible mechanism of fluoride activation of TBBI molecule at C1 position and sensing of carbon dioxide

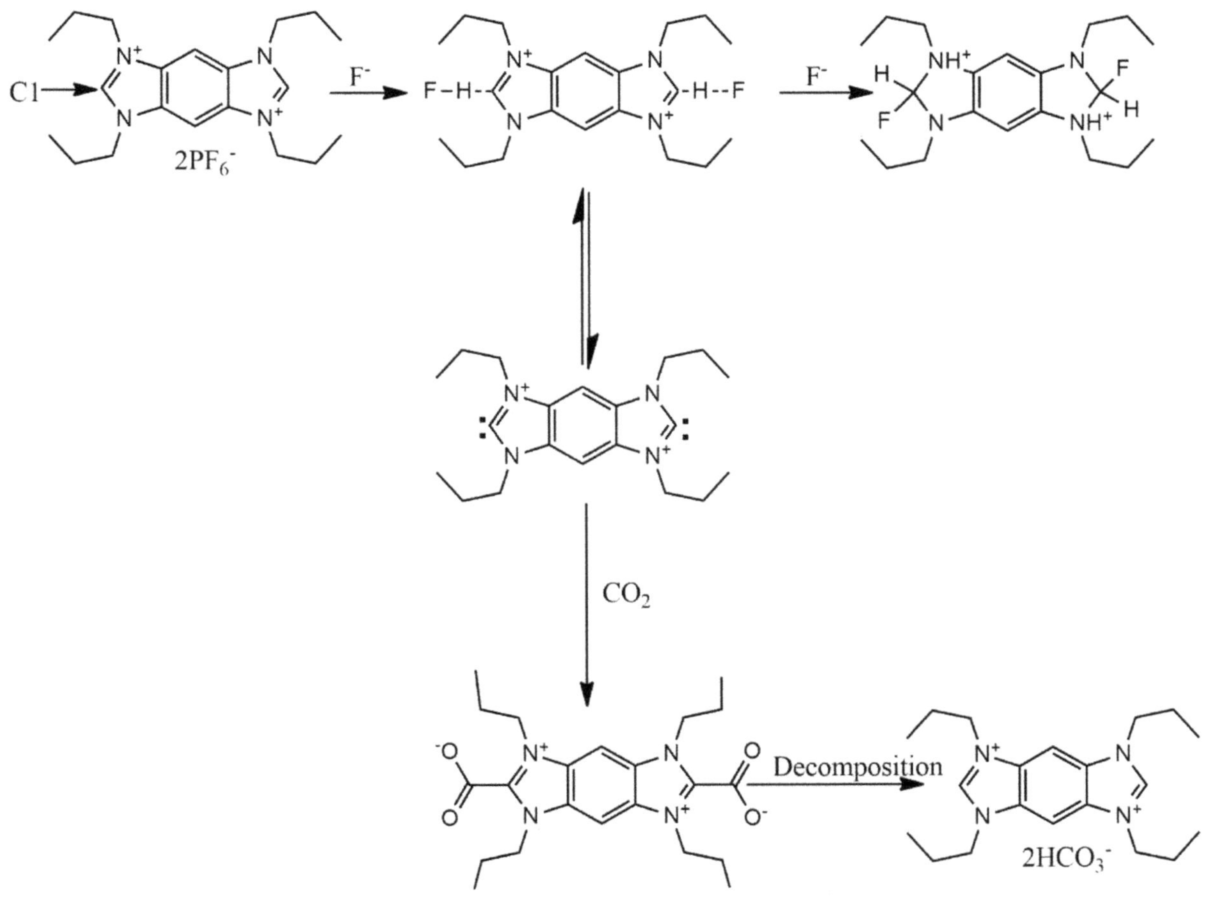

as 1.96×10^{-4} M and 4.45×10^{-3} M with CN^- activated sensor. For the other derivative, the calculated LOD is 8.44×10^{-6} M and 1.16×10^{-5} M.

Adding further, Ishida et al. (2013) synthesized N-fused aza-indacene-based (D10) (Figure 6) fluorogen through the condensation involving benzimidazole-carbinol, trifluoroacetic acid and indole. The N-fused aza-indacenes behaves as optical chemical sensors for CO_2 which is supposed to work through fluoride anion-mediated deprotonation. The fluorescence intensity of the sensor diminishes on addition of fluoride ion, indicating deprotonation of the sensor and this deprotonation event has occurred only on addition of fluoride ion not with other ions. When fluoride pre-treated sensor brought into contact with carbon dioxide a pronounced variation was observed in optical spectra. It was thought theoretically that this contact will aid in the retrieval of the fluorogen. This thought was further supported by the ^{1} H-NMR studies, the spectra of original sensor and that of the sensor after treatment with CO_2 corroborate well. The detection limit for CO_2 under these conditions was thus calculated to be 4.1 x 10^{-7} M.

Lee et al. (2015) prepared naphthalimide derivative, N-[2-(2-hydroxyethoxy)ethyl]-4,5-di{[(2-methylthio)ethyl]amino}-1,8-naphthalimide(D11) (Figure 6) as a carbon dioxide sensor activated by anion taking cognizance of previous studies. They introduced naphthalimide amine derivatives as sensors for carbon

Figure 6. Molecular structures of sensor D8, D9, D10, and D11

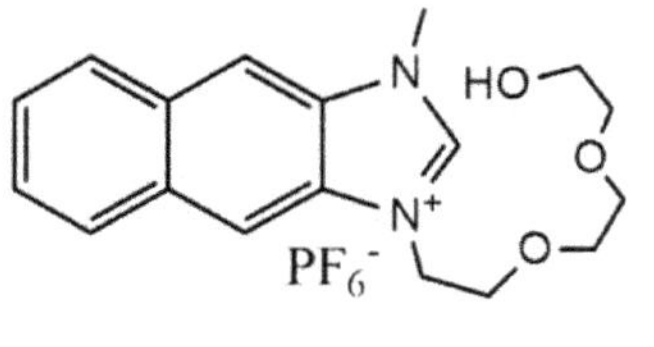

D8

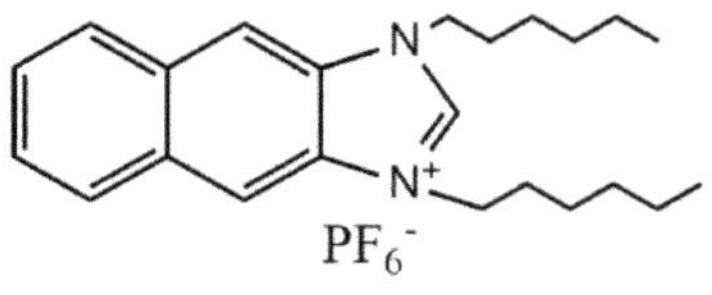

D9

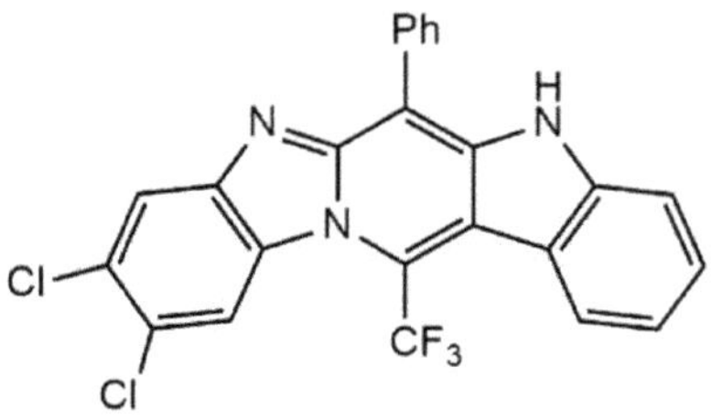

D10

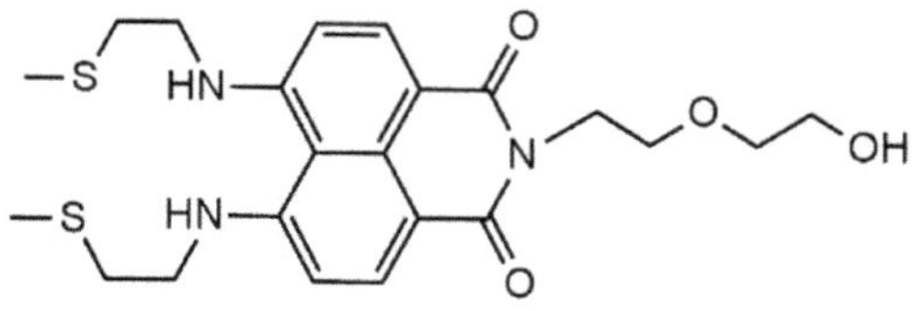

D11

dioxide gas following the fluoride activated deprotonation of amine functional group and consequential interaction with carbon dioxide. The molecule displayed selectivity towards fluoride and cyanide ions both fluorometrically and colorimetrically, which may be because of strong hydrogen bonding attraction of hydrogen of amine functional and anions. On addition of CO_2 the color and fluorescence of the molecule is retrieved and showed distinct "On-Off-On" fluorescence emission variations. The LOD of the anion activated system for carbon dioxide is calculated to be 2.04×10^{-7} M.

A category of chemical dyes known as squaraine dyes exhibits strong fluorescence and substantial absorption coefficient, usually in the red and NIR spectral regions. Considerable absorbance, fluorescence spectral features and ample fluorescence quantum efficiency makes them a suitable as signaling fragment in various chemical sensors and dosimeters (Mayerhöfer et al., 2012). Realizing the successful applicability of squaraine based chemosensors towards anions, Xia et al. (2015) thought to use anion (anion used here is $[Bu_4N]F$) activated squaraine based molecules for visible and reversible recognition of carbon dioxide gas. Between the two isomers (i.e. 1,2 or 1,3 derivatized) of squaraine, the hydrogen bond donor type capability of 1,3 substituted isomer is more as compared to other, which makes it a potent contender for hydrogen bond involving anion sensing. Considering this, the group has prepared unsymmetrical 1,3-squaraine (Figure 7) and investigated its sensory behavior for carbon dioxide in detail through UV-VIS spectrophotometry and ^{1}HNMR spectroscopy in dimethylsulphoxide medium. In presence of 20 folds of fluoride ions the sensor turned green in color owing to large bathochromic shift from 445 nm to 655 nm. When this green solution was kept in open air, the intensity of the signal at 655 showed decrement and the peak at 525 nm showed increment in intensity and green color turned back to purple. This observation suggests that atmospheric CO_2 gas responds to sensor. Hence, increasing concentration of CO_2 gas was given to the sensor solution pretreated with 20 folds of fluoride ions and the spectrum overlapped with that of 5 fold fluoride ions added solution. These features enable the sensing system to behave as a carbon dioxide and fluoride controlled "OFF-ON-OFF" molecular switch. The LOD of the molecule towards the gas was calculated as 15.6 ppm along with immediate reaction.

The plausible mechanistic pathway for the detection is given in the Figure 7. On F^- ion addition, the sensor SH_2 deprotonate following the equilibrium:

$SH_2 + F^- \leftrightarrows SH^- + HF$ (IX)

$SH^- + 2F^- \leftrightarrows S^{2-} + HF_2^-$ (X)

Green colored sulphide ions were highly unstable and reacts with carbon dioxide to form $S\text{-}COO^{2-}$ which immediately break down to SH^- ion and bicarbonate ions.

Another anion activated D-p-A type fluorophore was reported by Ali et al. (2016). It was prepared by connecting imidazole (donator D) with benzothiazole (Gainer A) by means of a phenyl (p) ring. It displays intramolecular charge transfer turn on fluorescent behavior. The photophysical responses of the fluorogen is studied in solvents of varying polarities and pH. The molecule displayed high selectivity towards F^- ions in presence of other ions in 20% DMSO water solvent system. Job's plot established 1: 1 bonding between the sensor and fluoride ion displaying notable binding stability and limit of detection (30 ppb). Fluoride bound sensor when exposed to carbon dioxide significant naked eye color variations are observed in which the initial color is retrieved along with corresponding revival of emission intensity. This behavior could be assigned to emergence of imidazolium hydrogen carbonate/ imidazolium ions/ salt originated due the reaction imidazolium ion with carbon dioxide gas. The studies revealed LOD as 100ppb;2.29mM).

Figure 7. Anion activated mechanistic pathway for carbon dioxide sensing by unsymmetrical 1,3-squaraine

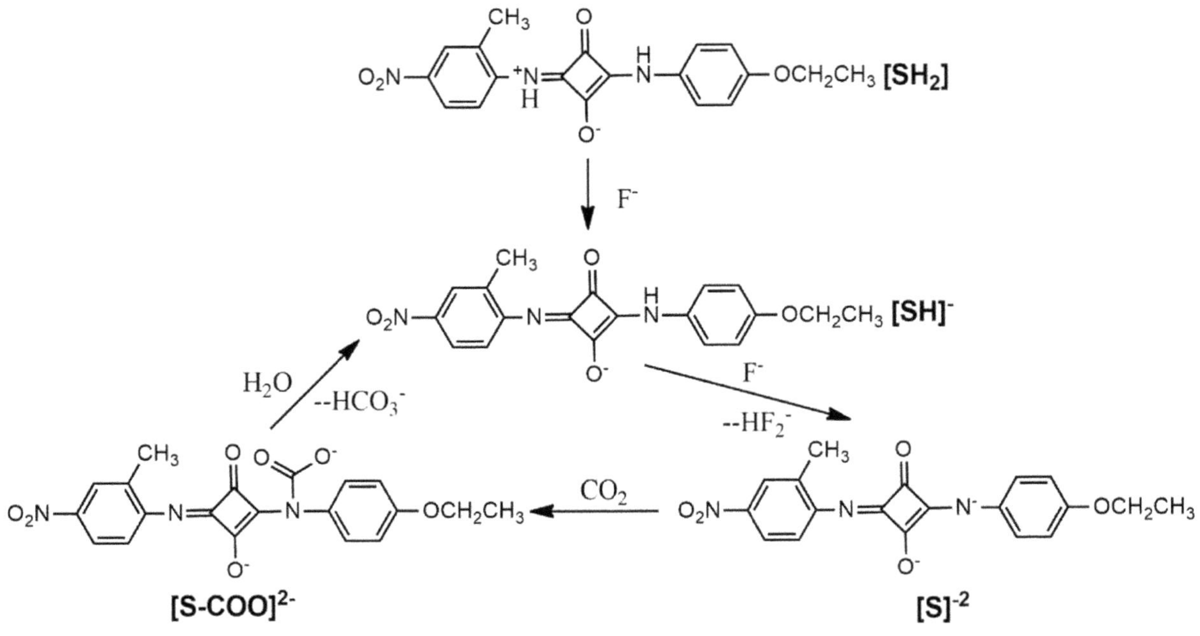

[Indolo[3,2-b]carbazole(ICZ) derivatives consist of two N-H moieties, and it has been reported that N-H moiety can be deprotonated by anion and the deprotonated product can react with carbon dioxide to form the adduct N-COO^-. Based on these cognitions, the ICZ is functionalized with phenyl groups to improve the solubility and 6,12-Diphenyl-5,11-dihydroindolo[3,2-b]carbazole(DP-ICZ) derivative is obtained (Chen et al., 2017). On addition of fluoride ions the color of the solution changed to clay bank and the obvious color changes also reflected in photoluminescence behavior. It displayed blue fluorescence at 450 nm which changed to faintly yellow (562 nm) in the presence of fluoride ions with a concomitant decrease in fluorescent intensity from 972 to 40. In presence of other competitive ions this effect is found to be absent which establishes the selectivity of the probe towards fluoride ion. Previous reports established that nucleophilic N is one among the functional moieties that can use infirm electrophilic character of carbon dioxide gas for recognition application. The N-H fragment of DP-ICZ deprotonates on reaction with fluoride ions and on subjection to carbon dioxide gas showed disappearance of fluorescent peak at 562 nm along with appearance of new peak at 450 nm that appeared as retrieval of DP-ICZ. The LOD was determined as 1.07 μM. The mechanism of sensing may be attributed to transition from DP-2F to DP-ICZ. The assumption was further supported by ^{1}H NMR spectra of sensor, fluoride activated sensor (DP-2F) changes to DP-ICZ on exposure of CO_2. On adding four equivalents of F^- followed by exposure to excessive gas in deuterated chloroform, the NH peak located around 8.10 ppm diminishes first and reappeared afterwards. This event indicated that on exposure to the gas, the molecule transforms from fluorinated sensor to DP-ICZ. To investigate further, deuterated chloroform was changed to DMSO-d_6 to solubilize the precipitates that were present in chloroform resulted in appearance of signal at 8.45 ppm. Previous reports established the appeared peak for HCO^{3-} anion. Therefore, the mechanistic pathway can be proposed to proceed through the formation of DP-2F and CO_2 to DP-ICZ and HCO^{3-}.

Intramolecular Hydrogen Bond and Colorimetric Dye Based Fluorogens for Carbon Dioxide Sensing

A hydrogen bond may be defined as intermolecular or intramolecular associative interactions between the two dipoles, when a highly electronegative atom containing lone pair of electrons is close to the hydrogen attached to an electronegative atom forming two dipoles. These bonds are as crucial as that of other forces like van der Waals interaction, non-covalent attractions and electrovalent bonding. These can be classified into two types: strong hydrogen bonds (energy is 12 kcal/mol) and standard hydrogen bonds (energy is 3-5 kcal/mol). Owing to their capability to stabilize the intermediates produced in the course of the reaction procedure, strong hydrogen bonds become imperative to many enzymatic and catalytic processes. To our interest, these interaction are also found to stabilize the intermediate formed on exposure to carbon dioxide and enable sensing of the analytical gas. Kang et al. (2015) reported a fluorophore for detection of carbon dioxide gas consisting of anthracene molecule appended to an amino acid. The probe is capable of stabilizing the carbamic acid formed during the formation of probe –CO_2 adduct via intramolecular hydrogen bonding. The amine moiety inherent in the chemical structures interact with carbon dioxide gas resulting in the formation of carbamic acid. The carbamic acid formed as mentioned above, act as proton contributor and carboxylate moiety of amino acid act as proton acceptor, thus resulting in the formation of hydrogen bonds. The emergence of these interactions leads to stabilization of carbamic acid thus making amino acid and functionalized amino acids a potential candidate for carbon dioxide gas sensing. The fluorescent emission for 1-aminomethylanthracenes moieties is diminished through photoinduced electron transfer by amine moiety towards anthracene.

But, on binding with metallic ions derivatized 1-aminomethylanthracene forms amide bond resulting in appreciation of fluorescent intensity, as result of blockage in the occurrence of PET. Taking these outcomes into cognizance, Kang et al. (2015), designed a fluorophore by appending anthracene moiety to sodium salt of amino acid as shown in the Figure 8. On exposure to carbon dioxide gas the fluoroprobe react to form carbamic acid bonded via hydrogen bonding with carboxylate ion, thus eventually blocking photoinduced electron transfer mechanism and induces fluorophoric increment of the adduct displayed in the Figure 8. The spectral information enables the researcher to estimate the detection limit of the probe for CO_2 to be ca. 2 ppm.

A series of effectual, solid-state probes for sensing carbon dioxide were prepared by Chatterji and Sen (2015), the probes consists of a gas receptive tertiary amino alcohol for example triethanol amine along with pH-indicator like cresol red, supported on aluminium oxide solid assistance. These probes demonstrated detectable visual signal even in the presence of minute quantity (ppm level) of environmental CO_2 and SO_2 without any interference from water molecules. The sensitivity possessed by the sensors was investigated with the signaling dyes bromothymol blue (BTB) or cresol red (CR), base (N,N'-dimethyl-2-methoxyethylamine, triethylamine, equimolar mixture of triethylamine and ethanol). These reaction mixture did not showed any change in color, demonstrating unequivocally that the formation of the alkylcarbonate moiety requires tertiary amino alcohol as skeleton. Keeping in view the observations, the researchers proposed the mechanism to be based on protonation-deprotonation equilibrium. Cresol red (CR) react with tertiary amino alcohol to interchange proton to form deprotonated anionic dye species which are purple in color. Amino alcohols when interact with carbon dioxide gas results in zwitterionic ammonium carbonate adduct. This hinders the proton removal from the signaler dye molecule by the amino alcohol, resulting yellow color of the protonated dye is visualized.

One more cresol based sensor, consisting of sensing mixture that comprise of the sensor sensitive to pH variations(cresol red) embedded on a gas-penetrable membrane, is synthesized by Kwanyoung et al. (2020). pH signaler when subjected to carbon dioxide demonstrated color variation, this variation is quantified by an RGB application. The shift in G and B values of the sensing mixture displayed noteworthy linear relationship with amount of carbon dioxide present in soil measured through non-dispersive infra-red (NDIR) probe enabling quantification of analyte concentration. Experiments through carbon

Figure 8. Mechanism of sensing of CO_2 by anthracence based fluorescent molecule via, stabilization of carbamic acid, formed by interaction of CO_2 with ligand, through hydrogen bonding

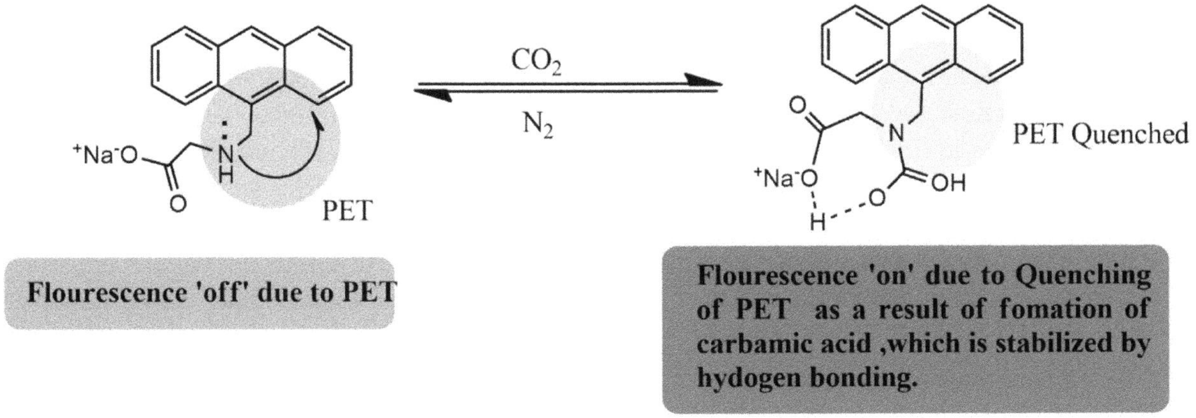

dioxide injection chamber revealed that the indicator has the ability to recognize soil carbon dioxide amount of 0.1 to 30% in a very short span of time. Real place investigation at a natural CO_2 opening and CO_2 leakage location revealed that this sensor can be utilized in measuring surficial CO_2 leakage and it offers a cost effective means of sensing along with the advantage of uncomplicated system installation.

Schutting et al. (2013) prepared Diketo-pyrrolo-pyrrole (DPP) based novel class of pH responsive sensors for carbon dioxide sensing and these were rendered dissolvable in organic solvents and in polymeric solvents through functionalizing dialkyl sulfonamide functional moieties. DPP and their derivatives are utilized in developing biophotonics and optoelectronics possessing less bandgap polymeric or copolymeric materials offering conjugation to be used in semiconductors. Qu et al. (2010) suggested a fluoride sensor that caused a shift to longer wavelength in wavelength of maximum absorption due to deprotonation of the lactam moiety of DPP chromophoric sensor. These observation suggested that DPP derivatives offers a promising potential for developing sensors for carbon dioxide as they are based on acid base equilibrium. The DPP based indicators synthesized by Schutting et al. (2013) display notably important absorption and emission spectral features in non-deprotonated or deprotonated forms that makes colorimetric and ratiometric measurements possible. The probe also offers modulatable sensitivity, appreciable optical variation even at less concentration of carbon dioxide and considerable photostability. The synthesized sensors coupled with tetraoctylammoniumhydroxide was planted on ethylcellulose to form plastic sensing entity for the gas under consideration. The probes sense CO_2 and showed color variation (blue when CO_2 is absent and pink when present) thus making colorimetric recognition possible. Noticeable shifts in the fluorescent intensity further support the color variation. For all the derivatives the fluorescent emissivity both in neutral and anionic form showed increment with increasing CO_2 amount. This increment of intensity for anionic form can be attributed to appreciable decrease in concentration of neutral form not the FRET. This may be because of spectral overlapping of emission spectra due to non-deprotonated species and absorption spectra due to anionic species formed by deprotonation.

Another diketo-pyrrolo-pyrrole (DPP) based sensor, 2-hydro-5-tert-butylbenzyl-3,6-bis(4-tert-butylphenyl)-pyrrolo[3,4-c]pyrrole-1,4-dione(DPPtBu^3) to sense carbon dioxide optically is synthesized following a facile one step synthetic route from 'Pigment Orange 73'. Lactam nitrogen of 'Pigment Orange 73' is derivatized with tert-butylbenzyl group. The signaling dye molecule can be easily solubilized in organic/polymeric solvents and displays pH-dependent absorption. Both the species also demonstrate notable fluorescent quantum efficiency (φ_{prot}0.86;φ_{deprot}0.66). Thus, allowing the sensing of the carbon dioxide gas colorimetrically and ratiometrically (fluorescence intensity). The emission from the two species sensor corroborate well with the signals obtained from green/red routes for RGB photographic system. This makes possible the visualization of carbon dioxide dispersal through an uncomplicated and cost efficient optical set-up. This sensing system demonstrates exceptional sensitivity and is especially auspicious for tracking carbon dioxide levels in the atmosphere (Schutting et al., 2014).

It has been documented that certain amines other than primary amines interact with carbon dioxide and form carbonate ionic liquids when alcohols are present or form of carbamate ionic liquids with one another. Consequentially, physical characteristics, which include polarity and viscosity showed significant variation to larger values. These variation in physical parameters could be reversed to original standards by desorbing carbon dioxide by bubbling nitrogen gas. These amine/alcohol combinations are therefore sometimes called "smart solvent systems". These variations in said physical parameters can be visualized by bare eyes with the assistance of indicator dyes. Charge transfer dyes (CT dye) assist in visualizing the variations in polarity. Fluorescent intensity decreases along with concomitant red shift with the increase in polarity. While, the viscosity variation can be seen using molecular rotor dyes (MR dye). Because of restriction of intramolecular rotation, the fluorescent emission caused by molecular rotor type pigments usually showed noticeable increment with the rise in solvent viscosity. Jin et al. (2016) dissolved both the dyes in amine/alcohol concoction. The fluorescence emission bands for the dyes do not overlap, the emission caused by CT dyes dominates in lower polarity and high viscosity solutions (the case before the interaction with carbon dioxide) however, on exposure to carbon dioxide the emission from CT dyes diminishes and emissions of MR dyes emerges and dominates in more polar and viscous solution (Figure 9). In particular, the group has taken combination of a Nile Red with allyl-2-cyano-3-(1,2,3,5,6,7-hexahydropyrido[3,2,1]quinolin-9-yl)acrylate(ACHQA) as the CT dye and MR dye, respectively. Equimolar combination of 1,8-diazabicyclo-[5.4.0]-undec-7-ene(DBU) with 1-propanol was the choice for smart solvent concoction. Nile Red and ACHQA in DBU/1-propanol medium emanated red light attributed to Nile Red, while the color subsequently moved towards green as the concentration of CO_2 is increased.

Aazaphthalocyanine dyes (AzaPc) have been demonstrated to exhibit favorable spectrum features, notable molar absorptivity (over 200 000 $L.mol^{-1}.cm^{-1}$) and appreciable quantum yields (up to 50%). These dyes are stable under the effect of radiant energy and easy derivatizable as per the requirement. Derivatizing with appropriate detection receptors these dyes can be made responsive to pH and metal ions.

By controlling the intramolecular charge transfer mechanism the fluorescence switching to on and off state can be modulated. These novel pH indicators responsive in basic pH range offers the promising potentiality for use in optical CO_2 sensors. This could help these sensors overcome some of the shortcomings of the most advanced ones. Lochman et al. (2017) prepared CO_2 sensors reliant on zinc azaphthalocyanine (Zn-AzaPc) signalers implanted on polyurethane polymeric frameworks. All the four sensors displayed no fluorescence in their deprotonated form (when carbon dioxide is not present) and switch to 'on' mode when exposed the gas under analysis. The receptor carrying one phenolate receptor is found to be most appropriate because of the exceptional brightest emission. Hydrophillic Hydrothane is used as matrix in place of ethylcellulose matrix in case of hydrophobic Zn-AzaPc sensors. Hydrothane 25 displayed exceptional outcomes, increasing fluorescence intensity twelve folds in off/ on stages, notably effective quantum yield (0.071) and repeatable reaction towards carbon dioxide ranging in 0 - 95 kPa, along with little dependability on temperature ranging in 15-35 °C. Crucially, sensitiveness rose dramatically when a base's alkyl chains were shortened, following the sequence TOAOH ˂ TBAOH ˂ TEAOH. This provides useful mechanism for extending the range of probes for variety of utilities.

Pfeifer et al. (2018) synthesized one more class of indicators based on derivatives of perylene molecule for optical recognition of pH and CO_2 gas. These molecular scaffolds showed appreciable luminescent features photostability and versatile chemical derivatization. The sensors can be prepared through a facile one step synthesis from Lumogen Orange. Distinguished fluorescence spectral profiles of these molecules for various forms made possible the colorimetric and ratiometric investigation of pH and carbon dioxide. The sensitivity towards carbon dioxide can be modified by varying the dye, amount of plasticizer and type of quaternary ammonium salt. The di-substituted perylene displays distinct features with two protonation/deprotonation equilibria capacitate recognition of analyte gas molecules in unprecedentedly broad dynamic range.

Nanomaterials Based Sensing Probes for Carbon Dioxide

The sensitivity of the sensing probe for the analyte is significantly influenced by the dimensions of the material. The unique and enhanced properties exhibited by materials with nanoscale dimensions may

Figure 9. Mechanism of sensing of CO_2 by the concoction of CT organic molecule and MR organic molecule

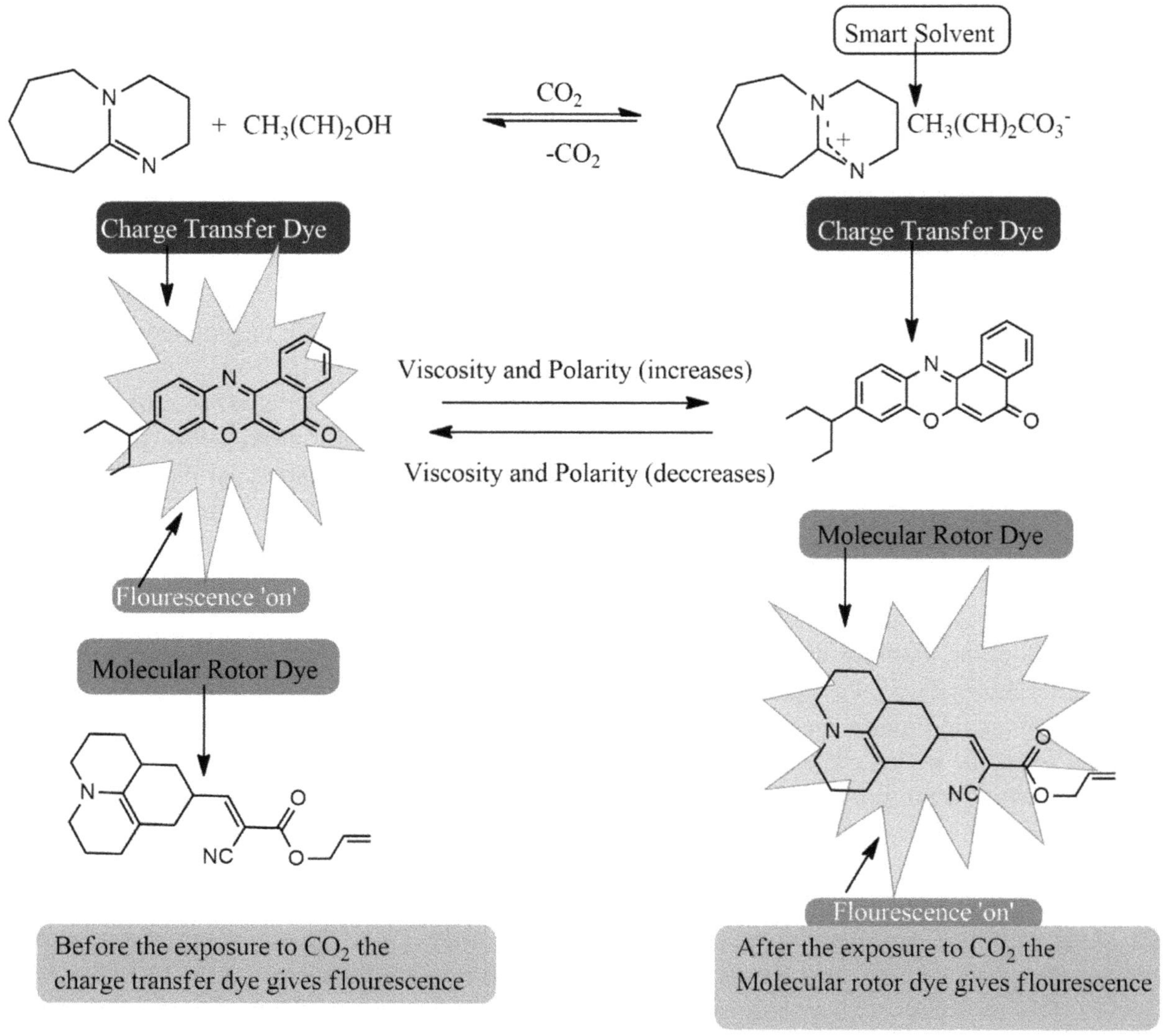

be ascribed to the enhancement of surface area at their disposal. The enhanced properties encompass size-dependent characteristics, increased catalytic activity, accelerated response and recovery, and surface energy. These attributes are pivotal in facilitating the enhanced sensing capabilities of sensors based on nanomaterials. As these materials own a larger surficial area per unit mass inevidently lead to increment in the surface energy. The enhanced surface energy improves cohesiveness of the material for the gas molecules and facilitates gas adsorption, thus enabling the sensing of the gas. Nano dimensions resulted in quantum confinement effect, which substantially influence the optical properties of material along with band structure and electronic features, thus increasing sensitivity towards environmental changes. Also, the larger surface and small dimensions accelerate gas diffusivity or adsorption, making the response and recovery times' quick.

Among the various forms of nanomaterial, burgeoning interest has been drawn towards carbon nanodots (CDs) on account of their extraordinary behavior in terms of ease of synthesis, chemically inert, unsusceptible to photobleaching, less toxic behavior, and satisfactory compatibility towards biotic system (Baker et al., 2010). Thus, enabling the carbon nanodots to act as sensors, catalysts, bioimaging entities and in drug delivery system (Zhao et al., 2011). Carbon nanodots normally have carboxylic acid or amine groups on the outer surficial area, which allows easy derivatization. Thus, functionalized carbon nanodots able to sense a particular entity can be formulated by derivatizing the surface with the receptors responsive towards analyte. Amino polymers like polyethyleneimine (PIE) are easily available and respond to carbon dioxide frequently. The amine-coated CDs produced by pyrolyzing a combination of carbon carrier and PEI are reportedly insensitive to CO_2, despite the fact that substituted PEI are coated as protecting layers on luminescent CDs. This insensitivity may be primarily because of the carbonization of carbon dots at elevated temperatures. Pei et al. (2017) prepared PEI coated carbon dots (PEI-CDs) by using three polymeric materials available in the market with molecular weight (Mn = 600, 1800, 10000 g/mol). The fluorescence spectral studies along with ^{13}C NMR studies were carried out to investigate CO_2-molecular switch 'on/off' type fluorescent behavior of the prepared PEI-CDs in water/DMSO solvent system, which were further employed to quantify the acidic gases like CO_2 and SO_2. The precipitates formed by the interaction of PEI-CD with carbon dioxide in dimethyl sulphoxide and water solution were gathered and ^{13}C NMR studies were performed. A peak located on 160.2 ppm appeared and attributed to HCO^{3-} anion, which evidenced the emergence of polyammonium bicarbonates precipitates. The pH of the solution showed decrement from 9.3 to 7.5 on attaining the equilibrium after the exposure to carbon dioxide gas. To further investigate the factors responsible for switching on/off of fluorescence, the variation caused by pH change on the fluorescence intensity of the system was investigated. To attain the purpose, hydrochloric acid was poured into the solution having PEI1.8K-CD (4 mg/ml) to maintain the pH value to 4.0, 5.0, 6.0, 7.0, 8.0 and 9.0 respectively, when the system is not exposed to carbon dioxide. Only a little change in fluorescence intensity was observed on varying pH from 7.5 to 9.0, which indicates that the pH change can only diminish/recover fluorescence intensity but not responsible for switching between the two modes. When carbon dioxide gas was exposed to the aqueous solution of PEI1.8K-CD and aqueous DMSO solution of PEI1.8K-CD at the same PEI-CD content of 4 mg/ml, no precipitates was observed for aqueous solution while for aqueous DMSO solution system polyammonium bicarbonate precipitates were formed. The formed precipitates were insoluble in the solution, it thus decreased the concentration of carbon dots and lead to switching off of the fluorescence. The results revealed that the emergence of precipitate of polyammoniumbicarbonates in water/DMSO medium engage in a vital role in switching the fluorescence modes on/off. Thus, the redox equilibrium between carbon dots and CO_2 followed by the production of precipitates (hydrophilic polyammoniumbicarbonates) are responsible for the fluorescence "on-off" outcomes. The reaction can be reversed on bubbling nitrogen gas.

Nemade & Wahguley (2014) fabricated chemically resistant carbon dioxide sensing probes comprised of graphene/aluminium oxide quantum dots. Composite material by varying percentage weight of graphene (20–80 wt%) and fixed one gram of aluminium oxide were prepared and sensing experiments were carried out. The outcomes of the investigations revealed that graphene/Al_2O_3-based sensors demonstrated notable identification which increases proportionally on incremental introduction of graphene.

The mechanistic pathway for recognition of gas is proportional to defect density caused by addition of grapheme onto the fixed amount of aluminium oxide. These defects offer vacancies for the atmospheric oxygen to be adsorbed. The adsorbed oxygen undergoes the reaction given below:

$O_2(g) \longrightarrow O_2$ (adsorbed) (XI)

O_2(adsorbed) + $e^- \longrightarrow O_2^-$ (adsorbed) (XII)

O_2^- (adsorbed) + $e^- \longrightarrow 2O^-$ (adsorbed) (XIII)

Recognition of carbon dioxide depends mainly upon the reaction of the surfacial molecules of probe and adsorbed oxide ions. Carbon dioxide is adsorbed by the bridging oxygens, resulting in emergence of surface carbonates, which eventually increases the resistivity of chemiresistor probe revealing the nature of probe to be n-type. 80 wt% graphene/Al_2O_3 composite found to have best among others in terms of sensor features like reaction time, temperature requirement and stability. The same research group also synthesized Graphene/Y_2O_3 (20 wt% graphene into 1 gram Y_2O_3) quantum dots composite chemiresistor type sensor for carbon dioxide gas. Graphene and the 20 weight percent graphene/Y_2O_3 composite exhibit good dependence on the CO_2 gas concentration (Nemade & Wahguley, 2013). The optimal sensing was calculated to be 1.08 for 20-wt% of graphene/ Y_2O_3 at 35 ppm of carbon dioxide amount. The mechanism for gas detection can be attributed to electron transfer reaction in which one of the reactant is oxidized while other is reduced. The existence of weak electrical attraction between the reactant entities (ions/ molecules) is the necessity for electron transfer process to proceed. The deviation of the reduction potential of donating and accepting reactants provides the driving free energy for the redox reaction. The detection pathway is similar to above composite that is the interaction of adsorbed oxygen ions and carbon dioxide. Due to the electron withdrawing nature of CO_2, it draw the electrons from the surface resulting in increment of electrical resistance, thus enabling the sensing.

The sensitivity of the sensor probe that works on the principle of interaction of surface molecules with entity to be analyzed, can be improved by the augmentation of the surface area. Electrospinning is one of the cost effective technique for preparing membranes with increased surface. In this method stronger static voltage is applied to generate an framework of short chained fibres (having diameter value from 10 to 1000 nm). Electrospun nanofibrous frameworks fabricated using this method are reported to possess 1 to 2 orders increased surface area as compared to regular thin films. This huge accessible surface area is anticipated to have the potential for exceptional sensitivity and quick reaction times. Aydogdu et al. (2011) and researchers prepared electrospun nanofibres using Poly(methyl methacrylate), ethyl cellulose, the plasticizer (dioctylphthalate, 8-hydroxypyrene-1,3,6-trisulfonicacidtrisodiumsalt and tetraoctylammoniumbromide. The probe displayed large sensitivity assigned to large surface area-to-volume ratio of the synthesized electrospun material. Stern-Volmer investigations revealed that these materials have 24 to 120 fold high sensitivities along with short response times and signal reversibility.

Fernández-Sánchez et al. (2007) described carbon dioxide responsive films by incorporating phenol dyes, α-naphtholphthalein (NAF), naphthol blue black (NBB) and calmagite (CMG), accompanied by phase-transfer agent, tetraoctylammonium hydroxide (TONOH), into metal-oxide nanoporous matrices (aluminum (AlOOH), silicon (SiO_2) and zirconium (ZrO_2) oxides.). The sensing films responded to CO_2 concentrations in the gas phase between 0.25% and 40% CO_2 (v/v) for NAF–TONOH, 4.1% and 30% CO_2 for NBB–TONOH and 0.6% and 40% CO2 for CMG–TONOH, with LOD of 0.25%, 4.1% and 0.6% CO2 (v/v) for NAF–TONOH, NBB–TONOH and CMG–TONOH, respectively. The film displayed appreciable dispersion and availability of signaler, resulting in fast response, reduced likelihood of agglomeration, the potential for gamma radiation sterilization, resistance towards chemicals and increased stability of the environment.

Nanoparticles based optical sensor consisting of 12-μm polystyrene (PS) film, upconverting nanoparticles (UCNPs; 40–100 nm in size) of the type $NaYF_4$:Yb,Er, and pH probe bromothymol blue (BTB) in its anionic (blue) form were synthesized by Ali et al. (2010). Selective permeability of the polystyrene for carbon dioxide over proton is the reason behind the choice. The UCNPs were exposed to 980-nm laser to result in green emission at 542 nm along with another red emission at 657 nm. The fluorescent intensity of the nanoparticles located at 542, 657 nm showed increment with the incremental exposure to carbon dioxide. The bromothymol blue (a sulfonate) and tetrabutylammonium cation (TBA) combines to result in an ion pair and tetraoctylammoniumhydroxide transforms BTB to phenoxide (blue) ionic species and forms buffer. This ensemble demonstrates a reaction time of ~10 s on increasing carbon dioxide to 1% in pure argon medium, with the reversal of system within 180 s, and the LOD of 0.11% for carbon dioxide.

Another CO_2-responsive nanocomposite consisting of polymer, poly(N-(3-amidino)-aniline) (PNAAN), coated gold NPs (AuNPs) prepared by the reduction of $HAuCl_4$ using N-(3- amidino)-aniline (NAAN) (Ma et al., 2016). The amidine functional moiety of PNAAN is easily protonable to form a hydrophilic amidinium group due to the presence of dissolved carbon dioxide. This results in swelling of PNAAN and causes the detachment of the polymer from gold nanoparticles which consequently aggregates the nanopartilces to demonstrate the visible color variation. The optical measurement revealed LOD to be 0.0024 hPa.

In addition to above, layered double hydroxide-based nanocomposite HPTS/NiFe-LDH (8-hydroxypyrene-1,3,6-trisulfonicacid trisodium, HPTS) was prepared by simple one-step hydrothermal method (Li et al., 2016). The nanocomposite displays no fluorescence but when exposed to CO_2 gas the carbonate ions formed were then enter the interlayer of NiFe-LDH and result in the release of anionic dye, consequently recovering fluorescence. The fluorescence intensity increases proportionally to the amount of carbon dioxide.

Recently, colorimetric assay including silver nanoparticles was prepared by employing two capping agents: first is thiomalic acid and second is maltol (Sheini, 2020). The system's function depends on the hydrolytic cleavage of urea into NH_3 and CO_2. These products react with nanoparticles resulting in aggregation. Between the two capping agents, the maltol capped with silver nanoparticles selectively responds to carbon dioxide and easily visualized in color changes from yellow to red. The linear range was 0.08 mg.dL^{-1}-220.0 mg.dL^{-1} for CO_2 and limits of detection was 0.06 mg.dL^{-}.

Li et al. (2021) synthesized a composite of tetraphenylethene (TPE) and gold nanoclusters (Au NCs) embedded onto disulfide functionalized hyperbranched poly(amidoamine) framework for recognition of dissolved CO_2 ($3.8 \times 10^{-4} - 1.9 \times 10^{-2}$ mol·L^{-1}) with LOD 7.8 μM and the reaction is reversible by bubbling nitrogen gas.

Sol Gel Based Sensors for Carbon Dioxide

Sol gel based sensors employ silica sol-gel immobilized matrix which ensnare the analyte-responsive probe while permitting the analyte to permeate through the matrix. Sol-gel and polymer materials are the two types of immobilization matrices that are most frequently utilized. The sol gel method basically includes two steps: hydrolysis followed by condensation reaction with a suitable metal alkoxide. As a result, a porous glass matrix is formed that is responsible for the entrapping of analyte sensitive reagent, thus restricting its leaching, however it allows ingression of analytical entity. The sol gel preparation process offers many tailorable features like pH, starting material type and amount, amount of water, curing temperature, possibility of modification of physicochemical properties of the sol gel and available potentiality of designing hybrid matrixes containing both hydrophobic and hydrophilic moieties so as to improve the capability of sensor. However, this versatile nature is sometimes found to be disadvantageous as culminating sol gel morphology is highly depended upon the processing conditions. Particularly speaking, the final morphological structure of the matter significantly impacted by temperature. Thus, for high temperature application polymers are preferred as compared to sol gel.

A significant benefit of these materials in comparison to polymeric materials is their adaptability. Additionally, these materials are able to deposit on a variety of solid support without any tedious process, through various methods. As compared to conventional methods, this chemical procedure proceed via single step- moderate temperature pathway that was first utilized to fabricate an extensive variety of materials, including glasses or ceramics, with improved quality and uniformity. The significant property of sol gel method includes potential for the preparation of materials that combine inorganic and organic components even at the molecular level. The organic component of the above mentioned combination will stay unchanged during the sol-gel treatment owing to gentle processing at low temperatures, which is exceedingly challenging to do by traditional methods. The uniform mixing of inorganic /organic components to form a single phase offers excellent options for customizing the end attributes of the material which also includes optical. Therefore, the materials obtained from sol-gel present itself as a potential option for optochemical probe. The sol gel process includes sequence of consecutive (and typically overlapping) procedures: (a) hydrolytic cleavage of reactants (b) condensation, (c) gel formation (d) ageing and (e) drying. The whole procedure results the emergence of combined matrix of inorganic /organic components proceeds via origination of colloidal suspension (called sol) followed by formation of interconnected framework in a continual liquid form called as gel. Hydrolysis step involves the reaction of precursor's alkoxy functionality and water to generate silanol functional moieties (≡Si–OH) followed by condensation which includes the reaction of silanol with silanol /alkoxy groups to form siloxane (≡Si–O–Si≡) with concomitant removal of H_2O or alcohol. The end product of the reaction procedure is the generation of sol, a fluidic phase of high viscosity comprised of polymeric compounds, aggregated materials and medium used for reaction. The sol can be molded to form gel or may attached on the substrate. With the increase in siloxane bonds the molecules result in aggregation, thus linking sol molecules leading to gelation. On attaining the gelling point only the viscosity of the system increases without any chemical reaction. The speed of the reaction and variation in physical properties may decrease on attaining solid form but still, the system remains active and the conditions under which it ages significantly impact the ultimate structure of the material. Ageing process includes generation of crosslinking, shrinking and hardening of the gel.

Lo and Chu, (2009) reported sol gel based carbon dioxide probe which includes n-octyltriethoxysilane (Octyl-triEOS)/tetraethylorthosilane(TEOS) with dispersed form of 1-hydroxy-3,6,8-pyrenetrisulfonic acidtrisodiumsalt(HPTS, PTS-), granules of silica and tetraoctylammonium hydroxide (TOAOH). The flourometric studies revealed that the comparative fluorescence integration of indicator showed decrement on incremental exposure to carbon dioxide and found to have notable sensitivity of 26 (ratio I_{N2} / I_{CO2}, I_{N2} and I_{CO2} are the fluorescent integration in nitrogen medium and carbon dioxide environments, respectively) with a direct proportionality to carbon dioxide concentration in the span of 0-100%. The sensing system also exhibited significant response time of 9.8s (when system switches to carbon dioxide atmosphere from nitrogen) while on transition to reverse mode it is 195.4 s.

Another optical fluorescent sol-gel sensors was synthesized by Dansby-Sparks et al. (2010) using redesigned silica-doped framework with 1-hydroxy-pyrene-3,6,8-trisulfonate(HPTS) organic dye with

LOD 80 ppm accompanying limit of quantification as 200 ppm for carbon dioxide. On exposure to the target gas the fluorescent intensity due to dye molecule got diminished, because of the presence of the carbon dioxide gas which causes the protonation of its anionic form.

Other group ion paired 1-hydroxypyrene-3,6,8-trisulfonicacid and cetyltrimethylammonium bromide, trapped within a hybrid sol–gel-based three dimensional cage comprised of n-propyltriethoxysilane and liphophilic organic base (Wencel et al., 2010). This ensemble demonstrated two pH-dependent variations in optical signals, thus offering dual excitation ratiometric sensing to quantify dissolved carbon dioxide. This measuring method unresponsive to amount of indicator, percolation/photobleaching of the fluorescent responsive molecule and instrumental alterations. The experiments displayed high degree of reproducibility, reversing ability, stabilizing ability with LOD of 35 ppb.

An additional class of gel other than sol gel are polymer hydrogels (PHGs), these are a three-dimensional viscoelastic matrix prepared by the interlinking of polymeric chains through various chemical and physical entanglements. These hydrophilic polymers form a network that expand in biological liquids without dissolution followed by soaking up a significant quantity of them without dissolving in them. They can absorb more than 400 times their initial weight, which is over 20% of their original weight, and even more.

Wang et al. (2019) prepared a CO_2 sensitive hydrogel from Schiff's base reaction between polyethyleneimine with significant branching (BPEI) and moderately oxygenated dextran (PO-Dex) via stacking layer on layer. The swelling characteristics of the gel was investigated using Fabry Perot fringes. The films, similar to other hydrogels, exhibit swelling when exposed to water. Furthermore, the introduction of CO_2 leads to a more significant expansion of the particles, as a result of the chemical reaction of CO_2 and the $-NH_2$ belonging to BPEI. The expansion caused by gas can be detected through observation of the displacement of the Fabry–Perot fringes. Consequently, the film can serve as sensing probe towards dissolved CO_2 following optical mode. The signal produced by the probe towards CO_2 is linear/reversible and quick unlike most hydrogels.

Recently, Zhang et al. (2023) reported colorimetric carbon dioxide (CO_2) optical sensing ensemble having polypropylene framework placed onto a papery layer of thickness (30 µm) comprised of polyurethane. The polypropylene membrane facilitated the diffusion of CO_2, resulting in pH alterations within the hydrogel that contained an indicator (capable of dissolving in lipids), a cation exchange resin, and amine as cation. The sensitivity of the system is efficiently modulated by adjusting the proportion of the indicator molecule which is lipophilic in nature and the cation exchange resin. Slight excess amount of cation exchanger allow better sensing of carbon dioxide even at lower concentration. Color of the hydrogel ensemble corresponds well to the concentration of carbon dioxide gas and thus enable its quantification.

ELECTROCHEMICAL CARBON DIOXIDE SENSORS

The beginning of the electrochemistry dates back to late 18th century with Galvani, Volta, and Nicholson. Since then the field of electrochemistry has found basic implications in various phenomena associated with regular life related processes such as corrosion, electroplating of metals, electrochemical treatment of drinking water, batteries, fuel cells, and electroanalytical sensors (Deng et al., 2008; Fu et al., 2008; Bergmann and Koparal, 2005; Laik et al., 2008; Robel et al., 2006; Xiang et al., 2007).

Electroanalytical chemistry has various applications which include environmental observation, quality control industrial sector, and biomedical investigation. Electroanalytical research has witnessed a mammoth uptrend in last few years with a range of development which include the fabrication of customized interfaces and monolayers of molecules (Braun et al., 1998; Wang et al., 2001), the linking of biological constituents with electrochemical transducer (Alfonta et al., 2001), and advancement in the basic understanding of voltammetric techniques. Electrochemical probes are attracting a significant attention for exploring chemical sensors.

Electroanalytical techniques are broadly classified as potentiometric and potentiostatic sensing. Current of electrochemical system is fixed at zero and potential established across the membrane is measured in potentiometry while in potentiostatic techniques electrode potential derives reactions involving electron-transfer and the current arising as a result is observed. The measured current is directly proportional to the rate of electrons across the interface of electrode and solution. Potentiostatic techniques are more popular due to their various distinct advantages.

Electroanalytical Techniques

Linear sweep voltammetry is simple electroanalytical technique where working electrode potential is changed with time in a linear manner in a particular direction (either from low positive potential to high positive potential or *vice versa*) at a constant scan rate and the resultant current is recorded with respect to applied potential or time.

Peak current obtained for a system which is reversible in linear sweep voltammetry is depicted by Randles-Sevčik equation:

$$i_p = [2.69 \times 10^5]\, n^{3/2} A D^{1/2} C \nu^{1/2}$$

here i_p is peak current in μA
n is the number of electrons
A is area of electrode surface in cm^2
D is diffusion coefficient of the species being oxidized or reduced in cm^2/s
C is concentration in mol/cm^3
ν is rate of scan in V/s

Figure 10. Potential-time waveform applied for linear scan voltammetry

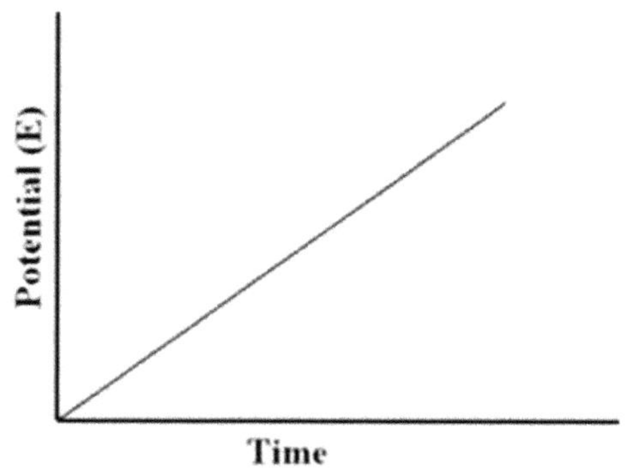

Variations in voltammetric patterns of linear sweep voltammograms for different types of electrode reactions have been discussed comprehensively by Nicholson and Shain (1964). The outcome of their observations proved out as milestone in qualitative electrochemistry.

Cyclic voltammetry is another electroanalytical technique of prime importance and is based on linear variation of the working electrode potential. The technique is basically a different version of linear sweep voltammetry with provision of reverse potential scan and is carried out by switching the direction of the scan at a particular potential. In cyclic voltammetry, working electrode potential is varied from a potential (E_1) to another potential (E_2) at a predefined scan rate (v); but unlike linear sweep voltammetry, at potential E_2 the course of potential sweep is retreated instead of being concluded. On re-attaining the original potential, E_1, the sweep cycle may be ceased, again retreated, or continued to new value E_3. Single or a greater number of cycles are employed depending on the information sought.

Cyclic voltammetry is considered as a powerful technique which gives important data about redox couple under investigation and allows simultaneous monitoring of oxidation as well as reduction processes. Hence, the technique is found to be very useful for investigating the redox behavior of biological molecules, organic or inorganic compounds with special emphasis on fundamental aspects of redox processes, electron transfer reactions, and understanding reaction intermediates. Application of cyclic voltammetry for analytical purposes is limited on account of the coupling of faradaic current with charging current.

Pulse voltammetry history accounts to the early work of Kemula which is centered on the concept of mechanical switch (Osteryoung and Wechter, 1989). In pulse technique, changes in potentials are discontinuous on the time scale of the experiment and sampling of the current is done only when the potential becomes constant for some-time. As a consequence, background current has been nullified, and reliable data can be collected at lower concentrations also. The credit for the basis of modern pulse voltammetric techniques goes to the early work of Barker and Jenkin (1952).

Differential pulse voltammetry is a very useful method for determination of organic as well as inorganic species. In differential pulse voltammetry, a pulse of constant height is superimposed on a linearly varying base potential applied on the working electrode.

The current is measured, just afore applying the pulse, and again just afore the pulse ends and the difference of the values are instrumentally recorded with respect to the potential. The height of resultant

Figure 11. Applied potential-time waveform for differential pulse voltammetry

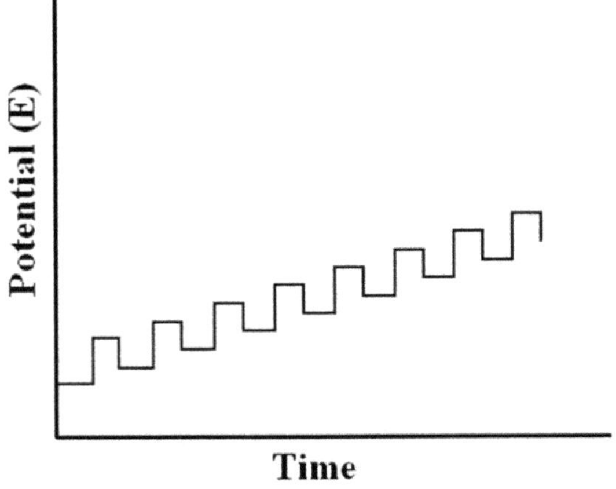

differential pulse voltammograms varies linearly with the concentration of the corresponding analytes. The technique is based on the fact that the charging and the faradaic currents declines with different rates after pulse application. The charging current decays quickly in comparison to faradaic current. Hence, recording current on conclusion of the pulse gives recording of the faradaic current, and charging current becomes negligible at this stage. Suppression of charging current helps in achieving lower detection limits (Bard and Faulkner, 2001). Differential pulse voltammetry gives better detection limits in comparison to linear sweep voltammetry and cyclic voltammetry. Due to its peak shaped nature, differential pulse voltammetry offers another advantage of well resolved peaks of closely spaced electrode processes. It makes possible to simultaneously monitor more than one electroactive species present in solution at a time.

Chronocoulometry employs square-wave voltage signal as an excitation signal and monitors charge as the function of time. The measured charge in a chronocoulometric experiment is assigned to contribution from (a) electrolysis of active specie, (b) electrolysis of active specie adsorbed on the surface of electrode, and (c) charging of the double-layer to new potential. Strength of the technique stems from the simple separation of adsorption and double-layer components from the diffusional component by plotting Q *versus* $t^{1/2}$, which makes it the most lucrative technique to determine surface excess of an electroactive species.

Modified Electrodes and Sensors

Electrochemical sensors include all those sensors which employ electrodes as detector device and detect the presence of an analyte by producing electrical signal which can be correlated to the concentration of an analyte. An electrochemical gas sensor specifies the gas molecules of interest and the information is converted into a suitable measurable property. Electrochemical sensors for gases comprise of gas sensing electrodes which are based on pH electrodes, and can detect gases which forms acidic or basic solutions. A gas sensor comprises of an ion-selective electrode which is engulfed by a thin film of an intermediate electrolyte solution and covered by a gas permeable membrane. Gas sensor for carbon dioxide was developed by Severinghaus and Bradley (1957). It is based on the principle that carbon dioxide gas decompose into bicarbonate and a proton which changes pH of the solution. However, in

Figure 12. Potential-time waveform for chronocoulmetry.

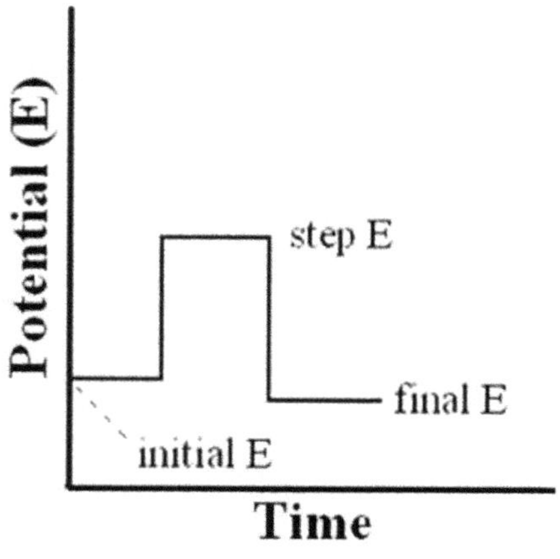

present methodology, CO_2 was not sensed directly and presence of other volatile compounds and gases can also impact the pH of the electrolyte. Possibilities of several types of electrochemical sensors using wide variety of materials have been explored. A screen printed electrode was developed by Ostrick et al. (2000) for CO_2 detection. A limitation of this electrode is its sensitivity to humidity. As far as sensitivity and selectivity is concerned; gold, silver or platinum electrodes are considered better while economical electrodes like vitreous carbon electrodes are capable of separating overlapping signals in blood gas analysis (Hrnc̆ı́r̆ova et al., 2000). Zhou et al. (2001) have described working of multicomponent gas sensor based on transient electrochemistry principle. Schwandt et al. (2018) developed a solid state electrochemical sensor for CO_2 gas. Their work is unique as the individual sensor component can be prepared individually and can be assembled into the full sensor later. Atifi et al. (2018) reported conversion of CO_2 to CO at thin film Bi electrode which is accelerated by using suitable ionic liquid. Linear sweep voltammetry was the electrochemical technique employed for this purpose.

Struzik et al. (2018) reported potentiometric CO_2 sensors based on $Li_7La_3Zr_2O_{12}$ as conducting solid electrolytes, and the sensing activity is assigned to the cyclic redox process. The reported method has a distinct advantage of quick response time and sensing ability at relatively low temperature of 320 °C. Dorner et al. (2023) extensively studied the electrochemical reduction of CO_2 at silver electrodes using cyclic voltammetry. They found that selectivity for CO_2 consumption can be improved with increase in electrode rotation rate. A summary of electrochemical sensors for CO_2 detection has been tabulated in Table 1.

CONCLUSION

In this book chapter, the fluorescent and electrochemical sensing probes for sensing carbon dioxide have been systematically summarized. This chapter provides a methodical overview of sensors for the

Table 1. Electrochemical sensors for carbon dioxide sensing

Material used	Method	Electrolyte	Ref.
Glass electrode	Potentiometry	$NaHCO_3$-NaCl	Severinghaus and Bradley (1957)
$BaCO_3$	Potentiometry	--	Ostrick et al. (2000)
Reticulated vitreous carbon	Voltammetry	solid polymer electrolyte	Hrnc̆ı́r̆ova et al., 2000
Teflon & acetylene black	Voltammetry	--	Zhou et al. (2001)
Na_2CO_3	Potentiometry	Na-β/β”-alumina and Na_2SO_4	Schwandt et al. (2018)
Bi/ILs having 1,8-diazabicyclo[5.4.0] undec-7-ene	Controlled-potential electrolysis	MeCN-based electrolytes	Atifi et al. (2018)
Li_2CO_3–Au	Potentiometry	$(Li_{6.75}La_3Zr_{1.75}Ta_{0.25}O_{12})$	Struzik et al. (2018)
Silver rotating disk electrode	Cyclic voltammetry	0.1M $KHCO_3$	Dorner et al. (2023)
nano-SnO_2/graphene on GC electrode	Voltammetry	0.1 M $NaHCO_3$	Jhang et al. (2014)
Cu modified boron doped diamond electrode	Linear Sweep voltammetry	0.5 MKOH	Jiwanti et al. (2018)
SnO_2/MWCNT on carbon paper	Linear Sweep Voltammetry and Chronoamperometry	0.5 M $NaHCO_3$	Bashir et al. (2016)
NiO on MWCNT with paper	Linear Sweep Voltammetry and Chronoamperometry	0.5 M $NaHCO_3$	Bashir et al. (2015)

recognition of carbon dioxide by highlighting the design concept, detection mechanistic pathway, characteristics, and sensing abilities in terms of limit of detection.

While much progress has been made in this area of study, we believe that two key factors should be taken into account when developing novel optical and electrochemical sensors in future. Firstly, to improve the applicability of the sensor for biotic systems, the sensors should be designed and synthesized offering stability, compatibility to biological arrangement, and dissolvability in aqueous medium. Secondly, the majority of the documented sensors just sense the analyte and do not adsorb/absorb the gas. Evidently, the adsorption/absorption of CO_2 is more required as compared to its mere detection. Therefore, it is especially important to develop the sensor that can detect, separate, and adsorb/ absorb carbon dioxide. In summary, the substantial roles that carbon dioxide plays in different fields, such as the chemical, environmental, therapeutic, and other industries, it is certain that the fluorescent and electrochemical sensors will receive a lot of recognition in future.

REFERENCES

Alfonta, L., Singh, A. K., & Willner, I. (2001). Liposomes labeled with biotin and horseradish peroxidase: A probe for the enhanced amplification of antigen-antibody or oligonucleotide-DNA sensing processes by the precipitation of an insoluble product on electrodes. *Analytical Chemistry*, *73*(1), 91–102. doi:10.1021/ac000819v PMID:11195517

Ali, R., Razi, S. S., Gupta, R. C., Dwivedi, S. K., & Misra, A. (2016). An efficient ICT based fluorescent turn-on dyad for selective detection of fluoride and carbon dioxide. *New Journal of Chemistry*, *40*(1), 162–170. doi:10.1039/C5NJ01920F

Ali, R., Saleh, S. M., Meier, R. J., Azab, H. A., Abdelgawad, I. I., & Wolfbeis, O. S. (2010). Upconverting nanoparticle based optical sensor for carbon dioxide. *Sensors and Actuators. B, Chemical*, *150*(1), 126–131. doi:10.1016/j.snb.2010.07.031

Atifi, A., Boyce, D. W., Dimeglio, J. L., & Rosenthal, J. (2018). Directing the outcome of CO_2 reduction at bismuth cathodes using varied ionic liquid promoters. *ACS Catalysis*, *8*(4), 2857–2863. doi:10.1021/acscatal.7b03433 PMID:30984470

Ausdall, B. R., Glass, J. L., Wiggins, K. M., Aarif, A. M., & Louie, J. J. (2009). A systematic investigation of factors influencing the decarboxylation of imidazolium carboxylates. *The Journal of Organic Chemistry*, *74*(20), 7935–7942. doi:10.1021/jo901791k PMID:19775141

Aydogdu, S., Ertekin, K., Suslu, A., Ozdemir, M., Celik, E., & Cocen, U. (2011). Optical CO_2 Sensing with Ionic Liquid Doped Electrospun Nanofibers. *Journal of Fluorescence*, *21*(2), 607–613. doi:10.1007/s10895-010-0748-4 PMID:20945079

Baker, S. N., & Baker, G. A. (2010). Luminescent carbon nanodots: Emergent nanolights. *Angewandte Chemie International Edition*, *49*(38), 6726–6744. doi:10.1002/anie.200906623 PMID:20687055

Bard, A. J., & Faulkner, L. R. (2001). *Electrochemical Methods: Fundamentals and Applications*. John Wiley & Sons.

Barker, G. C., & Jenkin, I. L. (1952). Square-wave polarography. *Analyst*, *77*(920), 685–696. doi:10.1039/an9527700685

Bashir, S., Hossain, S. S., Rahman, S. U., Ahmed, S., Al-Ahmed, A., & Hossain, M. M. (2016). Electrocatalytic reduction of carbon dioxide on SnO_2/MWCNT in aqueous electrolyte solution. *Journal of CO_2 Utilisation*, *16*, 346–353. doi:10.1016/j.jcou.2016.09.002

Bashir, S. M., Hossain, S. S., Rahman, S., Ahmed, S., & Hossain, M. M. (2015). NiO/MWCNT Catalysts for Electrochemical Reduction of CO_2. *Electrocatalysis (New York)*, *6*(6), 544–553. doi:10.1007/s12678-015-0270-1

Bergmann, H., & Koparal, S. (2005). The formation of chlorine dioxide in the electrochemical treatment of drinking water for disinfection. *Electrochimica Acta*, *50*(25-26), 5218–5228. doi:10.1016/j.electacta.2005.01.061

Braun, E., Eichen, Y., Sivan, U., & Yoseph, G. B. (1998). DNA-templated assembly and electrode attachment of a conducting silver wire. *Nature*, *391*(6669), 775–778. doi:10.1038/35826 PMID:9486645

Chatterjee, C., & Sen, A. (2015). Sensitive colorimetric sensors for visual detection of carbon dioxide and sulfur dioxide. *Journal of Materials Chemistry. A, Materials for Energy and Sustainability*, *3*(10), 5642–5647. doi:10.1039/C4TA06321J

Chen, D., Wang, H., Dong, L., Liu, P., Zhang, Y., Shi, J., Feng, X., Zhi, J., Tong, B., & Dong, Y. (2016). The fluorescent bioprobe with aggregation-induced emission features for monitoring to carbon dioxide generation rate in single living cell and early identification of cancer cells. *Biomaterials*, *103*, 67–74. doi:10.1016/j.biomaterials.2016.06.055 PMID:27372422

Chen, S., Yu, H., Zhao, C., Hu, R., Zhu, J., & Li, L. (2017). Indolo[3,2-b]carbazole derivative as a fluorescent probe for fluoride ion and carbon dioxide detections. *Sensors and Actuators. B, Chemical*, *250*, 591–600. doi:10.1016/j.snb.2017.05.012

Chu, C., & Lo, Y. (2009). Highly sensitive and linear optical fiber carbon dioxide sensor based on sol–gel matrix doped with silica particles and HPTS. *Sensors and Actuators. B, Chemical*, *143*(1), 205–210. doi:10.1016/j.snb.2009.09.019

Dansby-Sparks, R. N., Jin, J., Mechery, S. J., Sampathkumaran, U., Owen, T. W., Yu, B. D., Goswami, K., Hong, K., Grant, J., & Xue, Z. L. (2010). Fluorescent-dye-doped sol−gel sensor for highly sensitive carbon dioxide gas detection below atmospheric concentrations. *Analytical Chemistry*, *82*(2), 593–600. doi:10.1021/ac901890r PMID:20038093

Deng, H., Nanjo, H., Qian, P., Xia, Z., Ishikawa, I., & Suzuki, T. M. (2008). Corrosion prevention of iron with novel organic inhibitor of hydroxamic acid and UV irradiation. *Electrochimica Acta*, *53*(6), 2972–2983. doi:10.1016/j.electacta.2007.11.008

Dorner, I., Röse, P., & Krewer, U. (2023). Dynamic vs. Stationary Analysis of Electrochemical Carbon Dioxide Reduction: Profound Differences in Local States. *ChemElectroChem*, *10*(24), e202300387. doi:10.1002/celc.202300387

Fernández-Sánchez, J. F., Cannas, R., Spichiger, S., Steiger, R., & Spichiger-Keller, U. E. (2007). Optical CO_2-sensing layers for clinical application based on pH-sensitive indicators incorporated into nanoscopic metal-oxide supports. *Sensors and Actuators. B, Chemical*, *128*(1), 145–153. doi:10.1016/j.snb.2007.05.042

Finn, C., Schnittger, S., Yellowlees, L. J., & Love, J. B. (2012). Molecular approaches to the electrochemical reduction of carbon dioxide. *Chemical Communications*, *48*(10), 1392–1399. doi:10.1039/C1CC15393E PMID:22116300

Fu, J., Cherevko, S., & Chung, C. H. (2008). Electroplating of metal nanotubes and nanowires in a high aspect-ratio nanotemplate. *Electrochemistry Communications*, *10*(4), 514–518. doi:10.1016/j.elecom.2008.01.015

Gravenstein, N., & Jaffe, M. B. (2021). Capnography, Anesthesia Equipment (Third Ed.). Principles and Applications (pp. 239-252). Cambridge University Press

Guo, Z., Song, N. R., Moon, J. M., Kim, M., Jun, E. J., Choi, J., Lee, J. Y., Bielawski, C. W., Sessler, J. L., & Yoon, J. (2012). A Benzobisimidazolium-Based Fluorescent and Colorimetric Chemosensor for CO_2. *Journal of the American Chemical Society*, *134*(43), 17846–17849. doi:10.1021/ja306891c PMID:22931227

Hrncˇı'rˇova, P., Opekar, F., & Sˇtulik, K. (2000). Amperometric solid-state NO_2 sensor with a solid polymer electrolyte and a reticulated vitreous carbon indicator electrode. *Sensors and Actuators. B, Chemical*, *69*(1-2), 199–204. doi:10.1016/S0925-4005(00)00540-2

Ishida, M., Kim, P., Choi, J., Yoon, J., Kim, D., & Sessler, J. L. (2013). Benzimidazole-embedded N-fused aza-indacenes: Synthesis and deprotonation-assisted optical detection of carbon dioxide. *Chemical Communications*, *49*(62), 6950–6952. doi:10.1039/c3cc43938k PMID:23811989

Jang, M., Kang, S., & Han, M. S. (2019). A simple turn-on fluorescent chemosensor for CO_2 based on aggregation-induced emission: Application as a CO_2 absorbent screening method. *Dyes and Pigments*, *162*, 978–983. doi:10.1016/j.dyepig.2018.11.031

Jin, Y. J., Moon, B. C., & Kwak, G. (2016). Colorimetric fluorescence response to carbon dioxide using charge transfer dye and molecular rotor dye in smart solvent system. *Dyes and Pigments*, *132*, 270–273. doi:10.1016/j.dyepig.2016.05.003

Jiwanti, P. K., Natsui, K., Nakata, K., & Einaga, Y. (2018). The electrochemical production of C2/C3 species from carbon dioxide on copper-modified boron-doped diamond electrodes. *Electrochimica Acta*, *266*, 414–419. doi:10.1016/j.electacta.2018.02.041

Kang, S., Kim, J., Park, J. H., Ahn, C. K., Rhee, C. H., & Han, M. S. (2015). Intra-molecular hydrogen bonding stabilization based-fluorescent chemosensor for CO_2: Application to screen relative activities of CO_2 absorbents. *Dyes and Pigments*, *123*, 125–131. doi:10.1016/j.dyepig.2015.07.033

Kaur, K., & Baral, M. (2014). Synthesis of imine-naphthol tripodal ligand and study of its coordination behaviour towards Fe (III), Al (III), and Cr (III) metal ions. *Bioinorganic Chemistry and Applications*. doi:10.1155/2014/915457

Khandare, D. G., Joshi, H., Banerjee, M., Majik, M. S., & Chatterjee, A. (2015). Fluorescence turn-on chemosensor for the detection of dissolved CO_2 based on ion-induced aggregation of tetraphenylethylene derivative. *Analytical Chemistry*, *87*(21), 10871–10877. doi:10.1021/acs.analchem.5b02339 PMID:26458016

Ko, K., Lee, J., & Chung, H. (2020). Highly efficient colorimetric CO_2 sensors for monitoring CO_2 leakage from carbon capture and storage sites. *The Science of the Total Environment*, *729*, 138786. doi:10.1016/j.scitotenv.2020.138786 PMID:32380324

Kwon, N., Baek, G., Swamy, K. M. K., Lee, M., Xu, Q., Kim, Y., Kim, S. J., & Yoon, J. (2009). Naphthoimidazolium based ratiometric fluorescent probes for F^- and CN^-, and anion-activated CO2 sensing. *Dyes and Pigments*, *171*, 107679. doi:10.1016/j.dyepig.2019.107679

Laik, B., Eude, L., Ramos, J.-P. P., Cojocaru, C. S., Pribat, D., & Rouvière, E. (2008). Silicon nanowires as negative electrode for lithium-ion microbatteries. *Electrochimica Acta*, *53*(17), 5528–5532. doi:10.1016/j.electacta.2008.02.114

Lakowicz, J. R. (Ed.), *Topics in Fluorescence Spectroscopy* (pp. 119–161). Springer.

Lee, M., Jo, S., Lee, D., Xu, Z., & Yoon, J. (2015). New Naphthalimide Derivative as a Selective Fluorescent and Colorimetric Sensor for Fluoride, Cyanide and CO_2. *Dyes and Pigments*, *120*, 288–292. doi:10.1016/j.dyepig.2015.04.029

Li, C., Yang, Q., Zhang, T., Lv, Z., Wang, Y., & Chen, Y. (2021). A hybrid CO_2 ratiometric fluorescence sensor synergizing tetraphenylethene and gold nanoclusters relying on disulfide functionalized hyperbranched poly(amido amine). *Sensors and Actuators. B, Chemical*, *346*, 130513. doi:10.1016/j.snb.2021.130513

Li, H., Su, X., Bai, C., Xu, Y., Pei, Z., & Sun, S. (2016). Detection of carbon dioxide with a novel HPTS/NiFe-LDH nanocomposite. *Sensors and Actuators. B, Chemical*, *225*, 109–114. doi:10.1016/j.snb.2015.11.007

Lochman, L., Zimcik, P., Klimant, I., Novakova, V., & Borisov, S. M. (2017). Red-emitting CO_2 sensors with tunable dynamic range based on pH-sensitive azaphthalocyanine indicators. *Sensors and Actuators. B, Chemical*, *246*, 1100–1107. doi:10.1016/j.snb.2016.10.135

Loudet, A., & Burgess, K. (2007). BODIPY dyes and their derivatives: Syntheses and spectroscopic properties. *Chemical Reviews*, *107*(11), 4891–4932. doi:10.1021/cr078381n PMID:17924696

Luo, J., Xie, Z., Lam, W. Y. J., Cheng, L., Chen, H., Qiu, C., Kwok, S. H., Zhan, X., Liu, Y., Zhu, D., & Tang, Z. B. (2001). Aggregation-induced emission of 1-methyl-1,2,3,4,5-pentaphenylsilole. *Chemical Communications*, (18), 1740–1741. doi:10.1039/b105159h PMID:12240292

Ma, Y., Promthaveepong, K., & Li, N. (2016). CO_2-Responsive Polymer-Functionalized Au Nanoparticles for CO_2 Sensor. *Analytical Chemistry*, *88*(16), 8289–8293. doi:10.1021/acs.analchem.6b02133 PMID:27459645

Ma, Y., Zeng, Y., Liang, H., Ho, C. H., Zhao, Q., Huang, W., & Wong, W. Y. (2015). A water-soluble tetraphenylethene based probe for luminescent carbon dioxide detection and its biological application. *Journal of Materials Chemistry. C, Materials for Optical and Electronic Devices*, *3*(45), 11850–1185. doi:10.1039/C5TC03327F

Mai, Z., Li, H., Gao, Y., Niu, Y., Li, Y., Rooij, N. F. D., Umar, A., Al-Assiri, M. S., Wang, Y., & Zhou, G. (2020). Synergy of CO_2-response and aggregation induced emission in a small molecule: Renewable liquid and solid CO_2 chemosensors with high sensitivity and visibility. *Analyst*, *145*(10), 3528–3534. doi:10.1039/D0AN00189A PMID:32190881

Mayerhöfer, U., Fimmel, B., & Wüthner, F. (2012). Bright Near-Infrared Fluorophores Based on Squaraines by Unexpected Halogen Effects. *Angewandte Chemie International Edition*, *51*(1), 164–167. doi:10.1002/anie.201107176 PMID:22105993

Mills, A. (2009). Optical Sensors for Carbon Dioxide and Their Applications. In: Baraton, MI. (Ed.) Sensors for Environment, Health and Security. NATO Science for Peace and Security Series C: Environmental Security (pp. 347-370). Springer. doi:10.1007/978-1-4020-9009-7_23

Mills, A., & Hodgen, S. (2005). Fluorescent Carbon Dioxide Indicators. In C. D. Geddes & J. R. Lakowicz (Eds.), *Topics in Fluorescence Spectroscopy, Advanced Concepts in Fluorescence Sensing, Part A: Small Molecule Sensing* (pp. 119–161). Springer.

Mishra, R. K., Vijayakumar, S., Mal, A., Karunakaran, V., Janardhanan, J. C., Maiti, K. K., Praveen, V. K., & Ajayaghosh, A. (2019). Bimodal detection of carbon dioxide using a fluorescent molecular aggregates. *Chemical Communications*, *55*(43), 6046–6049. doi:10.1039/C9CC01564G PMID:31065654

Neethirajan, S., Jayas, D. S., & Sadistap, S. (2009). Carbon dioxide (CO_2) sensors for the agri-food industry-a review. *Food and Bioprocess Technology*, *2*(2), 115–121. doi:10.1007/s11947-008-0154-y

Nemade, K. R., & Waghuley, S. A. (2013). Carbon dioxide gas sensing application of graphene/Y_2O_3 quantum dots composite. *International Journal of Modern Physics. Conference Series*, *22*, 380–384. doi:10.1142/S2010194513010404

Nemade, K. R., & Waghuley, S. A. (2014). Highly responsive carbon dioxide sensing by graphene/Al_2O_3 quantum dots composites at low operable temperature. *Indian Journal of Physics and Proceedings of the Indian Association for the Cultivation of Science*, *88*(6), 577–583. doi:10.1007/s12648-014-0454-1

Nicholson, R. S., & Shain, I. (1964). Theory of stationary electrode polarography. Single scan and cyclic methods applied to reversible, irreversible, and kinetic Systems. *Analytical Chemistry*, *36*(4), 706–723. doi:10.1021/ac60210a007

Osteryoung, J., & Wechter, C. (1989). *Development of Pulse Polarography and Voltammetry* (J. T. Stock & M. E. Orna, Eds.). American Chemical Society. doi:10.1021/bk-1989-0390.ch025

Ostrick, B., Fleischer, M., Meixner, H., & Kohl, C. D. (2000). Investigation of the reaction mechanisms in work function type sensors at room temperature by studies of the cross-sensitivity to oxygen and water: The carbonate-carbon dioxide system. *Sensors and Actuators. B, Chemical*, *68*(1-3), 197–202. doi:10.1016/S0925-4005(00)00429-9

Pan, Z. H., Luo, G. G., Zhou, J. W., Xia, J. X., Fang, K., & Wu, R. B. (2014). A simple BODIPY-aniline-based fluorescent chemosensor as multiple logic operations for the detection of pH and CO2 gas. *Dalton Transactions (Cambridge, England)*, *43*(22), 8499–8507. doi:10.1039/C4DT00395K PMID:24756338

Pei, X., Xiong, D., Fan, J., Li, Z., Wang, H., & Wang, J. (2017). Highly efficient fluorescence switching of carbon nanodots by CO_2. *Carbon*, *117*, 147–153. doi:10.1016/j.carbon.2017.02.090

Pfeifer, D., Klimant, I., & Borisov, S. M. (2018). Ultrabright red-emitting photostable perylene bisimide dyes: New indicators for ratiometric sensing of high pH or carbon dioxide. *European Journal of Chemistry*, *24*(42), 10711–10720. doi:10.1002/chem.201800867 PMID:29738607

Puligundla, P., Jung, J., & Ko, S. (2012). Carbon dioxide sensors for intelligent food packaging applications. *Food Control*, *25*(1), 328–333. doi:10.1016/j.foodcont.2011.10.043

Qazi, H. H., Mohammad, A. B. B., & Akram, M. (2012). Recent Progress in Optical Chemical Sensors. *Sensors (Basel)*, *12*(12), 16522–16556. doi:10.3390/s121216522 PMID:23443392

Qu, Y., Hua, J., & Tian, H. (2010). Colorimetric and Ratiometric Red Fluorescent Chemosensor for Fluoride Ion Based on Diketopyrrolopyrrole. *Organic Letters*, *12*(15), 3320–3323. doi:10.1021/ol101081m PMID:20590106

Robel, I., Girishkumar, G., Bunker, B. A., Kamat, P. V., & Vinodgopal, K. (2006). Structural changes and catalytic activity of platinum nanoparticles supported on C_{60} and carbon nanotube films during the operation of direct methanol fuel cells. *Applied Physics Letters*, *88*(7), 073113/1–073113, 3. doi:10.1063/1.2177354

Schutting, S., Borisov, S. M., & Klimant, I. (2013). Diketo-Pyrrolo-Pyrrole dyes as new colorimetric and fluorescent pH indicators for optical carbon dioxide sensors. *Analytical Chemistry*, *85*(6), 3271–3279. doi:10.1021/ac303595v PMID:23421943

Schutting, S., Jokic, T., Strobl, M., Borisov, S. M., Beer, D. D., & Klimant, I. (2015). NIR optical carbon dioxide sensors based on highly photostable dihydroxy-aza-BODIPY dyes. *Journal of Materials Chemistry. C, Materials for Optical and Electronic Devices*, *3*(21), 5474–5483. doi:10.1039/C5TC00346F

Schutting, S., Klimant, I., Beer, D. K., & Borisov, S. M. (2014). New highly fluorescent pH indicator for ratiometric RGB imaging of pCO2. *Methods and Applications in Fluorescence*, *2*(2), 02400. doi:10.1088/2050-6120/2/2/024001 PMID:29148465

Schwandt, C., Kumar, R. V., & Hills, M. P. (2018). Solid state electrochemical gas sensor for the quantitative determination of carbon dioxide. *Sensors and Actuators. B, Chemical*, *265*, 27–34. doi:10.1016/j.snb.2018.03.012

Severinghaus, J. W., & Bradley, A. F. (1957). Electrodes for blood pO_2 and pCO_2 determination. *Journal of Applied Physiology*, *13*(3), 515–520. doi:10.1152/jappl.1958.13.3.515 PMID:13587443

Sheini, A. (2020). A paper-based device for the colorimetric determination of ammonia and carbon dioxide using thiomalic acid and maltol functionalized silver nanoparticles: Application to the enzymatic determination of urea in saliva and blood. *Mikrochimica Acta*, *187*(10), 565. doi:10.1007/s00604-020-04553-8 PMID:32920692

Struzik, M., Garbayo, I., Pfenninger, R., & Rupp, J. L. M. (2018). A Simple and Fast Electrochemical CO_2 Sensor Based on $Li_7La_3Zr_2O_{12}$ for Environmental Monitoring. *Advanced Materials*, *30*(44), 1804098. doi:10.1002/adma.201804098 PMID:30238512

Sun, J., Ye, B., Xia, G., Zhao, X., & Wang, H. (2016). A colorimetric and fluorescent chemosensor for the highly sensitive detection of CO_2 gas: Experiment and DFT calculation. *Sensors and Actuators. B, Chemical*, *233*, 76–82. doi:10.1016/j.snb.2016.04.052

Tian, T., Chen, X., Li, H., Wang, Y., Guo, L., & Jiang, L. (2013). Amidine-based fluorescent chemosensor with high applicability for detection of CO_2: A facile way to "see" CO_2. *Analyst*, *138*(4), 991–994. doi:10.1039/C2AN36401H PMID:23259156

Wang, H., Chen, D., Zhang, Y., Liu, P., Shi, J., Feng, X., Tong, B., & Dong, Y. (2015). A fluorescent probe with an aggregation enhanced emission feature for real-time monitoring of low carbon dioxide levels. *Journal of Materials Chemistry. C, Materials for Optical and Electronic Devices*, *3*(29), 7621–7626. doi:10.1039/C5TC01280E

Wang, J., Xu, D., Kawde, A. N., & Polsky, R. (2001). Metal nanoparticle-based electrochemical stripping potentiometric detection of DNA hybridization. *Analytical Chemistry*, *73*(22), 5576–5581. doi:10.1021/ac0107148 PMID:11816590

Wang, R., Zhang, M., Guan, Y., Chen, M., & Zhang, Y. (2019). A CO_2-responsive hydrogel film for optical sensing of dissolved CO_2. *Soft Matter*, *15*(30), 6107–6115. doi:10.1039/C9SM00958B PMID:31282902

Wencel, D., Abel, T., & McDonagh, C. (2014). Optical Chemical pH Sensors. *Analytical Chemistry*, *86*(1), 15–29. doi:10.1021/ac4035168 PMID:24180284

Wencel, D., Moore, J., Stevenson, N., & McDonagh, C. (2010). Ratiometric fluorescence-based dissolved carbon dioxide sensor for use in environmental monitoring applications. *Analytical and Bioanalytical Chemistry*, *398*(5), 1899–1907. doi:10.1007/s00216-010-4165-y PMID:20827465

Xia, G., Liu, Y., Ye, B., Sun, J., & Wang, H. (2015). A Squaraine-based Colorimetric and F^- Dependent Chemosensor for Recyclable CO_2 Gas Detection: Highly Sensitive Off-On-Off Response. *Chemical Communications*, *51*(72), 13802–13805. doi:10.1039/C5CC04755B PMID:26235137

Xiang, Y., Xie, M., Bash, R., Chen, J. J. L., & Wang, J. (2007). Ultrasensitive label-free aptamer-based electronic detection. *Angewandte Chemie International Edition*, *46*(47), 9054–9056. doi:10.1002/anie.200703242 PMID:17957666

Xu, Z., Kim, S. K., & Yoon, J. (2010). Revisit to imidazolium receptors for the recognition of anions: Highlighted research during 2006–2009. *Chemical Society Reviews*, *39*(5), 1457–1466. doi:10.1039/b918937h PMID:20419201

Yang, L. H., & Wang, H. M. (2014). Recent advances in carbon dioxide capture, fixation, and activation by using n-heterocyclic carbenes. *ChemSusChem*, *7*(4), 962–998. doi:10.1002/cssc.201301131 PMID:24644039

Zhang, S., Kang, P., & Meyer, T. J. (2014). Nanostructured tin catalysts for selective electrochemical reduction of carbon dioxide to formate. *Journal of the American Chemical Society*, *136*(5), 1734–1737. doi:10.1021/ja4113885 PMID:24417470

Zhang, Y., Du, X., Zhai, J., & Xie, X. (2023). Tunable colorimetric carbon dioxide sensor based on ion-exchanger- and chromoionophore-doped hydrogel. *Analysis & Sensing*, *3*(6), e202300032. doi:10.1002/anse.202300032

Zhao, H. X., Liu, L. Q., Liu, Z. D., Wang, Y., Zhao, X. J., & Huang, C. Z. (2011). Highly selective detection of phosphate in very complicated matrixes with an off-on fluorescent probe of europium-adjusted carbon dots. *Chemical Communications*, *47*(9), 2604–2606. doi:10.1039/c0cc04399k PMID:21234476

Zhou, Z. B., Feng, L. D., Liu, W. J., & Wu, Z. G. (2001). New approaches for developing transient electrochemical multi-component gas sensors. *Sensors and Actuators. B, Chemical*, *76*(1-3), 605–609. doi:10.1016/S0925-4005(01)00654-2

Zhua, J., Jiaa, P., Lia, N., Tanb, S., Huangb, J., & Xu, L. (2018). Small-molecule fluorescent probes for the detection of carbon dioxide. *Chinese Chemical Letters*, *29*(10), 1445–1450. doi:10.1016/j.cclet.2018.09.002

Ziebart, C., Federsel, C., Anbarasan, P., Jackstell, R., Baumann, W., Spannenberg, A., & Beller, M. (2014). Well-defined iron catalyst for improved hydrogenation of carbon dioxide and bicarbonate. *Journal of the American Chemical Society*, *134*(51), 20701–20704. doi:10.1021/ja307924a PMID:23171468

Chapter 5
Water Pollutants, Sensor Types, and Their Advantages and Challenges

Sanjeevi Ramakrishnan
Nims Institute of Allied Medical Science and Technology, Nims University, Jaipur, India

Prashantkumar Sathvara
https://orcid.org/0000-0002-3548-5584
Nims Institute of Allied Medical Science and Technology, Nims University, Jaipur, India

Sandeep Tripathi
Nims Institute of Allied Medical Science and Technology, Nims University, Jaipur, India

Anuradha Jayaraman
Nims Institute of Allied Medical Science and Technology, Nims University, Jaipur, India

ABSTRACT

Water pollution is a global crisis impacting ecosystems, health, and economies. This chapter explores strategies to combat it, stressing advanced water quality sensors' vital role. It scrutinizes pollutants, emphasizing modern sensor tech's importance in ensuring water safety. Tackling pollution is crucial for biodiversity, human health, and clean water access. Pollutants include heavy metals, chemicals, pathogens, and sediments, requiring precise monitoring by sensors using various technologies. They offer real-time detection and response, covering chemical, biological, physical, remote sensing, and IoT-enabled sensors. Challenges like maintenance persist, requiring protocols and training. Collaboration and sensor tech are pivotal in ensuring cleaner water. This chapter highlights technology's role in managing water quality, emphasizing innovation for safeguarding this vital resource.

INTRODUCTION

Water pollution is indeed a critical global environmental challenge with far-reaching impacts on ecosystems, human health, and economies worldwide. Here's a breakdown of how water pollution affects

DOI: 10.4018/979-8-3693-1930-7.ch005

each of these areas like in ecosystem Water pollution can have devastating effects on aquatic ecosystems. Contaminants such as heavy metals, pesticides, and industrial chemicals can accumulate in water bodies, harming aquatic plants, animals, and microorganisms (Lin et al., 2022a). Pollution can disrupt food chains, alter habitats, and lead to declines in biodiversity. Eutrophication, caused by excessive nutrient runoff, can result in oxygen depletion and the formation of harmful algal blooms, further degrading aquatic ecosystems (Anwar Abdelrahman Aly et al., 2016). Water pollution poses significant risks to human health. Contaminated water can transmit pathogens and cause waterborne diseases such as cholera, typhoid fever, and dysentery. Exposure to pollutants like heavy metals, pesticides, and industrial chemicals through contaminated water sources can lead to various health problems, including neurological disorders, cancer, and reproductive issues (Manoiu et al., 2022). Vulnerable populations, such as children, pregnant women, and communities lacking access to clean water and sanitation facilities, are particularly at risk. The economic consequences of water pollution are substantial. Contaminated water sources can render drinking water supplies unsafe, leading to increased healthcare costs due to waterborne illnesses and the need for water treatment infrastructure upgrades (Radu et al., 2022). Pollution-related damage to fisheries, aquaculture, and recreational water bodies can result in lost revenue and livelihoods for communities dependent on these resources. Additionally, industries may face regulatory fines and cleanup expenses for environmental contamination incidents, impacting their profitability and competitiveness (Sur et al., 2022).

Addressing water pollution requires coordinated efforts at local, national, and global levels to implement effective pollution prevention and control measures (Karunanidhi et al., 2021). This includes adopting pollution reduction strategies, investing in wastewater treatment infrastructure, promoting sustainable agricultural practices, and enforcing environmental regulations. Public awareness, education, and community engagement are also essential for fostering responsible water stewardship and ensuring the sustainable management of water resources for future generations.

VARIOUS CATEGORIES OF WATER POLLUTANTS

Water pollutants can be categorized into various groups based on their sources, characteristics, and impacts on aquatic ecosystems and human health. Heavy Metals including lead, mercury, cadmium, arsenic, and chromium, can accumulate in water bodies and bioaccumulate in aquatic organisms, posing risks to human health and ecosystems (Ahamad et al., 2020). Organic Compounds such as pesticides, herbicides, industrial chemicals, solvents, and pharmaceuticals, can leach into water sources from agricultural runoff, industrial discharges, and wastewater effluents, affecting water quality and aquatic organisms (Dwivedi, 2017). Nutrients such as nitrogen and phosphorus, can lead to eutrophication and harmful algal blooms when present in excessive amounts, degrading water quality, oxygen levels, and aquatic habitats (Fischer et al., 2012). Persistent Organic Pollutants including polychlorinated biphenyls, polycyclic aromatic hydrocarbons, and chlorinated pesticides, which are resistant to degradation, bioaccumulate in aquatic organisms and pose long-term risks to humans and environmental health (Gambhir et al., 2012). Chlorinated Compounds such as chlorine, chloramines, and chlorinated solvents, are used for disinfection and industrial processes but can form disinfection by-products when reacting with organic matter in water, some of which are carcinogenic or toxic (Kordbacheh & Heidari, 2023).

Some pathogens including bacteria, viruses, protozoa, and parasites, can contaminate water sources through sewage discharges, agricultural runoff, and animal waste, causing waterborne diseases such as cholera, typhoid fever, gastroenteritis, and hepatitis. Algal Toxins produced by harmful algal blooms and cyanobacteria, including microcystins, saxitoxins, and cylindrospermopsin, can contaminate drinking water sources, posing risks to human health, livestock, and aquatic organisms (Fischer et al., 2012).

Physical Pollutants like Suspended Solids including sediments, silt, and particulate matter, can cloud water, reduce water clarity, and smother benthic habitats when suspended in high concentrations, affecting aquatic ecosystems and habitat quality (Gambhir et al., 2012). Sedimentation resulting from soil erosion, construction activities, and land development, can degrade water quality, impair aquatic habitats, and increase turbidity, sedimentation, and nutrient loading in water bodies (Verma & Ratan, 2020). Thermal Pollution resulting from the discharge of heated water from industrial processes, power plants, and urban runoff, can raise water temperatures, disrupt aquatic ecosystems, and decrease dissolved oxygen levels, affecting fish populations and biodiversity. Radionuclides including isotopes of uranium, radium, cesium, and strontium, can enter water sources through natural processes, mining activities, nuclear accidents, and nuclear waste disposal, posing risks to human health and the environment due to their radioactive properties (Sanjeevi, 2011). These categories of water pollutants interact with each other and with natural environmental processes, influencing water quality, ecosystem health, and human well-being.

WATER POLLUTION EFFECTS IN ECOSYSTEM

Water pollution can have profound and multifaceted impacts on ecosystems, affecting various components and processes within aquatic environments: Pollutants introduced into water bodies can alter their chemical composition, leading to changes in pH levels, nutrient concentrations, and dissolved oxygen levels (Arenas-Sánchez et al., 2016). For example, industrial effluents containing acids or bases can disrupt the natural pH balance of water, affecting the survival and reproduction of aquatic organisms adapted to specific pH ranges (Bourdeau & Treshow, 1978). Excessive nutrient runoff from agricultural activities or sewage discharges can result in eutrophication, causing algae to bloom and consume oxygen upon decomposition, leading to hypoxic or anoxic conditions harmful to aquatic life. Many pollutants, such as heavy metals, pesticides, and persistent organic pollutants (POPs), have the potential to bio-accumulate in aquatic organisms (Chen et al., 2019). Bioaccumulation occurs when organisms absorb pollutants from their environment at a rate faster than they can excrete or metabolize them, leading to the accumulation of toxins in their tissues. Bio-magnification refers to the increase in pollutant concentrations at higher trophic levels of the food chain, as predators consume contaminated prey (Chen et al., 2019). This can result in high levels of pollutants in apex predators, posing risks to their health and reproductive success. Water pollution can degrade aquatic habitats essential for the survival of many species. Sedimentation, resulting from soil erosion due to deforestation or land development, can smother benthic habitats like coral reefs, sea-grass beds, and spawning grounds, reducing their productivity and biodiversity. Toxic pollutants can also directly damage habitat structures, such as coral skeletons or aquatic vegetation, affecting the availability of shelter, food, and breeding sites for aquatic organisms (Fatima et al., 2020). Pollution-induced changes in water quality and habitat conditions can disrupt aquatic food webs, altering species interactions and community dynamics. For example, declines in primary producers like phytoplankton due to eutrophication can affect the abundance and distribution of herbivores and primary consumers. This ripple effect can propagate through the food chain, impacting higher trophic levels and ultimately leading to shifts in species composition and biodiversity (Håkanson & Bryhn, 1999). The cumulative impacts of water pollution on water quality, habitats, and food webs can lead to the loss of biodiversity in aquatic ecosystems. Species adapted to specific environmental conditions may decline or disappear due to pollution-induced stressors, habitat degradation, or competition with invasive species. Biodiversity loss can weaken ecosystem resilience, making ecosystems more vulnerable to further environmental disturbances and reducing their ability to provide valuable ecosystem services, such as water purification, nutrient cycling, and climate regulation.

WATER POLLUTION EFFECT ON HUMAN HEALTH

Water pollution poses significant risks to human health through various pathways of exposure. Contaminated water sources can harbor a wide range of pathogens, including bacteria, viruses, protozoa, and parasites, which can cause waterborne diseases. Examples include cholera, typhoid fever, dysentery, giardiasis, and cryptosporidiosis (Malik et al., 2020). These diseases are typically spread through the ingestion of contaminated water or food prepared with contaminated water. Symptoms can range from mild gastrointestinal discomfort to severe illness and, in some cases, death, particularly among vulnerable populations such as children, the elderly, and individuals with weakened immune systems. Water pollution can introduce various chemical contaminants into drinking water sources, posing health risks to human populations (Halder & Islam, 2015). Heavy metals such as lead, mercury, arsenic, and cadmium can leach into water sources from industrial discharges, mining activities, and natural geological processes. Pesticides, herbicides, and industrial chemicals can also contaminate water supplies through agricultural runoff, urban runoff, and industrial effluents (Haseena et al., 2017; Ramakrishnan & Jayaraman, 2019). Chronic exposure to these chemical contaminants through drinking water consumption can lead to adverse health effects, including neurological disorders, developmental delays, cancer, reproductive problems, and organ damage. Some pollutants present in water bodies have the potential to bioaccumulate in aquatic organisms and biomagnify through the food chain, increasing human exposure risks. For example, methylmercury, a highly toxic form of mercury, bioaccumulates in fish tissue and can reach high concentrations in predatory fish species commonly consumed by humans (Inyinbor Adejumoke et al., 2018). Chronic consumption of contaminated fish can result in mercury poisoning, leading to neurological impairments, cardiovascular problems, and developmental disorders, particularly in fetuses and young children. Excessive nutrient runoff from agricultural activities and wastewater discharges can promote the growth of harmful algal blooms (HABs) in water bodies. Some species of algae produce toxins known as cyanotoxins, which can contaminate drinking water supplies and pose health risks to humans and animals. Exposure to cyanotoxins through ingestion, inhalation, or skin contact can cause a range of health effects, including gastrointestinal illness, liver damage, respiratory problems, and neurological symptoms (Lin et al., 2022b). Water pollution incidents, such as chemical spills or contamination of water treatment plants, can disrupt water and sanitation services, compromising access to safe drinking water and sanitation facilities. Communities affected by such incidents may experience shortages of clean water for drinking, cooking, and hygiene, increasing the risk of waterborne diseases and sanitation-related health problems (R. Qadri & Faiq, 2020; Ranjan et al., n.d.). Vulnerable populations, including those living in low-income areas or disaster-prone regions, are particularly susceptible to the health impacts of disrupted water and sanitation services. Addressing water pollution and protecting human health requires comprehensive strategies that focus on pollution prevention, source control, water quality monitoring, and effective water treatment and sanitation infrastructure (Sanjeevi, 2011; Sanjeevi et al., 2017). Public awareness, education, and community engagement are also essential for promoting water stewardship practices and ensuring access to clean water for all.

WATER POLLUTION IMPACT ON ECONOMIC GROWTH

The economic impacts of water pollution can be substantial and wide-ranging, affecting various sectors and aspects of economic activity. Water pollution-related illnesses can impose significant healthcare costs on individuals, communities, and healthcare systems (Zhang et al., 2017). Treating waterborne diseases, such as cholera, typhoid fever, and gastrointestinal infections, requires medical attention, medications, hospitalization, and sometimes long-term care. These healthcare expenses can strain household budgets, burden public health systems, and reduce productivity due to illness-related absenteeism from work or school. Contaminated water sources require treatment to meet safe drinking water standards, which can incur substantial costs for water utilities and municipalities (Reddy & Behera, 2006). Water treatment processes such as filtration, disinfection, and chemical treatment are necessary to remove or neutralize pollutants and pathogens from drinking water supplies. Investing in water treatment infrastructure upgrades, maintenance, and operational costs can impose financial burdens on water providers and may lead to increased water tariffs for consumers (Orubu & Omotor, 2011). Water pollution can degrade the quality of recreational water bodies, such as beaches, lakes, and rivers, reducing their attractiveness to tourists and outdoor enthusiasts. Polluted water bodies may be subject to swim advisories, beach closures, or fishing bans due to health concerns, resulting in lost revenue for tourism-dependent businesses, including hotels, restaurants, marinas, and recreational outfitters (Muyibi et al., 2008). Declines in tourism and recreational activities can harm local economies and employment opportunities in coastal and inland communities. Water pollution can have detrimental effects on fisheries and aquaculture operations, leading to declines in fish stocks, shellfish harvests, and aquaculture yields (Cai et al., 2020). Contaminants such as heavy metals, pesticides, and toxins from harmful algal blooms can accumulate in aquatic organisms, rendering them unsafe for human consumption and damaging commercial fisheries and aquaculture enterprises (Li & Li, 2021; R. Sanjeevi et al., 2022). Economic losses from reduced fishery yields, fishery closures, and market disruptions can affect livelihoods and income for fishermen, fish farmers, and seafood processors. Water pollution can diminish the value of waterfront properties and real estate investments in areas with polluted water bodies. Contaminated water sources can detract from the aesthetic appeal and recreational appeal of waterfront properties, leading to decreased demand and lower property values (Juma et al., 2014). Homeowners, businesses, and developers may experience financial losses and reduced returns on investment in polluted waterfront areas, affecting property tax revenues and municipal budgets. Water pollution can impair agricultural productivity and soil fertility through contamination of irrigation water sources and agricultural runoff. Excessive nutrient runoff, pesticides, and industrial chemicals can accumulate in soil and water, affecting crop yields, quality, and marketability. Farmers may incur additional costs for water treatment, soil remediation, and crop loss mitigation measures, impacting farm profitability and agricultural competitiveness (Juma et al., 2014; Sathvara et al., 2023). Water pollution-related restrictions on irrigation water use can exacerbate water scarcity issues in agricultural regions, affecting food production and rural economies (Choi et al., 2015).

ADVANCED WATER QUALITY IN MONITORING AND ADDRESSING

Advanced water quality monitoring and addressing techniques encompass a range of innovative technologies and approaches aimed at enhancing the effectiveness, efficiency, and timeliness of water quality assessment and management (Park et al., 2020). Remote sensing technologies, including satellite imagery, aerial drones, and unmanned aerial vehicles (UAVs), are increasingly used for monitoring water quality over large spatial scales. These technologies can provide real-time or near real-time data on parameters such as water clarity, chlorophyll concentration, turbidity, and algal blooms, allowing for the timely detection of pollution events and environmental changes. Deploying networks of sensors and monitoring devices in water bodies enables continuous, high-resolution monitoring of water quality parameters (Mishra et al., 2016; Olatinwo & Joubert, 2019). These sensors can measure various physical, chemical, and biological indicators, such as temperature, pH, dissolved oxygen, nutrients, pollutants, and microbial contaminants. Advanced sensor networks can provide comprehensive datasets for analyzing spatial and temporal variations in water quality and identifying pollution sources and trends (Olatinwo & Joubert, 2018). Advanced data analytics techniques, including machine learning algorithms, statistical analyses, and numerical modeling, are used to process and interpret large volumes of water quality data. These tools can identify patterns, correlations, and anomalies in water quality datasets, predict future water quality conditions, and simulate the impacts of pollution sources and management strategies (Abbasi et al., 2013; O'Grady et al., 2021). Data-driven approaches enhance decision-making for water quality management and facilitate targeted interventions to address pollution hotspots. Real-time monitoring systems enable the continuous surveillance of water quality parameters and the rapid detection of pollution events. Citizen science initiatives engage the public in monitoring and addressing water quality issues, leveraging the collective efforts of volunteers, community groups, and citizen scientists (Kumar et al., 2022; Prashantkumar B. Sathvara et al., 2023). Crowdsourcing platforms and mobile applications enable citizens to collect water quality data, report pollution sightings, and participate in environmental monitoring efforts. Citizen-generated data supplements traditional monitoring programs, enhances spatial coverage, and fosters community awareness and stewardship of water resources. Ongoing advancements in sensor technology, nanotechnology, microfluidics, and biosensors are driving the development of novel water quality monitoring tools and techniques. Miniaturized, low-cost sensors, wearable devices, and smartphone-based technologies are expanding access to water quality monitoring capabilities, particularly in resource-constrained or remote regions (El-Shafeiy et al., 2023). Emerging technologies hold promise for improving the affordability, portability, and accessibility of water quality monitoring solutions, democratizing access to environmental data, and empowering communities to address water pollution challenges. By harnessing advanced monitoring technologies, data analytics, and collaborative approaches, stakeholders can enhance their capacity to monitor, assess, and respond to water quality issues effectively, ultimately contributing to the sustainable management and protection of freshwater resources and ecosystems.

ADVANCED WATER QUALITY SENSORS

Advanced water quality sensors encompass a variety of cutting-edge technologies designed to detect and measure a wide range of parameters with high accuracy, sensitivity, and efficiency. These sensors play a crucial role in environmental monitoring, water resource management, and pollution control efforts.

1. **Optical Sensors:** Optical sensors utilize light-based techniques to measure various water quality parameters. Fluorescence sensors can detect organic matter, chlorophyll-a, and other fluorophores, providing insights into nutrient levels and algal biomass. Turbidity sensors use light scattering to quantify suspended solids and particulate matter in water, indicating water clarity and sediment levels. Additionally, optical sensors can measure parameters such as dissolved oxygen, pH, and chemical contaminants through spectroscopic methods. It plays a crucial role in water quality monitoring. These sensors convert light into electronic signals, measuring incident light intensity and transforming it into readable data (Ankitkumar B Rathod et al., 2023; Parambil et al., 2022). In water quality assessment, optical sensors detect interactions of light with particles or dissolved constituents. For instance, dissolved constituents like nitrate and organic matter convert absorbed light into other forms of energy. UV/Vis and fluorescence spectroscopy are common optical methods for in situ monitoring. Glass-fiber-optic turbidity sensors use optical fibers to detect changes in turbidity caused by suspended particles in water. Overall, optical sensors contribute to real-time monitoring and environmental protection by assessing water quality.
2. **Electrochemical Sensors:** Electrochemical sensors play a significant role in detecting chemical ions, molecules, and pathogens in water and other applications. These sensors offer several advantages, including sensitivity, portability, speed, affordability, and suitability for online and in-situ measurements compared to other methods. They can detect compounds that undergo specific transformations within a potential window, making them versatile for multiple ion detections (Kanoun et al., 2021). Electrochemical sensors rely on electrochemical reactions to detect and quantify specific ions, gases, and pollutants in water. Ion-selective electrodes (ISEs) are commonly used to measure concentrations of ions such as chloride, nitrate, ammonia, and heavy metals. pH sensors utilize glass or membrane electrodes to measure hydrogen ion activity in water, providing insights into water acidity or alkalinity. Electrochemical sensors are known for their rapid response times, high sensitivity, and suitability for in-situ monitoring applications.
3. **Biosensors:** Biosensors play a pivotal role in water quality detection, offering simple and accessible solutions for water management. Unlike traditional sensors, biosensors utilize natural microbes and enzymes engineered to respond to specific toxins. When exposed to contaminants, these biosensors produce chemical reactions, emit light, or even sound. Biosensors utilize biological components such as enzymes, antibodies, or microorganisms to detect target analyses in water samples. Enzyme-based biosensors can detect specific pollutants or contaminants through enzymatic reactions, offering high selectivity and specificity. Immune-sensors utilize antigen-antibody interactions to detect pathogens, toxins, or chemical contaminants in water. Microbial biosensors employ living

Figure 1. Optical sensors working principal

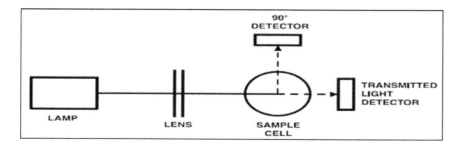

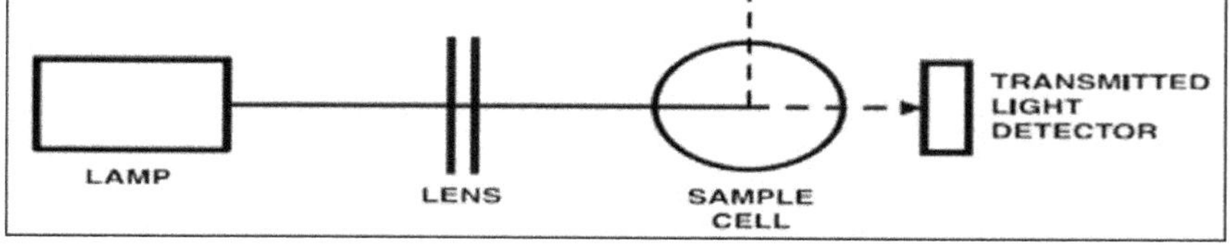

Figure 2. Electrochemical sensor working principal

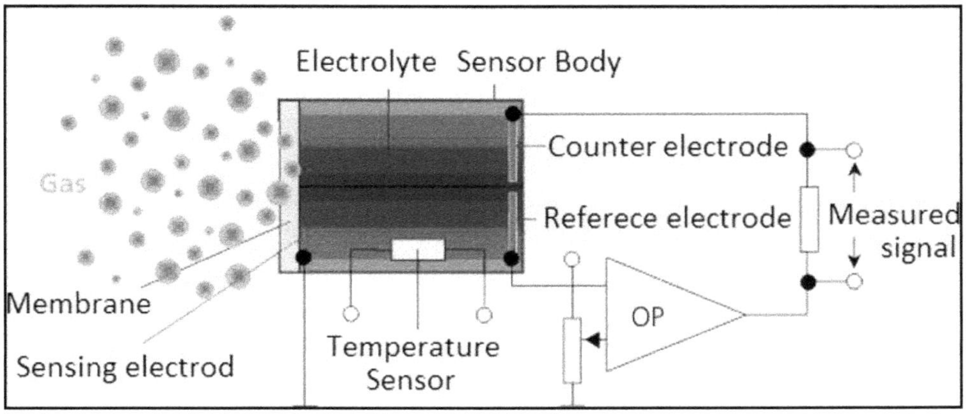

microorganisms to assess water quality based on metabolic activity or inhibition responses (Mustafa et al., 2017). Biosensors offer the advantage of biological recognition elements, enabling sensitive and selective detection of target analytes in complex environmental samples.

4. **Acoustic Sensors**: Acoustic sensors are essential tools for water quality analysis. These sensors utilize sound waves to determine various water properties. Sonar-based acoustic sensors accurately measure water depth, playing a crucial role in hydrographic surveys, underwater mapping, and managing water resources. Additionally, they are permanently installed in water distribution networks for leak detection, automatically recording noise samples from water flow within pipes. By analyzing these samples, leaks, and anomalies can be detected promptly, minimizing water loss. Acoustic sensors also contribute to environmental monitoring by assessing underwater noise levels, which impact marine ecosystems. Understanding noise pollution helps protect aquatic life and maintain ecological balance. Furthermore, acoustic sensors provide valuable data for mapping underwater terrain, identifying submerged objects, and assessing seafloor conditions, crucial for navigation safety and environmental studies. Acoustic sensors use sound waves or ultrasonic waves

Figure 3. Biosensor working principal

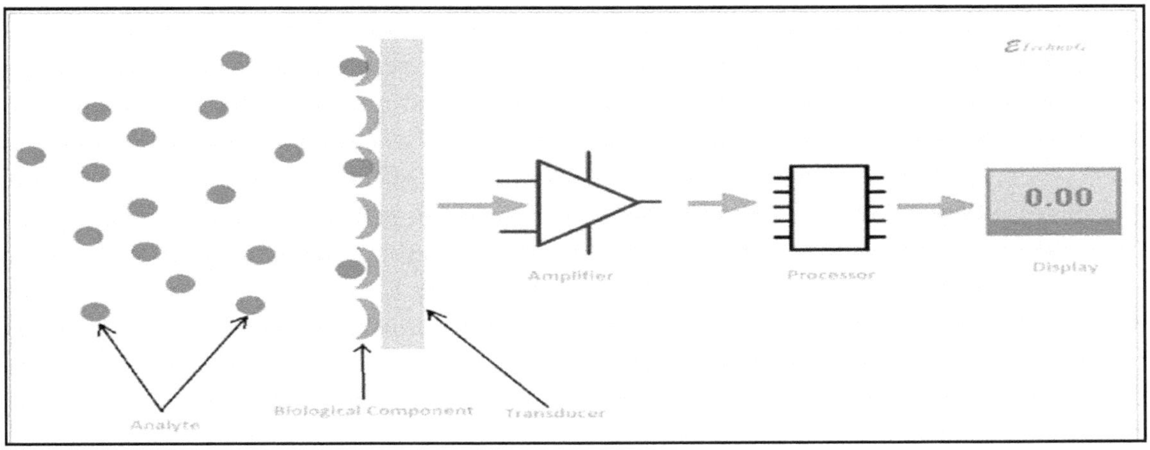

to characterize water properties and detect changes in water quality. Acoustic Doppler sensors can measure water velocity, flow rates, and sediment transport in rivers, streams, and estuaries, providing insights into hydrodynamic processes and sedimentation dynamics (An et al., 2021). Acoustic spectroscopy techniques can assess water quality parameters such as suspended solids, particle size distribution, and bubble concentration based on acoustic backscatter or attenuation measurements.

5. **Nanotechnology-based Sensors:** Nanotechnology-based sensors hold immense promise for water quality monitoring. These sensors, designed with nanomaterials, offer high efficiency, multiplex functionality, and flexibility. While many Nano sensors can achieve these goals, further development is needed to create user-friendly tools capable of detecting analyses in previously inaccessible locations. Water quality monitoring faces challenges due to matrix variability, complex compositions, and low pollutant concentrations. Nanomaterial-enabled sensor platforms promise ultralow multiplex detection and rapid analysis times. Although excitement surrounds nano-enabled sensors, only a few have made it to the market. Nevertheless, their potential impact remains significant, providing widespread and potentially low-cost monitoring of chemicals, microbes, and other analyses in drinking water. Addressing global water challenges requires harnessing the full potential of nanotechnology-enabled sensors (Sanjeevi et al., 2017; Vikesland, 2018). Nanotechnology-based sensors leverage nanomaterials and nanostructures to enhance sensitivity, selectivity, and performance in water quality monitoring applications. Nanomaterials such as carbon nanotubes, graphene, and metal nanoparticles exhibit unique properties that can be exploited for sensing purposes (Nasture et al., 2022). Nano-scale sensors can detect trace levels of contaminants, pathogens, and pollutants in water with high sensitivity and specificity, offering potential for miniaturization, portability, and integration into autonomous monitoring platforms.
6. **Smart Sensors and IoT Integration:** Smart sensors integrated with the Internet of Things (IoT) play a pivotal role in advancing water quality monitoring. These intelligent devices enhance our ability to assess and manage crucial components of daily human needs, especially in agriculture and daily life. Researchers propose IoT-based systems equipped with wireless sensor networks and water quality sensors. These sensors include turbidity, conductivity, temperature, pH, and oxidation-reduction potential sensors. Fuzzy logic models are implemented to predict local water contamination risks. Such systems enable real-time monitoring and management of water quality. While smart sensors offer immense potential, challenges remain. These include ensuring sensor accuracy, addressing data privacy concerns, and developing user-friendly interfaces. However, the benefits of real-time monitoring, early detection of anomalies, and efficient resource utilization make IoT-based water quality systems indispensable (Anwar Abdelrahman Aly et al., 2016; Zulkifli et al., 2022). Smart sensors equipped with data logging, wireless communication, and Internet of Things (IoT) connectivity capabilities enable real-time monitoring and remote data transmission from field locations to central databases or cloud-based platforms. Integrated sensor networks and IoT systems facilitate continuous monitoring of multiple water quality parameters across large spatial scales, enabling data-driven decision-making, early warning systems, and adaptive management strategies for water resource management and pollution control (Bhardwaj et al., 2022).

These advanced water quality sensors contribute to an enhanced understanding of aquatic ecosystems, early detection of pollution events, and informed decision-making for sustainable water management and environmental protection efforts. Continued innovation and technological advancements in sensor

Table 1. Water quality parameter, sensor, advantages, and disadvantages

Water Quality Parameter	Sensor Type	Advantages	Disadvantages	References
Colored Dissolved Organic Matter (CDOM)	Spectrophotometry	- Rapid measurements. - Non-destructive.	- Influenced by water turbidity. - Limited depth penetration.	
Secchi Disk Depth (SDD)	Visual observation	- Simple and low-cost. - Provides a quick estimate of water Clarity.	- Subjective interpretation. - Limited precision.	
Turbidity	Turbidity sensors	- Continuous monitoring. - Reflects suspended solids levels.	- Calibration needed. - Affected by sensor fouling.	
Total Suspended Sediments (TSS)	Turbidity sensors	- Quantifies sediment load. - Real-time data.	- Limited accuracy in high turbidity conditions.	
Water Temperature (WT)	Temperature sensors	- Easy to deploy. - Real-time monitoring.	- Prone to drift. - Affected by solar radiation.	
Total Phosphorus (TP)	Spectrophotometry	- Measures nutrient levels. - Non-destructive.	- Requires reagents for analysis. - Interference from other compounds.	
Sea Surface Salinity (SSS)	Conductivity sensors	- Reflects salinity variations. - Real-time data.	- Calibration needed. - Affected by fouling.	
Dissolved Oxygen (DO)	Optical sensors	- Vital for aquatic life. - Continuous monitoring.	- Affected by temperature and pressure. - Requires calibration.	
Biochemical Oxygen Demand (BOD)	BOD sensors	- Indicates organic pollution. - Real-time assessment.	- Requires incubation period. - Interference from other substances.	
Chemical Oxygen Demand (COD)	COD sensors	- Measures organic and inorganic pollutants. - Quick results.	- Requires reagents. - Interference from other compounds.	

design, fabrication, and deployment will further advance capabilities for water quality monitoring and address emerging challenges in water resource sustainability and pollution mitigation.

The Capabilities of Modern Sensors and Their Significance in Achieving the Mission of Cleaner and Safer Waters

Modern sensors possess advanced capabilities that play a pivotal role in achieving the mission of cleaner and safer waters by enabling efficient, accurate, and timely monitoring of water quality parameters.

1. High Sensitivity: Modern sensors exhibit high sensitivity, allowing them to detect low concentrations of contaminants and pollutants in water. This capability is essential for early detection of pollution events and monitoring compliance with water quality standards, enabling prompt intervention and corrective actions to prevent or mitigate adverse impacts on aquatic ecosystems and human health.
2. Real-time Monitoring: Many modern sensors enable real-time or near real-time monitoring of water quality parameters, providing continuous data streams that capture temporal variations and dynamics in water quality. Real-time monitoring facilitates rapid detection of pollution incidents, timely response to changing environmental conditions, and adaptive management strategies for maintaining water quality within acceptable limits.

3. Multi-parameter Measurement: Advanced sensors are capable of measuring multiple water quality parameters simultaneously, offering comprehensive insights into the chemical, physical, and biological characteristics of water bodies. Multi-parameter sensors streamline monitoring efforts, reduce the need for multiple instruments or sampling campaigns, and provide integrated datasets for holistic assessments of water quality and ecosystem health.
4. Remote Monitoring: Modern sensors are equipped with wireless communication capabilities, enabling remote monitoring of water quality in remote or inaccessible locations. Remote monitoring systems transmit data to centralized databases or cloud-based platforms, allowing stakeholders to access real-time monitoring data, receive alerts and notifications, and make informed decisions regarding water quality management and pollution control measures.
5. Accuracy and Precision: Advanced sensor technologies offer high accuracy and precision in measuring water quality parameters, minimizing measurement errors and uncertainties. Accurate sensor data enhance the reliability and credibility of monitoring results, supporting evidence-based decision-making and regulatory compliance efforts to protect water resources and public health.
6. Autonomous Operation: Some modern sensors are designed for autonomous or unmanned operation, capable of self-calibration, self-diagnosis, and long-term deployment in field environments. Autonomous sensors reduce the need for manual intervention, maintenance, and data collection, optimizing resources and personnel allocation for water quality monitoring programs.
7. Cost-effectiveness: Advances in sensor technology have led to the development of cost-effective monitoring solutions, including low-cost sensors, miniaturized devices, and scalable sensor networks. Cost-effective sensors increase accessibility to water quality monitoring capabilities, particularly in resource-constrained or underserved regions, and facilitate community-based monitoring initiatives and citizen science programs.
8. Data Integration and Analysis: Modern sensors are often integrated with data logging, storage, and analysis functionalities, facilitating the processing, interpretation, and visualization of monitoring data. Integrated sensor networks and data management systems enable seamless data integration, trend analysis, and spatial mapping of water quality parameters, supporting evidence-based decision-making and long-term planning for water resource management and pollution control.

MISSION TO COMBAT WATER POLLUTION

A mission to combat water pollution involves a multifaceted approach aimed at reducing, mitigating, and preventing the introduction of pollutants into water bodies, restoring degraded ecosystems, and promoting sustainable water management practices (Wang et al., 2016). In water pollution prevention implementing measures to prevent pollution at the source is crucial for reducing the influx of contaminants into water bodies. This includes regulatory measures to control industrial discharges, agricultural runoff, and urban storm water runoff, as well as promoting pollution prevention practices among businesses, industries, and individuals. Establishing comprehensive water quality monitoring programs to assess the status and trends of water quality parameters in rivers, lakes, estuaries, and coastal waters is essential. Continuous monitoring using advanced sensor technologies enables early detection of pollution events, identification of pollution sources, and data-driven decision-making for water quality management (Rene et al., 2019). Enacting and enforcing robust regulatory frameworks and water quality standards is necessary to protect water resources and ensure compliance with pollution control measures. Regulatory agencies play a critical role in monitoring compliance, issuing permits, enforcing environmental regulations, and imposing penalties for violations of water quality standards. Investing in wastewater treatment infrastructure and technologies to treat domestic, industrial, and agricultural effluents is essential for removing pollutants before they are discharged into water bodies. Upgrading and expanding wastewater treatment plants, implementing advanced treatment processes, and promoting water reuse and recycling initiatives contribute to reducing pollution and conserving freshwater resources. Restoring degraded aquatic ecosystems, such as wetlands, riparian zones, and mangrove forests, helps improve water quality, enhance biodiversity, and provide habitat for aquatic flora and fauna. Ecosystem restoration projects involve habitat rehabilitation, reforestation, sediment remediation, and reintroduction of native species to restore ecological functions and services. Increasing public awareness and fostering community engagement are essential for promoting responsible water stewardship practices and behavior change. Educational campaigns, outreach initiatives, and citizen science programs empower individuals, communities, and stakeholders to take proactive measures to prevent pollution, conserve water resources, and protect aquatic ecosystems (Anwar Abdelrahman Aly et al., 2016; Lees, 1994). Addressing trans-boundary water pollution requires international cooperation and collaboration among neighboring countries sharing water resources. Bilateral and multilateral agreements, joint monitoring programs, and collaborative initiatives facilitate data sharing, coordinated management, and mutual support for addressing shared water pollution challenges. Harnessing innovation and technology, including advanced sensor technologies, data analytics, remote sensing, and artificial intelligence, enhances the effectiveness and efficiency of water pollution monitoring, management, and remediation efforts. Investing in research and development, fostering partnerships between academia, industry, and government, and promoting technology transfer facilitate the adoption of innovative solutions for combating water pollution.

By implementing a comprehensive and coordinated approach encompassing these components, a mission to combat water pollution can achieve significant progress in safeguarding water resources, protecting aquatic ecosystems, and ensuring access to clean and safe water for all. Such efforts contribute to sustainable development, environmental conservation, and the well-being of present and future generations.

To Preserve Aquatic Ecosystems and Protect Biodiversity by Reducing Water Pollution

Preserving aquatic ecosystems and protecting biodiversity by reducing water pollution requires concerted efforts across multiple fronts.

1. Source Reduction and Pollution Prevention:
    - Implement pollution prevention measures to reduce the release of pollutants into water bodies.
    - Enforce regulations and incentivize industries to adopt cleaner production practices and technologies.
    - Promote sustainable agriculture practices, such as precision farming, integrated pest management, and soil conservation, to minimize nutrient runoff and pesticide contamination.
    - Encourage the use of eco-friendly products and materials to reduce the generation of pollutants in households and businesses.

2. Wastewater Treatment and Management:
    - Upgrade and expand wastewater treatment infrastructure to improve the quality of treated effluents.
    - Implement advanced treatment processes, such as tertiary treatment and disinfection, to remove contaminants like nutrients, pathogens, and emerging pollutants.
    - Implement decentralized wastewater treatment systems and constructed wetlands to treat sewage and storm water runoff at the source.
    - Promote water reuse and recycling initiatives to reduce the discharge of treated effluents into water bodies and alleviate pressure on freshwater resources.
3. Ecosystem Restoration and Conservation:
    - Restore and rehabilitate degraded aquatic ecosystems, including wetlands, riparian zones, coral reefs, and estuaries, to enhance biodiversity and ecosystem services.
    - Implement habitat conservation measures to protect critical habitats for aquatic flora and fauna, including endangered species.
    - Establish marine protected areas (MPAs) and biodiversity conservation zones to safeguard vulnerable ecosystems and promote sustainable fisheries management.
    - Implement sustainable aquaculture practices to minimize environmental impacts and preserve wild fish stocks.
4. Integrated Water Resources Management:
    - Adopt integrated water resources management (IWRM) approaches to balance competing water demands and optimize water allocation for ecological, economic, and social needs.
    - Promote watershed management initiatives to address non-point source pollution and land-use impacts on water quality.
    - Foster collaboration among stakeholders, including government agencies, local communities, NGOs, and private sectors, to develop and implement holistic water management plans.
    - Incorporate ecosystem-based approaches into water resource planning and decision-making processes to maintain the health and resilience of aquatic ecosystems.
5. Public Awareness and Education:
    - Raise public awareness about the importance of preserving aquatic ecosystems and protecting biodiversity.
    - Educate communities about the impacts of water pollution on human health, ecosystems, and the economy.
    - Engage stakeholders through outreach programs, educational campaigns, and citizen science initiatives to promote participation in water conservation and pollution prevention efforts.
    - Foster environmental literacy and empower individuals to take action to protect water resources through sustainable behaviors and practices.
6. Research and Innovation:
    - Invest in research and development to advance knowledge and technologies for water pollution monitoring, management, and remediation.
    - Support interdisciplinary research collaborations to address complex water pollution challenges and develop innovative solutions.
    - Foster technology transfer and knowledge exchange to facilitate the adoption of best practices and technologies for water quality improvement.

- Promote the development of green technologies and nature-based solutions for water pollution control, such as phytoremediation, bio-filtration, and ecological engineering.

By implementing these strategies in a coordinated and collaborative manner, it is possible to reduce water pollution effectively, preserve aquatic ecosystems, and protect biodiversity for the benefit of present and future generations.

WATER POLLUTION HAS DETRIMENTAL EFFECTS ON AQUATIC LIFE, ENDANGERING SPECIES AND ECOSYSTEMS

Pollutants such as heavy metals, pesticides, industrial chemicals, and oil spills can be highly toxic to aquatic organisms. These pollutants disrupt biological processes, impair vital functions, and cause physiological damage, leading to illness, deformities, reproductive impairments, and mortality among aquatic species (Ogidi & Akpan, 2022). Certain pollutants can bioaccumulate in the tissues of aquatic organisms, meaning they accumulate in higher concentrations as they move up the food chain. Predatory species at the top of the food chain, such as large fish or marine mammals, can accumulate high levels of pollutants through biomagnification, increasing their susceptibility to adverse health effects and reproductive issues. Water pollution can degrade aquatic habitats critical for the survival and reproduction of many species (Moyle & Leidy, 1992). Sedimentation, nutrient runoff, and chemical contamination can smother benthic habitats, destroy coral reefs, degrade wetlands, and eliminate spawning grounds, reducing available habitat and altering ecosystem structure and function. Excessive nutrient runoff from agricultural activities and sewage discharges can lead to eutrophication, a condition characterized by nutrient enrichment and algal overgrowth in water bodies. Harmful algal blooms (HABs) can form, releasing toxins that poison aquatic organisms, deplete oxygen levels, and disrupt food webs, leading to mass mortality events and ecosystem degradation. Water pollution can interfere with the reproductive success and development of aquatic species. Endocrine-disrupting chemicals (EDCs) can disrupt hormonal balance, causing reproductive abnormalities, feminization of male organisms, and decreased fertility rates. Pollutants can also interfere with embryo development, larval growth, and juvenile survival, impairing population recruitment and resilience. Chronic exposure to water pollution can lead to declines in species richness, abundance, and genetic diversity in aquatic ecosystems. Pollution-sensitive species may disappear from contaminated habitats, leading to shifts in species composition and ecosystem structure (Malmqvist & Rundle, 2002). Biodiversity loss weakens ecosystem resilience, making ecosystems more vulnerable to additional stressors and less capable of providing essential ecosystem services. Acid rain and acid mine drainage can lower pH levels in freshwater ecosystems, causing acidification and impairing the ability of aquatic organisms to regulate their internal pH. Acidification can harm fish, amphibians, mollusks, and other aquatic organisms, affecting their survival, growth, reproduction, and physiological functions. Elevated carbon dioxide levels in the atmosphere lead to increased absorption of carbon dioxide by oceans, resulting in ocean acidification (Häder et al., 2020). Water pollution damages aquatic life, particularly species with calcium carbonate shells, like corals and mollusks, imperiling their survival. Urgent action is vital for ecosystem restoration and pollution control.

IMPACT OF POLLUTED WATER ON HUMAN HEALTH AND THE FUNDAMENTAL NEED FOR ACCESS TO CLEAN AND SAFE DRINKING WATER

Polluted water poses significant risks to human health, emphasizing the fundamental need for access to clean and safe drinking water.

Contaminated water sources can harbor a variety of pathogens, including bacteria, viruses, protozoa, and parasites, which can cause waterborne diseases such as cholera, typhoid fever, dysentery, giardiasis, and cryptosporidiosis. Ingestion of contaminated water or food prepared with contaminated water can lead to gastrointestinal illnesses ranging from mild diarrhea to severe dehydration and death, particularly among vulnerable populations such as children, the elderly, and individuals with weakened immune systems (Schwarzenbach et al., 2010). Water pollution introduces chemicals like heavy metals, pesticides, and industrial substances into drinking water, causing health issues like neurological disorders, cancer, and reproductive problems. Agricultural runoff can also spur harmful algal blooms (Khan et al., 2013). Cyanotoxins from algae can contaminate water, causing gastrointestinal illness, liver damage, respiratory issues, and neurological symptoms. Polluted water breeds disease vectors, spreading malaria, dengue, Zika, and schistosomiasis (Fawell & Nieuwenhuijsen, 2003). Water pollution threatens food security by tainting freshwater used in agriculture and fishing, leading to foodborne illnesses and nutritional deficiencies. Reduced fish yields and biodiversity hinder nutrient access, affecting physical and cognitive development, especially in children. Waterborne diseases incur social and economic costs, perpetuating poverty and inequality. Universal access to clean water is crucial for human health and sustainable development.

Advantages and Challenges of Advanced Water Quality Sensors

Advanced water quality sensors offer several advantages for monitoring and managing water resources effectively, but they also come with certain challenges (Kruse, 2018).

Advantages:

1. High Accuracy and Precision: Advanced sensors can provide highly accurate and precise measurements of various water quality parameters, enabling reliable monitoring of changes in water quality over time and across different locations.
2. Real-Time Monitoring: Many advanced sensors offer real-time or near real-time monitoring capabilities, allowing for rapid detection of pollution events, timely response to water quality fluctuations, and proactive management of water resources.
3. Multi-parameter Capability: Some advanced sensors are capable of measuring multiple water quality parameters simultaneously, providing comprehensive insights into the chemical, physical, and biological characteristics of water bodies with a single device.
4. Remote Monitoring: Advanced sensors equipped with wireless communication capabilities enable remote monitoring of water quality in remote or inaccessible locations, facilitating data collection and analysis across large spatial scales.
5. Data Integration and Analysis: Advanced sensors are often integrated with data logging, storage, and analysis functionalities, enabling seamless integration of monitoring data, trend analysis, and visualization of water quality parameters for informed decision-making.

Figure 4. Advantage of water quality sensors

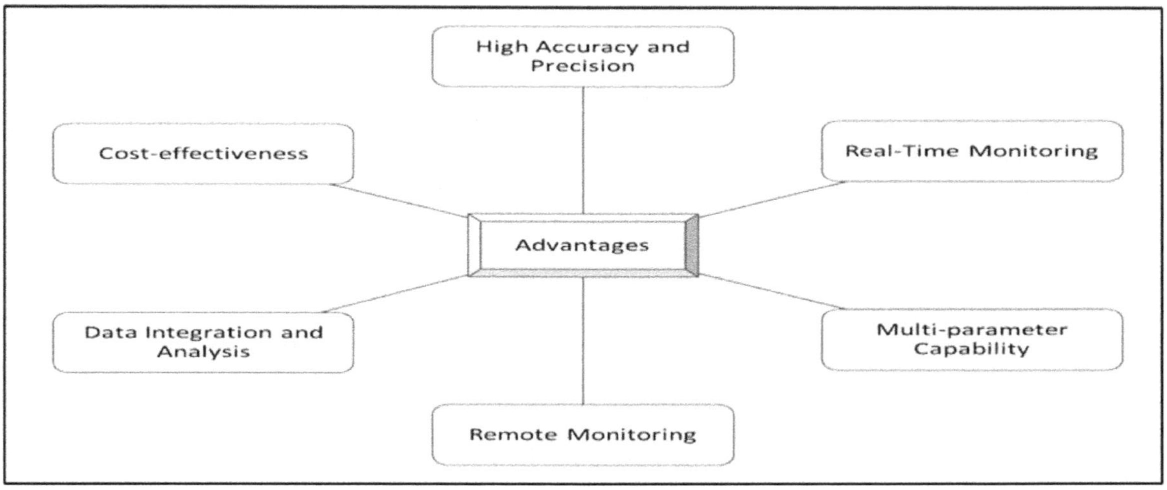

6. Cost-effectiveness: Advances in sensor technology have led to the development of cost-effective monitoring solutions, including low-cost sensors and scalable sensor networks, which increase accessibility to water quality monitoring capabilities, particularly in resource-constrained or underserved regions.

CHALLENGES

Advanced sensors require regular calibration and maintenance to ensure accurate and reliable measurements. Calibration drift, fouling, and sensor degradation over time can affect sensor performance and data quality, necessitating ongoing attention and resources for maintenance. Some advanced sensors may be complex to operate and require technical expertise for installation, calibration, and troubleshooting. Limited technical capacity and expertise among end-users can pose challenges to the widespread adoption and effective use of advanced sensor technologies. Environmental factors such as temperature variations, fouling, biofouling, and turbidity can affect sensor performance and data accuracy (Jan et al., 2021). Managing sensor interference and environmental impacts on measurements entails careful calibration adjustments. Handling large sensor data volumes and interpreting complex datasets poses challenges, especially for non-specialists. Effective data management and analysis are vital for informed decision-making. Standardization efforts are crucial for data reliability and interoperability, while privacy and security concerns must be addressed in deploying advanced sensor networks for water quality monitoring. Despite challenges, advanced sensors offer significant benefits, requiring continued innovation and collaboration for maximizing their potential in safeguarding water resources and public health.

CRITICAL ENVIRONMENTAL ISSUE

Indeed, understanding the broader context of water quality is crucial for addressing this critical environmental issue effectively. Water quality encompasses a complex interplay of physical, chemical, and biological factors that influence the health and integrity of aquatic ecosystems, as well as human well-being. Here are some key aspects of the broader context of water quality and the technological solutions available to address this issue:

Water quality is influenced by both natural processes and human activities. Natural factors such as weathering, erosion, and biological processes interact with anthropogenic sources of pollution, including industrial discharges, agricultural runoff, urbanization, and wastewater effluents. Understanding the sources, pathways, and impacts of pollutants on water quality is essential for developing effective management strategies (Shoushtarian & Negahban-Azar, 2020). Water quality plays a critical role in supporting healthy aquatic ecosystems and biodiversity. Clean and healthy water bodies provide essential habitats for a diverse range of aquatic species, including fish, invertebrates, plants, and microorganisms (Prăvălie, 2016). Maintaining water quality is essential for preserving ecosystem services such as water purification, nutrient cycling, flood regulation, and habitat provision. Access to clean and safe drinking water is essential for human health and well-being. Contaminated water sources can pose risks to human health through waterborne diseases, chemical exposures, and other health impacts. Ensuring access to clean water for drinking, sanitation, and hygiene is fundamental for preventing water-related illnesses and promoting public health.

Regulatory frameworks and water quality standards play a crucial role in protecting water resources and public health. Governments and regulatory agencies establish standards and guidelines for water quality parameters such as drinking water quality, recreational water quality, and environmental quality. Compliance with regulatory requirements helps mitigate pollution and ensure the safety of water supplies for human consumption and ecosystem health. Technological innovations offer valuable tools and solutions for monitoring, assessing, and managing water quality. Advanced sensor technologies, remote sensing platforms, data analytics, and modeling tools enable real-time monitoring of water quality parameters, early detection of pollution events, and informed decision-making for pollution control measures (Pimentel et al., 2004). Nature-based solutions, such as constructed wetlands, riparian buffers, and green infrastructure, offer cost-effective approaches to improving water quality and enhancing ecosystem resilience. Adopting integrated water management approaches is essential for addressing water quality challenges comprehensively. Integrated water management considers the interconnectedness of water resources, land use, ecosystems, and human activities, taking into account social, economic, and environmental factors. Collaborative governance, stakeholder engagement, and adaptive management strategies are key elements of integrated water management approaches (Alcamo & Henrichs, 2002). By gaining insights into the broader context of water quality and leveraging technological solutions, stakeholders can work together to address this critical environmental issue effectively. By protecting and restoring water quality, we can ensure the sustainability of water resources for current and future generations, safeguarding both ecosystems and human well-being.

CONCLUSION

In conclusion, addressing water quality is a multifaceted challenge that requires a comprehensive understanding of its broader context, including ecological, social, economic, and technological dimensions. Water quality is intricately linked to the health of aquatic ecosystems, biodiversity, human well-being, and sustainable development. Pollution from both natural and anthropogenic sources poses significant threats to water quality, leading to adverse impacts on ecosystems, public health, and socio-economic activities.

However, technological solutions offer valuable tools and approaches for monitoring, assessing, and managing water quality effectively. Advanced sensor technologies, remote sensing platforms, data analytics, and modeling tools enable real-time monitoring, early detection of pollution events, and informed decision-making for pollution control measures. Additionally, nature-based solutions and integrated water management approaches provide cost-effective strategies for improving water quality, enhancing ecosystem resilience, and promoting sustainable water use.

To address water quality challenges successfully, collaboration and cooperation among stakeholders are essential. Governments, regulatory agencies, communities, industries, academia, and non-governmental organizations must work together to implement robust regulatory frameworks, promote pollution prevention measures, and invest in sustainable water management practices. By safeguarding water quality, we can protect aquatic ecosystems, preserve biodiversity, ensure access to clean and safe drinking water, and promote the well-being of present and future generations.

In summary, addressing water quality is not only a matter of environmental conservation but also a fundamental requirement for sustainable development and human prosperity. By recognizing the importance of water quality within its broader context and adopting holistic approaches and technological solutions, we can achieve cleaner and safer waters for the benefit of both ecosystems and societies worldwide.

REFERENCES

Abbasi, T., Sanjeevi, R., Anuradha, J., & Abbasi, S. A. (2013). Impact of Al 3+ on sludge granulation in UASB reactor. *Indian Journal of Biotechnology*, *12*(2), 254–259.

Ahamad, A., Madhav, S., Singh, A. K., Kumar, A., & Singh, P. (2020). Types of Water Pollutants: Conventional and Emerging. In D. Pooja, P. Kumar, P. Singh, & S. Patil (Eds.), *Sensors in Water Pollutants Monitoring: Role of Material* (pp. 21–41). Springer Singapore. doi:10.1007/978-981-15-0671-0_3

Alcamo, J., & Henrichs, T. (2002). Critical regions: A model-based estimation of world water resources sensitive to global changes. *Aquatic Sciences*, *64*(4), 352–362. doi:10.1007/PL00012591

Aly, A. A., Al-Omran, A. M., Sallam, A. S., Al-Wabel, M. I., & Al-Shayaa, M. S. (2016). Vegetation cover change detection and assessment in arid environment using multi-temporal remote sensing images and ecosystem management approach. *Solid Earth*, *7*(2), 713–725. doi:10.5194/se-7-713-2016

An, J., Ra, H., Youn, C., & Kim, K. (2021). Experimental Results of Underwater Acoustic Communication with Nonlinear Frequency Modulation Waveform. *Sensors (Basel)*, *21*(21), 21. doi:10.3390/s21217194 PMID:34770501

Ankitkumar,, B., Rathod, A. J., Prashantkumar B., Tripathi, S., & Sanjeevi, R. (2023). Vegetational Change Detection Using Machine Learning in GIS Technique: A Case Study from Jamnagar (Gujarat). *Journal of Data Acquisition and Processing*, *38*(1), 1046–1061. doi:10.5281/zenodo.7700655

Arenas-Sánchez, A., Rico, A., & Vighi, M. (2016). Effects of water scarcity and chemical pollution in aquatic ecosystems: State of the art. *The Science of the Total Environment*, *572*, 390–403. doi:10.1016/j.scitotenv.2016.07.211 PMID:27513735

Bhardwaj, A., Dagar, V., Khan, M. O., Aggarwal, A., Alvarado, R., Kumar, M., Irfan, M., & Proshad, R. (2022). Smart IoT and Machine Learning-based Framework for Water Quality Assessment and Device Component Monitoring. *Environmental Science and Pollution Research International*, *29*(30), 46018–46036. doi:10.1007/s11356-022-19014-3 PMID:35165843

Bourdeau, P., & Treshow, M. (1978). Ecosystem response to pollution. *Prin Ciples Ecotoxicol*, *5*, 79–88.

Cai, H., Mei, Y., Chen, J., Wu, Z., Lan, L., & Zhu, D. (2020). An analysis of the relation between water pollution and economic growth in China by considering the contemporaneous correlation of water pollutants. *Journal of Cleaner Production*, *276*, 122783. doi:10.1016/j.jclepro.2020.122783

Chen, B., Wang, M., Duan, M., Ma, X., Hong, J., Xie, F., Zhang, R., & Li, X. (2019). In search of key: Protecting human health and the ecosystem from water pollution in China. *Journal of Cleaner Production*, *228*, 101–111. doi:10.1016/j.jclepro.2019.04.228

Choi, J., Hearne, R., Lee, K., & Roberts, D. (2015). The relation between water pollution and economic growth using the environmental Kuznets curve: A case study in South Korea. *Water International*, *40*(3), 499–512. doi:10.1080/02508060.2015.1036387

Dwivedi, A. K. (2017). Researches in water pollution: A review. *International Research Journal of Natural and Applied Sciences*, *4*(1), 118–142.

El-Shafeiy, E., Alsabaan, M., Ibrahem, M. I., & Elwahsh, H. (2023). Real-Time Anomaly Detection for Water Quality Sensor Monitoring Based on Multivariate Deep Learning Technique. *Sensors (Basel)*, *23*(20), 8613. doi:10.3390/s23208613 PMID:37896705

Fatima, S., Muzammal, M., Rehman, A., Rustam, S. A., Shehzadi, Z., Mehmood, A., & Waqar, M. (2020). Water pollution on heavy metals and its effects on fishes. *International Journal of Fisheries and Aquatic Studies*, *8*(3), 6–14.

Fawell, J., & Nieuwenhuijsen, M. J. (2003). Contaminants in drinking water: Environmental pollution and health. *British Medical Bulletin*, *68*(1), 199–208. doi:10.1093/bmb/ldg027 PMID:14757718

Fischer, K., Fries, E., Körner, W., Schmalz, C., & Zwiener, C. (2012). New developments in the trace analysis of organic water pollutants. *Applied Microbiology and Biotechnology*, *94*(1), 11–28. doi:10.1007/s00253-012-3929-z PMID:22358315

Gambhir, R. S., Kapoor, V., Nirola, A., Sohi, R., & Bansal, V. (2012). Water Pollution: Impact of Pollutants and New Promising Techniques in Purification Process. *Journal of Human Ecology (Delhi, India)*, *37*(2), 103–109. doi:10.1080/09709274.2012.11906453

Häder, D.-P., Banaszak, A. T., Villafañe, V. E., Narvarte, M. A., González, R. A., & Helbling, E. W. (2020). Anthropogenic pollution of aquatic ecosystems: Emerging problems with global implications. *The Science of the Total Environment*, *713*, 136586. doi:10.1016/j.scitotenv.2020.136586 PMID:31955090

Håkanson, L., & Bryhn, A. (1999). Water pollution. *Backhuys Publ, Leiden*. https://www.researchgate.net/profile/Lars-Hakanson-2/publication/236024084_WATER_POLLUTION_-_methods_and_criteria_to_rank_model_and_remediate_chemical_threats_to_aquatic_ecosystems/links/02e7e515d7425f30ed000000/WATER-POLLUTION-methods-and-criteria-to-rank-model-and-remediate-chemical-threats-to-aquatic-ecosystems.pdf

Halder, J. N., & Islam, M. N. (2015). Water pollution and its impact on the human health. *Journal of Environment and Human*, *2*(1), 36–46. doi:10.15764/EH.2015.01005

Haseena, M., Malik, M. F., Javed, A., Arshad, S., Asif, N., Zulfiqar, S., & Hanif, J. (2017). Water pollution and human health. *Environmental Risk Assessment and Remediation, 1*(3). https://eastafricaschoolserver.org/content/_public/Environment/Teaching%20Resources/Environment%20and%20Sustainability/Water-pollution-and-human-health.pdf

Inyinbor Adejumoke, A., Adebesin Babatunde, O., Oluyori Abimbola, P., Adelani Akande Tabitha, A., Dada Adewumi, O., & Oreofe Toyin, A. (2018). Water pollution: Effects, prevention, and climatic impact. *Water Challenges of an Urbanizing World*, *33*, 33–47.

Jan, F., Min-Allah, N., & Düştegör, D. (2021). Iot based smart water quality monitoring: Recent techniques, trends and challenges for domestic applications. *Water (Basel)*, *13*(13), 1729. doi:10.3390/w13131729

Juma, D. W., Wang, H., & Li, F. (2014). Impacts of population growth and economic development on water quality of a lake: Case study of Lake Victoria Kenya water. *Environmental Science and Pollution Research International*, *21*(8), 5737–5746. doi:10.1007/s11356-014-2524-5 PMID:24442964

Kanoun, O., Lazarević-Pašti, T., Pašti, I., Nasraoui, S., Talbi, M., Brahem, A., Adiraju, A., Sheremet, E., Rodriguez, R. D., Ben Ali, M., & Al-Hamry, A. (2021). A Review of Nanocomposite-Modified Electrochemical Sensors for Water Quality Monitoring. *Sensors (Basel)*, *21*(12), 12. doi:10.3390/s21124131 PMID:34208587

Karunanidhi, D., Subramani, T., Roy, P. D., & Li, H. (2021). Impact of groundwater contamination on human health. *Environmental Geochemistry and Health*, *43*(2), 643–647. doi:10.1007/s10653-021-00824-2 PMID:33486701

Khan, S., Shahnaz, M., Jehan, N., Rehman, S., Shah, M. T., & Din, I. (2013). Drinking water quality and human health risk in Charsadda district, Pakistan. *Journal of Cleaner Production*, *60*, 93–101. doi:10.1016/j.jclepro.2012.02.016

Kordbacheh, F., & Heidari, G. (2023). Water pollutants and approaches for their removal. *Materials Chemistry Horizons*, *2*(2), 139–153.

Kruse, P. (2018). Review on water quality sensors. *Journal of Physics. D, Applied Physics*, *51*(20), 203002. doi:10.1088/1361-6463/aabb93

Kumar, T., Naik, S., & Jujjavarappu, S. E. (2022). A critical review on early-warning electrochemical system on microbial fuel cell-based biosensor for on-site water quality monitoring. *Chemosphere*, *291*, 133098. doi:10.1016/j.chemosphere.2021.133098 PMID:34848233

Lees, P. (1994). *Combating water pollution.* Cabid Digital Library. https://www.cabidigitallibrary.org/doi/full/10.5555/19951802017

Li, C., & Li, G. (2021). Impact of China's water pollution on agricultural economic growth: An empirical analysis based on a dynamic spatial panel lag model. *Environmental Science and Pollution Research International*, *28*(6), 6956–6965. doi:10.1007/s11356-020-11079-2 PMID:33025434

Lin, L., Yang, H., & Xu, X. (2022a). Effects of Water Pollution on Human Health and Disease Heterogeneity: A Review. *Frontiers in Environmental Science*, *10*, 880246. https://www.frontiersin.org/articles/10.3389/fenvs.2022.880246. doi:10.3389/fenvs.2022.880246

Lin, L., Yang, H., & Xu, X. (2022b). Effects of water pollution on human health and disease heterogeneity: A review. *Frontiers in Environmental Science*, *10*, 880246. doi:10.3389/fenvs.2022.880246

Malik, D. S., Sharma, A. K., Sharma, A. K., Thakur, R., & Sharma, M. (2020). A review on impact of water pollution on freshwater fish species and their aquatic environment. *Advances in Environmental Pollution Management: Wastewater Impacts and Treatment Technologies*, *1*, 10–28. doi:10.26832/aesa-2020-aepm-02

Malmqvist, B., & Rundle, S. (2002). Threats to the running water ecosystems of the world. *Environmental Conservation*, *29*(2), 134–153. doi:10.1017/S0376892902000097

Manoiu, V.-M., Craciun, A.-I., Kubiak-Wójcicka, K., Antonescu, M., & Olariu, B. (2022). An Eco-Study for a Feasible Project: "Torun and Its Vistula Stretch—An Important Green Navigation Spot on a Blue Inland Waterway.". *Water (Basel)*, *14*(19), 19. doi:10.3390/w14193034

Mishra, S., Anuradha, J., Tripathi, S., & Kumar, S. (2016). In vitro antioxidant and antimicrobial efficacy of Triphala constituents: Emblica officinalis, Terminalia belerica and Terminalia chebula. *Journal of Pharmacognosy and Phytochemistry*, *5*(6), 273–277.

Moyle, P. B., & Leidy, R. A. (1992). Loss of Biodiversity in Aquatic Ecosystems: Evidence from Fish Faunas. In P. L. Fiedler & S. K. Jain (Eds.), *Conservation Biology* (pp. 127–169). Springer US. doi:10.1007/978-1-4684-6426-9_6

Mustafa, F., Hassan, R. Y. A., & Andreescu, S. (2017). Multifunctional Nanotechnology-Enabled Sensors for Rapid Capture and Detection of Pathogens. *Sensors (Basel)*, *17*(9), 9. doi:10.3390/s17092121 PMID:28914769

Muyibi, S. A., Ambali, A. R., & Eissa, G. S. (2008). The Impact of Economic Development on Water Pollution: Trends and Policy Actions in Malaysia. *Water Resources Management*, *22*(4), 485–508. doi:10.1007/s11269-007-9174-z

Nasture, A.-M., Ionete, E. I., Lungu, F. A., Spiridon, S. I., & Patularu, L. G. (2022). Water Quality Carbon Nanotube-Based Sensors Technological Barriers and Late Research Trends: A Bibliometric Analysis. *Chemosensors (Basel, Switzerland)*, *10*(5), 5. doi:10.3390/chemosensors10050161

O'Grady, J., Zhang, D., O'Connor, N., & Regan, F. (2021). A comprehensive review of catchment water quality monitoring using a tiered framework of integrated sensing technologies. *The Science of the Total Environment, 765*, 142766. doi:10.1016/j.scitotenv.2020.142766 PMID:33092838

Ogidi, O. I., & Akpan, U. M. (2022). Aquatic Biodiversity Loss: Impacts of Pollution and Anthropogenic Activities and Strategies for Conservation. In S. Chibueze Izah (Ed.), *Biodiversity in Africa: Potentials, Threats and Conservation* (Vol. 29, pp. 421–448). Springer Nature Singapore. doi:10.1007/978-981-19-3326-4_16

Olatinwo, S. O., & Joubert, T.-H. (2018). Energy efficient solutions in wireless sensor systems for water quality monitoring: A review. *IEEE Sensors Journal, 19*(5), 1596–1625. doi:10.1109/JSEN.2018.2882424

Olatinwo, S. O., & Joubert, T.-H. (2019). Enabling communication networks for water quality monitoring applications: A survey. *IEEE Access : Practical Innovations, Open Solutions, 7*, 100332–100362. doi:10.1109/ACCESS.2019.2904945

Orubu, C. O., & Omotor, D. G. (2011). Environmental quality and economic growth: Searching for environmental Kuznets curves for air and water pollutants in Africa. *Energy Policy, 39*(7), 4178–4188. doi:10.1016/j.enpol.2011.04.025

Parambil, A. R. U. (2022). Water-soluble optical sensors: Keys to detect aluminium in biological environment. *RSC Advances, 12*(22), 13950–13970. doi:10.1039/D2RA01222G PMID:35558844

Park, J., Kim, K. T., & Lee, W. H. (2020). Recent advances in information and communications technology (ICT) and sensor technology for monitoring water quality. *Water (Basel), 12*(2), 510. doi:10.3390/w12020510

Pimentel, D., Berger, B., Filiberto, D., Newton, M., Wolfe, B., Karabinakis, E., Clark, S., Poon, E., Abbett, E., & Nandagopal, S. (2004). Water resources: Agricultural and environmental issues. *Bioscience, 54*(10), 909–918. doi:10.1641/0006-3568(2004)054[0909:WRAAEI]2.0.CO;2

Prashantkumar, B. Sathvara, J. Anuradha, Sandeep Tripathi, & R. Sanjeevi. (2023). Impact of climate change and its importance on human performance. In Insights on Impact of Climate Change and Adaptation of Biodiversity (1st ed., pp. 1–9). KD Publication.

Prăvălie, R. (2016). Drylands extent and environmental issues. A global approach. *Earth-Science Reviews, 161*, 259–278. doi:10.1016/j.earscirev.2016.08.003

Qadri, R., & Faiq, M. A. (2020). Freshwater Pollution: Effects on Aquatic Life and Human Health. In H. Qadri, R. A. Bhat, M. A. Mehmood, & G. H. Dar (Eds.), *Fresh Water Pollution Dynamics and Remediation* (pp. 15–26). Springer Singapore. doi:10.1007/978-981-13-8277-2_2

Radu, C., Manoiu, V.-M., Kubiak-Wójcicka, K., Avram, E., Beteringhe, A., & Craciun, A.-I. (2022). Romanian Danube River Hydrocarbon Pollution in 2011–2021. *Water (Basel), 14*(19), 19. doi:10.3390/w14193156

Ramakrishnan, S., & Jayaraman, A. (2019). Pesticide contaminated drinking water and health effects on pregnant women and children. In *Handbook of research on the adverse effects of pesticide pollution in aquatic ecosystems* (pp. 123–136). IGI Global. https://www.igi-global.com/chapter/pesticide-contaminated-drinking-water-and-health-effects-on-pregnant-women-and-children/213500 doi:10.4018/978-1-5225-6111-8.ch007

Ranjan, H., Sanjeevi, R., Vardhini, S., Tripathi, S., & Anuradha, J. (n.d.). *Biogenic Production of Silver Nanoparticles Utilizing an Arid Weed (Saccharum munja Roxb.) and Evaluation of its Antioxidant and Antimicrobial Activities.*

Reddy, V. R., & Behera, B. (2006). Impact of water pollution on rural communities: An economic analysis. *Ecological Economics*, *58*(3), 520–537. doi:10.1016/j.ecolecon.2005.07.025

Rene, E. R., Shu, L., & Jegatheesan, V. (2019). *Appropriate technologies to combat water pollution.* Challenges in Environmental Science and Engineering (CESE-2017), Kunming, China. https://www.cabidigitallibrary.org/doi/full/10.5555/20210094724

Sanjeevi, R. (2011). *Studies on the treatment of low-strength wastewaters with upflow anaerobic sludge blanket (UASB) reactor: With emphasis on granulation studies.* Centre for Pollution Control and Environmental Engineering.

Sanjeevi, R., & Ankitkumar, B., Rathod, Prashantkumar B. Sathvara, Aviral Tripathi, J. Anuradha, & Tripathi, S. (2022). Vegetational Cartography Analysis Utilizing Multi-Temporal Ndvi Data Series: A Case Study From Rajkot District (Gujarat), India. *Tianjin Daxue Xuebao (Ziran Kexue Yu Gongcheng Jishu Ban)/ Journal of Tianjin University Science and Technology, 55*(4), 490–497. https://doi.org/doi:10.17605/OSF.IO/UGJYM

Sanjeevi, R., Haruna, M., Tripathi, S., Singh, B., & Jayaraman, A. (2017). Impacts of Global Carbon Foot Print on the Marine Environment. *International Journal of Engineering Research & Technology (Ahmedabad)*, *5*, 51–54.

Sathvara, P. B., Anuradha, J., Sanjeevi, R., Tripathi, S., & Rathod, A. B. (2023). Spatial Analysis of Carbon Sequestration Mapping Using Remote Sensing and Satellite Image Processing. In Multimodal Biometric and Machine Learning Technologies (pp. 71–83). Springer. doi:10.1002/9781119785491.ch4

Schwarzenbach, R. P., Egli, T., Hofstetter, T. B., Von Gunten, U., & Wehrli, B. (2010). Global Water Pollution and Human Health. *Annual Review of Environment and Resources*, *35*(1), 109–136. doi:10.1146/annurev-environ-100809-125342

Shoushtarian, F., & Negahban-Azar, M. (2020). Worldwide regulations and guidelines for agricultural water reuse: A critical review. *Water (Basel)*, *12*(4), 971. doi:10.3390/w12040971

Sur, I. M., Moldovan, A., Micle, V., & Polyak, E. T. (2022). Assessment of Surface Water Quality in the Baia Mare Area, Romania. *Water (Basel)*, *14*(19), 19. Advance online publication. doi:10.3390/w14193118

Verma, P., & Ratan, J. K. (2020). Assessment of the negative effects of various inorganic water pollutants on the biosphere—An overview. *Inorganic Pollutants in Water*, 73–96.

Vikesland, P. J. (2018). Nanosensors for water quality monitoring. *Nature Nanotechnology*, *13*(8), 8. doi:10.1038/s41565-018-0209-9 PMID:30082808

Wang, Y., Mukherjee, M., Wu, D., & Wu, X. (2016). Combating river pollution in China and India: Policy measures and governance challenges. *Water Policy*, *18*(S1), 122–137. doi:10.2166/wp.2016.008

Zhang, C., Wang, Y., Song, X., Kubota, J., He, Y., Tojo, J., & Zhu, X. (2017). An integrated specification for the nexus of water pollution and economic growth in China: Panel cointegration, long-run causality and environmental Kuznets curve. *The Science of the Total Environment*, *609*, 319–328. doi:10.1016/j.scitotenv.2017.07.107 PMID:28753507

Zulkifli, C. Z., Garfan, S., Talal, M., Alamoodi, A. H., Alamleh, A., Ahmaro, I. Y. Y., Sulaiman, S., Ibrahim, A. B., Zaidan, B. B., Ismail, A. R., Albahri, O. S., Albahri, A. S., Soon, C. F., Harun, N. H., & Chiang, H. H. (2022). IoT-Based Water Monitoring Systems: A Systematic Review. *Water (Basel)*, *14*(22), 22. doi:10.3390/w14223621

Chapter 6
Sensors for Monitoring Water Pollutants

Surjit Singha
https://orcid.org/0000-0002-5730-8677
Kristu Jayanti College (Autonomous), India

ABSTRACT

This chapter presents an overview of water pollutants and sensor technologies for monitoring them. The chapter emphasizes detection and quantification techniques while discussing chemical, physical, and biological contaminants in surface and groundwater. In addition to examining real-time monitoring advancements, this study delves into critical sensors, including spectroscopic, electrochemical, biosensor, and remote sensing technologies that are emerging, lab-on-a-chip, and nanomaterials. An analysis is conducted on the prospects of water pollutant sensors that progressively improve sensitivity, selectivity, and cost-effectiveness. This extensive evaluation enhances comprehension and resolution of water pollution issues while advocating for sustainable water management strategies that benefit ecosystems and human health.

INTRODUCTION

Water pollution is a substantial peril to ecosystems and human health on a global scale. The detrimental consequences of water body contamination extend to the equilibrium of aquatic ecosystems and the purity of potable water, ultimately resulting in extensive environmental deterioration. The surveillance of water contaminants is of the utmost importance to guarantee adherence to regulations, enable practical remediation endeavours, and protect valuable water resources (Liu et al., 2022).

Researchers and engineers have developed and implemented various sensor technologies that detect and quantify contaminants in water sources to meet this urgent demand. The capabilities of these sensors are extensive, ranging from the detection of chemical pollutants to the evaluation of physical parameters and the monitoring of biological indicators. Using these cutting-edge sensors, scholars and practitioners in the environmental field can acquire significant knowledge concerning the condition of

DOI: 10.4018/979-8-3693-1930-7.ch006

aquatic ecosystems and formulate well-informed choices concerning approaches to alleviate pollution (Shahra & Wu, 2020).

This chapter aims to thoroughly examine the significant categories of water pollutants encountered in diverse settings and investigate the state-of-the-art sensor technologies used to detect and monitor these pollutants. By reviewing the most recent developments in sensor technology and their implementations in water quality evaluation, this chapter aims to make a scholarly contribution to the ongoing dialogue surrounding water pollution and the promotion of sustainable water resource management.

Water pollution is an urgent ecological concern with wide-ranging consequences for biodiversity, ecosystems, and human health. The presence of diverse pollutants—chemicals, pathogens, and physical debris—in water bodies poses a significant risk to providing potable water and causes disturbances in aquatic ecosystems. To address water pollution, it is necessary to conduct exhaustive monitoring to identify sources, evaluate hazards, and implement effective mitigation strategies. Sophisticated sensor technologies are paramount in these surveillance endeavours as they facilitate instantaneously identifying and measuring contaminants within water systems (Ma et al., 2023).

Water pollutant sensors are rooted in engineering, chemistry, environmental science, and physics. The subject matter encompasses comprehension of the conduct of contaminants within aquatic ecosystems, such as their origins, methods of transport, and interactions with both biotic and abiotic elements. Furthermore, sensor technologies employ spectroscopy, electrochemistry, biology, and remote sensing principles to identify and assess water samples' impurities. Theoretical models and concepts guide the design, development, and deployment of sensor systems used in water quality monitoring. These frameworks enable the precise measurement and interpretation of data.

The framework for water pollutant sensors incorporates the conceptualization of sensor technologies in environmental monitoring and management. Identification of critical elements, including sensor types, detection mechanisms, data processing algorithms, and decision support systems, is required. The conceptual framework incorporates additional elements that impact sensor performance, such as environmental conditions, sensor calibration, and data validation. By harmonizing these elements within a unified structure, scholars and professionals alike can proficiently devise and execute sensor-driven surveillance approaches to tackle the complexities of water pollution.

This study's objective is a comprehensive overview of sensor technologies for monitoring water pollutants. The objective is to investigate the advancements, applications, and underlying principles of spectroscopic, electrochemical, biosensors, and remote sensing technologies for water quality evaluation. New developments in microfluidics, nanomaterials, and real-time monitoring networks are being examined as part of the research. It will help researchers understand what is coming next in water pollutant sensing.

This research includes various sensor technologies that surveil chemical, physical, and biological contaminants in water systems. The subject encompasses substantial water contaminants, such as thermal pollution, heavy metals, pesticides, pathogens, and sediment. In addition, the study examines the applications of sensor technology in surface water, groundwater, and effluent treatment systems. The study not only examines well-established sensor technologies but also investigates emergent advancements and potential improvements to the capabilities of monitoring water quality.

MAJOR TYPES OF WATER POLLUTANTS

Water pollution is an intricate and widespread menace that jeopardizes the health of humans and ecosystems. It comprises a multitude of contaminants that may originate from diverse pathways and sources. Chemical contaminants from metropolitan development, agricultural operations, and industrial processes collectively contribute significantly to water pollution. Heavy metals, including lead, mercury, cadmium, and arsenic, are released into aquatic environments via atmospheric deposition, mining, and industrial effluents. These substances accumulate in the environment and persist for extended periods, presenting significant health hazards. Commonly employed in agriculture to augment crop productivity, pesticides and herbicides have the potential to seep into surface and groundwater via discharge and leaching processes. It can harm aquatic organisms and expose humans to health hazards if they consume contaminated water or marine food sources. Moreover, pharmaceutical substances such as analgesics, hormones, and antibiotics enter water systems via effluent discharges. The compounds mentioned above present challenges due to their extensive application, prolonged persistence, potential environmental harm, and contribution to the proliferation of antibiotic resistance (Akhtar et al., 2021; Herath et al., 2022).

Physical pollutants also play a role in the deterioration of water quality, alongside chemical contaminants. These pollutants modify the physical attributes of aquatic ecosystems and worsen ecological pressures. Sedimentation caused by construction activities, deforestation, and soil erosion can disrupt marine habitats, suffocate benthic organisms, and impair water clarity, thereby endangering the health and biodiversity of an ecosystem. In addition, urban areas, power plants, and factories can contribute to thermal pollution, which can increase water temperatures, decrease dissolved oxygen levels, and disrupt thermal stratification patterns. This phenomenon may induce detrimental consequences for aquatic organisms, including alterations in their metabolic rates, reproductive difficulties, and increased susceptibility to diseases and parasites (Gonzalez et al., 2023).

Biological pollutants, including microorganisms and pathogens, present supplementary obstacles to water quality and human health. Bacterial pathogens such as Escherichia coli and Salmonella, which frequently arise from human and animal faecal contamination, can potentially induce waterborne diseases and present substantial hazards to public health, especially in regions where sanitation and hygiene standards are insufficient. Parasites such as Giardia and Cryptosporidium, in addition to waterborne viruses including norovirus, hepatitis A virus, and rotavirus, can cause gastrointestinal infections and other health complications. These are particularly hazardous for infants, older people, and individuals with compromised immune systems (Gall et al., 2015). A comprehensive strategy is required to address the myriad issues arising from water pollution. This strategy should incorporate cutting-edge sensor technologies that enable real-time monitoring and early detection of contaminants, stringent regulations ensuring adherence to pollution control protocols and compliance with water quality standards, and active engagement and participation of the community to foster consciousness, promote accountability, and mobilize collective efforts towards sustainable water management practices. By embracing this comprehensive methodology, stakeholders can engage in cooperative efforts to alleviate the consequences of water pollution, maintain the ecological soundness of aquatic ecosystems, and ensure the general public's well-being for current and future cohorts.

SENSOR TECHNOLOGIES FOR WATER POLLUTANTS

Ensuring the health of ecosystems and taking preventative measures against water pollutants requires the implementation of sophisticated sensor technologies that can precisely identify contaminants in various environmental settings. Spectroscopic sensors use spectroscopy techniques, such as UV-Vis, infrared (FTIR), and fluorescence spectroscopy, to study how matter and electromagnetic radiation interact. These sensors' outstanding sensitivity and specificity facilitate identifying a wide range of contaminants, such as dissolved gases, organic compounds, and heavy metals. According to Naimaee et al. (2024), spectroscopic sensors give environmental scientists and analysts detailed chemical profiles that let them know exactly how much certain pollutants are in water samples. The ability of electrochemical sensors to detect alterations in electrical characteristics caused by chemical reactions occurring at the surfaces of electrodes is vital. In water samples, electrochemical sensors are proficient at detecting ions, redox-active species, and gases. Due to their portability, swift response times, and low detection limits, these devices are indispensable for field-based monitoring applications. Their adaptability and dependability enable prompt intervention and remediation initiatives in the event of contamination incidents, making real-time monitoring of critical water quality parameters possible (Kanoun et al., 2021).

Biosensors are an innovative method for detecting pollutants that utilize biological recognition elements in conjunction with transducer technologies, such as antibodies or enzymes. With minimized interference from matrix components, these sensors provide specificity and selectivity in detecting target pollutants beyond all comparisons. Putting biosensors into portable devices makes on-site monitoring easier. It gives us helpful information about water quality so we can act and use proactive management strategies to lower the risk of contamination (Gavrilaş et al., 2022). Remote sensing technologies, which include aerial and satellite platforms, bring about a significant paradigm shift in the surveillance and evaluation of water quality on a large scale. Remote sensing techniques employ multispectral, thermal infrared, and LiDAR to produce maps that accurately depict critical water quality indicators, including turbidity, chlorophyll concentration, and temperature gradients. According to Yang et al. (2022), these technologies provide decision-makers with crucial knowledge regarding temporal and spatial trends in water quality. It enables them to allocate resources and make well-informed decisions supporting sustainable water management practices.

Using remote sensing technologies, spectroscopic sensors, electrochemical sensors, biosensors, and biosensors, environmental scientists and water resource managers can improve their ability to track water pollutants effectively, find where the pollution comes from, and implement targeted plans to clean it up. Sensor technologies of this nature are essential in promoting sustainable water management, preserving the vitality of aquatic ecosystems, and shielding the general public from the detrimental consequences of water pollution.

EMERGING SENSING TECHNOLOGIES

Emerging sensing technologies are revolutionizing environmental monitoring, offering innovative solutions to enhance pollution detection efforts' efficiency, accuracy, and timeliness. Nanomaterials-based sensors can find contaminants even at deficient concentrations with unmatched sensitivity and selectivity using the unique properties of materials like carbon nanotubes, graphene, and metal nanoparticles. This capability detects various pollutants, from heavy metals to organic compounds, making them indispensable for water quality monitoring. Additionally, their compact size and portability enable their deployment in diverse environmental settings, including remote or inaccessible locations, thereby extending the reach of monitoring efforts (Willner & Vikesland, 2018).

Microfluidic and lab-on-a-chip sensors represent another promising avenue in environmental sensing. These sensors integrate multiple analytical functions onto compact platforms, allowing for rapid and efficient detection of water pollutants in real-time. With reduced sample volume requirements and fast analysis times, these sensors are well-suited for on-site monitoring applications, providing high precision and enabling timely intervention measures. Their versatility and efficiency make them invaluable tools for monitoring water quality parameters in various environmental contexts (Kapoor et al., 2020). Real-time monitoring networks, powered by advancements in sensor technology and data analytics, offer continuous surveillance of water quality parameters across large spatial scales. By collecting real-time data on critical indicators such as pH, dissolved oxygen, temperature, and nutrient concentrations, these networks enable prompt responses to emerging threats, supporting proactive management strategies for sustainable water resource management. Integrating remote sensing technologies, IoT devices, and machine learning algorithms further enhances the capabilities of these networks, enabling predictive modelling and adaptive management strategies to anticipate and address potential issues before they escalate (Sejdiu et al., 2022). Emerging sensing technologies offer promising opportunities to revolutionize water quality monitoring and pollution detection efforts. By providing enhanced detection, analysis, and response capabilities, these technologies protect and conserve water resources, ensuring their availability and quality for current and future generations.

FUTURE OUTLOOK FOR WATER POLLUTANT SENSORS

Water pollutant sensors will undergo significant advancements shortly, transforming environmental monitoring. Enhanced sensitivity and selectivity will facilitate the identification of a broader range of contaminants at lower concentrations as materials science, nanotechnology, and signal processing techniques continue to advance. It will contribute to an enhanced comprehension of the dynamics of water quality and facilitate the timely identification of pollution.

Portability and field testing will be of the utmost importance to meet the growing need for platforms that are simple to implement and can be utilized by individuals without specialized knowledge in diverse environmental conditions. A strong emphasis on accessibility will facilitate prompt responses to pollution incidents, enable expedited on-site monitoring, and support the development of proactive management strategies.

The democratization of access to sophisticated sensor technologies will benefit resource-constrained communities and developing regions through reduced costs and maintenance obligations. Incorporating self-calibration and self-cleaning mechanisms into these sensors will reduce the burden of upkeep and maintenance while ensuring their long-term performance and dependability.

In addition, integrating AI algorithms and networked systems with water pollutant sensors will revolutionize water quality monitoring. The integration of these components will facilitate autonomous operation, real-time data analysis, and well-informed decision-making, ultimately resulting in enhanced strategies for detecting and mitigating pollution. By promoting adaptive management approaches, these developments will contribute to preserving freshwater ecosystems for present and future generations and aid in the sustainable management of water resources.

Incorporating artificial intelligence (AI), selectivity, sensitivity, portability, and cost-effectiveness presents significant potential in augmenting our capacity to protect freshwater resources. These advancements will be instrumental in tackling the issue of water pollution and guaranteeing the long-term viability of water management methodologies. Ensuring the health of ecosystems and preventative measures against water pollutants requires the implementation of sophisticated sensor technologies that can precisely identify contaminants in various environmental settings. Spectroscopic sensors employ spectroscopy principles, including UV-Vis, infrared (FTIR), and fluorescence spectroscopy, to analyze the interactions between matter and electromagnetic radiation. These sensors' outstanding sensitivity and specificity facilitate identifying a wide range of contaminants, such as dissolved gases, organic compounds, and heavy metals. Spectroscopic sensors enable analysts and environmental experts to precisely determine the concentrations of particular pollutants in water samples through comprehensive chemical profiles (Naimaee et al., 2024). The ability of electrochemical sensors to detect alterations in electrical characteristics caused by chemical reactions occurring at the surfaces of electrodes is vital. In water samples, electrochemical sensors are proficient at detecting ions, redox-active species, and gases. Due to their portability, swift response times, and low detection limits, these devices are indispensable for field-based monitoring applications. Real-time monitoring of critical water quality parameters is made possible by their adaptability and dependability; this enables prompt intervention and remediation initiatives in the event of contamination incidents (Kanoun et al., 2021).

Biosensors are an innovative method for detecting pollutants that utilize biological recognition elements in conjunction with transducer technologies, such as antibodies or enzymes. With minimized interference from matrix components, these sensors provide specificity and selectivity in detecting target pollutants beyond all comparisons. The integration of biosensors into portable devices facilitates on-site monitoring, allowing for the provision of actionable insights into water quality conditions and the implementation of proactive management strategies aimed at mitigating contamination risks (Gavrilaş et al., 2022). Remote sensing technologies, which include aerial and satellite platforms, bring about a significant paradigm shift in the surveillance and evaluation of water quality on a large scale. Remote sensing techniques employ multispectral, thermal infrared, and LiDAR to produce maps that accurately depict critical water quality indicators, including turbidity, chlorophyll concentration, and temperature gradients. According to Yang et al. (2022), these technologies provide decision-makers with crucial knowledge regarding temporal and spatial trends in water quality. It enables them to allocate resources and make well-informed decisions supporting sustainable water management practices.

By utilizing remote sensing technologies, spectroscopic sensors, electrochemical sensors, biosensors, and biosensors, environmental scientists and water resource managers can augment their capacity to monitor water pollutants efficiently, detect the origins of contamination, and execute focused mitigation strategies. Sensor technologies of this nature are essential in promoting sustainable water management, preserving the vitality of aquatic ecosystems, and shielding the general public from the detrimental consequences of water pollution.

DISCUSSION

The discourse on water pollutant monitoring sensors incorporates many critical facets essential for comprehending the present state of affairs and the potential of environmental monitoring. This segment examines the implications of the previously discussed sensor technologies, including their practical implementations, obstacles, and potential directions for future progress. Implementing sophisticated sensor technologies, such as electrochemical, spectroscopic, biosensor, and remote sensing systems, signifies a substantial progression in water quality monitoring. These sensors provide improved functionalities for detecting contaminants with exceptional sensitivity, specificity, and efficacy. Real-time data on critical water quality parameters enables policymakers, environmental experts, and researchers to make well-informed judgments concerning strategies to mitigate pollution and manage resources. Spectroscopic sensors use spectroscopy principles to make finding and measuring chemical compounds in water samples easier by looking at their unique spectral signatures. This technological advancement provides a swift and destructive approach to water sample analysis, enabling the identification of dissolved gases, heavy metals, and organic compounds. Researchers must overcome obstacles such as interference from matrix components and calibration requirements to ensure precise and dependable measurements. The ability of electrochemical sensors to detect alterations in electrical properties caused by chemical reactions occurring at electrode surfaces is crucial. These sensors' portability, quick response times, and low detection limits make them ideal for field-based monitoring applications. Notwithstanding their extensive implementation, sensor drift, corrosion, and calibration variability continue to be subjects of ongoing investigation and advancement.

Biosensors can find pollutants accurately and selectively using biological recognition elements and transducer technologies. The prospective applications of these sensors in portable devices for on-site monitoring include the detection of waterborne pathogens and contaminants in real time. Before their full potential in environmental monitoring applications can be realized, it is necessary to resolve sensor stability, storage life, and reproducibility challenges. Remote sensing technologies, which encompass aerial and satellite platforms, bring about a paradigm shift in the surveillance and evaluation of water quality on a large scale. These technologies facilitate the creation of maps that accurately represent water quality indicators spatially precisely. It aids in making well-informed decisions and allocating resources supporting sustainable water management practices. However, obstacles, including data processing, interpretation complexity, and spatial and temporal resolution limitations, necessitate additional development and research.

Water pollutant sensors are expected to progress in portability, cost-effectiveness, selectivity, and sensitivity with an eye toward the future. Real-time monitoring networks, microfluidic and lab-on-a-chip sensors, and nanomaterial-based sensors are all examples of emerging technologies that can potentially significantly transform environmental monitoring endeavours. By resolving obstacles, including sensor drift, calibration variability, and the intricacy of data interpretation, these technologies can augment our capacity to oversee, identify, and alleviate water pollution. In doing so, they will ultimately aid in conserving freshwater resources and safeguarding human health and ecosystems.

The discourse underscores the significance of sensor technologies in effectively tackling the complex issues associated with water contamination. By capitalizing on progress in sensor technology, materials science, and data analytics, interested parties can augment their ability to monitor and manage the environment efficiently. Academia, industry, government agencies, and communities must work together to foster innovation, overcome technological barriers, and encourage the broad implementation of sensor technologies in sustainable water resource management.

CONCLUSION

The surveillance and evaluation of water contaminants are critical to protect human well-being, maintain ecological integrity, and guarantee the long-term viability of water supplies. This chapter has comprehensively examined notable water pollutants and the sensor technologies employed for their detection and monitoring. The overview includes chemical, physical, and biological contaminants in both surface and groundwater. In addition to remote sensing technologies, spectroscopic, electrochemical, and biosensors, we have examined emerging trends in lab-on-a-chip, real-time monitoring, and nanomaterials. Upon contemplation of the future, it becomes evident that water contaminant sensors that are progressively more sensitive, cost-effective, portable, and selective could be enhanced by incorporating artificial intelligence. By capitalizing on these developments, interested parties can improve their capacity to oversee, identify, and alleviate water pollution, thereby fostering sustainable water management methodologies and guaranteeing the ongoing accessibility and excellence of water resources for present and future cohorts.

REFERENCES

Akhtar, N., Ishak, M. Z., Bhawani, S. A., & Umar, K. (2021). Various natural and anthropogenic factors responsible for water quality degradation: A review. *Water (Basel)*, *13*(19), 2660. doi:10.3390/w13192660

Chidiac, S., Najjar, P. E., Ouaini, N., Rayess, Y. E., & Azzi, D. E. (2023). A comprehensive review of water quality indices (WQIs): History, models, attempts and perspectives. *Reviews in Environmental Science and Biotechnology*, *22*(2), 349–395. doi:10.1007/s11157-023-09650-7 PMID:37234131

Gall, A. M., Mariñas, B. J., Lu, Y., & Shisler, J. L. (2015). Waterborne viruses: A barrier to safe drinking water. *PLoS Pathogens*, *11*(6), e1004867. doi:10.1371/journal.ppat.1004867 PMID:26110535

Gonzalez, L., McCallum, A., Kent, D., Rathnayaka, C., & Fairweather, H. (2023). A review of sedimentation rates in freshwater reservoirs: Recent changes and causative factors. *Aquatic Sciences*, *85*(2), 60. doi:10.1007/s00027-023-00960-0

Herath, I. K., Wu, S., Ma, M., & Huang, P. (2022). Heavy metal toxicity, ecological risk assessment, and pollution sources in a hydropower reservoir. *Environmental Science and Pollution Research International*, *29*(22), 32929–32946. doi:10.1007/s11356-022-18525-3 PMID:35020150

Kanoun, O., Lazarević-Pašti, T., Pašti, I. A., Nasraoui, S., Talbi, M., Brahem, A., Adiraju, A., Sheremet, E., Rodriguez, R. D., Ali, M. B., & Al-Hamry, A. (2021). A review of Nanocomposite-Modified Electrochemical Sensors for Water Quality monitoring. *Sensors (Basel)*, *21*(12), 4131. doi:10.3390/s21124131 PMID:34208587

Kapoor, A., Balasubramanian, S., Ponnuchamy, M., Vaishampayan, V., & Sivaraman, P. (2020). Lab-on-a-chip devices for water quality monitoring. In Nanotechnology in the life sciences (pp. 455–469). Springer. doi:10.1007/978-3-030-45116-5_15

Liu, L., Yang, H., & Xu, X. (2022). Effects of water pollution on Human Health and Disease Heterogeneity: A review. *Frontiers in Environmental Science*, *10*, 880246. doi:10.3389/fenvs.2022.880246

Ma, T., Zhang, D., Li, X., Hu, Y., Zhang, L., Zhu, Z., Sun, X., Lan, Z., & Guo, W. (2023). Hyperspectral remote sensing technology for water quality monitoring: Knowledge graph analysis and Frontier trend. *Frontiers in Environmental Science*, *11*, 1133325. doi:10.3389/fenvs.2023.1133325

Naimaee, R., Kiani, A., Jarahizadeh, S., Asadollah, S. B. H. S., Melgarejo, P., & Jódar-Abellán, A. (2024). Long-term water quality monitoring: Using satellite images for temporal and spatial monitoring of thermal pollution in water resources. *Sustainability (Basel)*, *16*(2), 646. doi:10.3390/su16020646

Sejdiu, B., Ismaili, F., & Ahmedi, L. (2022). A Real-Time Semantic Annotation to the Sensor Stream Data for the Water Quality Monitoring. *SN COMPUT. SCI.*, *3*, 254. doi:10.1007/s42979-022-01145-6

Shahra, E. Q., & Wu, W. (2020). Water contaminants detection using sensor placement approach in smart water networks. *Journal of Ambient Intelligence and Humanized Computing*, *14*(5), 4971–4986. doi:10.1007/s12652-020-02262-x

Willner, M. R., & Vikesland, P. J. (2018). Nanomaterial-enabled sensors for environmental contaminants. *Journal of Nanobiotechnology*, *16*(1), 95. doi:10.1186/s12951-018-0419-1 PMID:30466465

Yang, H., Kong, J., Hu, H., Du, Y., Gao, M., & Chen, F. (2022). A review of Remote sensing for water quality Retrieval: Progress and challenges. *Remote Sensing (Basel)*, *14*(8), 1770. doi:10.3390/rs14081770

KEYWORDS AND DEFINITIONS

Biosensors: Sensors incorporating biological recognition elements and transducer technologies to detect target analytes, offering high specificity and selectivity.

Electrochemical Sensors: Devices that detect changes in electrical properties resulting from chemical reactions, commonly used to measure ions or redox-active species in water.

Nanomaterials: Materials with dimensions on the nanoscale, such as carbon nanotubes or metal nanoparticles, are utilized for their unique properties in sensor development.

Real-Time Monitoring: Continuous environmental parameter monitoring provides instantaneous data collection and analysis for timely decision-making and intervention.

Remote Sensing: Techniques employing satellite or aerial platforms to collect data on Earth's surface, used in water quality assessment and environmental monitoring.

Sensor Technologies: Tools and systems designed to detect and measure the environment's physical, chemical, or biological parameters.

Spectroscopic Sensors: Instruments that analyze the interaction between matter and electromagnetic radiation to identify and quantify substances based on their unique spectral signatures.

Water Pollutants: Substances introduced into water bodies that degrade water quality, including chemical, physical, and biological contaminants.

Chapter 7
Smart Sensors for Water Quality Monitoring Using IoT

Kumud
D.S. College, Aligarh, India

ABSTRACT

Due to the increasing contamination and pollution of drinking water, water pollution has emerged as a major concern in recent years. Infectious illnesses spread by contaminated water have a domino effect on ecological life cycles. Early detection of water contamination allows for the implementation of appropriate solutions, therefore preventing potentially disastrous circumstances. It is important to monitor the water quality in real-time to ensure a steady supply of clean water. Improvements in sensor technology, connectivity, and the internet of things (IoT) have led to a rise in the importance of smart solutions for water pollution monitoring. This chapter presents a comprehensive overview of recent developments in the field of smart water pollution monitoring systems. An efficient and cost-effective smart water quality monitoring system that continuously checks quality indicators is proposed in this research. After running the model on three different water samples, the parameters are sent to the server in the cloud for further processing.

INTRODUCTION

Anthropogenic and trophic disruption of aquatic ecosystems (including wetlands, rivers, and lakes) have been on the rise in recent years. The role of aquatic ecosystems as both a reservoir and a route for human-caused pollution is growing. Our inability, unwillingness, and incapacity to sufficiently monitor ambient water quality on a large scale means that its precise scope and consequences are yet unknown. An increase in worldwide freshwater monitoring is necessary to halt this trend and learn how to clean up our waterways so they can support future generations. In order to manage freshwater ecosystems sustainably, it is essential to monitor water quality in aquatic ecosystems as a result of human activities and future climate change. Notably, there are new possibilities for improved modeling and monitoring brought about by the water sector's quick digitalization (Muinul et al., 2014). Natural and artificial abnormalities may be detected early on with the use of enhanced water quality monitoring, which is both

DOI: 10.4018/979-8-3693-1930-7.ch007

innovative and cost-effective. Sewage intrusion, pollution, agricultural practices, and surface runoff are among the many human-caused and climate-related threats to freshwater ecosystems in the modern era. In the end, these stresses lead to a decline in water quality and a deleterious effect on aquatic ecosystems. In order to respond quickly to negative changes in freshwater ecosystems, it is crucial to constantly monitor aquatic ecosystems. Ponds, rivers, lakes, seas, and oceans are all susceptible to water pollution when harmful substances enter, dissolve in, or settle to the bottom of these bodies of water. Water quality and purity are both negatively impacted by pollution (Jianhua et al., 2015). The excessive use of chemicals and other impurities makes it very difficult to guarantee water that is both clean and safe to drink. Industrial waste discharge and municipal sewage are two of the most common sources of water contamination, however there are many more potential causes. Pollutants that seep into water bodies from many sources, such as soils, the atmosphere (via rain or groundwater systems), or both, are known as secondary causes of pollution. The majority of the materials found in soils and groundwater are the byproducts of current farming techniques and poorly managed industrial waste. Microorganisms, chemicals, parasites, drugs, insecticides, and plastics are among the most significant contaminants of water. Though they may be difficult to detect as pollutants, these substances will not necessarily change the water's hue. So, in order to find out how good the water is, scientists take a little sample from these water sources and look at marine life.

The health, ecology, and economics are all negatively impacted by water quality declines. The pollution of aquatic ecosystems is causing impact on economic growth and exacerbating poverty in various ways to different states. The GDP growth of the will fall by one-third if the biological oxygen demand, a metric for organic pollution in water, exceeds the threshold. The abundance of contaminants, the most of which are produced by humans, makes it difficult to guarantee that water is safe to drink. The over-depletion of natural resources is the primary driver of water quality issues. Water contamination has been further exacerbated by the fast-paced industrialization and increased focus on agricultural expansion, together with innovative technologies, chemicals, and the lack of enforcement of regulations. The uneven distribution of precipitation might worsen the situation at times. The quality of water is affected by individual habits as well (Central Ground Water Board, 2017).

Sewage overflow, industrial effluent, agricultural run-off, and urban run-off are all examples of non-point sources of pollution that degrade water quality. Droughts and floods, as well as consumers' lack of information and training, are other causes of water pollution. Involvement from users is essential for water quality maintenance, as is consideration of other factors such as sanitation, hygiene, storage, and disposal.

Disease, mortality, and societal and economic stagnation are all results of water that isn't up to par. In 2017, water-related illnesses claimed the lives of almost 5 million people globally (Water Resource Information System of India). Rainwater has the potential to wash away pesticides and fertilizers applied by farms, eventually ending up in rivers. Rivers and lakes also get industrial waste. Toxic amounts of these contaminants build up in the food chain, where they kill animals, fish, and birds. Additionally, chemical companies dump their effluent into waterways. A lot of factories get their water for running machines or cooling them down from rivers. Water temperature changes disrupt aquatic ecosystems by reducing dissolved oxygen levels (Central Ground Water Board, 2017). Water quality monitoring is crucial due to all of the aforementioned reasons.

In order to identify present conditions, establish trends, etc., water quality monitoring involves collecting information at specific sites and at regular intervals (Niel et al., 2016). The primary goals of conducting water quality monitoring online are to assess important water quality parameters to detect

Figure 1. Author's adaption
(Adopted from: Allen et al., 2022)

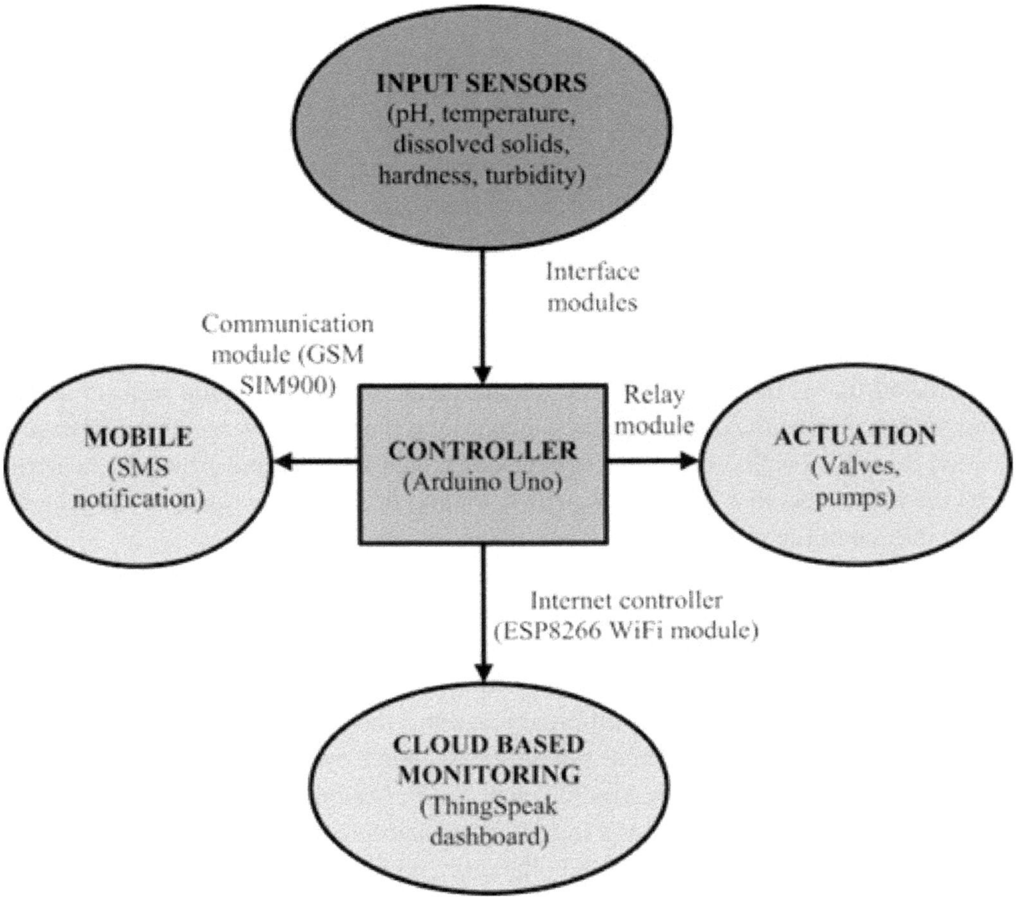

parameter variations; and to offer early warning of potential dangers. The monitoring system also analyses the data in real-time and suggests fixes based on what it finds.

There are two goals to this paper. One is to compile a comprehensive literature review on smart water quality monitoring including topics such as latest research, communication technologies, sensor kinds, and applications. The second objective is to provide a simplified smart water quality monitoring system that is both affordable and easy to use. This system will use a controller that has an integrated Wi-Fi module to track metrics including conductivity, turbidity, and pH. An alert feature is also a part of the system, which notifies the user when water quality metrics deviate from the norm.

REVIEW OF EXISTING LITERATURE

The following are the three primary subsystems that have been identified:

The data management subsystem consists of the application that accesses the cloud storage environment for data and presents the same to the end user. The data transmission subsystem is comprised of a wireless communication device that is equipped with built-in security mechanisms. This device is responsible for transmitting the data from the controller to the cloud which stores the data. The data collection subsystem is comprised of multi-parameter sensors and a wireless communication device that may be optionally used to transfer the information gathered from the sensors to the controller. The controller is responsible for collecting and processing the data.

When looking at the block diagram, the sensors are located at the very bottom. For the purpose of monitoring water quality indicators, there are several sensors available. These sensors are inserted into the water that is going to be tested, which refers to either water that is stored or water that is flowing. Through the use of an optional wireless communication device, sensors are able to transform the physical parameter into an equivalent quantifiable electrical amount. This electric quantity is then sent to controllers as an input. The controller's primary responsibility is to get the information from the sensor, do any necessary processing on it, and then transmit it to the application via making use of the communications method that is most suitable. The requirements of the application need to be taken into consideration while selecting the communication technology and the parameters that will be monitored. This application has functionality for managing data, doing data analysis, and providing an alarm system depending on the parameters that are being watched. A more in-depth discussion of the prior work that was completed in each of the subsystems is presented in this part.

Consumption of domestic water is meant for human consumption, including for the purposes of drinking and cooking. According to the Central Ground Water Board (2017), the Bureau of Indian Standards gives information on the allowed limits of pollutants such as aluminium, ammonia, iron, zinc, and other similar compounds. Traditional methods of measuring water quality involve the collection of water samples by hand from a number of different locations, the storage of those samples in a central location, and the subsequent testing of those samples in a laboratory (Thinagaran et al., 2015; Vinod & Sushama, 2016; Pandian & Mala, 2015; Azedine et al., 2000; Offiong et al., 2014). These methods are not regarded to be successful since there is a lack of information on the water's quality that is available in real time, there is a delay in the detection of pollutants, and there is no solution that is cost effective. Therefore, the need of continuous online water quality monitoring is brought to light in a number of studies (Xiuli et al., 2011; Sathish et al., 2016).

For applications involving lake and sea water, intelligent water quality measures have been taken into consideration. Such applications need the utilization of distributed wireless sensor networks in order to monitor the parameters across a more extensive region and transmit the data that is monitored to a centralized controller via the utilization of wireless communication. Chlorophyll (Francesco et al., 2015), DO content (Christie et al., 2014; Anthony et al., 2014), and temperature (Peng et al., 2009; Francesco et al., 2015; Christie et al., 2014) are some of the parameters that are often monitored by apps of this kind.

For the purpose of ensuring the healthy development of aquatic organisms, aquaculture facilities need to be equipped with water quality monitoring and forecasting systems (Goib et al., 2015; Gerson et al., 2012; Xiuna et al. 10). Using Arduino microcontrollers, the authors of the study (Gerson et al., 2012) have constructed biosensors in order to detect changes in animal behavior that are caused by pollution in aquatic environments. There is a possibility that the aberrant behavior of animals is an indicator that the water is affected by pollution. A smart water quality monitoring system that makes use of artificial neural networks has been suggested by the authors of the paper (Xiuna et al., 2010) in order to make predictions about water quality. For a period of twenty-two months, extensive experiments have been carried out at a local area network that is completely isolated, and the data has been sent to the internet by using CDMA technology. In the context of the management of distributed wireless sensor networks

(WSN), water quality monitoring in distribution systems presents a number of challenges. Eliades et al. (2014) made a presentation in which they discussed a water distribution network that was designed to monitor the quantity of chlorine. (Ruan & Tang, 2011) suggests that solar-enabled distributed WSN might be used for the purpose of monitoring parameters such as pH, turbidity, and oxygen density. For the purpose of monitoring water at various locations in real time, an architecture that is comprised of sensor nodes that are equipped with solar cells and a base station is used. There are a number of benefits associated with the approach that is presented in the research, including flexibility, minimal carbon emission, and low power use. Using additional sensors for monitoring air temperature and relative humidity, Mitar et al. (2016) suggest a combined system for assessing water and air quality. This system would be more accurate than the current approach.

MONITORING OF THE PARAMETERS

It has been determined, on the basis of extensive experimental evaluation carried out by the USEPA, that the chemical and biological contaminants that are utilized have an impact on a variety of water parameters that are monitored. These parameters include turbidity, ORP, EC, and pH. According to Theofanis et al. (2014), it is possible to infer the quality of the water provided that the parameters of the water are monitored and changes in those parameters are detected.Because it determines whether the water is basic or acidic, the pH of the water is one of the most significant aspects to consider while conducting an investigation into the quality of the water. It is possible for the eyes, skin, and mucous membranes to get irritated when using water with a pH of 11 or higher. According to Niel et al. (2016), the corrosive impact of acidic water (also known as water with a pH of 4 or below) may also induce irritation. It is essential for aquaculture facilities to take measurements of DO since this parameter is responsible for determining whether or not a particular species is able to live in the water source in question. A material's ORP is a measurement of the degree to which it is capable of oxidizing or reducing another chemical. When utilizing an ORP meter, the measurement of ORP is done in millivolts (mv). The ORP is positive for both tap water and bottled water. The concentration of such particles that are suspended in water is referred to as turbidity. Conductivity is a measure that may be used to determine the amount of contaminants that are present in water; the level of conductivity decreases as the water becomes cleaner. Conductivity is also closely related with TDS in many instances. However, this is not always the case.

Communication Between Sensors and Controller

Either directly via the use of the UART protocol or remotely through the use of the Zigbee protocol, sensors are linked to the controller. In wireless networks, ZigBee is a technology that allows for the transport of data. In addition to having a low energy usage, it is suited for use in multichannel control systems, alarm systems, and lighting control. For low-rate WPANs, ZigBee is built on the physical layer and media access control that are established in IEEE standard 802.15.4. The Zigbee protocol is used in smart water quality systems for the purpose of communication between sensor nodes and the controller. This is the case when the sensors are situated in distant locations that are not in close proximity to the controller. Direct connection between the controller and the sensors is the ideal method for providing in-pipe home monitoring (Tomoaki et al., 2016). A WSN system for monitoring water quality has been created by the authors. UART is used to establish connections between the transmission module and the sensors. This is accomplished via the use of the Internet connection and the 3G mobile network in order to communicate with the outside world of the sensor nodes. There is a water quality monitoring system that has been suggested by the authors of (Theofanis et al., 2014) for the purpose of in-pipe monitoring and the evaluation of water quality on flies. There are sensor nodes that are put in the pipelines that deliver water to the residences of consumers.

Transmission of Information Between the Controller and the Data Storage

Long-range communication technologies, such as Internet and 3G, are used in order to facilitate communication between the controller and the centralized data storage. The purpose of some of the earlier efforts is to send a text message to the user informing them about the water quality. An extra SIM card is required for the GPRS module that is linked with the controller in these kinds of systems (Wei et al., 2012). One of the disadvantages of such systems is that they incur extra costs for making use of SIM cards. Additionally, the user's premises do not have the capacity to store and retrieve significant amounts of data. IoT enabled solutions have recently gained relevance. The authors of the article "Alessio et al., 2016" provide a survey on the extensive variety of applications that may be achieved via the use of cloud computing and the Internet of Things.

The IoT is a relatively new communication paradigm in which objects that are used in everyday life are outfitted with microcontrollers and transceivers for digital communication. This will enable the objects to communicate with one another and with users, thereby transforming them into an essential component of the Internet (Andrea et al., 2014). An external Wi-Fi module is linked to the controller in the research conducted by Vijayakumar and Ramya (2015), Thinagaran et al. (2015), and Mitar et al. (2016). This establishes a connection between the controller and the closest Wi-Fi hotspot, which in turn allows the controller to establish a connection to the Internet cloud.

Utilized Sensors

A number of sensors are now available for purchase for the purpose of monitoring water quality. These sensors are used in the research conducted by Thinagaran et al. (2015), Vinod and Sushama (2016), and Niel et al. (2016). Among the works that have been published in the literature are examples of sensors that have been built for increased usability. (Tomoaki et al., 2016) makes use of a sensor node of the buoy type that has been manufactured for the purpose of parameter monitoring. In addition to a power module and a transmission module, the sensor that was manufactured also has a solar cell and a Li-ion battery. The research carried out by Mitar et al. (2016) makes use of a thick film pH resistive sensor that is produced in-house using TiO2. Additionally, the output of this sensor module may be directly linked to the microcontroller, eliminating the need for extra signal processing devices. The authors of the study (Theofanis et al., 2014) have created a turbidity sensor that is accurate, inexpensive in cost, and simple to use. This sensor is intended for continuous monitoring of turbidity in pipes.

Theofanis et al. (2014), Peng et al. (2009), Xiuna et al. (2010), Francesco et al. (2015), and Azedine et al. (2000) have all provided extensive data analysis and information processing. These examples may be found in the aforementioned publications. According to Haroon and Anthony (2016), a hierarchical routing method has been developed with the purpose of lowering the communication overhead and increasing the life time of WSN that are appropriate for monitoring river and lake water. A study of the Smart Water Grid system that incorporates ICT is presented in the publication that was published by the

Public Utilities Board of Singapore in 2016. (Woon et al., 2016) presents a discussion on an integrated management model that encompasses the whole water cycle, from the sources to the taps, with the goal of ensuring the consistency, safety, and efficiency of water.

POWER CONSUMPTION RELATED ISSUES

As a result of the fact that Internet of Things apps are most likely to run on batteries, power consumption is a significant limitation for these applications. One of the most significant sources of power consumption is the transmission of data. There are two phases of data transfer that take place for applications such as an intelligent water quality monitoring system. One kind of communication is that which occurs between the controller and the sensors, while the other type of communication occurs between the controller and the application. According to Al-Fuqaha et al. (2015) and Ray (2016), It has been determined by Shuker et al. (2016) that Wi-Fi is not an appropriate method for communication between sensor nodes and the controller due to the significant amount of power that is lost and wasted. As far as our literature review is concerned, every single study has used the zigbee protocol for the purpose of communication between the controller and the sensor nodes.

A home water quality monitoring system is the focus of the work that is being proposed. It is presumed that the sensors are linked at the in-pipe level. The user premises are the location where the controller and the sensors are put together as a single module. For this reason, the controller is directly linked to the sensors' devices. For applications such as monitoring the water in lakes, rivers, and the ocean, the controller and the sensors are separated by a significant distance throughout the process. The use of short-range communication protocols, such as Zigbee, is recommended in such circumstances. Wi-Fi is an attractive option for the purpose of facilitating communication between the controller and the application. When using these various short-range protocols, the sensor nodes are able to interact with the controller with relative ease. However, when attempting to link the system to the Internet, it is necessary to have an adapter of some kind that is capable of communicating with both the sensors and the Internet. The increased hardware overhead is caused by this. The aforementioned issue does not emerge while using Wi-Fi since there is already an infrastructure that has been constructed and is in operation at this point. The fact that the Wi-Fi standard was developed for personal computers and laptops, which have a totally different power need than battery-operated smart items, is one of the limitations of the technology. As a result, manufacturers have begun manufacturing Wi-Fi devices that possess low power consumption. The CC3200 is an example of an embedded low-power Wi-Fi device that has a major emphasis on power management and extending the battery life of its products. The microcontroller is run in one of the four power modes, which are Hibernate, Low Power Deep Sleep mode, Sleep mode, and Active mode (Texas instrument CC3200 Simple Link, 2017). This reduction in power consumption is accomplished by operating the microcontroller in one of these modes.

Thomas et al., 2016 examined the power consumption of a standalone microcontroller with Zigbee and BLE modules, as well as a controller that had an integrated Wi-Fi device. When compared to standalone microcontrollers, it has been shown that devices that have Wi-Fi integrated in utilize less power. This was discovered based on the results of the experiments. Because of the additional power consumption that occurs during the process of establishing and disestablishing a connection during transmission in standalone devices, this is the reason. When using the Wi-Fi integrated controller, the Wi-Fi module enters a sleep state, after which it maintains the connections that it has previously established. Since this is the case, it is not necessary to create a fresh connection each time the Wi-Fi module becomes active. The usage of electricity is significantly decreased as a result of this. A comparison of the CC3200 with the microcontroller and embedded boards that have been utilized in the research that has been published (Al-Fuqaha et al., 2015; Ray, 2016).

UTILIZATION OF WIRELESS SENSOR NETWORKS

In the most recent decades, WSNs have been created and are progressively being used for the purpose of monitoring water quality by researchers, municipalities, and commercial firms alike. According to Taperello et al. (2017), the characteristics of a general WSN system architecture include data gathering, transmission, processing, storage, and redistribution. These sensor probes have progressed to the point that they are now capable of measuring a wide variety of physicochemical parameters, including conductivity, turbidity, DO, and pH (Marcé et al., 2016). Data collecting is accomplished by the use of a network of in-situ sensor(s) as well as a certain sampling frequency. The data that has been collected is then sent to the central monitoring hub using various technologies. These technologies often include cellular networks like GSM, as well as more recent networks like ZigBee and even Wi-Fi. As the IoT continues to attract attention from researchers and expands in terms of commercialization, it is anticipated that these networks will continue to evolve. Processing, storing, and analyzing the data are all possible after it has been sent. Technologies such as these make it possible to perform remote continuous real-time monitoring and visualisation of water-body quality parameters at fixed locations (Tapparello et al., 2017). Additionally, it has been discovered that these technologies provide a more accurate description of water-bodies in comparison to manual methods, which enables the comprehension of biogeochemical processes (Kirchner et al., 2004; Ivanovsky et al., 2016).

WSNs provide a great deal of flexibility in terms of the frequency at which water quality parameters may be monitored. This is in contrast to systems that monitor water quality parameters at low frequencies, which are prone to ambiguity (Birgand et al., 2013). This is especially important for flashy streams, where timing sampling with peak flow can be difficult to achieve. Khalil and Ouarda (2009) conducted a review in which they discovered that the use of multiparameter sondes and the much finer possible temporal sampling resolutions were able to capture transient events that were likely to have been missed by grab sampling. However, despite the fact that high-frequency data collecting does not involve an excessive amount of expense, there are limitations to the transport and storage of data in systems that do not have automated telemetry (Chappell et al., 2017). A monitoring frequency that is too high may also return redundant information (Khalil and Ouarda, 2009), which can lead to an increase in the possibility for noise in the data. Additionally, it can increase the demand for power, which can lead to an increase in the capital cost of installing a renewable source (such as a solar panel) in the event that mains energy is unavailable. In light of these factors, it is necessary to locate a frequency that strikes a balance between the examination of each physicochemical parameter.

CONCLUSION

The pollution of water poses a significant risk to any nation since it harms people's health, the economy, and the biodiversity of the environment. Presented in this article are the causes and consequences of water pollution, as well as a detailed examination of several techniques of water quality monitoring and a discussion of an effective approach for monitoring water quality that is based on the IoT. Despite the fact that there have been a great number of successful intelligent water quality monitoring systems, the research field continues to be particularly difficult. This article provides an overview of the current work that has been carried out by researchers in order to create water quality monitoring systems intelligent, low powered, and highly efficient. The goal of these systems is to ensure that monitoring will be continuous and that alarms and notifications will be forwarded to the relevant authorities for further processing. In addition to being easy to use and economical, the model that was designed is also versatile. Three different samples of water are put through a series of tests, and the findings allow for the determination of whether or not the water is suitable for consumption. The IoT is used in the process of developing a system for the monitoring and management of water quality in real time for a water treatment facility. Through the use of a wireless sensor network, the system is able to deliver immediate results and enables real-time monitoring and control. In addition, a simulation model and a prototype that is operational are constructed in order to illustrate how the system functions according to a variety of field variables. The quality of the water is evaluated at each and every step of the treatment process, and any deviations from the WHO standard will result in the implementation of an independent control measure. The monitoring and regulation of water quality in real time contributes to an improvement in water security. It eliminates the need of conventional techniques of water filtration, which are unable to monitor and manage water quality parameters from the beginning stages of the process. In addition to this, it eliminates the need for human contact, which eliminates the time that is wasted when water samples are submitted to the laboratory for analytical testing as well as the expense of testing offline. This product makes a contribution to the prevention of the spread of diseases that are transmitted by water. As a recommendation for a future directive, it is suggested that the most recent sensors be used for the purpose of detecting a variety of other quality metrics, that wireless communication standards be utilized for improved communication, and that the Internet of Things be utilized to develop a more effective system for monitoring water quality, and that the water resources be made safe by prompt action.

LIST OF ABBREVIATIONS

BLE: Bluetooth Low Energy
DO: Dissolved Oxygen
EC: Electrical Conductivity
GDP: Groos Domestic Product
GPRS: General Packet Radio Service
ICT: Information and Communication Technology
IEEE: Institute of Electrical and Electronics Engineers
IoT: Internet of Things
ORP: Oxidation-Reduction Potential
UART: Universal Asynchronous Receiver / Transmitter
WPANs: Wireless Personal Area Networks

REFERENCES

Al-Fuqaha, A., Guizani, M., Mohammadi, M., Aledhari, M., & Ayyash, M. (2015). Internet of Things: A Survey on Enabling Technologies, Protocols, and Applications. *IEEE Communications Surveys and Tutorials*, *17*(4), 2347–2376. doi:10.1109/COMST.2015.2444095

Alessio, B., Walter, D., Valerio, P., & Antonio, P. (2016). Integration of Cloud computing and Internet of Things: A survey. *Future Generation Computer Systems*, *56*, 684–700. doi:10.1016/j.future.2015.09.021

Andrea, Z., Nicola, B., Angelo, C., Lorenzo, V., & Michele, Z. (2014). Internet of Things for Smart Cities. *IEEE Internet of Things Journal*, *1*(1), 22–32. doi:10.1109/JIOT.2014.2306328

Anthony, F., Aloys, N., Hector, J., Maria, C., Albino, J., & Samuel, B. (2014). Wireless Sensor Networks for Water Quality Monitoring and Control within Lake Victoria Basin: Prototype Development. *Wirel Sens Netw*, *6*(12), 281–290. doi:10.4236/wsn.2014.612027

Azedine, C., Antoine, G., Patrick, B., & Michel, M. (2000). Water quality monitoring using a smart sensing system. *Measurement*, 219–224.

Biljana, L., Risteska, S., & Kire, V. T. (2017). A review of Internet of Things for smart home: Challenges and solutions. *Journal of Cleaner Production*, *140*(3), 1454–1464.

Birgand, F., Appelboom, T. W., Chescheir, G. M., & Skaggs, R. W. (2013). Estimating nitrogen, phosphorus, and carbon fluxes in forested and mixed-use watersheds of the lower coastal plain of North Carolina: Uncertainties associated with infrequent sampling. *Transactions of the ASABE*, *54*, 2099–2110. doi:10.13031/2013.40668

Bushra, R., & Mubashir, H. R. (2016). Applications of wireless sensor networks for urban areas: A survey. *Journal of Network and Computer Applications*, *60*, 192–219. doi:10.1016/j.jnca.2015.09.008

Chafa, A. T., Chirinda, G. P., & Matope, S. (2022). Design of a real–time water quality monitoring and control system using Internet of Things (IoT). *Cogent Engineering*, *9*(1), 1. doi:10.1080/23311916.2022.2143054

Chappell, N. A., Jones, T. D., & Tych, W. (2017). Sampling frequency for water quality variables in streams: Systems analysis to quantify minimum monitoring rates. *Water Research*, *123*, 49–57. doi:10.1016/j.watres.2017.06.047 PMID:28647587

Christie, R., Mallory, C., Jared, L., & Alan, M. (2014) Remote Delay Tolerant Water Quality Monitoring. In: IEEE global humanitarian technology conference. IEEE.

Eliades, D., Lambrou, T., Panayiotou, C., & Polycarpou, M. (2014) Contamination Event Detection in Water Distribution Systems using a Model-Based Approach. In: *16th Conference on Water Distribution System Analysis*. 14–17 July 2014 10.1016/j.proeng.2014.11.229

Francesco, A., Filippo, A., Carlo, G. C., & Anna, M. L. (2015). A Smart Sensor Network for Sea Water Quality Monitoring. *IEEE Sensors Journal*, *15*(5), 2514–2522. doi:10.1109/JSEN.2014.2360816

Gerson, G., Christopher, B., Stephen, M., & Richard, O. (2012) Real-time Detection of Water Pollution using Biosensors and Live Animal Behavior Models, In: 6th eResearch Australasia Conference, 28 Oct –1 Nov 2012

Goib, W., Yudi, Y., Dewa, P., Iqbal, S., & Dadin, M. (2015). Integrated online water quality monitoring. In: *International conference on smart sensors and application*. IEEE.

Haroon, M., & Anthony, S. (2016) Towards Monitoring the Water Quality Using Hierarchal Routing Protocol for Wireless Sensor Networks. In: *7th International Conference on Emerging Ubiquitous Systems and Pervasive Networks*. IEEE.

Jayti, B., & Patoliya, J. (2016). IoT based water quality monitoring system. In: *Proc of 49th IRF Int Conf.* IEEE.

Jianhua, D., Guoyin, W., Huyong, Y., Ji, X., & Xuerui, Z. (2015). A survey of smart water quality monitoring system. *Environmental Science and Pollution Research International*, *22*(7), 4893–4906. doi:10.1007/s11356-014-4026-x PMID:25561262

Khalil, B., & Ouarda, T. B. M. J. (2009). Statistical approaches used to assess and redesign surface water-quality-monitoring networks. *Journal of Environmental Monitoring*, *2009*(11), 1915–1929. doi:10.1039/b909521g PMID:19890548

Kirchner, J. W., Feng, X., Neal, C., & Robson, A. J. (2004). The fine structure of water-quality dynamics: The (high-frequency) wave of the future. *Hydrological Processes*, *18*(7), 1353–1359. doi:10.1002/hyp.5537

Li, L. (2014) Software development for water quality's monitoring centre of wireless sensor network. *Computer Modeling New Tech,* 132–136.

Li, Z., Wang, K., & Liu, B. (2013). Sensor-Network based Intelligent Water Quality Monitoring and Control. *International Journal of Advanced Research in Computer Engineering and Technology*, *2*(4), 1659–1662.

Mahmoud, S., & Mahmoud, A. (2016). A Study of Efficient Power Consumption Wireless Communication Techniques/ Modules for Internet of Things (IoT). *Applications Advances in Internet of Things*, *6*(2), 19–29. doi:10.4236/ait.2016.62002

Marcé, R., George, G., Buscarinu, P., Deidda, M., Dunalska, J., De Eyto, E., Flaim, G., Grossart, H.-P., Istvanovics, V., Lenhardt, M., Moreno-Ostos, E., Obrador, B., Ostrovsky, I., Pierson, D. C., Potužák, J., Poikane, S., Rinke, K., Rodríguez-Mozaz, S., Staehr, P. A., & Jennings, E. (2016). Automatic high frequency monitoring for improved lake and reservoir management. *Environmental Science & Technology*, *50*(20), 10780–10794. doi:10.1021/acs.est.6b01604 PMID:27597444

Muinul, H., Syed, I., Alex, F., Homayoun, N., Rehan, S., Manuel, R., & Mina, H. (2014). Online Drinking Water Quality Monitoring: Review on Available and Emerging Technologies. *Critical Reviews in Environmental Science and Technology*, *44*(12), 1370–1421. doi:10.1080/10643389.2013.781936

Niel, A. C., Reza, M., & Lakshmi, N. (2016). Design of Smart Sensors for Real-Time Water Quality Monitoring. *IEEE Access : Practical Innovations, Open Solutions*, *4*, 3975–3990. doi:10.1109/ACCESS.2016.2592958

Offiong, N., Abdullahi, S., Chile, B., Raji, H., & Nweze, N. (2014). Real Time Monitoring Of Urban Water Systems for Developing Countries. *IOSR Journal of Computer Engineering*, *16*(3), 11–14. doi:10.9790/0661-16321114

Pandian, D. R., & Mala, K. (2015). Smart Device to monitor water quality to avoid pollution in IoT environment. *Int J Emerging Tech Comput Sci Electron*, *12*(2), 120–125.

Peng, J., Hongbo, X., Zhiye, H., & Zheming, W. (2009). Design of a Water Environment Monitoring System Based on Wireless Sensor Networks. *Journal of Sensors*, *9*(8), 6411–6434. doi:10.3390/s90806411 PMID:22454592

Poonam,, J. (2016). IoT Based Water Quality Monitoring. *Int J Modern Trends Eng Res*, *3*(4), 746–750.

Public Utilities Board Singapore (PUB). (2016). Managing the water distribution network with a Smart Water Grid. *Smart Water International Journal.*

Ray, P. P. (2016). (in press). A survey on Internet of Things architectures. *Journal of King Saud University. Computer and Information Sciences.*

Ruan, Y., & Tang, Y. (2011) A water quality monitoring system based on wireless sensor network & solar power supply. In: *2011 IEEE International Conference on Cyber Technology in Automation, Control, and Intelligent System*, (pp. 20–23). IEEE.

Sathish, K., Sarojini, M., & Pandu, R. (2016). IoT based real time monitoring of water quality. *Int J Prof Eng Stud*, *VII*(5), 174–179.

Tapparello, C., Abdulai, J. D., Katsriku, F. A., Heinzelman, W., & Adu-Manu, K. S. (2017). Water quality monitoring using wireless sensor networks. *ACM Transactions on Sensor Networks*, *13*, 1–41.

Theofanis, P. L., Christos, C. A., Christos, G. P., & Marios, M. P. (2014). A Low-Cost Sensor Network for Real-Time Monitoring and Contamination Detection in Drinking Water Distribution Systems. *IEEE Ssensors J*, *14*(8), 2014.

Thinagaran, P., Nasir, S., & Leong, C. Y. (2015) Internet of Things (IoT) enabled water monitoring system. In: *4th IEEE global conference on consumer electronics*. IEEE.

Thomas, D., Wilkie, E., & Irvine, J. (2016). Comparison of Power Consumption of Wi-Fi Inbuilt Internet of Things Device with Bluetooth Low Energy. *Intl J Comput Electrical Automation Control Inf Eng*, *10*(10), 1837–1840.

Tomoaki, K., Masashi, M., Akihiro, M., Akihiro, M., & Sang, L. (2016) A wireless sensor network platform for water quality monitoring. In: *IEEE Sensors*.

Ubidots. (2017). *IoT platform*. Ubidots. https://ubidots.com/.

Vijayakumar, N., & Ramya, R. (2015) The real time monitoring of water quality in IoT environment. In: international conference on circuits, power and computing technologies. IEEE.

Vinod, R., & Sushama, S. (2016) Wireless acquisition system for water quality monitoring. In: *Conference on advances in signal processing*. IEEE.

Water Resource Information System of India. (2017). http://www.india-wris.nrsc.gov.in/wrpinfo/index.php?title=River_Water_Quality_Monitoring. Accessed 2 May 2017

Wei, D., Liu, P., & Lu, B. (2012). Water Quality Automatic Monitoring System Based on GPRS Data Communications. In: *International conference on modern hydraulic engineering*. IEEE.

Xin W, Longquan M, & Huizhong Y (2011). Online Water Monitoring System Based on ZigBee and GPRS. *Adv Control Eng Inform Sci.*

Xiuli, M., Xo, H., Shuiyuan, X., Qiong, L., Qiong, L., & Chunhua, T. (2011) Continuous, online monitoring and analysis in large water distribution networks. In: *International Conference on Data Engineering*. IEEE.

Xiuna, Z., Daoliang, L., Dongxian, H., Jianqin, W., Daokun, M., & Feifei, L. (2010). A remote wireless system for water quality online monitoring in intensive fish culture. *Computers Electronics Agriculture*.

Chapter 8
Sensors in the Marine Environments and Pollutant Identifications and Controversies

Minakshi Harod
Devi Ahilya Vishwavidyalaya, India

ABSTRACT

The quality of marine environments is influenced by a range of anthropogenic and natural hazards, which may adversely affect human health, living resources, and the general ecosystem. The most common anthropogenic wastes found in marine environments are dredged spoils, sewage, and industrial and municipal discharges. These wastes generally contain a wide range of pollutants, notably heavy metals, petroleum hydrocarbons, polycyclic aromatic hydrocarbons, and others. Real-time measurements of pollutants, toxins, and pathogens across a range of spatial scales are required to adequately monitor these hazards, manage the consequences, and to understand the processes governing their magnitude and distribution. Significant technological advancements have been made in recent years for the detection and analysis of such marine hazards. This chapter aims to review the availability and application of sensor technology for the detection of marine hazards and for observing marine ecosystem status.

INTRODUCTION

Marine debris detected as marine pollution involves solid materials, plastics and bags that are unintentionally or intentionally has been disposed in the ocean. Marine ecosystems act as natural sink of greenhouse gases and carbon dioxide. Anthropogenic activities are seen to severely pollute the oceans for last few centuries. Ocean is seen to contain enormous seawater with many ecosystem regulations, with global based heat cycles, climate systems and carbon cycles. Marine pollution is caused due to the discharge of the sewage within rivers and excessive dumping of pesticides as well as fertilizers is the main cause of marine pollution (Clark et al., 1989). Pollutants like oil, plastic, chemicals that are toxic, radioactive

DOI: 10.4018/979-8-3693-1930-7.ch008

wastes are seen as the main pollutants in oceans. Long term protection as well as ocean management needs to be done in order to save the oceans from being polluted. Approximately 8 million tons plastic are seen to contribute 80% of the total ocean based pollution (Jambeck et al., 2015). In the light of the growing concerns remote sensors has been found to be usable and potential for monitoring environmental based issues in marine environment. In this regard, this chapter provides a wider knowledge about the sensors with the classification of the sensors that are used to prevent pollution in the ocean. In addition, the identification of the pollution using those sensors and their implication for the environment will be discussed. Plastic is the main cause of pollution in oceans while other pollutants are non-point-based sources, industrial chemicals, fertilizer runoffs and other sewage. This pollution is extremely harmful that kills ecosystems, affects human life and threatens wildlife also. This causes damage in the marine ecosystem and that creates a knock-on effect on wildlife populations. By getting entangled with fishing gear 100000 marine animals dies in each year (Simmonds and Peter 2012).

As in year 2023 estimated 170 trillion particles of plastics in the ocean is seen which is roughly 21250 plastics. This indicates a surprising fact that plastic is one of the main sources of pollutant in ocean ecosystem. International union for conservation of nature detected that 14 million tons plastic ends up every year in the ocean making 80% marine debris. As per the study of university of Cadiz 10 plastic products made up to 75% of the polluting items within ocean. Worst part from the lot states that plastic bags take 20 years for decomposition and stays there harming the marine animals causing pollution.

Ocean bound plastic as the name suggests generates 80% of the marine pollution and as per NIH meso-plastics, macro plastics and micro plastics makes litters in ocean and are the leading source of marine pollution.

INTRODUCTION TO OCEAN SENSORS

Oceanography has modern electronic tools that allow humans to investigate deep down analysis and monitor the highest depths in the ocean. Electronic techniques with low energy, delicate, and accurate sensors collect vital information, and have helped investigators to record and measure the environment of the deep sea. Transportable submersible robots and underwater drones with intelligent sensors are used for observing the temperature of the water, auto compass bearing, and depth, pitch, and roll, turns, have been created. For technological improvements, underwater drones can drive around freely nowadays, quickly, and more smoothly. Artificial intelligence and robots have been developed to investigate the oceans where people cannot reach. Human-like bodies in these robots with touch sensors can take out a combination of works in the deep of the sea. Sensors are the tools used to get information of the ocean. Modern instruments operating electronically in oceanography helped humans to explore the deeper and hidden facts underneath the ocean and sensors are among them.

In order to reduce this impact sensing the ocean with sensors has become a must. To study ocean pollution for example ocean chemists are likely to use sensors for detection of the synthetic compounds like plastics, petroleum, automobile exhaust, sewage runoff, fertilizers and chemicals.

Today due to rapid advances in biotechnology, nanotechnology and micro technology cost effective precision-based sensors has been introduced. It creates an observatory-based platform having unprecedented access to underwater vehicles that sends data after observation to the sender. In 2003 WHOI ocean life institute and deep ocean exploration institute along with national science foundation and office of naval research introduced one workshop *"The Next Generation of in situ Biological and Chemical*

Sensors in the Ocean." This created a roadmap to sensors where through quantum leaps observation of oceans is done. These sensors not only detected the temperature and salinity of the ocean but also helped in detecting the percentage of dangerous pollutants that are present in the ocean. However, between 2013 and 2016 European researchers developed innovative techniques to access pollutants. The BRAAVOO team prepared one specific sensor where chop sensors access the presence of pollutants through light. More light means more contamination in the water.

UNDERWATER SENSOR NETWORKS

Adhoc network operating wireless has the capability to monitor the oceans and marine environment. The proposal details an underwater wireless network where autonomous underwater vehicles (AUVs), unmanned undersea vehicles (UUVs) and buoys to receive data from sensors and periodically send back to a base station (Kerry Taylor-Smith 2018).

Ocean based sensors detects radiance and through imagery resolutions the plastics and other pollutants that are harming the oceans got detected. This creates nodes that communicate with neighbors with shorter range of acoustic communication. That acts as early warner that can prevent catastrophic events and protects the oceans. By the use of remote sensors detection of some properties of the ocean like is done:

- Reflectance
- Temperature
- Colour
- Roughness

The pollutant gets detected when any of the above property changes for example oil spills in the ocean dampens the surface of the water and so lower roughness-based signal means oil is there in huge quantity in the ocean.

TECHNOLOGIES TO DETECT SENSOR-BASED IDENTIFICATION OF POLLUTANTS

A broad type of sensor devices is operated in the sea, including benthic flow meters, dissolved oxygen sensors, mass spectrometers, hydrophones, resistivity probes, conductivity-temperature-depth, and turbulent-flow current meters. Remote sensors are used to capture the electromagnetic interaction with the water. Checking total absorption, non-algal pigments, and suspended sediments present in the water gets detected by the help of these sensors. These sensors, for identification of pollution, can be categorized into two types which are airborne and spaceborne (Ogola et al., 2010). Some other sensors are also utilized to detect water-based pollution for example dissolved oxygen sensors, benthic flow meters, hydrophones, mass spectrometers, resistivity probes, turbulent flow current meters, and conductivity-temperature depth. The Airborne sensors are utilized to detect water pollution, and for operating the Airborne sensors aircraft are used which fly at relatively lower altitudes. For spill surveillance airborne sensors are (Theo., 2008):

- Side-looking airborne radar
- Microwave radiometer
- Laser fluor sensor

Based on the satellite sensors water quality and other parameters are checked so that spills percentage can be checked.

The aircraft usually fly a few hundred meters to a few kilometers above the surface thus providing a higher level of accuracy and details regarding the data collected. For real-time monitoring of oil and chemical spills airborne data is particularly useful. The types of airborne sensors for pollution detection are listed below:

a) *Infrared/ultraviolet line scans (IR/UVLS)*

IR/UV Line scanner is regarded as a standard tool in oil spill remote sensing. IR/UV line sensors are capable of simultaneously mapping the total extent of oil spills as well as the areas where there are intermediate and large oil thicknesses. This sensor helps in resolving the issue of mapping the very thick and thin oil layers in the water surface. The IR/ULVS sensor is highly applicable in imaging remote sensors for thermal mapping applications and basic sensors for the automatic creation of the thematic map for the oil seen. The scanner is characterized by a unique optical design, excellent maintainability, and a broad operating temperature range.

b) *Laser floor sensors*

Laser floor sensors are capable of detecting oil and complicated petroleum products in marine, coastal, and terrestrial environments. The major capabilities of this sensor involve the ability to discriminate between oiled and various naturally occurring non-oiled substances such as seaweed. Laser door sensors reflect uniqueness by identifying the primary characteristics of oil, namely the unique oil fluorescence spectral signature whereas generic sensors depend upon the secondary characteristics of oil (Carl E Brown, 2011).

c) ***Space-based sensors***

Space borne sensors are used to cover extensive and remote areas for monitoring the water quality. Optical space borne sensors are utilized for marine monitoring and this operates in the sun synchronous orbit. The only difference is that the GOCI sensor, which is specifically designed for marine monitoring, is placed in the geostationary orbit. Spatial coverage of the space-bound sensors can vary from tens to hundreds of kilometers and the temporal frequency can vary from hourly to weekly monitoring (Hafeez, 2018).

Large algorithms have been developed for getting the water quality information such as the primary productivity, Chl-a variability,SS, total suspended solids (TSS), CDOM etc (Hafeez, 2018). The Orb View 2 was launched at 1 August 1997 with a spectral brand of 8 nm whose marine perimeter access was Chl-a. SeaWiFS made a simple but elegant measurement regarding the greenery of the earth from the period between 1998 -2001. The measurement provided the scientist an array regarding the capability of the planets to support life. The following data also enable the scientist to obtain a benchmark for studying the effect of the planet's response to a changing environment.

d) *Remote sensors*

Chlorophyll and algal blooms are the main planktons that form the base of the marine life. Accelerated growth of the algae mainly blocks the sunlight and too much dissolved oxygen that it takes is hazardous for marine life and this are checked by the remote sensors (O'Reilly, 2000). Based on the above figure it is seen that spectral response and level of chlorophyll concentration is checked. By the use of above sensors, the detection of pollutants are done and in later chapters elaborative discussion will be done.

CONTROVERSIES SURROUNDING POLLUTANT IDENTIFICATION AND MONITORING

Various controversies arise due to pollutant identification and monitoring through sensors. Despite the relevance of the sensors in oceans some challenges in the form of controversies are still seen. Purchase and maintenance of the sensors are the biggest controversy that is seen in sensor usage. Difficulty in maintenance of sensors in harsh conditions remains in the top list of controversy that will be discussed.

Reliability and accuracy: Main controversy that arises due to sensor useis how much reliable and accurate information it gives. As per the critics this technology may work on lower precisions, without proper diagnostics and faulty based alarm data can give wrong information to the radar and that can lead to wrong results on percentage of pollution calculation. Other error-based factors like fouling, calibration drift, and interference with the other substances in the water.

Accessibility and cost: Modern and innovative sensor-based technology has advanced more now and are seen to be less accessible among all countries. Lesser equality in access to all may lead to use of less effective sensors that could result in late detection or wrong detection of pollutants that has the potential to disparities in data collection-based efforts.

Detection Limits: Most top-rated controversy is seen in this aspect where sensors cannot detect accurate information especially in lower concentration. As per the critique limit of detection (LOD) is stated as the lowest concentration of sample analyte within 95% probability. In other words underestimation of the pollution levels mainly masks the true source of environmental harm. Wrong and timely detection constraints make this a poor and faulty instrument.

Validation and standardization: Lack of protocol-based standardization in terms of calibration, data interpretation and sensor deployment again arouse as a source of controversy. As per the argument done by critics without proper and consistent standards it becomes challenging to compare data that gets collected from sensors and this results in wrong measurement.

Data security and privacy: With the increased proliferation of sensor-based networks raised concerns on security and privacy has been seen as the main issue. Controversies like unauthorized access of sensors, wrong usage and tampered usage of sensors. It is seen that underwater sensor-based nodes gets installed in dangerous and harsh conditions that resulted in network attacks. External and internal attacks like high error rates, higher latency in propagation, computational based limitations in capability and limited bandwidth hampers the smooth operations of sensors.

Temporal and spatial resolution: Another controversy states concerns on temporal and spatial resolution for sensing data. With satellite sensors broader coverage of larger areas can be done but spatial resolution is seen to be restricted for small scale and localized based pollution events. Controversy that arises here states that frequency in the satellite passing by can capture only temporal variations within pollution levels.

Sensitivity to environmental conditions: The main challenges faced here is the environmental conditions that hampers the image generation of the sensors. Cloud and heavy fog can reduce the accuracy of the data as due to cloud proper picture can't be taken and that hampers the reliability and this has been raised as another controversy by the researchers. Atmospheric and water turbidity affects the reliability and quality of data. Data biasness and errors leads to errors in accurate monitoring of data.

Proper addressing of controversies involves collaboration with remote sensors-based scientists, policy makers, and local communities involve robust use of remote sensing technologies. Efforts for addressing ethical and privacy-based concerns have been seen subjective to further discussion.

SUMMARY

This chapter presented an overview of sensors that are used in ocean for detection of pollutants.

- It explained the common pollutants of the marine environment Ocean bound plastics, Ocean bound plastic, non-point-based sources, industrial chemicals, fertilizer runoffs and other sewage.
- Different type of sensors like autonomous underwater vehicles (AUVs), unmanned undersea vehicles (UUVs) Infrared/ultraviolet line scan (IR/UVLS), Laser floor sensors and) Space-based sensors has been found
- It has been learnt that several controversies like reliability and accuracy, detection limits, validation and standardization, temporal and spatial resolution and sensitivity to environmental conditions has made the sensor usage faulty and biased.

It has been concluded that by the use of sensors the detection of pollutants has become easier although there are many controversies still sensors have made the detection of oil spills, chemicals and sewage runoffs easier. This detection through remote and oceanographic sensors shows that measurements and detection earlier can help in reducing the harmful impacts of hazardous pollutants in oceans.

REFERENCES

Al-Dossari, M., Awasthi, S. K., Mohamed, A. M., Abd El-Gawaad, N. S., Sabra, W., & Aly, A. H. (2022). Bio-alcohol sensor based on one-dimensional photonic crystals for detection of organic materials in wastewater. *Materials (Basel)*, *15*(11), 4012. doi:10.3390/ma15114012 PMID:35683310

Bai, C., Zhang, H., Zeng, L., Zhao, X., & Ma, L. (2020). Inductive magnetic nanoparticle sensor based on microfluidic chip oil detection technology. *Micromachines*, *11*(2), 183. doi:10.3390/mi11020183 PMID:32050692

Bellou, N., Gambardella, C., Karantzalos, K., Monteiro, J. G., Canning-Clode, J., Kemna, S., Arrieta-Giron, C. A., & Lemmen, C. (2021). Global assessment of innovative solutions to tackle marine litter. *Nature Sustainability*, *4*(6), 516–524. doi:10.1038/s41893-021-00726-2

Fattah, S., Gani, A., Ahmedy, I., Idris, M. Y. I., & Targio Hashem, I. A. (2020). A survey on underwater wireless sensor networks: Requirements, taxonomy, recent advances, and open research challenges. *Sensors (Basel)*, *20*(18), 5393. doi:10.3390/s20185393 PMID:32967124

Ford, H. V., Jones, N. H., Davies, A. J., Godley, B. J., Jambeck, J. R., Napper, I. E., Suckling, C. C., Williams, G. J., Woodall, L. C., & Koldewey, H. J. (2022). The fundamental links between climate change and marine plastic pollution. *The Science of the Total Environment*, *806*, 150392. doi:10.1016/j.scitotenv.2021.150392 PMID:34583073

Gola, K. K., & Gupta, B. (2020). Underwater sensor networks:'Comparative analysis on applications, deployment and routing techniques'. *IET Communications*, *14*(17), 2859–2870. doi:10.1049/iet-com.2019.1171

Hojjati-Najafabadi, A., Mansoorianfar, M., Liang, T., Shahin, K., & Karimi-Maleh, H. (2022). A review on magnetic sensors for monitoring of hazardous pollutants in water resources. *The Science of the Total Environment*, *824*, 153844. doi:10.1016/j.scitotenv.2022.153844 PMID:35176366

Khan, H., Hassan, S. A., & Jung, H. (2020). On underwater wireless sensor networks routing protocols: A review. *IEEE Sensors Journal*, *20*(18), 10371–10386. doi:10.1109/JSEN.2020.2994199

Kumar, P. S. (2020). *Modern treatment strategies for marine pollution*. Elsevier.

Lin, M., & Yang, C. (2020). Ocean observation technologies: A review. *Chinese Journal of Mechanical Engineering*, *33*(1), 1–18. doi:10.1186/s10033-020-00449-z

Luo, J., Yang, Y., Wang, Z., & Chen, Y. (2021). Localization algorithm for underwater sensor network: A review. *IEEE Internet of Things Journal*, *8*(17), 13126–13144. doi:10.1109/JIOT.2021.3081918

Mahrad, B. E., Newton, A., Icely, J. D., Kacimi, I., Abalansa, S., & Snoussi, M. (2020). Contribution of remote sensing technologies to a holistic coastal and marine environmental management framework: A review. *Remote Sensing (Basel)*, *12*(14), 2313. doi:10.3390/rs12142313

Marra, G., Fairweather, D. M., Kamalov, V., Gaynor, P., Cantono, M., Mulholland, S., Baptie, B., Castellanos, J. C., Vagenas, G., Gaudron, J. O., Kronjäger, J., Hill, I. R., Schioppo, M., Barbeito Edreira, I., Burrows, K. A., Clivati, C., Calonico, D., & Curtis, A. (2022). Optical interferometry–based array of seafloor environmental sensors using a transoceanic submarine cable. *Science*, *376*(6595), 874–879. doi:10.1126/science.abo1939 PMID:35587960

Painting, S. J., Collingridge, K. A., Durand, D., Grémare, A., Créach, V., Arvanitidis, C., & Bernard, G. (2020). Marine monitoring in Europe: Is it adequate to address environmental threats and pressures? *Ocean Science*, *16*(1), 235–252. doi:10.5194/os-16-235-2020

Pal, A., Campagnaro, F., Ashraf, K., Rahman, M. R., Ashok, A., & Guo, H. (2022). Communication for Underwater Sensor Networks: A Comprehensive Summary. *ACM Transactions on Sensor Networks*, *19*(1), 1–44. doi:10.1145/3546827

Rizzato, S., Leo, A., Monteduro, A. G., Chiriacò, M. S., Primiceri, E., Sirsi, F., Milone, A., & Maruccio, G. (2020). Advances in the development of innovative sensor platforms for field analysis. *Micromachines*, *11*(5), 491. doi:10.3390/mi11050491 PMID:32403362

Salgado-Hernanz, P. M., Bauzà, J., Alomar, C., Compa, M., Romero, L., & Deudero, S. (2021). Assessment of marine litter through remote sensing: Recent approaches and future goals. *Marine Pollution Bulletin*, *168*, 112347. doi:10.1016/j.marpolbul.2021.112347 PMID:33901907

Salgado-Hernanz, P. M., Bauzà, J., Alomar, C., Compa, M., Romero, L., & Deudero, S. (2021). Assessment of marine litter through remote sensing: Recent approaches and future goals. *Marine Pollution Bulletin*, *168*, 112347. doi:10.1016/j.marpolbul.2021.112347 PMID:33901907

Sana, S. S., Dogiparthi, L. K., Gangadhar, L., Chakravorty, A., & Abhishek, N. (2020). Effects of microplastics and nanoplastics on marine environment and human health. *Environmental Science and Pollution Research International*, *27*(36), 44743–44756. doi:10.1007/s11356-020-10573-x PMID:32876819

Shen, L., Wang, Y., Liu, K., Yang, Z., Shi, X., Yang, X., & Jing, K. (2020). Synergistic path planning of multi-UAVs for air pollution detection of ships in ports. *Transportation Research Part E, Logistics and Transportation Review*, *144*, 102128. doi:10.1016/j.tre.2020.102128

Sørensen, A. J., Ludvigsen, M., Norgren, P., Ødegård, Ø., & Cottier, F. (2020). Sensor-Carrying Platforms. POLAR NIGHT Marine Ecology: Life and Light in the Dead of Night, 241-275. Springer. doi:10.1007/978-3-030-33208-2_9

Sun, K., Cui, W., & Chen, C. (2021). Review of underwater sensing technologies and applications. *Sensors (Basel)*, *21*(23), 7849. doi:10.3390/s21237849 PMID:34883851

Thakur, A., & Kumar, A. (2022). Recent advances on rapid detection and remediation of environmental pollutants utilizing nanomaterials-based (bio) sensors. *The Science of the Total Environment*, *834*, 155219. doi:10.1016/j.scitotenv.2022.155219 PMID:35421493

Thushari, G. G. N., & Senevirathna, J. D. M. (2020). Plastic pollution in the marine environment. *Heliyon*, *6*(8), e04709. doi:10.1016/j.heliyon.2020.e04709 PMID:32923712

Whitt, C., Pearlman, J., Polagye, B., Caimi, F., Muller-Karger, F., Copping, A., Spence, H., Madhusudhana, S., Kirkwood, W., Grosjean, L., Fiaz, B. M., Singh, S., Singh, S., Manalang, D., Gupta, A. S., Maguer, A., Buck, J. J. H., Marouchos, A., Atmanand, M. A., & Khalsa, S. J. (2020). Future vision for autonomous ocean observations. *Frontiers in Marine Science*, *7*, 697. doi:10.3389/fmars.2020.00697

Yusuf, A., Sodiq, A., Giwa, A., Eke, J., Pikuda, O., Eniola, J. O., Ajiwokewu, B., Sambudi, N. S., & Bilad, M. R. (2022). Updated review on microplastics in water, their occurrence, detection, measurement, environmental pollution, and the need for regulatory standards. *Environmental Pollution*, *292*, 118421. doi:10.1016/j.envpol.2021.118421 PMID:34756874

Zou, X., Ji, Y., Li, H., Wang, Z., Shi, L., Zhang, S., Wang, T., & Gong, Z. (2021). Recent advances of environmental pollutants detection via paper-based sensing strategy. *Luminescence*, *36*(8), 1818–1836. doi:10.1002/bio.4130 PMID:34342392

Chapter 9
Removal of Heavy Metals From Waste Water Using Different Biosensors

Kamal Kishore
https://orcid.org/0000-0002-7870-4142
Department of Chemistry & Biochemistry, Eternal University, India

Yogesh Kumar Walia
Department of Chemistry, Career Point University, India

ABSTRACT

Recently, much effort has been made to reach to an effective strategy for wastewater monitoring. Several pieces of evidence support the special role of biosensors in plans for the administration of water resources. Concerning this fact, there are some technical and practical limitations and complications, which should be overcome to develop more efficient and commercial applicable biosensors. To achieve this goal for the detection of a broad range of wastewater pollutants, it is necessary to design novel sensing systems with larger detection range and capability for the simultaneous detection of several compounds. Additionally, the limit of detection in the lower concentration range should be possible, and also biosensor should have long-storage stability. This chapter explores the various ways by which heavy metals can be removed from wastewater. Different biosensors are under investigation that can be used to remove different pollutants form different ecosystems. This will help to solve the problem of water pollution and will also help to reduce human health impact.

INTRODUCTION

Although heavy metals are essential for life on Earth, they pose a threat when they build up in living things. Lead, nickel, cadmium, chromium, arsenic, and mercury are among the most common heavy metals found in environmental contamination. Various activities, both natural and man-made, emit cadmium into the atmosphere, and its exposure to animals and people varies. Industrial waste, surface runoff, and

DOI: 10.4018/979-8-3693-1930-7.ch009

absorption all contribute to cadmium contamination of aquatic ecosystems. The metal cadmium may be toxic to humans if they consume it in the form of food, air, or water. cadmium lacks any properties that would be beneficial to plant development or metabolic activities.

The biosphere may include mercury, a very dangerous heavy metal. Additionally, it has grown into a significant air pollutant as a result of human activities and is becoming increasingly common in the air. The very toxic compound methylmercury is formed when mercury reacts with sediments in water (Ansari, 2017). The methylmercury contaminates human bodies as it passes through the food chain from marine organisms, such as shellfish and fish, that ingest pathogenic bacteria. as absorbed into the human body, it causes a variety of neurological disorders as it enters the circulation (Olaniran et al., 2008).

A non-biodegradable metal, lead occurs naturally in very minute amounts. Lead levels in the atmosphere are steadily rising due to human activities including mining, manufacturing, and burning fossil fuels. Lead poses health risks to people when exposure levels above the recommended safe threshold. Kids are more likely to have lead poisoning to begin with, and dust that contains environmental lead may make their condition worse (Zhou et al., 2017). Naturally occurring in various oxidation levels, manganese is the most prevalent of the hazardous heavy metals. Manganese oxides are released into the atmosphere when the petrol additive methylcyclopentadienyl manganese tricarbonyl (MMT) is burned. Consuming too much manganese is quite hazardous, even though it is necessary for many physiological processes (Kummu et al, 2016).

The metal chromium is both poisonous and carcinogenic. In comparison to chromium (VI), chromium (III) is not as dangerous. In the course of manufacturing, they are able to interconvert to one another. Because chromium (III) is less poisonous than chromium (VI), its conversion to chromium (VI) is less damaging to the environment. Numerous businesses utilise chromium, which is harmful to local climates.

Traditional analytical approaches for environmental contaminant monitoring are time-consuming, costly, and need specialised reagents. These methods include gas chromatography and high performance liquid chromatography, as well as capillary electrophoresis and mass spectrometry. To address the deteriorating state of the environment, there is an urgent need for more sensitive, less costly, quicker, simpler to use, and portable biosensing technology. These techniques would be ideal for monitoring contaminants that are harmful to ecosystems and individuals. Traditional techniques are worthless for in-situ measurements, such as in the event of an unintentional pesticide release or acute poisoning, since tiny, lightweight, and rapid equipment, such as environmental monitoring biosensors, is required. Nanotechnology is critical in the creation of successful biosensing devices; modern biosensors improve analytical performance, such as sensitivity and limit of detection, by adding nanoparticles and unique nanocomposites into their designs. Gold nanostructures, for example, offer a large surface area and excellent electron mediation capabilities, making them an ideal basis for enzyme immobilisation matrices. Furthermore, even in vivo, gold nanoparticles showed little citotoxicity and great biocompatibility. As a result, one present and future technical challenge in environmental science is how to increase the number of commercially available biosensors for in-situ measurements by enhancing the analytical performance of sensing systems through the inherent contributions of nanotechnology and biotechnology (Geng et al., 20-08).

Biosensors integrate chemical and physical sensing methods into a single analytical sensor. Their effectiveness stems from the fact that they have a physical or chemical immobilization mechanism that brings the biological and physicochemical components into close proximity with one another. At its core, a biological element is a receptor (bioreceptor), which allows it to interact with the medium of interest in order to identify a certain analyte. A physicochemical transducer takes the reaction that happens at the interface between analytes and bioreceptors and turns it into a quantifiable signal that can be processed and shown as values. The immobilization of the biological substance in close proximity to the transducer is necessary for accurate biosensor functioning. This immobilization may be achieved by chemical attachment or physical entrapment (Lyngberg et al., 1999). Since bio receptor molecules are to be employed for measurements several times, only minimal numbers are needed. Because of its versatility in detecting a broad variety of analytes, including contaminants, microbes, fungi, pharmaceuticals, and food additives, biosensors have seen fast and diverse development over the last few decades. The wide range of industries that may benefit from these characteristics includes robotics, security and defence, agriculture, forensic chemistry, pharmaceuticals, industry, environmental monitoring, medicine, food, and more. Biosensors are most useful when used to activities that are unique to that field. Their value in the food business was shown when it came to controlling quality and safety. They were able to distinguish between natural and fake ingredients, keep an eye on fermentation, and more. Control procedures are where they are most useful in industry. Their usage is highly recommended in the fields of drug development, clinical and medical sciences, and the quick detection of viruses and substances that cause a wide range of disorders, including cancer (Kokkinos, and Economou, 2017).

ELECTROCHEMICAL BIOSENSORS

Because they were the most often reported, electrochemical biosensors were the first to be developed and sold commercially. By generating or consuming ions or electrons, the chemical interaction between the biomolecule (bioreceptor) and the analyte (target) changes the electrical characteristics of the analyte solution (current, potential, etc.) in this biosensor type. The transducer can detect these variations in the analyte concentration in the sample solution since its electrochemical signal is proportional to it (Bachmann, and Schmid, 1999).

One benefit of electrochemical biosensors is their sensitivity even with very tiny sample volumes, and another is that they need very little sample preparation. Direct sample analysis is another option, opening the door to automation. Detections have issues with stability and repeatability. Biosensors based on (ISFETs), potentiometric, amperometric, and conductometric electrochemical sensors are the four main categories into which these devices fall. A customized electrochemical cell design is usually necessary when working with different measuring principles. A potentiometric biosensor works by placing an ion-selective membrane on top of an ion-selective electrode (ISE). This membrane acts as a selector, allowing the sensor to detect target ions even while other ions in the sample are interfering. These devices calculate the difference between the potentials of the working and reference electrodes with almost little current, which is a function of the analyte concentration. Amperometric biosensors dominate the electrochemical biosensor market. Compared to potentiometric biosensors, amperometric ones are slower and less sensitive, and they might be impacted by irrelevant electroactive species (Zhang et al., 2014).

The electrical conductivity of a sample solution measured between two electrodes is the basis of conduct metric biosensors, which are based on the biological process. Conduct metric biosensors are great for a lot of things as they aren't light-sensitive, they don't need a reference electrode, and they can be mass-produced utilizing inexpensive technology. The fourth category of electrochemical biosensors is those that can directly detect ions; these are biosensors built on (ISFETs). When the gate electrode comes into touch with the analyte solution, its potential changes due to changes in the sample's ion activity. Afterwards, the electric potential change is quantified (Saber, and Pişkin, 2003).

OPTICAL BIOSENSORS

When a bioreceptor interacts with a target analyte, optical biosensors measure the resulting changes in the input light's optical intensity, which are directly proportional to the analyte concentration in the sample. Being non-electrical biosensors, these optical devices have a number of benefits, including being able to be used in vivo, being simple, compact, and unaffected by electromagnetic interference. Biosensors may be either intrinsic or extrinsic, depending on their optical design. In intrinsic biosensors, the analyte is detected by means of the interaction between the incident light wave and a structure that permits it to pass via a wave guide or optical fibre. The optical fibre transmits the signal from an extrinsic biosensor, which uses light waves to react with the sample phase as they travel through it (Tagad et al., 2016).

Based on the observation that various analytes absorb light of different wavelengths, absorption-based biosensors are low-cost, straightforward instruments for determining analyte concentrations. The light may be sent from the source to the sample using a single optical fibre, and then from the sample to the detector using an additional cable. One kind of biosensor that uses light for detection is the surface plasmon resonance (SPR) sensor. This technique creates electromagnetic waves (plasmons) via excitation of electrons on a metal's surface at its interaction with a dielectric. Biomolecular interactions, such as the analyte's particular binding, may alter the material's refractive index close to the metal surface, which can have a profound effect on Plasmon propagation (Hayat and Marty, 2014).

By monitoring the wavelength shift in electromagnetic radiation, optical biosensors based on fluorescence may identify specific atoms or molecules. Alternatively, fluorescence energy transfer (FRET) or fluorescent markers may be used to conduct detection indirectly. Luminescent optical biosensors may be broadly classified into two categories: chemiluminescent and bioluminescent. Unlike fluorescence-based biosensors, these devices use an exothermic chemical process to get the target atoms or molecules to a triggered state. Upon returning to the ground state, the excited species create light with little heat. Bioluminescence is the result of this kind of chemical reaction taking place within living things (Zehani et al., 2015).

PIEZOELECTRIC BIOSENSORS

Devices that combine a bio recognition element with a piezoelectric material for transducing are known as piezoelectric biosensors. One of the most common types of materials with a piezoelectric effect is quartz crystal, which is widely used due to its abundance, high temperature tolerance, and chemical stability in water. This biosensor type relies on the piezoelectric material's capacity to undergo mechanical stress-induced electrical potential generation and electric field-induced elastic deformation as its fundamental measuring mechanism (Zehani et al., 2014).

THERMAL BIOSENSORS

Thermoelectric transducers detect changes in temperature that correlate with the concentration of analytes in thermal biosensors, which are also known as calorimetric or thermometric biosensors. Both thermistors and thermopiles are used as heat transducers in these apparatuses. Among the many benefits of thermal biosensors are their ability to detect reactions without labelling reactants, their lack of sensitivity to changes in sample electrochemistry or optical characteristics, and their lack of requirement for regular recalibration. Because enzyme-catalyzed reactions are exothermic, the majority of published research on this kind of sensor describes trials employing these sensors in conjunction with enzyme-based thermal biosensors (Yang et al., 2018).

BIOSENSORS FOR WASTEWATER MONITORING

Reusing wastewater and making informed decisions about treatment process design and operation are both impacted by the component concentrations in the wastewater. Due to the fact that wastewater contaminants may change over time and in different places, there has been a push to create new technology that can track these waste products in real-time and at low cost (Michael-Kordatou et al., 2015; Rehman et al., 2015).

Waste fluids, whether they are disposed of or reused, may have their various qualities measured using a battery of quality indicators. Many of these methods rely on laboratory procedures that are time-consuming and expensive due to the need to collect data often and treat it before analysis (Chong et al., 2013, Yang et al., 2015). Recently, there has been a great deal of effort put into creating biosensors that can identify environmental pollutants. One great thing about these sensors is that they can pick up toxins even in very little amounts in complex matrices, like wastewater. Biosensors are miniature devices that allow for the fabrication of portable sensors that can monitor wastewater on-site (Tsopela et al., 2016). This research organises different types of biosensors into four categories based on the processes they use for transduction: (i) electrochemical, (ii) optical, (iii) piezoelectric, and (iv) thermal biosensors. Molecularly imprinted polymer (MIP), immunochemical, non-enzymatic, whole-cell, and DNA elements are all part of these categories, which together include a wide range of biorecognition approaches. Recent years have seen a rapid expansion in the field of powerful biosensors based on nanomaterials. These sensors possess exceptional characteristics that allow them to detect chemical reagents and biological events. Existing techniques used to identify contaminants in wastewater were affected, and there was an increase in the number of initiative activities based on real-time biosensing of wastewater sources. There has also been a lot of study on nanotechnology-based wastewater treatment methods. References include Berekaa (2016), Cloete et al. (2010), and Henze et al. (2008).

Detecting Organic Substances With Biosensors

Anthropogenic and naturally occurring organic pollutants are present in effluents from a wide variety of sources, including households, farms, and businesses. Some of the organic molecules that may be discovered in industrial wastewaters include hydrocarbons, aromatic compounds, chlorine compounds, diphenylmethane, and surfactants. Polybrominated diphenyl ethers (PBDEs), polybrominated biphenyls (PBBs), and polychlorinated biphenyls (PCBs) are three examples of the persistent organic pollutants (POPs) found in wastewater, especially in industrial waste leachate (Weber et al. 2011). Agricultural wastewater sometimes contains harmful compounds, such as organic herbicides and pesticides, which are used to manage weeds, unwanted plants, and a variety of pests (Wauchope 1978). Many antimicrobial household cleaning solutions include organic compounds like triclosan, which, when discharged into wastewater and groundwater, might be harmful. Since these organic waste compounds may bind to endogenous hormone receptors as either agonists or antagonists, it is now necessary to identify endocrine disrupting chemicals (EDCs) in wastewater.

For the sake of both humans and the environment, it is crucial to accurately measure organic matter levels throughout wastewater treatment and water reclamation activities. There are certain issues with using organic components in wastewater treatment. As an example, they have the potential to foul membrane filtration systems and generate harmful by-products. It is usual practice to look at two factors when analysing the organic matter content of water. Biodegradable organic compound breakdown oxygen demand is quantified using the biochemical oxygen demand (BOD) index. There is also the matter of the chemical oxygen demand (COD), which shows the amount of oxygen that the water's chemical processes are using (Wacheux 1998). Traditional approaches to characterising BOD have relied on either short-term in-situ measurements or off-line monitoring for 5 or 7 days. When compared to the conventional method, biosensors detect these indices more quickly and accurately. The purpose of this section is to provide a synopsis of the most recent biosensing techniques for tracking organic matter in different wastewater resources, with a focus on transducing procedures.

OPTICAL BIOSENSORS FOR DETECTION OF ORGANIC MATERIALS

Some organic pollutants in wastewater, including triclosan, may be accurately identified using a combination of surface plasmon resonance (SPR) and molecular imprinting polymers (MIPs). Molecular imprinted-SPR nanosensors were created to accurately identify certain chemical substances in wastewater, such as triclosan, by simulating the activity of their biological receptors. Immunoanalytical techniques, which rely on fluorescently tagged antibodies having a strong affinity to the target component, are another prevalent mechanism used in optical biosensors for organic matter monitoring. Optical biosensors for organic contaminants often use this technique, called fluorimetry, for signal transduction. Groundwater samples have previously been used to track organic environmental contaminants such hormones, pesticides, endocrine disrupting chemicals, antibiotics, and sperm (Tschmelak et al. 2005). By monitoring variations in bioluminescence response, they demonstrated that this approach can detect COD in wastewater.

Biosensors That Detect Organic Compounds via Electrochemistry

Another option for detecting organic pollutants in wastewater is the use of electrochemical bio-receptors, which may include enzymes, antibodies, and even whole cells. When it comes to biosensors, ambulometric devices are considered superior. Nomngongo et al. (2012) used a biosensing device based on the horseradish peroxidase (HRP) bioreceptor to characterise persistent organic pollutants in wastewater samples using an amperometric transduction mechanism. Attaching HRPs electrostatically to a platinum electrode that had been treated with polyaniline (PANi) allowed for immobilization. To find the BOD in wastewater samples, a set of bespoke Clark-type oxygen sensors were used. Immobilised microorganisms on a bio-membrane linked to an amperometric sensor that monitors dissolved oxygen concentrations provide the fundamental basis of these systems. More specifically, Verma and Singh described a biosensor that used amperometric oxygen probes made of gold or silver attached to cellulose acetate membranes that had living bacteria immobilised on them. This instrument is perfect for assessing BOD in industrial wastewater because to its fast response time of around 7 minutes and detection limit of 1 mg/L (Verma and Singh 2012). In order to assess the biochemical oxygen demand (BOD) in wastewater in real-time,

Yamashita et al. looked into a novel electrochemical open-type biosensor. Anodes of the open type, in contrast to their closed-type counterparts, are constantly exposed to oxygen when placed in an aerated tank containing wastewater (Yamashita et al. 2016). This biosensor provides effective detection in both aerobic and anaerobic environments, thanks to a high correlation ($R2 > 0.9$) between the generated current and the loading of BOD up to 250 mg/L in wastewater. Using a potentiometric sensor based on metal-ion polymers, Kou et al. (2013) developed a biosensor with a lowered limit of detection (LOD) for neutral bisphenol A as part of their investigation on the nanomolar quantities of organic toxicants in wastewater. Several biosensors for water quality monitoring have been developed recently using microbial fuel cells (MFCs). Chemical oxygen demand (COD), biological oxygen demand (BOD), and dissolved oxygen (DO) levels in water samples may be found with these instruments, and they are also often used to identify and analyse different organic pollutants (Zhou et al. 2017). Anaerobic bacteria are housed in the anode chamber of these electro-biosensors, which is separated from the cathode chamber by a proton exchange membrane.

HOT-SENSOR DETECTORS FOR ORGANIC INGREDIENTS

By tracking the rate of thermal change in wastewater and water, Yao et al. developed a biosensing device that effectively monitors chemical oxygen demand (COD). The COD values were determined using a flow injection analysis equipment with periodic acid solutions as oxidants (Yao et al. 2014). The detection range for COD in water samples from different sources was found to be broad and the LOD to be low using this sensing technique. The findings of the dichromate technique, which is used for conventional COD monitoring, are associated with the COD levels detected in wastewater samples. In terms of speed, stability over time, and robustness, the calorimetric COD determination technique outperforms the conventional method in terms of detection efficacy.

BIOSENSORS FOR DETECTION OF HEAVY METALS (HMS)

According to Sayari et al. (2005), heavy metal pollution in wastewaters is often associated with many industrial effluents, including agricultural runoff, chemical manufacture, mining, cosmetics, nuclear waste, metal processing, and painting. For metabolic reactions to occur in living organisms, certain amounts of HMs such as copper, iron, manganese, molybdenum, nickel, selenium, and zinc are required. The non-essential HMs lead, chromium, cadmium, mercury, arsenic, and antimony are very dangerous and carcinogenic even in trace amounts. By attaching to bacteria and then climbing the food chain, they have the potential to cause a wide range of chronic diseases in humans (Jaishankar et al. 2014). Analytical methods such as chromatography, ultraviolet-visible spectroscopy, and atomic absorption spectroscopy have been widely used for the control of these pollutants. Time and money required for sample preparation and pre-concentration restrict the uses of the previously listed procedures, even if they are sensitive and selective enough (Gumpu et al. 2015). Because of their great sensitivity, simplicity, and observable signals, biosensors have recently been proposed as a remarkable alternative to traditional sensing systems. We detail many biosensors and nanobiosensors that can detect HMs in this research.

Optical Biosensors for Heavy Metal Detection

Several microbial biosensors that emit bioluminescence signals were created to work in environments with low levels of HMs (Jouanneau et al. 2012). Recombinant Escherichia coli strains producing fluorescent proteins are the backbone of bacterial biosensors. According to Raja and Selvam (2011) and Ravikumar et al. (2012), when these bacteria come into contact with HMs, they increase their synthesis of fluorescent proteins, which allows them to detect potentially dangerous quantities of HMs in wastewater samples. Biosensors based on enzymes rely on the inhibitory effects of HMs. Because of its distinct optical properties, which improve detection efficacy, porous silicon (PS) structures have seen significant application as enzyme immobilisation substrates for the monitoring of chromium ions in water and wastewater (Biswas et al. 2017). Two optical fibre biosensors based on immobilised urease and acid phosphatase have been used in earlier attempts to detect low levels of HMs. Additionally, a colorimetric bioactive paper sensor that quantifies concentrations of Ag+, Ni^{2+}, and Cr^{4+} in parts per million (ppm) was created by using β-galactosidase (β-GAL). You can't beat the speed and accuracy of this sensor.

In light of recent developments in DNA-based devices and nanotechnology, a promising strategy for multiplexed heavy metal detection is the development of highly sensitive nanobiosensors using fluorescence resonance energy transfer (FRET). Xudong and colleagues used this technology to construct a new FRET sensor. The nanobiosensor's receptor is a thin film of silica that has been covalently attached to DNAzymes, two quenchers, and quantum dots (QDs). Because of their photostability, they have the potential to improve the selectivity and specificity of the system. The quantum dots exhibited a change in the fluorescence signals reflected by them when subjected to various metal ions. Many HMs may be measured in water samples at once using multidimensional aptasensors that contain Au- or Ag-NPs (Song et al., 2016).

METAL DETECTION USING ELECTROCHEMICAL BIOSENSORS

Zhao et al. (2018) developed innovative electro-biochemical sensors named sediment microbial fuel cells (SMFCs), single-chamber batch-mode cube microbial fuel cells (CMFCs), and double-chamber microbial fuel cells (DMFCs) to monitor HMs in wastewater influent in real-time. Biosensors may change voltage signals by including immobilised electrogenic microorganisms within them. These biosensors show promise as sensitive wastewater quality monitors, especially for HMs like Cr^{6+}, Cd^{2+}, Cu^{2+}, and Zn^{2+} that have low detectable quantities (Tao et al. 2017).

According to Nguyen et al. (2018) and Song et al. (2018), electrochemical biosensors that use nanomaterials may significantly improve device performance. For example, Chen et al. showed that an electrochemical apt sensor could be made more sensitive by using gold nanoparticles (GNPs) in their investigation on Cu^{2+} detection (Chen et al. 2011b). The biosensing method depends on the direct formation of guanine wire structures, which are based on the production of strong and stable T-Hg2^{+-} T mismatches in the presence of Hg^{2+}. Triggering the electrochemical oxidation of 3,3',5,5'-tetramethylbenzidine (TMB) using the generated guanine nanowire on the surface of the gold electrode results in a notable enhancement of the electrical signal. The feasibility of detecting Cd^{2+} and Hg^{2+} simultaneously using a modified Au electrode including multi-walled carbon nanotube (MWCNT)/peptide was investigated using samples of industrial effluent. The peak-current density, reaction rate, and sensitivity of the modified MWCNT electrode were enhanced by adding peptides. The limited detection limits (LODs) of these systems are below the limits specified by the EPA for Cd^{2+} and Hg^{2+} in drinking water (Rahman et al. 2012). Ganjali et al. detailed a nanobiosensor, an electrochemical sensor that uses an L-cysteine functionalized glassy carbon (GC) electrode. It was used to improve the detection of Cd^{2+}, Pb^{2+}, Cu^{2+}, and Hg^{2+} ions in industrial effluents. The World Health Organisation has established standards for the measurement of metal ions, and this platform allows you to do so with acceptable LOD values for several ions at once. November 2011 was the year.

Biosensor That Detects Heavy Metals Using Piezoelectric Technology

Ion sensors that monitor different ions have recently been produced using AlGaN/GaN HEMTs. Unlike many other kinds of ion sensors, these devices may function without a reference electrode and are sensitive to surface charge properties. Researchers Asadnia et al. demonstrated that ionophore-coated PVC-based membranes on such devices can detect Hg^{2+} and Ca^{2+} ions in water (Asadnia et al., 2017).

Devices for the Microfluidic Analysis of Heavy Metals

The use of portable microfluidic devices to detect HMs in various fluids has recently shown remarkable efficiency, suggesting that these devices may find use in dark seas. An unique example of this is a cost-effective biosensor for lead concentration monitoring that was produced by integrating a microfluidic technique with an optical transducing technology. In order to detect levels as low as 0.094 nM of Cr(III) and Cr(VI) in different water samples, Peng et al. (2017) created a (MWCNTS). Another approach was to use a bio-inspiration technique to create a microelectromechanical system (MEMS) that could detect lead on-site by simulating the function of a shark's olfactory sensory system (Wang et al. 2016). Another microfluidic Pb^{2+} biosensor that Cropek and colleagues developed was a microchannel made of polymethylmethacrylate (PMMA) that was immobilised with lead-specific catalytic DNA (Dalavoy et al. 2008). Microfluidic devices based on field-effect transistors (FETs) have been developed in a number of recent research. A protein-functionalized rGO thin-film structure was used to construct the sensing microchannel, which was shown to be able to detect metal ions in a variety of water-based solutions (Sudibya et al. 2011). At room temperature, this gadget quickly reacts to Hg^{2+} inside microfluidic channels and is completely portable.

BIOSENSORS FOR DETECTION OF MICROORGANISMS

Significant global mortality and morbidity are caused by a wide range of aquatic microbial illnesses. Many various kinds of microbes, including bacteria, viruses, protozoa, algae, fungus, and yeasts, find sewage to be a perfect environment for development. They have a wide range of organismal compositions and are mostly discharged into wastewater from food processing and domestic activities. Microorganisms may be either autotrophic or heterotrophic, and the amounts of these types of microbes can change depending on where they come from. In contrast, Grøndahl-Rosado et al. (2014) found that the sickness rate in a community, which might be an indicator of public health concerns, is closely related to the concentration of parasites and harmful viruses. Furthermore, different microbial pathogens have distinct minimum infective doses. Traditional approaches to microbial pathogen detection rely on microscopic techniques or artificial media, which come with a host of technical drawbacks such lengthy processing times, limited sensitivity, and poor detection accuracy. As an example, rather of monitoring microbes in general, coliform tests are typically used to evaluate coliform bacteria in environmental samples. There has been a lot of focus on discovering more efficient ways to detect tiny levels of viruses in dark waters because of the aforementioned shortcomings of the traditional approaches.

EYE-SENSORS USED TO DETECT MICROBRANTS

Two gram-negative bacteria, E. coli and Listeria, as well as gram-positive bacteria, were detected using surface-enhanced Raman spectroscopy (SERS) molecular probes, which showed very fast detection periods of bacterial specific antigens (Xiao 2010). In order to identify E. coli O157, a H7 strain, in wastewater samples, Yildirim et al. (2014) created a portable optical fibre biosensor. A specific aptamer that was fluorescently labelled was used to accomplish this. By combining them with complementary DNA probes that are surface-immobilized on an optical electrode fibre, one may ascertain the quantity of free aptamers. The higher fluorescent signal could be associated with the decreased E. coli level in the samples that were analysed. Due to its sensitivity, portability, low cost, and speed, the developed device is perfect for on-site wastewater testing.

Researchers have shown that silver and gold nanoparticles, as reported in studies by Singh et al.(2009) and Tripp et al.(2008), work well as substrates for SERS pathogen biosensors. Another potential mechanism by which nanomaterials enable the detection of microbial DNA is via the use of metal nanoparticles. The need for intricate and costly fluorescent labelling might be rendered obsolete by the colorimetric alterations induced by hybridization of gold nanoparticles with single-stranded DNA probes. According to Jyoti et al. (2016), bacterial sensing equipment often use Au NPs to assess the minute concentrations of organisms in samples of dark water. In the presence of the target bacteria, the aptamers undergo a conformational shift, leading to the aggregation of AuNPs and, as a consequence, a visible change in colour from red to pinkish-purple. This approach can directly quantify the total bacterial load without the need for specialized equipment or pretreatment; it is also quick, specific, and easy to implement. There is also an interesting method that labels an antibody against E. coli O157: H7 using carboxyl functionalized graphene quantum dots (cf-GQDs) via the antibody's main amines.

ELECTROCHEMICAL BIOSENSORS FOR DETECTION OF MICROORGANISMS

Numerous portable instruments have been developed in recent years to amperometrically detect and track the electrooxidation and electroreduction processes of different microbes. The fact that almost all E. coli strains possess β-D-Glucuronidase (GLUase) activity has been well-documented. In their study, Rochelet et al. used a paminophenyl β-D-glucuronide substrate to observe E. coli GLUase activity in wastewater samples using disposable, one-time use carbon screen-printed electrodes. This approach may have potential uses in water and wastewater analysis, despite its low sensitivity (Rochelet et al. 2015).

Yemini et al. demonstrated a cheap, fast, and sensitive amperometric biosensor that could detect minuscule concentrations of Mycobacterium smegmatis and Bacillus cereus. The amperometric biosensor's general construction allows for the rapid evaluation of various species' particular enzymes on electrodes in under eight hours (Yemini et al. 2007). Borisova et al. (2018) found that this approach can quickly identify very tiny amounts of real bacteria. Despite the lack of common usage for potentiometric techniques in bacterial contamination detection, Ercole et al. showed that they could measure Escherichia coli, DH5a in water using a tiny, quick, inexpensive, and very sensitive potentiometric alternating biosensor (Ercole et al. 2002). To detect Staphylococcus aureus (S. aureus) in water samples, Abbaspour et al. (2015) proposed an electrochemical dual aptasensor. The detection process began with an aptamer that was biotin-streptavidin-adhered to magnetic beads. The sandwich immunosensing approach, which involves secondary aptamer-conjugated AgNPs, was used to report the presence of S. aureus. Finally, the acidic solution's dissolved AgNPs produced a differential pulse stripping voltammetric signal. S. aureus quantitation in the samples was shown indirectly by this signal.

A bio nanocomposite of polypyrole (PPy)/AuNP/MWCNT/Chitosan was used to modify the graphite electrode and create a sensitive voltammetric immunosensor. The technology allowed the immobilisation of an anti-E. coli O157: H7 monoclonal antibody thanks to the large surface area given by MWCNT and AuNPs. E. coli O157: H7 monitoring for environmental and food quality management was much enhanced by this modified electrode technique. Güner et al. (2017) noted that a major obstacle to using this biosensor in real-world samples was the antibodies' inability to regenerate for future detections. Due to its dual nature, the impedance biosensor relies on complex mathematical formulas for its prediction. Despite the possibility of label-free recognition and the simplification of sensor setup, no research has been conducted on microbe identification using impedance biosensors so far. Muhammad-Tahir and Alocilja (2003) detailed a one-time use impedimetric technique for detecting E. coli. Using a self-assembled monolayer of gold electrode and an anti-E. coli antibody, an impedance immunosensor was shown for the detection of E. coli in samples of river water. This particular sensor can keep tabs on effluent. When E. coli specifically bound to the electrode, it increased the electro-transfer resistance, and Geng et al. (2008) demonstrated a direct association between bacterial concentration and this effect.

THE ROLE OF NANOTECHNOLOGIES IN BIOSENSING FROM WASTEWATER

Biosensors built using nanomaterials allow for further component miniaturization, according to Lv et al. (2018). When it comes to monitoring the composition of wastewater, the tiny size of the sensor makes it possible to integrate numerous processes into one device, allowing for the design of real-time monitoring of different materials.

A decrease in the cost and labour intensity of biosensing equipment is complemented by the prospective uses of nanotechnology. Nano biosensors rely on the hypersensitivity of nanostructures to contaminants and the selectivity of interactions between tiny biomolecules used as bio recognition components to work (Ghasemzadeh et al. 2014). According to Kaittanis et al. (2010), nanoparticles are very suitable for environmental specimen monitoring due to their unique properties such as significant extinction coefficients, strong photo stability, luminescence, and catalytic capability. In addition, as stated by Lim et al. (2015), portable microbial biosensors may achieve a greater retention time when nanomaterials are used in conjunction with the freeze-drying process. The detection of harmful substances in water has recently come to rely on nanomaterials, which are notable for their tiny size and distinct physicochemical properties (Dasgupta et al. 2017).

Environmental monitoring has been a primary focus of recent advances in nanotechnology's incorporation with physiologically sensing systems (Justino et al., 2017; Reverté et al., 2016; Stoytcheva et al., 2018). Not long ago, there were a plethora of groundbreaking biosensors that used nanomaterials to enhance the precision and efficacy of contaminant detection in wastewater samples. Several nanobiosensors have been developed for the detection of organic components, heavy metals, and microorganisms in dark water resources. These include the ones mentioned earlier Several intriguing avenues for improving biosensors for environmental wastewater monitoring may be explored by combining nanotechnology with specific bio recognition mechanisms and different transducing techniques.

TRANSITION TO COMMERCIALIZATION

When it comes to environmental monitoring, there aren't as many commercial options as there are for medical biosensors. Commercial usage in wastewater treatment facilities is still a way off for the majority of the existing biosensors, which need substantial improvements in sensitivity, selectivity, functionality, etc. Here, biosensors may be enhanced to detect lower concentrations of analytes, especially in turbid samples like wastewaters, by including efficient bio-recognition components, such as whole microbes or biomolecules. Utilising a biosensor that quantifies biological oxygen demand (BOD) is the gold standard for assessing water and wastewater quality. In 1983, according to our study (Iranpour et al., 1997), the first BOD biosensor was offered to the market by Nisshin Denki (Electric) Co. Ltd. Jouanneau et al. (2014) states that a commercial BOD biosensor using electrochemically-microbial fuel cell technology was created by Korbi Co., Ltd. Anam et al. (2017) describe a novel biosensor approach that uses mediator-less microbial fuel cells to measure BOM in wastewater. This approach has the potential to enter the commercial market in the near future. Numerous commercial solutions for toxicant detection have been developed to quench the bioluminescence activity of luminous bacteria, such as *Vibrio fischeri* or *Photobacterium phosphoreum*. Various optic biosensors are available for use in detecting organic chemicals and heavy metals in different environments. One of the several commercially available surface plasmon resonance (SPR) biosensors for application in environmental sample monitoring is the β-SPR biosensor from SENSIA in Spain, which can detect Carbaryl. Spreeta 2000, developed by Texas Instruments Inc. using nanotechnology, is a cheap and portable surface plasmon resonance (SPR)-sensor that can detect a variety of biomolecule pollutants in soil or water (Chinowsky et al. 2003). While nanotechnology has played a significant role in biomolecular detection, label-free cantilever sensors, such as Biocom's VeriScan 3000, have been developed to monitor more than 100 distinct chemicals in environmental samples. Companies have developed a plethora of operational systems for measuring microorganisms in physiological and environmental fluids; these systems may be expanded to include wastewater samples.

Notable among these patents are a few that showed exceptional promise for commercialization. Two new bio sensing systems were developed and patented by Corrèa and colleagues to detect and remove copper nanoparticles from industrial settings. These systems make use of the decomposing biomass of *Rhodotorula mucilagin*, *Hypocrea lixii*, and *Trichoderma koningiopsis* . Additionally, a patent was awarded for the purpose of detecting methane, a byproduct of biodegradation, in order to monitor organic compounds in wastewater and water fluids. This innovation used a microbe that produces methane without oxygen in a bio-electrochemical system that includes a reactor with anodes and cathodes (Silver et al. 2017). In their 2017 publication, Dooley and Burns detailed a method for monitoring wastewater concentration that involves selectively sensing the quantity of dissolved oxygen using two accurate oxygen sensors. Indeed, a very sensitive patent was obtained not long ago for the recognition of different viruses and macromolecules at very low levels. A substrate made of single-walled carbon nanotubes (SWNT) and two electrodes were used to non-covalently attach particular bioreceptors (April et al. 2018).

CONCLUSION

The health, nutrition, energy, and financial well-being of a population are profoundly impacted by water quality. Traditional methods of defining recycled water have many drawbacks, including inefficiency, high costs, and lengthy processing times, in addition to the fast expanding field of wastewater reclamation. This is associated with the inability to provide financial capacity above traditional monitoring systems and their technological constraints. Biosensors, a potential game-changer in the field of freshwater resource management, are a viable option for tracking down environmental contaminants using real-time, on-site tactics.

Predicting the physicochemical parameters, such as pH, conductivity, and turbidity, of wastewaters becomes challenging due to the diverse compositions of these samples collected from various locations and times. It is possible that the expected sample parameters and the detection of the target components might be impacted by the complicated matrix of dark waters. A further crucial consideration is the stability of the immobilized biological materials (e.g., cells, antibodies, tissue, enzymes, etc.) throughout transport, storage, and use. A stable biosensor, according to some research (Brecht 2005; Kissinger 2005), should be able to detect changes in biological variables over a period of six months when exposed to ambient conditions.

It is necessary to compare and contrast the current methods with conventional analytical systems that monitor water and wastewater. For each, it is necessary to develop standards for quantitative and qualitative validation (Cárdenas and Valcárcel, 2005). There has been a lot of recent focus on developing a reliable method for wastewater monitoring. Several pieces of evidence point to biosensors as playing a unique role in water resource management strategies. To create more effective and commercially viable biosensors, it is necessary to address certain technological and practical constraints and challenges related to this reality. New sensing technologies that can identify more chemicals at once and have a wider detection range are required to accomplish this objective of detecting a wide variety of wastewater contaminants. The biosensor must also have the ability to detect at low concentrations and remain stable for long periods of time. One approach being considered to address the significant challenges of using biomolecules as recognition elements is the creation of recombinant fragments that are specific to a target and the site-directed immobilizations of antibodies and enzymes. This would allow for the rapid, inexpensive, and in-situ screening of organic contaminants in wastewater. Alternately, biosensors may have their sensitivity and accuracy enhanced by nanomaterial integration; miniaturization and multi-sensor array devices can be used to increase detection efficiency; and so on. While the majority of the published research has focused on HM ion detection (e.g., Hg^{2+}, Pb^{2+}, Cd^{2+}, and Cu^{2+}), there is growing enthusiasm for tracking other inorganic elements as well. Developing commercial bacterial biosensors to monitor wastewater may be a challenging process. A low detection limit, strain and species selectivity independent of pre-enrichment procedures, and the capacity to identify living and dead microbes independently are all crucial. For the in-situ application, a portable device should be used to tackle these challenges. The limit of detection and standard calibration curves for the aforementioned substances also vary across various water resources. Additional research is needed to gain a better understanding of the concepts involved in measuring toxicants in wastewater, specifically for organic and heavy metal pollutants, microorganisms, and groundwater and river water. This is because numerous studies have used biosensors to measure toxicants in these environments. New possibilities for improved detection performance may arise as a result of recent developments in nanotechnology, which hold tremendous promise for enhancing certain features of biosensors (Aragay et al. 2012).

REFERENCES

Abbaspour, A., Norouz-Sarvestani, F., Noori, A., & Soltani, N. (2015). Aptamer-conjugated silver nanoparticles for electrochemical dual-aptamer-based sandwich detection of staphylococcus aureus. *Biosensors & Bioelectronics*, *68*, 149–155. doi:10.1016/j.bios.2014.12.040 PMID:25562742

Anam, M., Yousaf, S., Sharafat, I., Zafar, Z., Ayaz, K., & Ali, N. (2017). Comparing natural and artificially designed bacterial consortia as biosensing elements for rapid non-specific detection of organic pollutant through microbial fuel cell. *International Journal of Electrochemical Science*, *12*(4), 2836–2851. doi:10.20964/2017.04.49

Ansari, S. (2017). Combination of molecularly imprinted polymers and carbon nanomaterials as a versatile biosensing tool in sample analysis: Recent applications and challenges. *Trends in Analytical Chemistry*, *93*, 134–151. doi:10.1016/j.trac.2017.05.015

April, G., Yildirim, N., Lee, J., Cho, H., & Busnaina, A. (2018). *Nanotube-based biosensor for pathogen detection*. Google Patents.

Aragay, G., Pino, F., & Merkoçi, A. (2012). Nanomaterials for sensing and destroying pesticides. *Chemical Reviews*, *112*(10), 5317–5338. doi:10.1021/cr300020c PMID:22897703

Asadnia, M., Myers, M., Umana-Membreno, G. A., Sanders, T. M., Mishra, U. K., Nener, B. D., Baker, M. V., & Parish, G. (2017). Ca2+ detection utilising AlGaN/GaN transistors with ion-selective polymer membranes. *Analytica Chimica Acta*, *987*, 105–110. doi:10.1016/j.aca.2017.07.066 PMID:28916033

Bachmann, T. T., & Schmid, R. D. (1999). A disposable multielectrode biosensor for rapid simultaneous detection of the insecticides paraoxon and carbofuran at high resolution. *Analytica Chimica Acta*, *401*(1), 95–103. doi:10.1016/S0003-2670(99)00513-9

Berekaa, M.M. (2016). Nanotechnology in wastewater treatment; influence of nanomaterials on microbial systems. *International journal of current microbiolgy and applied sciences* *5*(1), 713-726.

Biswas, P., Karn, A. K., Balasubramanian, P., & Kale, P. G. (2017). Biosensor for detection of dissolved chromium in potable water: A review. *Biosensors & Bioelectronics*, *94*, 589–604. doi:10.1016/j.bios.2017.03.043 PMID:28364706

Borisova, T., Kucherenko, D., Soldatkin, O., Kucherenko, I., Pastukhov, A., Nazarova, A., Galkin, M., Borysov, A., Krisanova, N., Soldatkin, A., & El'skaya, A. (2018). An amperometric glutamate biosensor for monitoring glutamate release from brain nerve terminals and in blood plasma. *Analytica Chimica Acta*, *1022*, 113–123. doi:10.1016/j.aca.2018.03.015 PMID:29729731

Bourgeois, W., Burgess, J. E., & Stuetz, R. M. (2001). On-line monitoring of wastewater quality: A review. *Journal of Chemical Technology and Biotechnology*, *76*(4), 337–348. doi:10.1002/jctb.393

Brecht, A. (2005). Multianalyte bioanalytical devices: Scientific potential and business requirements. *Analytical and Bioanalytical Chemistry*, *381*(5), 1025–1026. doi:10.1007/s00216-004-2912-7 PMID:15726339

Cárdenas, S., & Valcárcel, M. (2005). Analytical features in qualitative analysis. *Trends in Analytical Chemistry*, *24*(6), 477–487. doi:10.1016/j.trac.2005.03.006

Chinowsky, T., Quinn, J., Bartholomew, D., Kaiser, R., & Elkind, J. (2003). Performance of the Spreeta 2000 integrated surface plasmon resonance affinity sensor. *Sensors and Actuators. B, Chemical, 91*(1-3), 266–274. doi:10.1016/S0925-4005(03)00113-8

Chong, S. S., Aziz, A., & Harun, S. W. (2013). Fibre optic sensors for selected wastewater characteristics. *Sensors (Basel), 13*(7), 8640–8668. doi:10.3390/s130708640 PMID:23881131

Dalavoy, T. S., Wernette, D. P., Gong, M., Sweedler, J. V., Lu, Y., Flachsbart, B. R., Shannon, M. A., Bohn, P. W., & Cropek, D. M. (2008). Immobilization of DNAzyme catalytic beacons on PMMA for Pb 2+ detection. *Lab on a Chip, 8*(5), 786–793. doi:10.1039/b718624j PMID:18432350

Ganjali, M. R., Eshraghi, M. H., Ghadimi, S., & Mojtaba, S. (2011). Novel chromate sensor based on MWCNTs/nanosilica/ionic liouid/Eu complex/graphite as a new nano-composite and Its application for determination of chromate ion concentration in waste water of chromium electroplating. *International Journal of Electrochemical Science, 6*(3), 739–748. doi:10.1016/S1452-3981(23)15031-0

Geng, P., Zhang, X., Meng, W., Wang, Q., Zhang, W., Jin, L., Feng, Z., & Wu, Z. (2008). Self-assembled monolayers-based immunosensor for detection of Escherichia coli using electrochemical impedance spectroscopy. *Electrochimica Acta, 53*(14), 4663–4668. doi:10.1016/j.electacta.2008.01.037

Geng, P., Zhang, X., Meng, W., Wang, Q., Zhang, W., Jin, L., Feng, Z., & Wu, Z. (2008). Self-assembled monolayers-based immunosensor for detection of Escherichia coli using electrochemical impedance spectroscopy. *Electrochimica Acta, 53*(14), 4663–4668. doi:10.1016/j.electacta.2008.01.037

Grøndahl-Rosado, R. C., Tryland, I., Myrmel, M., Aanes, K. J., & Robertson, L. J. (2014). Detection of microbial pathogens and indicators in sewage effluent and river water during the temporary interruption of a wastewater treatment plant. *Water Quality, Exposure, and Health, 6*(3), 155–159. doi:10.1007/s12403-014-0121-y

Gumpu, M. B., Sethuraman, S., Krishnan, U. M., & Rayappan, J. B. B. (2015). A review on detection of heavy metal ions in water–An electrochemical approach. *Sensors and Actuators. B, Chemical, 213*, 515–533. doi:10.1016/j.snb.2015.02.122

Güner, A., Çevik, E., Şenel, M., & Alpsoy, L. (2017). An electrochemical immunosensor for sensitive detection of Escherichia coli O157: H7 by using chitosan, MWCNT, polypyrrole with gold nanoparticles hybrid sensing platform. *Food Chemistry, 229*, 358–365. doi:10.1016/j.foodchem.2017.02.083 PMID:28372186

Hayat, A., & Marty, J. L. (2014). Aptamer based electrochemical sensors for emerging environmental pollutants. *Frontiers in Chemistry, 2*, 41. doi:10.3389/fchem.2014.00041 PMID:25019067

Henze, M., Harremoes, P., la Cour Jansen, J., & Arvin, E. (2001). *Wastewater treatment: biological and chemical processes* (3rd ed.). Springer science & business media.

Iranpour, R., Straub, B., & Jugo, T. (1997). Real time BOD monitoring for wastewater process control. *Journal of Environmental Engineering, 123*(2), 154–159. doi:10.1061/(ASCE)0733-9372(1997)123:2(154)

Jaishankar, M., Tseten, T., Anbalagan, N., Mathew, B. B., & Beeregowda, K. N. (2014). Toxicity, mechanism and health effects of some heavy metals. *Interdisciplinary Toxicology*, *7*(2), 60–72. doi:10.2478/intox-2014-0009 PMID:26109881

Jouanneau, S., Durand, M. J., & Thouand, G. (2012). Online detection of metals in environmental samples: Comparing two concepts of bioluminescent bacterial biosensors. *Environmental Science & Technology*, *46*(21), 11979–11987. doi:10.1021/es3024918 PMID:22989292

Jyoti, A., Tomar, R. S., & Shanker, R. (2016). *Nanosensors for the detection of pathogenic bacteria. Nanoscience in food and agriculture 1*. Springer.

Kissinger, P. T. (2005). Biosensors—a perspective. Biosensors and bioelectronics 20(12), 2512-2516.

Kokkinos, C., & Economou, A. (2017). Emerging trends in biosensing using stripping voltammetric detection of metal-containing nanolabels–A review. *Analytica Chimica Acta*, *961*, 12–32. doi:10.1016/j.aca.2017.01.016 PMID:28224905

Kokkinos, C., & Economou, A. (2017). Emerging trends in biosensing using stripping voltammetric detection of metal-containing nanolabels–A review. *Analytica Chimica Acta*, *961*, 12–32. doi:10.1016/j.aca.2017.01.016 PMID:28224905

Kummu, M., Guillaume, J., de Moel, H., Eisner, S., Flörke, M., Porkka, M., Siebert, S., Veldkamp, T., & Ward, P. (2016). The world's road to water scarcity: Shortage and stress in the 20th century and pathways towards sustainability. *Scientific Reports*, *6*(1), 6. doi:10.1038/srep38495 PMID:27934888

Lyngberg, O., Stemke, D., Schottel, J., & Flickinger, M. (1999). A single-use luciferase-based mercury biosensor using Escherichia coli HB101 immobilized in a latex copolymer film. *Journal of Industrial Microbiology & Biotechnology*, *23*(1), 668–676. doi:10.1038/sj.jim.2900679 PMID:10455499

Lyngberg, O., Stemke, D., Schottel, J., & Flickinger, M. (1999). A single-use luciferase-based mercury biosensor using Escherichia coli HB101 immobilized in a latex copolymer film. *Journal of Industrial Microbiology & Biotechnology*, *23*(1), 668–676. doi:10.1038/sj.jim.2900679 PMID:10455499

Michael-Kordatou, I., Michael, C., Duan, X., He, X., Dionysiou, D., Mills, M., & Fatta-Kassinos, D. (2015). Dissolved effluent organic matter: characteristics and potential implications in wastewater treatment and reuse applications. *Water Research, 77.*

Muhammad-Tahir, Z., & Alocilja, E. C. (2003). A conductometric biosensor for biosecurity. *Biosensors & Bioelectronics*, *18*(5), 813–819. doi:10.1016/S0956-5663(03)00020-4 PMID:12706596

Nguyen, T.-T., Hwang, S.-Y., Vuong, N. M., Pham, Q.-T., Nghia, N. N., Kirtland, A., & Lee, Y.-I. (2018). Preparing cuprous oxide nanomaterials by electrochemical method for non-enzymatic glucose biosensor. *Nanotechnology*, *29*(20), 205501. doi:10.1088/1361-6528/aab229 PMID:29480163

Nomngongo, P. N., Ngila, J. C., Msagati, T. A., Gumbi, B. P., & Iwuoha, E. I. (2012). Determination of selected persistent organic pollutants in wastewater from landfill leachates, using an amperometric biosensor. *Physics and Chemistry of the Earth Parts A/B/C*, *50-52*, 252–261. doi:10.1016/j.pce.2012.08.001

Olaniran, A. O., Motebejane, R. M., & Pillay, B. (2008). Bacterial biosensors for rapid and effective monitoring of biodegradation of organic pollutants in wastewater effluents. *Journal of Environmental Monitoring*, *10*(7), 889–893. doi:10.1039/b805055d PMID:18688458

Peng, G., He, Q., Lu, Y., Huang, J., & Lin, J.-M. (2017). Flow injection microfluidic device with on-line fluorescent derivatization for the determination of Cr (III) and Cr (VI) in water samples after solid phase extraction. *Analytica Chimica Acta*, *955*, 58–66. doi:10.1016/j.aca.2016.11.057 PMID:28088281

Rahman, N. A., Yusof, N. A., Maamor, N. A. M., & Noor, S. M. M. (2012). Development of electrochemical sensor for simultaneous determination of Cd (II) and Hg (II) ion by exploiting newly synthesized cyclic dipeptide. *International Journal of Electrochemical Science*, *7*(1), 186–196. doi:10.1016/S1452-3981(23)13330-X

Raja, C. E., & Selvam, G. (2011). Construction of green fluorescent protein based bacterial biosensor for heavy metal remediation. *International Journal of Environmental Science and Technology*, *8*(4), 793–798. doi:10.1007/BF03326262

Rehman, U., Vesvikar, M., Maere, T., Guo, L., Vanrolleghem, P. A., & Nopens, I. (2015). Effect of sensor location on controller performance in a wastewater treatment plant. *Water Science and Technology*, *71*(5), 700–708. doi:10.2166/wst.2014.525 PMID:25768216

Rochelet, M., Solanas, S., Betelli, L., Chantemesse, B., Vienney, F., & Hartmann, A. (2015). Rapid amperometric detection of Escherichia coli in wastewater by measuring β-D glucuronidase activity with disposable carbon sensors. *Analytica Chimica Acta*, *892*, 160–166. doi:10.1016/j.aca.2015.08.023 PMID:26388487

Saber, R., & Pişkin, E. (2003). Investigation of complexation of immobilized metallothionein with Zn (II) and Cd (II) ions using piezoelectric crystals. *Biosensors & Bioelectronics*, *18*(8), 1039–1046. doi:10.1016/S0956-5663(02)00217-8 PMID:12782467

Sayari, A., Hamoudi, S., & Yang, Y. (2005). Applications of pore-expanded mesoporous silica. 1. Removal of heavy metal cations and organic pollutants from wastewater. *Chemistry of Materials*, *17*(1), 212–216. doi:10.1021/cm048393e

Silver, M., Buck, J., & Taylor, N. (2017). *Systems and devices for treating water, wastewater and other biodegradable matter*. Google Patents.

Singh, A. K., Senapati, D., Wang, S., Griffin, J., Neely, A., Candice, P., Naylor, K. M., Varisli, B., Kalluri, J. R., & Ray, P. C. (2009). Gold nanorod based selective identification of Escherichia coli bacteria using two- photon Rayleigh scattering spectroscopy. *ACS Nano*, *3*(7), 1906–1912. doi:10.1021/nn9005494 PMID:19572619

Song, L., Mao, K., Zhou, X., & Hu, J. (2016). A novel biosensor based on Au@ Ag core–shell nanoparticles for SERS detection of arsenic (III). *Talanta*, *146*, 285–290. doi:10.1016/j.talanta.2015.08.052 PMID:26695265

Sudibya, H. G., He, Q., Zhang, H., & Chen, P. (2011). Electrical detection of metal ions using field-effect transistors based on micropatterned reduced graphene oxide films. *ACS Nano*, *5*(3), 1990–1994. doi:10.1021/nn103043v PMID:21338084

Tagad, C. K., Kulkarni, A., Aiyer, R., Patil, D., & Sabharwal, S. G. (2016). A miniaturized optical biosensor for the detection of Hg2+ based on acid phosphatase inhibition. *Optik (Stuttgart)*, *127*(20), 8807–8811. doi:10.1016/j.ijleo.2016.06.123

Tagad, C. K., Kulkarni, A., Aiyer, R., Patil, D., & Sabharwal, S. G. (2016). A miniaturized optical biosensor for the detection of Hg2+ based on acid phosphatase inhibition. *Optik (Stuttgart)*, *127*(20), 8807–8811. doi:10.1016/j.ijleo.2016.06.123

Tao, X. (2017). *A double-microbial fuel cell heavy metals toxicity sensor*. 2nd International conference on environmental science and energy engineering (ICESEE), Beijing, China.

Tripp, R. A., Dluhy, R. A., & Zhao, Y. (2008). Novel nanostructures for SERS biosensing. *Nano Today*, *3*(3), 31–37. doi:10.1016/S1748-0132(08)70042-2

Tschmelak, J., Proll, G., & Gauglitz, G. (2005). Optical biosensor for pharmaceuticals, antibiotics, hormones, endocrine disrupting chemicals and pesticides in water: Assay optimization process for estrone as example. *Talanta*, *65*(2), 313–323. doi:10.1016/j.talanta.2004.07.011 PMID:18969801

Tsopela, A., Laborde, A., Salvagnac, L., Ventalon, V., Bedel-Pereira, E., Séguy, I., Temple-Boyer, P., Juneau, P., Izquierdo, R., & Launay, J. (2016). Development of a lab-on-chip electrochemical biosensor for water quality analysis based on microalgal photosynthesis. *Biosensors & Bioelectronics*, *79*, 568–573. doi:10.1016/j.bios.2015.12.050 PMID:26749098

Verma, N., Sharma, R., & Kumar, S. (2016). Advancement towards microfluidic approach to develop economical disposable optical biosensor for lead detection.

Wacheux, H. (1998). *Sensors for waste water: Many needs but financial and technical limitations. Monitoring of water quality*. Elsevier. doi:10.1016/B978-008043340-0/50018-7

Wauchope, R. (1978). The pesticide content of surface water draining from agricultural fields—A review. *Journal of Environmental Quality*, *7*(4), 459–472. doi:10.2134/jeq1978.00472425000700040001x

Weber, R., Watson, A., Forter, M., & Oliaei, F. (2011). Persistent organic pollutants and landfills-a review of past experiences and future challenges. *Waste Management & Research*, *29*(1), 107–121. doi:10.1177/0734242X10390730 PMID:21224404

Yamashita, T., Ookawa, N., Ishida, M., Kanamori, H., Sasaki, H., Katayose, Y., & Yokoyama, H. (2016). A novel open-type biosensor for the in-situ monitoring of biochemical oxygen demand in an aerobic environment. *Scientific Reports*, *6*(1), 38552. doi:10.1038/srep38552 PMID:27917947

Yang, L., Han, D. H., Lee, B.-M., & Hur, J. (2015). Characterizing treated wastewaters of different industries using clustered fluorescence EEM–PARAFAC and FT-IR spectroscopy: Implications for downstream impact and source identification. *Chemosphere*, *127*, 222–228. doi:10.1016/j.chemosphere.2015.02.028 PMID:25746920

Yao, N., Wang, J., Zhou, Y., (2014). Rapid determination of the chemical oxygen demand of water using a thermal biosensor. *Sensors (Basel)*, *14*(6), 9949–9960.

Yemini, M., Levi, Y., Yagil, E., & Rishpon, J. (2007). Specific electrochemical phage sensing for Bacillus cereus and Mycobacterium smegmatis. *Bioelectrochemistry (Amsterdam, Netherlands)*, *70*(1), 180–184. doi:10.1016/j.bioelechem.2006.03.014 PMID:16725377

Yu, D., Zhai, J., Liu, C., Zhang, X., Bai, L., Wang, Y., & Dong, S. (2017). Small microbial three-electrode cell based biosensor for online detection of acute water toxicity. *ACS Sensors*, *2*(11), 1637–1643.

Zehani, N., Fortgang, P., Lachgar, M. S., Baraket, A., Arab, M., Dzyadevych, S. V., & Kherrat, R. (2015). Highly sensitive electrochemical biosensor for bisphenol A detection based on a diazonium-functionalized boron-doped diamond electrode modified with a multi-walled carbon nanotube-tyrosinase hybrid film. *Biosensors & Bioelectronics*, *74*, 830–835. doi:10.1016/j.bios.2015.07.051 PMID:26232678

Zhan, S., Xu, H., Zhang, D., Xia, B., Zhan, X., Wang, L., Lv, J., & Zhou, P. (2015). Fluorescent detection of Hg2+ and Pb2+ using GeneFinder™ and an integrated functional nucleic acid. *Biosensors & Bioelectronics*, *72*, 95–99. doi:10.1016/j.bios.2015.04.021 PMID:25966463

Zhou, T., Han, H., Liu, P., Xiong, J., Tian, F., & Li, X. (2017). Microbial fuels cell-based biosensor for toxicity detection: A review. *Sensors (Basel)*, *17*(10), 2230. doi:10.3390/s17102230 PMID:28956857

Chapter 10
Sensors in the Oceans, Pollutant Identifications, and Controversies:
Radio-Isotopic Tracing of Pollutants in Marine Ecosystems and Controveries

Sanchari Biswas
https://orcid.org/0000-0002-1159-5826
Amity University, Kolkata, India

ABSTRACT

Oceans are the largest means of survival for millions of people and also the source of many life forms. Human activities have made the environmental conditions in marine habitats more dire for the last fifty years. The discharge of agricultural nutrients, heavy metals, and persistent organic pollutants (plastics, pesticides) threaten the coastal zones. Chemical compounds containing one or more radioisotope atoms are known as radiotracers, which are particularly useful for identifying and analysing pollutants as they can readily identify trace amounts of a particular radioisotope and short-lived isotope decays. It is thus important to identify such sources of contaminants by quantifying essential pollutants separately and gathering dependable information regarding their origin, movement, and ultimate destination. Nuclear and isotope techniques help in gathering such data. This book chapter gives an overview of the modern techniques available for probing the various contaminants across marine ecosystems and several drawbacks and controversies associated with the same.

INTRODUCTION

Primarily the source of survival for hundreds of millions of people, Oceans, are the greatest natural resource on Earth. It is also the source of most life forms. Unplanned coastal zone settlement and rapid industrialization were linked to ocean pollution, which was a major issue of the 20th century. The 21st

DOI: 10.4018/979-8-3693-1930-7.ch010

century still sees it as a problem. In coastal zones, heavy metals, persistent organic pollutants including pesticides, hazardous algal toxins, and agricultural runoff of fertilisers pose the most environmental threats. A constant threat to birds, marine life, and beaches comes from oil spills from ships and tankers (Betti et al., 2011).

Marine pollution is a transboundary problem that is frequently brought on by economic activity that occurs inland and alters the marine environment's ecology. Examples include sewage discharges into lakes and rivers that encourage the occurrence of harmful algal blooms (HABs), the use of chemicals in agriculture, air emissions from factories and cars, and numerous other phenomena occurring hundreds or thousands of kilometres away from the coast. The ecosystem of estuaries, bays, coastal waters, and even entire seas is impacted by these operations sooner or later, which affects the maritime industry's economy. Furthermore, a lot of pesticides that are outlawed in the majority of developed nations are still in use today, and their transboundary dispersion pathway has an impact on marine ecosystems all over the world. Aside from these issues, permitted nuclear plant that release off radioactive materials into rivers and coastal areas have contaminated the marine environment and have the potential to do so in the future. This has resulted in the ocean dumping of radioactive waste. To guarantee the sustainable use of marine ecosystems, it is imperative to quantify contaminants separately and gather dependable information regarding their origin, concentrations, movements, and ultimate destination. Isotope analysis offers a special source of data for tracking the movements of both nuclear and nonnuclear pollutants in the environment and determining their biological consequences (Betti et al., 2011).

For more than 150 years, radioisotopes have been employed in earth and environmental studies. They offer special instruments for the in-depth study of environmental processes at all scales, from the level of individual cells to that of oceanic basins. Through lab and field research, these nuclear approaches have been used to better understand coastal and marine ecosystems and how aquatic species react to various environmental stresses such as temperature, pH, nutrients, metals, organic anthropogenic pollutants, and biological toxins. Global marine problems that affect marine environments and impose different threats to the environment and economy include ocean warming, deoxygenation, plastic pollution, acidification of the ocean, prolonged and intense harmful algal blooms (HABs), and coastal contamination.

Isotopes and nuclear methods can be effectively used to analyse the various sources of pollution and the related diffusion mechanisms.

The ability to consistently evaluate the state of marine and coastal ecosystems and their potential responses to future disturbances might yield crucial insights for society in managing their marine environments sustainably. This book chapter contains information on emerging concerns that would benefit from present and novel radiotracer technologies, a summary of the historical use of radiotracers in these systems and an explanation of how current techniques of radioecological tracing can be modified for specific contemporary environmental issues. Additionally, the potential and problems associated with using radioecological tracers are discussed, along with ways to make the most of their use and significantly improve environmental managers' capacity to manage coastal and marine ecosystems using evidence.

The health and sustainability of coastal and marine ecosystems are threatened by a number of environmental issues that are caused by or made worse by a multitude of anthropogenic stressors that are made worse by a changing ocean and climate (Dwight et al., 2005). The marine habitats are being impacted by global marine challenges, which pose a variety of environmental and economic threats (Speers et al., 2016; Yagi, 2016; Creswell et al., 2020). These issues include ocean warming, deoxygenation, plastic pollution, ocean acidification, increased duration and severity of toxic harmful algal blooms (HABs), and coastal contamination (Beaumont et al., 2019). To manage their maritime environments sustainably, society can benefit greatly from having access to reliable knowledge about the state of coastal and marine ecosystems and how they can react to upcoming disturbances. The application of a variety of radioactive isotope tracers or radiotracers, in field settings (Fowler, 2011; Harmelin-Vivien et al., 2012) and controlled lab experiments (Stewart et al., 2008; Metian et al., 2019) has aided in learning about the impacts of the same in coastal marine environments.

Radioisotope analysis is a swift, cost-effective method of element analysis that allows for element experiments at trace quantities, far lower than those found in most natural water. Thus, this approach has proven helpful in determining the following characteristics of pollutants: bioavailability (the percentage of the total exposure that can be absorbed by biota); bioaccumulation (contaminant uptake from food and water into an organism); bioconcentration (the amount of contaminant present in the system of living organism contrasting with ambient water); and, primarily for organometallic compounds, biomagnification (the process through contaminant concentration levels in the tissues of a predator exceeds compared to its prey).

New insights into the rates and routes of absorption (such as bioaccumulation) and biomagnification of both radioactive and non-radioactive pollutants have been made possible by nuclear techniques. Similar to this, these methods have been applied to the quick identification and measurement of biochemical toxins in seafood, the evaluation of metabolic processes in the presence of rising ocean temperatures, and the assessment of the effects of persistent ocean acidification on a variety of calcifying organisms. Likewise, the reconstruction of Earth's geochemical evolution and the historical reconstruction of important environmental processes and rates have been made possible by isotopes.

This book chapter discusses determining the benefits of utilizing radioisotopes to address both established and emergent environmental processes in marine ecosystems. This book chapter has been divided into four parts for a constructive understanding of the following parts:

- an overview of the past applications and advantages of radiotracers used in marine and coastal environments.
- a section outlining the evolution viz. current radiotracer techniques
- a section listing novel, frequently interdisciplinary instruments and methods for addressing current maritime and coastal problems that could be helped by using radiotracers to evaluate the state of ecosystems
- a segment outlining the prospects and problems in this industry moving ahead

BRIEF HISTORY OF RADIO-ISOTOPIC TRACING OF MARINE POLLUTANTS

One of the most important factors in assessing the prevalence and seriousness of contaminants in the marine environment is identifying the sources of the contamination. The effects of human activities have worsened the environmental conditions in marine ecosystems during the last fifty years. In essence, a variety of waste and discharge from local, sub-regional, and regional enterprises and activities mix with global ocean currents to produce a worldwide dispersion of toxins. Nations need to implement environmental regulations that balance environmental protection on a local and global level with socio-economic development to stop these phenomena. However, the only way to effectively regulate the environment is to establish a clear connection between known processes or sources and the diffusion of contaminants. Analyses using stable carbon isotopes can be used to locate the sources of organic contaminants. A composite series of events culminates in a contaminant's stable isotopic composition in the environment. It is anticipated that chemicals made from various sources using essentially different techniques will have unique isotopic compositions that can be utilised to pinpoint the sources.

Knowledge of natural processes has improved as a result of the investigation and application of radioisotopes in environmental science. Using Pb isotope systematics (Patterson et al., 1955; Creswell et al., 2020) a timeline of Earth's differentiation and planetary evolution was developed. Over the last 60 years, as analytical techniques have been established and greatly improved, the utility of radioisotopes has continuously changed. Applications of radioisotopes have ranged from using products of legacy nuclear weapons testing to derive sedimentary age models to generating a pedagogy of evolution of the Earth using Pb isotope systematics. The Suess effect—which relies on radiocarbon (^{14}C) dendrochronology—became the main argument in favor of human-caused global warming and contributed to the measurement of CO_2 exchange rates between the ocean and atmosphere (Revelle and Suess, 1957; Swart et al., 2010). In the marine sciences, radionuclides—which can be man-made or natural—have been utilized to track numerous processes. Radionuclides are powerful timepieces because of their ability to be used as a clock to trace the rate of various processes due to their rate of radioactive decay, or loss. The development of nuclear energy for civil purposes and atmospheric nuclear weapons testing from 1945 to 1980 (peaking in 1963) had released several manmade radionuclides into the marine environment, including ^{3}H, ^{238}Pu, ^{137}Cs, ^{239}Pu, ^{90}Sr, ^{129} I, ^{240}Pu (Creswell et al., 2020).

The traditional three U-Th series radioactive decay chains were used more frequently to study marine processes almost simultaneously, as a result of advancements in analytical chemistry and technology (Benitez-Nelson et al., 2018). One of the first initiatives to use radionuclides to examine basin-scale processes was the GEOSECS (Geochemical Ocean Sections) program (Broecker and Peng, 1982). Since the early 1970s, recent silt has been dated and sedimentation rates have been calculated using natural and artificial fallout radionuclides (Appleby, 2008; Creswell et al., 2020). Sediment partition coefficients (Kds) and bioconcentration factors (BCFs) of long-lived radionuclides in marine sediments and organisms were compiled as a result of the many studies that started to characterize the sorption of radionuclides to marine sediment and the bioaccumulation of radionuclides in aquatic biota in the 1960s and 1970s (IAEA, 2004). In modelling efforts to assess the cycling and possible effects of radionuclides in marine ecosystems, these Kds and BCFs have been employed (ICRP, 2008; Creswell et al., 2020). Additionally, several research studies investigated the use of man-made and natural radionuclides to assess processes like sediment geochronology (Luoma and Rainbow, 2005; Baumann and Fisher, 2011), sediment plume dynamics, carbon flux, GEOTRACES/GEOSECS/JGOFS programs, ground water discharges and water mass ventilation (Creswell et al., 2020).

Additionally, stable water isotopes like δ^{18O} can be coupled with radiotracers like ^{222}Rn and ^{226}Ra to differentiate between fresh and saline groundwater discharge, as well as source terms of submarine groundwater discharge. This knowledge may prove important to groundwater resource managers responsible for generating potable water budgets and marine resource managers interested in providing contaminants and nutrients carried by SGD to coastal ecosystems (Rocha et al., 2016).

CURRENT TECHNIQUES OF RADIO-ISOTOPIC TRACING IN MARINE POLLUTION

The development of compound-specific stable isotope analysis (CSIA) is the most significant in analytical chemistry in recent years. This method, which measures specifically carbon isotopic content of individual chemicals within a complex mixture, is based on gas chromatography/isotope ratio mass spectrometry (GC/IRMS). Natural pollutants include chlorinated solvents, refined hydrocarbon elements, crude oils, PCBs (polychlorinated biphenyls), PAHs (polycyclic aromatic hydrocarbons and crude oils can all be individually identified by carbon-semi conductance analysis (CSIA).

The nitrogen and hydrogen isotopic composition of certain substances can be determined using GC/IRMS. A class of pollutants known as hydrocarbons and oil products have a complex and varied makeup. They affect living things in many ways, causing anything from physical and physicochemical harm to cancerous consequences. Based on statistics from 1988-1997, it is projected that 1,245,200 tonnes of oil enter the maritime environment annually from ships and other sea-based operations.

A study published in Science Advances and detailed in an IAEA stated that only approximately seventeen per cent of the plastic manufactured between 1950 and 2015 has been recycled and is still in use. Roughly twelve per cent have been burned, and the remaining sixty per cent have been dumped in landfills where they could potentially contaminate rivers, groundwater, and the ocean. The IAEA estimates that as 2025 proceeds, the chance is that one metric tonne of plastic in the oceans for every three metric tonnes of fish, might exist, and by 2050, there might be more plastic than fish. This prediction is based on current trends. The establishment of NUTEC Plastics aimed to support nations in incorporating isotope and nuclear methods to combat plastic pollution. It takes two approaches: (i) to present data based on science to evaluate and characterise marine microplastic contamination; and (ii) to show how ionising radiation can be used to convert plastic trash into useable resources. The IAEA claims that NUTEC Plastics will improve labs' capacity to investigate the effects of plastic pollution in marine and coastal ecosystems by using nuclear techniques to precisely track and measure the movement and effects of co-contaminants and microplastics (Joanne Liou, 2021).

AN OUTLINE OF OTHER AVAILABLE TECHNIQUES FOR DETERMINING CONTAMINANTS IN ENVIRONMENTAL SAMPLES

Table 1. Some commonly used techniques in the last decade for determining Radionuclide stressors

S.No	Detection Technique	Range of Applications	Author
1	Accelerator Mass Spectrometry- AMS, Inductively Coupled Plasma-Mass Spectrometry-ICP-MS, Resonance Ionization Mass Spectrometry-RIMS, Secondary Ion Mass Spectrometry-SIMS, Thermal Ionization Mass Spectrometry-TIMS	Radionuclide (^{129}I)	Povinec (2017)
2	Accelerator Mass Spectrometry- AMS	$^{240}Pu/^{239}Pu/^{241}Pu$	Garcia-Leon, M (2018)
3	ICP-MS coupling with linear quadrupole, TOF (Time of Flight), FTICR (Fourier transform ion cyclotron resonance)	$^{135}Cs/^{137}Cs$ ratios	Russell et al., (2015)
4	Combined with traditional radioanalytical techniques and mass spectrometric measurement techniques	^{99}Tc	Shi et al., (2012)

STRESSORS AND MULTIPLE CONTAMINANTS THAT INFLUENCE THE BIOAVAILABILITY OF METALS IN OCEANS

Bioavailability of Metals

There are two implications of "bioavailability" in the ecotoxicology of metals which can be categorized primarily as environmental and secondarily as toxicological. The metal that is accessible for absorption by living forms and incorporated into its metabolic processes is referred to as bioavailability in the environment. The percentage of a metal's concentration which thus gets absorbed and/or adsorbed (bodily) is known as toxicological bioavailability. When the absorbed fraction interacts with physiological sites and receptors essential to the body's metabolism, harmful effects follow (Rainbow and Luoma 2011).

Metal exposure in aquatic animals can happen through consumption or by direct absorption via the water. Feeding is the main way that predators, such as some fish, amphibian, and invertebrate species, and animals that scavenge, are exposed to and accumulate metals. The concentration of metals deposited in the tissues of exposed animals is often not connected with harmful effects, even though many research have sought to determine the toxicokinetics of metals. Metals can either permanently or temporarily chelate with different ligands inside the body, which prevents them from being accessible to disrupt metabolism (Wang 2013).

It has long been known that the physical and chemical speciation of metal pollutants can affect how bioavailable they are to aquatic species, such as animals and plants like phytoplankton. In relation to metals dissolved in freshwater and saltwater, this has been thoroughly researched (Paiva Magalhães et al., 2015). The assessment of metal contaminants' bioavailability in sediments has proven to be difficult, especially when it comes to the impact of metal speciation in sediments on benthic animal bioavailability. Even though the bioaccumulation of metals from sediments has been the subject of countless research (Wang et al., 1999). The speciation of metals, especially those associated to iron and manganese oxides, may vary seasonally due to the eutrophication of sediments, especially those found in coastal sediments. This might impact the bioavailability of these metals for organisms that live on the bottom.

As a result, estimating their harmful effects in the environment or bioaccumulated levels stands challenging. The use of biomonitoring instruments, such as biomarkers and bioindicators, has proven to be a more uncomplicated and reliable methodology in detecting environmental conditions, owing to the intricacy of generating toxicokinetic patterns.

The speciation of metals, particularly those associated with iron and manganese oxides, may vary annually due to the seasonal changes in redox conditions in sediments brought on by eutrophication, especially in coastal sediments. This variation can therefore impact the bioavailability of these metals for organisms that live on the bottom.

The Fukushima-Daiichi nuclear power plant accident raised the bioavailability of ^{137}Cs from contaminated soils off the coast of Japan. As a result, benthic fish in Japanese coastal waters have higher amounts of this pollutant than pelagic fish (Buesseler et al., 2017; Creswell et al., 2020). Although the speciation of sediment-bound Cs in those areas has not been thoroughly investigated, data indicates that Cs can accumulate in benthic food chains by way of assimilation in deposit-feeding worms, which then transfer Cs to fish or macroinvertebrates from contaminated sediments (Wang et al., 2016). The bioaccumulation of metals and radionuclides may also be impacted by physical changes in the world's oceans, such as the temperature and pH of the saltwater. It is yet unknown

Figure 1. The pathway and fate of contaminants entering the coastal ecosystem network

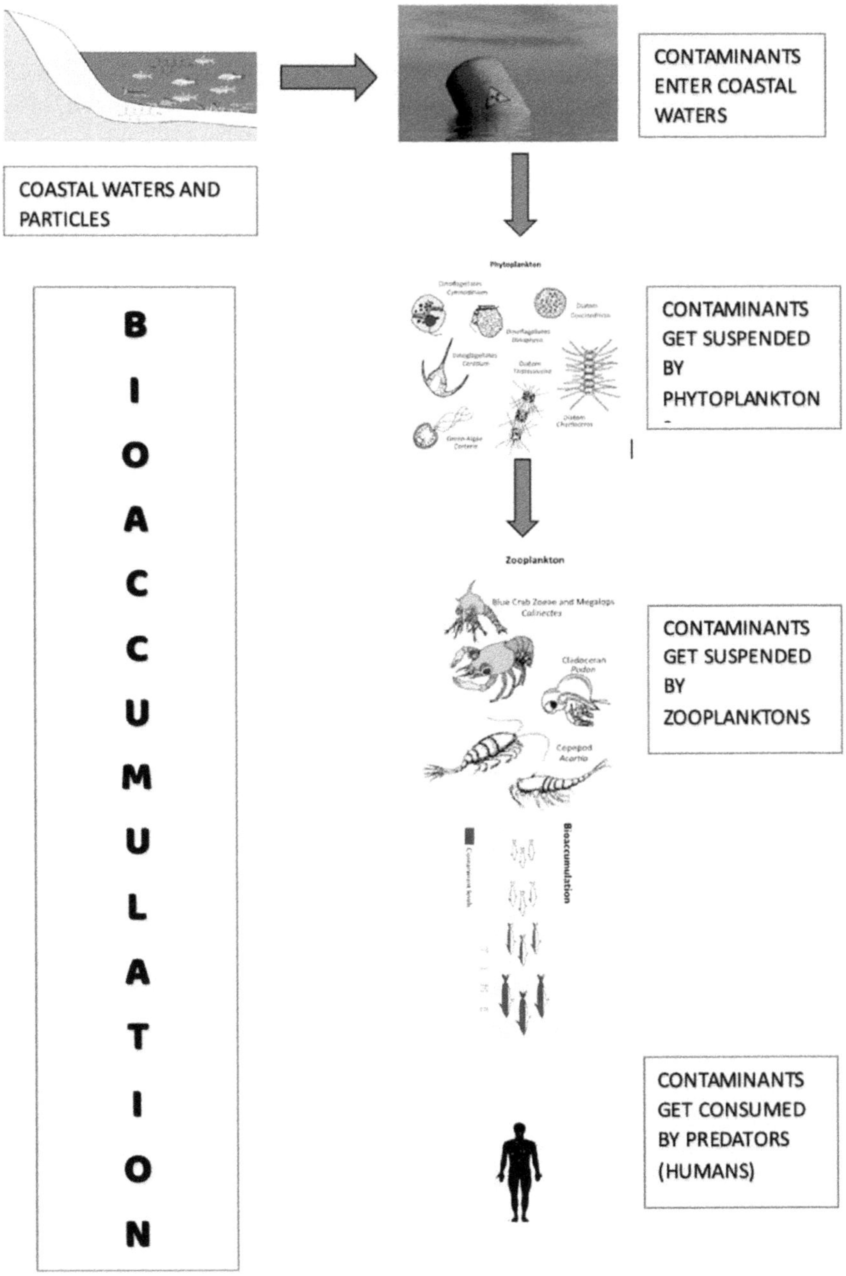

how much these physical characteristics might affect the bioavailability of contaminants. Although field conditions are difficult to duplicate in controlled laboratory sorption experiments, radiotracers provide a quick way to comprehend the parameters controlling contaminant partitioning between the phases of sediment, solution, and suspended particulate matter, which can help with field data interpretation (Payne, et al., 2004).

Hydrocarbons

A group of pollutants known as hydrocarbons and oil products have a complex and varied makeup. They affect living things in different ways, ranging from physicochemical and physical harm to carcinogenic consequences. Based on statistics from 1988-1997, it is projected that 1,245,200 tons of oil enter the maritime environment annually from ships and other sea-based operations (GESAMP, 2007). Most molecular fingerprint recognition was used in source apportionment studies of hydrocarbons in the environment until recently. Nonetheless, because of preferential compound losses or degradation, hydrocarbons being impacted by a series of processes in the marine environment viz. photooxidation, dissolution, biodegradation and evaporation, may change the initial hydrocarbon molecular profiles, raising the possibility of ambiguity when using molecular profiles in source identification. Numerous isotopic signals are produced during petroleum genesis, and these signals generally deviate greatly from the isotopic compositions of clean marine environments. It is possible to categorize crude oils, petroleum products, and tars using the distinctive carbon-13/carbon-12 ratios produced by the intricate isotopic fractionation patterns brought about by physical and biological processes.

Other stable isotopes have several forensic applications, but bulk carbon isotopes have received the majority of attention regarding source and correlation work. Isotopic abundances of sulfur, nitrogen, and deuterium also provide source and geological formation histories, including ratios unique to individual oil fields. It is possible to "fingerprint" oils spilled into the environment to identify the source or sources by using the distinctive isotopic ratios of these elements. The compound-specific isotope analysis of various elements within a typical and specific oil constitutes a distinctive indication of its origin and maturity, even though bulk measurements offer valuable information.

Forensic identification of gasoline and crude oil spills has been attempted using GC-IRMS in conjunction with currently available biomarker technologies. By comparing the isotopic makeup of specific hydrocarbons, it is possible, for instance, to establish a correlation between hydrocarbons spilled in aquatic habitats and their presumed source or sources. It is easy to spot differences in the isotopic makeup of specific compounds within a sample of gasoline or crude oil, as these differences correspond to the various sources of the chemicals.

In this context, weathering can lead to a notable loss of volatile hydrocarbons; nevertheless, non-volatile and semi-volatile molecules' $\delta^{13}C$ values remain unchanged and their isotopic profiles can be utilized to pinpoint and track the origin of an oil spill.

Crude oil hydrocarbons exhibit a wide compositional variety of hydrogen isotope ratios and are conserved during biodegradation (aerobic). Measurement of hydrogen isotope ratios (specifically in petroleum hydrocarbons) additionally, is a potent tool to determine the source of the pollutant. In the future, tracking the origins of pollution and keeping an eye on biodegradation should take into account data from both hydrogen and carbon isotopes.

Aside from stable isotopes, measurements of radiocarbon (carbon-14) may also reveal information on oil contamination. While organic matter derived from petroleum is free of carbon-14, all marine organic matter is labelled with carbon-14, which is created naturally in the atmosphere by nuclear weapons explosions as well as cosmic radiation. Therefore, a relevant quantitative indicator of the petroleum carbon contribution to the total marine organic matter occurs as a result of the absence of carbon-14 signal (in contaminants) that originates from fossil fuel.

Polycyclic Aromatic Hydrocarbons (PAHs)

The main ways that organic pollutants known as polycyclic aromatic hydrocarbons (PAHs) enter the marine environment are through oil spills and atmospheric deposition. They can originate from incomplete combustion of organic matter (such as coal, oil, or wood) or be a substantial component of crude oil (petrogenic origin). They may also have a diagnostic origin (biological and physiochemical transformation) from organic material that happens in sediments after deposition. Since some PAHs have the potential to cause cancer and mutagenesis, understanding their sources is crucial because of their eco-toxicological nature, which poses long-term health risks to nearshore marine systems and has an impact on social and commercial uses of marine resources as well as ecological processes and public health. When combined with molecular fingerprint analysis, the molecular stable carbon isotopic composition of PAHs by GC/IRMS is a potent method for researching the origins and environmental destiny of hydrocarbons in the contemporary environment.

For example, the isotope signals of the PAH products can be used to trace air pollutants from combustion source materials, such as soot from burning coal, natural gas, biomass, and vehicle exhausts, back to the source materials (Macko, 1994; Ballentine, 1996; Betti et al., 2011).

Persistent Organic Pollutants (POPs)

The majority of halogenated organic compounds fall under the category of persistent organic pollutants, which are known to bioaccumulate and have harmful and mutagenic consequences when they move up the food chain. To create baseline data for the identification of these anthropogenic contaminants in the future, a few studies have reported the carbon-13 compound specific isotope analysis (CSIA) of PCNs (polychlorinated naphthalene mixtures) and PCBs (some commercial polychlorinated biphenyls), such as Phenoclors, Aroclors and Kanechlors. Some bi-pyrrolic halogenated organic compounds have also been shown in recent investigations to be present throughout the world and building up in marine food webs (Powell, 1999: Reddy et al., 2004). Determining whether these substances are natural products or the result of industrial synthesis has proven to be challenging thus far.

A tracer used to identify between natural and manmade chemicals is called radiocarbon, or carbon-14. Since all recent natural products have modern or contemporary quantities of carbon-14, radiocarbon analysis can be utilized in this context to determine the origin of halogenated organic molecules. Conversely, there is no detectable carbon-14 in synthetic compounds made from petrochemical processes.

STABLE ISOTOPES FOR BIOREMEDIATION RESEARCH

One of the most significant approaches to lessen the harm caused by hydrocarbon spills and other marine oil spills is in situ bioremediation. On the other hand, precise comprehension and measurement of biotransformation processes are essential. When engineered remediation is used, quantifying intrinsic biodegradation may help lower the cost of site remediation. The natural abundance of stable isotopes of carbon, hydrogen, and oxygen—the elements necessary for the biodegradation processes—is used to track the occurrence of in situ biodegradation as well as the pathways, rates, and extent of biodegradation of fuel or chlorinated hydrocarbons. The natural abundance of stable isotopes of carbon, hydrogen, and oxygen—the elements necessary for the biodegradation processes—is used to track the occurrence of *in situ* biodegradation as well as the pathways, rates, and extent of biodegradation of fuel or chlorinated hydrocarbons.

Analysis of the isotopic compositions of (a) the degradation products, (b) the residual fractions of the contaminant, and (c) the dissolved inorganic carbon of the water can be used to monitor *in situ* biotransformation using stable isotopes because isotopic fractionation leads to a preferable degradation of chemical bonds with lighter compared to heavier isotopes (Sturchio et al., 2009). The characteristic carbon-13 isotope ratios of carbon dioxide are caused by the oxidation of organic materials, the solubility of inorganic carbon, or the breakdown of contaminating hydrocarbons. Their carbon-13 levels are valuable instruments for evaluating in situ biodegradation in intricate settings. Moreover, microbial processes are the main factors influencing the oxygen isotopic compositions of molecular oxygen, nitrate, and sulphate in complex systems; isotopic fractionation during microbial respiration results in a notable alteration in the δ^{18O} of the leftover molecules. Carbon dioxide comes from living things The isotopic compositions of CO^2 and O^2 work together to estimate microbial respiration rates and to differentiate between aerobic and anaerobic CO^2 generation.

MONITORING BIO-MAGNIFICATION OF POLLUTANTS USING STABLE ISOTOPES

Because they are hydrophobic, a large number of persistent organic pollutants (POPs) found in aquatic environments tend to accumulate in aquatic species. These compounds' pervasiveness and persistence have been connected to several negative environmental impacts, such as water, sediment, and aquatic food chain pollution. Exposure to these substances can cause a variety of health impacts, from immune system or nervous system issues to an increased risk of some cancers. Stable isotope ratio measurement of bio-elements, such as carbon and nitrogen, has made it easier to examine the biomagnification5 profiles of POPs, including PCBs, PAHs, Organotins, and trace elements across aquatic food webs within the past 20 years. In a food chain, the stable nitrogen isotope ratio ($\delta^{15}N$) typically rises by 3–4% for each trophic level. Therefore, the value of $\delta^{15}N$ can be used to determine each organism's trophic position within a food web. A food web's principal carbon sources can be identified using the stable carbon isotope ratio, $\delta^{13}C$, which is slightly enriched by around 1% at every trophic level.

The Application of Radioisotopes in Harmful Algae Blooms (HABS) to Identify Toxins

The production of toxins by some algae species, which can build up in seafood items (mostly fish and shellfish) and endanger human health, is one of the most concerning effects of HABs. Although it is unknown when hazardous algal blooms (HABs) begin and produce biotoxins, certain environmental factors (such as temperature, light, and nutrients) are probably necessary for the blooms to occur. Trace metals affect phytoplankton not only by toxicity and nutritional limitation, but also by increasing the production of toxins by HABs due to their bioavailability (Sunda, 2006; Creswell et al., 2020). The presence of pseudonitzschia and the neurotoxic domoic acid after iron fertilization studies or with the addition of copper are the most thoroughly researched (Trick et al., 2010; Creswell et al., 2020). Bloom dynamics may also be influenced by interactions between algae and bacteria, with siderophore-producing bacteria possibly providing iron to HAB species (Yarimizu et al., 2018).

Figure 2. Global patterns showing coastal phytoplankton and harmful (algal) blooms HABs across the world between 2003 and 2020
(Dai et al, 2023)

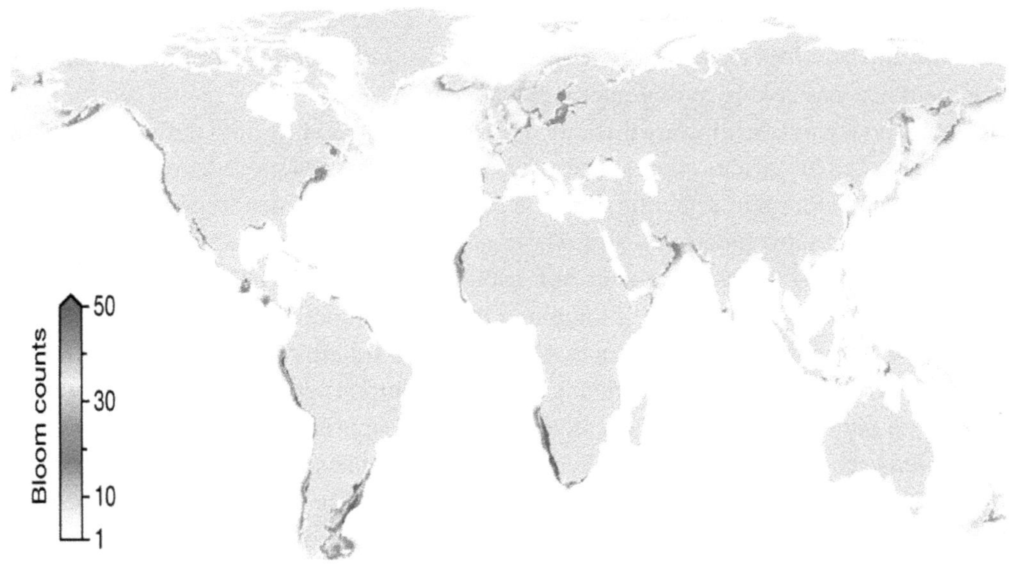

Figure 3. Global patterns showing Harmful Algal Information System (HAIS),
(https://data.hais.ioc-unesco.org)

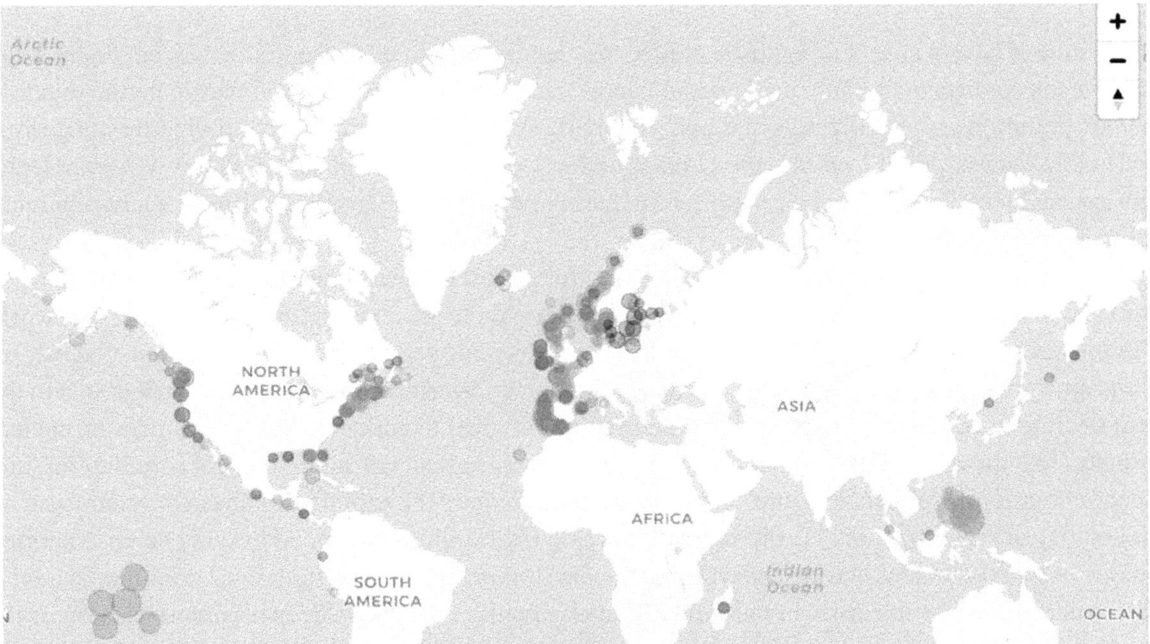

Trace elements including Li, Se, and Ni have been connected to HAB species in other research (Kudela et al., 2010). Using sensitive radiotracers, such as ^{59}Fe, allows tests with ecologically relevant amounts of compounds, potentially involved in biotoxin synthesis, initiating HAB and its physiology, given the possible connections between different trace metals and HAB toxin production. The exact identification of trace elements and possible localization of such elements within cells or in the surrounding phycosphere would be made possible by radiotracer techniques.

Toxin detection and quantification are difficult due to the heterogeneity both within and between toxin classes. Toxins produced by various algae species vary in type; Fig. III-4 provides examples. Many methods have been developed for toxin detection; they are generally divided into three categories: quantitative instrumental analysis, in vitro bioassays, and whole animal (in vivo) bioassays. The receptor binding assay is a helpful analytical method that uses radioisotopes (RBA). The extremely specific interaction between a toxin and a biological receptor is known as the toxin's function or medicinal activity, and this is the basis of the RBA approach. For instance, saxitoxins (STXs) are poisonous because they attach to and block sodium channels in specific human tissue types, impairing muscle function and ultimately resulting in death or paralysis. When doing a receptor binding experiment for STXs, the sodium channel makes the most sense choice of the receptor. Toxins that impact the sodium channel, such as PSP, NSP, and CFP toxins, have given rise to RBAs (Vandola et al., 1994; Betti et al., 2011). A liquid scintillator counter then measures how well the toxin molecules containing radionuclides, like tritium (^{3}H), attach to the receptor sites.

CONCLUSION

Isotope studies have a special diagnostic ability that can be used to understand the dangers being faced at the marine environment. Only nuclear and isotope techniques can be used to investigate the majority of major pollution issues, affecting the marine environment. These techniques provide the diagnostic and flexible data required to probe the contamination's origins, its originating history, environmental pathways, and its effect on the ecosystems. To make cost-effective mitigation decisions, such techniques need to be incorporated.

Although radioecological tracers have been used for almost 60 years, there is a serious risk that these methods will become obsolete at one point in time because there are a meagre number of scientists working in this area. This is caused in part by the stringent safety and regulatory criteria that must be met for the transportation, storage, use, and disposal of radioactive materials. It is also partly attributable to the recent decline in training in radiochemical and radiobiological techniques. The assessment of nuclear accidents (like the one at Fukushima), the siting and possible effects of future nuclear installations, and the decommissioning of outdated nuclear power plants all require the expertise of an era of scientists with radioecological specialization. Furthermore, to develop fresh approaches for addressing newly emerging problems with coastal and marine ecosystems, the environmental science field must continue to advise and instruct upcoming scientists in the field of radioecological tracers. A deeper comprehension of the state of coastal and marine ecosystems can be attained by fusing traditional environmental biological and chemical approaches with radioecological tracing techniques. When combined with newly emerging "omic" techniques (detecting genes, proteins, to mRNA), the advancement of radioecological tracer technologies should prioritize tackling specific environmental challenges, such as alterations in animal physiology in response to anthropogenic stressors and changing climate conditions.

ACKNOWLEDGEMENT

The author is thankful to the Department of Environment Sciences, AMITY Institute of Environmental Science Kolkata (AMITY University, Kolkata) for providing basic infrastructure, library, and technical support.

REFERENCES

Appleby, P. G. (2008). Three decades of dating recent sediments by fallout radionuclides: A review. *The Holocene*, *18*(1), 83–93. doi:10.1177/0959683607085598

Ballentine, D. C., Macko, S. A., Turekian, V. C., Gilhooly, W. P., & Martincigh, B. (1996). Compound specific isotope analysis of fatty acids and polycyclic aromatic hydrocarbons in aerosols: Implications for biomass burning. *Organic Geochemistry*, *25*(1-2), 97–104. doi:10.1016/S0146-6380(96)00110-6

Baumann, Z., and Fisher, N. S. (2011). Modeling metal bioaccumulation in a deposit-feeding polychaete from labile sediment fractions and from pore water. *Sci. Total Environ. 409*, 2607–2615. doi: . 03.009 doi:10.1016/j.scitotenv.2011

Beaumont, N. J., Aanesen, M., Austen, M. C., Börger, T., Clark, J. R., Cole, M., Hooper, T., Lindeque, P. K., Pascoe, C., & Wyles, K. J. (2019). Global ecological, social and economic impacts of marine plastic. *Marine Pollution Bulletin*, *142*, 189–195. doi:10.1016/j.marpolbul.2019.03.022 PMID:31232294

Benitez-Nelson, C. R., Buesseler, K., Dai, M., Aoyama, M., Casacuberta, N., Charmasson, S., Johnson, A., Godoy, J. M., Maderich, V., Masqué, P., Moore, W., Morris, P. J., & Smith, J. N. (2018). Radioactivity in the marine environment: Understanding the basics of radioactivity. Limnol. Oceanogr. *Limnology and Oceanography e-Lectures*, *8*(1), 1–58. doi:10.1002/loe2.10010

Betti, M., Boisson, F., Eriksson, M., Tolosa, I., & Vasileva, E. (2011). Isotope analysis for marine environmental studies. *International Journal of Mass Spectrometry*, *307*(1-3), 192–199. doi:10.1016/j.ijms.2011.03.008

Buesseler, K., Dai, M., Aoyama, M., Benitez-Nelson, C., Charmasson, S., Higley, K., Maderich, V., Masqué, P., Morris, P. J., Oughton, D., & Smith, J. N. (2017). Fukushima Daiichi–derived radionuclides in the ocean: Transport, fate, and impacts. *Annual Review of Marine Science*, *9*(1), 173–203. doi:10.1146/annurev-marine-010816-060733 PMID:27359052

Cresswell, T., Metian, M., Fisher, N. S., Charmasson, S., Hansman, R. L., Bam, W., Bock, C., & Swarzenski, P. W. (2020). Exploring new frontiers in marine radioisotope tracing–adapting to new opportunities and challenges. *Frontiers in Marine Science*, *7*, 406. doi:10.3389/fmars.2020.00406

de Paiva Magalhães, D., da Costa Marques, M., Baptista, D., & Buss, D. (2015). Metal bioavailability and toxicity in freshwaters. *Environmental Chemistry Letters*, *13*(1), 69–87. doi:10.1007/s10311-015-0491-9

Dwight, R. H., Fernandez, L. M., Baker, D. B., Semenza, J. C., & Olson, B. H. (2005). Estimating the economic burden from illnesses associated with recreational coastal water pollution—A case study in Orange County, California. *Journal of Environmental Management*, *76*(2), 95–103. doi:10.1016/j.jenvman.2004.11.017 PMID:15939121

Fowler, S. W. (2011). 210Po in the marine environment with emphasis on its behaviour within the biosphere. *Journal of Environmental Radioactivity*, *102*(5), 448–461. doi:10.1016/j.jenvrad.2010.10.008 PMID:21074911

Garcia-Leon, M. (2018). Accelerator Mass Spectrometry (AMS) in Radioecology. *Journal of Environmental Radioactivity*, *186*, 116–123. doi:10.1016/j.jenvrad.2017.06.023 PMID:28882579

Harmelin-Vivien, M., Bodiguel, X., Charmasson, S., Loizeau, V., & Mellon-Duval, C. (2012). Differential biomagnification of PCB, PBDE, Hg and radiocesium in the food web of the European hake from the NW Mediterranean. *Mar. Pollut. Bull., 64,* 974–983. doi: . 02.014 doi:10.1016/j.marpolbul.2012

IAEA. (2004). Sediment Distribution Coefficients and Concentration Factors for Biota in the Marine Environment. *Technical Report Series No. 422*. Vienna: International Atomic Energy Agency.

ICRP. (2008). Environmental Protection - the Concept and Use of Reference Animals and Plants. ICRP Publication 108. *Annals of the ICRP*, *38*, 1–242.

Kudela, R. M., Seeyave, S., & Cochlan, W. P. (2010). The role of nutrients in regulation and promotion of harmful algal blooms in upwelling systems. *Progress in Oceanography*, *85*(1-2), 122–135. doi:10.1016/j.pocean.2010.02.008

Liou, J. (2021). *NUTEC Plastics: Using Nuclear Technologies to Address Plastic pollution*. IAEA Office of Public Information and Communication.

Luoma, S. N., & Rainbow, P. S. (2005). Why is metal bioaccumulation so variable? Biodynamics as a unifying concept. *Environmental Science & Technology*, *39*(7), 1921–1931. doi:10.1021/es048947e PMID:15871220

Macko, S. A. (1994). Pollution studies using stable isotopes. *Stable isotopes in ecology and environmental science.*

Metian, M., Pouil, S., & Fowler, S. W. (2019). Radiocesium accumulation in aquatic organisms: a global synthesis from an experimentalist's perspective. *J. Environ. Radioact. 198*, 147–158. doi: . 11.013 doi:10.1016/j.jenvrad.2018

Patterson, C., Tilton, G., & Inghram, M. (1955). Age of the Earth. *Science*, *121*(3134), 69–75. doi:10.1126/science.121.3134.69 PMID:17782556

Payne, T. E., Hatje, V., Itakura, T., McOrist, G. D., & Russell, R. (2004). Radionuclide applications in laboratory studies of environmental surface reactions. *Journal of Environmental Radioactivity*, *76*(1-2), 237–251. doi:10.1016/j.jenvrad.2004.03.029 PMID:15245851

Povinec, P. P. (2017). Analysis of radionuclides at ultra-low levels: a comparison of low and high-energy mass spectrometry with gamma-spectrometry for radiopurity measurements. *Appl. Radiat. Isotopes, 126*, 26–30. doi: . apradiso.2017.01.029 doi:10.1016/j

Powell, C. L., & Doucette, G. J. (1999). A receptor binding assay for paralytic shellfish poisoning toxins: Recent advances and applications. *Natural Toxins*, *7*(6), 393–400. doi:10.1002/1522-7189(199911/12)7:6<393::AID-NT82>3.0.CO;2-C PMID:11122535

Reddy, C. M., Xu, L., O'Neil, G. W., Nelson, R. K., Eglinton, T. I., Faulkner, D. J., Norstrom, R., Ross, P. S., & Tittlemier, S. A. (2004). Radiocarbon evidence for a naturally produced, bioaccumulating halogenated organic compound. *Environmental Science & Technology*, *38*(7), 1992–1997. doi:10.1021/es030568i PMID:15112798

Rocha, C., Veiga-Pires, C., Scholten, J., Knoeller, K., Gröcke, D. R., Carvalho, L., Anibal, J., & Wilson, J. (2016). Assessing land–ocean connectivity via submarine groundwater discharge (SGD) in the Ria Formosa Lagoon (Portugal): Combining radon measurements and stable isotope hydrology. *Hydrology and Earth System Sciences*, *20*(8), 3077–3098. doi:10.5194/hess-20-3077-2016

Russell, B. C., Croudace, I. W., & Warwick, P. E. (2015). Determination of ^{135}Cs and ^{137}Cs in environmental samples: A review. *Analytica Chimica Acta*, *890*, 7–20. doi:10.1016/j.aca.2015.06.037 PMID:26347165

Shi, K., Hou, X., Roos, P., & Wu, W. (2012). Determination of technetium-99 in environmental samples: a review. *Anal. Chim. Acta 709*, 1–20. doi: . aca.2011.10.020 doi:10.1016/j

Speers, A. E., Besedin, E. Y., Palardy, J. E., and Moore, C. (2016). Impacts of climate change and ocean acidification on coral reef fisheries: an integrated ecological–economic model. *Ecol. Econ. 128*, 33–43. doi: .2016.04.012 doi:10.1016/j.ecolecon

Stewart, G., Fowler, S. W., & Fisher, N. S. (2008). In S. Krishnaswami & J. K. Cochran (Eds.), The bioaccumulation of U- and Th- series radionuclides in marine organisms. U-Th Series Nuclides in Aquatic Systems (pp. 269–305). Elsevier Science. doi:10.1016/S1569-4860(07)00008-3

Sturchio, N. C., Caffee, M., Beloso, A. D. Jr, Heraty, L. J., Böhlke, J. K., Hatzinger, P. B., & Dale, M. (2009). Chlorine-36 as a tracer of perchlorate origin. Environmental science & technology, 43(18), 6934-6938.

Sunda, W. G. (2006). Trace metals and harmful algal blooms. In E. Granéli & J. T. Turner (Eds.), Ecology of Harmful Algae (pp. 203–214). Springer. doi:10.1007/978-3-540-32210-8_16

Swart, P. K., Greer, L., Rosenheim, B. E., Moses, C. S., Waite, A. J., Winter, A., Dodge, R. E., & Helmle, K. (2010). The 13C Suess effect in scleractinian corals mirror changes in the anthropogenic CO2 inventory of the surface oceans. *Geophysical Research Letters*, *37*(5), L05604. doi:10.1029/2009GL041397

Trick, C. G., Bill, B. D., Cochlan, W. P., Wells, M. L., Trainer, V. L., & Pickell, L. D. (2010). Iron enrichment stimulates toxic diatom production in high- nitrate, low-chlorophyll areas. *Proceedings of the National Academy of Sciences of the United States of America*, *107*(13), 5887–5892. doi:10.1073/pnas.0910579107 PMID:20231473

Wang, C., Baumann, Z., Madigan, D. J., & Fisher, N. S. (2016). Contaminated marine sediments as a source of cesium radioisotopes for benthic fauna near Fukushima. *Environ. Sci. Technol. 50*, 10448–10455. doi: . 6b02984 doi:10.1021/acs.est

Wang, W.-X. (2013). Prediction of metal toxicity in aquatic organisms. *Chinese Science Bulletin*, *58*(2), 194–202. doi:10.1007/s11434-012-5403-9

Wang, W.-X., & Fisher, N. S. (1999). Delineating metal accumulation pathways for marine invertebrates. *The Science of the Total Environment*, *237-238*, 459–472. doi:10.1016/S0048-9697(99)00158-8

Yagi, N. (2016). Impacts of the nuclear power plant accident and the start of trial operations in fukushima fisheries. In T. M. Nakanishi & K. Tanoi (Eds.), Agricultural Implications of the Fukushima Nuclear Accident: The First Three Years (pp. 217–228). Springer Japan. doi:10.1007/978-4-431-55828-6_17

Yarimizu, K., Cruz-López, R., & Carrano, C. J. (2018). Iron and harmful algae blooms: Potential algal-bacterial mutualism between Lingulodinium polyedrum and Marinobacter algicola. *Frontiers in Marine Science*, *5*, 180. doi:10.3389/fmars.2018.00180

Chapter 11
Identification of Different Pollutants in Lotic and Lentic Ecosystems by Biosensors

Afaq Majid Wani
College of Forestry, India

Pritam Kumar Barman
https://orcid.org/0000-0002-5630-4965
College of Forestry, India

Satyendra Nath
College of Forestry, India

ABSTRACT

Environmental pollution is becoming a major global concern, especially about new pollutants, poisonous heavy metals, and other dangerous agents. Pollutants have a profound impact on ecosystems and present serious threats to the health of both the natural world and human communities. Water is one of the most important resources on the planet, since it is required for all species' survival and well-being. Surface water in an aquatic system is referred to as an inland water environment and is divided into lentic and lotic systems. In contrast to lotic water ecosystems, which share continuous habitats through the connection of many basins in unidirectional flow within the dendritic structure of river networks, lentic water ecosystems display discontinuous habitats as aquatic matrices inside the terrestrial system. The lentic water ecology is more similar to terrestrial waters than the lentic water ecosystem, which differs greatly. Aquatic ecosystems are diverse and, despite making up just a little of the planet's surface, are essential for several reasons.

INTRODUCTION

A specific environment, or biotope, and the species that inhabit it, referred to as the biocenosis, make up an ecosystem, also known as an ecological system (GWP/INBO, 2015). Numerous services provided by

DOI: 10.4018/979-8-3693-1930-7.ch011

the planet's ecosystems are essential to our everyday existence. Ecosystems on habitable planets can be broadly divided into two groups: terrestrial ecosystems that are based on land and aquatic ecosystems that are based on water. An ecosystem of or about land, as opposed to an ecosystem of water, is called a terrestrial ecosystem (Reddy et al., 2018). A subgroup of ecosystems where water plays a significant role is known as an aquatic environment (GWP/INBO, 2015). Ecosystems that rely mostly or partially on freshwater flooding are known as aquatic ecosystems (Reddy et al., 2018). All of the components of a water-based environment, both living and non-living, and their interactions together comprise an aquatic ecosystem (Reddy et al., 2018). Production, regulation, and organization are just a few of the many tasks that a healthy aquatic environment may undertake (GWP/INBO, 2015). Humans and water ecosystems have long been associated. 71% of the Earth's surface is generally covered by aquatic habitats. Unlike saltwater habitats, which make up three-fourths of the planet's surface, freshwater ecosystems only make up a small portion of the planet's surface. Nearly 70% of people on Earth reside in regions that are adjacent to bodies of water, such as lakes, rivers, and coasts, where civilizations have historically flourished (Reddy et al., 2018). Freshwaters are strongly linked to land use and climate. The sustainability of ecosystems and society depends on fresh water. The natural sciences (such as limnology, hydrology, and ecology), which highlight water quality, hydrological fluxes, and habitat quality as possible ecosystem services, have traditionally been used to study freshwaters. Freshwater ecosystem services include recreation, transportation via water, and the provision of clean water. Freshwater from streams, rivers, and lakes is necessary for human survival and provides essential functions. A varied collection of habitats depending on water, aquatic ecosystems contain significant biodiversity and offer numerous advantages to humans.

Since it is essential to the existence and well-being of all living beings, water is one of the most important resources on Earth and vital to all forms of life (Ahmed et al., 2017). Wetlands and other aquatic ecosystems offer homes for several kinds of plants as well as a place for animals to eat, rest, and reproduce (GWP/INBO, 2015). Aquatic ecosystems are abundant in biodiversity and support a wide range of species and habitats, benefiting the nation's economy and society in many ways. Many different species can be found in freshwater habitats on the surface of the Earth. A sizable portion of the biological variety of Earth is comprised of these aquatic species and the environments in which they coexist (Reddy et al., 2018). Freshwater that flows off land into rivers, lakes, and wetlands is what makes all land an essential component of an inland water ecosystem, as seen from the perspectives of ecology, hydrology, the environment, and socioeconomics. Water ecosystems on land are found in inland waters. Non-saline water, or water with very little or no salt content, is found in freshwater habitats. Within continental landmasses, the ecology of inland water is a complex of living things in free water. Within the Earth's surface, inland aquatic ecosystems make up less than 1%, but they are frequently among the most productive regions (Reddy et al., 2018). The study of all freshwater and saltwater aquatic systems, including wetlands, lakes, bogs, marshes, streams, ponds, rivers, reservoirs, oceans, and so on, with an emphasis on their physical, chemical, and biological properties, is known as limnology (Ziauddin, 2021). Lotic and lentic ecosystems are two different types of aquatic environments, each with its own population, special traits, and ecological dynamics. In addition to delivering a variety of ecosystem services, these ecosystems are essential for maintaining biodiversity and controlling regional temperatures. It is easier to recognize their importance and the need for their conservation when one is aware of their variations and roles.

The term "lotic" refers to flowing water that flows in a single direction, derived from the Greek word *"lavo"*, meaning "to wash" (Eramma et al., 2023). Water flowing through rivers, streams, tributaries, waterfalls, springs, canals, and creeks is a defining feature of lotic ecosystems. Their continuous movement modifies the habitat, affecting the way organisms are distributed and the ecosystem's physical makeup. The richness of their ecosystems was influenced by the size, pace, and volume of water that streams and rivers contain. Different species have different habitats at headwaters, which are usually small and swift-moving, compared to lower reaches, which are usually broader and slower-moving. There is a lot of variation among lotic habitats. Low salt content characterizes littoral environments. Water flows through lotic ecosystems from a source, like a spring or glacier, to a terminus, which could be an ocean, a bigger stream or river, or another kind of reservoir. In lotic ecosystems, gravity causes the surface water to flow downhill along the slopes. The majority of decisions in lotic ecosystems are made by the velocity and current of the flowing water because they create pathways and influence living things. In flowing rivers, the surface current is more noticeable than the bottom substrate. Because of this, the bottom substrate conditions resemble those of lentic environments. A path to the ocean is formed by the connections between numerous lotic habitats (Reddy et al., 2018). In the form of biological matter flowing strongly in one way, the lotic ecosystem is essentially an open system. However, the energy input of solar radiation and the close connection between streams and the nearby terrestrial ecosystems allow running waters to remain energetically viable.

The Latin word *"lentus"*, which means sluggish, is where the word "lentic" originates (Reddy et al., 2018). Standing, or comparatively still, water is a defining feature of lentic ecosystems, which include ponds, lakes, and wetlands. Lakes and other lentic environments are created by glaciers, volcanoes, tectonic plate movement, and occasionally by human activity. The area, depth, and nutrients of these bodies of water vary greatly, creating a variety of ecosystems that are home to a vast variety of creatures. Lakes are larger and can be classified as oligotrophic (low nutrients) or eutrophic (rich nutrients), depending on their nutrient content. Ponds, on the other hand, are the smallest among lentic ecosystems and are frequently shallow and temporary. However, bogs, swamps, marshes, and are all considered to be wetlands because of their saturation and the presence

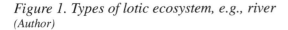

Figure 1. Types of lotic ecosystem, e.g., river
(Author)

of specific plant and animal species that have evolved to survive in wet environments. Compared to the lotic ecosystem, the lentic ecosystem has a significantly lower oxygen level due to its intense breakdown at the bottom and a small percentage of its surface area being in direct contact with the environment. A greater salt content is seen in certain lentic ecosystems, while freshwater environments have a lower salinity level. Little, transient ponds and massive lakes are examples of lentic habitats. A community of both abiotic and biotic interactions exists in lentic environments. Various creatures are supported by distinct layers in lentic habitats, which vary based on temperature and light levels. Lentic ecosystems are areas of standing water that are home to a wide range of creatures, such as fish, birds, frogs, and plants with rooted and floating leaves. A lentic ecology is the perfect home for frogs and ducks.

Both lotic and lentic environments support a wide variety of plants and animals (Padmanabha, 2017). While lotic habitats may be home to algae, mosses, insects, fish, amphibians, and mammals like otters and beavers that have adapted to the flowing water and the nearby land, lentic environments are frequently home to species like water lilies, cattails, frogs, turtles, fish, and various bird species. These ecosystems have intricated and interrelated dynamics. For example, in lentic ecosystems, the growth of plants and algae is influenced by nutrient levels, temperature fluctuations, and water depth. These factors also affect the availability of habitat and oxygen levels for other organisms. Temperature variations, sediment transport, and flow rate all influence the physical environment and the life cycles of the resident species in lotic ecosystems. Lotic and lentic ecosystems are severely impacted by human activity. These ecosystems are delicately balanced, but pollution from industrial waste, agricultural runoff, habitat degradation, and the introduction of alien species upset the equilibrium. Additionally affecting the integrity of these ecosystems are overfishing, damming, and modifications to natural river flows. For lotic and lentic ecosystems to remain healthy and preserved, conservation measures are essential. The negative consequences of human activity are lessened through programs including pollution control, wetland restoration, watershed preservation, and sustainable fishing methods. To protect these important

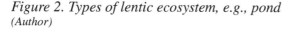

Figure 2. Types of lentic ecosystem, e.g., pond
(Author)

Table 1. Characteristics of lotic and lentic ecosystems

S. No.	Characteristics	Lotic Ecosystem	Lentic Ecosystem
1	Water source	Fed by springs, precipitation, rain, snowmelt, and lower-order streams and diffuse inputs	Surface water predominated, fed by higher-order streams; lotic waters from rivers, streams, and creeks flow into lakes and ponds in addition to ground water.
2	Width of water body	Comparatively limited	Comparatively wide
3	Depth of water body	Shallower	Deeper
4	Morphometry of water body	Longer, narrower, shallower, linear basin with a more intricate border	Deeper, round basin with a simpler border
5	Landscape position	Placement at natural flow	Lower in the watershed
6	Water flow	Consistent and precise guidance	Lacks a continuous, clear direction of flow
7	Duration of water retention and/or water residency	During a drought, there may be very little water, and the area would dry out, killing several creatures. Occasionally, there is a lot of water, such as after a strong storm.	Last longer and allow organisms to survive in spite of the reduced supply
8	Regional distribution and function	Determined by the geomorphology	Impacted by the local need for ecosystem services associated with dams
9	Permanence	Last several thousand years	Last several hundred to several thousand years
10	Adaptability of creatures	Need to become used to their outdated surroundings	Also need to adjust to their surroundings
11	Current velocity	High	Low
12	Dissolved oxygen content	High	Low
13	Salt content	Low	High
14	Speciation rates	Low	High
15	Geographical range size	High	Low
16	Species diversity	High	Low
17	Stability	High	Low

Source: Reddy et al., 2018, Eramma et al., 2023

aquatic habitats for future generations, public knowledge and participation in conservation initiatives are essential. Pollution in lotic and lentic environments must be closely monitored to keep aquatic ecosystems healthy and in balance.

The lotic and lentic environments provide distinct habitats for a wide variety of flora and fauna, and it is important to recognize the importance of pollutant monitoring in these regions (Sharma and Giri, 2018). Because water velocity directly affects the dispersion of contaminants, monitoring pollution in lotic ecosystems is essential. Introduced pollutants into these systems have the potential to move quickly downstream, impacting wider areas and a variety of ecosystems. Aquatic life and water quality are at risk from contaminants, including pesticides, fertilizers, and heavy metals, that are introduced via improper waste disposal, agricultural runoff, and industrial discharges. Frequent monitoring aids in the identification of pollution sources, the evaluation of the level of contamination, and the development of control and restoration plans for the environment. Because of their comparatively closed systems, which can cause pollutants to accumulate over time, it is imperative to monitor contaminants in these environments. Runoff from adjacent areas, urbanization, and industrial operations can contaminate these bodies of water with chemicals, fertilizers, and silt, which can cause algal blooms, eutrophication, and deterioration of water quality. To preserve the equilibrium of these ecosystems, it is helpful to study the ecological changes, pinpoint the sources of pollution, and put remedial measures in place.

To evaluate the general state of aquatic ecosystems, pollution in both lotic and lentic environments must be monitored. Knowing how contaminants affect biodiversity, the food chain, and the overall ecological balance is made easier with its assistance. The maintenance of water quality standards is ensured through routine monitoring. Assuring the water's suitability for drinking, farming, and recreation, it assists in identifying variations in pH levels, dissolved oxygen, nutrient levels, and the presence of hazardous materials. Many types of aquatic life are supported by lentic and lotic settings. The identification and mitigation of hazards, the prevention of habitat deterioration, and the preservation of biodiversity are all made possible by the monitoring of pollutants. Many activities are supported by these bodies of water, which also serve as sources of drinking water. The protection of human health from contaminants that may be harmful due to recreational or water-related activities requires continuous monitoring of pollutants. Using sustainable resource management techniques is made easier by having an understanding of how contaminants affect lotic and lentic settings. To stop pollution and encourage conservation, it helps stakeholders and policymakers make well-informed choices. To comprehend, mitigate, and prevent the harmful consequences of pollution on aquatic ecosystems, biodiversity, human health, and sustainable resource management, it is imperative to monitor pollutants in both lotic and lentic habitats. Informed decision-making is essential to protecting these essential water systems for coming generations, and it forms the basis for that effort. One of the most important factors in achieving environmentally sustainable development and reducing poverty in the context of global change is better governance that respects the environment. For a complete grasp of freshwater ecosystems, one must be able to identify, define, and comprehend their features. Thus, appreciation and knowledge can result in better management and preservation of all freshwater ecosystems, ensuring the sustainable use and provision of freshwater ecosystem services (Reddy et al., 2018).

In the field of biotechnology, the use of biosensors for industrial applications and environmental monitoring is important. Better environmental control is now required by law and public concern. More appropriate analytical techniques are also needed due to the growing number of analytes that need to be monitored. Samples must be submitted to a laboratory for testing to conduct a traditional "off-site" analysis. Conventional procedures are the most accurate with low limit of detection (LOD), but they require highly skilled individuals and are costly and time-consuming. The present trend of conducting field monitoring has increased the development of biosensors as novel analytical instruments capable of delivering affordable, quick, sensitive, and dependable measurements many of which are intended for on-site examination. Many biosensors use is being developed, such as environmental and bioprocess control, food quality management, agriculture, the military, and, most notably, medical applications. The majority of biosensor systems that are sold commercially are used in the pharmaceutical and clinical industries. As a result, this field has received the majority of research and development attention. Biosensors can deliver quick and precise information on contaminated sites for environmental control and monitoring. They provide additional benefits over existing analytical techniques, including the potential for portability, the ability to operate on-site, and the capacity to measure contaminants in intricate matrices with little sample preparation.

BIOSENSORS: PRINCIPLES, TYPES AND ROLE IN POLLUTANT DETECTION

The word "sensors" comes from the Latin word *"sentire"*, which implies perception. When it comes to the detection and removal of environmental contaminants found in tainted soil, water, and air, biotechnology is crucial (Ahmad and Kumar, 2020). Biotechnology is a significant factor in the treatment and monitoring of environmental toxins found in contaminated soil, water, and air. A biosensor is a biotechnology-based analytical method for pollutants that consists of components for a signal transducer that, while sensing contaminants, produces observable or quantifiable signals (Huang et al., 2023). A biosensor combines transducer and biological sensing elements, such as proteins, enzymes, nucleic acids, and microbes, to produce a signal proportionate to the measured analyte quantity. In a biosensor, the input transducer's job is to change the biological signal to produce a measured reaction. This response can be potential, current, or absorption through an electrochemical or optical process (Ahmad and Kumar, 2020; Nigam et al., 2015; Pakshirajan et al., 2015; Khanam et al., 2020, Kumar et al., 2020). Differential signals are created depending on whether the microbe is acting on the analyte or if it produces a biological recognition element. Enzymes, oligonucleotides, and antibodies are more frequently used as biosensing elements in biosensors than microorganisms. When compared to other biological elements, using microorganisms as a response element has advantages in terms of being more cost-effective and adaptable (Kumar et al., 2020). The types and guiding concepts of biosensors are as follows:

1. Biological recognition element: a biochemical shift is brought about by this component's specific interactions with the target analyte. Enzymes, antibodies, DNA/RNA, cells, or entire organisms could be involved.
2. Transducer: creates a measurable output (such as an electrical, optical, or mass-related signal) from the biochemical signal produced by the contact.
3. Signal processing unit: generates a quantitative result that expresses the concentration or existence of the target analyte by amplifying, processing, and translating the transducer output.

Compared to traditional systems and techniques, biosensors for environmental monitoring have several benefits, such as portability, miniaturization, and the capacity to measure a contaminant with a minimum number of samples. Several benefits, including real-time monitoring, downsizing, and improved selectivity and sensitivity, are associated with electrochemical biosensors. Furthermore, complex

Figure 3. Schematic diagram of a biosensor
(Gieva et al., 2014)

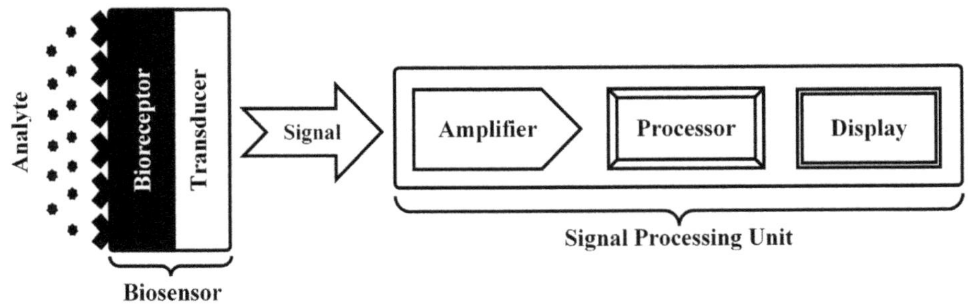

signalling components are not required because electrochemical reactions deliver electronic signals. This enables the creation of mobile systems for on-site environmental monitoring and clinical testing (Arduini et al., 2017). Advances in sensitivity, stability, selectivity, and their use in environmental monitoring have been emphasized in recent studies (Kaur et al., 2015; Justino et al., 2017; Felix and Angnes, 2018; Berberich et al., 2019; Gavrilaş et al., 2022). Biosensors are often categorized into multiple fundamental types based on signal transduction and biorecognition principles. Typically, they are categorized based on the transducer or the bio sensitive element. The biosensors' classification based on the transducer is illustrated in Figure 4.

There is an increase in potentially dangerous pollutants affecting the environment. Developing monitoring systems is essential for ongoing environmental monitoring. To do this, new technology and suitable operational procedures must be developed. The biosensors are suitable choices in the present scenario. By enabling the quick and accurate identification of several pollutants, toxins, and biological agents, biosensors are essential to environmental monitoring. In general, there are two types of environmental pollutants: inorganic and organic substances. Organic pollutants include pesticides, hormones, dioxins, polychlorinated biphenyls, phenols, bisphenol A (BPA), surfactants, linear alkylbenzene sulfonates, alkanes, polycyclic aromatic hydrocarbons, and antibiotics. Metals, inorganic phosphates, and nitrate are examples of inorganic contaminants (Marrazza, 2014; Goradel et al., 2018). Using appropriate biosensors, pollutants from the environment can be identified and monitored. Table 2 displays the types of biosensors used for organic and inorganic pollutant detection.

Figure 4. Classes and sub-classes of biosensors based on the type of transducer
(Gieva et al., 2014)

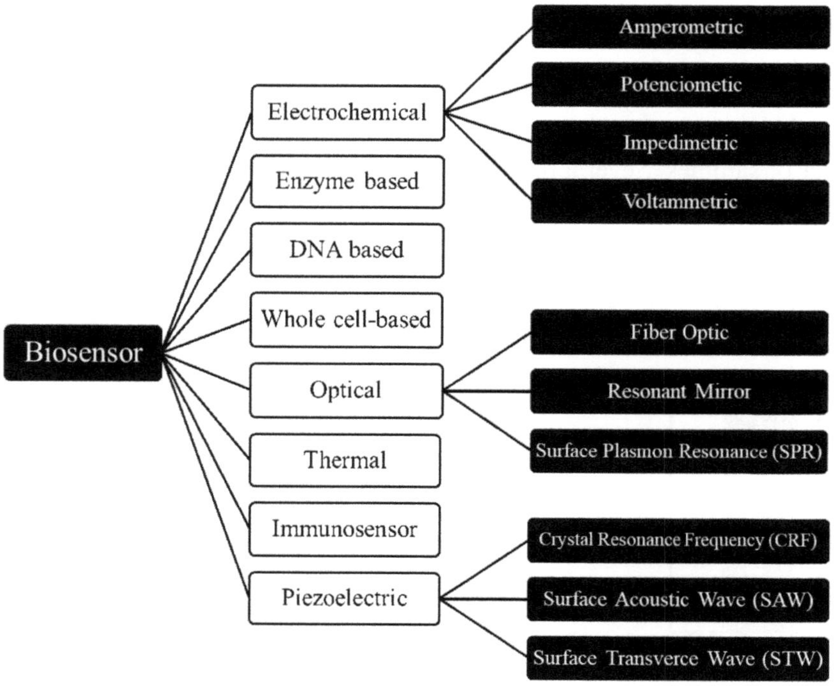

Table 2. Types of biosensors used for organic and inorganic pollutant detection

Target class	Biological sensing element	Physical transducer	Reference
Arsenite, Selenite	Microorganism (*E. coli*)	Optical	Ooi et al., 2015, Goradel et al., 2018
Atrazine,	Antibody	Piezoelectric	Jia et al., 2013
Carbaryl	Piezoelectric	Antibody	Wang et al., 2014
Catechol	Microorganism (Lactobacillus)	Electrochemical	Vogrinc et al., 2015
Copper	Enzyme (HRP)	Electrochemical	Moyo et al., 2014
Copper	Microorganism (yeast)	Optical	Vopálenská et al., 2015
Methyl Parathion	Microorganism (Flavobacterium)	Optical	Vogrinc et al., 2015
Serotonin	Microorganism (yeast)	Optical	Nakamura et al., 2015
Zinc, Cadmium	Microorganism (*E. coli*)	Optical	Vogrinc et al., 2015

Environmental quality assessment can greatly benefit from their usefulness due to their characteristics, including portability, specificity, and real-time monitoring. Biosensors are useful instruments for environmental monitoring and analysis because they have many advantages for pollution detection. In terms of pollutant detection, biosensors have the following major benefits:

1. High Sensitivity and Selectivity: Pollutants can be detected by biosensors with high sensitivity, even at low concentrations. Target pollutants can be selectively detected, reducing negative results, through the use of particular biological recognition elements like enzymes, antibodies, or whole cells.
2. Rapid Response Time: Rapid detection and reaction times are frequently offered by biosensors. In contrast to conventional analytical techniques, biosensors' biological components enable swift interactions with the target contaminants, enabling speedier and real-time monitoring.
3. Miniaturization and Portability: Pollution may be monitored in situ and on-site with biosensors because of their compact and portable design. Because of their small size, they may be easily deployed in various environmental conditions, making them appropriate for field applications.
4. Cost-Effectiveness: Biosensors can sometimes be more affordable than conventional analytical techniques. They are frequently easier to make with less expensive components and reagents. Savings are further increased by the fact that certain biosensors can be reused.
5. Ease of Use: The use of biosensors requires little technical knowledge and is intended to be user-friendly. This enables them to be used by non-experts, such as field workers and community members, enabling more extensive and broad-based environmental monitoring.
6. Multiplexing Capability: To monitor complicated environmental samples with a variety of pollutant profiles, biosensors can be designed to detect numerous pollutants at once. Monitoring efforts are more effective thanks to this multiplexing feature.
7. Biocompatibility: Because biosensors can be designed to detect numerous contaminants at once, complex environmental samples with a variety of pollutant profiles can be monitored. The efficiency of monitoring efforts is improved by this multiplexing feature.

8. Long-Term Stability: Many biosensors have strong long-term stability, making it possible to monitor continuously for lengthy periods. Understanding temporal fluctuations in pollution levels and trends depends on this characteristic.
9. Potential for Real-Time Monitoring: Real-time monitoring is possible with biosensors, which can provide ongoing information on the concentrations of pollutants. Being able to comprehend dynamic environmental processes and react quickly to emergent pollution events are critical skills.
10. Integration with Emerging Technologies: Environmental monitoring systems can be made more capable and effective by integrating biosensors with other cutting-edge technologies, such as data analytics and wireless communication.

SOURCES OF POLLUTANTS IN LOTIC AND LENTIC ECOSYSTEMS

The term "pollutant" refers to any material or agent released into the environment that harms ecosystems, living things, or the environment in general. In addition to man-made sources, including industrial operations, transportation, agriculture, and waste disposal, pollutants can also come from natural sources like volcanic eruptions or wildfires. These chemicals may negatively impact the air, water, soil, or other environmental components. They can exist in a variety of forms, such as solid particles, liquids, gases, or energy. Pollutants can have a negative effect on plants, animals, people, and the general equilibrium of ecosystems. Identification, monitoring, and mitigation of pollution sources and effects are necessary for effective pollution management and control to reduce environmental harm and advance sustainable practices. Contaminants such as pesticides, heavy metals, industrial chemicals, and pathogens are examples of water and soil pollutants. Water pollution can endanger human health, damage aquatic habitats, and lower the quality of drinking water. In addition to harming ecosystems and lowering agricultural productivity, soil pollution can contaminate food. When heated water from power plants or industrial activities is dumped into natural water bodies, it is referred to as thermal pollution. High temperatures have the potential to harm biodiversity and aquatic habitats. It is critical to remember that continual investigation and observation are essential to comprehending newly emerging contaminants and creating sensible environmental management plans.

The heavy metals represent some of the most the most dangerous pollution issues. Both the environment and human health are at risk, even at low doses. Lead, zinc, mercury, cadmium, and copper are the heavy metal contaminants that are most frequently found in the environment. The examination of heavy metals in the environment is the primary purpose of many biosensors that use bacteria as a sensing element. Certain types of bacteria serve as sensors and are resistant to a variety of heavy metals, including zinc, copper, tin, silver, mercury, and cobalt (Gieva et al., 2014). When treating different plants or fertilizing the soil, nitrates are frequently utilized as a protective agent. They are harmful to human health, have an adverse effect on hemoglobin, and may have irreversible effects. Pesticides are compounds or mixtures of substances used to keep pests away. Due to their long half-lives in the environment, pesticides are a concern for the environment. These have been shown through studies to persist in solids and sediments for extended periods. Pesticides are the most prevalent contaminants in the environment, present in soil, water, air, plants, and food. Using herbicides, particular undesirable weeds are eliminated without harming the intended crop. They may be carcinogenic and have highly variable toxicity. Although the effects are highly diverse, several herbicides have a deleterious effect on bird populations. Dioxins are environmental and persistent organic contaminants that are byproducts of several industrial processes.

As contaminating residues found in soil and water, dioxins are potentially hazardous substances that have a significant effect on the environment. Long-distance industrial dioxin emissions can be carried by wind, rivers, and sea currents. Dioxins can be continually recycled in the environment and have a long half-life, ranging from years to centuries.

Plants and microbes both create phenolic compounds, which can also be manufactured commercially. Phenolic structures are present in a large variety of environmental organic contaminants. Phenols and the derivatives they produce from the manufacturing of pharmaceuticals, polymers, paper, pulp, plastics, dyes, and oil refineries, among other things, have harmful effects in people, animals, and plants. Phenolic compounds alter the taste drinking water even at lower concentrations. Numerous contaminants have the potential to negatively harm the health and biodiversity of lotic and lentic ecosystems. Depending on the distinctive characteristics of these habitats and the adjacent land uses, pollutants in lotic and lentic ecosystems can differ.

LOTIC ECOSYSTEMS

1. Sedimentation:
   - Source: Soil erosion from construction sites, agriculture, and deforested areas.
   - Effect: Sedimentation can degrade water quality, impair habitats, and interfere with the reproductive processes of aquatic organisms.
2. Nutrient Pollution:
   - Source: Agricultural runoff, urban stormwater, and wastewater discharges.
   - Effect: Excessive nutrients may cause algal blooms, oxygen depletion, and changes in the composition of aquatic communities.
3. Industrial Discharges:
   - Source: Effluents from manufacturing processes.
   - Effect: Discharge of pollutants from industries can introduce toxic substances, heavy metals, and other contaminants into flowing water, affecting aquatic life.
4. Channelization and Altered Flow:
   - Source: Modifications to river channels for navigation, flood control, or agriculture.
   - Effect: Changes in natural flow patterns can disrupt habitats, alter sediment transport, and impact the health of lotic ecosystems.
5. Urban Runoff:
   - Source: Stormwater runoff from urban areas.
   - Effect: Urban runoff can affect aquatic ecosystems and water quality by introducing contaminants, including chemicals, heavy metals, and oil, into rivers.
6. Invasive Species:
   - Source: Introduction of non-native species.
   - Effect: In lotic ecosystems, invasive species can displace native species, modify food webs, and upset the ecological equilibrium.

LENTIC ECOSYSTEMS

1. Nutrient Pollution:
   - Source: Agricultural runoff, urban stormwater, and wastewater discharges.
   - Effect: Overnutrition (nitrogen and phosphorus) can cause eutrophication, which can harm aquatic life by creating algal blooms and oxygen depletion.
2. Sedimentation:
   - Source: soil erosion in deforested areas, agricultural areas, and construction sites.
   - Effect: Aquatic environments can be harmed by sedimentation, which can veil the water and reduce light penetration. Pesticides and fertilizers are among the associated contaminants that it can carry.
3. Toxic Chemicals:
   - Source: Industrial discharges, agricultural runoff, and urban runoff.
   - Effect: Pesticides, heavy metals, and other hazardous substances have the potential to damage aquatic life, interfere with food chains, and build up in sediments.
4. Pathogens:
   - Source: Untreated sewage, stormwater runoff, and agricultural runoff.
   - Effect: Water contamination by pathogens like bacteria and viruses can endanger aquatic life as well as human health.
5. Thermal Pollution:
   - Source: Discharges of heated water from industrial processes and power plants.
   - Effect: Warm water can have detrimental effects on aquatic ecosystems by decreasing the solubility of oxygen and altering the metabolic rates of organisms.
6. Floating Debris:
   - Source: Improper waste disposal and urban runoff.
   - Effect: Floating debris, including plastics and other materials, can degrade water quality, entangle wildlife, and disrupt aquatic ecosystems.

To monitor pesticides and herbicides, biosensors are essential. Furthermore, these substances might accumulate in grains and vegetables, impacting human health and development. By changing the fertility of the soil and eliminating numerous beneficial insects and bacteria, these chemicals can also modify the ecosystem during their operations. This can have a negative influence on biodiversity (Ahmad and Kumar, 2020). It is crucial to remember that the presence and effects of pollutants can differ significantly depending on the unique qualities of each water body, the land used in their catchment areas, and local environmental laws. In lotic and lentic ecosystems, regular monitoring and management techniques are crucial to reducing the negative effects of pollution.

APPLICATION OF BIOSENSORS FOR POLLUTANT DETECTION IN LOTIC AND LENTIC ECOSYSTEMS

Biosensors play a crucial role in monitoring and managing lotic and lentic ecosystems. Biosensors contribute to biodiversity and ecological studies by providing data on key environmental parameters. This information aids in understanding the relationships between environmental conditions and the diversity of aquatic ecosystems. The continuous development of biosensor technology enhances their capabilities for environmental monitoring, making them valuable tools for maintaining and preserving the health of lentic ecosystems. Their applications contribute to the assessment of water quality, detection of pollutants, and understanding ecological dynamics. The following are some important uses for biosensors in lotic and lentic environments.

Pesticides

Pesticides are extensively used in agriculture to improve yields and productivity, which has resulted in their widespread presence in the environment (Verma and Bhardwaj 2015). Pesticides are often present in natural streams since they are extensively used in agriculture. Pesticide concentration restrictions in certain environmental waterways have been imposed by the European Community due to concerns about their toxicity and long-term environmental persistence. A limit of 0.1 µg/l for individual pesticides and 0.5 µg/l for total pesticides has been imposed by Directive 98/83/EC concerning the safety of water for human consumption (Rodríguez-Mozaz et al., 2004). Chemical structure is used to categorize pesticides into five groups: inorganic pesticides, carbamate, synthetic pyrethroids, organochlorine (Atrazine), and organophosphate (Parathion) (Verma and Bhardwaj 2015). Organochlorine pesticides negatively impact fish reproduction (Vigneshvar et al., 2016). Among the most pervasive pesticides due to its slow breakdown, atrazine has been found in water bodies all over the world and is an endocrine-disrupting chemical (EDC) that affects fish, amphibians, reptiles, and mammals' ability to develop effectively (Scognamiglio et al., 2019).

Several approaches have been used in recent studies to quantify pesticides in water using biosensors. Liu et al., (2017) achieved an LOD of 7.53 µg/L for 2,4-D, a popular herbicide, by using fluorescence biosensors to detect it. Gosset et al., (2019) tested three distinct kinds of algae and three different pesticides; the best outcome they obtained was an LOD of 10 µg/L. Duan et al., (2020) tested the liquid crystal resonator biosensor using two analytes; the LOD was achieved at 1 pg/mL for dimethoate and 0.1 pg/mL for fenobucarb. Utilizing the latest interferometer technology, Ramirez-Priego et al., (2021) achieves an LOD of 0.29 ng/mL. In accordance with Fang et al., (2020), who employed immunosensors for the last five years to detect pesticides, they reported that immunosensors work well for detecting a variety of matrices because of their high sensitivity, ease of use, simplicity, and wide linear range.

The majority of current research on the detection of pesticides in lotic and lentic ecosystems has employed either fluorescence or interferometers. Interferometers have low LODs, and high sensitivity. A resonator that could detect even lower amounts has been developed recently (Herrera-Domínguez et al., 2023).

Table 3. Types of biosensors used for pesticides detection

Analyte	Biological sensing element	Physical transducer	Source	Reference
Atrazine	MIP	Fluorescence	Lake water	Liu et al., 2011
Atrazine	Antibody	Grating couplers	River water	Gao et al., 2017
Fenobucarb Dimethoate	Enzyme	Resonator	River water	Duan et al., 2020

Heavy Metals and Toxins

Because heavy metals can bioaccumulate in living things, particularly marine organisms, and because their levels in the environment are rising, it is crucial to find any residues of these metals, which include Cu, Cd, Hg, and Zn. Strong acids are typically used to evaluate metal content after digestion. Ion chromatography, inductively coupled plasma, and polarography are frequently employed analytical methods. Ion-selective electrodes can also be used to measure heavy metals. Being able to react just to the accessible fraction of metal ions is one benefit of whole-cell sensors. Recombinant bacterial sensors have been developed and employed to determine the presence of a certain metal. Chemical analysis techniques and nonspecific toxicity biosensors are less sensitive than specific biosensors, which are based on inducible promoters linked to reporter genes (Rodríguez-Mozaz et al., 2004).

Industrial operations such as mining raise the number of heavy metals. Because they are not biodegradable, heavy metals gradually build up in the environment. ROS are produced by almost all heavy metals, which transmit their harmful effects (Gutiérrez et al., 2015). Heavy metals like cadmium, cobalt, copper, and nickel can be found using glucose oxidase (GO) inhibition as a biosensor. The activity of a reporter gene driven by an inducible promoter serves as the basis for optical biosensors that detect metals. The reporter signal level rises with the pollutant concentration in this technique, known as "turn-on assay". The most often used reporter genes in biosensor systems are GFP, luciferase (*luc*), and β-galactosidase (*lacZ*) (Gutiérrez et al., 2015; Goradel et al., 2018). Gieva et al., 2014 reported that optical and electrochemical biosensors are the most commonly utilized for heavy metals detection.

The collection of substances known as toxins is highly diverse and can impact several biochemical processes, including ion transport, transmitter release, membrane function, and DNA and protein synthesis. In many cases, it is necessary to comprehend the finer points of a toxin's molecular method of action and site of action. A sulfur-oxidizing bacterium (SOB)-based biosensor was used in the online detection of heavy metals and other harmful compounds in water, as reported by Eom et al., (2019). In harmless water, sulfuric acid (H_2SO_4) is produced when elemental sulfur (S_8) is oxidized by SOB, lowering the pH of the water while raising its electrical conductivity (EC). In 2019, Chouler et al., developed an MFC-based biosensor that uses a microbial fuel cell to detect toxic compounds in water on-site. The MFCs are easy to use and create a bio-electrochemical sensor that is sensitive, portable, and reasonably priced by using microalgae.

Pharmaceuticals

One of the current concerns about water contamination is improper excretion of hospital wastes. Commonly founded pharmaceuticals in the water are likes of antihypertensive drugs, beta-blockers, analgesics/anti-inflammatories, antibiotics and psychiatric drugs. Therefore, detection of these compounds is essential to prevent life threating effects on human and other living organisms. Enzyme-based biosensors using peroxidases, laccases, and tyrosinases are main biosensors for the detection of pharmaceuticals (Rebollar-Pérez et al., 2015). When pharmaceutical substances are excreted or their residues are improperly disposed of, they end up in the environment. Similar to pesticides, these substances are persistent and build up in the environment; some even affect the development of some species of algae and microalgae and have estrogenic effects on particular creatures (Herrera-Domínguez et al., 2023).

Pharmaceuticals in water can be found using SPR and LSPR, two widely used techniques (Herrera-Domínguez et al., 2023). To detect the antibiotic ciprofloxacin, Luo et al., (2016) developed an SPR

Table 4. Optical biosensors and electrochemical biosensors used for toxic heavy metals detection

Sensing material	Contaminant	Limit of detection	Working range	Detection time
Optical sensors				
RGO	Hg^{2+}	1 nM	1–28 nM	tens of seconds
SWCNT (no probe)	Hg^{2+}	10 nM	10 nM to 1 mM	few seconds
Au NP/RGO	Hg^{2+}	25 nM	25 nM to 14.2 μM	few seconds
Au NP/RGO	Pb^{2+}	10 nM	10 nM to 10 μM	few seconds
SiNW	Pb^{2+}	1 nM	1–104 nM	few seconds
Au NP	Pb^{2+}	3 nM	3 nM to 1 μM	6 minutes
Au NP	Hg^{2+}	9.9 nM	9.9–600 nM	10 minutes
Au NP	Hg^{2+}	5 nM	50 nM to 10 μM	10 minutes
Au NP	Hg^{2+}	1 nM	1 nM to 1 mM	15 minutes
GO QD	Pb^{2+}	0.09 nM	0.1–1000 nM	20 minutes
SWCNT	*E. coli* DH5a	3×10^3 CFU mL^{-1}	3×10^3–1×10^6 CFU mL^{-1}	20 minutes
RGO	*E. coli* O157:H7	803 CFU mL^{-1}	803–10^7 CFU mL^{-1}	25 minutes
Au NP	Pb^{2+}	100 nM	0.1–50 μM	25 minutes
Graphene	*E. coli* K12	10 CFU mL^{-1}	10–10^5 CFU mL^{-1}	30 minutes
CNT	Cd^{2+}	88 nM	88 nM to 8.8 μM	30 minutes
Electrochemical sensors				
$MgSiO_3$	Pb^{2+} (0.1 M NaAc–HAc)	0.247 nM	0.1–1.0 μM	tenths of seconds
Fe_3O_4/RTIL	As^{3+} (acetate buffer)	0.01 nM	13.3–133 nM	few minutes
Nanosized Co.	H_2PO_4 – (KH_2PO_4 solution)	-	10^{-5} to 10^{-2} M	1 minute or less
Au	As^{3+} (1 M HCl)	0.26 nM	0.26–195 nM	100 seconds
Au–Pt NP	Hg^{2+} (1 M HCl)	0.04 nM	0.04–10 nM	100 seconds
Au NP/CNT	Hg^{2+} (0.1 M $HClO_4$)	0.3 nM	0.5 nM to 1.25 μM	2 minutes
Carbon NP	Hg^{2+} (1 M HCl)	4.95 nM	4.95–49.5 nM	2 minutes
CNT	Pb^{2+} (1 M HCl)	0.96 nM	9.6–480 nM	180 seconds
MWCNT/GO	Pb^{2+} (0.1 M NaAc–HAc)	0.96 nM	0.96–144 nM	3 minutes
Bi–CNT	Pb^{2+} (0.1 M acetate buffer)	6.24 nM	9.6–480 nM	300 seconds
Graphene/nafion	Pb^{2+} (0.1 M acetate buffer)	0.096 nM	2.4–240 nM	300 seconds
Nanosized hydroxyapatite	Pb^{2+} (0.2 M HAc-NaAc)	1 nM	5.0 nM to 0.8 μM	10 minutes
Graphene nanodots	Cu^{2+} (ammonium acetate solution)	9 nM	9 nM to 4 μM	15 minutes

Source: Hernandez-Vargas *et al.,* 2018.

biosensor using a molecularly printed polymer. It was estimated that this biosensor has an LOD of 0.08 μg/L. One of the most remarkable developments is the biosensor created by Shrivastav et al., (2015), which combines fiber optics with SPR and LSPR. The biosensor in this particular case has been made up of an optical fiber with a portion of its core exposed. With tetracycline as the target chemical, they obtained an LOD of 0.97 μg/L.

The most recent attempt to use fluorescence to detect antibiotics, Huang et al., (2021) were able to quantify ciprofloxacin by obtaining an LOD of 6 μM. Another substance that has been identified by fluorescence is diclofenac. Schirmer et al., (2019) developed this biosensor using yeast cells that illuminate when the drug is present. They obtained an LOD of 10 μM. Weber et al., (2017) developed an alternative method for antibiotic detection that involved the quantification of penicillin using an interferometer with a minimum concentration test of 0.25 μg/mL.

E. coli

The primary pathogen-transmission vector is surface water. In contaminated, treated, and untreated waterways, bacteria, viruses, and other microbes are frequently detected and represent a global public health concern. Since controlling diseases from these sources can benefit from proper monitoring of the water supply for the presence of pathogens, new technologies such as biosensors have been developed to provide fast identification of contamination by microorganisms at the source and in real-time, whereas conventional analytical methods take days or weeks to produce a result. The traditional approach to quantitative microbiological investigation involves counting visible microbial colonies for a wide range of prokaryotic and eukaryotic bacteria. A single culture cell in a sample can grow into an obvious colony, which is one of the benefits of colony-counting tests' straightforward methods and great sensitivity. However, the method hasn't advanced much since colony-counting tests were created in the past (Woldu, 2022).

Peixoto et al., (2019) developed bioactive paper sensors for real-time water quality monitoring. This study describes a technique for the very sensitive and specific multiplexed detection of *E. coli* utilizing a lab-on-paper test strip (bioactive paper) that is based on the activity of intracellular enzymes (β-glucuronidase (GUS) or β-galactosidase (B-GAL). Roda et al., (2019) used lectin and a porous silicon substrate to detect *E. coli* and Staphylococcus aureus, and they were able to acquire an LOD of 103 cells/mL. An optical biosensor was created by Jung and Lee (2016) for automated, continuous monitoring of the establishment and expansion of microbial colonies in water. The device may use high-resolution sub-pixel sweeping microscopy to dynamically identify individual microcolonies.

Biochemical Oxygen Demand

The oxygen needed to neutralize organic wastes over five days at 20°C is known as biochemical oxygen demand (BOD), or BOD_5. This metric is frequently used to determine the amount of biodegradable organic material present in water (Abdelghani and Jaffrezic-Renault, 2001). The traditional BOD test has a few advantages, including the ability to measure the majority of wastewater samples universally

Table 5. Types of biosensors used for pharmaceuticals detection

Analyte	Biological sensing element	Physical transducer	Source	Reference
Ciprofloxacin	MIP	Fluorescence	River water	Huang et al., 2021
Sulfadimine	Antibody	Fluorescence	Lake water	Liu et al., 2014
17β-estradiol	ER hERα	SPR	Pond water	Liu et al., 2021
Tetracycline	Aptamer and Antibody	SPR/LSPR	River water	Kim et al., 2017

and the lack of expensive equipment requirements. However, because of its time-consuming nature, it is not appropriate for online process monitoring. As a result, creating a different approach that may get around the traditional BOD test's drawback is imperative. Biosensor-based techniques could accomplish quick BOD determination.

Early on in the development of the new technology, biosensors were created to measure BOD. An amperometric sensor for measuring dissolved oxygen was developed in 1962 by Clark and Lyons. A self-powered BOD biosensor for signal-frequency-based water quality monitoring was developed by Pasternak et al., (2017) using an electroactive microbe. This biosensor is self-powered and has a five-month autonomy period because it is made from electroactive microorganisms. More research on the reliability of MFC-based biosensors is necessary and frequently goes unreported, as reported by Cui et al., in 2019. The catalysts used in MFC are self-renewing microorganisms. However, bacteria may adapt efficiently to modifications in their environment during long-term operation. Consequently, bacteria with a high extracellular electron transfer rate may be screened to enhance MFC-based biosensors, as this will negatively impact the biosensors' sensitivity, selectivity, and repeatability.

Endocrine Disrupting Compounds

Endocrine-disrupting chemicals (EDCs) have drawn more attention to themselves as pollutants found in municipal wastewater. According to recent studies, these substances are frequently not effectively removed in WWTPs and may have detrimental effects on the ecosystem (Woldu, 2022). Rapini & Marrazza, (2017) used aptamer-based electrochemical biosensors to evaluate water quality. Certain EDCs are tested for toxicity using a two-stage, multi-channel mini-bioreactor that is based on genetically modified bioluminescent bacteria (Shi et al., 2014). For determining EDCs, other biosensing techniques such as cell proliferation, luciferase induction, ligand binding, antigen-antibody interactions, or vitellogenin induction are available (Woldu, 2022).

Organic Compounds

Industrial supplies, fuels, detergents, personal care items, and derivatives made of plastic are all included in this category. Phenolic and halogenated chemicals are among them. Their detrimental impacts on aquatic species and even humans draw particular attention to them (Herrera-Domínguez et al, 2023). A biosensor based on an enzymatic membrane was proposed by Shahar et al., (2019) to detect organohalide, a halogenated compound, using an optical fiber reflectometer. An LOD of 0.908 mg/L was attained using this technique. BPA is one of the compounds that SPR biosensors have identified. Xue et al., (2019) employed functionalized gold nanoparticles using an inhibitory format; as a result, they acquired an LOD of 5.2 pg/mL. BPA is used in the making of plastics and can operate as an endocrine disruptor. Förster Resonance Energy Transfer (FRET) is an additional method of using fluorescence as an analytical tool. This method is based on how energy is transferred in biological systems. A biosensor for the detection of BPA was created by combining FRET with graphene; the outcome of this effort was an LOD of 0.1 ng/mL (Gupta and Wood, 2017). Cheng et al., (2022) developed a biosensor that measures variations in an immunoassay's fluorescence using a smartphone. The data is processed in an app that allows real-time tracking of the measurement. Using antibodies tagged with the Cy5.5 dye, the immunoassay was used to detect BPA in water samples from a lake and tap water, with an LOD of 0.1 nM for free Cy5.5.

Table 6. Types of biosensors used for organic compounds detection

Analyte	Biological sensing element	Physical transducer	Source	Reference
BPA	DNA	Fluorescence	River water	Gupta and wood, 2017
BPA	Antibody	Fluorescence	Lake water	Cheng et al., 2022
Dichloroethane	Enzyme	Fiber optic	River water	Shahar et al., 2019

CHALLENGES AND FUTURE DIRECTIONS

Biosensors are incredibly promising tools that can be used to quickly, easily, affordably, and reliably screen a large number of real samples. These devices' inherent qualities make them applicable to a wide range of industries, but the biological, agricultural, and environmental sectors seem to benefit the most from their frequent use in repetitive analysis. For the identification of contaminants in lotic and lentic environments, biosensors present a promising alternative. However, there are a few difficulties related to these environments in their implementation:

Lotic Ecosystems

1. Flow variation: Constant water flow in rivers and streams can alter the stability and positioning of sensors. It can be difficult to maintain sensor locations in quickly moving waters to provide reliable data.
2. Biofouling: An important problem is biofouling, or the build-up of organic materials on sensor surfaces. It may impede the functionality of the sensor, decreasing accuracy and requiring regular cleaning or maintenance.
3. Sensitivity to environmental conditions: There can be variations in turbidity, pH, and temperature in lotic ecosystems. The accuracy and dependability of sensors may be impacted by their sensitivity to these changes.
4. Limited access: Installing, maintaining, and retrieving sensors for calibration or repair can be challenging in lotic ecosystems because of their remote and frequently inaccessible nature.

Lentic Ecosystems

1. Stratification: Layers of differing environmental conditions are frequently found in lakes and ponds. As water depths vary, so do the temperature, oxygen content, and pollution dispersion. These variations must be taken into consideration by sensors.
2. Sediment interference: Sediment build-up at still water bodies' bottoms may prevent direct contact with the water or change the chemical composition being measured, which can both have an impact on sensor accuracy.
3. Biological interference: In lentic ecosystems, interactions between sensors and aquatic life can cause biofouling or even modify the way the sensors' function.
4. Spatial variability: Selecting sensor locations that effectively represent overall water quality can be difficult due to the variation in the spatial distribution of contaminants in lakes and ponds.

Biologists, environmental scientists, engineers, and technology developers must work with multidisciplinary teams to overcome these challenges. The development of advanced sensor technologies, persistent calibration techniques, and innovative deployment approaches is a continual effort to mitigate these concerns and augment the efficacy of biosensors in the surveillance and detection of contaminants in lotic and lentic environments. Biosensors for environmental monitoring have several advantages over conventional environmental analysis techniques, including low cost, low technical expertise and sample pre-treatment requirements, feasibility for on-site use, energy savings, and non-use of hazardous compounds (Thavarajah et al., 2020). The field of biosensor technology is rapidly developing, and there are several possible developments and future paths that might be taken to improve analyte identification and monitoring. The following are some crucial areas for future development:

1. Nanotechnology Integration: The sensitivity, selectivity, and response times of biosensors can be improved by integrating nanomaterials, such as nanoparticles, nanotubes, and nanowires. Nanomaterials can help create biosensor platforms that are more sensitive and effective by offering a large surface area for immobilizing biological recognition elements.
2. 3D Printing Technology: Complex and customized constructions can be manufactured with 3D printing. 3D printing can be used in biosensor development to build intricate designs that increase biological component immobilization and boost sensitivity and performance.
3. Flexible and Wearable Biosensors: Biosensors that adapt to the outlines of surrounding surfaces or the human body may be developed as a result of developments in flexible and wearable electronics. Particularly for health-related applications and environmental exposure evaluations, these biosensors may provide continuous, non-invasive monitoring capabilities.
4. Integration with Mobile Devices: Biosensors can offer intuitive interfaces for data collection, processing, and transfer when they easily connect with mobile devices, including smartphones and tablets. This integration allows for real-time monitoring in a variety of scenarios and improves accessibility.
5. Artificial Intelligence (AI) and Machine Learning: Biosensors can handle data more efficiently if AI and machine-learning algorithms are integrated. These technologies can improve the accuracy and dependability of biosensor-based measurements by helping with pattern identification, data interpretation, and the creation of predictive models.
6. Energy Harvesting: Developing biosensors that use energy harvesting methods, such as piezoelectric or triboelectric materials, can enable self-powered sensor systems. In remote or resource-constrained situations, this would lessen reliance on external power sources and improve the sustainability of biosensor deployments.
7. Single-Cell Analysis: Technological developments in biosensors may make single-cell analysis possible, enabling the identification and observation of biological substances at the cellular level. Understanding complex biological processes and using this expertise for medical diagnostics would be especially beneficial.
8. Multi-Analyte Detection Platforms: Using integrated sensor arrays, future biosensors might concentrate on the simultaneous identification of several analytes. The capacity to identify many analytes at once is essential for the thorough monitoring of complex samples, including biological fluids or environmental waters.

9. Biosensors for Biomarker Discovery: When it comes to finding biomarkers for a variety of illnesses and environmental circumstances, biosensors can be extremely important. The discovery of novel biomarkers and the creation of biosensors for early illness detection and monitoring could result from further study in this field.
10. Synthetic Biology Integration: By utilizing synthetic biology techniques, biosensors can be developed with more functions, increased target specificity, and increased stability. Additionally, biological components with specific qualities for use in biosensor applications can be created with the aid of synthetic biology technologies.
11. *In vivo* Biosensors: Advances in *in vivo* biosensors could make it possible to continuously monitor the physiological characteristics of living things. These biosensors could be applied to illness management, personalized therapy, and the comprehension of dynamic biological processes.

Future biosensor technology is expected to be driven by interdisciplinary research, scientific-engineer collaboration, and advances in materials science and bioengineering. These factors will likely result in more advanced, dependable, and adaptable sensing platforms for better identification and monitoring across a variety of fields.

SUMMARY

Emerging remediation technologies and governance techniques are being developed with a sustainable future in mind, acknowledging the growing demand for more socially, economically, and environmentally responsible societies. Pollution severely impacts ecosystems, which has a debilitating effect on economic growth. Pollutants also threaten human life and the lives of other living things. Biotechnology places a lot of importance on the usage of biosensors in industrial settings and for environmental monitoring. Building monitoring systems are essential for ongoing environmental monitoring. To do this, new technology and suitable field procedures must be developed. Since biosensors perform better than any other diagnostic instrument on the market regarding sensitivity and selectivity, they will quickly become indispensable analytical tools. Numerous biosensors were previously developed in research labs, and a sizable body of literature has been written about them. Although biosensors have been created for many uses, environmental analysis applications have demonstrated the greatest potential for biosensor development in the future. As a result, these techniques for identifying environmental contaminants must be improved. Although biosensors have many benefits, it is vital to remember that they also have drawbacks, including issues with repeatability, standardization, and possible interference from intricate sample matrices. These issues are still being researched, and new developments in technology are helping to overcome these obstacles and raise the general efficacy of biosensors in pollution detection. Lotic and lentic ecosystems, in summary, are complex and varied aquatic environments, each with its own distinct characteristics and ecological dynamics. Realizing the complexity of these ecosystems and their many benefits to humans and wildlife alike emphasizes how crucial conservation efforts are to ensuring their sustainability.

REFERENCES

Abdelghani, A., & Jaffrezic-Renault, N. (2001). SPR fibre sensor sensitised by fluorosiloxane polymers. *Sensors and Actuators. B, Chemical*, *74*(1–3), 117–123. doi:10.1016/S0925-4005(00)00720-6

Ahmad, R. G., & Kumar, V. (2020). Microorganism Based Biosensors to Detect Soil Pollutants. *Plant Archives*, *20*(2), 2509–2516.

Ahmed, M. H., El-Hamed, N. N. B. A., & Shalby, N. I. (2017). Impact of Physico-chemical parameters on composition and diversity of zooplankton Community in Nozha Hydrodrome, Alexandria, Egypt. *Egyptian Journal of Aquatic Biology and Fisheries*, *21*(1), 49–62. doi:10.21608/ejabf.2017.2382

Arduini, F., Cinti, S., Scognamiglio, V., Moscone, D., & Palleschi, G. (2017). How cutting-edge technologies impact the design of electrochemical (bio) sensors for environmental analysis. A review. *Analytica Chimica Acta*, *22*(959), 15–42. doi:10.1016/j.aca.2016.12.035 PMID:28159104

Berberich, J. A., Li, T., & Sahle-Demessie, E. (2019). Chapter 11 - Biosensors for Monitoring Water Pollutants: A Case Study with Arsenic in Groundwater. *Separation Science and Technology*, *11*, 285–328. doi:10.1016/B978-0-12-815730-5.00011-9

Cheng, Y., Wang, H., Zhuo, Y., Song, D., Li, C., Zhu, A., & Long, F. (2022). Reusable Smartphone-Facilitated Mobile Fluorescence Biosensor for Rapid and Sensitive On-Site Quantitative Detection of Trace Pollutants. *Biosensors & Bioelectronics*, *199*, 113863. doi:10.1016/j.bios.2021.113863 PMID:34894557

Chouler, J., Monti, M. D., Morgan, W., Cameron, P. J., & Lorenzo, M. D. (2019). A photosynthetic toxicity biosensor for water. *Electrochimica Acta*, *309*, 392–401. doi:10.1016/j.electacta.2019.04.061

Cui, Y., Lai, B., & Tang, X. (2019). Microbial Fuel Cell-Based Biosensors. *Biosensors (Basel)*, *9*(3), 92. doi:10.3390/bios9030092 PMID:31340591

Duan, R., Hao, X., Li, Y., & Li, H. (2020). Detection of Acetylcholinesterase and Its Inhibitors by Liquid Crystal Biosensor Based on Whispering Gallery Mode. *Sensors and Actuators. B, Chemical*, *308*, 127672. doi:10.1016/j.snb.2020.127672

Eom, H., Hwang, J., Hassan, S. H. A., Joo, J. H., Hur, J. H., Chon, K., Jeon, B.-H., Song, Y.-C., Chae, K.-J., & Oh, S.-E. (2019). Rapid detection of heavy metal-induced toxicity in water using a fed-batch sulfur-oxidizing bacteria (SOB) bioreactor. *Journal of Microbiological Methods*, *161*, 35–42. doi:10.1016/j.mimet.2019.04.007 PMID:30978364

Eramma, N., Lalita, H. M., Satishgouda, S., Jyothi, S. R., Venkatesh, C. N., & Patil, S. J. (2023). Zooplankton Productivity Evaluation of Lentic and Lotic Ecosystem. *Intech Open*, 1-16. doi:10.5772/intechopen.107020

Fang, L., Liao, X., Jia, B., Shi, L., Kang, L., Zhou, L., & Kong, W. (2020). Recent progress in immunosensors for pesticides. *Biosensors & Bioelectronics*, *164*, 1–57. doi:10.1016/j.bios.2020.112255 PMID:32479338

Felix, F. S., & Angnes, L. (2018). Electrochemical immunosensors-A powerful tool for analytical applications. *Biosensors & Bioelectronics*, *102*, 470–478. doi:10.1016/j.bios.2017.11.029 PMID:29182930

Gao, H., Generelli, S., & Heitger, F. (2017). Online Monitoring the Water Contaminations with Optical Biosensor. *Proceedings*, *1*(4), 522. doi:10.3390/proceedings1040522

Gavrilaş, S., Ursachi, C. Ş., Perţa-Crişan, S., & Munteanu, F.-D. (2022). Recent Trends in Biosensors for Environmental Quality Monitoring. *Sensors (Basel)*, *22*(4), 1513. doi:10.3390/s22041513 PMID:35214408

Gieva, E., Nikolov, G. T., & Nikolova, B. (2014). Biosensors For Environmental Monitoring. *Challenges in Higher Education & Research*, *12*, 123–127.

Goradel, N. H., Mirzaei, H., Sahebkar, A., Poursadeghiyan, M., Masoudifar, A., Malekshahi, Z. V., & Negahdari, B. (2018). Biosensors for the Detection of Environmental and Urban Pollutions. *Journal of Cellular Biochemistry*, *119*(1), 207–212. doi:10.1002/jcb.26030 PMID:28383805

Gosset, A., Oestreicher, V., Perullini, M., Bilmes, S. A., Jobbágy, M., Dulhoste, S., Bayard, R., & Durrieu, C. (2019). Optimization of Sensors Based on Encapsulated Algae for Pesticide Detection in Water. *Analytical Methods*, *11*(48), 6193–6203. doi:10.1039/C9AY02145K

Gupta, S., & Wood, R. (2017). Development of FRET biosensor based on aptamer/functionalized graphene for ultrasensitive detection of bisphenol a and discrimination from analogs. *Nano-Structures & Nano-Objects*, *10*, 131–140. doi:10.1016/j.nanoso.2017.03.013

Gutiérrez, J. C., Amaro, F., & Martín-González, A. (2015). Heavy metal whole-cell biosensors using eukaryotic microorganisms: An updated critical review. *Frontiers in Microbiology*, *6*(48), 1–8. doi:10.3389/fmicb.2015.00048 PMID:25750637

GWP (Global Water Partnership)/INBO (International Network of Basin Organizations). (2015). *The Handbook for Management and Restoration of Aquatic Ecosystems in River and Lake Basins*, 1-100. GWP. https://www.gwp.org/globalassets/global/toolbox/references/a-handbook-for-management-and-restoration-of-aquatic-ecosystems-in-river-and-lake-basins-no.3-2015.pdf

Hernandez-Vargas, G., Sosa-Hernández, J. E., Saldarriaga-Hernandez, S., Villalba-Rodríguez, A. M., Parra-Saldivar, R., & Iqbal, H. M. N. (2018). Electrochemical Biosensors: A Solution to Pollution Detection with Reference to Environmental Contaminants. *Biosensors (Basel)*, *8*(2), 1–21. doi:10.3390/bios8020029 PMID:29587374

Herrera-Domínguez, M., Morales-Luna, G., Mahlknecht, J., Cheng, Q., Aguilar-Hernández, I., & Ornelas-Soto, N. (2023). Optical Biosensors and Their Applications for the Detection of Water Pollutants. *Biosensors (Basel)*, *13*(3), 370. doi:10.3390/bios13030370 PMID:36979582

Huang, Q.-D., Lv, C.-H., Yuan, X.-L., He, M., Lai, J.-P., & Sun, H. (2021). A Novel Fluorescent Optical Fiber Sensor for Highly Selective Detection of Antibiotic Ciprofloxacin Based on Replaceable Molecularly Imprinted Nanoparticles Composite Hydrogel Detector. *Sensors and Actuators. B, Chemical*, *328*(1-2), 129000. doi:10.1016/j.snb.2020.129000

Jia, K., Adam, P. M., & Ionescu, R. E. (2013). Sequential acoustic detection of atrazine herbicide and carbofuran insecticide using a single micro-structured gold quartz crystal microbalance. *Sensors and Actuators. B, Chemical*, *188*, 400–404. doi:10.1016/j.snb.2013.07.033

Jung, J. H., & Lee, J. E. (2016). Real-time bacterial microcolony counting using on-chip microscopy. *Scientific Reports*, *6*(1), 21473. doi:10.1038/srep21473 PMID:26902822

Justino, C. I. L., Duarte, A. C., & Rocha-Santos, T. A. P. (2017). Recent Progress in Biosensors for Environmental Monitoring: A Review. *Sensors (Basel)*, *17*(12), 2918. doi:10.3390/s17122918 PMID:29244756

Kaur, H., Kumar, R., Babu, J. N., & Mittal, S. (2015). Advances in arsenic biosensor development–A comprehensive review. *Biosensors & Bioelectronics*, *63*, 533–545. doi:10.1016/j.bios.2014.08.003 PMID:25150780

Khanam, Z., Gupta, S., & Verma, A. (2020). Endophytic fungi-based biosensors for environmental contaminants-A perspective. *South African Journal of Botany*, *134*, 401–406. doi:10.1016/j.sajb.2020.08.007

Kim, S., & Lee, H. J. (2017). Gold Nanostar Enhanced Surface Plasmon Resonance Detection of an Antibiotic at Attomolar Concentrations via an Aptamer-Antibody Sandwich Assay. *Analytical Chemistry*, *89*(12), 6624–6630. doi:10.1021/acs.analchem.7b00779 PMID:28520392

Kumar, H., Kumari, N., & Sharma, R. (2020). Nanocomposites (conducting polymer and nanoparticles) based electrochemical biosensor for the detection of environment pollutant: Its issues and challenges. *Environmental Impact Assessment Review*, *85*(12), 106438. Advance online publication. doi:10.1016/j.eiar.2020.106438

Liu, L., Zhang, X., Zhu, Q., Li, K., Lu, Y., Zhou, X., & Guo, T. (2021). Ultrasensitive Detection of Endocrine Disruptors via Superfine Plasmonic Spectral Combs. *Light, Science & Applications*, *10*(1), 1–14. doi:10.1038/s41377-021-00618-2 PMID:34493704

Liu, L., Zhou, X., Lu, M., Zhang, M., Yang, C., Ma, R., Memon, A. G., Shi, H., & Qian, Y. (2017). An Array Fluorescent Biosensor Based on Planar Waveguide for Multi-Analyte Determination in Water Samples. *Sensors and Actuators. B, Chemical*, *240*, 107–113. doi:10.1016/j.snb.2016.08.118

Liu, L., Zhou, X., Xu, W., Song, B., & Shi, H. (2014). Highly Sensitive Detection of Sulfadimidine in Water and Dairy Products by Means of an Evanescent Wave Optical Biosensor. *RSC Advances*, *4*(104), 60227–60233. doi:10.1039/C4RA10501J

Liu, R., Guan, G., Wang, S., & Zhang, Z. (2011). Core-Shell Nanostructured Molecular Imprinting Fluorescent Chemosensor for Selective Detection of Atrazine Herbicide. *Analyst*, *136*(1), 184–190. doi:10.1039/C0AN00447B PMID:20886153

Luo, Q., Yu, N., Shi, C., Wang, X., & Wu, J. (2016). Surface plasmon resonance sensor for antibiotics detection based on photo-initiated polymerization molecularly imprinted array. *Talanta*, *161*, 797–803. doi:10.1016/j.talanta.2016.09.049 PMID:27769483

Marrazza, G. (2014). Piezoelectric biosensors for organophosphate and carbamate pesticides: A review. *Biosensors (Basel)*, *4*(3), 301–317. doi:10.3390/bios4030301 PMID:25587424

Moyo, M., Okonkwo, J. O., & Agyei, N. M. (2014). An amperometric biosensor based on horseradish peroxidase immobilized onto maize tassel-multi-walled carbon nanotubes modified glassy carbon electrode for determination of heavy metal ions in aqueous solution. *Enzyme and Microbial Technology*, *56*, 28–34. doi:10.1016/j.enzmictec.2013.12.014 PMID:24564899

Nakamura, Y., Ishii, J., & Kondo, A. (2015). Applications of yeast-based signaling sensor for characterization of antagonist and analysis of site-directed mutants of the human serotonin 1A receptor. *Biotechnology and Bioengineering*, *112*(9), 1906–1915. doi:10.1002/bit.25597 PMID:25850571

Nigam, V. K., & Shukla, P. (2015). Enzyme based biosensors for detection of environmental pollutants-A review. *Journal of Microbiology and Biotechnology*, *25*(11), 1773–1781. doi:10.4014/jmb.1504.04010 PMID:26165317

Nnachi, R. C., Sui, N., Ke, B., Luo, Z., Bhalla, N., He, D., & Yang, Z. (2022). Biosensors for rapid detection of bacterial pathogens in water, food and environment. *Environment International*, *166*, 1–20. doi:10.1016/j.envint.2022.107357 PMID:35777116

Ooi, L., Heng, L. Y., & Mori, I. C. (2015). A high-throughput oxidative stress biosensor based on Escherichia coli $roGFP_2$ cells immobilized in a *k*-carrageenan matrix. *Sensors (Basel)*, *15*(2), 2354–2368. doi:10.3390/s150202354 PMID:25621608

Padmanabha, B. (2017). Comparative study on the hydrographical status in the lentic and lotic ecosystems. *Global Journal of Ecology, 2*(1), 015-018. doi:10.17352/gje.000005

Pakshirajan, K., Rene, E. R., & Ramesh, A. (2015). Biotechnology in environmental monitoring and pollution abatement. *BioMed Research International*, *1-3*, 1–3. doi:10.1155/2015/963803 PMID:26526980

Pasternak, G., Greenman, J., & Ieropoulos, I. (2017). Self-powered, autonomous Biological Oxygen Demand biosensor for online water quality monitoring. *Sensors and Actuators. B, Chemical*, *244*, 815–822. doi:10.1016/j.snb.2017.01.019 PMID:28579695

Peixoto, P. S., Machado, A., Oliveira, H. P., Bordalo, A., & Segundo, M. (2019), Paper-Based Biosensors for Analysis of Water, Biosensors for Environmental Monitoring, *Intech Open*, 1-15. doi:10.5772/intechopen.84131

Ramirez-Priego, P., Estévez, M.-C., Díaz-Luisravelo, H. J., Manclús, J. J., Montoya, Á., & Lechuga, L. M. (2021). Real-Time Monitoring of Fenitrothion in Water Samples Using a Silicon Nanophotonic Biosensor. *Analytica Chimica Acta*, *1152*, 338276. doi:10.1016/j.aca.2021.338276 PMID:33648644

Rapini, R., & Marrazza, G. (2017). Electrochemical aptasensors for contaminants detection in food and environment: Recent advances. *Bioelectrochemistry (Amsterdam, Netherlands)*, *118*, 47–61. doi:10.1016/j.bioelechem.2017.07.004 PMID:28715665

Rebollar-Pérez, G., Campos-Terán, J., Ornelas-Soto, N., Méndez-Albores, A., & Torres, E. (2015). Biosensors based on oxidative enzymes for detection of environmental pollutants. *Biocatalysis*, *1*(1), 118–129. doi:10.1515/boca-2015-0010

Reddy, M. T., Sivaraj, N., Venkateswaran, K., Pandravada, S. R., Sunil, N., & Dikshit, N. (2018). Classification, Characterization and Comparison of Aquatic Ecosystems in the Landscape of Adilabad District, Telangana, Deccan Region, India. *OAlib*, *5*(4), 1–111. doi:10.4236/oalib.1104459

Roda, A., Zangheri, M., Calabria, D., Mirasoli, M., Caliceti, C., Quintavalla, A., Lombardo, M., Trombini, C., & Simoni, P. (2019). A Simple Smartphone-Based Thermochemiluminescent Immunosensor for Valproic Acid Detection Using 1,2-Dioxetane Analogue-Doped Nanoparticles as a Label. *Sensors and Actuators. B, Chemical*, *279*, 327–333. doi:10.1016/j.snb.2018.10.012

Rodríguez-Mozaz, S., Marco, M.-P., Alda, M. J. L. D., & Barceló, D. (2004). Biosensors for environmental applications: Future development trends. *Pure and Applied Chemistry*, *76*(4), 723–752. doi:10.1351/pac200476040723

Schirmer, C., Posseckardt, J., Schröder, M., Gläser, M., Howitz, S., Scharff, W., & Mertig, M. (2019). Portable and Low-Cost Biosensor towards on-Site Detection of Diclofenac in Wastewater. *Talanta*, *203*, 242–247. doi:10.1016/j.talanta.2019.05.058 PMID:31202333

Scognamiglio, V., Antonacci, A., Arduini, F., Moscone, D., Campos, E. V. R., Fraceto, L. F., & Palleschi, G. (2019). An Eco-Designed Paper Based Algal Biosensor for Nano formulated Herbicide Optical Detection. *Journal of Hazardous Materials*, *373*, 483–492. doi:10.1016/j.jhazmat.2019.03.082 PMID:30947038

Shahar, H., Tan, L. L., Ta, G. C., & Heng, L. Y. (2019). Optical Enzymatic Biosensor Membrane for Rapid in Situ Detection of Organohalide in Water Samples. *Microchemical Journal*, *146*, 41–48. doi:10.1016/j.microc.2018.12.052

Sharma, P., & Giri, A. (2018). Productivity evaluation of lotic and lentic water body in Himachal Pradesh, India. *MOJ Ecology & Environmental Sciences*, *3*(5), 311–317. doi:10.15406/mojes.2018.03.00105

Shi, X., Zhou, J. L., Zhao, H., Hou, L., & Yang, Y. (2014). Application of passive sampling in assessing the occurrence and risk of antibiotics and endocrine disrupting chemicals in the Yangtze Estuary, China. *Chemosphere*, *111*, 344–351. doi:10.1016/j.chemosphere.2014.03.139 PMID:24997938

Shrivastav, A. M., Mishra, S. K., & Gupta, B. D. (2015). Localized and Propagating Surface Plasmon Resonance Based Fiber Optic Sensor for the Detection of Tetracycline Using Molecular Imprinting. *Materials Research Express*, *2*(3), 35007. doi:10.1088/2053-1591/2/3/035007

Thavarajah, W., Verosloff, M. S., Jung, J., Alam, K. K., Miller, J. D., Jewett, M. C., Young, S. L., & Lucks, J. B. (2020). A primer on emerging field-deployable synthetic biology tools for global water quality monitoring. *NPJ Clean Water*, *3*(1), 1–10. doi:10.1038/s41545-020-0064-8 PMID:34267944

Verma, N., & Bhardwaj, A. (2015). Biosensor technology for pesticides-a review. *Applied Biochemistry and Biotechnology*, *175*(6), 3093–3119. doi:10.1007/s12010-015-1489-2 PMID:25595494

Vigneshvar, S., Sudhakumari, C. C., Senthilkumaran, B., & Prakash, H. (2016). Recent Advances in Biosensor Technology for Potential Applications-An Overview. *Frontiers in Bioengineering and Biotechnology*, *4*(11), 1–9. doi:10.3389/fbioe.2016.00011 PMID:26909346

Vogrinc, D., Vodovnik, M., & Marinsek-Logar, R. (2015). Microbial biosensors for environmental monitoring. *Acta Agriculturae Slovenica*, *106*(2), 67–75. doi:10.14720/aas.2015.106.2.1

Vopálenská, I., Váchová, L., & Palková, Z. (2015). New biosensor for detection of copper ions in water based on immobilized genetically modified yeast cells. *Biosensors & Bioelectronics*, *72*, 160–167. doi:10.1016/j.bios.2015.05.006 PMID:25982723

Wang, J., Liu, W., Chen, D., Xu, Y., & Zhang, L.-y. (2014). A micro-machined thin film electro-acoustic biosensor for detection of pesticide residuals. *Journal of Zhejiang University SCIENCE C*, *15*(5), 383–389. doi:10.1631/jzus.C1300289

Weber, P., Vogler, J., & Gauglitz, G. (2017). Development of an Optical Biosensor for the Detection of Antibiotics in the Environment. *Proceedings of the Society for Photo-Instrumentation Engineers*, *10231*, 102312L. doi:10.1117/12.2267467

Woldu, A. (2022). Biosensors and its applications in Water Quality Monitoring. *International Journal of Scientific and Engineering Research*, *13*(5), 12–29. Retrieved January 8th, 2024, from https://www.ijser.org/researchpaper/Biosensors-and-its-applications-in-Water-Quality-Monitoring.pdf

Xue, C. S., Erika, G., & Jiří, H. (2019). Surface Plasmon Resonance Biosensor for the Ultrasensitive Detection of Bisphenol A. *Analytical and Bioanalytical Chemistry*, *411*(22), 5655–5658. doi:10.1007/s00216-019-01996-8 PMID:31254055

ZiauddinG. (2021). Study of Limnological status of two selected floodplain wetlands of West Bengal. Research Square, 1-68. doi:10.21203/rs.3.rs-443104/v1

KEY TERMS AND DEFINITIONS

Biosensor: A device that combines biological material and transducers, usually in intimate association, to signal the presence of a particular substance or group of substances.

Ecosystem: A functional ecological unit in which the biological, physical, and chemical components of the environment interact.

Environment: The term "environment" refers to the surrounding area and includes all of the factors that influence an organism's life. The environment encompasses the interactions between water, air, and land, as well as their impact on humans, other living organisms, and property.

Lentic: Still water system, *i.e.*, ponds and lakes.

Lotic: Running waters, such as rivers and streams.

Pollutant: The components, chemicals, and particles that cause pollution are called pollutants; it is widely known that exposure to these substances can have negative impacts on human health as well as plant health.

LIST OF ABBREVIATIONS

% : Percent
µg : Microgram
µM : Micrometre
2,4-D : 2,4-dichlorophenoxyacetic acid
3D : Three-dimensional
AI : Artificial intelligence
BOD : Biochemical oxygen demand
BPA : Bisphenol A
Cd : Cadmium
CNT : Carbon Nanotube
Cu : Copper
DNA : Deoxyribonucleic acid
E. coli : Escherichia coli
EC : Electrical conductivity
EDC : Endocrine-disrupting chemical
GFP : Green fluorescent protein
GO : Glucose oxidase
Hg : Mercury
L : Liter
LOD : Limit of detection
LSPR : Localized surface plasmon resonance
M : Meter
MFC : Microbial fuel cell
mL : Milliliter
mM : Millimeter
ng : Nanogram
nM : Nanometer
pg : Picogram
pH : Potential of Hydrogen
RGO : Reduced graphene oxide
RNA : Ribonucleic acid
ROS : Reactive oxygen species
SOB : Sulfur oxidizing bacteria
SPR : Surface plasmon resonance
SWCNT : Single-wall carbon Nanotube
WWTPs : Wastewater treatment plants
Zn : Zinc
α : Alpha
β : Beta

Chapter 12
Limnology and the Science of Biosensors

Hemendra Wala
Devi Ahilya Vishwavidyalaya, India

ABSTRACT

Increasing concern about levels of pollution in the aquatic environment has led to the adoption of a number of preventive measures to assist in maintaining the quality of water bodies. The development of new user-friendly, portable, and low-cost bioanalytical methods is the focus of research, and biosensors are in the forefront of these research works. Biosensors have various prospective and existing applications in the detection of contaminants in the aquatic environment by transducing a signal. Biosensors are able to detect a wide range of analytes in complex matrices and have proven a great potential in environment monitoring, clinical diagnostics and food analysis Hence, the aim of this work is to provide a description of the state of the art about the development and application of biosensors to detect contaminants in freshwater ecosystems.

INTRODUCTION

Limnology is the study of inland water bodies, including ponds, lakes, rivers, and streams, with interactions between environmental factors. Limnology plays a primary part in the use of water and distribution, and nature habitat security. Limnologists work in reservoir management and lake, river, and stream protection, wildlife and fish enhancement, water pollution control, and artificial wetland construction. Limnology is the study of inland waters, streams, rivers, wetlands, groundwater, and reservoirs as systems of ecology related to their waste inlets and the environment. A biosensor is a tool that calculates chemical and biological reactions by developing signals symmetrical to the engagement of an analyte in the response. Despite protecting only 0.01% of the world's entire surface, they deliver essential services to the ecosystem, such as water, energy, and food provisions to billions of individuals. The oceans include 97.3% of all sources of water in the world. Accordingly, the ratio of freshwater sources is just 2.7% of all the sources of freshwater, and just 2.9% of it is open to people in the state of freshwater rivers, groundwater, and lakes **(Nakamura, 2018)**. The analyte connects to

DOI: 10.4018/979-8-3693-1930-7.ch012

the substance of biology, including a determined analyte, which causes countable electric responses. Electrochemical biosensors are easy tools that operate as bio electrodes and are used to calculate electric present conductance and ionic changes. Biosensor is an integrated receptor-transducer device that can convert biological responses into electrical signals.

Tropic State Index

The health and water quality of the oceans depend on their nutrient concentrations and water clarity. It includes three parts, mainly eutrophic, mesotrophic, and oligotrophic, which are determined by the level of nitrogen, prosperous, and chlorophyll concentration.

The Seasonal Cycle: Stratification and Mixing

In ponds and lakes moderately deep, the water undergoes mixing and thermal satisfaction's seasonable cycle. Warm water is lighter than cool water, which indicates that in summer., the surface water warms and becomes lighter, hovering above the heavier cool water that drops to the bottom(**Moore *et al.*, 2009**). Cold winds and air cool the water at the surface, which becomes heavier and drops in the column of the water, blending with the ground, water that stays cool all summer. In the winter season, when ice appears on the surface of the lake, the water below stays diverse but with slight temperature variations. Cold water poses instantly below the ice, while water straight, roughly 4 degrees Celsius, stays at the bottom of the lake. The water from dissolving ice is colder than the lake water. Hence,, this heavier, colder water drops to the bottom of the lake, and the surface water warms, directing to summer stratification of the column of the water.

Figure 1. Connecticut DEEP trophic categories

Category	T.P. (ppb)	T. Nitrogen (ppb)	Secchi Depth (m)	Chlorophyll *a* (ppb)
Oligotrophic	0 -- 10	2 -- 200	6 +	0 -- 2
Oligo-mesotrophic	10 -- 15	200 -- 300	4 -- 6	2 -- 5
Mesotrophic	15 -- 25	200 -- 500	3 -- 4	5 -- 10
Meso-eutrophic	25 -- 30	500 -- 600	2 -- 3	10 -- 15
Eutrophic	30 -- 50	600 -- 1000	1 -- 2	15 -- 30
Highly eutrophic	50 +	1000 +	0 -- 1	30 +

OVERVIEW OF FRESHWATER ECOSYSTEMS AND THEIR CHARACTERISTICS

Freshwater ecosystems are a subset of the world's aquatic ecosystems, and they can be contracted with oceanic ecosystems, which include a more extensive salt content—limnologists who are studying freshwater ecosystems. Freshwater ecosystems are interactive techniques in biotic species in their development and adaptation. It connected biological productivity, energy flows, and nutrient cycling between plants, inland marine microbial, as well as animal communities combined with their atmosphere **(Boehm *et al.,* 2017)**. The organic matter and nutrient range of drainage water from the area of the catchment is changed in the stream and wetland littoral. In addition to terrestrial features, water carries down gradient to and within the reservoir or lake itself. An ecosystem of freshwater includes a lot of minerals and nutrients, and it is small-salinized when likened to different ecosystems. The freshwater habitat experiences a temperature of 30 to 71° F during summers and 35 to 45° F in winter. The three kinds of freshwater ecosystems are lotic, wetlands, and lentic. Lotic ecosystems are fast-moving water like rivers and streams, and wetlands ecosystems are areas where the ground is inundated or saturated for a part of the time. Lentic ecosystems are gradually moving water like lakes, ponds, and pools.

Freshwater Ecosystems Characteristics

The ecosystem of freshwater is a habitat for different animal and plant species. The main reason is that it is particularly wealthy in words of minerals and nutrition. The ecosystem of freshwater is slightly salty, unlike the ocean ecosystem. The ecosystem's temperature changes depending on the same elements like season, depth, and location from the surface of the water. In the summer, the freshwater ecosystem's temperature typically varies from 30- 70 degrees Fahrenheit. However, in the winter the temperature varies from 35-45 degrees Fahrenheit. The shape and size of the freshwater ecosystems vary depending on the water bodies' depth, area covered, and location **(Laffoley *et al.,* 2020)**. The ecosystem of fresh-water contains residues at the ground. In gentle-flowing still-water bodies and freshwater bodies the residues stay in place. The ecosystem of freshwater delivers a suitable atmosphere for different species of fauna and flora.

Importance of Water Quality Monitoring in Limnology

Water quality monitoring involves the repeated study of water quality at specific locations, which helps to identify trends and take action to intercept and remediate any adverse effects caused by human activity on the marine environment. Parameters such as dissolved oxygen, ORP, pH, turbidity, temperature, chloride, ammonia, nitrate, and algae are continuously monitored to ensure water quality **(Lin et al. 2022)**. Monitoring the quality of ocean water (OWQ) provides a better understanding of the ocean's water quality as well as the near shore atmosphere. OWQ measurements can include biological, chemical, and physical characteristics of marine waters. Low values of OWQ measurements indicate an unhealthy ecosystem. Ocean water quality refers to the existence or absence of any pollutants in sea waters. More essential pollutants are heavy metals, plastics, nutrients, oil, thermal pollution, and sedimentation. Monitoring of water quality depends on taking appropriate measurements of seawater. In oceans, there are 5.25 trillion plastic waste calculated, 4 billion microfibers in every km^2, and 269,000 tons float live under the surface. In the ecosystems, there are 70% of waste sinks present, 15% land, and 15% float on the beaches. 8.3 million tons of plastics are scrapped in the ocean yearly **(Topp *et al.,* 2020).**

OCEAN ACIDIFICATION IN THE SALISH SEA

It has just become a rapidly growing danger to many exposed locations in the Salish Sea. It is the incremental growth in acidity in the earth's seas because of their carbon dioxide (CO2) uptakes from the environment. Scientists calculate that the seas have found roughly 1/3 of carbon dioxide, which is human-produced over the last 250 years, and it has shown a 30% growth in acidity. Ocean acidification is an international problem caused by the international growth in atmospheric CO2, influenced instantly not only by carbon dioxide's local emissions and different greenhouse gases but also by sources of freshwater **(Falkenberg *et al.,* 2020).** For example, organic carbon and nutrients from ground runoff can show improved algae growth and the algae mortality cycle. The growth in photosynthesizing algae boosts CO2 consumption and oxygen production in the waters at the surface of the coast during the day, but at night algae deliver CO2 during respiration.

Moreover, as algae sink and die, bacteria spoil the other organic issue that comes to the sea bottom, which can occasionally generate hypoxic situations. The Salish Sea is inherently more acidic than further waters, primarily due to increased rates of nutrient upwelling, so actually, small differences in pH can create a danger to ocean animals in the area. Acidic waters make it hard for oysters and various shellfish to construct their calcium carbonate bodies while in the stage of larval. The effect of ocean acidification on shellfish has fundamental economic developments **(Sallée *et al.,* 2021).** Gathering shellfish has been a rule in many coastal Indigenous societies since the immemorial period. However, as sea acidity boosts and permits shellfish reduction, the flow of learning from seniors to childhood becomes disrupted. "The Governor's Marine Resources Advisory Council" was created in "Washinton State" to handle the danger of sea acidification in 2013. "The Washington State Legislature" has since funded about $10 million in the research of sea acidification, modelling, and monitoring, with the help of the board. It has helped the shellfish enterprise start adjusting their methods to enhance shellfish survival. It has started promotions to wastewater remedy plants, which will facilitate local nutrient intakes that contribute to acidification.

What is Being Done About It?

A prohibition on plastic bags in "Washinton State" was marked into regulation in 2020, and the regulation came into effect in 2021. A prohibition on the one-time use of plastics will also be implemented in Canada before 2021 (**Cantonati*et al.* 2020**). These measures will help decrease the quantity of microplastics in the sea. Promotions to the "Annacis Island Wastewater Treatment Plant in Delta, B.C." will even take place starting in 2020. This promotion will enhance the power of the works to treat better wastewater and will fix and substitute older regions of the works.

DEFINITION AND PRINCIPLES OF BIOSENSORS

Biosensors are tools that are used for environmental monitoring and can be used to evaluate biological or chemical mixtures of water. These are machines that are used to evaluate the information of living organisms' early responses to environmental stressors and to provide signals on ocean ecosystem harm and pathology due to both natural and artificial pollution **(Navaand Leoni, 2021).**

The biosensor is a type of detection system using typical biomolecule recognition factors, significantly interacting with ocean toxins and discovering the positively sensitive detection of exact signal transformation. Due to the profitability, autorotation, and miniaturization of related tools, biosensors have an appropriate application option for online monitoring and in-field detection **(Koehnken *et al.*, 2020)**. The sensor is formed in an empty carbon electrode changed with carbon microspheres and palladium particles and can reach a precision between 96% ± 1% and 105% ± 3%. A specific biosensor includes a transducer, a display, an analyte, and a bioreceptor.

TYPES OF BIOSENSORS AND THEIR APPLICATIONS

Ocean pollutant perseverance and biosensors present many benefits with specificity, quick response times, and high sensitivity. Some kinds of biosensors are typically used for noticing pollutants in seas with their applications which are

Immunosensors

Antigens and antibodies are immobilized on the surface of the sensor to grab specific contaminants, including a complex of antigen-antibody. Helpful in detecting pollutant levels like pesticides, toxins, and metals, in seawater. **(Majors *et al.*, 2020)**.

Enzyme Based Biosensors

Enzymes like peroxidases, hydrolases, and oxidases are immobilized on the surface of the transducer. Applications contain the detection of organic pollutants like phenols, pesticides, and petroleum hydrocarbons in seawater. **(Da Silva *et al.*, 2022).**

DNA-Based Biosensors

DNA oligonucleotides or strands are immobilized on the surface of the sensor, and differences in their design or exchanges with target contaminants are detected. Good for noticing typical DNA arrangements of genetically or pathogens changed organisms in ocean samples. **(Suslova and Grebenshchikova, 2020)**.

Whole Cell Biosensors

Living cells, such as algae, yeast, and bacteria, are employed as feeling elements. Applications range from observing the seawater's toxicity due to heavy metals, organic pollutants, and pesticides to noticing pathogens in oceanic environments.

Nanomaterial-Based Biosensors

Nanomaterials, such as graphene, nanoparticles, and carbon nanotubes, are operated to improve selectivity and sensitivity.

Aptamer-Based Biosensors

Aptamers, molecules of single-stranded DNA or RNA, are established to bind expressly to target contaminants.

Optical Biosensors

optical properties such as surface plasmon resonance (SPR) and absorbance are operated to notice differences generated by pollutant binding. Good for real-time observation of pollutants like chemical contaminants, microplastics, and oil spills in seawater.

These biosensors discover applications in different areas of ocean pollution observation, including evaluating the effect of human movements, detecting pollutants in aquaculture, and protecting ocean ecosystems and general health **(Villalobos *et al.,* 2020)**. Additionally, progressions in remote sensing, nanotechnology, and miniaturization are directed to the growth of transportable and independent biosensor designs appropriate for in-situ monitoring of sea pollutants.

Pollution of heavy metals is often removed as an anthropogenic result, and activities of industry and their corruption threaten the environment and human health. Elements like Cu, Mn, Zn, and Fe, then highly toxic As, Hg, Cr, Cd, Pb, etc, are extremely immune to biodegradation. Heavy metals are enchanted into the atmosphere, particularly water sources, and efficiently fascinated by living organisms **(Baronas et al. 2021)**. For their apparent toxicity, these heavy metals are essential for observing the atmosphere. Observing water contamination is vital for environmental protection and the deterrence of conditions. For the collection in the atmosphere and nature, such as animals and plants, heavy metals create a threat over a long time. Many techniques have been created to notice their engagements and existence in the matrix of the environment. Specifically, biosensors can notice the heavy metal levels and resolve how much corruption they caus. Biosensors can efficiently notice the existence of heavy metals to control and operate water quality and safety **(Sezginturk, 2020)**. The whole-cell biosensors represent a future strategy for the detection of heavy metals and are required for sensitivity and selectivity. Furthermore, biosensors can be combined into pollutants direction and operated to forecast chemical pollution in the atmosphere.

ORGANIC POLLUTANTS MONITORING

Anthropologic activities confirm that organic contaminants pollute the natural atmosphere. A broad range of pollutants develops from various household, agricultural, and industrial activities. Agricultural waste pesticides and organic herbicides include toxic mixes mostly discovered in wastewater, and the mixtures are broadly used to extract pests' unwanted vegetation and weeds. "Persistent organic pollutants (POPs)", including "polybrominated biphenyls (PBBs)", "phthalate esters (PAEs)", "polybrominated diphenyl ethers (PBDEs)", "polychlorinated biphenyls (PCBs)", and other dangerous contaminants exist in industrial wastewater **(Alin *et al.,* 2023).** Observing the particular organic topic in wastewater is an essential element of the environment and human health, especially water reclamation methods and wastewater treatment. Biosensor applications for noticing organic pollutants in wastewater receive the quickest and most precise effects compared to different traditional methods.

Wastewater quality monitoring and environmental viability

Biosensors offer stimulating solutions for observing sea pollutants, but many challenges continue, along with appearing locations for future growth. There are some challenges and future directions in the biosensors for seas:

Durability and Stability: Biosensors deployed in sea atmospheres must resist harsh situations such as temperature variations, biofouling, and salinity. Producing full protective coatings and sensor materials to provide long-term equilibrium and implementation is crucial.

Sensitivity and Selectivity: Enhancing the selectivity and sensitivity of biosensors to notice low attention of contaminants amidst the complicated matrix of seawater remains a challenge. Future investigations should focus on improving the particularity of biological distinction elements and producing creative signal amplification methods **(Tetyana *et al.,* 2021).**

Biofouling Mitigation: Biofouling, the collection of algae, microorganisms, and different ocean organisms on the surface of the sensor, can interfere with sensor implementation over time. Research measures may concentrate on creating antifouling materials and techniques to resolve this issue.

Real-Time Monitoring: Improvements in data processing strategies and sensor technologies are required to allow real-time observation of sea pollutants. Wireless transmission and small sensing abilities can enable continuous data communication and analysis.

Emerging contaminants: Continuously observing and noticing emerging contaminants in seawater, such as pharmaceuticals, personal care products, and microplastics, creates new challenges that need creative sensor techniques and detection procedures **(Bhattarai and Hameed, 2020)**.

Managing these challenges and increasing the abilities of biosensors for seas will need interdisciplinary collaborations concerning researchers from fields like biology, materials science, oceanography, chemistry, and engineering. Moreover, expanded funding help and assets in ocean technologies for monitoring are essential for driving the creation and summarizing of research results into practical explanations for conservation efforts and environmental management.

SUMMARY

This chapter offered an overview of Limnology and the Science of Biosensors of the ocean.

- It presents an overview of the freshwater ecosystem and its characteristics. It also represents the importance of water quality monitoring in limnology and Ocean Acidification in the Salish Sea.
- Different types of biosensors like immunosensors, whole Cell Biosensors, Optical Biosensors, DNA-Based Biosensors, and Enzyme based biosensors have been found.
- The application of biosensors in limnology, wastewater quality monitoring, and environmental viability are also discussed.

It has been concluded that the use of biosensors to detect ocean pollutants is easier. This detection via biosensors sensors indicates that proportions and detection earlier can support reducing the damaging effects of dangerous pollutants in seas.

REFERENCES

Alin, S. R., Newton, J. A., Feely, R. A., Greeley, D., Curry, B., Herndon, J., & Warner, M. (2023). A decade-long cruise time-series (2008–2018) of physical and biogeochemical conditions in the southern Salish Sea, North America. *Earth System Science Data Discussions*, *2023*, 1–37.

Baronas, R., Ivanauskas, F., & Kulys, J. (2021). *Mathematical modeling of biosensors*. Springer International Publishing. doi:10.1007/978-3-030-65505-1

Bhattarai, P., & Hameed, S. (2020). Basics of biosensors and nanobiosensors. Nanobiosensors: From Design to Applications.

Boehm, A. B., Ismail, N. S., Sassoubre, L. M., & Andruszkiewicz, E. A. (2017). Oceans in peril: Grand challenges in applied water quality research for the 21st century. *Environmental Engineering Science*, *34*(1), 3–15. doi:10.1089/ees.2015.0252

Cantonati, M., Poikane, S., Pringle, C. M., Stevens, L. E., Turak, E., Heino, J., Richardson, J. S., Bolpagni, R., Borrini, A., Cid, N., Čtvrtlíková, M., Galassi, Hájek, Hawes, Levkov, Naselli-Flores, Saber, Cicco, Fiasca, & Znachor. (2020). Characteristics, main impacts, and stewardship of natural and artificial freshwater environments: Consequences for biodiversity conservation. *Water (Basel)*, *12*(1), 260. doi:10.3390/w12010260

da Silva, F. L., Fushita, Â. T., da Cunha-Santino, M. B., & Bianchini, I. Jr. (2022). Adopting basic quality tools and landscape analysis for applied limnology: An approach for freshwater reservoir management. *Sustainable Water Resources Management*, *8*(3), 65. doi:10.1007/s40899-022-00655-8

Falkenberg, L. J., Bellerby, R. G., Connell, S. D., Fleming, L. E., Maycock, B., Russell, B. D., Sullivan, F. J., & Dupont, S. (2020). Ocean acidification and human health. *International Journal of Environmental Research and Public Health*, *17*(12), 4563. doi:10.3390/ijerph17124563 PMID:32599924

Koehnken, L., Rintoul, M. S., Goichot, M., Tickner, D., Loftus, A. C., & Acreman, M. C. (2020). Impacts of riverine sand mining on freshwater ecosystems: A review of the scientific evidence and guidance for future research. *River Research and Applications*, *36*(3), 362–370. doi:10.1002/rra.3586

Laffoley, D., Baxter, J. M., Amon, D. J., Currie, D. E., Downs, C. A., Hall-Spencer, J. M., Harden-Davies, H., Page, R., Reid, C. P., Roberts, C. M., Rogers, A., Thiele, T., Sheppard, C. R. C., Sumaila, R. U., & Woodall, L. C. (2020). Eight urgent, fundamental and simultaneous steps needed to restore ocean health, and the consequences for humanity and the planet of inaction or delay. *Aquatic Conservation*, *30*(1), 194–208. doi:10.1002/aqc.3182

Lin, L., Yang, H., & Xu, X. (2022). Effects of water pollution on human health and disease heterogeneity: A review. *Frontiers in Environmental Science*, *10*, 880246. doi:10.3389/fenvs.2022.880246

Moore, T. S., Mullaugh, K. M., Holyoke, R. R., Madison, A. S., Yücel, M., & Luther, G. W. III. (2009). Marine chemical technology and sensors for marine waters: Potentials and limits. *Annual Review of Marine Science*, *1*(1), 91–115. doi:10.1146/annurev.marine.010908.163817 PMID:21141031

Nakamura, H. (2018). Current status of water environment and their microbial biosensor techniques–Part I: Current data of water environment and recent studies on water quality investigations in Japan, and new possibility of microbial biosensor techniques. *Analytical and Bioanalytical Chemistry*, *410*(17), 3953–3965. doi:10.1007/s00216-018-0923-z PMID:29470662

Nava, V., & Leoni, B. (2021). A critical review of interactions between microplastics, microalgae and aquatic ecosystem function. *Water Research*, *188*, 116476. doi:10.1016/j.watres.2020.116476 PMID:33038716

Sallée, J. B., Pellichero, V., Akhoudas, C., Pauthenet, E., Vignes, L., Schmidtko, S., Garabato, A. N., Sutherland, P., & Kuusela, M. (2021). Summertime increases in upper-ocean stratification and mixed-layer depth. *Nature*, *591*(7851), 592–598. doi:10.1038/s41586-021-03303-x PMID:33762764

Sezgintürk, M. K. (2020). Introduction to commercial biosensors. In *Commercial Biosensors and Their Applications* (pp. 1–28). Elsevier. doi:10.1016/B978-0-12-818592-6.00001-3

Suslova, M. Y., & Grebenshchikova, V. I. (2020). Water quality monitoring of the Angara River source. *Limnology and Freshwater Biology*, (4), 1040–1041. doi:10.31951/2658-3518-2020-A-4-1040

Tetyana, P., Shumbula, P. M., & Njengele-Tetyana, Z. 2021. Biosensors: design, development and applications. In Nanopores. IntechOpen. doi:10.5772/intechopen.97576

Topp, S. N., Pavelsky, T. M., Jensen, D., Simard, M., & Ross, M. R. (2020). Research trends in the use of remote sensing for inland water quality science: Moving towards multidisciplinary applications. *Water (Basel)*, *12*(1), 169. doi:10.3390/w12010169

Vick-Majors, T.J., Michaud, A.B., Skidmore, M.L., Turetta, C., Barbante, C., Christner, B.C., Dore, J.E., Christianson, K., Mitchell, A.C., Achberger, A.M. and Mikucki, J.A., 2020. Biogeochemical connectivity between freshwater ecosystems beneath the West Antarctic Ice Sheet and the sub-ice marine environment. *Global Biogeochemical Cycles, 34*(3).

Villalobos, C., Love, B. A., & Olson, M. B. (2020). Ocean acidification and ocean warming effects on Pacific Herring (Clupea pallasi) early life stages. *Frontiers in Marine Science*, *7*, 597899. doi:10.3389/fmars.2020.597899

Chapter 13
Biosensors for Environmental Monitoring

Rajni Gautam
K.R. Mangalam University, India

ABSTRACT

Environmental monitoring is essential to safeguard our planet's ecosystems, public health, and natural resources. Biosensors have emerged as powerful tools for assessing environmental parameters due to their sensitivity, specificity, and versatility. From the detection of pollutants to the monitoring of water and air quality, biosensors offer a wide array of applications that contribute to comprehensive environmental assessment. In this chapter, the principles, applications, and significance of biosensors in the context of environmental monitoring are explored in detail. Future prospects and challenges are discussed as well.

INTRODUCTION

Biosensors are analytical devices that play a pivotal role in environmental monitoring, offering the capability to detect and quantify specific analytes in the environment. These biosensors function through a combination of biological recognition elements and transducers, which work synergistically to provide accurate and reliable data in environmental monitoring applications (Huang et al., 2023) as shown in Figure 1.

Biosensors employ biological recognition elements such as enzymes, antibodies, DNA, or whole cells. Based on the target analyte of interest in environmental monitoring, these components have been carefully chosen. For example, enzymes are frequently used in biosensors to catalyze particular interactions with molecules of interest or environmental contaminants, including pollutants in the air or water. Conversely, antibodies are useful for identifying diseases or allergens in a variety of environmental samples due to their high specificity in identifying certain antigens. In order to help with the assessment of biodiversity and the existence of genetically modified species, DNA-based identification elements are able to identify genetic material from microorganisms in soil or water. The physiological reactions of whole cells, usually bacterial strains or microorganisms, to environmental changes are monitored,

DOI: 10.4018/979-8-3693-1930-7.ch013

Figure 1. Components of a biosensor

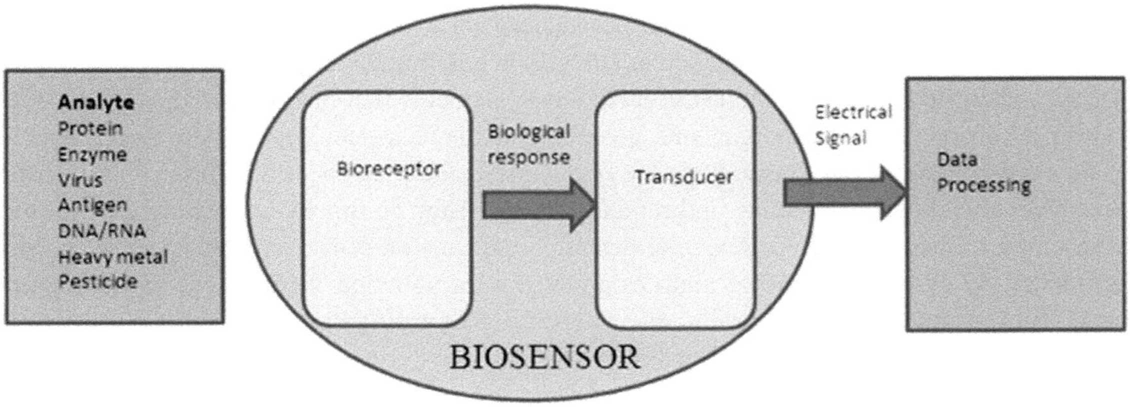

which makes them adaptable for detecting a variety of analytes in environmental samples (Geng et al., 2017; Justino et al., 2017).

Another essential part of biosensors are transducers, which turn the biological response produced by the recognition element into a signal that can be measured. Transducers are frequently made with optical or electrical systems interfaces when used in environmental monitoring applications For example, electrochemical transducers can produce an electrical signal in response to the interaction between the recognition element and the analyte. To identify the presence of the target analyte, optical transducers may rely on variations in light absorption, fluorescence, or luminescence (Pohanka, 2018). These transduction processes are necessary for distant sensing, real-time data provision, and continuous monitoring of pollutants and environmental conditions.

The sensitivity and specificity of biosensors are of utmost importance when detecting trace amounts of environmental contaminants accurately. High sensitivity allows for the detection of analytes at low concentrations, which is crucial when monitoring pollutants or microorganisms in the environment. Specificity ensures that the biosensor responds only to the target analyte, minimizing false-positive results (Hashem et al., 2018). Achieving these qualities involves optimizing both the biological recognition elements and the transduction mechanisms. Selection of the appropriate recognition element with high affinity for the target analyte and engineering transducers for optimal signal amplification are key considerations to enhance sensitivity and specificity (Justino et al., 2017; Pohanka et al., 2018).

In the food industry, bisensors are integral for monitoring both temperature and humidity levels in storage facilities (Kryuk et al., 2023), ensuring optimal conditions for food preservation and safety. In the pharmaceutical industry, they play a crucial role in quality control by simultaneously measuring parameters like pH and conductivity during drug manufacturing processes, ensuring product consistency and efficacy (Polshettiwar et al., 2021). In biomedicine and healthcare (Mohankumar et al., 2021), bisensors are utilized for real-time monitoring of physiological parameters such as glucose levels and blood pressure, enabling precise diagnosis and personalized treatment regimens. In industrial settings, bisensors contribute to process optimization by monitoring variables like pressure and flow rate simultaneously, enhancing efficiency and product quality in manufacturing operations (Kišija et al., 2020). These applications demonstrate the versatility and importance of bisensors in enabling multifaceted monitoring and control across various sectors.

Biological recognition components and transducers are used by biosensors for environmental monitoring to identify and measure particular analytes in environmental samples. Enzymes, antibodies, DNA, and entire cells are examples of recognition elements that offer selectivity by interacting with specific substances or microorganisms, guaranteeing the biosensor's capacity to target particular environmental contaminants. Transducers are frequently made to connect with optical or electronic systems for real-time data collection and remote sensing. They do this by converting the biological reaction into a measurable signal. Accurate detection depends on sensitivity and specificity, which are optimized by carefully choosing and designing the transduction mechanisms and recognition elements. Biosensors are still essential to environmental monitoring because they help us recognize environmental problems and find solutions.

TYPES OF BIOSENSORS FOR ENVIRONMENTAL MONITORING

Enzymatic Biosensors

Enzymatic biosensors, a prominent category of biosensors, are widely employed in environmental monitoring due to their capacity to detect a diverse range of environmental parameters. These biosensors are instrumental in assessing parameters such as biochemical oxygen demand, heavy metals, and organic pollutants, making them indispensable tools for environmental researchers and regulators.

One of the key features that render enzymatic biosensors particularly valuable for environmental monitoring is their ability to capitalize on enzymatic reactions. Enzymes, as biological recognition elements, exhibit high specificity for particular substrates. When applied to environmental analysis, enzymes catalyze reactions with the target analytes, converting them into detectable signals. This catalytic activity is crucial for the accurate and selective measurement of various environmental parameters. For instance, in monitoring BOD, enzymes like glucose oxidase are used to catalyze the oxidation of organic compounds, generating a measurable electrochemical signal (Radhakrishnan et al., 2022). In this context, the enzyme's activity is directly proportional to the concentration of organic material in the water, providing a quantitative assessment of water pollution.

Additionally, enzymatic biosensors are widely utilized in the detection of heavy metals in environmental samples. Heavy metals, such as lead, cadmium, and mercury, can have severe ecological and health impacts when present in elevated concentrations. Enzymatic biosensors, leveraging the specificity of enzymes for certain metal ions, are capable of detecting these contaminants at low levels. For instance, the enzyme alkaline phosphatase can be used in biosensors to detect heavy metal ions by inhibiting its enzymatic activity in the presence of these ions (Mishra et al., 2018).

Moreover, enzymatic biosensors are essential tools for monitoring organic pollutants in environmental samples. These pollutants can originate from industrial discharges, agricultural runoff, or various human activities, and they can pose significant risks to aquatic ecosystems and human health. Enzymes such as acetylcholinesterase or tyrosinase have been employed to detect specific organic pollutants through enzymatic reactions, producing quantifiable signals proportional to the pollutant concentration (Rai et al., 2016; Li et al., 2023).

Immunological Biosensors

Immunological biosensors represent a critical category of biosensors that are extensively applied in environmental monitoring. These sensors play a pivotal role in recognizing and quantifying specific environmental antigens, including pathogens, toxins, and allergens. Their application is particularly significant in the monitoring of water and air quality, where the identification and quantification of these specific agents are essential for ensuring the safety of ecosystems and public health.

The core principle of immunological biosensors involves the use of antibodies as the biological recognition element. Antibodies exhibit an exceptional degree of specificity in binding to antigens. When applied to environmental monitoring, immunological biosensors can specifically recognize and quantify a range of environmental antigens, offering a high level of selectivity in the process.

In the context of water quality monitoring, immunological biosensors are instrumental in the detection of waterborne pathogens. Pathogenic microorganisms, such as bacteria and viruses, can pose significant threats to human health when present in water sources. Immunological biosensors, equipped with antibodies designed to target specific pathogens, enable the rapid and precise identification of these microorganisms. This is crucial for timely responses to potential disease outbreaks and ensuring safe drinking water (Zhao et al., 2023).

Immunological biosensors also have vital applications in air quality monitoring, particularly in detecting allergens and toxins. Allergens can be found in the air and can lead to respiratory issues, making their detection important for public health. Toxins, such as those produced by mold or harmful airborne chemicals, can pose health risks when inhaled. Immunological biosensors equipped with antibodies specific to these allergens or toxins can accurately detect and quantify their presence, providing data for assessing air quality and mitigating potential health hazards (Aghababai et al., 2022).

These biosensors allow for the accurate identification and measurement of pathogens, poisons, allergens, and other pertinent chemicals by utilizing the high specificity of antibodies for specific environmental antigens. Protecting public health, maintaining ecosystem health, and mitigating possible environmental hazards all depend on this competence.

DNA-Based Biosensors

DNA-based biosensors have gained significant recognition in environmental monitoring due to their unique capability to identify genetic material from environmental microorganisms. These biosensors have proven invaluable in a variety of applications, including the detection of pathogens, genetically modified organisms (GMOs), and the assessment of biodiversity.

One of the primary applications of DNA-based biosensors in environmental monitoring is the detection of pathogens. Pathogenic microorganisms, whether in water sources, soil, or air, can pose serious health risks to humans, animals, and ecosystems. DNA-based biosensors can target specific DNA sequences unique to these pathogens. When environmental samples are tested using these biosensors, they can accurately identify the presence of pathogenic DNA. This capability enables early and precise detection of potential disease outbreaks and helps guide public health measures (Ali et al., 2021).

Another crucial application of DNA-based biosensors is the detection of genetically modified organisms (GMOs). The introduction of GMOs into the environment, whether intentionally or unintentionally, can have ecological and health implications. DNA-based biosensors, by targeting specific DNA sequences characteristic of GMOs, provide a reliable means of identifying their presence in the environment. This is vital for regulatory compliance, biodiversity preservation, and risk assessment (Bennett et al., 2019).

Additionally, DNA-based biosensors are used in biodiversity assessment. Environmental DNA (eDNA) is genetic material shed by various organisms into their surroundings. By analyzing eDNA using DNA-based biosensors, researchers can gain insights into the presence and diversity of species in a given ecosystem. This non-invasive approach has revolutionized biodiversity assessment, allowing for the monitoring of aquatic and terrestrial ecosystems without disturbing the organisms themselves (Deiner et al., 2017).

Whole-Cell Biosensors

Whole-cell biosensors represent a cutting-edge and versatile category of biosensors used in environmental monitoring. They utilize live microorganisms, such as bacteria or yeast, as the biological recognition elements, and these organisms are engineered to monitor changes in their physiology in response to exposure to environmental contaminants. These sensors have the unique advantage of being adaptable to detect a wide range of analytes in various environmental samples.

The fundamental principle of whole-cell biosensors is based on the fact that microorganisms respond to the presence of specific analytes by triggering changes in their physiology (Yagi, 2007). These changes can include alterations in gene expression, enzyme activity, or metabolic pathways. By introducing genetic modifications, researchers can tailor the response of these microorganisms to target analytes of interest. This engineering process allows for the creation of biosensors with remarkable specificity and sensitivity, making them suitable for the detection of a wide array of contaminants in the environment (Van der Meer & Belkin, 2010).

One of the key strengths of whole-cell biosensors is their versatility. These sensors can be designed to detect a broad spectrum of environmental analytes, including heavy metals, organic pollutants, toxins, and various chemicals. For example, when engineered to respond to heavy metals, whole-cell biosensors can exhibit changes in fluorescence, bioluminescence, or electrical conductance when exposed to these contaminants (Roggo et al., 2017).

Whole-cell biosensors are particularly valuable in environmental monitoring scenarios where there is a need to assess the presence of multiple analytes simultaneously. For instance, in the context of monitoring water quality, these sensors can be designed to detect various pollutants, providing comprehensive data for assessing the overall environmental health. Their ability to produce real-time, in-situ data makes them indispensable for field applications, where timely responses to environmental changes are crucial.

Optical and Electrochemical Biosensors

Optical and electrochemical biosensors are two major classes of biosensors that find extensive application in the realm of environmental monitoring. These biosensors are capable of detecting a wide range of environmental parameters, including pH, heavy metals, and dissolved gases. Their attributes, such as rapid response times and compatibility with data transmission systems, make them ideal for in-situ and remote monitoring applications.

Optical biosensors rely on the measurement of light-related properties to detect changes in the environment. They offer exceptional sensitivity and are particularly useful for monitoring various environmental parameters (Sharma & Sharma, 2023), (Chen & Wang, 2020).Optical biosensors are frequently used for pH measurements in environmental samples. They function by detecting changes in the absorption, fluorescence, or luminescence properties of indicator dyes in response to alterations in pH. These sensors are valuable for assessing water quality and soil conditions. They can also be engineered to detect heavy metals through the use of specific fluorescent dyes or quantum dots. When heavy metals bind to these recognition elements, they induce changes in the optical properties, allowing for their sensitive and selective detection. Certain optical biosensors can also be designed to detect dissolved gases, including oxygen and carbon dioxide. These sensors utilize luminescent indicators to monitor changes in gas concentrations, making them suitable for applications in environmental science and aquatic research.

Electrochemical biosensors, on the other hand, are based on the measurement of electrical signals generated by biochemical reactions at the sensor's surface. These sensors are known for their high sensitivity, rapid response, and applicability to various environmental parameters (Ariyant et al., 2020),(Cesewski et al., 2020), (Kaya et al., 2021) . Electrochemical biosensors are widely used to monitor pH levels and ion concentrations in environmental samples. These sensors rely on pH-sensitive or ion-selective electrodes that produce electrical signals in response to changes in the sample's ionic composition. They are critical for understanding water quality and soil conditions. They can detect heavy metals through specific redox reactions that occur when the metal ions interact with the recognition elements on the sensor's surface. This capability is essential for assessing contamination levels in water sources. Electrochemical biosensors are also used for monitoring dissolved gases like oxygen, which is crucial for assessing water quality and the health of aquatic ecosystems. These sensors often utilize amperometric or potentiometric techniques to measure gas concentrations.

Both optical and electrochemical biosensors have the distinct advantage of being well-suited for in-situ and remote monitoring. They can be integrated with data transmission systems, providing real-time data, and can be deployed in field applications to assess and respond to environmental changes promptly. This makes them invaluable tools for safeguarding the environment and ensuring the health and sustainability of ecosystems.

APPLICATIONS OF BIOSENSORS IN ENVIRONMENTAL MONITORING

Water Quality Monitoring

Water quality monitoring is a critical aspect of environmental management and public health. Biosensors have emerged as indispensable tools for the assessment of water quality in a variety of settings, including natural bodies of water, wastewater treatment plants, and industrial facilities. These biosensors offer a rapid and accurate means of detecting a wide range of contaminants, ensuring compliance with environmental regulations and safeguarding the health of ecosystems and human populations (Adekunle et al., 2021), (Laad & Ghule, 2023).

Detection of Heavy Metals

Biosensors are frequently used to monitor water quality for heavy metal contaminants. Heavy metals such as lead, cadmium, and mercury can enter water sources through industrial discharges, agricultural runoff, and other sources. Biosensors designed to detect heavy metals employ specific biological recognition elements, such as enzymes or antibodies, that interact with these contaminants. When the heavy metals bind to the recognition element, they induce a change in the sensor's signal, which can be measured quantitatively. This ability to detect heavy metals is crucial for assessing water quality, ensuring regulatory compliance, and protecting aquatic ecosystems (Karthik et al., 2022), (Velusamy, et al., 2022).

Detection of Microorganisms

Water quality monitoring also involves the detection of microorganisms, including bacteria, viruses, and protozoa, which can pose health risks if present in high concentrations. Biosensors designed for microorganism detection use biological recognition elements, often antibodies or DNA probes, which are specific to the target microorganism. These biosensors can provide rapid and sensitive measurements of microbial contamination, enabling the timely response to potential outbreaks of waterborne diseases. They are essential for ensuring the safety of drinking water and recreational water sources (Nayak et al., 2009), (Wang et al., 2017).

Detection of Organic Pollutants

Organic pollutants, such as pesticides, industrial chemicals, and pharmaceuticals, can also enter water sources and impact water quality. Biosensors are well-suited for the detection of these organic contaminants. Enzymatic biosensors, for example, can be engineered to catalyze specific reactions with organic compounds, generating signals proportional to their concentration. This enables the accurate measurement of organic pollutant levels in water samples. It is crucial for assessing the impact of human activities on aquatic ecosystems and the safety of water supplies (Zhao et al., 2021).

Air Quality Monitoring

Air quality monitoring is a crucial component of environmental protection and public health. Biosensors have emerged as essential tools for monitoring air quality by enabling the detection of a wide range of airborne pollutants, including gases, particulate matter, and biological agents. Real-time data generated by these biosensors is invaluable for assessing air pollution and implementing mitigation strategies.

Detection of Gases

One of the primary applications of biosensors in air quality monitoring is the detection of gases. Various gases, including common air pollutants such as carbon monoxide (CO), nitrogen dioxide (NO2), sulfur dioxide (SO2), and volatile organic compounds (VOCs), can have adverse effects on human health and the environment. Biosensors, equipped with specific recognition elements that interact with these gases, provide rapid and sensitive measurements. When these gases bind to the recognition elements, they induce changes in the sensor's signal, which can be quantified. This real-time data allows for the timely assessment of air quality and the implementation of measures to reduce exposure to harmful gases (Sokhi et al., 2022), (Usman et al., 2023).

Particulate Matter Monitoring

Biosensors are also utilized to monitor particulate matter (PM) in the air. PM includes tiny solid particles and liquid droplets suspended in the air and is categorized based on size, with PM2.5 (particles with a diameter of 2.5 micrometers or smaller) and PM10 (particles with a diameter of 10 micrometers or smaller) being of particular concern due to their potential health impacts. Biosensors, often employing immunological recognition elements, can detect specific proteins or antigens associated with PM. This information helps in assessing the concentration of PM in the air and its potential health effects. Real-time PM data is essential for implementing air quality control measures, particularly in urban areas (Nieckarz et al., 2023).

Biological Agent Detection

In addition to chemical pollutants, biosensors can be engineered to detect biological agents in the air, such as bacteria, viruses, and allergens. These agents can have implications for public health, particularly in indoor environments or healthcare facilities. Biosensors equipped with antibodies or DNA probes specific to these agents enable the rapid identification and quantification of biological contaminants in the air. This capability is vital for implementing infection control measures and ensuring the safety of indoor air quality (Jakupciak et al., 2009).

Soil and Agriculture

In agriculture, biosensors have become invaluable tools for monitoring soil health and optimizing crop management practices. These sensors are designed to detect various parameters related to soil quality, nutrient levels, and the presence of contaminants, ultimately contributing to sustainable and environmentally friendly agricultural practices.

Soil Health Monitoring

Biosensors play a vital role in assessing soil health, which is essential for maintaining soil fertility and productivity. Soil health parameters, such as pH, electrical conductivity, and organic matter content, can be measured using biosensors (Archeka et al., 2022), (Mandal, et al., 2020). These sensors often employ enzymatic or microbial recognition elements that react with specific soil components. Monitoring these parameters helps farmers and agricultural researchers make informed decisions about soil management practices, including soil amendment and fertilization.

Nutrient Level Detection

Nutrient monitoring is a crucial aspect of agriculture, as it allows for precise nutrient management and reduces the risk of over-fertilization (Sempionatto, et al., 2020). Biosensors can detect essential nutrients such as nitrogen (N), phosphorus (P), and potassium (K) in the soil. By providing real-time data on nutrient concentrations, these sensors enable farmers to adjust their fertilizer application rates based on the specific needs of their crops. This not only maximizes crop yield but also minimizes nutrient runoff into water bodies, mitigating environmental pollution.

Contaminant Detection

Agricultural biosensors are also employed to detect contaminants in the soil, such as pesticides, heavy metals, and organic pollutants. Pesticide biosensors, for instance, utilize enzymes or antibodies to detect the presence of specific pesticides in soil samples (Zhao, et al., 2015). This information allows farmers to assess the impact of pesticide use on soil quality and make informed decisions about pesticide application (Xie et al., 2022).

FUTURE PROSPECTS AND CHALLENGES

Biosensor technology has made significant strides in various fields, from healthcare to environmental monitoring. Looking ahead, several prospects and challenges are likely to shape the future of biosensors:

Miniaturization and Integration

Miniaturization of biosensors will continue, enabling the development of portable and wearable devices for real-time monitoring. Integration with other technologies, such as microfluidics, nanomaterials, and wireless communication, will further enhance the capabilities of biosensors (Liu et al., 2020). Miniaturization raises challenges in terms of signal amplification and robustness, as smaller sensors may be more susceptible to interference and require improved signal processing techniques .

Multi-Parameter Sensing

The ability to measure multiple parameters simultaneously is a promising avenue. Multi-parameter biosensors will offer a comprehensive view of complex systems, be it in healthcare diagnostics or environmental monitoring, leading to more accurate and actionable data (Pérez-López et al., 2011). Designing biosensors that can efficiently detect and differentiate multiple analytes without cross-reactivity remains a challenge. Ensuring the reliability and specificity of multi-parameter sensors is essential (Choi et al., 2019).

Data Management and Analysis

Advanced data analysis techniques, including machine learning and artificial intelligence, will be integrated with biosensors to extract meaningful insights from complex datasets. This will enhance the predictive and diagnostic capabilities of biosensors (Hasanzadeh et al., 2017). Handling, storing, and securing large volumes of data generated by biosensors can be challenging. Data privacy and ethical concerns will also need to be addressed (Fan et al., 2023).

Standardization

Standardization in biosensor development will ensure the reproducibility and reliability of results. Internationally recognized standards can facilitate regulatory approvals and promote wider adoption of biosensor technology (Liu et al., 2021). Establishing global standards that encompass the diverse range of biosensors and applications can be complex. Coordination among regulatory agencies and research communities is needed.

CONCLUSION

Biosensors have become essential instruments for environmental monitoring, enabling us to monitor and alleviate the effects of pollutants, pollutants, and climate change on Earth. Their ongoing advancement and integration with state-of-the-art technology portend a time when we may rely on real-time, high-quality data to support biodiversity conservation, environmental protection, and the sustainability of our natural resource base. This chapter gives an extensive review of working phenomena, types and applications of biosensors for environmental applications. Future prospects and challenges are also discussed to pave the path for better, cleaner environment.

REFERENCES

Adekunle, A., Rickwood, C., & Tartakovsky, B. (2021). On-line monitoring of water quality with a floating microbial fuel cell biosensor: Field test results. *Ecotoxicology (London, England)*, *30*(5), 851–862. doi:10.1007/s10646-021-02409-2 PMID:33851335

Aghababai Beni, A., & Jabbari, H. (2022). *Nanomaterials for Environmental Applications. Results in Engineering*. Elsevier B.V., doi:10.1016/j.rineng.2022.100467

Ali, M. R., Bacchu, M. S., Setu, M. A. A., Akter, S., Hasan, M. N., Chowdhury, F. T., Rahman, M. M., Ahommed, M. S., & Khan, M. Z. H. (2021). Development of an advanced DNA biosensor for pathogenic Vibrio cholerae detection in real sample. *Biosensors & Bioelectronics*, *188*, 113338. Advance online publication. doi:10.1016/j.bios.2021.113338 PMID:34030094

Archeka, C. N., Neelam, Kusum, & Hooda, V. (2022). Nanobiosensors for Monitoring Soil and Water Health. In Nanotechnology in Agriculture and Environmental Science (pp. 183–202). CRC Press. doi:10.1201/9781003323945-15

Ariyanti, D., Iswantini, D., Sugita, P., Nurhidayat, N., Effendi, H., Ghozali, A. A., & Kurniawan, Y. S. (2020). Highly Sensitive Phenol Biosensor Utilizing Selected Bacillus Biofilm Through an Electrochemical Method. *Makara Journal of Science*, *24*(1), 24–30. doi:10.7454/mss.v24i1.11726

Bennett, A. B., Chi-Ham, C., Barrows, G., Sexton, S., & Zilberman, D. (2013). Agricultural biotechnology: Economics, environment, ethics, and the future. *Annual Review of Environment and Resources*, *38*(1), 249–279. doi:10.1146/annurev-environ-050912-124612

Cesewski, E., & Johnson, B. N. (2020). *Electrochemical biosensors for pathogen detection. Biosensors and Bioelectronics*. Elsevier Ltd., doi:10.1016/j.bios.2020.112214

Chen, C., & Wang, J. (2020). *Optical biosensors: An exhaustive and comprehensive review. Analyst*. Royal Society of Chemistry., doi:10.1039/C9AN01998G

Choi, J. R., Yong, K. W., Choi, J. Y., & Cowie, A. C. (2019, February 2). Emerging point-of-care technologies for food safety analysis. *Sensors (Switzerland)*. MDPI AG. doi:10.3390/s19040817

Deiner, K., Bik, H. M., Mächler, E., Seymour, M., Lacoursière-Roussel, A., Altermatt, F., & Bernatchez, L. (2017). *Environmental DNA metabarcoding: Transforming how we survey animal and plant communities. Molecular Ecology*. Blackwell Publishing Ltd. doi:10.1111/mec.14350

Fan, C., Hu, K., Yuan, Y., & Li, Y. (2023). A data-driven analysis of global research trends in medical image: A survey. *Neurocomputing*, *518*, 308–320. doi:10.1016/j.neucom.2022.10.047

Geng, Z., Zhang, X., Fan, Z., Lv, X., Su, Y., & Chen, H. (2017). *Recent progress in optical biosensors based on smartphone platforms. Sensors*. MDPI AG. doi:10.3390/s17112449

Hasanzadeh, M., & Shadjou, N. (2017). *Advanced nanomaterials for use in electrochemical and optical immunoassays of carcinoembryonic antigen. A review*. Microchimica Acta. Springer-Verlag Wien. doi:10.1007/s00604-016-2066-2

Hashem, A., Hossain, M. A. M., Marlinda, A. R., Mamun, M. A., Simarani, K., & Johan, M. R. (2021). *Nanomaterials based electrochemical nucleic acid biosensors for environmental monitoring: A review*. Applied Surface Science Advances. Elsevier B.V. doi:10.1016/j.apsadv.2021.100064

Huang, C. W., Lin, C., Nguyen, M. K., Hussain, A., Bui, X. T., & Ngo, H. H. (2023). *A review of biosensor for environmental monitoring: principle, application, and corresponding achievement of sustainable development goals. Bioengineered. NLM*. Medline. doi:10.1080/21655979.2022.2095089

Jakupciak, J. P., & Colwell, R. R. (2009). Biological agent detection technologies. *Molecular Ecology Resources*, *9*(s1, SUPPL. 1), 51–57. doi:10.1111/j.1755-0998.2009.02632.x PMID:21564964

Justino, C. I. L., Duarte, A. C., & Rocha-Santos, T. A. P. (2017). *Recent progress in biosensors for environmental monitoring: A review. Sensors*. MDPI AG. doi:10.3390/s17122918

Karthik, V., Karuna, B., Kumar, P. S., Saravanan, A., & Hemavathy, R. V. (2022). Development of lab-on-chip biosensor for the detection of toxic heavy metals: A review. *Chemosphere*, *299*, 134427. doi:10.1016/j.chemosphere.2022.134427 PMID:35358561

Kaya, H. O., Cetin, A. E., Azimzadeh, M., & Topkaya, S. N. (2021). *Pathogen detection with electrochemical biosensors: Advantages, challenges and future perspectives. Journal of Electroanalytical Chemistry*. Elsevier B.V., doi:10.1016/j.jelechem.2021.114989

Kišija, E., Osmanović, D., Nuhić, J., & Cifrić, S. (2020). Review of biosensors in industrial process control. In I. F. M. B. E. Proceedings (Ed.), Vol. 73, pp. 687–694). Springer Verlag., doi:10.1007/978-3-030-17971-7_103

Kryuk, R., Mukhsin, M.-Z., Kurbanova, M., & Kryuk, V. (2023). Biosensors in Food Industry. *BIO Web of Conferences*. IEEE. 10.1051/bioconf/20236401006

Laad, M., & Ghule, B. (2023). *Removal of toxic contaminants from drinking water using biosensors: A systematic review*. Groundwater for Sustainable Development. Elsevier B.V., doi:10.1016/j.gsd.2022.100888

Li, J., Chang, H., Zhang, N., He, Y., Zhang, D., Liu, B., & Fang, Y. (2023). *Recent advances in enzyme inhibition based-electrochemical biosensors for pharmaceutical and environmental analysis*. Talanta. Elsevier B.V., doi:10.1016/j.talanta.2022.124092

Liu, D., Wang, J., Wu, L., Huang, Y., Zhang, Y., Zhu, M., & Yang, C. (2020). *Trends in miniaturized biosensors for point-of-care testing. TrAC - Trends in Analytical Chemistry*. Elsevier B.V., doi:10.1016/j.trac.2019.115701

Liu, G. (2021). Grand Challenges in Biosensors and Biomolecular Electronics. *Frontiers in Bioengineering and Biotechnology*, *9*, 707615. doi:10.3389/fbioe.2021.707615 PMID:34422782

Mandal, N., Adhikary, S., & Rakshit, R. (2020). Nanobiosensors: Recent Developments in Soil Health Assessment. In Soil Analysis: Recent Trends and Applications (pp. 285–304). Springer Singapore. doi:10.1007/978-981-15-2039-6_15

Mishra, R. K., Rhouati, A., Bueno, D., Anwar, M. W., Shahid, S. A., Sharma, V., & Hayat, A. (2018). Design of a portable luminescence bio-tool for on-site analysis of heavy metals in water samples. *International Journal of Environmental Analytical Chemistry*, *98*(12), 1081–1094. doi:10.1080/03067319.2018.1521395

Mohankumar, P., Ajayan, J., Mohanraj, T., & Yasodharan, R. (2021). Recent developments in biosensors for healthcare and biomedical applications: A review. Measurement. *Measurement*, *167*, 108293. doi:10.1016/j.measurement.2020.108293

Nayak, M., Kotian, A., Marathe, S., & Chakravortty, D. (2009). Detection of microorganisms using biosensors-A smarter way towards detection techniques. *Biosensors & Bioelectronics*, *25*(4), 661–667. doi:10.1016/j.bios.2009.08.037 PMID:19782558

Nieckarz, Z., Pawlak, K., Baran, A., Wieczorek, J., Grzyb, J., & Plata, P. (2023). The concentration of particulate matter in the barn air and its influence on the content of heavy metals in milk. *Scientific Reports*, *13*(1), 10626. doi:10.1038/s41598-023-37567-2 PMID:37391588

Onyancha, R. B., Ukhurebor, K. E., Aigbe, U. O., Osibote, O. A., Kusuma, H. S., Darmokoesoemo, H., & Balogun, V. A. (2021). *A systematic review on the detection and monitoring of toxic gases using carbon nanotube-based biosensors*. Sensing and Bio-Sensing Research. Elsevier B.V., doi:10.1016/j.sbsr.2021.100463

Pérez-López, B., & Merkoçi, A. (2011). Nanomaterials based biosensors for food analysis applications. *Trends in Food Science & Technology*, *22*(11), 625–639. doi:10.1016/j.tifs.2011.04.001

Pohanka, M. (2018). *Overview of piezoelectric biosensors, immunosensors and DNA sensors and their applications*. Materials. MDPI AG., doi:10.3390/ma11030448

Polshettiwar, S. A., Deshmukh, C. D., Wani, M. S., Baheti, A. M., Bompilwar, E., Choudhari, S., Jambhekar, D., & Tagalpallewar, A. (2021). Recent Trends on Biosensors in Healthcare and Pharmaceuticals: An Overview. *International Journal of Pharmaceutical Investigation*, *11*(2), 131–136. doi:10.5530/ijpi.2021.2.25

Radhakrishnan, S., Lakshmy, S., Santhosh, S., Kalarikkal, N., Chakraborty, B., & Rout, C. S. (2022). *Recent Developments and Future Perspective on Electrochemical Glucose Sensors Based on 2D Materials*. Biosensors. MDPI., doi:10.3390/bios12070467

Rai, M., Ingle, A. P., Birla, S., Yadav, A., & Santos, C. A. D. (2016). *Strategic role of selected noble metal nanoparticles in medicine. Critical Reviews in Microbiology*. Taylor and Francis Ltd., doi:10.3109/1040841X.2015.1018131

Roggo, C., & van der Meer, J. R. (2017). *Miniaturized and integrated whole cell living bacterial sensors in field applicable autonomous devices. Current Opinion in Biotechnology*. Elsevier Ltd., doi:10.1016/j.copbio.2016.11.023

Sempionatto, J. R., Khorshed, A. A., Ahmed, A., De Loyola, E., Silva, A. N., Barfidokht, A., Yin, L., & Wang, J. (2020). Epidermal Enzymatic Biosensors for Sweat Vitamin C: Toward Personalized Nutrition. *ACS Sensors*, *5*(6), 1804–1813. doi:10.1021/acssensors.0c00604 PMID:32366089

Sharma, K., & Sharma, M. (2023). *Optical biosensors for environmental monitoring: Recent advances and future perspectives in bacterial detection. Environmental Research*. Academic Press Inc., doi:10.1016/j.envres.2023.116826

SokhiR. S.MoussiopoulosN.BaklanovA.BartzisJ.CollI.FinardiS.KukkonenJ. (2022). Advances in air quality research - current and emerging challenges. Atmospheric Chemistry and Physics. Copernicus GmbH. doi:10.5194/acp-22-4615-2022

Usman, F., Ghazali, K. H., Muda, R., Dennis, J. O., Ibnaouf, K. H., Aldaghri, O. A., & Jose, R. (2023). *Detection of Kidney Complications Relevant Concentrations of Ammonia Gas Using Plasmonic Biosensors: A Review*. Chemosensors. MDPI. doi:10.3390/chemosensors11020119

Van Der Meer, J. R., & Belkin, S. (2010). Where microbiology meets microengineering: Design and applications of reporter bacteria. *Nature Reviews. Microbiology*, *8*(7), 511–522. doi:10.1038/nrmicro2392 PMID:20514043

Velusamy, K., Periyasamy, S., Kumar, P. S., Rangasamy, G., Nisha Pauline, J. M., Ramaraju, P., & Nguyen Vo, D. V. (2022). Biosensor for heavy metals detection in wastewater: A review. *Food and Chemical Toxicology*, *168*, 113307. doi:10.1016/j.fct.2022.113307 PMID:35917955

Wang, C., Madiyar, F., Yu, C., & Li, J. (2017). Detection of extremely low concentration waterborne pathogen using a multiplexing self-referencing SERS microfluidic biosensor. *Journal of Biological Engineering*, *11*(1), 9. doi:10.1186/s13036-017-0051-x PMID:28289439

Xie, M., Zhao, F., Zhang, Y., Xiong, Y., & Han, S. (2022). *Recent advances in aptamer-based optical and electrochemical biosensors for detection of pesticides and veterinary drugs. Food Control*. Elsevier Ltd. doi:10.1016/j.foodcont.2021.108399

Yagi, K. (2007). Applications of whole-cell bacterial sensors in biotechnology and environmental science. *Applied Microbiology and Biotechnology*, *73*(6), 1251–1258. doi:10.1007/s00253-006-0718-6 PMID:17111136

Zhao, K., Veksha, A., Ge, L., & Lisak, G. (2021). Near real-time analysis of para-cresol in wastewater with a laccase-carbon nanotube-based biosensor. *Chemosphere*, *269*, 128699. doi:10.1016/j.chemosphere.2020.128699 PMID:33121813

Zhao, Y., Huang, F., Wang, W., Gao, R., Fan, L., Wang, A., & Gao, S. H. (2023). *Application of high-throughput sequencing technologies and analytical tools for pathogen detection in urban water systems: Progress and future perspectives. Science of the Total Environment*. Elsevier B.V., doi:10.1016/j.scitotenv.2023.165867

Chapter 14
Advances in Sensor Technologies for Detecting Soil Pollution

Ranjit Singha
https://orcid.org/0000-0002-3541-8752
Christ University, India

Surjit Singha
https://orcid.org/0000-0002-5730-8677
Kristu Jayanti College (Autonomous), India

V. Muthu Ruben
https://orcid.org/0009-0006-7723-8596
Christ University, India

Alphonsa Diana Haokip
https://orcid.org/0000-0003-2578-0114
Christ King High School, India

ABSTRACT

The present chapter elucidates progressions in the surveillance of soil pollution, with a specific emphasis on integrated systems and sensor technologies. Future trends (e.g., enhanced selectivity, regulatory adoption), deployment platforms (field-deployable, wireless networks), and sensor types (electrochemical, optical, and biosensors) are discussed. Increasing sensitivity and specificity, facilitating on-site, real-time analysis, and integrating sensing with remediation strategies are priorities. The discourse highlights the revolutionary capacity that soil pollution sensors possess to propel environmental monitoring and management forward. Collaboration among stakeholders is critical for successfully implementing sensor-based approaches and driving innovation.

INTRODUCTION

DOI: 10.4018/979-8-3693-1930-7.ch014

Globally, soil contamination by inorganic and organic contaminants is a significant environmental concern. Due to industrial activities and improper waste disposal, soils have become contaminated with numerous hazardous substances, such as emerging contaminants, pesticides, heavy metals, and hydrocarbons. Ingestion or direct exposure to these soil contaminants may result in severe health hazards for humans, including diminished soil fertility, groundwater and crop contamination, and accumulation in the food chain. So, quickly, cheaply, and accurately finding soil contaminants is very important for describing sites, keeping track of contamination levels, and judging how well cleanup efforts are working (He & Cai, 2021; Zaghloul et al., 2019). Gas chromatography and mass spectrometry are conventional analytical methods for measuring soil pollutants, are costly, time-consuming, and offer limited spatial information. Recent developments in sensor technologies hold significant potential for enhanced in-situ and real-time monitoring and detection of soil pollution. These sensors can identify and quantify contaminants in soil samples using sensing mechanisms, including optical, electrochemical, biological, and physical processes. These devices are well-suited for field applications and continuous monitoring due to their portability, affordability, rapid response times, and high sensitivity (Abdulraheem et al., 2023; Nadporozhskaya et al., 2022).

This chapter examines recent advancements in sensor technologies for detecting soil pollution, focusing on their operating principles, performance characteristics, applications, and challenges. This study aims to look into what could be done by connecting sensors to wireless communication networks, data acquisition systems, and geographic information systems (GIS) to make it easier to map out soil contamination and keep an eye on it from afar. Furthermore, the paper will examine the potential of sensor networks and drones to augment the resolution and spatial coverage of monitoring initiatives about soil pollution. The chapter endeavours to underscore the significance of sensor technologies in furthering our comprehension of the dynamics of soil contamination and bolstering the efficacy of environmental management approaches. Various human activities, including industrial operations, agricultural methods, and inadequate refuse management, cause substantial soil pollution, a significant global environmental issue. Soil contamination by various substances, such as organic pollutants, pesticides, and heavy metals, can harm ecosystem health, agricultural output, and human welfare. The assessment of contamination levels, identification of pollutant sources, and implementation of remediation strategies to mitigate environmental impacts all rely heavily on the efficacy of soil pollution monitoring (Cui et al., 2021). Theoretical underpinnings for soil pollution monitoring comprise engineering, chemistry, environmental science, and biology principles. It involves comprehending the mechanisms by which pollutants degrade in soil matrices, including adsorption, leaching, and microbial degradation. Theoretical models and concepts from these disciplines guide the design and development of sensor technologies, data analysis techniques, and remediation strategies to tackle soil pollution challenges.

Soil pollution monitoring integrates data analytics, sensor technologies, and environmental management strategies. Sensors equipped to identify a diverse array of contaminants in soil samples, along with data processing algorithms that decipher sensor data and pinpoint areas of high pollution, are utilized in the process. Monitoring soil contamination requires knowledge to formulate environmental management strategies, such as land use planning and remediation techniques, to alleviate environmental hazards and advance sustainable land management approaches. Monitoring soil pollution to determine the source of pollutants, identify the extent and severity of contamination in soil environments, and direct remediation efforts to safeguard human health and ecosystems. Soil pollution monitoring involves developing and deploying sensor technologies to detect pollutants, integrating sensor data with environmental models to assess risks, and implementing remediation and prevention management strategies. The objective is to develop an all-encompassing structure for proactively managing soil contamination that considers human health and environmental sustainability.

TYPES OF SOIL POLLUTANTS AND THEIR ENVIRONMENTAL IMPACTS

Soil pollution is caused by various contaminants, each with distinct environmental consequences. Industrial operations, agricultural methodologies, and urban effluent frequently contribute to the infiltration of heavy metals, including arsenic, cadmium, lead, and mercury, into the soil. These metals pose significant hazards to both ecosystems and human health. They are capable of causing soil structure degradation, fertility reduction, and microbial inhibition, all of which contribute to stunted plant growth. Moreover, the infiltration of heavy metals into groundwater can contaminate water sources and give rise to critical health concerns. Additionally, the bioaccumulation of heavy metals along the food chain elevates toxicity levels for organisms occupying higher trophic levels (Liu et al., 2022; Nnaji et al., 2023). The extensive use of pesticides for pest control in agriculture and public health exacerbates soil pollution complications. Surface waters are contaminated by runoff and leaching of pesticides, which have detrimental effects on aquatic life. In addition, using pesticides can potentially disturb the communities and biodiversity of soil microbes, impacting the cycling of nutrients and the overall functioning of ecosystems (Raffa & Chiampo, 2021; Sun et al., 2018).

Urban runoff, petroleum hydrocarbons discharged from oil accidents, and industrial operations significantly contaminate soil and groundwater. These compounds exhibit prolonged persistence in soil, thereby presenting hazards to terrestrial ecosystems and organisms. Significant implications may ensue regarding biodiversity (Chen et al., 2023; Stepanova et al., 2022).

The presence of emerging contaminants, such as pharmaceuticals and personal care products, in soil health is a growing concern due to its novelty and growing significance. These substances demonstrate ecotoxicity, can disrupt the endocrine system, and tend to accumulate and persist in the environment. Constant oversight and effective administration are imperative to confront this escalating issue (Bayabil et al., 2022; Pérez-Sirvent & Bech, 2023).

Soil pollutants induce toxicity in humans via diverse mechanisms, resulting in acute poisoning, chronic diseases, and developmental abnormalities. Furthermore, the distribution and mobility of soil pollutants are determined by their intricate fate and transport processes, which are impacted by soil properties, climate, and land use practices. Implementing comprehensive mitigation and environmental preservation strategies to tackle soil pollution is imperative. It is essential to comprehend the intricacies of soil pollution dynamics to protect the well-being of ecosystems and human populations.

CONVENTIONAL ANALYTICAL TECHNIQUES FOR SOIL POLLUTION MONITORING

Traditional analytical methods are fundamental to the surveillance of soil pollution as they provide accurate and consistent methods for identifying and measuring pollutants. The procedure commences with carefully gathering samples and employing augers or corers to guarantee that all samples are present throughout the area of interest. After that, the samples undergo a series of preparatory procedures, including dehydration, sieving, grinding, and digestion, to remove any impurities efficiently before the subsequent analysis. Chromatographic methodologies, including high-performance liquid chromatography (HPLC) and gas chromatography (GC), play a critical role in the separation and quantification of organic pollutants present in soil samples. In heavy metal analysis, spectroscopic methods such as inductively coupled plasma mass spectrometry (ICP-MS) and atomic absorption spectroscopy (AAS) are indispensable.

Soil contamination monitoring can be made more accessible with immunoassay methods like enzyme-linked immunosorbent assay (ELISA), which quickly identifies specific impurities and expands the range of tests that can be used (Eugenio et al., 2020; Singh et al., 2019). However, several obstacles impede the widespread implementation of these methodologies, including the requirement for expensive equipment, specialized expertise, and labour-intensive sample preparation processes. It frequently restricts their availability, especially in settings with limited resources where regular monitoring is critical. However, it is still imperative to utilize these traditional methods to evaluate environmental hazards precisely and devise efficient remediation approaches (Eugenio et al., 2020; Luo et al., 2023). To mitigate these constraints and improve the long-term viability of soil pollution monitoring initiatives, it is critical to expedite the development of alternative techniques that are quicker, less expensive, and easier for users to implement. Allocating resources towards developing novel technologies and methodologies guarantees the ongoing preservation of environmental health and the effectiveness and efficiency of soil pollution monitoring initiatives.

ELECTROCHEMICAL SENSORS FOR HEAVY METALS AND ORGANIC POLLUTANTS

Electrochemical sensors have become indispensable in detecting and monitoring organic pollutants and heavy metals in soil. This is due to their exceptional sensitivity, quick response times, and capability to conduct measurements on-site. Voltammetric sensors stand out among the primary categories of electrochemical sensors because they rely on measuring the current generated by redox reactions at the sensor's surface. Square wave voltammetry (SWV), cyclic voltammetry (CV), and differential pulse voltammetry (DPV) are all standard methods that can be used to measure minimal amounts of pollutants. It is because they are compassionate and have low detection limits. New developments in this field include adding nanomaterials like graphene and carbon nanotubes, which improve sensitivity and selectivity, and making sensor platforms smaller to make them easier to carry and use in the field (Baranwal et al., 2022; Venkateswara Reddy et al., 2023, p. 30). Potentiometric sensors measure potential differences between a reference electrode and an indicator electrode in response to changes in analyte concentration. It has been observed that ion-selective electrodes (ISEs) frequently detect heavy metals in soil samples. Although these sensors provide advantages such as affordability, user-friendliness, and simplicity, their detection limits may be higher than those of voltammetric sensors. Quite recently, there have been significant steps forward in creating wireless and small sensor systems designed for remote monitoring. New materials have also been used to change electrodes to make them more stable and selective (Baumbauer et al., 2022; Zhai et al., 2022).

Amplimetric biosensors are a crucial variety. Biological recognition elements on the sensor's surface selectively locate target analytes. Enzymatic biosensors have demonstrated the ability to detect organic pollutants in soil samples. The biological recognition element's affinity for the target analyte determines these sensors' substantial specificity and sensitivity. Improvements in microfabrication techniques have enabled the fabrication of more compact and stable biosensor platforms. Furthermore, incorporating nanomaterials has enhanced their sensitivity and capacity for detecting a more comprehensive range of substances (Islam et al., 2023; Kratasyuk et al., 2021). Electrochemical sensors rely on their selectivity, reproducibility, and detection limits. While most voltammetric sensors have low detection limits, some can identify pollutants at concentrations as low as sub-ppb. Surface modification techniques such as molecularly imprinted polymers or selective coatings on electrode surfaces can increase selectivity. The reliability of monitoring is contingent upon reproducibility, and recent developments in sensor fabrication and calibration techniques have enhanced the reproducibility of electrochemical measurements.

Utilizing nanomaterials and microfabrication techniques has significantly improved the functionality and practicality of electrochemical sensors. The utilization of microfabrication methods such as photolithography and microfluidics has enabled the creation of sensor platforms that are more compact, portable, and capable of seamless integration. Furthermore, incorporating nanomaterials—including graphene, metal nanoparticles, and carbon nanotubes—has improved sensors' sensitivity, selectivity, and stability. This has facilitated the advancement of multi-analyte sensor arrays, which can simultaneously detect multiple pollutants in soil samples. Electrochemical sensors offer adaptable and pragmatic solutions for monitoring soil contamination, and their effectiveness in environmental monitoring has significantly improved recently.

OPTICAL SENSORS BASED ON FLUORESCENCE, COLORIMETRY, AND SPECTROSCOPY

The advent of optical sensors utilizing fluorescence, colourimetry, and spectroscopy signifies a significant advancement in monitoring soil contamination. These sensors' remarkable sensitivity, selectivity, and portability primarily contribute to this progress. These sophisticated sensing technologies provide an array of functionalities that empower environmental agencies and researchers to identify and measure soil pollutants with unprecedented efficiency. For example, fluorescence sensors utilize the distinctive characteristic of fluorophores, which is their ability to emit light when excited. This enables the precise detection of specific analytes. Quantum dots and metal-organic frameworks (MOFs) have become frontrunners in this field because of their adaptable characteristics, which permit customized reactions to particular pollutants. Scientists can fabricate fluorescence sensors with improved sensitivity and selectivity using these materials. As a result, these instruments have become indispensable in identifying diverse contaminants present in soil samples.

In contrast, colourimetric sensors detect the presence of analytes through discernible colour changes. Researchers frequently functionalize gold nanoparticles with particular ligands to build numerous colourimetric sensing platforms. These nanoparticles enable the swift and dependable detection of target contaminants. The simplicity and user-friendliness of this method render it appropriate for on-site analysis without intricate instrumentation.

Surface-enhanced Raman spectroscopy (SERS) is an innovative technology for monitoring soil contamination. Nanostructured metal surfaces are used to boost Raman scattering signals in SERS, which gives it a great mix of sensitivity and specificity. This allows for the precise detection and measurement of trace-level contaminants present in soil samples. The attribute mentioned above renders SERS notably advantageous in identifying low-concentration pollutants, including specific organic compounds and heavy metals (Li et al., 2013; Yang et al., 2021).

This new combination of smartphone scanners and miniature spectrometers has also changed how on-site analysis is done, making it possible for scientists to quickly check for soil contamination without using expensive lab equipment. This development enables stakeholders to gather data more efficiently and cost-effectively by improving accessibility and scalability in soil pollution monitoring initiatives (Tobiszewski & Vakh, 2023). Optical sensing technologies provide adaptable and functional solutions for detecting soil pollutants, thereby making substantial strides in environmental monitoring. These technologies establish a solid groundwork for sustainable ecological stewardship by facilitating the prompt and accurate identification of contaminants, vital for protecting ecosystems and human populations. Electrochemical sensors have become indispensable in detecting and monitoring organic pollutants and heavy metals in soil. This is due to their exceptional sensitivity, quick response times, and capability to conduct measurements on-site. Voltammetric sensors stand out among the primary categories of electrochemical sensors because they rely on measuring the current generated by redox reactions at the sensor's surface. Square wave voltammetry (SWV), cyclic voltammetry (CV), and differential pulse voltammetry (DPV) are all standard methods that can be used to measure tiny amounts of pollutants. This is because they are susceptible and have low detection limits. New developments in this field include adding nanomaterials like graphene and carbon nanotubes, which improve sensitivity and selectivity, and making sensor platforms smaller to make them easier to carry and use in the field (Baranwal et al., 2022; Venkateswara Reddy et al., 2023, p. 30).

Potentiometric sensors measure potential differences between a reference electrode and an indicator electrode in response to changes in analyte concentration. It has been observed that ion-selective electrodes (ISEs) frequently detect heavy metals in soil samples. Although these sensors provide advantages such as affordability, user-friendliness, and simplicity, their detection limits may be higher than those of voltammetric sensors. Quite recently, there have been significant steps forward in creating wireless and small sensor systems designed for remote monitoring. New materials have also been used to change electrodes to make them more stable and selective (Baumbauer et al., 2022; Zhai et al., 2022). Additionally, amplimetric biosensors are a crucial variety. Biological recognition elements on the sensor's surface selectively locate target analytes. Enzymatic biosensors have demonstrated the ability to detect organic pollutants in soil samples. The biological recognition element's affinity for the target analyte determines these sensors' substantial specificity and sensitivity. Improvements in microfabrication techniques have enabled the fabrication of more compact and stable biosensor platforms. Furthermore, incorporating nanomaterials has enhanced their sensitivity and capacity for detecting a more comprehensive range of substances (Islam et al., 2023; Kratasyuk et al., 2021).

Electrochemical sensors rely on their selectivity, reproducibility, and detection limits. While most voltammetric sensors have low detection limits, some can identify pollutants at concentrations as low as sub-ppb. Surface modification techniques such as molecularly imprinted polymers or selective coatings on electrode surfaces can increase selectivity. The reliability of monitoring is contingent upon reproducibility, and recent developments in sensor fabrication and calibration techniques have enhanced the reproducibility of electrochemical measurements. Utilizing nanomaterials and microfabrication techniques has significantly improved the functionality and practicality of electrochemical sensors. The utilization of microfabrication methods such as photolithography and microfluidics has enabled the creation of sensor platforms that are more compact, portable, and capable of seamless integration. Furthermore, incorporating nanomaterials—including graphene, metal nanoparticles, and carbon nanotubes—has improved sensors' sensitivity, selectivity, and stability. This has facilitated the advancement of multi-analyte sensor arrays, which can simultaneously detect multiple pollutants in soil samples. Electrochemical sensors offer adaptable and pragmatic solutions for monitoring soil contamination, and their effectiveness in environmental monitoring has significantly improved recently.

BIOSENSORS UTILIZING BACTERIA, ENZYMES, ANTIBODIES, AND APTAMERS

The utilization of biosensors comprising microbes, enzymes, antibodies, and aptamers has significantly transformed the domain of soil pollution monitoring due to their exceptional sensitivity, specificity, and versatility attributes. (Bae et al., 2020) These cutting-edge instruments have surpassed conventional analytical techniques by offering a comprehensive method for identifying and measuring various contaminants in soil environments. To detect target analytes, whole-cell biosensors utilize the metabolic activity of living cells, such as yeast or bacteria. By integrating reporter genes, these biosensors facilitate continuous monitoring of contaminants, such as pesticides, organic compounds, and heavy metals. By capitalizing on the intrinsic biological mechanisms of microorganisms, it is possible to achieve precise and sensitive identification, even in intricate environmental matrices. Enzyme-based biosensors detect pollutants in soil by utilizing the catalytic properties of enzymes such as oxidases and dehydrogenases. Due to their remarkable sensitivity and specificity, these biosensors are essential for accurately identifying contaminants. The discernible signals generated by the enzymatic reactions of these biosensors offer valuable information regarding the concentrations and dynamics of pollutants in soil environments.

Leveraging the specific binding affinity between target analytes and antibodies, immunoassays are an additional potent instrument for monitoring soil contamination. Adaptable to contaminants, including heavy metals, pesticides, and organic pollutants, immunoassays feature exceptional specificity and breadth. Immunoassays offer dependable and robust detection capabilities by utilizing the immune system's inherent capacity to identify and bind to particular molecules. In the context of biosensors, deoxyribonucleic acid (DNA/RNA) aptamers function as recognition elements by binding with high affinity and specificity to target molecules. The aptamers provide the benefit of easy synthesis and modification, allowing for a wide range of applications in monitoring soil contamination. Aptamer-based biosensors, which take advantage of the unique characteristics of nucleic acid molecules, are an effective method for detecting contaminants in soil (Flores-Contreras et al., 2022; McConnell et al., 2020). As a group, these biosensors symbolize a fundamental change in environmental monitoring, as they permit the precise and efficient detection of soil pollutants in real-time and at the location of the incident. Their implementation signifies a substantial progression in ecological preservation and management endeavours, guaranteeing the safeguarding of ecosystems and human health while proactively mitigating soil contamination. Using the interplay between biology and technology, biosensors facilitate the establishment of a more sustainable ecological guardianship system and an improved planet for subsequent generations.

INTEGRATED SENSOR SYSTEMS AND NETWORKS FOR REAL-TIME ANALYSIS

Soil pollution monitoring has been revolutionized by implementing integrated sensor systems and networks, which provide invaluable insights and real-time analysis of environmental contamination. By precisely manipulating and analyzing small volumes of fluid, microfluidics and lab-on-a-chip devices have revolutionized the field, enabling rapid and high-throughput capabilities ideal for on-site soil pollution monitoring. These devices would allow scientists to conduct a wide range of analytical procedures in a compacted form, which decreases the amount of sample and reagent required for the analysis and improves its speed and sensitivity. Additionally, the portability of these instruments allows for immediate analysis at the collection location, eradicating the necessity for sample transportation and deployment in the field (Smolka et al., 2016). Installing wireless sensor networks (WSNs) is essential for continuously collecting data on environmental conditions across vast geographical areas. WSNs, which consist of spatially dispersed autonomous sensors, utilize wireless communication to monitor ecological parameters, soil properties, and pollutant concentrations across enormous tracts of land. Through real-time data transmission and collection, WSNs afford environmental agencies and researchers a holistic comprehension of the dynamics of soil pollution. This empowers them to implement management strategies and intervene opportunely (Le et al., 2023; Lloret et al., 2021).

Remote sensing technologies, including aerial drones and satellite imagery, are crucial in surveilling soil contamination through acquiring high-resolution spatial data. These technological advancements enable scientists to identify alterations in soil characteristics, vegetation vitality, and land utilization patterns linked to pollution. In contrast to satellite imagery, which provides a broad spatial scope, aerial drones offer precise, localized data. Abdulraheem et al. (2023) describe them as complementary instruments that monitor hotspots of soil contamination and evaluate environmental impacts on a large and small scale, respectively. Proximal soil sensing is an approach that employs sensor-equipped mobile platforms to collect comprehensive data on soil characteristics and contaminants at the surface. These platforms, which may consist of portable devices or tractors, facilitate researchers' direct data collection from the soil surface. This capability permits the accurate characterization of pollution levels and soil conditions. Proximal soil sensing techniques yield significant data that can be utilized to discern areas with high pollution concentrations, track temporal trends, and determine the order of importance for remediation endeavours. GIS is an exceptionally potent instrument that facilitates integrating and examining data derived from diverse origins, such as soil sampling, sensor networks, and remote sensing. Through the spatial visualization of soil pollution data and the superimposition of pertinent information, including topography and land use patterns, GIS enables environmental managers and researchers to precisely delineate and identify areas of high pollution. By conducting this spatial analysis, well-informed decisions can be made and targeted remediation strategies can be implemented to prevent soil pollution, protect ecosystems, and ensure human health. Integrated sensor systems and advanced analytics enable proactive soil pollution monitoring management strategies. Technology advancements such as real-time data collection, spatial visualization, and actionable insights improve environmental stewardship and significantly contribute to preserving ecosystems and human health.

TRENDS AND FUTURE OUTLOOK ON SOIL POLLUTION SENSORS

Researchers are currently focusing on critical aspects such as field deployability, selectivity, and sensitivity in the ongoing developments of soil pollution sensors, which hold great promise for their future. Advancements in materials, specifically molecularly imprinted polymers and nanomaterials, possess significant promise for enhancing the performance of sensors. Nanomaterials have unique properties, such as a high surface area to volume ratio and surface chemistry that can be changed. These properties can make sensors better at finding specific pollutants while also making it less likely that they will react with background contaminants. Instead, molecularly imprinted polymers have a recognition element that is molecularly selective and specific to a given analyte. This ensures the analyte is correctly identified even when many other things are happening around it. Advancements in signal amplification methodologies and algorithms for processing sensor signals are poised to enhance the sensitivity of soil pollution sensors. By substantially amplifying the detection signal, signal amplification techniques such as enzymatic signal amplification and nanoparticle-based signal enhancement permit the detection of pollutants at even lower concentrations. Sophisticated signal processing algorithms such as machine learning and artificial intelligence enhance the precision and dependability of pollutant detection by efficiently examining intricate sensor data.

The need for real-time, on-site soil contamination monitoring motivates the trend toward field-deployable sensor platforms, particularly in remote or resource-constrained regions. Portable and rugged sensor devices, like integrated sensor networks and mobile devices, make it easy and inexpensive for stakeholders to assess soil pollution thoroughly. These field-deployable platforms enable stakeholders, including environmental agencies, researchers, and others, to promptly collect practical data that can be utilized to make informed decisions and proactively manage pollution. In addition, endeavours to establish uniformity in sensor methodologies and technologies are vital to guaranteeing extensive implementation and regulatory approval. Standardized protocols for sensor calibration, validation, and data interpretation will ensure that measurements made with sensors are reliable, comparable, and repeatable across various platforms and applications. Regulatory bodies increasingly acknowledge the significance of sensor-based monitoring in environmental assessment and management. As a result, they have established guidelines and frameworks to streamline the incorporation of sensor data into regulatory decision-making procedures.

Integrating sensing technologies with remediation strategies is encouraging for effectively addressing soil pollution issues. Adaptive management strategies and sensor-guided remediation techniques using sensor-based feedback could help cleanup efforts by focusing on specific polluted areas and monitoring how healthy treatments are always working. An integrated approach can enhance pollution remediation efforts by executing them more efficiently and cost-effectively while reducing environmental impact and protecting ecosystems and human health. Standardization, regulatory acceptance, selectivity, sensitivity, field deployability, and integration with remediation technologies will improve proactive soil pollution monitoring and management. Regulatory agencies, industry stakeholders, and researchers will collaborate to propel the successful integration of sensor-based methodologies for soil pollution management. By applying state-of-the-art sensor technologies and interdisciplinary cooperation, it is possible to tackle soil pollution issues efficiently while positively contributing to environmental sustainability and human welfare.

DISCUSSION

When deliberating on soil pollution sensors, it is critical to recognize the complex and diverse characteristics of soil pollution and the imperative for all-encompassing monitoring approaches. Even though it is essential to make improvements in selectivity, sensitivity, and field deployability, the problem of soil contamination is very complicated and needs a whole-systems approach that looks at contaminant sources, transport mechanisms, and ecosystem interactions, among other things. An essential element to contemplate is the fluidity of soil contamination, which is susceptible to temporal and spatial fluctuations attributable to influences such as alterations in land use, weather patterns, and human activities. Hence, it is critical to establish sensor networks that can continuously monitor and collect data across extensive geographical regions to comprehensively capture the dynamics of soil pollution. By incorporating information from these networks into models, we can gain significant insights into contamination sources by predicting the pathway and destiny of pollutants. Community engagement and citizen science are essential components of soil pollution monitoring initiatives. Granting training programs and sensor technologies to local communities facilitates their active engagement in data collection and decision-making procedures, cultivating a sense of ownership and responsibility towards environmental stewardship. Citizen-generated data has the potential to supplement official monitoring endeavours by bolstering soil pollution assessments with supplementary spatial and temporal coverage.

Furthermore, as we contemplate the future, data analytics and machine learning progress presents the potential for extracting significant insights from extensive quantities of sensor data. Using these technologies, it is possible to discern correlations, patterns, and trends in soil pollution data that might need to be more readily discernible using conventional analytical techniques. This data-driven approach enables the implementation of proactive management strategies, including early warning systems for contamination events or targeted remediation efforts in high-risk areas. We can better understand ecosystem resilience and health by integrating soil pollution sensors into broader environmental monitoring systems, such as those used for monitoring air or water quality. By situating soil pollution within interconnected environmental systems, we can develop more effective strategies for promoting sustainable land management practices and mitigating pollution impacts. Soil pollution sensor technology is crucial in combating environmental issues and safeguarding ecosystems and human well-being. We can lay the foundation for a future that is more resilient and sustainable through the adoption of a holistic approach to monitoring and management, the promotion of collaboration, and the embracement of innovation.

CONCLUSION

Soil pollution is a substantial environmental issue with wide-ranging consequences for ecosystems and human well-being. Throughout this discussion, we have examined numerous facets of soil pollution monitoring, including cutting-edge sensor technologies. Various advanced instruments, including electrochemical and optical sensors and others, provide unparalleled sensitivity, selectivity, and portability levels. As a result, these tools enable stakeholders to detect and mitigate soil contamination with greater efficacy. Additionally, integrating sensor data with remediation strategies enhances the potential for proactively tackling soil pollution challenges. In anticipation of forthcoming developments, ongoing scientific inquiry, technological progress, and cooperative initiatives will be indispensable for guaranteeing environmentally responsible soil management and stewardship.

REFERENCES

Abdulraheem, M. I., Zhang, W., Li, S., Moshayedi, A. J., Farooque, A. A., & Hu, J. (2023). Advancement of remote sensing for soil measurements and applications: A comprehensive review. *Sustainability (Basel), 15*(21), 15444. doi:10.3390/su152115444

Bae, J. W., Seo, H. B., Belkin, S., & Gu, M. B. (2020). An optical detection module-based biosensor using fortified bacterial beads for soil toxicity assessment. *Analytical and Bioanalytical Chemistry, 412*(14), 3373–3381. doi:10.1007/s00216-020-02469-z PMID:32072206

Baranwal, J., Barse, B., Gatto, G., Broncová, G., & Kumar, A. (2022). Electrochemical sensors and their applications: A review. *Chemosensors (Basel, Switzerland), 10*(9), 363. doi:10.3390/chemosensors10090363

Baumbauer, C., Goodrich, P., Payne, M. E., Anthony, T. L., Beckstoffer, C., Toor, A., Silver, W. L., & Arias, A. C. (2022). Printed potentiometric nitrate sensors for use in soil. *Sensors (Basel), 22*(11), 4095. doi:10.3390/s22114095 PMID:35684715

Bayabil, H. K., Teshome, F. T., & Li, Y. (2022). Emerging contaminants in soil and water. *Frontiers in Environmental Science, 10*, 873499. doi:10.3389/fenvs.2022.873499

Chen, G., Ye, M., Ma, B., & Ren, Y. (2023). Responses of petroleum contamination at different sites to soil physicochemical properties and indigenous microbial communities. *Water, Air, and Soil Pollution, 234*(8), 494. doi:10.1007/s11270-023-06523-1

Cui, H., Zhou, J., Li, Z., & Gu, C. (2021). Editorial: Soil and sediment pollution, processes and remediation. Frontiers in Environmental Science. doi:10.3389/fenvs.2021.822355

Eugenio, N. R., Naidu, R., & Colombo, C. (2020). Global approaches to assessing, monitoring, mapping, and remedying soil pollution. *Environmental Monitoring and Assessment, 192*(9), 601. doi:10.1007/s10661-020-08537-2 PMID:32857292

Flores-Contreras, E. A., González-González, R. B., Gonzalez-González, E., Melchor-Martínez, E. M., Parra-Saldívar, R., & Iqbal, H. M. (2022). Detection of emerging pollutants using AptAMer-Based biosensors: Recent advances, challenges, and outlook. *Biosensors (Basel), 12*(12), 1078. doi:10.3390/bios12121078 PMID:36551045

He, Y., & Cai, P. (2021). Special issue on soil pollution, control, and remediation. *Soil Ecology Letters, 3*(3), 167–168. doi:10.1007/s42832-021-0110-6

Islam, M. S., Sazawa, K., Sugawara, K., & Kuramitz, H. (2023). Electrochemical biosensor for evaluation of environmental pollutants toxicity. *Environments (Basel, Switzerland), 10*(4), 63. doi:10.3390/environments10040063

Kratasyuk, V. A., Kolosova, E. M., Sutormin, O. S., Lonshakova-Mukina, V. I., Baygin, M. M., Rimatskaya, N. V., Sukovataya, I. E., & Шпедт, A. A. (2021). Software for matching standard activity enzyme biosensors for soil pollution analysis. *Sensors (Basel), 21*(3), 1017. doi:10.3390/s21031017 PMID:33540862

Le, M. T., Pham, C. D., Nguyen, T. P. T., Nguyen, T. L., Nguyen, Q., Hoang, N. B., & Nghiem, L. D. (2023). Wireless powered moisture sensors for smart agriculture and pollution prevention: Opportunities, challenges, and future outlook. *Current Pollution Reports*, *9*(4), 646–659. doi:10.1007/s40726-023-00286-3

Li, D., Zhai, W., Li, Y., & Long, Y. (2013). Recent progress in surface-enhanced Raman spectroscopy for the detection of environmental pollutants. *Mikrochimica Acta*, *181*(1–2), 23–43. doi:10.1007/s00604-013-1115-3

Liu, Y., Cao, X., Hu, Y., & Cheng, H. (2022). Pollution, risk and transfer of heavy metals in soil and rice: A case study in a typical industrialized region in South China. *Sustainability (Basel)*, *14*(16), 10225. doi:10.3390/su141610225

Luo, M., Liu, X., Legesse, N., Liu, Y., Wu, S., Han, F. X., & Ma, Y. (2023). Evaluation of agricultural non-point source pollution: A review. *Water, Air, and Soil Pollution*, *234*(10), 657. doi:10.1007/s11270-023-06686-x

McConnell, E. M., Nguyen, J., & Li, Y. (2020). Aptamer-based biosensors for environmental monitoring. Frontiers in Chemistry. doi:10.3389/fchem.2020.00434

Nnaji, N. D., Onyeaka, H., Miri, T., & Ugwa, C. (2023). Bioaccumulation for heavy metal removal: A review. *SN Applied Sciences*, *5*(5), 125. doi:10.1007/s42452-023-05351-6

Pérez-Sirvent, C., & Bech, J. (2023). Spatial assessment of soil and plant contamination. *Environmental Geochemistry and Health*, *45*(12), 8823–8827. doi:10.1007/s10653-023-01760-z PMID:37973774

Raffa, C. M., & Chiampo, F. (2021). Bioremediation of Agricultural Soils Polluted with Pesticides: A Review. *Bioengineering (Basel, Switzerland)*, *8*(7), 92. doi:10.3390/bioengineering8070092 PMID:34356199

Singh, G., Sinam, G., Kriti, K., Pandey, M., Kumari, B., & Kulsoom, M. (2019). Soil pollution by fluoride in India: Distribution, chemistry and analytical methods. Springer. doi:10.1007/978-981-13-6358-0_12

Smolka, M., Puchberger-Enengl, D., Bipoun, M., Klasa, A., Kiczkajlo, M., Śmiechowski, W., Sowiński, P., Krutzler, C., Keplinger, F., & Vellekoop, M. J. (2016). A mobile lab-on-a-chip device for on-site soil nutrient analysis. *Precision Agriculture*, *18*(2), 152–168. doi:10.1007/s11119-016-9452-y

Stepanova, A. Y., Gladkov, E. A., Osipova, E. S., Gladkova, O. V., & Терешонок, Д. В. (2022). Bioremediation of soil from petroleum contamination. *Processes (Basel, Switzerland)*, *10*(6), 1224. doi:10.3390/pr10061224

Sun, S., Sidhu, V., Rong, Y., & Zheng, Y. (2018). Pesticide pollution in agricultural soils and sustainable remediation methods: A review. *Current Pollution Reports*, *4*(3), 240–250. doi:10.1007/s40726-018-0092-x

Tobiszewski, M., & Vakh, C. (2023). Analytical applications of smartphones for agricultural soil analysis. *Analytical and Bioanalytical Chemistry*, *415*(18), 3703–3715. doi:10.1007/s00216-023-04558-1 PMID:36790460

Yang, Y., Creedon, N., O'Riordan, A., & Lovera, P. (2021). Surface enhanced Raman Spectroscopy: Applications in agriculture and food safety. *Photonics*, *8*(12), 568. doi:10.3390/photonics8120568

Zaghloul, A., Saber, M., & Abd-El-Hady, M. (2019). Physical indicators for pollution detection in terrestrial and aquatic ecosystems. *Bulletin of the National Research Center*, *43*(1), 120. doi:10.1186/s42269-019-0162-2

Zhai, J., Luo, B., Li, A., Dong, H., Jin, X., & Wang, X. (2022). Unlocking all-solid ion selective electrodes: Prospects in crop detection. *Sensors (Basel)*, *22*(15), 5541. doi:10.3390/s22155541 PMID:35898054

KEY TERMS AND DEFINITIONS

Biosensors: Biological sensing devices that employ biological recognition elements, such as enzymes or antibodies, to detect and quantify target analytes, often used in environmental monitoring applications.

Electrochemical Sensors: Instruments that detect analytes by measuring changes in electrical properties, commonly used in soil pollution monitoring for their sensitivity and selectivity.

Emerging Contaminants: Previously unrecognized pollutants or chemicals of concern are becoming increasingly prevalent in the environment, posing potential risks to ecosystems and human health.

Environmental Sensors: Devices designed to detect and measure physical, chemical, or biological parameters in the environment to monitor pollution levels or environmental conditions.

Optical Sensors: Sensors that utilize light-based techniques, such as fluorescence or absorption, to detect and quantify analytes, offering advantages in sensitivity and real-time analysis.

Soil Pollution: Soil contamination by hazardous substances, such as heavy metals or pesticides, harms ecosystems and human health.

Chapter 15
Sensors for Waste Management

Ankur Bhardwaj
https://orcid.org/0000-0003-0687-8810
Department of Life Sciences, Shri Vaishnav Institute of Science, India

Surendra Prakash Gupta
https://orcid.org/0000-0002-4873-6346
Department of Life Sciences, Shri Vaishnav Institute of Science, India

ABSTRACT

The globe is becoming more and more urbanised and industrialised, making waste management an urgent worldwide challenge. Traditional waste management methods have proven to be insufficient in addressing the challenges posed by increasing waste volumes, environmental concerns, and resource scarcity. One potential answer to these problems is the use of sensor technologies in waste management (WM) systems. It explores the various types of sensors used in waste management applications, ranging from simple bin-level sensors to advanced technologies such as remote sensing and IoT-based systems. The chapter also discusses the key advantages of sensor-driven waste management, including improved efficiency, cost reduction, and enhanced environmental sustainability. As WM continues to evolve in response to the demands of the 21st century, this chapter underscores the pivotal role that sensor technologies play in revolutionizing the industry, a glimpse into a more sustainable and efficient future for WM practices worldwide.

INTRODUCTION

In an era where environmental sustainability is paramount, WM stands as a critical battleground. The burgeoning global population coupled with rapid urbanization has amplified the challenges associated with waste disposal and recycling. Traditional WM methods often fall short in effectively addressing these challenges, leading to environmental degradation, public health risks, and economic inefficiencies (Mondal et al., 2023). However, the advent of sensor technology offers a promising avenue for revolutionizing WM practices. By integrating sensors into various aspects of WM systems, we can achieve heightened efficiency, optimize resource allocation, and mitigate environmental impacts.

DOI: 10.4018/979-8-3693-1930-7.ch015

Sensors play a pivotal role in modern WM by providing real-time (RT) data and insights across the entire WM cycle. From collection and sorting to processing and recycling, sensors enable precise monitoring and management, thereby optimizing each stage for maximum efficiency. One of the primary areas where sensors make a significant impact is in waste collection. Traditional waste collection methods often follow predetermined schedules, leading to inefficient routes and unnecessary fuel consumption. However, sensor-equipped waste bins can autonomously signal when they reach capacity, enabling waste collection vehicles to prioritize their routes (Szpilko et al., 2023). This dynamic approach not only reduces fuel consumption and emissions but also minimizes the likelihood of overflowing bins, thus enhancing public hygiene and aesthetics.

Moreover, sensors can facilitate waste sorting by accurately identifying recyclable materials within the waste stream. Advanced sensor technologies, such as near-infrared spectroscopy and hyperspectral imaging, can swiftly differentiate between various types of materials, allowing for automated sorting processes (Nanda and Berruti, 2021). By streamlining sorting operations, these sensors improve the efficacy of recycling facilities, thereby increasing the quantity of recyclable materials diverted from landfills.

Furthermore, sensors contribute to the optimization of waste processing facilities by monitoring factors such as temperature, humidity, and gas emissions. By continuously assessing these parameters, sensors can detect anomalies and malfunctions, enabling prompt intervention to prevent operational disruptions or environmental hazards (Mohanty et al., 2022(a)). Additionally, sensors can aid in the extraction of valuable resources from waste streams, such as methane from organic waste or metals from electronic scrap, thus promoting resource recovery and circular economy principles.

In addition to operational efficiencies, sensor technology also enhances the transparency and accountability of WM practices. By providing RT data on waste generation, collection, and disposal, sensors enable stakeholders, including government agencies, municipalities, and WM companies, to make informed decisions and track progress towards sustainability goals. Moreover, the integration of sensors with blockchain technology offers tamper-proof data integrity, ensuring the traceability and authenticity of waste-related transactions (Bahramian et al., 2023).

However, there are several obstacles standing in the way of the broad use of sensor technologies in WM, even with their promise for transformation. The most significant of these is the initial outlay of funds needed to install sensor infrastructure and integrate it with already-in-place WM systems. Additionally, concerns regarding data privacy, cybersecurity, and interoperability must be addressed to ensure the seamless operation and acceptance of sensor-equipped solutions (Martikkala et al., 2023).

Using sensor technology to its full potential throughout the WM cycle, we can move towards a future where waste is viewed not as a burden but as a valuable resource to be managed intelligently and responsibly. Through collaboration between policymakers, industry stakeholders, and technology innovators, we can realize the full potential of sensors in revolutionizing WM practices for the benefit of current and future generations.

SENSORS IN WM

Sensors are devices capable of detecting and measuring various parameters, such as fill levels, temperature, humidity, and even hazardous gas emissions, in real-time. When applied to WM, sensors have the potential to revolutionize the industry in several ways:

(a) **Fill Level Monitoring:** One of the primary applications of sensors in WM is the RT monitoring of waste container fill levels. These sensors are typically installed in trash bins, dumpsters, and recycling containers. By continuously assessing the fill levels, waste collection companies can optimize collection routes, reducing unnecessary trips to empty partially filled containers. This not only saves fuel and reduces emissions but also minimizes wear and tear on collection vehicles (Pardini et al., 2020).
(b) **Predictive Maintenance:** Sensors can also be used for the predictive maintenance of waste collection trucks and equipment. By monitoring the performance and condition of vehicles and machinery, maintenance issues can be identified early, preventing breakdowns and costly repairs (Theissler et al., 2021).
(c) **Environmental Monitoring:** Sensors can detect hazardous substances and emissions from waste sites, helping to ensure compliance with environmental regulations. This is essential for safeguarding surrounding populations and reducing the negative environmental effects of trash disposal (Fatimah et al., 2020).
(d) **Resource Recovery:** In recycling facilities, sensors play a critical role in sorting and separating recyclable materials efficiently. Advanced optical sensors can identify and sort diverse types of materials, such as plastics, metals, and paper, facilitating resource recovery and reducing landfill waste (Krechetov et al., 2019).
(e) **Data Analytics:** The data generated by sensors can be analysed to gain valuable insights into waste generation patterns, trends, and seasonal variations. This data-driven approach enables WM companies to make informed decisions, allocate resources efficiently, and implement targeted waste reduction initiatives (Jiang et al., 2020).
(f) **Smart WM Systems**: The integration of sensors into WM systems allows for the development of "smart" WM networks. These systems use data from sensors to optimize collection routes in real-time, reducing fuel consumption and greenhouse gas emissions. They can also provide residents with RT information on waste collection schedules and encourage responsible waste disposal practices (Pardini et al., 2020).
(g) **Public Engagement:** Sensors can facilitate greater public engagement in WM. Mobile apps and online platforms can provide residents with information on the nearest recycling centres, collection points, and guidelines for responsible waste disposal (Ramírez-Moreno et al., 2021).

While sensors offer numerous advantages in WM, their adoption is not without challenges. These include the initial investment cost, data management and privacy concerns, sensor maintenance, and the need for infrastructure upgrades (Joshi et al., 2022).

Sensors have the potential to revolutionize WM by making it more efficient, cost-effective, and environmentally friendly. By providing RT data and insights, sensors empower WM companies and municipalities to optimize their operations, reduce environmental impact, and engage communities in responsible waste disposal practices (Munir et al., 2023). As technology continues to advance, the integration of sensors into WM systems will likely become increasingly commonplace, contributing to a more sustainable and efficient WM ecosystem.

CHALLENGES IN WASTE MANAGEMENT: THE ROLE OF SENSORS

Waste management is a critical aspect of maintaining environmental sustainability and public health. As urban populations continue to grow and waste generation increases, the challenges in WM become more complex. One innovative solution that has gained prominence in recent years is the use of sensors to optimize waste collection and disposal processes (Martikkala et al., 2023). While sensors hold great promise in transforming WM, they also come with their own set of challenges.

(a) **Cost of Implementation:** One of the primary challenges in adopting sensor-based WM systems is the initial cost of implementation. Installing sensors in waste bins, collection trucks, and waste processing facilities can be expensive. Municipalities and WM companies often face budget constraints that may hinder their ability to invest in sensor technology (Vishnu et al., 2022; Fiorillo and Merkaj, 2024).

(b) **Data Management and Integration:** Sensor-generated data needed to be effectively managed and integrated into existing WM systems. This involves the advances in robust data infrastructure and analytics capabilities. Ensuring data security and privacy is also essential, as waste-related information can be sensitive (Arshi and Mondal, 2023).

(c) **Sensor Reliability:** Sensors are exposed to harsh environmental conditions, such as extreme temperatures and humidity, which can affect their reliability and accuracy. Regular maintenance and calibration are necessary to ensure sensors continue to function correctly. Failures or inaccuracies in sensor data can disrupt waste collection schedules and lead to inefficiencies.

(d) **Interoperability:** In many cases, WM systems involve various stakeholders, including municipalities, waste collection companies, recycling facilities, and more. Ensuring that sensors and data systems are compatible across different entities can be a challenge (Phonchi-Tshekiso et al., 2020). Lack of interoperability can lead to data silos and hinder the optimization of WM processes.

(e) **Scalability:** Implementing sensor-based WM on a small scale may be manageable, but scaling up to cover entire cities or regions is a significant challenge. It requires substantial infrastructure development, coordination among multiple parties, and the allocation of resources for widespread sensor deployment (D'Amico et al., 2020).

(f) **Community Acceptance:** The introduction of sensors in waste bins and collection trucks may raise concerns about privacy and surveillance among residents. Ensuring community acceptance and addressing privacy concerns is crucial for the successful adoption of sensor-based WM systems (Pardini et al., 2020; Pelonero et al., 2020).

(g) **Environmental Impact:** While sensor technology can improve WM efficiency, the production and disposal of sensors themselves may have environmental consequences. One issue that requires consideration is making sure that the procedures used in the fabrication of sensors are sustainable and that they are disposed of properly after their lives are over (Kundariya et al., 2021).

(h) **Technological Advancements:** Rapid advancements in sensor technology can make existing systems quickly obsolete. Staying up-to-date with the latest sensor innovations and integrating them into WM processes can be challenging for municipalities and WM companies (Das et al., 2019).

(i) **Data Analysis and Decision-Making:** Collecting vast amounts of data from sensors is only useful if it can be effectively analysed and turned into actionable insights. Developing the capability to analyse sensor data and make informed decisions based on that data is a challenge that organizations must address (Sliusar et al., 2022).

Since sensors hold tremendous potential to revolutionize WM by making it more efficient and environmentally friendly, there are several challenges that must be overcome. These challenges include cost, data management, sensor reliability, interoperability, scalability, community acceptance, environmental impact, technological advancements, and data analysis (Milson and Mehmet, 2023; Allioui and Mourdi, 2023). Addressing these challenges will be essential for the successful implementation of sensor-based WM systems and for achieving sustainable WM practices in the future.

SENSOR TECHNOLOGIES FOR WASTE MANAGEMENT

Waste management is a global challenge that requires innovative solutions to handle the ever-increasing volume of waste generated by societies. Sensor technologies have emerged as a powerful tool in modern WM, enabling more efficient and sustainable practices across the entire WM lifecycle (Bibri, 2018). These sensor technologies play a pivotal role in optimizing waste collection, improving recycling processes, and minimizing environmental impact (Table 1).

1. Fill Level Sensors

Fill level sensors are among the most widely adopted sensor technologies in WM. Installed in trash bins, dumpsters, and recycling containers, these sensors continuously monitor the fill levels of waste receptacles. By collecting RT data, waste collection companies can optimize their routes, ensuring that collection trucks only visit containers that are nearing full capacity (Premgi, 2019). This not only reduces operational costs but also cuts down on fuel consumption and greenhouse gas emissions.

2. Weight Sensors

Weight sensors are used to measure the weight of waste containers, enabling WM companies to accurately assess the amount of waste collected. This data is invaluable for billing purposes and monitoring waste generation trends (Wang et al., 2021). Weight sensors can be integrated into collection trucks, automated sorting systems, and waste compactors.

3. Environmental Sensors

Environmental sensors are employed to monitor air quality and detect hazardous substances emitted from waste sites and landfills. They help in ensuring compliance with environmental regulations and safeguarding the health of nearby communities. These sensors can detect pollutants such as methane, sulphur dioxide, and volatile organic compounds, providing early warnings in case of potential environmental hazards (Munir et al., 2019).

4. Optical and Infrared Sensors

Optical and infrared sensors are used in recycling facilities to sort and separate various types of recyclable materials efficiently. These sensors can identify and sort materials such as plastics, metals, paper, and glass based on their optical properties. By automating the sorting process, recycling facilities can enhance resource recovery and reduce contamination in recycled materials (Gundupalli et al., 2017; Senturk et al., 2021).

5. Temperature Sensors

Temperature sensors are employed to monitor the temperature within waste piles, particularly in composting and bioconversion processes. Maintaining the right temperature is crucial for the efficient decomposition of organic waste and preventing the release of harmful emissions (Duarte et al., 2017).

6. RFID and Barcode Sensors

RFID (Radio-Frequency Identification) and barcode sensors are used for tracking and identifying waste containers and their contents. These technologies facilitate better inventory management, enhance traceability, and enable more accurate tracking of waste materials from collection to disposal (Mostaccio et al., 2023).

7. GPS and Geographic Information Systems (GIS)

GPS and GIS technologies are integrated with sensors in WM to track the RT location of collection trucks, monitor routes, and optimize collection schedules. These technologies help minimize fuel consumption, reduce vehicle wear and tear, and improve overall operational efficiency (Hayat, 2023; Singh et al., 2024).

8. Smart WM Systems

Smart WM systems combine various sensor technologies with data analytics and IoT (Internet of Things) platforms. These systems provide RT insights into waste collection, enable route optimization, and enhance decision-making for WM companies (Kannan et al., 2024). They often include user-friendly interfaces for both WM professionals and residents, promoting community engagement and responsible waste disposal practices.

While sensor technologies have revolutionized WM, challenges remain, such as the initial investment cost, data security and privacy concerns, sensor maintenance, and the need for infrastructure upgrades. However, the benefits of these technologies, including reduced operational costs, minimized environmental impact, and improved resource recovery, make them a compelling choice for modern WM systems.

Sensor technologies have ushered in a new era of efficiency and sustainability in WM. By providing RT data and insights, these sensors empower WM companies to optimize their operations, reduce environmental harm, and engage communities in responsible waste disposal practices. As sensor technology continues to evolve, its integration into WM systems is expected to become increasingly prevalent, ultimately aiding more sustainable and efficient WM ecosystem (Farjana et al., 2023).

Table 1. Sensor technologies used in waste management, along with their typical applications and functions

S. No	Sensor Technology	Application	Function
1	Fill Level Sensors	Waste Containers	Monitor and report fill levels of bins for optimized collection routes.
2	Weight Sensors	Collection Trucks, Waste Bins	Measure the weight of waste materials for billing and tracking.
3	Environmental Sensors	Waste Sites, Landfills	Detect and monitor emissions, ensuring compliance with environmental regulations.
4	Optical and Infrared Sensors	Recycling Facilities	Identify and sort recyclable materials based on their optical properties.
5	Temperature Sensors	Composting, Bioconversion Sites	Monitor temperature for efficient decomposition of organic waste.
6	RFID and Barcode Sensors	Waste Containers, Materials	Track and identify waste containers and contents for inventory management.
7	GPS and GIS Technologies	Collection Trucks, Routes	Track vehicle locations, monitor routes and optimize collection schedules.
8	Smart Waste Management (SWM) Systems	Integrated Systems	Combine various sensors, data analytics and IoT for RT WM.

Source: (Bharadwaj et al., 2016; Anagnostopoulos et al., 2017; Nižetić et al., 2019; Wang et al., 2021; Sosunova and Porras, 2022).

SENSOR APPLICATIONS IN WASTE COMPOSITION ANALYSIS

Effective WM is a demanding global challenge in the 21st century, urged by the need to reduce environmental pollution, conserve resources, and promote sustainable practices (Jin et al., 2023). Waste composition analysis is a crucial component of this effort, as it helps policymakers, waste managers, and researchers understand the composition of waste streams, identify recyclable materials, and develop strategies for waste reduction and recycling. Sensors play a pivotal role in waste composition analysis, enabling accurate and efficient data collection and analysis (Piskulova and Gorbanyov, 2023).

1. Sorting and Segregation

Sensors are extensively used in automated waste sorting and segregation processes. In recycling facilities, sensors such as infrared (IR), near-infrared (NIR), and X-ray sensors are employed to identify and sort different materials based on their spectral properties. NIR sensors, for example, can distinguish between various plastics, paper, and glass, allowing for efficient separation and recycling. In addition to improving garbage sorting efficiency and accuracy, this automation also lessens the need for physical labour (Shirke et al., 2019).

2. Weight Measurement

Weight sensors are employed in waste bins and trucks to measure the weight of waste collected. These sensors provide valuable data for WM operations, enabling accurate billing for waste collection services and helping municipalities monitor waste generation trends. Weight data can also be used to optimize collection routes and reduce fuel consumption, contributing to cost savings and reduced carbon emissions (Chen et al., 2022).

3. Gas Sensors for Landfills

Landfills are a significant source of greenhouse gas emissions, primarily methane (CH_4) and carbon dioxide (CO_2). Gas sensors are utilized to monitor and control these emissions. Methane sensors can detect the presence and concentration of CH_4 in landfill gas, allowing for timely interventions to prevent gas leaks and reduce environmental impact (Dhall et al., 2021). Additionally, CO_2 sensors help assess the progress of landfill gas-to-energy projects, which convert methane into electricity.

4. pH and Temperature Sensors

In composting facilities, pH and temperature sensors are crucial for maintaining optimal conditions for the decomposition of organic waste. These sensors ensure that the composting process is efficient and produces high-quality compost, which can be used as a soil conditioner or fertilizer (Mengqi et al., 2021). Monitoring pH and temperature helps prevent issues such as odour problems and the production of harmful compounds.

5. Moisture Sensors

Moisture sensors are essential in waste composition analysis, especially for determining the moisture content of waste materials. High moisture content can make waste less suitable for recycling or energy recovery processes. These sensors help waste facilities manage moisture levels to ensure that waste is processed efficiently and effectively (Yin et al., 2021).

SENSORS FOR WASTE COLLECTION AND SORTING: REVOLUTIONIZING WASTE MANAGEMENT

Waste management is a global challenge that requires innovative solutions to reduce environmental impact, optimize resource utilization, and promote sustainability. Sensors have emerged as a game-changer in this field, offering enhanced capabilities for waste collection and sorting (Table 2). Here the pivotal role of sensors in transforming WM, from efficient waste collection to precise material sorting, ultimately contributing to a more sustainable future is explored (Szpilko et al., 2023).

1. Smart Waste Collection

Traditional waste collection methods often involve fixed schedules and routes, leading to inefficient resource allocation and increased environmental footprint due to unnecessary fuel consumption. Smart waste collection systems leverage sensors to address these challenges. These systems are equipped with various sensors, including ultrasonic level sensors, to monitor waste bin fill levels in real time (Lu et al., 2020).

Ultrasonic level sensors use sound waves to measure the depth of waste in bins. When a bin reaches a certain fill level, the sensor sends a signal to WM authorities or service providers, triggering timely collection (Silva et al., 2022). This approach minimizes the number of unnecessary collections, reducing costs and greenhouse gas discharges while optimizing resource allocation.

2. Compactor Sensors

Compactor sensors are frequently used in waste compactors, which are machines designed to compress waste materials to reduce volume. These sensors monitor the compaction process and help optimize compaction cycles for maximum efficiency. By ensuring that waste is compacted to its fullest potential, these sensors contribute to reduced transportation costs and decreased landfill usage (Kumar et al., 2023).

3. Sorting Sensors

In recycling facilities, sorting sensors are the backbone of material recovery processes. They identify and segregate recyclable materials from mixed waste streams. Common types of sorting sensors include:

(a) **Infrared (IR) and Near-Infrared (NIR) Sensors:** These sensors detect the spectral properties of materials, allowing for the identification and sorting of plastics, paper, glass, and metals based on their unique signatures. NIR sensors are highly versatile and can distinguish between various plastics, enhancing the efficiency of recycling processes (Araujo-Andrade et al., 2021).
(b) **Magnetic Sensors:** Magnetic sensors are used to separate ferrous metals (containing iron) from non-ferrous metals. They work by attracting ferrous materials with a magnetic field, while non-ferrous materials are unaffected (Brooks et al., 2019).
(c) **Eddy Current Sensors:** Eddy current sensors detect non-ferrous metals such as aluminium and copper by inducing electrical currents in these materials and measuring their response (Smith et al., 2019).
(d) **X-ray Sensors:** X-ray sensors are employed for advanced sorting applications, particularly in electronic waste recycling. They can identify different materials based on their density and atomic composition (Sterkens et al., 2021).

4. Environmental Sensors

Beyond waste collection and sorting, environmental sensors play essential role in monitoring the impact of WM operations. Gas sensors, for instance, detect and measure emissions from landfills, helping to mitigate harmful environmental effects. pH sensors and temperature sensors are used in composting facilities to maintain optimal conditions for decomposition (Ramírez-Moreno et al., 2021).

While the globe still struggles with the issues of environmental sustainability and waste generation, the role of sensors in collection and sorting will only grow in significance, contributing to more responsible and eco-friendly WM practices.

MONITORING AND CONTROL SYSTEMS: ROLE OF SENSORS

In today's rapidly urbanizing world, WM has become a serious challenge. The efficient collection, disposal, and recycling of waste are essential not only for environmental sustainability but also for public health and well-being. To address these challenges, monitoring and control systems with advanced sensor technologies are playing an increasingly pivotal role in optimizing WM processes (Yang et al., 2017).

Table 2. Sensors used in waste collection and sorting, along with potential future applications

S. No	Type of Sensor	Current Applications	Potential Future Applications
1	Ultrasonic Level	Smart waste bin fill monitoring in waste collection systems.	Intelligent waste compaction in smart cities.
2	Weight	Weighing waste in collection trucks and disposal facilities.	RT waste composition analysis for waste reduction and recycling optimization.
3	Compactor	Monitoring compaction in waste compactors to reduce volume.	Integration with AI for predictive maintenance and energy optimization.
4	Infrared (IR) and Near-Infrared (NIR)	Material sorting in recycling facilities based on spectral properties.	Enhanced recycling systems with improved material recognition and separation.
5	Magnetic	Separating ferrous metals from non-ferrous metals in recycling facilities.	Enhanced sorting of challenging waste streams, like electronic waste.
6	Eddy Current	Detecting non-ferrous metals like aluminium and copper.	Advanced resource recovery from complex waste streams.
7	X-ray	Advanced material sorting in recycling, especially electronic waste recycling.	Sorting for rare materials, trace contaminants, or hazardous materials.
8	Gas	Monitoring landfill gas emissions for environmental protection.	Early detection and prevention of gas leaks and better gas-to-energy conversion strategies.
9	pH and Temperature	Maintaining optimal conditions for composting facilities.	Precision control of compost processes for biogas production and soil improvement.
10	RFID and Barcode	Tracking and tracing waste containers and recyclable materials.	Enhanced waste tracking in circular economy systems.

Source: Araujo-Andrade et al., 2021; Brooks et al., 2019; Smith et al., 2019; Sterkens et al., 2021; Ramírez-Moreno et al., 2021.

Sensors are at the heart of modern WM systems, providing valuable RT data that enables municipalities and WM companies to make informed decisions and streamline their operations. These sensors come in various forms, from simple bin fill-level sensors to sophisticated environmental monitoring devices. Sensors revolutionizing WM are as follows:

(a) **Fill-Level Sensors:** One of the most basic yet crucial sensor applications in WM is fill-level sensors. These sensors are placed inside waste bins and containers to monitor their fill levels. By continuously measuring the waste volume, waste collection routes can be optimized. This leads to significant cost reductions in terms of fuel consumption, labour, and vehicle maintenance (Vishnu et al., 2022).

(b) **Environmental Sensors:** Beyond monitoring waste containers, environmental sensors are used to assess air quality, detect hazardous materials, and monitor emissions from waste processing facilities. These sensors help ensure compliance with environmental regulations and provide early warning (EW) systems for potential hazards, protecting both the environment and the health of nearby communities (Yekeen et al., 2020).

(c) **Sorting and Recycling Sensors**: In recycling facilities, sensors are employed to separate and classify discrete types of materials inevitably. Optical sensors, for example, can identify and sort various plastics, metals, and paper products based on their composition and characteristics. This automation not only increases the efficiency of recycling but also enhances the quality of recycled materials (Gundupalli et al., 2017).

(d) **GPS and RFID Technology:** Global Positioning System (GPS) and Radio-Frequency Identification (RFID) technology are often integrated into waste collection vehicles and bins. These technologies provide RT tracking and data collection capabilities, allowing WM companies to supervise the movement of their assets, optimize collection routes, and enhance overall operational efficiency (Arebey et al., 2010; Hannan et al., 2015).

(e) **Data Analytics and Remote Monitoring:** Sensor data is invaluable for WM companies when it is processed through advanced analytics. By analyzing historical and RT data, decision-makers can identify trends, predict maintenance needs, and optimize resource allocation. Remote monitoring systems also allow for RT visibility into WM operations, enabling rapid response to issues as they arise (Munir et al., 2023).

(f) **Smart Bins and IoT Integration:** The concept of smart bins has gained traction in recent years. These bins are equipped with sensors and IoT (Internet of Things) technology, allowing them to communicate with central systems. Smart bins can send alerts when they are full, help plan collection schedules more efficiently, and even encourage responsible waste disposal behaviour among users (Allioui and Mourdi, 2023).

(g) **Cost Savings and Sustainability:** Perhaps one of the most significant benefits of sensor-driven monitoring and control systems in WM is the promise for cost savings and environmental sustainability (Mohanty et al., 2022(b)). By reducing unnecessary collection trips, optimizing routes, and improving recycling rates, WM becomes more economically viable and environmentally friendly (Das et al., 2019).

Sensors enable RT monitoring, data collection, and informed decision-making, leading to cost savings, improved operational efficiency, and a reduced environmental footprint. As technology continues to advance, the integration of sensors and data analytics into WM processes will only become more sophisticated, helping us address the growing challenges of WM in an increasingly urbanized world.

CASE STUDIES AND EXAMPLES: TRANSFORMING THE TRASH INDUSTRY

Modern urban living requires effective trash management, and the use of sensors and technology is changing the way garbage is handled. WM techniques may be made more effective and sustainable by utilising the RT data and insights these sensors give. Here are some case studies and examples of sensors in WM:

(a) Smart Bins and Automated Collection

Barcelona, Spain: Barcelona has implemented an extensive network of smart waste bins equipped with fill-level sensors. These sensors monitor the waste levels inside the bins and communicate this information to waste collection teams. As a result, collection routes are optimized, reducing unnecessary pickups and decreasing fuel consumption (Camero et al., 2019). This smart system has led to significant cost savings and a more efficient waste collection process.

(b) Recycling Sorting Facilities

San Jose, California: The city of San Jose operates a state-of-the-art Material Recovery Facility (MRF) equipped with advanced sensor technology. Automated sorting machines use sensors to identify and separate different types of recyclables, such as paper, plastics, and metals, with remarkable precision (Cimpan et al., 2015). This technology not only reduces the need for manual labor but also ensures higher recycling rates by efficiently sorting materials.

(c) Landfill Gas Monitoring

Los Angeles County, California: The Puente Hills Landfill, one of the largest landfills in the United States, utilizes gas monitoring sensors. These sensors continuously monitor methane levels within the landfill. When methane levels exceed certain thresholds, gas extraction systems are activated to capture and convert the gas into electricity (Yadav et al., 2019). This process both reduces greenhouse gas emissions and generates renewable energy.

(d) Underground Waste Containers

Amsterdam, Netherlands: Amsterdam has adopted an underground waste container system equipped with sensors. These containers feature compaction systems and sensors that notify waste collectors when they need emptying. By eliminating the need for traditional above-ground bins, this technology enhances the city's aesthetic appeal while improving WM efficiency (Hancke and Hancke, 2013).

(e) GPS Tracking for Collection Vehicles

New York City, USA: The New York City Department of Sanitation has equipped its waste collection fleet with GPS tracking sensors. These sensors enable RT tracking of collection routes and vehicle activities. With this data, collection routes can be optimized to reduce fuel consumption, lower carbon emissions, and upgrade overall operational proficiency (Isik et al., 2021).

(f) E-Waste Recycling

Belgium: Electronic waste (e-waste) contains valuable and hazardous materials that require specialized recycling processes. Sensors are used to automatically identify and sort e-waste, ensuring the recovery of precious metals and components while safely managing toxic materials. Facilities like Recupel, a non-profit organisation that collects and processes used electrical and electronic appliances and light bulbs) in Belgium demonstrate the efficient use of sensor technology in e-waste recycling (Verstricht et al., 2022).

(g) Waste Composition Analysis

Austin, Texas: Waste composition analysis is crucial for informed waste reduction and recycling efforts. Sensors are employed to analyse the composition of waste streams, identifying types and quantities of materials present. This information helps authorities tailor recycling and waste reduction programs effectively (Sahadewa et al., 2014). The waste characterization study in Austin, Texas, relies on sensor technology to guide its WM strategies.

These case studies illustrate how sensor technology is helping cities and regions worldwide reduce costs, improve sustainability, and minimize the environmental impact of WM. As technology continues to advance, we can expect even more innovative solutions to emerge, further revolutionizing the WM industry.

FUTURE TRENDS AND EMERGING TECHNOLOGIES

Sensors embedded in waste bins, trucks, and sorting facilities will be interconnected, allowing RT data collection and analysis. This connectivity will enable predictive maintenance, route optimization, and better decision-making in waste collection.

Artificial intelligence and machine learning algorithms will become more sophisticated in predicting waste generation patterns. These technologies will help optimize collection routes, predict fill levels, and identify contamination in recycling streams. By analysing historical data, AI can help cities and WM companies make data-driven decisions.

Advancements in sensor technology will lead to smaller, more affordable sensors. These compact sensors can be deployed in various WM applications, including monitoring landfill gas emissions, detecting leaks in waste containment systems, and assessing waste composition. Miniaturization will make sensor deployment more accessible and cost-effective.

Blockchain technology can enhance transparency and traceability in WM. Each step of the waste journey, from generation to disposal or recycling, can be recorded on a blockchain. This ensures accurate tracking and reporting of waste volumes, recycling rates, and compliance with regulations.

Robotics will play a significant role in waste sorting and recycling. Advanced robotic systems equipped with sensors and computer vision technology can identify and sort various types of waste with high precision. This reduces the need for manual labour and improves recycling rates.

Sensors will be instrumental in optimizing waste-to-energy processes. The RT monitoring of waste characteristics, such as calorific value and moisture content, will enable better control of incineration and gasification processes, maximizing energy recovery and minimizing environmental impact.

The impact of WM operations on the environment will be tracked using sensors. This includes air and water quality monitoring near landfills, in addition to evaluating the environmental impact of waste processing technologies. These sensors will help ensure compliance with environmental regulations.

Smart packaging with embedded sensors will help consumers and WM organizations identify the composition and recyclability of packaging materials. This technology will promote responsible consumer choices and facilitate the recycling process.

Emerging technologies may enable decentralized WM systems. These systems could include on-site waste processing, such as small-scale recycling units or composting solutions, all equipped with sensors for monitoring and control.

Mobile apps with WM sensors will empower citizens to actively participate in waste reduction and recycling efforts. These apps can provide RT information on waste collection schedules, recycling guidelines, and nearby recycling facilities.

These innovations promise to make WM more efficient, reduce environmental impact, and encourage responsible waste disposal and recycling. As technology continues to advance, WM will become smarter, more responsive, and better equipped to address the growing challenges of waste generation in our rapidly urbanizing world.

CONCLUSION: SENSORS IN WASTE MANAGEMENT - A SMARTER, SUSTAINABLE FUTURE

The use of sensors in garbage management has brought in a new era of effectiveness, sustainability, and environmental responsibility. Throughout this chapter, we have explored the myriad ways in which sensors are transforming the WM industry, from smart bins and recycling sorting facilities to landfill gas monitoring and waste composition analysis. We have delved into case studies and examples, highlighting the tangible benefits that sensor technology has already brought to cities and regions worldwide. Furthermore, we have examined future trends and emerging technologies, providing a glimpse into the exciting developments that promise to shape the future of WM.

In the wake of unprecedented urbanization and population growth, the challenges posed by WM have grown exponentially. Traditional WM practices are no longer sufficient to meet the demands of our ever-expanding cities. This is where sensors have proven to be invaluable. They provide RT data, enhance operational efficiency, reduce environmental impact, and empower communities to take a proactive role in waste reduction and recycling.

One of the key takeaways from this chapter is the concept of "smart waste management." Smart WM systems, powered by sensors and data analytics, allow for more responsive and adaptive approaches to waste collection and disposal. For instance, smart bins equipped with fill-level sensors can optimize collection routes, minimizing unnecessary trips and reducing fuel consumption. This not only translates into cost savings for municipalities but also reduces carbon emissions, contributing to a greener, more sustainable future.

Recycling, a cornerstone of WM, has also benefited immensely from sensor technology. Automated sorting machines equipped with sensors can accurately identify and separate recyclable materials from the waste stream, increasing recycling rates and reducing contamination. This technology is vital for achieving circular economies where resources are conserved and reused efficiently.

Sensors have also proved their worth in monitoring landfill gas emissions, particularly methane, a potent greenhouse gas. By continuously tracking gas levels within landfills and triggering gas extraction systems when thresholds are exceeded, sensors mitigate the environmental impact of landfills while harnessing methane as a valuable energy source.

Looking to the future, we anticipate a profound transformation of WM practices. The emergence of the Internet of Things (IoT) will create a web of interconnected sensors, bins, vehicles, and processing facilities. This interconnectedness will enable predictive maintenance, route optimization, and data-driven decision-making on an unprecedented scale. Artificial intelligence and machine learning will refine waste generation predictions, further optimizing waste collection routes and schedules.

Sensors will also play a crucial role in ensuring the responsible disposal and recycling of electronic waste (e-waste). As technology continues to advance, e-waste will become an increasingly significant component of the waste stream. Advanced sensors will help identify and sort e-waste, recovering valuable materials while safely managing hazardous components.

Furthermore, blockchain technology will provide transparency and traceability in WM, bolstering accountability and compliance with regulations. Recycling and garbage sorting operations will be more effective thanks to robotics, automation, and smart packaging. Apps for public participation and decentralised trash management will enable communities to take an active role in waste reduction initiatives.

In conclusion, sensors in WM are not merely technological innovations; they signify a paradigm change in the direction of a more sustainable, environmentally conscious, and efficient future. In the midst of our efforts to solve the urgent problems associated with WM in an urbanising globe, sensors will continue to be at the forefront of innovation. They will enable us to reduce waste, conserve resources, lower emissions, and ultimately create cleaner, healthier communities for generations to come. By embracing sensor technology and fostering collaboration between governments, industries, and communities, we can pave the way for a smarter, more sustainable future in WM.

REFERENCES

Allioui, H., & Mourdi, Y. (2023). Exploring the full potentials of IoT for better financial growth and stability: A comprehensive survey. *Sensors (Basel)*, *23*(19), 8015. doi:10.3390/s23198015 PMID:37836845

Anagnostopoulos, T., Zaslavsky, A., Kolomvatsos, K., Medvedev, A., Amirian, P., Morley, J., & Hadjieftymiades, S. (2017). Challenges and opportunities of waste management in IoT-enabled smart cities: A survey. *IEEE Transactions on Sustainable Computing*, *2*(3), 275–289. doi:10.1109/TSUSC.2017.2691049

Araujo-Andrade, C., Bugnicourt, E., Philippet, L., Rodriguez-Turienzo, L., Nettleton, D., Hoffmann, L., & Schlummer, M. (2021). Review on the photonic techniques suitable for automatic monitoring of the composition of multi-materials wastes in view of their posterior recycling. *Waste Management & Research*, *39*(5), 631–651. doi:10.1177/0734242X21997908 PMID:33749390

Arebey, M., Hannan, M. A., Basri, H., Begum, R. A., & Abdullah, H. (2010). RFID and integrated technologies for solid waste bin monitoring system. In *Proceedings of the world congress on engineering* (Vol. 1, pp. 316-32). Research Gate.

Arshi, O., & Mondal, S. (2023). Advancements in sensors and actuators technologies for smart cities: A comprehensive review. *Smart Construction and Sustainable Cities*, *1*(1), 18. doi:10.1007/s44268-023-00022-2

Bahramian, M., Dereli, R. K., Zhao, W., Giberti, M., & Casey, E. (2023). Data to intelligence: The role of data-driven models in wastewater treatment. *Expert Systems with Applications*, *217*, 119453. doi:10.1016/j.eswa.2022.119453

Bharadwaj, A. S., Rego, R., & Chowdhury, A. (2016). IoT based solid waste management system: A conceptual approach with an architectural solution as a smart city application. In *2016 IEEE annual India conference (INDICON)* (pp. 1-6). IEEE. 10.1109/INDICON.2016.7839147

Bibri, S. E. (2018). The IoT for smart sustainable cities of the future: An analytical framework for sensor-based big data applications for environmental sustainability. *Sustainable Cities and Society*, *38*, 230–253. doi:10.1016/j.scs.2017.12.034

Brooks, L., Gaustad, G., Gesing, A., Mortvedt, T., & Freire, F. (2019). Ferrous and non-ferrous recycling: Challenges and potential technology solutions. *Waste Management (New York, N.Y.)*, *85*, 519–528. doi:10.1016/j.wasman.2018.12.043 PMID:30803607

Camero, A., Toutouh, J., Ferrer, J., & Alba, E. (2019). Waste generation prediction in smart cities through deep neuroevolution. In *Smart Cities: First Ibero-American Congress, ICSC-CITIES 2018,* (pp. 192-204). Springer International Publishing. 10.1007/978-3-030-12804-3_15

Chen, J., Lu, W., Yuan, L., Wu, Y., & Xue, F. (2022). Estimating construction waste truck payload volume using monocular vision. *Resources, Conservation and Recycling*, *177*, 106013. doi:10.1016/j.resconrec.2021.106013

Cimpan, C., Maul, A., Jansen, M., Pretz, T., & Wenzel, H. (2015). Central sorting and recovery of MSW recyclable materials: A review of technological state-of-the-art, cases, practice and implications for materials recycling. *Journal of Environmental Management*, *156*, 181–199. doi:10.1016/j.jenvman.2015.03.025 PMID:25845999

D'Amico, G., L'Abbate, P., Liao, W., Yigitcanlar, T., & Ioppolo, G. (2020). Understanding sensor cities: Insights from technology giant company driven smart urbanism practices. *Sensors (Basel)*, *20*(16), 4391. doi:10.3390/s20164391 PMID:32781671

Das, S., Lee, S. H., Kumar, P., Kim, K. H., Lee, S. S., & Bhattacharya, S. S. (2019). Solid waste management: Scope and the challenge of sustainability. *Journal of Cleaner Production*, *228*, 658–678. doi:10.1016/j.jclepro.2019.04.323

Dhall, S., Mehta, B. R., Tyagi, A. K., & Sood, K. (2021). A review on environmental gas sensors: Materials and technologies. *Sensors International*, *2*, 100116. doi:10.1016/j.sintl.2021.100116

Duarte, L., Teodoro, A. C., Gonçalves, J. A., Ribeiro, J., Flores, D., Lopez-Gil, A., Dominguez-Lopez, A., Angulo-Vinuesa, X., Martin-Lopez, S., & Gonzalez-Herraez, M. (2017). Distributed temperature measurement in a self-burning coal waste pile through a GIS open source desktop application. *ISPRS International Journal of Geo-Information*, *6*(3), 87. doi:10.3390/ijgi6030087

Farjana, M., Fahad, A. B., Alam, S. E., & Islam, M. M. (2023). An iot-and cloud-based e-waste management system for resource reclamation with a data-driven decision-making process. *IoT*, *4*(3), 202–220. doi:10.3390/iot4030011

Fatimah, Y. A., Govindan, K., Murniningsih, R., & Setiawan, A. (2020). Industry 4.0 based sustainable circular economy approach for smart waste management system to achieve sustainable development goals: A case study of Indonesia. *Journal of Cleaner Production*, *269*, 122263. doi:10.1016/j.jclepro.2020.122263

Fiorillo, F., & Merkaj, E. (2024). Municipal strategies, fiscal incentives and co-production in urban waste management. *Socio-Economic Planning Sciences*, *101817*, 101817. doi:10.1016/j.seps.2024.101817

Gundupalli, S. P., Hait, S., & Thakur, A. (2017). A review on automated sorting of source-separated municipal solid waste for recycling. *Waste Management (New York, N.Y.)*, *60*, 56–74. doi:10.1016/j.wasman.2016.09.015 PMID:27663707

Hancke, G. P., & Hancke, G. P. Jr. (2013). The role of advanced sensing in smart cities. *Sensors (Basel)*, *13*(1), 393–425. doi:10.3390/s130100393 PMID:23271603

Hannan, M. A., Al Mamun, M. A., Hussain, A., Basri, H., & Begum, R. A. (2015). A review on technologies and their usage in solid waste monitoring and management systems: Issues and challenges. *Waste Management (New York, N.Y.)*, *43*, 509–523. doi:10.1016/j.wasman.2015.05.033 PMID:26072186

Hayat, P. (2023). Integration of advanced technologies in urban waste management. In *Advancements in Urban Environmental Studies: Application of Geospatial Technology and Artificial Intelligence in Urban Studies* (pp. 397–418). Springer International Publishing. doi:10.1007/978-3-031-21587-2_23

Isik, M., Dodder, R., & Kaplan, P. O. (2021). Transportation emissions scenarios for New York City under different carbon intensities of electricity and electric vehicle adoption rates. *Nature Energy*, *6*(1), 92–104. doi:10.1038/s41560-020-00740-2 PMID:34804594

Jiang, P., Van Fan, Y., Zhou, J., Zheng, M., Liu, X., & Klemeš, J. J. (2020). Data-driven analytical framework for waste-dumping behaviour analysis to facilitate policy regulations. *Waste Management (New York, N.Y.)*, *103*, 285–295. doi:10.1016/j.wasman.2019.12.041 PMID:31911375

Jin, L., Sun, X., Ren, H., & Huang, H. (2023). Biological filtration for wastewater treatment in the 21st century: A data-driven analysis of hotspots, challenges and prospects. *The Science of the Total Environment*, *855*, 158951. doi:10.1016/j.scitotenv.2022.158951 PMID:36155035

Joshi, L. M., Bharti, R. K., & Singh, R. (2022). Internet of things and machine learning-based approaches in the urban solid waste management: Trends, challenges, and future directions. *Expert Systems: International Journal of Knowledge Engineering and Neural Networks*, *39*(5), e12865. doi:10.1111/exsy.12865

Kannan, D., Khademolqorani, S., Janatyan, N., & Alavi, S. (2024). Smart waste management 4.0: The transition from a systematic review to an integrated framework. *Waste Management (New York, N.Y.)*, *174*, 1–14. doi:10.1016/j.wasman.2023.08.041 PMID:37742441

Krechetov, I. V., Skvortsov, A. A., Poselsky, I. A., Paltsev, S. A., Lavrikov, P. S., & Korotkovs, V. (2019). Implementation of automated lines for sorting and recycling household waste as an important goal of environmental protection. *Journal of Environmental Management and Tourism*, *9*(8), 1805–1812. doi:10.14505//jemt.v9.8(32).21

Kumar, A., Verma, S. K., Sharma, K., & Mendiburu, A. Z. (2023). Design and analysis of heat melt refuse compactor for solid waste with E-control mechanism. *Environmental Challenges*, *100740*, 100740. doi:10.1016/j.envc.2023.100740

Kundariya, N., Mohanty, S. S., Varjani, S., Ngo, H. H., Wong, J. W., Taherzadeh, M. J., Chang, J. S., Ng, H. Y., Kim, S. H., & Bui, X. T. (2021). A review on integrated approaches for municipal solid waste for environmental and economical relevance: Monitoring tools, technologies, and strategic innovations. *Bioresource Technology*, *342*, 125982. doi:10.1016/j.biortech.2021.125982 PMID:34592615

Lu, X., Pu, X., & Han, X. (2020). Sustainable smart waste classification and collection system: A bi-objective modeling and optimization approach. *Journal of Cleaner Production*, *276*, 124183. doi:10.1016/j.jclepro.2020.124183

Martikkala, A., Mayanti, B., Helo, P., Lobov, A., & Ituarte, I. F. (2023). Smart textile waste collection system–Dynamic route optimization with IoT. *Journal of Environmental Management*, *335*, 117548. doi:10.1016/j.jenvman.2023.117548 PMID:36871359

Mengqi, Z., Shi, A., Ajmal, M., Ye, L., & Awais, M. (2021). Comprehensive review on agricultural waste utilization and high-temperature fermentation and composting. *Biomass Conversion and Biorefinery*, 1–24. doi:10.1007/s13399-021-01438-5

Milson, S., & Mehmet, A. (2023). *Microfluidic Sensors for Environmental Monitoring: From Lab to Field Applications*. EasyChair. easychair.org/publications/preprint/6Spww

Mohanty, A., Mohanty, S. K., Jena, B., Mohapatra, A. G., Rashid, A. N., Khanna, A., & Gupta, D. (2022). (b)). Identification and evaluation of the effective criteria for detection of congestion in a smart city. *IET Communications*, *16*(5), 560–570. doi:10.1049/cmu2.12344

Mohanty, S., Saha, S., Santra, G. H., & Kumari, A. (2022). (a)). Future perspective of solid waste management strategy in India. In *Handbook of Solid Waste Management: Sustainability through Circular Economy* (pp. 191–226). Springer Nature Singapore. doi:10.1007/978-981-16-4230-2_10

Mondal, P., Nandan, A., Ajithkumar, S., Siddiqui, N. A., Raja, S., Kola, A. K., & Balakrishnan, D. (2023). Sustainable application of nanoparticles in wastewater treatment: Fate, current trend & paradigm shift. *Environmental Research*, *116071*, 116071. doi:10.1016/j.envres.2023.116071 PMID:37209979

Mostaccio, A., Bianco, G. M., Marrocco, G., & Occhiuzzi, C. (2023). RFID Technology for Food Industry 4.0: A Review of Solutions and Applications. *IEEE Journal of Radio Frequency Identification*, *7*, 145–157. doi:10.1109/JRFID.2023.3278722

Munir, M. T., Li, B., Naqvi, M., & Nizami, A. S. (2023). Green loops and clean skies: Optimizing municipal solid waste management using data science for a circular economy. *Environmental Research*, *117786*. doi:10.1016/j.envres.2023.117786 PMID:38036215

Nanda, S., & Berruti, F. (2021). Municipal solid waste management and landfilling technologies: A review. *Environmental Chemistry Letters*, *19*(2), 1433–1456. doi:10.1007/s10311-020-01100-y

Nižetić, S., Djilali, N., Papadopoulos, A., & Rodrigues, J. J. (2019). Smart technologies for promotion of energy efficiency, utilization of sustainable resources and waste management. *Journal of Cleaner Production*, *231*, 565–591. doi:10.1016/j.jclepro.2019.04.397

Pardini, K., Rodrigues, J. J., Diallo, O., Das, A. K., de Albuquerque, V. H. C., & Kozlov, S. A. (2020). A smart waste management solution geared towards citizens. *Sensors (Basel)*, *20*(8), 2380. doi:10.3390/s20082380 PMID:32331464

Pelonero, L., Fornaia, A., & Tramontana, E. (2020). From smart city to smart citizen: rewarding waste recycle by designing a data-centric iot based garbage collection service. In *2020 IEEE International Conference on Smart Computing (SMARTCOMP)* (pp. 380-385). IEEE. 10.1109/SMARTCOMP50058.2020.00081

Phonchi-Tshekiso, N. D., Mmopelwa, G., & Chanda, R. (2020). From public to private solid waste management: Stakeholders' perspectives on private-public solid waste management in Lobatse, Botswana. *Zhongguo Renkou Ziyuan Yu Huanjing*, *18*(1), 42–48. doi:10.1016/j.cjpre.2021.04.015

Piskulova, N., & Gorbanyov, V. (2023). Global Challenges: Environment. In *World Economy and International Business: Theories, Trends, and Challenges* (pp. 213–233). Springer International Publishing., doi:10.1007/978-3-031-20328-2_11

Premgi, A., Martins, F., & Domingos, D. (2019). An infrared-based sensor to measure the filling level of a waste bin. In *2019 International Conference in Engineering Applications (ICEA)* (pp. 1-6). IEEE. 10.1109/CEAP.2019.8883303

Ramírez-Moreno, M. A., Keshtkar, S., Padilla-Reyes, D. A., Ramos-López, E., García-Martínez, M., Hernández-Luna, M. C., Mogro, A. E., Mahlknecht, J., Huertas, J. I., Peimbert-García, R. E., Ramírez-Mendoza, R. A., Mangini, A. M., Roccotelli, M., Pérez-Henríquez, B. L., Mukhopadhyay, S. C., & Lozoya-Santos, J. J. (2021). Sensors for sustainable smart cities: A review. *Applied Sciences (Basel, Switzerland)*, *11*(17), 8198. doi:10.3390/app11178198

Sahadewa, A., Zekkos, D., Woods, R. D., Stokoe, K. H. II, & Matasovic, N. (2014). In-situ assessment of the dynamic properties of municipal solid waste at a landfill in Texas. *Soil Dynamics and Earthquake Engineering*, *65*, 303–313. doi:10.1016/j.soildyn.2014.04.004

Senturk, S. F., Gulmez, H. K., Gul, M. F., & Kirci, P. (2021). Detection and separation of transparent objects from recyclable materials with sensors. In *International Conference on Advanced Network Technologies and Intelligent Computing* (pp. 73-81). Cham: Springer International Publishing. 10.1007/978-3-030-96040-7_6

Shirke, S. I., Ithape, S., Lungase, S., & Mohare, M. (2019). Automation of smart waste management using IoT. *International Research Journal of Engineering and Technology, 6*(6), 414-419. IRJET-V6I615320190820-94746-xkw9ee-libre.pdf

Silva, A. S., Brito, T., de Tuesta, J. L. D., Lima, J., Pereira, A. I., Silva, A. M., & Gomes, H. T. (2022, October). Node Assembly for Waste Level Measurement: Embrace the Smart City. In *International Conference on Optimization, Learning Algorithms and Applications* (pp. 604-619). Cham: Springer International Publishing. 10.1007/978-3-031-23236-7_42

Singh, D., Dikshit, A. K., & Kumar, S. (2024). Smart technological options in collection and transportation of municipal solid waste in urban areas: A mini review. *Waste Management & Research*, *42*(1), 3–15. doi:10.1177/0734242X231175816 PMID:37246550

Sliusar, N., Filkin, T., Huber-Humer, M., & Ritzkowski, M. (2022). Drone technology in municipal solid waste management and landfilling: A comprehensive review. *Waste Management (New York, N.Y.)*, *139*, 1–16. doi:10.1016/j.wasman.2021.12.006 PMID:34923184

Smith, Y. R., Nagel, J. R., & Rajamani, R. K. (2019). Eddy current separation for recovery of non-ferrous metallic particles: A comprehensive review. *Minerals Engineering*, *133*, 149–159. doi:10.1016/j.mineng.2018.12.025

Sosunova, I., & Porras, J. (2022). IoT-enabled smart waste management systems for smart cities: A systematic review. *IEEE Access : Practical Innovations, Open Solutions*, *10*, 73326–73363. doi:10.1109/ACCESS.2022.3188308

Sterkens, W., Diaz-Romero, D., Goedemé, T., Dewulf, W., & Peeters, J. R. (2021). Detection and recognition of batteries on X-Ray images of waste electrical and electronic equipment using deep learning. *Resources, Conservation and Recycling*, *168*, 105246. doi:10.1016/j.resconrec.2020.105246

Szpilko, D., de la Torre Gallegos, A., Jimenez Naharro, F., Rzepka, A., & Remiszewska, A. (2023). Waste Management in the Smart City: Current Practices and Future Directions. *Resources*, *12*(10), 115. doi:10.3390/resources12100115

Theissler, A., Pérez-Velázquez, J., Kettelgerdes, M., & Elger, G. (2021). Predictive maintenance enabled by machine learning: Use cases and challenges in the automotive industry. *Reliability Engineering & System Safety*, *215*, 107864. doi:10.1016/j.ress.2021.107864

Verstricht, J., Li, X. L., Leonard, D., & Van Geet, M. (2022). Assessment of sensor performance in the context of geological radwaste disposal—A first case study in the Belgian URL HADES. *Geomechanics for Energy and the Environment*, *32*, 100296. doi:10.1016/j.gete.2021.100296

Vishnu, S., Ramson, S. J., Rukmini, M. S. S., & Abu-Mahfouz, A. M. (2022). Sensor-based solid waste handling systems: A survey. *Sensors (Basel)*, *22*(6), 2340. doi:10.3390/s22062340 PMID:35336511

Wang, C., Qin, J., Qu, C., Ran, X., Liu, C., & Chen, B. (2021). A smart municipal waste management system based on deep-learning and Internet of Things. *Waste Management (New York, N.Y.)*, *135*, 20–29. doi:10.1016/j.wasman.2021.08.028 PMID:34461487

Yadav, V., Duren, R., Mueller, K., Verhulst, K. R., Nehrkorn, T., Kim, J., Weiss, R. F., Keeling, R., Sander, S., Fischer, M. L., Newman, S., Falk, M., Kuwayama, T., Hopkins, F., Rafiq, T., Whetstone, J., & Miller, C. (2019). Spatio-temporally resolved methane fluxes from the Los Angeles Megacity. *Journal of Geophysical Research. Atmospheres*, *124*(9), 5131–5148. doi:10.1029/2018JD030062

Yang, H., Ma, M., Thompson, J. R., & Flower, R. J. (2017). Waste management, informal recycling, environmental pollution and public health. *J Epidemiol Community Health*. jech.bmj.com/content/72/3/237

Yekeen, S., Balogun, A., & Aina, Y. (2020). Early warning systems and geospatial tools: managing disasters for urban sustainability. In *Sustainable Cities and Communities* (pp. 129–141). Springer International Publishing., doi:10.1007/978-3-319-95717-3_103

Yin, H., Cao, Y., Marelli, B., Zeng, X., Mason, A. J., & Cao, C. (2021). Soil sensors and plant wearables for smart and precision agriculture. *Advanced Materials*, *33*(20), 2007764. doi:10.1002/adma.202007764 PMID:33829545

KEYWORDS AND DEFINITIONS

Environmental Sensor Networks: Interconnected sensors gathering environmental data for monitoring and analysis.

IoT in Waste Management: Internet-connected systems optimizing waste collection and processing.

Real-Time Waste Analytics: Instantaneous analysis of waste data for informed decision-making.

Sensor-Enabled Waste Bins: Waste receptacles equipped with sensors for monitoring.

Smart Waste Management: Utilizing technology for efficient and optimized waste handling.

Waste Sensors: Devices detecting and measuring waste levels or compositions.

Waste Tracking Technology: Systems tracing waste from generation to disposal.

Chapter 16
Environmental Sensors: Safeguarding the Ecosystem by Monitoring Sanitary Pad Disposal

Deepa V. Jose
Christ University, India

ABSTRACT

This chapter focuses on the applications of environmental sensors in general and their role in identifying and addressing the issues related to the improper disposal of sanitary pads, which is a growing concern. It also gives an overview of the pollutants associated with it, and the role that environmental sensors can play in mitigating this problem. By harnessing the power of advanced sensing technologies, we can gain a better understanding of the environmental impact of sanitary pad disposal and work towards sustainable solutions. This chapter aims to provide valuable insights and guidance for researchers and practitioners working to create a cleaner and healthier environment and generate self-awareness for individuals in safeguarding ecosystem.

INTRODUCTION

Sanitary hygiene is a fundamental aspect of human health and well-being. However sanitary pads, essential for women's menstrual health, have become ubiquitous in India. However, the improper disposal of these pads poses a severe environmental problem. The vast majority of sanitary pads are made from non-biodegradable materials like plastics and synthetic fibres, which can take centuries to decompose. When these pads are discarded irresponsibly, they clog waterways, litter streets, and contribute to the growing issue of plastic pollution in the environment (Qu et al.,2023) sensor.

Within the Indian landscape, critical concerns revolve around the pollution of rivers and water bodies, congestion within sewer systems, and the pressing challenges posed by open dumping sites (Ahmad,2021; Elledge et al.,2018)). These are urgent matter of concerns to be addressed seriously to ensure further complications. Wherever proper sanitary pad disposals mechanisms are not in place, it is widely observed that women dispose the used sanitary pads in rivers and other water

DOI: 10.4018/979-8-3693-1930-7.ch016

bodies and even dispose them along with other waste without segregating them. This not only have adverse effects on the aquatic ecosystems but also contaminates the nearby water sources which will be depended by many communities for drinking and agriculture purposes. The polluted water can lead to various health issues for the communities relying on these sources (Patel et al.2020) in the long run. They might also create serious health concerns for the people who does the waste management. Another issue is related to the dumping of used sanitary pads in toilets. Urban and suburban areas often experience clogging of drainage systems due to the flushing of sanitary napkins into the toilets. This leads to expensive and disruptive maintenance work, which ultimately burdens municipal authorities and taxpayers. Leaving dirty napkins in public places on rural areas is a significant health risk as it draws out wandering creatures and promotes the spread of diseases (Gadekar,2023). The problem is most pronounced in groups with lower levels of proper waste disposal facilities, which are often marginalized.

In India, the issue of disposing of sanitary pads is significant due to the country's large population and potential health and hygiene issues. This requires urgent action. Improper disposal of sanitary napkins threatens public health because it can cause the spread of diseases and infections. For women and girls, access to hygienic menstrual products and safe disposal methods is integral to their well-being (Melaku et al.2023). The non-biodegradable nature of most sanitary pads means that they remain in the environment for an extended period. This contributes significantly to plastic pollution and its associated environmental degradation. The social stigma surrounding menstruation further complicates this issue. Women and girls often feel ashamed or embarrassed to discuss proper disposal methods, which perpetuates the problem. The burden of dealing with clogged sewer systems and managing open dumping sites places a strain on already stretched municipal infrastructure and resources.

This chapter endeavours to explore the contemporary landscape of sanitary pad disposal within the different context. Its aim is to ascertain prevalent challenges, anticipate future issues, and proactively devise viable solution strategies for addressing this concern. The chapter is organised into different sections which addresses various concerns. It mentions the role of environmental sensors in general and the organizations that produce these sensors and effectively use them for monitoring different aspects. Then the methods and practices of sanitary pad disposal, the environmental pollutants associated with improper disposal and statistics and facts about the scale of the problem are stated. An overview of environmental sensing technologies, the types of sensors used for monitoring environmental pollutants and the importance of real-time data collection and monitoring is mentioned. It also gives a description of the various pollutants and contaminants associated with sanitary pads and highlight the challenges of detecting these pollutants through traditional means. Section 6 present case studies and examples of how environmental sensors have been used in similar pollution monitoring scenarios and discuss the potential of these sensors in sanitary pad disposal contexts. Potential of Environmental sensors in Sanitary Pad Disposal is discussed in Section 7. Section 8 deals with the proposal for potential solutions and mitigation strategies for the issues related to sanitary pad disposal and how environmental sensors can inform and improve these solutions. Case Studies and Success Stories which showcase real-world examples of successful implementations of environmental sensors in addressing sanitary pad disposal problems is mentioned in Section 9. Section 10 discuss the Future Directions and Research Opportunities.

THE ROLE OF ENVIRONMENTAL SENSORS

Environmental sensors are devices that detect and measure various parameters in the environment. Examples include Temperature Sensor, Humidity Sensor, Air Quality Sensor, Light Sensor (Photodetector), Pressure Sensor, Gas Sensor, Sound Sensor (Microphone) etc. Temperature sensors measure the degree of hotness or coldness of an object or environment. Thermocouples, resistance temperature detectors (RTDs), and thermistors are commonly used sensors (Al Mamun et al., 2019). They work based on the principle that temperature changes affect the electrical properties of the material they are made of (e.g., resistance change in thermistors or voltage change in thermocouples). Humidity sensors measure the amount of moisture present in the air. Capacitive, resistive, and thermal conductivity sensors are used for humidity measurement. They operate by detecting changes in electrical conductivity, capacitance, or thermal properties due to the presence of moisture. The air quality sensors measure various pollutants present in the air, such as particulate matter, volatile organic compounds (VOCs), carbon monoxide (CO), nitrogen dioxide (NO_2), etc. Different technologies like electrochemical cells, metal oxide semiconductors, optical detection, and particulate matter sensors are used to detect and quantify these pollutants. Light sensors measure the intensity of light in an environment. Photodiodes, phototransistors, and photoresistors (LDR - Light Dependent Resistor) are commonly used. They work by converting light energy into electrical signals, where the electrical output is proportional to the intensity of incident light. Pressure sensors measure force applied by a fluid (liquid or gas) on its surface. Various types include piezoelectric, capacitive, and strain gauge sensors (Rhouati et al.,2022;Riza et al.2020). They operate by detecting changes in capacitance, resistance, or mechanical deformation due to pressure changes. Gas sensors detect the presence and concentration of specific gases in the environment (Burgués et al.,2022). Electrochemical, semiconductor, and infrared sensors are common types. They work based on interactions between gases and the sensing material, which causes changes in electrical conductivity, optical properties, or chemical reactions. Sound sensors capture sound waves and convert them into electrical signals. Microphones, which are a type of sound sensor, use diaphragms or other elements that vibrate in response to sound waves, generating corresponding electrical signals. Figures 1 depicts a temperature and humidity tracking box which is commonly used for environmental monitoring green houses, smart homes etc.

These sensors are integral to various applications like weather monitoring, indoor air quality assessment, industrial automation, smart homes, healthcare, and more, enabling the collection of data crucial for decision-making and understanding the environment's conditions. Environmental sensors have the potential to play a pivotal role in addressing the challenge of sanitary pad disposal in India. These sensors can be integrated into waste collection systems, sewage networks, and public spaces to monitor and manage the issue effectively. For real-time monitoring the environmental sensors can detect the presence of non-biodegradable waste in sewer systems, enabling immediate action to prevent clogs and overflows. Smart bins equipped with sensors can identify and sort sanitary waste, directing it to appropriate disposal channels and ensuring responsible disposal (Mor & Ravindra, 2023;Naher & Ahsan, 2023). Environmental sensors can also be used to monitor public spaces for littering and illegal dumping, providing data to raise awareness and enforce regulations (Joe et al.2018;Javaid et al.,2021). The data collected by these sensors can help local authorities make informed decisions about waste management, allowing for more efficient and sustainable solutions.

Figure 1. AEM1000 environmental monitoring box

Environmental sensors are devices designed to detect and measure various environmental parameters (Chapman et al., 2020). Table 1 gives the examples of the same, including the name of the integrated circuit (IC), their use, and type. These sensors are widely used in various applications like weather stations, environmental monitoring, home automation, and industrial control systems (Demrozi et al.2020). Each sensor has its specific application based on its sensing capability and the parameter it measures (De Sousa et al.,2023).

Several Indian companies specialize in the production and development of environmental sensors, catering to a wide range of applications such as air quality monitoring, weather forecasting, agriculture, and industrial automation. Table 2 gives an overview of some of the major Indian companies in this field. These companies are involved in either the direct production of environmental sensors or the development of systems that incorporate these sensors. Their products are used in a variety of sectors including industrial, commercial, healthcare, and agriculture.

Environmental sensors are used in various aspects of our daily lives. Table 3 gives some examples of specific organizations and locations where environmental sensors are prominently used. These organizations are leaders in their respective fields and utilize environmental sensors to enhance their operations, research, and services in specific locations around the world.

The global environmental sensor market is experiencing significant growth and is influenced by a variety of factors, including the increasing use of these sensors in various sectors, advancements in technology, and governmental regulations on pollution control. However, the market also faces challenges such as lack of awareness and budgetary constraints. As of 2023, the global environmental sensor market is expected to grow from \$1.59 billion in 2022 to \$1.77 billion in 2023, demonstrating a compound annual growth rate (CAGR) of 11.2%. Looking ahead, the market is projected to reach around \$2.59 billion by 2027, maintaining a CAGR of approximately 10.1%. The surge is fuelled by various factors, including the growth of smart cities, rising demand in different industries, and the development of IoT and cloud-based services.

Environmental Sensors

Table 1. Examples for environmental sensors

Sl No	Sensor	IC Name	Use	Type
1	DHT22 (AM2302) - Temperature and Humidity Sensor	DHT22 or AM2302	Measures ambient temperature and humidity.	Capacitive humidity sensing and thermistor-based temperature sensing
2	BMP280 - Barometric Pressure Sensor	BMP280	Measures atmospheric pressure, useful in weather forecasting and altitude estimation	Piezoresistive pressure sensor
3	MQ-135 - Air Quality Sensor	MQ-135	Detects various gases like ammonia, nitrogen oxides, alcohols, aromatic compounds, sulphide, and smoke	Gas sensor, chemoreceptor
4	DS18B20 - Waterproof Temperature Sensor	DS18B20	Measures temperature in wet or underwater environments	Digital temperature sensor, one-wire interface
5	TSL2561 - Luminosity Sensor	TSL2561	Measures ambient light intensity, mimicking the human eye response	Digital light sensor
6	MPU-6050 - Gyroscope and Accelerometer	MPU-6050	Measures angular rate and acceleration. Useful in orientation and motion detection	MEMS (Micro-Electro-Mechanical Systems) sensor
7	CCS811 - VOC and CO2 Sensor	CCS811	Measures indoor air quality by detecting volatile organic compounds (VOCs) and equivalent CO2 levels	Metal oxide gas sensor
8	Rain Sensor Module	Often a simple conductive board without a specific IC	Detects rainwater or water level presence	Conductive sensor
9	Soil Moisture Sensor	Often generic, with no specific IC	Measures the moisture level in soil, aiding in agriculture and gardening	Capacitive or resistive soil moisture sensor
10	MAX6675 - Thermocouple Temperature Sensor	MAX6675	Measures high temperatures, often used in conjunction with a K-type thermocouple	Thermocouple-to-digital converter

The global market is mainly concentrated in Asia Pacific, which is an industrial hub with China as its largest producer. The market in this region is expected to grow significantly, fuelled by factors such as industrialization, water scarcity, and air pollution control. However, the North American market is expected to surpass the Asia-Pacific region in terms of market share during the forecast period due to the technological development of the local countries. Government-led environmental initiatives are expected to drive the market's expansion in Europe, the Middle East, and Africa.

The main applications of these sensors are the monitoring and recording of environmental parameters such as humidity, temperature and air quality, which are important in industry, smart homes, consumer electronics and other devices. The main types of environmental sensors are temperature, humidity, air quality, water quality, integral, gas, chemical, smoke, ultraviolet (UV), and soil moisture sensors. Major players in the market are making product innovations and strategic acquisitions to maintain and strengthen their market position. For example, Zebra Technologies Corporation launched a new line of environmental sensors in April 2023 focused on the food, pharmaceutical and health sectors. Similarly, Interlink Electronics Inc. acquired SPEC Sensors and KWJ Engineering in December 2022, expanding its product range in gas and ambient air quality sensors. These developments show a vibrant and evolving market where environmental sensors are an increasingly important part of various aspects of modern life, from smart city development to industrial and agricultural applications. The Market Data Forecast report discussed the growing concern for environmental protection across the

Table 2. Indian companies using environmental sensors

Sl No	Company Name	Focus Area
1	Eureka Forbes Ltd	Known for their air quality monitoring products, Eureka Forbes offers a range of sensors and systems for detecting various air pollutants
2	Samriddhi Automation Pvt. Ltd. (Sparsh CCTV)	While primarily known for security and surveillance products, they also provide environmental sensors, particularly for industrial applications
3	Ambetronics Engineers Pvt. Ltd	They manufacture and export a wide array of electronic instruments and systems, including gas detection and environmental monitoring systems
4	Phoenix Sensors	Specializes in producing a variety of sensors including those for environmental monitoring, such as temperature, humidity, and pressure sensors
5	Oizom Instruments Pvt. Ltd	Focuses on smart environmental monitoring. They offer solutions like air quality monitors, dust monitors, odor monitors, and weather monitoring systems
6	Napino Auto & Electronics Ltd	While they are primarily focused on automotive electronics, they also venture into the development of sensors, including environmental sensors.
7	Robert Bosch Engineering and Business Solutions Private Limited	Part of the global Bosch group, this Indian subsidiary is involved in the development of various sensors, possibly including environmental sensors, as part of their wide range of electronic and technical services
8	Sensinova	Known for a range of sensor solutions, including motion sensors and environmental sensors for smart and automated systems
9	SEDEMAC Mechatronics	Specializes in mechatronics solutions and may include environmental sensors in their product range, particularly for automotive and industrial applications
10	Ripples IOT Pvt Ltd	Provides IoT solutions which include environmental monitoring for industries and healthcare facilities

Table 3. Organizations using environmental sensors

Sl No	Purpose	Organisation	Place
1	Weather Monitoring and Forecasting	National Oceanic and Atmospheric Administration (NOAA)	United States, with weather monitoring stations and equipment deployed nationwide
2	Air Quality Monitoring	Central Pollution Control Board (CPCB) in India	Major cities across India, including Delhi, Mumbai, and Bangalore, where air quality monitoring stations are installed
3	Smart Agriculture and Soil Monitoring	The United States Department of Agriculture (USDA)	Across various agricultural research sites and farms in the United States, particularly in the Midwest, known for its extensive agricultural activities
4	Home Automation and Smart Buildings	Siemens Building Technologies	Implemented in smart buildings and modern housing complexes worldwide, particularly in urban areas like New York City, London, and Singapore
5	Environmental Research and Conservation	World Wide Fund for Nature (WWF)	In various ecological conservation areas globally, notably in the Amazon Rainforest, the Arctic, and the Great Barrier Reef.
6	Industrial Emission Monitoring	Environmental Protection Agency (EPA) in the United States	Industrial zones across the United States, particularly in areas with high industrial activity like Texas and California
7	Water Quality Monitoring	European Environment Agency (EEA)	Across various water bodies in Europe, including the Danube River and the Baltic Sea
8	Healthcare and Hospital Environments	National Health Service (NHS) in the United Kingdom	In hospitals and healthcare facilities across the UK, particularly in major cities like London and Manchester
9	Urban Planning and Development	Urban Redevelopment Authority (URA) of Singapore	In urban areas of Singapore for monitoring environmental conditions and planning urban development projects

globe, the impact of IoT on the development of environmentally sensitive technology and the impact of government initiatives on market growth. Challenges faced by the market such as lack of awareness and budget constraints were also discussed. Allied Market Research has provided detailed statistics on the environmental sensors market, including its size, share and industry forecast to 2030. This resource highlighted the widespread use of environmental sensors in various sectors, market value in 2020 and forecasts to 2030. Research and Markets provided a comprehensive overview of the environmental sensors market, including its expected growth from 2022-2023 and forecasts to 2027. The source also presented the types of environmental sensors and their applications in various industries and the strategic movements of the most important market participants. Disposal of sanitary napkins in India is not only a public health problem, but also a major environmental problem. It is important to understand the urgency of solving this problem and the potential of environmental sensors to manage it. Sustainable solutions and awareness are critical to mitigating environmental impacts and promoting responsible disposal practices (Kaur et al., 2018), which ultimately improve women's well-being and the health of the planet. It is high time India took collective action to tackle this problem and create a more hygienic and sustainable future for all.

SANITARY PADS: AN ENVIRONMENTAL CHALLENGE

Indispensable for women and hygiene, sanitary napkins have had a significant impact on the lives of women and people around the world. However, the improper disposal of sanitary napkins has become a growing environmental problem, along with the many benefits they provide. Awareness of sanitary pad methods and practices, environmental contamination associated with improper disposal (Ghosh et al., 2020; Harrison and Tyson, 2022) and statistics and facts about the extent of this problem are mandatory to address emerging issues. With these sanitary napkins are traditionally disposed of by wrapping them in paper or plastic and throwing them in the trash. However, this method is not without problems. Sanitary napkins are often non-biodegradable and can take hundreds of years to decompose, adding to landfill and posing a serious environmental threat. Moreover, the lack of proper waste management infrastructure in many areas means that sanitary napkins are often discarded in open areas, rivers or even flushed toilets, clogging drainage systems and further degrading the environment. Improper disposal of sanitary napkins results in many environmental pollutants that negatively affect both terrestrial and aquatic ecosystems. The main environmental problems associated with improper disposal are soil, water, air pollution and related health risks. Sanitary napkins contain plastic parts and other non-biodegradable materials that remain in the environment for a long time. When deposited in landfills or open areas, they add to soil and soil pollution. Flushing sanitary napkins in the toilet can cause blockages in drainage systems, which can lead to contamination of overflow channels and waterways. Plastic and synthetic materials in pillows can release harmful chemicals and microplastics into the water, which harms aquatic life. Incineration of sanitary napkins as a disposal method can release toxic fumes and airborne pollutants into the atmosphere, further degrading air quality. Improper disposal practices can also endanger the health of waste collectors and sanitation workers who come into contact with contaminated materials. The scale of the problem of inappropriate disposal of sanitary napkins is significant and alarming. Problems related to the disposal of sanitary napkins are not limited to developing countries. Even in countries with advanced waste management systems, there are still problems in ensuring proper disposal. Here are some statistics and facts that highlight the importance of this environmental problem: According to a study by the Menstrual Hygiene Association, about 80% of India's sanitary waste is disposed of improperly, causing environmental pollution. According to a UK study, more than 4.6 billion single-use sanitary products are used each year, placing a significant burden on the environment. The Ellen MacArthur Foundation reports that plastics used in sanitary napkins and tampons, which are often disposed of incorrectly, contribute to plastic pollution in the oceans.Sanitary napkins are an environmental problem that can no longer be ignored. To reduce the environmental damage of these important hygiene products, it is necessary to adopt responsible waste management practices. Recommending the use of biodegradable and environmentally friendly hygiene products, improving waste management systems and raising awareness about the correct disposal of sanitary napkins are important steps towards a more sustainable and environmentally friendly future. By solving this problem, we can protect both the health of our planet and the well-being of women around the world.

ENVIRONMENTAL SENSORS: TECHNOLOGY AND CHARACTERISTICS

Environmental sensing technologies have advanced significantly in recent years (Tajik et al., 2021; Thakur and Kumar, 2022), providing invaluable tools for monitoring and managing our planet and health. These technologies include a wide range of sensors and data collection methods designed to measure various environmental parameters. They play an important role in understanding, mitigating and preventing the negative effects of pollution, climate change and other environmental problems. One of the key components of environmental sensing technology is the array of sensors used to monitor environmental contaminants. These sensors are designed to detect and measure a wide range of pollutants such as air pollution, water quality parameters, soil pollutants, and more (Khanmohammadi et al., 2020). They come in many forms, from simple handheld devices to sophisticated automated systems. For example, air quality sensors can measure levels of gases such as carbon dioxide and ozone (Dhall et al., 2021), while water quality sensors can detect parameters such as pH, turbidity, and chemical concentrations. These sensors provide important information that informs decision makers, scientists and the public about the state of the environment.Real-time data collection and monitoring has become increasingly critical in our efforts to effectively address environmental issues. Timely availability of environmental data enables a rapid response to pollution incidents and evaluation of the effectiveness of environmental policies and measures. For example, using real-time air quality sensors, cities can provide health information during fog or forest fires. In addition, they contribute to the development of early warning systems for natural disasters and can help monitor the movement of pollutants, such as oil spills, to minimize their impact on ecosystems and human health. This immediate access to information improves our ability to make informed decisions and take proactive measures to protect the environment (Kumunda et al., 2021; Lazarević et al., 2023). Environmentally sensitive technologies not only provide important information to solve current problems, but also support long-term environmental monitoring and research. By collecting continuous and reliable data over long periods of time, scientists can identify trends and patterns that aid our understanding of environmental change. This information is essential to develop sustainable policies and strategies to address emerging challenges such as climate change. Continuous monitoring also serves as a valuable learning tool, raising awareness of environmental issues and encouraging responsible behaviour by individuals, industries and governments. Environmental sensors and sensor technologies are essential to keeping our planet safe. These technologies use a variety of sensors to monitor environmental pollutants (Du et al. 2023), which provide real-time data collection and monitoring capabilities. Such advances are critical to mitigating the effects of pollution and climate change because they provide information needed for informed decision-making and long-term environmental research. As we constantly face environmental challenges, the development and application of environmental sensors remains at the forefront of our efforts to protect our environment and ensure a sustainable future.

IDENTIFYING POLLUTANTS IN SANITARY PAD DISPOSAL

Sanitary pads, essential for women's health and hygiene worldwide, present significant environmental and health challenges attributable to the non-biodegradable materials employed in their production and disposal. A critical step in addressing these concerns involves the identification of pollutants associated with sanitary pad disposal. Common contaminants encompass non-biodegradable plastics like polyethylene and polypropylene, superabsorbent polymers such as sodium polyacrylate, and chemical additives like synthetic fragrances, dyes, and adhesives containing volatile organic compounds. Furthermore, used sanitary pads may host harmful bacteria and microorganisms, elevating health risks for individuals managing the waste. Recognizing and comprehending these pollutants is indispensable for formulating effective strategies to mitigate the environmental and health impacts of sanitary pad disposal (Peter & Abhitha, 2021; Aguilar et al., 2020).

In light of these challenges, it becomes evident that a comprehensive approach is required to address the multifaceted issues associated with sanitary pad disposal. This involves not only acknowledging the ecological consequences of non-biodegradable materials but also considering the potential health risks posed by chemical additives and microbial contaminants. By fostering awareness and promoting sustainable alternatives, society can contribute to the development of environmentally friendly menstrual hygiene practices, thus mitigating the adverse impacts of sanitary pad disposal on both ecological systems and human health.

CHALLENGES OF DETECTING POLLUTANTS IN SANITARY PADS

Analysing the environmental impact of sanitary pads poses several challenges due to their complex composition, characterized by a combination of multiple materials. Identifying and quantifying pollutants becomes a daunting task as traditional chemical analysis methods are often time-consuming and necessitate intricate sample preparation. Compounding the issue, pollutants may be present in low concentrations, rendering conventional techniques optimized for higher concentrations less effective. The heterogeneous nature of the waste stream generated by the varied usage and disposal conditions of sanitary pads further complicates the detection of pollutants in this matrix. Moreover, certain pollutants, such as microorganisms or volatile organic compounds (VOCs), may demand non-destructive analysis methods to preserve sample integrity, contrasting with traditional methods that involve destructive sample preparation. Additionally, the absence of standardized testing protocols hampers the assessment of environmental impacts, making it challenging to compare results across different studies and regions. Many advanced analytical methods for detecting pollutants are expensive and may not be readily available, especially in resource-constrained settings. The disposal of sanitary pads can be a sensitive issue, and obtaining samples for analysis may raise privacy and ethical concerns, particularly if personal information is involved.

To address these challenges, ongoing research is focused on developing more efficient and sensitive analytical methods, as well as promoting the development and use of eco-friendly materials in menstrual hygiene products to mitigate their environmental and health impacts.

The critical concern of environmental pollution, emphasizing its potential irreversible harm to genetic, nervous, and circulatory systems was highlighted by Yang (2023). Luminescent metal-organic frameworks (LMOFs) are spotlighted for their selectivity, sensitivity, and recyclability in detecting pollutants. The paper provides a systematic overview of MOF-based luminescent sensors, addressing the gap in their application to detect various environmental pollutants, including radioactive and heavy metal ions. Another study by Yuan et al. (2023) discuss the pervasive spread of emerging contaminants due to human industrial practices, posing threats to health and ecology. They introduce optical fiber microfluidic coupled sensors as a promising solution, offering controlled microfluidics for precise and stable detection of environmental contaminants. The paper comprehensively reviews sensor characteristics, methodological principles, and applications, envisioning the future role of these sensors in environmental detection. Zhang et al. (2021) examine the promise of sensors in addressing global environmental challenges through pollutant detection, emphasizing the need for advancements in sensitivity, efficiency, simplicity, and miniaturization. The review critically analyses current sensing strategies to guide the future evolution of sensor technology for environmental applications. Zhou et al. (2023) focus on the reproductive toxicity of Per- and polyfluoroalkyl substances (PFAS) in personal hygiene products. Their study reveals the presence of PFAS in products such as paper diapers, sanitary pads, and menstrual cups. Paper diapers exhibit the highest concentrations, raising concerns about potential exposure and environmental emissions of PFAS from these products. The study emphasizes a previously undisclosed exposure pathway of PFAS through personal hygiene items and underscores potential health implications. Additionally, raising awareness about proper disposal and recycling methods is essential to reduce the environmental footprint of sanitary pads.

APPLICATIONS OF ENVIRONMENTAL SENSORS

Environmental sensors (Liang et al.,2020; Liu et al.,2022; Li et al,2022) play a crucial role in monitoring and mitigating pollution across various scenarios. Exploring diverse realms of environmental monitoring, several case studies showcase the practical applications of sensor technologies. In the realm of air quality, urban areas leverage sensors to track pollutants like PM2.5, PM10, NO2, SO2, and O3. A notable example is Beijing, where a sensor network provides real-time air quality data, empowering the government's pollution control measures and enhancing public awareness. Transitioning to water quality, the Chesapeake Bay Program in the U.S. deploys sensors to continuously monitor parameters such as dissolved oxygen, pH, turbidity, and nutrient levels. This approach facilitates the evaluation of water pollution levels, identification of pollution sources, and the implementation of targeted corrective actions (Pooja et al., 2020).

Expanding the scope to industrial emissions, companies employ sensors to ensure compliance with environmental regulations. The Sulphur Dioxide Emission Reduction Project in India exemplifies this, utilizing continuous emission monitoring systems (CEMS) to track SO2 emissions from power plants, thereby contributing to efforts aimed at reducing air pollution. Shifting focus to noise pollution, urban areas benefit from sensors that measure noise levels and identify noisy zones. London's "The Noise App" illustrates this concept, allowing residents to use smartphones for recording and reporting noise pollution incidents, actively participating in collective noise control initiatives. Lastly, in the agricultural domain, soil sensors play a pivotal role in monitoring moisture content, temperature, and nutrient levels. These sensors empower farmers to optimize irrigation and fertilization practices, effectively mitigating soil pollution arising from excess nutrients.

POTENTIAL OF ENVIRONMENTAL SENSORS IN SANITARY PAD DISPOSAL

Environmental sensors play a crucial role in addressing the environmental challenges associated with sanitary pad disposal, contributing to pollution in water bodies and landfills. In various pollution monitoring scenarios, these sensors have demonstrated success by providing real-time data for decision-making and pollution control. In the context of sanitary pad disposal, they offer potential improvements in waste management, compliance monitoring, and reducing the environmental impact of sanitary pad waste.

One application of environmental sensors is bin level monitoring, where sensors placed in sanitary pad disposal bins can track fill levels. Alerts can be sent to waste management teams when bins reach capacity, ensuring timely collection and preventing overflowing bins, which can lead to littering and environmental pollution. Additionally, sensors combined with RFID or barcoding technology facilitate waste sorting and tracking, aiding in proper disposal, recycling, or safe incineration, thus reducing the environmental impact.

Compliance monitoring is another key application, as sensors can ensure adherence to sanitary waste disposal regulations. For instance, in public restrooms, sensors can monitor disposal practices, providing data for awareness campaigns or enforcement measures. Environmental sensors also contribute to the Environmental Impact Assessment (EIA) process by strategically placed measurements near disposal sites, including landfills and incinerators. These sensors assess air quality, soil composition, and water quality, providing essential information for making informed decisions regarding the environmental impact of sanitary pad disposal.

Environmental sensors are instrumental in air quality assessment, measuring pollutants released during disposal and incineration, such as VOCs and particulate matter. This accurate data helps understand the impact of emissions on nearby ecosystems and public health, informing assessments and protective measures. Similarly, sensors in water bodies detect changes in water quality due to leachate or runoff, offering critical data for safeguarding aquatic ecosystems and community water supplies.

Soil analysis near disposal sites is facilitated by sensors, providing crucial data for assessing the impact on soil ecosystems. This includes potential harm to plant life, disruptions to nutrient cycles, and impacts on soil organism health. Biodiversity and habitat monitoring benefit from environmental sensors, offering comprehensive data on how sanitary pad disposal affects local ecosystems, informing conservation efforts.

Moreover, environmental sensors contribute to assessing public health implications by measuring pollutants in the air, water, and soil. This data helps understand exposure pathways and health hazards, enabling the formulation of protective measures and regulations. Data-driven decision-making is empowered by environmental sensors, providing accurate and comprehensive information essential for developing effective regulations and minimizing negative impacts on ecosystems and human populations.

The continuous collection of long-term data is emphasized, as some consequences may only become apparent years after disposal activities cease. In conclusion, the data collected by environmental sensors are critical for conducting an EIA of sanitary pad disposal, enabling informed decisions and actions for the well-being of ecosystems and human populations.

SOLUTIONS AND MITIGATION

Proper disposal of sanitary pads is a significant environmental and public health concern, involving waste management, hygiene, and environmental impact. Solutions and mitigation strategies that can be considered are mentioned below.

Promoting awareness and education on proper disposal through community programs, schools, and social media campaigns. Environmental sensors can monitor changes in disposal habits, assessing the success of awareness initiatives.

Encouraging the use of biodegradable sanitary pads to reduce environmental impact. Sensors can monitor decomposition rates in landfills or compost facilities, providing insights into their sustainability.

Ensuring waste management facilities, such as incinerators and sanitary landfill sites, are available and equipped to handle sanitary pad disposal safely. Sensors monitor facility efficiency and capacity, preventing overloading and ensuring safe operation.

Installing dedicated disposal bins for sanitary pads in public and private spaces, with sensors monitoring fill levels. Real-time data helps schedule timely disposal and prevent overflow.

Investing in waste-to-energy conversion technologies for safe incineration. Sensors monitor emissions to ensure compliance with environmental standards and prevent harmful pollutants.

Promoting recycling programs for specific components of sanitary pads, such as plastic backings. Sensors track the recycling process, ensuring efficiency and eco-friendly use of recycled materials.

Encouraging the development and use of environmentally friendly materials for sanitary pads. Sensors monitor the environmental impact, aiding the transition to more sustainable practices.

Empowering communities to use environmental sensors for monitoring and reporting improper disposal. Community-driven sensors provide real-time data, enabling authorities to respond promptly.

Implementing and enforcing regulations on manufacturers and distributors, including labeling requirements and extended producer responsibility. Sensors help monitor compliance and track the environmental impact of regulations. Environmental sensors, including air quality sensors, landfill gas detectors, and waste monitoring systems, play a crucial role in informing and improving solutions to sanitary pad disposal issues. They provide essential data for making informed decisions, optimizing waste management processes, and assessing environmental impact, ultimately leading to more sustainable practices.

FUTURE DIRECTIONS AND RESEARCH OPPORTUNITIES

The field of environmental observations developed rapidly in response to concerns about climate change, pollution, and the sustainable management of natural resources. Technological innovations enable accurate and scalable data collection on various environmental parameters. Future R&D opportunities in this area include:Miniaturization and Sensor Networks: Development of smaller, cost-effective, and energy-efficient sensors for large-scale use in sensor networks. These networks can provide high-resolution data on air quality, temperature, humidity and soil conditions.Advanced Data Analytics: Process increasing environmental sensor data using advanced analytics, including machine learning and artificial intelligence, to efficiently process, analyze and interpret data and identify trends, anomalies and patterns.Integration with the Internet of Things and Smart Cities: Explore the seamless integration of environmentally sensitive technology into urban infrastructure for real-time monitoring and adaptive resource management as smart cities and the Internet of Things (IoT) evolve. Improved environmental forecasts: Improve models and tools for predicting environmental change, including weather forecasting, long-term climate modelling, and prediction of natural disasters such as hurricanes, fires, and floods. Ocean and Space Research: Extend environmental monitoring beyond Earth by exploring innovations in underwater sensor technology to monitor ocean health and study extreme environments, and space-based sensors to study Earth's atmosphere and beyond. Biodiversity Monitoring: Develop sensors and data collection methods for biodiversity monitoring and conservation, including acoustic sensors for wildlife monitoring, cameras for wildlife monitoring, and environmental DNA (eDNA) technologies for species identification and tracking. Environmental Health and Epidemiology: Explore how environmental monitoring can contribute to public health, such as using sensors to detect air and water quality to control the spread of diseases and epidemics. Environmental Ethics and Policy: Explore the ethical implications of large-scale environmental control, addressing privacy, data ownership and potential abuse. Develop rules and regulations that balance the benefits of surveillance with the protection of individual rights and privacy. Renewable energy and sustainable development: Sensor technology research to improve the efficiency and sustainability of renewable energy sources, to optimize energy production and storage, and to support the development of environmentally friendly technologies. Interdisciplinary Collaboration: Encourages collaboration among environmental scientists, engineers, data scientists, and social scientists for multidisciplinary research that addresses complex environmental problems and develops comprehensive solutions. Community Engagement and Citizen Science: Engaging the public in environmental monitoring through citizen science initiatives using crowdsourced data to improve the reach and accuracy of environmental monitoring efforts. Carbon Sequestration and Mitigation Technologies: Develop sensors and monitoring systems to track carbon sequestration efforts and evaluate their effectiveness in mitigating climate change.

CONCLUSION

The field of environmental sensing is ripe with opportunities for innovation and research. As technology continues to advance, there is enormous potential for environmental sensors to play a pivotal role in our efforts to understand, mitigate, and adapt to environmental changes. These innovations are crucial for achieving a sustainable and resilient future.

This book chapter aims to shed light on the critical issue of sanitary pad disposal and the role of environmental sensors in identifying and mitigating the associated problems. By providing a comprehensive overview of the topic, including case studies and potential solutions, it will serve as a valuable resource for researchers, practitioners, and policymakers working towards a cleaner and more sustainable environment.

REFERENCES

Aguilar-Pérez, K. M., Heya, M. S., Parra-Saldívar, R., & Iqbal, H. M. (2020). Nano-biomaterials in-focus as sensing/detection cues for environmental pollutants. *Case Studies in Chemical and Environmental Engineering*, *2*, 100055. doi:10.1016/j.cscee.2020.100055

Ahmad, O. (2021, March 8). *As India breaks the taboo over sanitary pads, an environmental crisis mounts.* The Third Pole. https://www.thethirdpole.net/en/culture/sanitary-pads-environmental-crisis/

Al Mamun, M. A., & Yuce, M. R. (2019). Sensors and systems for wearable environmental monitoring toward IoT-enabled applications: A review. *IEEE Sensors Journal*, *19*(18), 7771–7788. doi:10.1109/JSEN.2019.2919352

Allied Market Research. (n.d.). *Environmental Sensor Market Size, Share, Competitive Landscape and Trend Analysis Report by Type (Humidity, Temperature, Pressure, Gas, and Others) and End User (Industrial, Residential, Commercial, Automotive, Government & Public Utilities and Other): Global Opportunity Analysis and Industry Forecast*. Allied Market Research. https://www.alliedmarketresearch.com/environmental-sensors-market-A12896

Burgués, J., & Marco, S. (2020). Environmental chemical sensing using small drones: A review. *The Science of the Total Environment*, *748*, 141172. doi:10.1016/j.scitotenv.2020.141172 PMID:32805561

Chapman, J., Truong, V. K., Elbourne, A., Gangadoo, S., Cheeseman, S., Rajapaksha, P., Latham, K., Crawford, R. J., & Cozzolino, D. (2020). Combining chemometrics and sensors: Toward new applications in monitoring and environmental analysis. *Chemical Reviews*, *120*(13), 6048–6069. doi:10.1021/acs.chemrev.9b00616 PMID:32364371

Clean India Journal. (2023, October 20). The sustainable route to sanitary napkin disposal. *Clean India Journal.* https://www.cleanindiajournal.com/the-sustainable-route-to-sanitary-napkin-disposal/

De Sousa, J. F. Junior, Columbus, S., Hammouche, J., Ramachandran, K., Daoudi, K., & Gaidi, M. (2023). Engineered micro-pyramids functionalized with silver nanoarrays as excellent cost-effective SERS chemosensors for multi-hazardous pollutants detection. *Applied Surface Science*, *613*, 156092. doi:10.1016/j.apsusc.2022.156092

Demrozi, F., Pravadelli, G., Bihorac, A., & Rashidi, P. (2020). Human activity recognition using inertial, physiological and environmental sensors: A comprehensive survey. *IEEE Access : Practical Innovations, Open Solutions*, *8*, 210816–210836. doi:10.1109/ACCESS.2020.3037715 PMID:33344100

Dhall, S., Mehta, B. R., Tyagi, A. K., & Sood, K. (2021). A review on environmental gas sensors: Materials and technologies. *Sensors International*, *2*, 100116. doi:10.1016/j.sintl.2021.100116

Du, Y., Yu, D. G., & Yi, T. (2023). Electrospun nanofibers as chemosensors for detecting environmental pollutants: A review. *Chemosensors (Basel, Switzerland)*, *11*(4), 208. doi:10.3390/chemosensors11040208

Elledge, M., Muralidharan, A., Parker, A., Ravndal, K. T., Siddiqui, M., Toolaram, A. P., & Woodward, K. (2018). Menstrual Hygiene Management and Waste Disposal in Low and Middle Income Countries—A Review of the Literature. *International Journal of Environmental Research and Public Health*, *15*(11), 2562. doi:10.3390/ijerph15112562 PMID:30445767

Gadekar, U. (2023). Hygiene Practices and Community Well-being in a Rural Setting: The Case of Mhawlewadi Village. *International Journal of Engineering and Management Research*, *13*(4), 99–104.

Ghosh, I., Rakholia, D., Shah, K., Bhatt, D., & Das, M. N. (2020). Environmental Perspective on Menstrual Hygiene Management Along with the Movement towards Biodegradability: A Mini-Review. *Journal of Biomedical Research & Environmental Sciences*, *1*(5), 122–126. doi:10.37871/jels1129

Harrison, M., & Tyson, N. (2022). Menstruation: Environmental impact and need for global health equity. *International Journal of Gynaecology and Obstetrics: the Official Organ of the International Federation of Gynaecology and Obstetrics*, *160*(2), 378–382. doi:10.1002/ijgo.14311 PMID:35781656

He, Q., Wang, B., Liang, J., Liu, J., Liang, B., Li, G., Long, Y., Zhang, G., & Liu, H. (2023). Research on the construction of portable electrochemical sensors for environmental compounds quality monitoring. *Materials Today. Advances*, *17*, 100340. doi:10.1016/j.mtadv.2022.100340

Javaid, M., Haleem, A., Rab, S., Singh, R. P., & Suman, R. (2021). Sensors for daily life: A review. *Sensors International*, *2*, 100121. doi:10.1016/j.sintl.2021.100121

Joe, H. E., Yun, H., Jo, S. H., Jun, M. B., & Min, B. K. (2018). A review on optical fiber sensors for environmental monitoring. *International journal of precision engineering and manufacturing-green technology, 5*, 173-191.

Kaur, R., Kaur, K., & Kaur, R. (2018). Menstrual Hygiene, Management, and Waste Disposal: Practices and challenges faced by Girls/Women of Developing Countries. *Journal of Environmental and Public Health*, *2018*, 1–9. doi:10.1155/2018/1730964 PMID:29675047

Khanmohammadi, A., Jalili Ghazizadeh, A., Hashemi, P., Afkhami, A., Arduini, F., & Bagheri, H. (2020). An overview to electrochemical biosensors and sensors for the detection of environmental contaminants. *Journal of the Indian Chemical Society*, *17*, 2429–2447.

Kumar, B., Singh, J., Mittal, S., & Singh, H. (2023). The Indian perspective on the harmful substances found in sanitary napkins and their effects on the environment and human health. *Environmental Science and Pollution Research International*. doi:10.1007/s11356-023-26739-2 PMID:37022541

Kumunda, C., Adekunle, A. S., Mamba, B. B., Hlongwa, N. W., & Nkambule, T. T. (2021). Electrochemical detection of environmental pollutants based on graphene derivatives: A review. *Frontiers in Materials*, *7*, 616787. doi:10.3389/fmats.2020.616787

Lazarević-Pašti, T., Tasić, T., Milanković, V., & Potkonjak, N. (2023). Molecularly Imprinted Plasmonic-Based Sensors for Environmental Contaminants—Current State and Future Perspectives. *Chemosensors (Basel, Switzerland)*, *11*(1), 35. doi:10.3390/chemosensors11010035

Li, L., Zou, J., Han, Y., Liao, Z., Lu, P., Nezamzadeh-Ejhieh, A., Liu, J., & Peng, Y. (2022). Recent advances in Al (iii)/In (iii)-based MOFs for the detection of pollutants. *New Journal of Chemistry*, *46*(41), 19577–19592. doi:10.1039/D2NJ03419K

Liang, J., Zulkifli, M. Y., Choy, S., Li, Y., Gao, M., Kong, B., Yun, J., & Liang, K. (2020). Metal–organic framework–plant nanobiohybrids as living sensors for on-site environmental pollutant detection. *Environmental Science & Technology*, *54*(18), 11356–11364. doi:10.1021/acs.est.0c04688 PMID:32794698

Liu, B., Zhuang, J., & Wei, G. (2020). Recent advances in the design of colorimetric sensors for environmental monitoring. *Environmental Science. Nano*, *7*(8), 2195–2213. doi:10.1039/D0EN00449A

Liu, Y., Xue, Q., Chang, C., Wang, R., Liu, Z., & He, L. (2022). Recent progress regarding electrochemical sensors for the detection of typical pollutants in water environments. *Analytical Sciences*, *38*(1), 55–70. doi:10.2116/analsci.21SAR12 PMID:35287206

Ltd, R. a. M. (n.d.). *Environmental Sensor Global Market Report 2023*. Research and Markets Ltd 2024. https://www.researchandmarkets.com/reports/5880229/environmental-sensor-global-market-report

Melaku, A., Addis, T., Mengistie, B., Kanno, G. G., Adane, M., Kelly-Quinn, M., Ketema, S., Hailu, T., Bedada, D., & Ambelu, A. (2023). Menstrual hygiene management practices and determinants among schoolgirls in Addis Ababa, Ethiopia: The urgency of tackling bottlenecks-Water and sanitation services. *Heliyon*, *9*(5), e15893. doi:10.1016/j.heliyon.2023.e15893 PMID:37180900

Mor, S., & Ravindra, K. (2023). Municipal solid waste landfills in lower-and middle-income countries: Environmental impacts, challenges and sustainable management practices. *Process Safety and Environmental Protection*, *174*, 510–530. doi:10.1016/j.psep.2023.04.014

Naher, L., & Ahsan, M. F. (2023). Solid Waste Disposal Scenario of Three Ladies' Halls in the University of Chittagong, Chittagong, Bangladesh. *American Journal of Agricultural Science, Engineering, and Technology*, *7*(2), 1–6. doi:10.54536/ajaset.v7i2.1295

Online, E. (2023, October 2). Why menstrual waste is proving to be a "big bloody mess" in India. *The Economic Times*. https://economictimes.indiatimes.com/news/india/the-menstrual-waste-conundrum-in-india-and-why-its-proving-to-be-a-big-bloody-mess-for-sustainability/articleshow/104106414.cms

Panda, S. K., Kherani, N. A., Debata, S., & Singh, D. P. (2023). Bubble propelled micro/nano motors: A robust platform for the detection of environmental pollutants and biosensing. *Materials Advances*, *4*(6), 1460–1480. doi:10.1039/D2MA00798C

Park, C. J., Barakat, R., Ulanov, A., Li, Z., Lin, P.-C., Chiu, K., Zhou, S., Pérez, P. E., Lee, J., Flaws, J. A., & Ko, C. (2019). Sanitary pads and diapers contain higher phthalate contents than those in common commercial plastic products. *Reproductive Toxicology (Elmsford, N.Y.)*, *84*, 114–121. doi:10.1016/j.reprotox.2019.01.005 PMID:30659930

Patel, B. R., Noroozifar, M., & Kerman, K. (2020). Nanocomposite-based sensors for voltammetric detection of hazardous phenolic pollutants in water. *Journal of the Electrochemical Society*, *167*(3), 037568. doi:10.1149/1945-7111/ab71fa

Peter, A., & Abhitha, K. (2021). Menstrual Cup: A replacement to sanitary pads for a plastic free periods. *Materials Today: Proceedings*, *47*, 5199–5202. doi:10.1016/j.matpr.2021.05.527

Pooja, D., Kumar, P., Singh, P., & Patil, S. (Eds.). (2020). *Sensors in water pollutants monitoring: role of material* (p. 320). Springer. doi:10.1007/978-981-15-0671-0

Qu, K., Hu, X., & Li, Q. (2023). Electrochemical environmental pollutant detection enabled by waste tangerine peel-derived biochar. *Diamond and Related Materials*, *131*, 109617. doi:10.1016/j.diamond.2022.109617

Rhouati, A., Berkani, M., Vasseghian, Y., & Golzadeh, N. (2022). MXene-based electrochemical sensors for detection of environmental pollutants: A comprehensive review. *Chemosphere*, *291*, 132921. doi:10.1016/j.chemosphere.2021.132921 PMID:34798114

Riza, M. A., Go, Y. I., Harun, S. W., & Maier, R. R. (2020). FBG sensors for environmental and biochemical applications—A review. *IEEE Sensors Journal*, *20*(14), 7614–7627. doi:10.1109/JSEN.2020.2982446

Saral Designs. (2022, July 22). *Sanitary Napkins and its Environmental impact - Part 1*. Saral Designs. https://www.saraldesigns.in/uncategorized/sanitary-napkins-and-its-environmental-impact-part-1/

Sohrabi, H., Hemmati, A., Majidi, M. R., Eyvazi, S., Jahanban-Esfahlan, A., Baradaran, B., Adlpour-Azar, R., Mokhtarzadeh, A., & de la Guardia, M. (2021). Recent advances on portable sensing and biosensing assays applied for detection of main chemical and biological pollutant agents in water samples: A critical review. *Trends in Analytical Chemistry*, *143*, 116344. doi:10.1016/j.trac.2021.116344

Tajik, S., Beitollahi, H., Nejad, F. G., Dourandish, Z., Khalilzadeh, M. A., Jang, H. W., Venditti, R. A., Varma, R. S., & Shokouhimehr, M. (2021). Recent developments in polymer nanocomposite-based electrochemical sensors for detecting environmental pollutants. *Industrial & Engineering Chemistry Research*, *60*(3), 1112–1136. doi:10.1021/acs.iecr.0c04952 PMID:35340740

Thakur, A., & Kumar, A. (2022). Recent advances on rapid detection and remediation of environmental pollutants utilizing nanomaterials-based (bio) sensors. *The Science of the Total Environment*, *834*, 155219. doi:10.1016/j.scitotenv.2022.155219 PMID:35421493

Wu, H., Li, J. H., Yang, W. C., Wen, T., He, J., Gao, Y. Y., ... & Yang, W. C. (2023). Nonmetal-doped quantum dot-based fluorescence sensing facilitates the monitoring of environmental contaminants. *Trends in Environmental Analytical Chemistry*, e00218.

Yang, G. L., Jiang, X. L., Xu, H., & Zhao, B. (2021). Applications of MOFs as luminescent sensors for environmental pollutants. *Small*, *17*(22), 2005327. doi:10.1002/smll.202005327 PMID:33634574

Yang, S., Sarkar, S., Xie, X., Li, D., & Chen, J. (2023). Application of optical hydrogels in environmental sensing. *Energy & Environmental Materials*, 12646. doi:10.1002/eem2.12646

Yuan, Y., Jia, H., Xu, D., & Wang, J. (2023). Novel method in emerging environmental contaminants detection: Fiber optic sensors based on microfluidic chips. *The Science of the Total Environment*, *857*, 159563. doi:10.1016/j.scitotenv.2022.159563 PMID:36265627

Zhang, Y., Zhu, Y., Zeng, Z., Zeng, G., Xiao, R., Wang, Y., Hu, Y., Tang, L., & Feng, C. (2021). Sensors for the environmental pollutant detection: Are we already there? *Coordination Chemistry Reviews*, *431*, 213681. doi:10.1016/j.ccr.2020.213681

Zhou, Y., Lin, X., Xing, Y., Zhang, X., Lee, H. K., & Huang, Z. (2023). Per-and Polyfluoroalkyl Substances in Personal Hygiene Products: The Implications for Human Exposure and Emission to the Environment. *Environmental Science & Technology*.

Chapter 17
Bioindicators of Environmental Pollution With Emphasis to Wetlands of Kashmir

Kounsar Jan
Government Degree College Thindim Kreeri, India

Javid Majeed Wani
Government Degree College Dangiwacha, India

ABSTRACT

The use of bioindicators has grown in recent years, and they have provided a wealth of valuable data that has improved water resource management. One way to measure the quality of an environment is by looking at how well a species (or group of species) can adapt to different kinds of chemical, physical, and biological stresses. A further benefit of bioindicators is their capacity to detect the indirect biotic impacts of contaminants, a feat that is not accomplished by many physical or chemical tests. When used as bioindicators, the varying degrees of stress that various aquatic species can withstand might provide light on the nature of a given environmental problem. Zooplankton species such as Branchionus sp., Molina sp., Keratella cochlearis, Daphnia sp., and Cyclopus sp., as well as phytoplankton species such as Euglena viridis, Oscillatoria limosa, Nitzschia palea, and Scenedesmus quadricauda, are indicators of water pollution. The goal of this study is to showcase some new plankton research that focuses on their potential and uses as bioindicators of water quality.

INTRODUCTION

One way to measure environmental quality and its changes over time is via bioindicators, which might be biological processes, species, or communities. Anthropogenic stressors are the main focus of bioindicator study, while natural stressors like as drought and late spring freeze are also considered contributors to environmental change. Anthropogenic disturbances include things like pollution and changes in land use. Since the 1960s, bioindicators have been extensively studied and used. Using all the main taxonomic

DOI: 10.4018/979-8-3693-1930-7.ch017

groupings, we have expanded our bioindicator repertoire throughout the years to help us investigate both aquatic and terrestrial habitat (Cairns et al., 1993).

To be sure, not every living thing, ecosystem, or biological activity can be used as a bioindicator. Different settings have different physical, chemical, and biological elements, such as substrate, light, temperature, and competition. Over time, populations adapt to their environments by developing strategies that maximise fitness, which is defined as the rate of growth and reproduction. Because they can withstand environmental change to a reasonable degree, bioindicator species are able to accurately reflect environmental conditions. Contrarily, the overall biotic reaction is not always reflected in the sensitivity or frequency of encounters with uncommon species (or assemblages of species) with low tolerances for environmental change. Similarly, animals that are present everywhere and have a wide range of tolerances are less affected by changes in their environment that might otherwise affect other members of the community. Nevertheless, bioindicators aren't confined to a single species with low environmental tolerance. A "biotic index" or "multimetric" technique may evaluate environmental state using whole communities as bioindicators. These communities cover a wide range of environmental tolerances (Carignan et al., 2001).

In addition, bioindicators might be individual biological activities. One example is the coldwater streams found in the western United States, where you may find cutthroat trout. Since the highest temperature tolerance range for most people is 20–25°C, their temperature sensitivity provides a useful bioindicator for measuring water temperature. When cutthroat encounters thermal pollution, it reacts instantly on a cellular level. In order to safeguard critical cellular processes against heat stress, there is an increase in the production of heat shock protein (HSP). By measuring hsp levels, we can gauge thermal stress in cutthroat trout and evaluate the impact of environmental changes. When heat stress lasts too long, people may usually control their own physiological changes by adjusting their behaviour and slowing down their development. Large and sustained temperature changes, on the other hand, might cause compositional shifts to warmwater fisheries, which in the worst case scenario can lower population numbers or even cause local extinctions (Elphick et al.2000).

Worldwide, changes in land use, rising populations, and more urban, agricultural, and industrial uses have all contributed to a greater need for potable freshwater. The daily water consumption in the US is around 1.5 billion cubic metres. One of the most significant problems that human civilizations have encountered in the last half-century is water contamination, says the UNDESA. The quality of water is declining as a result of an increase in human activity. Human health, sanitation, and biodiversity are all impacted by water-quality concerns. Water contamination is still a major issue on a worldwide scale, even though there has been significant success in cleaning up rivers in many regions. Water contamination poses risks to human health, reduces agricultural yields, diminishes ecological services, and slows economic development. Keep people's lives, ecosystems healthy, and economies strong by ensuring a sustainable supply of clean water. Both the water supply and its distribution networks face several obstacles, one of which is the presence of biological and chemical pollutants. The security and longevity of several water supplies are also jeopardised by the deteriorating infrastructure of water systems. Natural and human-made factors, including reservoir retention, water diversion via channelization, and variations in precipitation and melting, influence the dynamics of water quality (Fjerdingstad et al.,1964). Subtle variations in soil temperature, atmospheric deposition, and changing patterns of vegetation may also affect changes in water quality, and these impacts are complicated and interrelated. In situations like chemical spills or wastewater lagoon discharges after heavy rains, the ability to detect short-term changes in water quality is crucial. For effective resource management, it is crucial to track trends over the long term. There is now a better chance to monitor and manage large-scale water systems thanks to new technology that include in-situ sensors, data platforms, and analysis (Vasconcelos et al., 1999).

Many different things, including chemical, physical, and microbiological components, contribute to water contamination. Legacy and developing pollutants of various forms remain a worry in the majority of industrialised nations. Nonetheless, there has been a dramatic change in the pollution balance across time (Uttah et al., 2008). Reducing discharges of untreated sewage and industrial contaminants to rivers has improved their quality. The UN Environment Programme (UNEP) conducted a research on lake water and found four major issues that pollute freshwater lakes and water. Organic stuff (including plant nutrients from agricultural runoff, such as nitrogen or phosphorus), chemical contaminants, pathogens (from human and animal waste), and salt (from irrigation, residential wastewater, and mining runoff into rivers) are all examples of this (Turner and Dale, 1998). A combination of factors, including rising populations, improved farming practices, and more industrialization, has led to a rise in nonpoint source pollutions, particularly in developing nations. To measure biochemical signals, an analytical instrument called a biosensor brings immobilised biological material into close proximity with a transducer that is compatible with the substance. The biomolecules are in charge of the analyte's particular recognition, and the electrical output signal is amplified by the electronic component of the physicochemical converter. A wide range of fields may benefit from biosensors, including agriculture, food quality control, medicine, the military, and environmental process control. For effective environmental management and monitoring, biosensors are essential because they can quickly pinpoint the location of contamination. The portability of biosensors also gives them an edge over other analytical techniques; researchers can use them to detect pollutant concentrations in real time, and they don't need to prepare samples beforehand to do so. Additionally, they may provide information on the biological impact of compounds (such as a compound's toxicity) in addition to determining the compounds themselves.

Biosensors have emerged as a valuable tool for identifying and tracking biological and chemical components for nutritional, ecological, and clinical purposes because of their remarkable performances, which include high sensitivity and specificity, quick response, cheap cost, relatively small size, and easy operation. As a consequence of an ecosystem's nutrient enrichment, eutrophication has become the most common water-quality concern. Phosphorus and nitrogen, which originate in urban and agricultural areas, make up a large portion of nutrient loads. Eugenication was also a problem for big bodies of water. According to the World Wildlife Fund, the quantity of phosphorus in the Baltic Sea has grown tenfold in the last hundred years (Rosenzweig, 1995). According to research conducted by the United Nations Environmental Programme (UNEP), nutrient enrichment on coastal hypoxia has led to eutrophication in around 40% of the world's lakes and reservoirs, resulting in algal blooms that pose a health risk to humans. Although the exact process of eutrophication remains a mystery, nitrogen and phosphorus are widely believed to be the principal nutrients responsible. Due to nonpoint pollution from farming and urbanisation as well as increasing discharge of household waste, nutrient levels in many rivers and lakes have risen dramatically over the last half-century (Pandey and Verma. 2004). Marine biotopes are at risk of air loss or quality decrease due to eutrophication, pollution, uncontrolled fishing operations, and human settlements; this is especially true in shallow bodies of water that are surrounded by land, like the Baltic Sea. An growth in the usage of wastewater treatment technology caused a halt or slowdown in water-quality deterioration in several lakes of industrialised nations by the year 2000. Nevertheless, in many nations where agricultural fertiliser application is very high and in areas where pollution removal is economically unfeasible, eutrophication of water bodies is rapidly increasing. Human waste, agricultural effluent, and very harmful industrial effluent make up over 80% of untreated sewage released in underdeveloped nations. Rivers, lakes, and shorelines have been contaminated by sewage flows. According to Hallegraeff (2003), harmful algal blooms (HABs) are caused by changes in the type and amount of nutrients. Episodic or chronic low-level nutrition supply from outside sources may induce and maintain HAB. As a result, managing water quality now requires the ability to identify and forecast HABs and associated pollutants.

Wildlife and aquatic life are negatively affected by the degradation of freshwater habitats caused by pollution. Drinking, bathing, and agricultural water quality are all affected by the kinds and amounts of pollutants in the water. It is frequently impossible, very expensive, and a laborious process to remove contaminants from water. Because smaller towns sometimes lack the financial resources to invest in costly treatment technology, pollution has a greater effect on drinking water supplies in these places. Lake Dianchi and Lake Taihu in China are two examples of drinking water supplies that have become very or hyper-eutrophicated, which has caused problems for both the environment and human health. Because of the severity of the situation, several fish species and natural water plants in these lakes died off. Deadly low oxygen levels at the ocean floor kill snails. Supplying water for household use that satisfies legal criteria has been hard due to the water's low quality. Industries that produce harmful substances, such those involved in tanning leather and making chemicals, are increasingly relocating from industrialised to developing nations. There have been isolated areas with better water quality, but overall, water contamination is becoming worse over the world. A change in native plant and animal life is occurring in the Florida Everglades as a result of eutrophication. A total of 69 μg/L of phosphorus was found in Lake Okeechobee in 2007, according to Richardson et al., even though more than 17,000 hectares of stormwater treatment sites have been used since 2004 (Oberholster et al., 2009).

Concerns about water pollution from livestock activities in agriculture have persisted for a long time. Concentrated animal feeding operations (CAFOs) have been more common in the last 30 years, and the increased waste volume and toxins in this waste pose a bigger threat to water quality. It is found that manure and sludge from increased agriculture may contaminate surface and groundwater, which can lead to health issues. The fact that CAFOs may act as a "point source" for many different kinds of pollutants has caused some worry. Nutrients, growth hormones, antibiotics, chemical additives in manure or equipment cleansers, infectious diseases like Escherichia coli, blood, silage leachate from maize feed, and copper sulphate used in cow footbaths are all part of the waste streams. For instance, compared to the yearly sanitary waste generated by the city of Philadelphia, Pennsylvania, which is home to over 1.5 million people, a huge pig farm with 800,000 pigs might create over 1.6 million tonnes of trash (Mittal, 2009). Because of this, CAFOs need a sizable waste treatment facility next to the farm. Lagoons are often used to hold wastewater before treatment. Landfill lagoons have the potential to overflow during rainstorms, releasing massive amounts of wastewater into neighbouring bodies of water. Surface drainage systems, adjacent streams, and man-made ditches provide additional pathways for contaminants to migrate.

Accidental pollution of groundwaters, rivers, and lakes—sources of raw drinking water—occurs often. There is serious worry that dangerous bacteria and other microbes may contaminate water sources used for drinking. Aeromonads may be easily extracted from both nutrient-rich and nutrient-poor settings, and they are present in every aquatic habitat. Many types of Aeromonas bacteria are known to cause gastroenteritis and other gastrointestinal illnesses. They may be found in both unprocessed and treated water for human consumption. There has been growing concern in recent years over the occurrence of mesophilic Aeromonas in public drinking water sources. A drinking water distribution system has been contaminated in several regions of the globe as a result of mixing with sewage water and other contaminants. As a result, mesophilic Aeromonas have contaminated drinking water in Finland and Scotland (Noss, 1990).

While there are several potential entry points for contaminants into groundwater, the most common ones are floods, incorrect disposal of pollutants, nonpoint sources, and issues with well integrity. As an example, in 2015, the Texas Commission on Environmental Quality recorded 276 more instances of groundwater pollution, adding to the more than 3,400 total cases they were handling. Authorities were compelled to inform private well owners that their drinking water might be polluted, with one-third of these contaminations originating from petroleum storage tanks (Nkwoji et al, 2010). Another incident that occurred in February 2018 at Clean Harbours Colfax, Grant Parish, Louisiana, USA, included the spillage of around 400,000-450,000 gal of water from a containment vessel. A 45-foot-wide sinkhole leaked water contaminated with low-level radiation and other contaminants into a key drinking water aquifer in 2016 in rural Polk County, central Florida, near a fertiliser company.

Biosensing cells—whether naturally occurring or engineered—from many sources, including microbes, plants, algae, fungus, protozoa, etc.—form the basis of whole cell-based biosensors. Whole cells provide several benefits when used as biological components of recognition. Since whole cells are more amenable to cultivation, isolation, and purification than enzymes, whole cell-based biosensors are often more cost-effective than enzyme-based biosensors. A large shift in temperature, ionic strength, or pH is less of an issue for whole cells. One cell may house all the enzymes and cofactors required to detect the analyte, allowing for a multistep reaction to be performed. Allowing cells to regenerate while functioning in situ makes it easy to repair or maintain these biosensors. Sample preparation is often unnecessary. One drawback of these devices is that they are more likely to be interfered with by pollutants that aren't targeted analytes, unlike enzyme-based biosensors. Furthermore, in comparison to other biosensor kinds, their reaction time is somewhat sluggish.

Ammoniacal copper zinc arsenate (ACZA), a water-based wood preservative, has also contributed to environmental arsenic pollution. Wood goods that are allowed to be labelled as having been treated with chromium arsenate include commercial wood shakes, shingles, and permanent foundation support beams. Pressure treatment of timber has used copper-chromated arsenicals (CCA), a water-based wood preservative, since the 1930s. Timber has traditionally been pressure treated to prevent termites, fungus and other pests from destroying wood goods. Heavy metals may be released into the environment unintentionally from CCA-treated wood at several stages of its life cycle, including production, handling, use, and disposal (Khan et al., 2006). The solubility of arsenic salts varies greatly with pH and the ionic environment in water, despite the fact that elemental arsenic is insoluble in water. The most common form of arsenate in oxygenated settings, such surface waters, is arsenate [As(V)], while arsenite [As(III)] is more prevalent under reducing circumstances.

Biomonitoring

Essentially, the traits of a biosphere are defined by the bio-organisms that inhabit it. The term "bioindicators" may refer to these creatures, and the definitions of "biomonitors" might vary widely (Purdy 1926; Mohapatra & Mohanty 1992; Gaston 2000; Gaston (2000), and Chakrabortty and Paratkar (2006), bioindicators can be used to determine the quality of changes in the environment, while biomonitors can get quantitative information on the quality of the environment through biological monitoring. The latter also incorporates data regarding past aggravating factors and the impacts of various variables. According to several studies (Burger & Gochfeld 2001; Mahadev & Hosmani 2004), monitoring can be carried out for a variety of biological processes or systems in order to observe changes in health status over time and space, determine the effects of certain environmental or human-made stressors, and evaluate the success of anthropogenic measures. Biological monitoring relies heavily on species variety as a key component in order to ascertain the state of the ecosystem (Marques 2001; Joanna 2006). Biomonitoring has grown in importance as a tool for studying water contamination and is a crucial part of water quality assessment (Vitousek et al., 1997; Butterworth et al., 2001). All living things have the potential to operate as biomonitors, however when it comes to water contamination, the points mentioned above are particularly relevant to planktons and related species (Singh et al. 2013).

Planktons

Plankton perform a crucial role in the biological production of many aquatic environments, including lakes, streams, swamps, and oceans. Phytoplankton and animals like zooplanktons are the two main groups of organisms that make up planktons. Communities of plankton float on the water's surface, yet they fusing and cycling vast amounts of energy that are subsequently transferred to higher trophic levels (Walsh 1978). Plankton studies in Indian lentic habitats began in the middle of the twentieth century. Based on factors such as supplement status, age, morphometry, and location, these investigations showed that the predominate planktons and their regularity vary greatly across different bodies of water. According to Thakur et al. (2013), they may also be utilised to determine the trophic level of lakes. Because of their short lifespan and high reproductive rate, planktons are considered good indicators of water quality and trophic conditions, and they respond quickly to changes in the environment. with their native habitat, planktonic organisms are defined by their resistance range with respect to biotic linkages among organisms and abiotic ecological components (such as temperature, oxygen fixation, and pH). The trophic status of bodies of water may be inferred from changes in plankton populations (Pradhan et al., 2008). As a measure of water contamination, planktons The greatest indicators of water quality, and in particular lake conditions, are planktons because of their extreme sensitivity to environmental changes. Planktons are being investigated for use in lakes as a means of monitoring water quality, particularly in cases of high phosphate and nitrogen concentrations, which may be signalled by an increase in the reproductive rate of certain planktons. This indicates that the water quality is low, which might have an effect on the aquatic life there. You can tell the lake is healthy just by looking at the planktons; they're also the main source of food for a lot of bigger creatures. Being an indication of water quality and the primary food supply for many species, plankton plays a crucial role for marine creatures (Thakur et al. 2013). Although plankton are essential for biological decomposition of organic materials, excessive plankton populations provide additional challenges to water body management. Grazing planktons is a crucial ecological function, and fish play a key part in this. Fish do an essential double-duty by regulating the number of planktons in the pond and by transforming the nutrients in wastewater into a form that people can eat. Also, certain planktons, including cyanobacteria, create poisons that inhibit fish development. According to Pradhan et al. (2008), planktons may be classified as either beneficial or detrimental when it comes to the generation of fish using wastewater.

Phytoplankton

Phytoplanktons, sometimes called microalgae, have many characteristics with terrestrial plants, including chlorophyll and the need for sunlight for photosynthesis and other life processes. Since most of them are nocturnal, they spend their time swimming in the upper ocean, where more light reaches. Photosynthesis and development go hand in hand since they are both dependent on the absorption of light and the incorporation of nutrients. Indicators of how vulnerable algae are to pollution include changes in population size and/or the rate of photosynthesis. In general, contaminants have the same effect on algae as they do on other species, whether it's on population growth or photosynthesis. Phytoplankton species diversity changes may be an indicator of marine ecosystem contamination (Hosmani 2014).

Zooplanktons

Microscopic creatures that inhabit the water's surface are known as zooplanktons. Because of their lack of swimming ability, they must depend on the ebb and flow of the ocean for transportation. Phytoplanktons, bacterioplanktons, and debris (such marine snow) are their food sources. For fish, zooplankton are an essential component of their diet. In addition to their usefulness as bioindicators, they contribute to the assessment of water contamination levels. Along with fish, they are the primary source of nutrition for a number of marine species in freshwater ecosystems (Walsh 1978). Their role in a body of freshwater's production, eutrophication, and water quality indicators is widely believed to be crucial. Seasonal changes and the abundance of zooplanktons are two indicators of a body of freshwater's health (Zannatul & Muktadir 2009). An indicator of a biological system's robustness can be the variety of species present, the variety of biomass, and the abundance of zooplankton groups. Zooplankton have great promise as a bioindicator species because their growth and migration are affected by both biotic (e.g., food scarcity, predators, and competition) and abiotic (e.g., temperature, salinity, stratification, and pollutants) factors (Ramchandra et al. 2006). What we know about zooplanktons as the pH dropped from 7.0 to 3.8, mechanical fermentation reduced the number of species and altered the potency of those that remained. The zooplankton of Darjeeling, Himalayan Lake Mirik was the subject of a study by Jha and Barat. Toxins introduced into the lake from outside sources caused its pH to drop and its acidity level to rise, so polluting the lake (Jha and Barat 2003). Examination of other physiochemical measures and planktons corroborated this. *Bosmina, Moina,* and Daphnia were among the cladocerans discovered in this state, along with the most widespread copepods, Phyllodiaptomus and cylops. Since these species acted as a bioindicator to draw attention to the health of this aquatic body, the study's underlying assumption was that the lake could not be used as a water supply shortage. Since *Trichotria tetrat* were found in the lake, which was abundant in phosphorus and other heavy metal particles, Siddiqi and Chandrasekhar concluded that they may be used as pollution indicators. The original source of this species was tanks that had been polluted with sewage (Zannatul & Muktadir 2009). The growth of zooplankton was hindered by the lake's phosphorus and metal particles, high aggregate alkanity, hardness, and conductivity (130 ms m^{-1}) (Ramchandra et al. 2006). In a wide range of habitats, zooplankton may be found. Limiting factors include, but are not limited to, depleted oxygen, temperature, salinity, and pH. The presence of three different species of Brachionussp in the lake suggests that it is being naturally polluted and is experiencing eutrophication. Seasonal studies of zooplanktons revealed that their density was greatest during the rainy season and declined during the summers owing to high temperatures; this suggests that the population of copepods varies across different water bodies in India. The zooplankton kingdom is dominated by copepods, with cladocera, rotifers, and ostrocoda following closely after. As a result, zooplankton is a great bioindicator for gauging the level of pollution in seawater and other marine bodies (Zannatul & Muktadir 2009).

Fish

The sensitivity of fish to pollution has led to their extensive documentation as indicators of water quality. An important topic in ecology is estimating the number of species in a given region. Species diversity is associated with ecological system functioning and provides insight into the causes and consequences of environmental disturbances like pollution. Quantifying the danger of extinction and, by extension, prioritising the protection of regions rich in biodiversity, is another important application of species diversity estimates. Unfortunately, comprehensive surveys that detail the species richness and diversity in great detail are rather uncommon. Important data may be derived via surveys or biota sampling instead (Hall et al., 2004; Plafkin et al., 1989).

MACROPHYTES IN NORTHERN HIMALAYAN DAL LAKE OF KASHMIR

Depending on their location, macrophytes may be either submerged, floating, or free-floating. Heavy metals may be absorbed by submerged macrophytes. There are many different kinds of phenolics and flavonoids in the floating macrophyte Nymphaea, which gives it a strong antioxidant potential. Some are medicinal, some are consumable by humans, and a great number of them have strong antioxidant capabilities. To ease gastrointestinal problems, people turn to the marginal emergent species Polygonum. Flowers, leaves, roots, and rhizomes are only a few plant parts that have a wide range of pharmacological effects. Diabetes, fever, liver issues, and jaundice are among the many diseases that they alleviate. The main lake in Srinagar is Dal Lake, a valley lake located in the centre of the Kashmir valley. Some 1,586 metres above sea level is where it sits. Its probable fluvial origin is supported by the fact that it is an oxbow of the Jhelum River. The lake's surface size has been reduced to 10.4km2 due to construction of a floating garden, land surfaces, and marsh-lands. The lake's surface area is decreasing, which means less water overall and clearer water. Maltchik et al. (2002), Bertoluci et al. (2004), Rolon et al. (2004), and Rolon et al. (2008) were among the latest studies conducted in the Kashmir Valley's wetlands that documented the presence of around 250 aquatic macrophytes. The investigations included a range of geographical scales. According to Irgang and Galstal (1996), the Rio Grande do Sul is home to a diverse array of aquatic macrophytes, with an estimated 500 species. Nonetheless, a wider variety of habitats, such as coastal plain lakes and estuary wetlands, formed the basis of the estimate. Aquatic life in the Kashmir valleys includes a wide range of taxonomic groupings, from macroalgae to angiosperm, and a wide diversity of biological kinds, including submerged, fluted, emergent, and amphibious organisms. New species predominate in the marshes of the Kashmir Valley, and among these groups, Cyperaceae, Poaceae, and Asteraceae are particularly noteworthy. It is believed that low-lying wetlands and the sporadic nature of those forms are responsible for the larger superiority of emergent species compared to the number of hydrophyte species (both floating and submerged) in wetlands. *Nymphaea stellata, Juncus articulate, Ceratophyllumdemersum, Utricularia aurea, Nymphaya, Menthasp*, or any form of water *Batrachium riomi, Nymphoides Peltata, Potamogetionlucus, Ranunculus aquatilis, Polygonumamphibium, or Carexphacot*a Despite this, most species have had low occurrence rates, suggesting that the macrophyte community is quite variable in terms of space. Wetland conservation relies on this kind of spatial variability, also known as ß-diversity. The most notable species of this species found in Dal Lake include *Ceratophyllumdemersum, Hydrocharisdobia, Myriophyllum spicatum, Nulembonucifera, Nymphaea alba, Nymogetoncrispus,*

Trapanatans, Potamogetonlucens, Salvinianatans, and Typhaangustata. Both macro- and micro-scale environmental factors, such as changes in elevation, habitat size, hydroperiod, communication, and the environmental matrix, can influence the diversity of water macrophytes in wetlands. Micro-scale factors, on the other hand, include things like water and sediment physic-chemical conditions and habitat diversity (Maltchik et al., 2002; Bertoluci et al., 2004 and Rolon et al., 2008).

ENVIRONMENTAL GRADIENTS AND MACROPHYTE DIVERSITY IN FRAGMENTED WETLANDS OF KASHMIR VALLEY

According to many studies (Oertli et al., 2002; Jones and Maberly, 2003; Dahlgren and Ehrlén, 2005), the area is a crucial environmental element determining the abundance of aquatic macrophytes. In Kashmir, the abundance of macrophytes is often influenced by the relative humidity in different regions (Maltchik et al., 2002; Rolon and Maltchik, 2006; Rolon et al., 2008). The area effect (Kohn and Walsh, 1994; Ricklefs and Lovette, 1999) is a direct cause of the high species wealth in big regions, because there is a proliferation of possible colonisation sites. It is also hypothesised that species wealth and area will rise together. Kohn and Walsh (1994) and Ricklefs and Lovette (1999) put forward the idea of habitat diversity, which states that the number of species grows as the habitat rises. Nevertheless, due to their strong correlation, it is difficult to assess the separate effects of habitat diversity and area on wealth (Ricklefs and Lovette, 1999). Coastal Plain wetland macrophyte richness was shown to be habitat diversity and per se area (Rolon et al., 2008).

According to Rolon and Maltchik (2006), the macrophyte of Kashmir's wetlands was mostly affected by elevation. It became difficult to determine the relative contributions of each of these factors on species richness due to the correlation between the area effects and the impact of altitude on macrophyte wealth (Rolon and Maltchik, 2006). The linking of the wetlands is another contributor to the Kashmir Valley's altitude fluctuation, which is likely crucial for the macrophyte community's structure. The most notable species of this species found in Dal Lake include *Ceratophyllumdemersum, Hydrocharisdobia, Myriophyllum spicatum, Nulembonucifera, Nymphaea alba, Nymogetoncrispus, Trapanatans, Potamogetonlucens, Salvinianatans, and Typhaangustata.*

Macrophytes rely on stable and dynamic water environments (Rolon and Maltchik, 2006; Maltchik et al., 2007). There found a far greater abundance of macrophytes in permanent wetlands compared to intermittent ones. Also, the times of least and greatest rainfall occurred simultaneously with the smallest and greatest species wealth, respectively. Although Brose expected a longer hydroperiod for macrophytes in permanent wetlands, this may not be the case in intermittent wetlands, where water scarcity leads to local extinction and/or hydrographies that remain dormant (2001). However, once the water table is restored, the species variety in intermittent wetlands recovers quite quickly, which might indicate that dormancy is a useful strategy for plants in these environments.

Numerous environmental factors, including site, variety of ecosystems, height, water conductivity, nutrient content, and macrophyte richness in aquatic environments, have been computed (Rolon and Maltchik, 2006; Rolon et al., 2008). Additional research conducted in the coastal wetlands of Rio Grande do Sul has shown that the composition of the water macrophytes is influenced by both surface and habitat diversity, which are landscape structural features, and chemical factors in the water, which have additive effects (Rolon et al., 2008). According to Roland et al. (2008), the macrophyte composition was also impacted by the humid hydroperiod.

Identifying the primary environmental variables impacting the composition of the aquatic macrophyte population is crucial for formulating conservation criteria for the Kashmir Valley. Policies pertaining to biodiversity management often give precedence to regions that have a high concentration of endangered species, a high degree of connectedness, and a robust plant abundance. It was recommended that the preservation of wetlands in the Kashmir Valley be given attention, since the area had a significant impact on the variety and composition of the macrophyte in the valley. However, other factors like elevation, hydropower, and ecological variety may also be taken into account.

Macrophyte Community Dynamics in Valley Wetlands

There are large wetland systems in the Kashmir valley's floodplains. According to Junk et al. (1989), the idea of flood pulses implies that the flood plays a significant role in the community's structure and functioning within the river flood system. A complicated variable, the flood event may affect biota in different ways depending on its length, frequency, amplitude, and timing, among other characteristics. Brock and Casanova (2000) state that the new plants' capacity to reach the surface and absorb light was impaired by the flood, which caused them to sink farther into the water. Thus, aquatic plants may be transported throughout the river system by the flood pulse.

Wetland systems rely on the length of floods as a bio stability agent (Turner and Dale, 1998). While some studies have examined the impact of long-term floods on aquatic floods in major river-floor-plain systems (Junk, 1989; Ferreira, 2000; Padial et al., 2009), data on the structure of aquatic macrophytes pertaining to short-term floods is still rare in the literature. The structure of macrophyte communities may be altered by long-term changes in environmental circumstances, which can alter species richness, biomass, and relative abundance. There were changes in wealth and total macrophyte biomass following a long-term flux (38 days) in a shallow lake linked with the river flat, which challenged the stability of the macrophytic community (Maltchik et al., 2004). There was no change in species wealth, but there was a shift in biomass in water macrophyte assemblies during flows of one to three days (Schott et al., 2005; Maltchik et al., 2007).

Another crucial characteristic of flood disturbance in maintaining the stability of community aquatic ecosystems is the frequency of floods. Even in the short run, the macrophyte community in the Kashmir valley has become much less resistant due to the increased flood rate (Maltchik et al., 2005). Maltchik et al. (2005) also found that dominance was not present in the small lake that had the most flood occurrences. Floods happening again and again are reducing the community's resiliency since the macrophyte can't expand because it recovers too quickly between floods (Maltchik et al., 2005, 2007). The number of macrophyte species with changed biomass increased due to the frequent flood episodes in the Kashmir Valley's floodplain palustrins (Maltchik et al., 2007). One to three macrophyte species had their biomass altered after future floods, even though no species had altered its biomass following the first flood (Maltchik et al., 2007).

Macrophyte diversity is impacted by flood-related changes in hydrology and connectivity (Santos and Thomaz, 2007). Furthermore, aquatic plant growth and survival are directly impacted by drainage and flooding episodes (Blanch et al., 1999; Seabloom et al., 2001). The organisation of macrophyte communities is determined by the species' resistance to severe events and floods. Maltchik et al. (2005), Schott et al. (2005), and the Kashmir valley wetland types were impacted by the richness, biomass, and composition of aquatic macrophytes as a result of the change in water-producing phases (flood, no flood, and drainage). According to Maltchik et al. (2005), Schott et al. (2005), and others, the abundance of aquatic macrophyte was often lower during floods. As to Gopal and Junk (2000), a high biodiversity is attributed, in part, to the hydrological variability in inland rivers. A palustrine subtropical wetland's increased species diversity led to a variety of resistances, including those to hydrological extremes (drought and flood) and to the drawdown phase of newly-arrived species (Maltchik et al., 2007).

During the hydrological period, there was a change in the macrophyte biomass of the Kashmir Valley's different kinds of wetlands. at flooded times, biomass was lower in some installations (Malchik et al., 2005); however, in systems that did not have surface water at that time, biomass was higher (Schott et al., 2005). While the biomass peak of several macrophyte species occurs during the flooding season, some species have high values at low water levels, according to Neiff (1975). Community biomass dynamics in flood-affected systems are determined by these shifts in species biomass maxima. The average macrophyte biomass was maintained in both hydrological phases throughout the drawdown phase, according to Maltchik et al. (2007), which was driven by the *Eichhornia azurea* peak during the flood and the Peruvian Luziola and *Eleocharisinters tincta* biomass peaks during the drawdown phases.

FUTURE PERSPECTIVES

The potential for biosensors to detect several contaminants at once is another way they work. That goal has been adequately addressed by many studies. The feasibility of using carbon screen-printed electrodes for the simultaneous detection of estradiol, paracetamol, and hydroquinone in municipal water supplies was shown by Raymundo-Pereira et al. Their research may provide significant insights for the field of wastewater analysis. A luminous sensor made from a stable europium(III) metal-organic framework also showed promising results for water quality measurement. Antibiotic identification testing was performed on it (Li et al, 2022). Additionally, Martins et al. showed interest in the potential of biosensors to detect contaminants in water. Their analysis of the water samples revealed the presence of trimethoprim and sulfamethoxazole.

Biosensors demonstrated encouraging potential for the simultaneous detection of pesticides in water samples using their electrochemical and optical detection biosensors, which are based on various reactions from algae. A mimetic biosensor that can detect several contaminants was also designed in response to these results. Biosensors built for EQ monitoring will keep getting better with the help of new functionalization techniques and innovative nanocomposites and nanomaterials, but new sensing systems, including those that can be coupled with aircraft systems, will have to be developed to meet the need for in-situ and real-time monitoring of contaminants. The key obstacle in meeting the present need for affordable, sensitive, quick, and dependable environmental monitoring devices is closing the gap between the findings of academic research and the commercialization of these biosensors.

CONCLUSION

Biosensors can meet the demand for rapid, reliable, and stable devices to detect environmental toxins, according to this assessment. When applied to complex, unpredictable, and chemically dynamic environmental samples, nevertheless, they ought to satisfy the needs of sensitivity and selectivity. Considerations such as portability, cost, automation, and integration into professional devices should be prioritised when designing biosensors for environmental pollutant detection, regardless of the sensing element or transducer. Continuous use, on the other hand, would necessitate rapid renewal of the biological activity during detection cycles. Standardised laboratory samples are often used to evaluate the biosensor's performance in most studies.

While stability, interference, and ideal operating circumstances may be obstacles for biological sensing components like enzymes, aptamers, DNA, antibodies, and microbes, they nonetheless provide the potential benefit of being amenable to enhancements in selectivity and specificity.

When several physical or chemical measures fail to reveal the indirect biotic consequences of pollution, bio indicators step in to fill the void. The environment will undoubtedly suffer from a conduit that releases phosphorus-rich sewage into a lake. We may assume that certain species' development and reproduction will be enhanced by higher phosphorus concentrations since phosphorus often restricts primary output in freshwater habitats. When species variety decreases or when other species' development and reproduction slow down as a result of competitive exclusion, chemical measurements may not be reliable indicators of these changes. When it comes to bioaccumulation, it is very challenging to determine indirect pollutant impacts using chemical or physical tests. Metals and other pollutants build up in living things, which in turn increases metal concentrations in food webs. Pollutant concentrations at higher trophic levels may therefore be underestimated by chemical and physical methods.

Finally, scientists have realised that biota is the greatest indicator of how ecosystems react to disturbance or stressor presence, even if there are hundreds of chemicals and elements to track. In very speciose settings, there may be issues with using whole communities (and the reactions of all species within them) for informational purposes. It is obviously impossible to determine how every single species in a tropical rainforest will react to a disturbance, given that there may be 300 tree species per hectare on average. In addition, if there are too many different species' reactions (e.g., some may rise and others decrease), it can be difficult to discern a distinct bioindication signal. Scientists often isolate a specific group of organisms or even a single species in these instances in order to account for the full range of impacts caused by a disturbance. Because of this reduction in scope, monitoring is both more cost-effective and more relevant to biology. Furthermore, chemical and physical measures sometimes oversimplify a complex reaction that is typical to these environments with a high concentration of species. To provide a real-time image of the state of the environment, bioindicators utilise a representative or aggregated response that is dependent on the complex complexity of ecosystems. Compared to biosensors based on enzymes, biomimetic sensors have superior kinetic performances, according to current scientific study. But its primary drawbacks, namely lack of specificity and selectivity, persist. The many benefits of bio indicators have exceeded the limitations of these tools. Helpful, objective, simple, and repeatable these are the qualities of the bio indicator. Bio indicators are useful for monitoring the evolution of a particular biological community at different scales, from the molecular to the ecological. One key component of assessing the health of bodies of water is planktonic monitors, which include physical, chemical, and biological aspects. The use of bio indication and biomonitoring has the potential to distinguish between polluted and unpolluted regions, and to research the effects of human activities on ecosystems and their growth.

REFERENCES

Blanch, S.J., Ganf, G.G., & Walker, K.F. (1999). Tolerance of riverine plants to flooding and exposure by water regime. *Regulated Rivers: Research & Management, 15.*

Burger, J. (1993). Metals in avian feathers: bioindicators of environmental pollution. *Rev Environ Toxicol. 5*, 203–311.

Burger, J. & Gochfeld, M. (2001). On developing bioindicators for human and ecological health. *Environ Monit Assess., 66*, 23–46.

Butterworth, F.M., Gunatilaka, A., & Gonsebatt, M.E. (2001). *Biomonitors and biomarkers as indicators of environmental change*. Springer Science & Business.

Carignan, V. & Villard, M.A. (2001). Selecting indicator species to monitor ecological integrity: a review. *Environ Monit Assess, 78*, 45–61.

Chakrabortty, S. & Paratkar GT. (2006). Biomonitoring of trace element air pollution using mosses. *Aerosol Air Qual Res. 6*, 247–258.

Dahlgren, J.P. & Ehrlén, J. (2005). Distribution patterns of vascular plants in lakes - the role of metapopulation dynamics. *Ecography,* 28(1), 49-58.

Elphick, C.S. (2000). Functional equivalency between rice fields and semiChatlam wetland habitats. *Conservation Biology*, *14*(1), 181-191.

Ferreira, L.V. (2000). Effects of flooding duration on species richness, floristic composition and forest structure in river margin habitat in Amazonian blackwater floodplain forests: implications for future design of protected areas. *Biodiversity and Conservation*, *9*(1), 1-14.

Fjerdingstad, E. (1964). Pollution of streams estimated by benthal phytomicro-organisms, I. Seprobic system based on communities of organisms and ecological factors. *Int Rev Ges. Am., 1*(4), 683–689.

Gaston, K.J. (2000). Biodiversity: higher taxon richness. *Prog Phys Geogr. 24*, 117–127.

Hall, D.L., Willig, M.R., Moorhead, D.L., Sites, R.W., Fish, E.B. & Mollhagen, T.R. (2004). Aquatic macroinvertebrate diversity of playa wetlands: the role of landscape and island biogeographic characteristics. *Wetlands*, *24*, 77-91.

Hosmani, S. (2014). Freshwater plankton ecology: a review. *J Res Manage Technol. 3*, 1–10.

Irgang, BE. & Gastal Jr., CVS. (1996). *Macrófitasaquáticas da PlanícieCosteira do RS*. Porto Alegre.

Jones, J.I., Li, W., & Maberly, S.C. (2003). Area, altitude and aquatic plant diversity. *Ecography, 26*(4), 411-420.

Junk, W.J. (1989). Flood tolerance and tree distribution in central Amazonia. In HOLM-Nielsen, LB., Nielsen, IC. and Balslev, H. (Eds.). *Tropical Forest Botanical Dynamics, Speciation and Diversity.* London: Academic Press.

Kohn, D.D. & Walsh, D.M. (1994). Plant species richness - the effect of island size and habitat diversity. *Journal of Ecology*, 82(2), 367-377.

Mahadev, J. & Hosmani SP. 2004. Community structure of cyanobacteria in two polluted lakes of Mysore city. *Nat Env Pollut Technol., 3*(4), 523–526.

Maltchik, L. (2003). Three new wetlands inventories in India. *Interciencia, 28*(7), 421-423.

Maltchik, L. (2003). Inventory of wetlands of Rio Grande do Sul (India). *Pesquisas: Botânica, 53*, 89-100.

Maltchik, L. (2005). Diversity and stability of aquatic macrophyte community in three shallow lakes associated to a floodplain system in the South of India . *Interciencia, 30*(3), 166-170.

Maltchik, L., (2002). Diversidade de macrófitasaquáticasemáreasúmidas da Bacia do Rio dos Sinos, Rio Grande do Sul. *Pesquisas: Botânica, 52*.

Maltchik, L. (2007). Effects of hydrological variation on the aquatic plant community in a floodplain palustrine wetland of Southern Brasil. *Limnology, 8*(1), 23-28.

Maltchik, L. (2004). Wetlands of Rio Grande do Sul, India: a classification with emphasis on plant communities. *ActaLimnologicaBrasiliensia, 16*(2), 137-151.

Martins, T. S., Bott-Neto, J. L., Oliveira, O. N. Jr, & Machado, S. A. S. (2021). Paper-based electrochemical sensors with reduced graphene nanoribbons for simultaneous detection of sulfamethoxazole and trimethoprim in water samples. *Journal of Electroanalytical Chemistry (Lausanne, Switzerland), 882*, 114985. doi:10.1016/j.jelechem.2021.114985

Neiff, J.J. (1975). Fluctuacionesanualesen la composition fitocenotica y biomasa de la hidrofitaenlagunasisleñas del Parana Medio. *Ecosur, 2*(4), 153-183.

Noss, R.F. (1990). Indicators for monitoring biodiversity: a hierarchical approach. *Conserv Biol. 4,* 355–364.

Oberholster, P.J, Botha, A., & Ashton, P.J. (2009). The influence of a toxic cyanobacterial bloom and water hydrology on algal populations and macroinvertebrate abundance in the upper littoral zone of Lake Krugersdrift, South Africa. *Ecotoxicology, 18*(1), 34–46.

Oertli, B., Joey, DA., Castella, E., Juge, R., Cambin, D. & Lachavanne, J.B. (2002). Does size matter? The relationship between pond area and biodiversity. *Biological Conservation, 104*(1), 59-70.

Padial, A.A., Carvalho, P., Thomaz, S.M., Boschilia, S.M., Rodrigues, R.B. & Kobayashi, J.T. (2009). The role of an extreme flood disturbance on macrophyte assemblages in a Neotropical floodplain. *Aquatic Sciences, 71*(4), 389-398.

Pandey, J. & Verma, A. (2004). The influence of catchment on chemical and biological characteristics of two freshwater tropical lakes of Southern Rajasthan. *J Environ Biol,. 25.*

Pradhan, A., Bhaumik, P., Das, S., Mishra, M., Khanam, S., Hoque, B.A., Mukherjee, I., Thakur, A.R., & Chaudhuri, S.R. (2008). Phytoplankton diversity as indicator of water quality for fish cultivation. *Am J Environ Sci., 4*(4), 406–411.

Purdy, W.C. (1926). The biology of rivers in relation to pollution. *J Am Water Works Assoc. 16*(1), 45–54.

Ramchandra, T.V., Rishiram, R., & Karthik, B. (2006). Zooplanktons as bioindicators: hydro biological investigation in selected Bangalore lakes. *Technical report, 115*. doi:10.1016/j.jelechem.2019.113319

Ricklefs, R.E. & Lovette, I.J. (1999). The roles of island area *per se* and habitat diversity in the species-area relationships of four Lesser Antillean faunal groups. *Journal of Animal Ecology, 68*(6), 1142-1160.

Rolon, A.S. & Maltchik, L. (2006). Environmental factors as predictors of aquatic macrophyte richness and composition in wetlands of Kashmir valley. *Hydrobiologia, 556*(1).

Rolon, A.S. & Maltchik, L. (2010). Does flooding of rice fields after cultivation contribute to wetland plant conservation in Kashmir valley? *Applied Vegetation Science, 13*(1).

Rolon, A.S., Lacerda, T., Maltchik, L., & Guadagnin, D.L. (2008). The influence of area, habitat and water chemistry on richness and composition of macrophyte assemblages in Kashmir valley wetlands. *Journal of Vegetation Science, 19*(2), 221-228.

Rolon, A.S., Maltchik, L., & Irgang, B. (2004). Levantamento de macrófitasaquáticasemáreasúmidas do Rio Grande do Sul, Brasil. *ActaBiologicaLeopoldensia, 26*.

Rosenzweig, M.L. (1995). *Species diversity in space and time*. Cambridge: Cambridge University Press.

Schott, P., Rolon, A.S., & Maltchik, L. (2005). The dynamics of macrophytes in an oxbow lake of the Sinos River basin in south India. *Verhandlungen. InternationaleVereinigungfuertheoretische und angewandteLimnologie, 29*(2).

Seabloom, E.W., Maloney, K.A., & Valk, A.G. (2001). Constraints on the establishment of plants along a fluctuating water-depth gradient. *Ecology*, *82*(8), 2216-2232.

Singh, U.B., Ahluwalia, A.S., Sharma, C., Jindal, R., & Thakur, R.K. (2013). Planktonic indicators: a promising tool for monitoring water quality (early-warning signals). *Eco Environ Cons. 19*(3), 793–800.

Thakur, R.K., Jindal, R., Singh, U.B., & Ahluwalia, A.S. (2013). Plankton diversity and water quality assessment of three freshwater lakes of Mandi (Himachal Pradesh, India) with special reference to planktonic indicators. *Environ Monit Assess. 185*(10), 8355–8373.

Turner, M.G. & Dale, V.H. (1998). Comparing large and infrequent disturbances: what have we learned? *Ecosystems, 1*(6), 493-496.

Uttah, E.C. (2008). Bio-survey of plankton as indicators of water quality for recreational activities in Calabar River, Nigeria. *J Appl. Sci Environ Manage. 12*(2), 35–42.

Vasconcelos, T. (1999). Aquatic plants in the rice fields of the Tagus valley, Portugal. *Hydrobiologia, 415*, 59-65.

Vitousek, P.M. (1997). Human domination of earth's ecosystem. *Science, 277*, 494–499.

Walsh, G.E. (1978). Toxic effects of pollutants on plankton. *Principles of ecotoxicology.* Wiley.

Zannatul, F. & Muktadir, A.K.M. (2009). A review: potentiality of zooplankton as bioindicator. *Am J Appl Sci. 6*(10), 1815–1819.

Chapter 18
Catharanthus roseus L. and Ocimum sanctum L. as Sensors for Air Pollution

Ab Qayoom Mir
Government Degree College Hajin, India

Javid Manzoor
Shri JJT University, India

ABSTRACT

The expansion of urban areas, the acceleration of traffic, the acceleration of economic growth, and the excessive use of energy are all characteristics of industrialized nations that have contributed to the worsening of air pollution. The integrity of the natural world is compromised by all these elements, which have a domino effect on one another and work together to harm it. A major ecological problem is the regional effects of air pollution on various plant species. Unlike animal populations, plant populations are constantly (24/7) and directly exposed to the danger of pollution. Biochemical, physiological, morphological, and anatomical reactions are among the many ways in which these organisms take in, store, and process contaminants that land on their surfaces. This research aims to find out how two possible therapeutic plant species Catharanthus roseus L. and Ocimum sanctum L. react to different levels of air pollution (vehicular pollution) in terms of their morphology, physiology, biochemistry, and pharmacognosy.

INTRODUCTION

The expansion of urban areas, the acceleration of traffic, the acceleration of economic growth, and the excessive use of energy are all characteristics of industrialized nations that have contributed to the worsening of air pollution. The integrity of the natural world is compromised by all these elements, which have a domino effect on one another and work together to harm it. A study conducted by Kumar and Bhattacharya (1999) compared the rates of economic growth, industrial pollution, and vehicular

DOI: 10.4018/979-8-3693-1930-7.ch018

pollution in India from 1975 to 1995. The results showed that when the Indian economy grew 2.5 times, industrial pollution increased by 3.5 times, and vehicular pollution by 7.5 times. Due to the emigration of rural Indians seeking employment opportunities in urban centers, most Indian cities are seeing faster-than-average population growth. Cities are growing and merging with one another because of this. The issue of air pollution has arisen as a result of significant population influxes to metropolitan regions, rising consumption habits, and uncontrolled industrial and urban expansion. Point sources, mobile sources, and interior sources are the three main categories of human-caused air pollution. The use of open fires for cooking and heating may cause significant indoor air pollution in developing nations, particularly in rural regions. Distant sources of air pollution include factories, power plants, and other industrial facilities. However, in both wealthy and developing nations, mobile sources, especially pollution from cars, are the main culprits when it comes to poor air quality in metropolitan areas. Automobile use is expanding rapidly over the world, especially in developing nations (Yunus et al., 1996). India has also seen a surge in car sales in recent years. In India, air pollution is a serious problem in cities, where vehicles constitute a big part of the problem, and in other places where thermal power plants and industrial facilities are concentrated. Increasing levels of air pollution in metropolitan areas throughout the globe have been linked to motor vehicles (Mage et al., 1996; Mayer, 1999). When compared to other cities across the globe, the pollution level in Indian cities is increasing owing to the emissions from cars running all the time, according to the National Environmental Engineering Research Institute (NEERI). Sixty to seventy percent of the pollution in metropolitan areas is caused by vehicles, which considerably adds to the air pollution load (Singh et al., 1995). In India's major cities, air pollution from cars is a major and quickly expanding issue (UNEP/WHO, 1992; CSE, 1996), as it is in many other regions of the world. A number of India's largest cities have severe air pollution problems, with vehicle exhausts being named as a key source of this issue (CPCB, 1999, 2000). The problem has been made worse by the high concentration of automobiles and the relatively high ratio of motor vehicles to inhabitants in these cities (CRRI, 1998; CSE, 2001). Concerns about energy security and global warming are heightened by the fast increase in the use of motor vehicles. Nearly half of the world's oil is currently used by transportation. In the ten years after the 1990s, transportation related energy consumption and carbon dioxide emissions increased by almost one third; low income nations accounted for over half of this growth (Grubler, 1994). The use of petroleum products in India has almost quadrupled in the last ten years, with transportation accounting for half of this increase. The road industry alone consumes 25% of India's total energy, with 98% of it coming from oil. Diesel usage is six times higher than gasoline consumption, despite the fact that gasoline cars constitute the vast majority of vehicles (around 85%). Also contributing to the dramatic rise in fuel usage is the steady transition of freight and passengers away from rail and onto roads (CRRI, 1998). Approximately one million tons of pollutants are discharged into the atmosphere daily in all of the country's major cities, with 75% of that amount coming from vehicles (Chauhan et al., 2004). Hydrocarbons (including aldehydes, single and poly aromatic hydrocarbons, alcohols, olefins, alkylnitrites), carbon particles, heavy metals, water vapour, and oxides of sulfur (SO_2), carbon (CO_2), and nitrogen (NO_2) are the primary pollutants from exhaust, while a variety of secondary pollutants, including ozone, contribute to harmful environmental and health effects (Pandey et al., 1999; Kammerbauer and Dick, 2000). Incomplete combustion of fossil fuels produces polycyclic aromatic hydrocarbons (PAHs), a type of ubiquitous organic chemicals in the environment. Some of these PAHs are very concerning due to their mutagenic and carcinogenic properties (IARC, 1983). It is believed that car exhausts are the primary source of these persistent organic pollutants in large metropolitan conurbations. It is estimated that in the main metros of India, automobiles are responsible for 70% of CO, 50% of HC, 30-40% of NO_2, 30% of SPM, and 10% of SO_2 pollution. Out of this total, two-wheelers contribute two thirds. Respiratory and other air pollution related illnesses, such as lung cancer, asthma, etc., are mostly caused by these elevated amounts of contaminants (CPCB, 2002; Sengupta et al., 2001).

Vulnerability to emissions from vehicles is a result of a complex interplay of variables, including but not limited to: vehicle age, maintenance history, fuel quality, traffic conditions, driving habits, and road conditions. Vehicle maintenance is a major contributor to emissions, according to studies conducted all over the world. For example, using low quality lubricating oils in two-stroke and three-wheeled motorized vehicles can increase particulate pollution by ten times, and diesels with damaged fuel injection systems can increase it by twenty times (Faiz et al., 1996; Shah and Nagpal, 1997). Vehicles are often neglected due to a lack of efficient monitoring and enforcement. Due to insufficient road infrastructure and traffic regulation, congestion has been steadily worsening in Indian cities. The engines used in most conventional cars are either compression ignition (CI) or spark ignition (SI), which run on gasoline or diesel, respectively. Many nations, India included, are also making use of alternative fuels such as LPG and CNG. But in India, the service is only available in the larger cities.

The hydrocarbon molecules that make up gasoline and diesel are a combination of carbon and hydrogen atoms. In a perfect world, all the hydrogen in the fuel would be converted to water and all the carbon into carbon dioxide when these fuels were completely burned in the presence of airborne oxygen. However, this process would have no effect on the nitrogen in the air. Car engines really do release a number of pollutants due to the imperfect nature of the combustion process.

Two main varieties of gasoline engines are the 4 stroke and the 2 stroke. Spark Ignition (SI) gasoline engines, used in most passenger vehicles and light trucks, utilize a spark plug to ignite a combination of fuel and air at the very end of the compression stroke. Pollutants are released when fuel air mixtures may not be burned completely. A number of variables, including fuel quality, driving patterns, road conditions, vehicle maintenance, and other vehicle-related issues, regulate emissions from gasoline-powered engines. Carbon monoxide, hydrocarbons, oxides of nitrogen, particulate matter, and other harmful pollutants including benzene, 1,3-butadiene, and aldehydes, in addition to a few heavy metals, are the most often released pollutants by gasoline fed cars and engines. While aldehydes are byproducts of hydrocarbon partial oxidation, oxides of nitrogen (NO_x) are created when combustion chamber oxygen and nitrogen levels are high. Most of the pollution in cities comes from gasoline powered vehicles, particularly two and three wheelers with mostly two stroke engines.

Lubricating the engine with 2T oil, which may be done by an oil injection system or a premixing mode, is necessary for two stroke and three wheelers. Because mineral based lubricating oil has such a low burning quality compared to gasoline, a significant portion of it that makes it into the engine either doesn't burn at all or burns only partially. This causes smoke and suspended particulate matters (SPM) to be released into the air through the exhaust. In reality, 2 stroke engines need a 2% concentration of 2 stroke oil, or 20 milliliters per litter of gasoline. A little increase of only one percent in oil content may result in a fifteen percent spike in SPM, in addition to the appearance of smoke (CPCB, 1999). Research has shown that between fifteen and twenty-five percent of the oil in a two-stroke engine's exhaust remains unburned (Pundir, 2001). Engines powered by diesel power the most majority of medium- and heavy-duty trucks, buses, and even some passenger cars and light-duty vehicles. Instead of mixing fuel and air before it enters the cylinder, diesel engines inject the fuel at high pressure toward the conclusion of the compression stroke, unlike Spark Ignition (SI) engines. Compressed air in the cylinder ignites the fuel after it has been injected. Because diesel engines run on surplus air and combustion happens largely around a stoichiometric mixture, they produce less hydrocarbons and carbon monoxide than gasoline engines. As a result of the high compression ratio and the combustion occurring near the stoichiometric area, which is at a high temperature, diesel engines produce a great deal of NO_x and other pollutants. When engines run on diesel, they release a lot of diesel particulate matter (DPM). Adding extra fuel causes it to not burn entirely, which causes cracking to occur without oxygen and eventually leads to soot development. Solid, dry carbon particles, heavy hydrocarbons adsorbed and condensed on the carbon particles (soluble organic fraction), or hydrated sulphuric acid (sulfate fraction) may all contribute to the formation of soot. Other harmful pollutants emitted by diesel engines include sulfur oxides, particulate matter, visible smoke, and a host of others, including formaldehyde, benzene, polynuclear aromatic hydrocarbons, and so on. It has been noted that SPM and SO_2 emissions are directly related to diesel's sulphur content (CPCB, 1999). Because of the alarming rise in car exhaust-related illnesses like hypertension, bronchial asthma, malignancies, and more, it is quite concerning that SPM levels in cities over the threshold limits are on the rise. The following factors may contribute to the presence of heavy metals in vehicle exhaust: i) slower vehicles experience more engine wear and tear and produce more particulate matter, including dust, than faster vehicles; ii) vehicles that stop at traffic signals contribute to air pollution; and iii) streets near high rise buildings have less effective ventilation and more pollutant dispersion. There has been a significant uptick in research on the elemental makeup of airborne particulate matter. High concentrations of metals in gasoline and diesel exhaust have been detected, which may be related to traffic (Que Hee, 1994). Zn mostly comes from worn tires, Cu from brake dust, and Fe and Zn from vehicle separation and fluid leaks (Ball et al., 1991). Particulate matter often enters Earth's atmosphere from two sources: first, vehicle exhaust, and second, the Earth's crust. Their ability to float in the air for extended periods of time makes them ideal vehicles for transporting and redistributing metals across the Earth's ecosystem. Their final removal from Earth's system is due to either dry or wet deposition. This means that the atmospheric cycling of the vast majority of metals, irrespective of their source, is closely tied to the destiny of SPMs (Kim et al., 2002). While several researchers from other parts of the globe have collected enough data on environmental trace element distributions (Kretzs Chamar et al., 1980; Carreras and Pignata, 2001), very little is known about India's metropolitan regions.

The locally detrimental air pollution consequences of the rapidly expanding motor vehicle and other energy intensive activities in Indian cities are significant for both regional and global reasons. Air pollution from cars may have consequences on a local, regional, and worldwide scale. Local affects include smoke influencing visibility, ambient air, noise, and other similar factors. Regional effects include smog and acidification. Finally, global warming is an example of a global problem. Soil, water, and air are all directly affected by pollution, along with all the biotic and abiotic characteristics of the ecosystem, including plants and animals. People in urban areas and the surrounding regions are more likely to experience a range of health issues due to air pollution. Because of this, it has gained a lot of attention from scientists and the general public. The fact that vehicle emissions are so close to ground level, where humans breathe, makes it very difficult, if not impossible, to escape vehicular pollution. The symptoms of automobile pollution, which include coughing, nausea, eye irritation, headache, bronchial issues, and impaired vision, contribute to the increased mortality and morbidity rates. An alarming surge in cases of respiratory allergies has been linked to the fast and noticeable worsening of air pollution in metropolitan areas (Hill, 1996). Diesel exhaust has a strong cancer causing potential, according to scientific data from worldwide research. Long term exposure to 1 $\mu g\ m^{-3}$ of diesel exhaust has the potential to cause 300 more incidences of lung cancer per million individuals (CARB, 1998). Living close to heavily populated roads is linked to more asthma hospitalizations, worse lung function, more wheezing episodes, and more severe allergic rhinitis, according to similar research (Diaz-Sanchez et al., 2003). Because air pollution may harm sperm, it can cause mutations, which in turn can cause birth abnormalities or miscarriages, in addition to bronchial or respiratory problems. A higher incidence of heart attacks may be seen among populations living in highly polluted regions.

Air pollution impacts not just people and other animals, but also plants. A major ecological problem is the regional effects of air pollution on various plant species. Unlike animal populations, plant populations are constantly (24/7) and directly exposed to the danger of pollution. Their morphological, biochemical, anatomical, physiological, and other reactions are varied, and they soak up, store, and incorporate the contaminants that hit their surfaces. Despite the fact that a great deal of work has been done in recent decades to establish how air pollution affects plants (Sharma and Tyre, 1973; Yunus et al., 1982; Khan et al., 1990; Jahan and Iqbal, 1992; Sheu, 1994; Carreres et al., 1996; Kammerbauer and Dick, 2000; Li, 2003; Hijano, 2005; Pandey, 2005; Rajput and Agrawal, 2005; Agarwal et al., 2006; Verma and Singh, 2006; Wahid, 2006), most of these studies have been conducted in vitro or to a lesser degree on avenue trees. Specifically, there has been very little research on the effects of vehicle pollution on medicinal plants, despite the fact that these plants are a major supply of medicines for both traditional medicine and modern pharmaceuticals. Air pollution is one of the elements that control the therapeutic efficacy of crude pharmaceuticals.

Additionally, a number of techniques have been used to evaluate the degree of damage to vegetation in its natural habitat and to investigate the interactions between plants and pollutants. To measure the quantifiable effects of contaminants on plants while they develop in their native environments, in situ approaches are often used. While conducting such experiments in natural settings has its benefits (Pandey and Pandey, 1994), there are also several drawbacks, such as the fact that plant material is heterogeneous and environmental conditions can vary greatly, making it hard to pinpoint exactly which environmental factor(s) govern plant responses. Particularly important among the many elements are edaphic (Farley and Fitter, 1999) and biotic (Narayan et al., 1994) regulation. Transplant studies, also known as "Transfer Experiment Studies," provide a more straightforward method by planting uniformly sized seedlings of the same age group in similar edaphic circumstances and irrigating them in the same way. It is important to maintain genetic consistency while propagating plants, hence common stock is used (Chaphekar, 1995). One way to influence the root environment and biotic variables is by using pot cultured transplants (Chaphekar, 2000). It offers a more realistic experience with the air around us, which improves our ability to measure the impact of the atmosphere and makes it easier to compare different locations. (Pandey and Pandey, 1996).

This is the background against which the current investigation into the effects of urban air pollution on medicinal plants in Lucknow was planned and executed. The environmental quality of Lucknow has been negatively impacted by the vehicular population, which grew from 6,79,326 in March 2004 to 9,04,831 in March 2007, a growth rate of 24.92%. *Catharanthus roseus L. and Ocimum sanctum L.* were chosen as the study's plant subjects because of their widespread distribution in Lucknow and their immense therapeutic value.

Invisible, flavorless, and odorless air is a gas mixture that includes 78% nitrogen and 21% oxygen. The remaining 1% is made up of nitrogen and sulfur oxides as well as water vapors, as well as trace quantities of ammonia, ozone, organic matter, salts, and suspended solid particles. This vast expanse of air is thought to be about 1018 kg in total mass. This distinctive atmospheric composition, which makes the planet livable and able to sustain life in its current form, was supposedly created some 4.5 billion years ago, according to evolutionary theory. The atmosphere has played a significant role in shaping plant and animal life on Earth from the dawn of life itself. A number of atmospheric chemical components and other physical variables have become manifestations of this impact, which may have both positive and negative effects (Posthumus, 1998). The chemical climatology of our planet is controlled by both natural and human made phenomena. The second one is a major worry on a global, national, and even regional scale since, unless humans drastically modify our atmosphere, it will be able to maintain a stable equilibrium (Lovelock, 1987). The atmospheric composition and complexity have been altered due to the addition of several contaminants brought about by human activity, which is regrettable and stems from a variety of technological breakthroughs. The growth and wellbeing of people are promoted by these activities, but at the cost of environmental contamination caused by the creation and discharge of undesired chemicals and substances.

The problem of air pollution has been there since the beginning of time, impacting both industrialized and developing nations. Since emissions from man-made sources first appeared, the existence of man-made air pollution has been well-documented. Numerous industrial processes, the biodegradation of waste products, some agricultural and forestry activities, and the combustion of fossil fuels in power plants and engines all contribute to the atmospheric emission of hundreds of distinct chemicals. Because of their low concentrations or lack of toxicity to biological systems, the majority of these substances have little to no impact on the environment. On the other hand, there are a few that are known to cause serious harm to humans, as well as to ecosystems in both their natural and manipulated forms (Barnes et al., 1996). But for quite some time, natural air pollution from things like swamps, lightning, forest fires, and volcanoes has been affecting weather patterns and plant life. Researchers have been looking at how air pollution affects plants for about 160 years (Stockhardt, 1850). Objecting to the soot damage inflicted on the old Roman temples, the Roman poet Horace records the first concerns about atmosphere in his works (Brimblecombe, 1982).

As far as the urban ecology is concerned, air pollution is one of the most detrimental effects of human operations. As a result of chemical reactions triggered by sunlight and a wide range of weather circumstances, gaseous and particle emissions into the atmosphere may cause harm to a wide range of organisms, including humans. There are a lot of different types of air pollutants in cities. The most common ones include sulfur oxides, nitrogen oxides, carbon oxides, carbon particles, heavy metals, hydrocarbons, and a host of secondary pollutants like ozone and peroxyacylnitrate (PAN). Pollutants have a wide range of negative effects on plants, and scientists have made incredible strides in documenting these effects in recent years (Sharma and Tyree, 1975; Yunus et al., 1979, 1982; Khan et al., 1990; Jahan and Iqbal, 1992; Marth, 1995; Singh et al., 1995; Carreras et al., 1996; Altaf, 1997; Shchberbakov et al., 2001; Li, 2003; Hijano et al., 2005; Pandey, 2005; Rajput and Agrawal, 2005; Verma and Singh, 2006; Wahid, 2006). Symptoms, changes in photosynthesis, metabolic and gaseous exchange activities, and tissue damage are all outcomes of the varied ways in which these pollutants impact plants, even at low concentrations (Posthumus, 1983). Numerous variables, including exposure duration, genetic composition, phenological phases, and dosage (pollutant concentration), determine the extent of the effect (Heck et al., 1965). Light, humidity, temperature, water availability, and mineral content are among environmental factors that might influence plant responses to harmful gasses. Several elements, like the source's proportion of emissions, wind direction and speed, geography, etc., affect the quantities of pollutants in the air, which in turn define the impact's intensity.

Pollutant Infusion

The process by which air contaminants are efficiently taken up by plants via their stomata goes like this: (1) Gas enters the leaf through stomata in its gas-phase; (2) Gas within the leaf's air space dissolves in the water on the surface of the plant cell; (3) Gas within the cell, in its liquid-phase, diffuses into the cells according to its concentration gradient; and (4) Gas within the cell undergoes metabolism or decomposition, thus sustaining a gas concentration gradient between the atmosphere and the interior of the plant cell (Omasa and Endo, 2001). According to Yunus et al. (1979), stomata size, quantity, and degree of opening significantly affect the entrance and effect of pollutants on plants. The sensitivity of plants to air pollution is thought to be significantly impacted by stomata, which control the entrance of contaminants into the plant. The cuticle is another potential entry point for contaminants entering plants, alongside the stomatal pathway (Omasa and Endo, 2001).

Oxides of sulfur, which include sulfur monoxide (SO), sulfur dioxide (SO_2), sulfur trioxide (SO_3), sulfur tetra oxide (SO_4), sulfur sesquioxide (S_2O_3), and sulfur heptoxide (S_2O_7), are the most abundant air pollutants. These compounds are primarily formed when inorganic sulphides and sulfur-containing organic compounds found in coal and oil are burned. When it comes to sulfur oxides, SO_2 is by far the most prevalent air pollutant (Gupta and Prakash, 1994). Plants are able to absorb sulfur dioxide from the air because they act as sinks. Because SO_2 is highly soluble and rapidly dissociates in the cell sap, its internal (mesophyll) resistance is low, which means that its diffusion via the stomata determines the foliar absorption. The process of oxidation to SO_4 may be either enzymatic or non-enzymatic. After that, it is moved into the vacuole, where its remobilization seems to be difficult (De Kok, 1990; De Kok and Tausz, 2001). In addition to the dissociation of ingested SO2 and the subsequent reactivity of the generated sulfite with cellular components, acidification of tissue and cells may also contribute to SO_2 toxicity. The effect of sulphur dioxide on plant activity is puzzling since plants may absorb and utilize the gas for both harmful and nutritional purposes (De Kok, 1990; De Kok et al., 1998; De Kok and Tausz, 2001; Deepak and Agrawal, 2001). The sulfur assimilation pathway may take in foliarly absorbed SO_2 either directly or after oxidation to SO_4. From there, it can be reduced to sulfite, which can be used as a nutrient, or it can be integrated into cysteine and organic sulfur compounds. (The works of De Kok (1990), Tausz et al. (1998), and De Kok and Tausz (2001)). Nevertheless, when the quantity of sulfur dioxide in the air reaches levels that are harmful to plants, the aforementioned processes are overwhelmed, and distinctive symptoms manifest, which may be either obvious (morphological) or subtle (biochemical or physiological). various environmental elements, such as the time of year, the availability of nutrients and water, and the presence of various contaminants in gaseous, particulate, and soluble forms, may have either complementary or antagonistic impacts on the development of any kind of injury. When compared to NO_x, sulfur dioxide may pose a greater danger to ecosystems and plant life. The relative phytotoxicity of SO_2 was found to be 2.0 to 2.6 times greater than an equally high concentration of NO_2 in the air, according to a quantitative study of the effects of SO_2 and NO_x on Norway Spruce conducted by Slovik (1996).

Even though nitrogen in the air is a gas with no discernible chemical properties, it undergoes a significant atmospheric event when fossil fuels and biomass are burned: the thermogenic oxidation of nitrogen to form nitrogen oxides (NO_x) ($NO + NO_2$). Products with negative impacts on biological systems are considered pollutants in addition to being chemically active. Several in situ and ex situ experiments have investigated the phytotoxic reactions of these gases and other nitrogenous compounds in the environment, including N_2O_5, HNO_3, HNO_2, NO_3-, and peroxyacylnitrate (PAN), sometimes known as odd nitrogen compounds (NOy) (Roberts, 1995). However, NO_2 has been the most investigated NO component, perhaps due to its role as a precursor molecule for the majority of NO components and O_3, another oxidant that is harmful to plants. Additionally, NO_2 is thought to be more hazardous to plants than NO. Because it forms nitrate and nitrite ions when dissolved in water, it is easily absorbed and transported across the moist interstitial spaces of cells. Comparable to SO_2 entrance, NO_2 enters leaves via the cuticle at a greater rate due to reduced cuticular resistance to NO_2 compared to SO_2 and O_3. Most nitrogen oxides (NOx) in the air come from burning fuels. Direct foliar depositions of NO_2 or indirect depositions in rainfall or soil are two ways in which this gas might enter the plant system. Species of plant, genes, concentration of NO_2, and other variables regulate its entry into the leaf via the stomata (Srivastava et al., 1975). When this happens, it's called translocation, and it happens all across the plant (Yoneyma et al., 1980). Whether it's via the leaves or the roots, plants may take NO_2 from the air or the soil when it dissolves in rainwater as HNO_3 and HNO_2.

Despite the discovery of many pathways by which NO_2 may potentially disrupt typical plant processes, the core mechanism of NO_2 phytotoxicity remains unknown. Several pieces of evidence point to a correlation between NO_{2-} accumulation and the obvious damage (Nouchi, 2002). According to Shimazaki et al. (1992), NO_2- accumulated in leaves from air NO_2 inhibited photosynthesis and produced reactive oxygen species (ROS), which may have contributed to the visual damage caused by NO_2. By altering lipid composition, inactivating enzymes, damaging DNA, and even triggering cell death, these species and free radicals may modify the functioning of membranes. But plants have developed a complex antioxidant defense system that includes low-molecular-weight components like glutathione and ascorbate as well as enzymatic components like SOD, POD, and CAT. These components detoxify O_{2-} and H_2O_2, respectively, and prevent the formation of HO radicals (Mittler, 2002). Plants, particularly those growing in soils lacking in nitrogen, may benefit from a low concentration of NO_2 in terms of their physiological and growth traits. Reason being, ambient NO_2 undergoes a series of reductions in the cell, beginning with NR in the cytosol, followed by nitrite reductase in the chloroplasts, and eventually, amino acids are formed from NH^{4+} (Kondo and Saji, 1992). Since NO_2 and SO_2 often coexist in urban settings, it would be prudent to investigate the combined impact of these two pollutants on plants there.

EFFECTS ON VEGETATION

Morphological Aspects

Air pollution is known to have negative impacts on plants, and the key components that contribute to this problem, such as ozone, sulfur dioxide, and nitrogen dioxide, are well known. The impact of pollutant gas mixtures on various plant species' growth and development may be additive, synergistic, or antagonistic, according to studies. A number of studies have examined the cumulative and sequential impacts of different pollutant combinations. It has long been known that plants are negatively impacted by air pollution, which in turn reduces their development and production (Ashmore and Marshall, 1999; Ashmore et al., 1988). There is consensus that higher plants can experience damage in polluted atmospheres with SO_2 concentrations above 25 pphm (parts per million hundred by volume) (Thomas, 1961; Daines, 1968; Guderian and Van Haut, 1970), and that even four weeks of exposure to 11 pphm SO_2 can reduce yields by half (Ashenden and Mansfield, 1977; Ashenden, 1978). Pollutants can cause a broad range of morphological, physiological, and biochemical changes in plants.

Researchers have shown that some species may experience noticeable leaf damage after being exposed to SO_2 and NO_2 for as little as four hours. It is possible that these gases have an additive or synergistic effect (Tingey et al., 1971). When studying the effects of SO_2 and NO_2 on pea seedlings' enzymatic and physiological levels, Wellburn et al. (1976) discovered that the two compounds worked together. At concentrations ranging from 10 to 25 parts per million of both gases, Bull and Mansfield (1974) found that net photosynthesis was additively inhibited. Ashenden (1979b) conducted an intriguing twenty-week study to find out how different concentrations of SO2 and NO_2 affected the growth of two species: *Dactylis glomerata L. and Poa pratensis L.* The results showed that both the SO_2 + NO_2 and SO_2 alone treatments significantly reduced the leaf area and all the dry weight fractions of the plants. Both species showed decreased tiller and leaf production in the SO_2 + NO_2 treatment, but only *P. pratensis* showed this effect when SO_2 was the only ingredient. Nevertheless, when sprayed alone, NO_2 had no impact on D. glomerata growth but significantly decreased *P. pratensis* leaf area and all assessed dry weight fractions. The same holds true for *Phaseolus vulgaris L.*; when applied separately, 10 pphm SO_2 and 10 pphm NO_2 increased transpiration rates for a short duration, but when applied together, they reduced (Ashenden, 1979a). In a separate investigation, Ashenden and Williams (1980) subjected *Phleum pratense L.* and *Lolium multifolium L* 6.8 pphm (194 µg m^{-3}) of SO_2 and 6.8 pphm (139 µg m^{-3}) of NO_2, which were treated individually and together. Both species yields of leaves and tillers, as well as their leaf area and all dry weight fractions, were drastically reduced when exposed to SO_2 and NO_2. Applying SO_2 alone had minimal effect on both species, but significantly reduced the amount of leaves and tillers produced by *Phleum pratense* and all dry weight fractions measured for *Lolium multifolium*, except for the green leaves dry weight fraction. To confirm the impact on development and growth, reddish *Raphanus sativus L.*cv. *cherry Belle* plants were subjected to 0.8 µl l^{-1} NO_2 and 0.8 µl l^{-1} SO_2 in both day and nighttime exposures. When compared to controls exposed to charcoal filtered air, growth was unaffected by sequential exposure to the two pollutants, but growth was significantly reduced when exposed to both pollutants at once (Hogsett, 1984). The cumulative impact of SO_2, NO_2, O_3, and other environmental conditions affecting the performance of four wheat cultivars was investigated experimentally by Ashmore et al. (1988). They found that SO_2, NO_2, and, to a lesser degree, O_3 significantly affected vegetative and reproductive development parameters according to a multiple regression analysis. The impact of low levels of O_3, SO_2, and NO_2 on the development and yield of spring rape plants in containers was studied by Adaros et al. (1991) both individually and in all potential combinations. It is supported by the fact that environmental conditions dictate the degree of impact since the single impacts of O_3 on growth and yield metrics were largely negative and the size of these effects varied with season. O_3 decreased plant dry weight by 11.3% to 18.6% and seed output by 11.4% to 26.9%, but a medium level of SO_2 increased pod weight by 33%. The yield decreased by 12.3% due to the higher concentrations (88 µg m^{-3}). Based on the observed substantial interactive effects, it can be concluded that SO_2 and NO_2 had largely good benefits when used alone, but had antagonistic interactions when combined and, in particular, when combined with O_3, since these pollutants reduced the negative impacts of O_3. The yield was significantly affected by the antagonistic impact of SO_2 on O_3 or NO_2. Even while 56 µg m^{-3} SO_2 improved yield by 9.9% when compared to the control group, it worsened yield loss due to O_3 by 16.18% to 21.4 percent and decreased yield stimulation from NO_2 by 11.8 percent to 4.2 percent. Using ambient and charcoal filtered air, Wahid et al. (1995) showed that two winter wheat cultivars had a 46% and 38% decrease in grain production, respectively, in an open top chamber research in Lahore, Pakistan. In addition, Maggs et al. (1995) demonstrated that wheat and rice yield metrics in Lahore were significantly reduced at yearly mean concentrations of 20–25 ppb of nitrogen dioxide (NO_2) and 6 hrs. mean concentrations of ozone (O_3) that reached 60 ppb in specific months. *Lolium perenne L.* (Perennial rye grass) and *Agrostis capillaries L.* (common bent-grass) were subjected to mixtures of gaseous contaminants and acid mists during 18 weeks and 22 weeks, respectively, by Ashenden et al. (1996). For the gaseous treatments, we had (a) air filtered through charcoal, (b) 40 ppb SO_2 + 40 ppb NO_2, (c) 40 ppb O_3 with peaks of 2x3 h at 80 ppb and 1x1 h at 110 ppb O_3, and (d) a mixture of (b) and (c) as well as 6 mm per week of solution at pH 2.5, 3.5, 4.5, and 5.6 for the mist treatments. The dry weight of *L. perenne* plants was significantly reduced by all gaseous treatments, with the exception of the SO_2 + NO_2 + O_3 treatment, which had a smaller impact than the additive effects of O_3 and SO_2 + NO_2. The development of *L. perenne* subjected to a pH of 2.5 did not differ in the charcoal filtered air and O_3 treatments compared to the less acidic mist, but this was not the case with SO_2 + NO_2 and SO_2 + NO_2 + O_3 gas treatments. The gas x mist treatments were more effective against A. capillaris plants, and the total shoot weights of plants cultivated in the gas exposure treatments were not significantly different from those grown in the charcoal filtered air. A. capillaris plants were simulated to grow in an environment with a pH of 2.5 and a lower concentration of acid mist in comparison to other gas treatments. For both species, exposure to any gaseous pollution treatment reduced the number of healthy shoots while increasing the number of dead ones. Another research found that Agrostis capillaris plants exposed to gaseous pollutants alone or in conjunction with wet nitrogen mist had leaves that progressively aged. Consistent with the trend of increasing damage over time, this matched up well with leaf dry weights. After 11, 13, and 15 weeks of exposure to pollutants, growth analysis showed that both the 20 ppb SO_2 + 20 ppb NO_2 and 40 ppb SO_2 + 40 ppb NO_2 treatments significantly reduced the leaf areas and dry weight of A. capillaris compared to the charcoal filtered controls and the 10 ppb SO_{2+} 10 ppb NO_2 treatment. According to Kupeinskiene (1997), the gaseous contaminants had a more detrimental impact on the shoots than the roots. These results show that the combined impacts of SO_{2+} NO_2 + O_3 or acid mist may be more harmful to plants than the sum of their separate components would indicate. Research by Shamsi et al. (2000) on the effects of several urban environmental factors, including SO2, NO_2, and O_3, on two crops wheat and rice found that both crops' yields were significantly reduced. Nevertheless, they attribute these decreased yields to the combined effects of NO_2 and O_3, as the concentrations of SO_2 were found to be very low during the research period. Agrawal (2003) found that when exposed to different levels of SO_2, NO_2, and O_3 at different urban sites, *Beta vulgaris, Triticum aestivum, Brassica compestris*, and *Vigna radiate* showed significantly reduced physiological characteristics, pigment content, and above-ground biomass compared to the least polluted sites. The development and production features of Helianthus annuus were shown to be significantly reduced as the dosage of SO_2 increased, when exposed to varied concentrations (142, 285, and 571 µg m^{-3}) of the pollutant (Siddiqui et al., 2004). While conducting a transplant research with wheat, Rajput and Agrawal (2005) found significant changes in a number of growth metrics, suggesting that urban air pollution affects sub-urban agriculture. Air pollution levels at various locations were positively correlated with the physiological traits, growth, and yield of transplanted wheat. Because of the significant impact on production, this research also demonstrated how urban air pollution impacts sub-urban agriculture. The same thing happened to three distinct wheat types in Pakistan when exposed to varying levels of urban air pollution, according to Wahid (2006). In a study conducted by Agrawal et al. (2006), mung beans were cultivated from seed to maturity in several regions of Varanasi, each with its unique pollution load. The results indicated a significant decrease in biomass accumulation and yield, which was strongly correlated with the pollution gradients. Using open top chambers, Rai et al. assessed the impacts of ambient gaseous air pollution on wheat (*Triticum aestivum L.*) grown in a suburban area located in the eastern Gangetic plain of India (2007). Filtered chambers (FCs), non-filtered chambers (NFCs), and open plots (OPs) were monitored for ambient concentrations of SO_2, NO_2, and O_3 every eight hours. We measured yield parameters at harvest after evaluating morphological, physiological, and biochemical characteristics throughout development. In the NFCs, the average SO_2 concentration was 8.4 ppb, NO_2 was 39.9 ppb, and O_3 was 40.1 ppb. As compared to NFCs, FCs had a 74.6% decrease in SO_2 concentrations, an 84.7% decrease in NO_2 concentrations, and a 90.4% decrease in O_3 concentrations. Photosynthetic rate, stomatal conductance, chlorophyll content, and Fv/Fm ratio were all greater in FC grown plants than in NFC and OP grown plants. Plants grown in FCs showed an improvement in morphological characteristics when compared to NFCs and OPs. There was a considerable improvement in plant yield in FCs as compared to NFCs (ventilated with ambient air) and OPs (grown in open spaces). Compared to FCs, NFCs significantly reduced plant development and biomass accumulation due to the detrimental impacts of NO_2 and O_3 on plant physiological and biochemical traits. Lower net assimilation per unit ground surface may result from less radiation absorption if leaf area is reduced.

Physio-Biochemical Aspects

The photosynthetic machinery of all phototropic organisms relies on several photosynthetic pigments. There are many kinds of chlorophyll found in higher plants, including chlorophyll a (the main pigment, which is yellow-green), chlorophyll b (blue-green), and accessory pigments. Light harvesting is linked to both the a and b pigments of chlorophyll. Also, according to Procházka et al. (1998), chlorophyll an is involved in the process of electric charge separation. Chlorophyll is protected from photo-oxidative degradation inside the chloroplast by carotenoids, which also function as antioxidants (Siefermann-Harms, 1987; Polle et al., 1992). However, when our bodies are under a lot of pressure, their usual defense mechanisms can't keep up, leading to cellular death and pigment deterioration (Senser et al., 1990). Many environmental conditions, including light intensity, water scarcity, pollution, and other pressures, have a significant impact on the concentrations of these pigments. Because air pollution is one of the external elements that might affect the levels of chlorophyll and carotenoids, many authors have shown that these compounds react to both internal and exterior stimuli. Chlorophyll depletion precedes carotenoids depletion (Sakaki et al., 1983), and an increase in chlorophyllase enzyme activity may be to blame for this decline in plant chlorophyll concentration (Mandal and Mukherji, 2000). According to Mandloi and Dubey (1988) and Rao and Dubey (1985), SO_2 is a key factor in lowering chlorophyll concentration. Acidic pollutants, such as SO_2, reduce chlorophyll concentration and lead to the synthesis of phaeophythin by acidification, as previously described by Rao and Leblance (1966). Metabolically active tissue of plants, mesophyll, receives sulfur dioxide via stomata diffusion and oxidizes it to sulfate ions. Mesophyll cell pigments are immediately affected by these ions in high concentrations, which may cause photo-oxidation, destruction, or pheophytinization (Rao and Leblance, 1966; Malhotra, 1977). According to Rao and Leblance (1966), when SO2 concentrations were greater, chlorophyll a was degraded into phaeophytin by exchanging Mg^{2+} ions in chlorophyll molecules, while chlorophyll b was degraded into chlorphyllide b by removing the phytol group. On the other hand, others claim that when chloroplasts are exposed to SO_2 in light, superoxide radicals are produced on the thylakoid membranes. These radicals then cause the chloroplasts to enlarge (Wellburn et al., 1972) or even disintegrate (Malhotra, 1976). Light catalyzes processes that oxidize carotenoids, producing epoxide, which is then reduced in the dark by an enzyme catalyzed reaction (Calvin, 1955). Such epoxide cycles indeed occur, and Krinsky (1966) verified their function in preventing photo-oxidation of chlorophylls. Several studies that have been conducted periodically have provided more evidence of this. In addition, Agrawal (1985) found that exposure to SO_2 and O_3 reduced the chlorophyll levels of some agricultural plants. According to Singh et al. (1988), when Dahlia rosea Cav. was exposed to SO_2, its photosynthetic pigments degraded. The most critical plant pigment for photosynthesis, chlorophyll a, is very sensitive to sulfur dioxide (SO_2), making it a negative influence on CO_2 fixation and ultimately stunting the plant's growth and development. Chlorophyll b is about two times as vulnerable to SO_2 while carotenoids are four times more so. The amount of chlorophyll in the air may be used as a measure of air pollution since it is a key component of photosynthetic activity (Pawar and Dubey, 1985). Khan et al. (1990) demonstrated a significant decrease in chlorophyll pigment in locations with greater air pollution compared to their control groups, in an experiment designed to confirm the impact of thermal power plant emissions on *Catharanthus roseus L.* They went on to say that this decrease was because chlorophyll was converted to phaeophytin, which altered its light spectrum properties, due to the SO_2 driven removal of Mg^{2+} ions by two hydrogen atoms from chlorophyll molecules. If *Catharanthus roseus L.* shows no outward signs of chlorophyll deficiency, it may be because its chlorophyll pigment production has slowed down (Shimazaki et al., 1980). Subsequently, Sharma et al. (1994) investigated how *Brassica campestris var.* responded to varying SO_2 concentrations. There was a marked drop in the chlorophyll concentration, as Krishna noticed. Chlorophyll an is more vulnerable to SO_2 pollution than the other photosynthetic pigments, and their findings reinforce this fact. Tripathi et al. (1999) found that compared to their control groups, *F.relegiosa, S. jambolana, A. indica, C. fistula,* and *M. indica* had significantly lower chlorophyll content and carotenoids due to NO_x pollution in the areas around silver refineries. Nighat *et al.* (2000) During research on how thermal power plant emissions affect some foliar features of *Ruellia tuberose L.*, it was found that plants grown in polluted areas had significantly less chlorophyll throughout the whole life cycle, from pre-flowering to blooming and beyond. For 20 days in a row, at an intensity of 8 hours per day, plants of Bel-W3 and seven commercial tobacco cultivars (*Nicotiana tabacum L.*) were subjected to two comparatively low ozone concentrations (90 or 135 ppb). Ozone has several negative effects on plants, including the development of necrotic and chlorotic patches, a quickening of the aging process in leaves, a decrease in chlorophyll, and a more severe degradation of chlorophyll a than chlorophyll b. Test tube experiments exposing leaf segments to high ozone concentrations (>1000 ppb) and in vitro ozone bubbling in chlorophyll extracts further corroborated chlorophyll a's greater sensitivity (Saitanis et al., 2001).

Reducing the photosynthetic pigment level of the plant correlates with the pollution load, according to Raina and Sharma (2003), who studied the effects of vehicle pollution on the micro morphological, anatomical, and chlorophyll contents of *Syzygium cumini L.* leaves. Similarly, in the suburbs of Varanasi, Agrawal et al. (2003) found that potted wheat plants exposed to ambient air containing mostly O_3 and NO_2 had a lower total chlorophyll content. Similarly, Siddiqui et al. (2004) found that varying doses of SO_2 significantly reduced the photosynthetic pigment (chl. a, chl. b, total chl., and carotenoids) in *Helianthus annuus*. Hemavathi and Jagannath (2004) also found that when grown in environments with varying levels of urban air pollution, the photosynthetic pigment content of *Peltophorum inerme* and *Azadaricta indica* decreased. Research by Kumari et al. (2005) examined the effects of vehicle exhaust pollution on various biochemical traits of roadside vegetation. They found that photosynthetic pigments in *Ficus relegiosa, Ricinus communis,* and *Carica papaya* were significantly lower in the roadside vegetation compared to the control group. Under increasing pollution loads, the chlorophyll content was reduced in *Carissa canadas L., Cassia fistula L.*, and *Psidium guajava L.* as well (Pandey, 2005). According to Verma et al. (2006), *Ipomea pes-tigridis L.* shows a substantial decrease in chlorophyll and carotenoid concentration when exposed to air pollution. Verma and Singh (2006) also found that when auto-pollution levels increased, the chlorophyll and carotenoid contents of roadside plants (*Ficus relegiosa L.* and *Thevetia nerefolia L.*) degraded. Using open top chambers, Rai et al. (2007) examined the effects of gaseous air pollution on wheat (*Triticum aestivum L.* var. HUW-234) grown in a suburban area in India's eastern Gangetic plain. Photosynthetic rate, stomatal conductance, chlorophyll content, and Fv/Fm ratio were all greater in FC-grown plants than in NFC and OP grown plants. Following this, Joshi and Swami (2007) used physiological markers such as chlorophyll a and b, total chlorophyll, carotenoids, ascorbic acid, pH, and relative water content to examine how four economically significant tree species Mango (*Mangifera indica*), Eucalyptus citriodora, Sagon (*Tectona grandis*), and Sal (*Shorea robusta*) reacted to pollution from cars on roadside. Leaf samples taken from trees along roadsides that were exposed to vehicle exhausts showed a significant change in all of these characteristics when compared to the control group. At the same time, Wali et al. (2007) found that field-grown marigold (*Calendula officinalis L.)* plants react differently to different concentrations of SO_2 (0.5, 1.0, and 2.0 ppm) stress during pre-flowering, flowering, and post-flowering stages of plant development. Under high SO_2 stress, chlorophyll and carotenoids are significantly reduced at all stages, but the lowest dose of SO_2 actually stimulates growth.

It is well-known that air pollution cause proteins and other physiologically significant molecules to degrade, leading to the production of malondialdehyde (Mudd, 1982). According to Foyer et al. (1994), plants that are exposed to SO_2 produce more free radicals, which can lead to impaired protein metabolism. This is because activated oxygen species change the structure of cellular proteins (Pacifici and Davies, 1990) and make them more vulnerable to proteolysis (Stadtmann and Oliver, 1991). Multiple researchers have found evidence of a decline in plant foliar protein content. The protein concentration of *Catharanthus roseus L.* has decreased significantly, but the free amino acid content has increased. published in the year 1990 by Khan et al. Some researchers have proposed that reduced photosynthesis (Constantinidou and Kozlowiski, 1979), blocked protein synthesis, or increased protein breakdown (Robe and Kreeb, 1980) might be responsible for the lower protein concentration. Degradation of existing protein molecules or enhanced de novo protein synthesis is shown by the reduction in protein content followed by an increase in free amino acid content. Thirty days after exposure to 99 ppb O_3 for two hours, Agrawal and Agrawal (1990) found a 39% decrease in Vicia faba and a 6.8% decrease in Cicer arietinum, respectively. every day. The protein content of soybean cv. was found to drop by 23% according to Verma and Agrawal (1996). for four hours with JS-72–44 treated at 0.15 ppm SO_2. for six weeks, five days a week. Protein hydrolysis or a reduction in protein synthesis have been proposed as the mechanisms by which SO_2 reduces protein content (Carlson and Bazzaz, 1985). Over the course of eight hours, Deepak and Agrawal (2001) exposed two soybean cultivars, Glycine max cv. Bragg and PK 472, to high levels of CO_2 (600 $\mu l\ l^{-1}$) and/or SO_2 (0.06 $\mu l\ l^{-1}$), respectively. under field circumstances to evaluate the response to SO_2 exposure resulting from CO_2 enrichment, from germination to grain maturity in open top chambers. Plant growth, biomass, yield, foliar starch, and protein content were all negatively affected in both soybean cultivars when exposed to SO_2 alone. A decrease in protein content was seen exclusively in PK 472 when subjected to CO_2 enrichment. It has been shown that plants grown in environments with increased CO_2 levels have a lower concentration of soluble proteins (Webber et al., 1994). Soybean cv. foliar protein content did not very much. There has also been a report of Kent at 700 µl l^{-1} CO_2 (Havelka et al., 1984). It seems that protein damage caused by SO_2 is considerably reduced in a CO_2 enriched environment, as protein maintenance under combined therapy is comparable to the control. It has been shown that wheat plants treated with O_3 exhibit CO_2 induced protein protection (Rao et al., 1995). *Ricinus communis L.* plants were studied by Kammerbauer and Dick (2000). was compared with controls maintained in a location free of direct motor vehicle emissions after being exposed to urban traffic exhaust emissions for 5 months. There were no outward signs of injury, but there were notable variations in terms of a number of physiological variables. After 2 months and 5 months, respectively, soluble leaf protein levels in the plants exposed to pollution were 32% and 14% lower than in the control plants. The increased levels of nitrogen oxides, carbon monoxide, and carbon dioxide in city air are likely to stimulate protein synthesis. Plants can tolerate low levels of CO_2 in the air. According to Bidwell and Bebee (1974), CO that leaves may absorb is transformed into CO_2, which is then added to serine, glycine, and eventually sucrose. Zeevart (1976) found that tomato plants increased their protein content and amino acid synthesis from CO_2 when given gaseous NO_2. Fluckiger et al. (1978b) also discovered that juvenile birches exposed to highway exhaust pollution stimulated amino acid production. Previous research by Kammerbauer et al. (1987) found that spruce trees along highway borders had reduced CO_2 absorption via stomatal conductance and photosynthetic capability. Thus, it is reasonable to assume that exhaust fumes are inhibiting protein production, however this does not rule out protein breakdown. Urban air pollution also causes protein breakdown in wheat transplants (Rajput and Agrawal, 2005). In their study, Tiwari et al. (2006) examined the effects of air pollution on carrot plants. The plants in the non-filtered chambers showed a significant decrease in protein concentration, while the plants in the filtered chambers were grown in open top chambers (OTCs) ventilated with either ambient air or charcoal filtered air. The experiment was conducted at a suburban site in Varanasi, India. The foliar protein content was found to be significantly lower at sites that received the highest amounts of pollution, according to a subsequent study by Verma and Singh (2006). They attributed this to either the breakdown of existing protein molecules or reduced de novo protein synthesis, building on previous work by Iqbal et al. (2000). The plants studied in this study were *Ficus relegiosa L.* and *Thevetia nerefolia L.*, which are found along roadsides.

According to Darrall (1989), Foyer et al. (1994), and Sandermann (1996), among other major and widely-spread air pollutants, SO_2, NO_2, and O_3 may cause oxidative stress in plants and negatively impact several physiological processes. When plants are exposed to oxidative stressors, they undergo a cascade of physiological and biochemical changes that contribute to their antioxidant defense mechanisms, some of which include enzymes and others that do not (Sharma and Davis, 1997). All cells, whether they are under stress or not, create reactive oxygen species (ROS). Plants have elaborate defensive mechanisms that deal with reactive oxygen species (ROS), which include mitigating their production and eliminating them when necessary. Oxygen production and removal are equal in relaxed circumstances. On the other hand, the immune system might become overwhelmed when faced with elevated ROS production during times of stress. However, plants have developed a sophisticated defense mechanism against free radicals called an antioxidant defense system. This system includes both low molecular weight components like glutathione and ascorbate and enzymatic components like SOD, POD, and CAT. The former two components detoxify oxygen radicals and H_2O_2, respectively, and the latter three prevent the formation of HO radicals (Mittler, 2002). According to Alscher et al. (2002), the superoxide dismutases (SODs) are the first defensive mechanisms that cells have against reactive oxygen species (ROS). Research indicates that SOD has the potential to shield cells from ROS-induced harm. H_2O_2 may diffuse quickly across cell membranes and is the most persistent ROS. Plants use hydrogen peroxide for two purposes: first, at low concentrations it triggers adaptive signaling, which in turn builds tolerance to different abiotic stresses (Karpinski et al., 1999; Dat et al., 2000). Second, at high concentrations it orchestrates programmed cell death (Lamb and Dixon, 1997; Alvarez et al., 1998). Superoxide dismutases, catalases, and peroxidases all have their activity amplified in response to oxidative stress (Hippeli and Elstner, 1996; Sharma and Davis, 1997). Peroxidase activity in plants has been proposed as a marker to assess urban air pollution (Puccinelli et al., 1998; Kammerbauer and Dick, 2000) and has been discovered to be a sensitive indicator of pollutant exposure (Keller, 1974; Curtis et al., 1976; Sarkar et al., 1986) among these antioxidant enzymes. Also, superoxide dismutase's conflicting function in plants' reactions to air pollution needs further investigation. Pollutant exposure increased superoxide dismutase activity in sugar beets (Dixon et al., 1995), pine and spruce (Tandy et al., 1989), and snap beans (Lee and Bennett, 1982). Chloroplasts and mitochondria in plants often create O_2 and H_2O_2 as part of their aerobic metabolism and in response to ambient oxidative stress (Sharma and Davis, 1997). To protect themselves against reactive oxygen species, plants have developed a number of molecular defensive mechanisms. Superoxide dismutases, for instance, convert O^{2-} to H_2O_2 and O_2. According to Harris (1992) and Sharma and Davis (1997), peroxidases and catalases proceed to further degrade H_2O_2 into water and oxygen. Hence, it is suggested that plants respond to oxidative stress caused by air pollutants by increasing the activity of antioxidant enzymes (Hippeli and Elstner, 1996; Sharma and Davis, 1997). The effects of acute levels of nitrogen dioxide (NO_2) on *Brassica campestris* seedlings were studied by Chun-yan et al. (2007) in a plant development chamber. The researchers also looked at whether plants might be pretreated with hydrogen peroxide (H_2O_2) to reduce the damage caused by NO_2. In a controlled setting, *B. campestris* plants that were 28 days old were subjected to varying concentrations of NO2 (0.25, 0.5, 1.0, and 2.0μl l^{-1}, respectively) during 24 hours after being sprayed with a 10 mmol l-1 H_2O_2 aqueous solution (equivalent to around 1.0 mg H_2O_2 per plant). The air that was filtered using charcoal was used as a control. Immediate post-exposure measurements were taken of ascorbate (ASA), malondialdehyde (MDA), total chlorophyll, photosynthetic rate, stomatal conductance, nitrate, nitrate reductase (NR), and plant biomass. Compared to the control, plants exposed to moderate doses of NO_2 (e.g., 0.25 μl l^{-1}) benefited from the exposure and had an increase in the dry weight of the aboveground part. However, plants exposed to high concentrations of NO_2 (e.g., 0.5 μl l^{-1} or higher) had a decrease in plant biomass and total chlorophyll. Furthermore, significant increases in the activity of superoxide dismutase (SOD) and NR were seen at NO_2 concentrations of 0.5 μl l^{-1} or above.

Pharmacognostic Aspects

Alkaloids, terpenes, cyanogenic glycosides, phenolics, and countless other secondary metabolites are produced by vascular plants. It is believed that many of these chemicals have no role in either the energy release mechanisms or the construction of structural components of cells. When it comes to protecting plants against herbivores, pests, and diseases, several secondary metabolites play crucial roles. Furthermore, their accumulation and synthesis differ between plant taxonomies (Harborne, 1999; Bourgaud et al., 2001; Ossipov et al., 2001). When it comes to the connections between plants and their environments, secondary metabolites specifically phenolic chemicals like flavonoids and phenols are very important (Haslam, 1989; Rhodes, 1994). Phenolic chemicals have a multi-functional role in terrestrial plants and are found all over the place (Harborne, 1997). Among their many roles, phenolics aid in wound healing after injury, are involved in plants' defensive responses to air pollution (Fluckiger et al., 1978a; Karolewski and Giertych, 1995), and protect plants from herbivorous insects and diseases (Hahlbroch and Scheel, 1989; Nicholson and Hammerschmidt, 1992; Dixion and Paiva, 1995; Hartley and Jones, 1997). Everything with an aromatic ring and one or more hydroxyls is considered a phenolic chemical (Waterman and Mole, 1994). There are a number of internal and external elements that influence foliar phenolic concentrations; one of the most important external ones is air pollution. According to Zobel (1996), Kanoun et al. (2001), and Loponen et al. (2001), secondary metabolite composition may be altered by air pollution, both quantitatively and qualitatively. According to Giertych et al. (1999), plants exposed to various harmful contaminants showed an increase in the levels of phenolic compounds. Researchers Kanoun et al. (2001) found that fumigating *Phaseolus vulgaris* cv. *Nerina* with ozone at 65-85 ppb significantly altered the accumulation of foliar phenolics in plants subjected to chronic stress. Glucose, fructose, and soluble phenol levels were shown to be lowered in *Pinus sylvestris* and *Picea abies* seedlings exposed to SO_2, according to Kainulainen et al. (1995). Agrawal and Deepak (2003) found that total soluble sugars, starch, and total phenolics were significantly higher under CO_2 and $CO_2 + SO_2$ exposures in an open top chamber study that evaluated the physiological and biochemical effects of two wheat cultivars (*Triticum aestivum L.* cv. *Malviya* 234 and HP1209). The foliar phenol contents of Aleppo pine (*Pinus halepensis Mill.*) needles were studied by Pasqualini et al. (2003). The needles were gathered from six different locations that were impacted by different types of air pollutants, including NO, NO_2, NO_x, O_3, and SO_2. Exposure to sulfur dioxide increased total phenol content, but exposure to nitrogen oxide pollution decreased it. One possible explanation for the negative connection between total phenolics and nitrogen oxides is that these pollutants have a favorable effect on nitrate reductase activity (Krywult et al., 1996). Several investigations have shown that there is a negative link between the quantities of nitrogen and phenolic compounds in the needles or leaves of different Pinus species, and this enzyme helps with nitrogen assimilation (Giertych et al., 1999). While sulphur dioxide and ozone reduced quantities of gallic acid and vanillin, respectively, exposure to nitrogen oxide pollution raised concentrations of p-coumaric acid, syringic acid, and 4-hydroxybenzoic acid. All phenolic compounds begin with cinnamic acid, which is produced by the shikimic acid pathway. P-Coumaric acid is produced via aromatic hydroxylation of cinnamic acid. One of the primary pathways to benzoic acids is the b-oxydation and aromatic hydroxylation of cinnamic acid side chains. Protocatechuic acid and vanillic acid are the products of a second aromatic hydroxylation followed by methylation. The p-coumaric acid routes may also lead to vanillic acid. According to Torssell (1981), syringic acid may be produced by methylating vanillic acid. There may be a hierarchy of sensitivity among the enzymes involved in these biosynthetic pathways with respect to certain contaminants (Loponen et al., 2001). Consequently, the effects of pollutants on secondary metabolism in plants are species-specific and chemically dependent. This is due to the fact that different plants have different metabolic routes for secondary chemicals. The enzymes chalcone synthetase and phenylalanine ammonia lyase, which control the production of phenolic compounds, are known to have their activity changed by exposure to ozone. O_3 fumigation did not affect the amounts of phenolic compounds in *Populus tremuloides* and *Betula papyrifera*, according to Lindroth et al. (2001). When *P. halepensis* needles accumulate both total and simple phenols, it indicates that the shikimate pathway is activated due to air pollution. Agrawal et al. (2005) conducted a field research in the suburbs of Allahabad, a city in a dry tropical region of India, to determine the effect of O_3 on mung bean plants (*Vigna radiata L.* var. *Malviya Jyoti*). The purpose of the evaluation was to determine if ethylene diurea (EDU) was suitable for this purpose. EDU has anti-ozonant properties and is a synthetic compound. Throughout the course of the experiment, the average monthly concentration of O_3 ranged from 64 to 69 μgm^{-3}. When it came to photosynthetic pigments, soluble proteins, ascorbic acid, and phenol levels, EDU application had a significant positive impact. The levels of pigments, protein, and ascorbic acid in the leaf of plants that were treated with EDU were greater than those of plants that were not, whereas the opposite was true for the phenol content. This would imply that by mitigating the detrimental effects of O_3 on proteins, its use decreased the production of phenolic chemicals. According to Tiwari et al. (2006), who used open top chambers (OTCs) to study the effects of air pollution on carrot (Dacus carota var. Pusa Kesar) plants, total phenolics increased significantly when the plants were stressed by air pollution. Previous research has also shown that exposure to pollutants may stimulate total phenolics (Howell, 1974). A study conducted by Rai et al. (2007) attempted to assess the effects of ambient gaseous air pollution on wheat (*Triticum aestivum L.*) using open top chambers. They found that total phenolic contents increased under air pollution stress, which could lead to a decrease in carbon fixation and ATP synthesis as well as an increase in respiration and chloroplast disintegration (Howell, 1974). The effect of ozone on phenols has been the subject of very few investigations (Howell, 1970; Louguet et al., 1989; Langebartels et al., 1990; Kainulainen et al., 1994). The enzyme activity that intervenes in phenol metabolism is affected by high ozone levels, as stated by Howell (1970). The presence of more phenolic compounds in plants exposed to ozone was shown by Langebartels et al. (1990). High levels of ozone were not associated with total phenol concentrations, however, as shown by Kainulainen et al. (1994). There are conflicting findings about the effect of ozone on total phenol concentrations, which may vary by species. In a study conducted by Katoh et al. (1989), it was shown that when *Cryptomeria japonica* was exposed to O_3, the quantities of soluble phenols, glucose, and shikimic acid all dropped. Wheat transplants stored in various locations exposed to varied concentrations of pollutants showed a considerable decrease in phenolic content, according to research by Rajput and Agrawal (2005).

Flavonoids, a type of low molecular weight phenolic chemicals found in many different plant species, have many different biological activities, and are essential in many ways in which plants interact with their environments (Shirley, 1996). Plant organs, tissues, and developmental variables all affect the content of phenolic chemicals such flavonoids (Bohm, 1987). Phytochemical adaptation to both biotic and abiotic conditions is what Dixon and Pavia mean when they say that flavonoid changes occur (1995). Bohm (1987), Tomas-Barberan et al. (1988), Midiwo et al. (1990), Cuadra et al. (1997), Cooper-Driver and Bhattacharya (1998), Lalova (1998), Markham et al. (1998), Penuelas et al. (1996), Simmonds (1998), Chaves et al. (2001), Sallem et al. (2001), and Robles et al. (2003) provide extensive data demonstrating that various biotic and abiotic factors, such as drought, atmospheric pollutants, phytopathogens, and insect deterrent, affect flavonoid synthesis. In spite of the abundance of literature documenting intraspecific flavonoid variation, very little is known about how flavonoids react to different levels of environmental pollution (Loponen et al., 1998; Robles et al., 2003; Nikolova and Ivancheva, 2005). The effect of an altitude gradient and environmental pollution on flavonoid aglycones that have accumulated from the outside was studied by Nikolova and Ivancheva (2005). Apigenin and quercetin 3,7,3'-trimethyl ether contents in *Artemisia vulgaris L.* and *Veronica chamaedrys L.,* respectively, were determined. There were significant variations in the quantity of quercetin 3,7,3'-trimethyl across *A. vulgaris* populations collected from environments with various types of contamination. Populations in environments contaminated by industry had a high quercetin 3, 7, 3'-trimethyl ether concentration. Pollution has been shown to raise levels of phenolic compounds and flavonoids in trees, while the exact mechanisms behind this phenomenon remain unknown (Loponen et al., 1997, 1998; Giertych et al., 1999). Accumulation of methylated flavonoids, according to Chaves et al. (1997), increases plants' overall stress tolerance and prevents water loss.

With a wide variety of structural forms, biosynthesis processes, and pharmacological actions, alkaloids are among the most varied groups of secondary metabolites identified in living organisms. These days, most people agree that secondary metabolites are important for an organism's continued existence. It is believed that alkaloids are an intricate chemical defense mechanism in plants. (Harborne, 1999; Bourgaud et al., 2001; Ossipov et al., 2001) The production and accumulation of secondary metabolites differ among plant taxonomies. External and internal stressors both have an impact on their concentrations (Senoussi et al., 2007; Jaleel et al., 2007; Misra and Gupta, 2006; Qureshi et al., 2004). The composition of secondary metabolites may undergo both qualitative and quantitative changes as a result of air pollution (Zobel, 1996; Kanoun et al., 2001; Lopanen et al., 2001). The production and accumulation of alkaloids in plants are thought to be heavily influenced by the availability of nitrogen, given that alkaloids are nitrogenous molecules. Tobacco, lupines, barley, Datura, Atropa, and Papaver are among the medicinal and non-medicinal plants whose alkaloids are amplified when nitrogen is added (Waller and Nowacki, 1979). It has been shown that indole alkaloid synthesis may be stimulated in cultures of Catharanthus roseus cell suspensions by nitric oxide (Xu and Dong, 2005).

CONCLUSION

The problem of air pollution has been there since the beginning of time, impacting both industrialized and developing nations. Since emissions from man-made sources first appeared, the existence of man-made air pollution has been well documented. Numerous industrial processes, the biodegradation of waste products, some agricultural and forestry activities, and the combustion of fossil fuels in power plants and engines all contribute to the atmospheric emission of hundreds of distinct chemicals. Low quantities or lack of toxicity to biological systems mean that most of these chemicals have little to no impact on the environment. On the other hand, there are a few that are known to have devastating impacts on human health and on flora and fauna in both wild and cultivated environments. A major ecological problem is the regional effects of air pollution on various plant species. Unlike animal populations, plant populations are constantly (24/7) and directly exposed to the danger of pollution. Their morphological, biochemical, anatomical, physiological, and other reactions are varied, and they soak up, store, and incorporate the contaminants that hit their surfaces.

According to the majority of research, there is a positive correlation between pollution load and the majority of morphological traits of various plant proteins and enzymes (Gavrila et al., 2022) Plants such as *Ocimum sanctum* and *Catharanthus roseus* have also shown this behavior. Exposure to vehicle pollution was shown to decrease the height, girth, and number of leaves of both plants. Both plants' chlorophyll, carotenoids, and protein contents dropped sharply as pollution levels rose. Both plants' carotenoids content decreased significantly when pollution levels rose, but *O. sanctum's* dropped the most.

While both plants' color and protein content increased due to vehicle pollution, their cysteine and proline levels increased significantly. The non-protein thiol content of both plants increased significantly in response to traffic pollution, with *O. sanctum* exhibiting the highest accumulation. Therefore, by observing the aforementioned characteristics, researchers may use these two plants as air pollution monitors.

REFERENCES

Adaros, G., Weigel, H. J., & Jager, H. J. (1991). Single and interactive effects of low levels of O_3, SO_2 and NO_2 on the growth and yield of spring rape. *Environmental Pollution*, *4*(4), 269–286. doi:10.1016/0269-7491(91)90002-E PMID:15092095

Agrawal, M. (1985). *Plant factors as indicator of SO_2 and O_3 pollutants. Proc. International Symposium on Biological Monitoring of the State Environment (Bio- indicator).* Indian National Science Academy.

Agrawal, M., & Agrawal, S. B. (1990). Effects of ozone exposure on enzymes and metabolites of nitrogen metabolism. *Scientia Horticulturae*, *43*(1-2), 169–177. doi:10.1016/0304-4238(90)90048-J

Agrawal, M., & Deepak, S. S. (2003). Physiological and biochemical response of two cultivars of wheat to elevated levels of CO_2 and SO_2 singly and in combination. *Environmental Pollution*, *121*(2), 189–197. doi:10.1016/S0269-7491(02)00222-1 PMID:12521107

Agrawal, M., Singh, B., Agrawal, S. B., Bell, J. N. B., & Marshall, F. (2006). The effect of air pollution on yield and quality of Mung bean grown in Peri-urban areas of Varanasi. *Water, Air, and Soil Pollution*, *169*(1-4), 239–254. doi:10.1007/s11270-006-2237-6

Agrawal, M., Singh, B., Rajput, M., Marshall, F., & Bell, J. N. B. (2003). Effect of air pollution on peri-urban agriculture: A case study. *Environmental Pollution*, *126*(3), 323–329. doi:10.1016/S0269-7491(03)00245-8 PMID:12963293

Agrawal, S. B., Singh, A., & Rathore, D. (2005). Role of ethylene diurea (EDU) in assessing impact of ozone on *Vigna radiata* L. plants in a suburban area of Allahabad (India). *Chemosphere*, *61*(2), 218–228. doi:10.1016/j.chemosphere.2005.01.087 PMID:16168745

Alscher, R. G., Erturk, N., & Heath, L. S. (2002). Role of superoxide dismutases (SODs) in controlling oxidative stress in plants. *Journal of Experimental Botany*, *53*(372), 1331–1341. doi:10.1093/jexbot/53.372.1331 PMID:11997379

Altaf, W. J. (1997). Effect of motorway traffic emissions on roadside wild-plants in Saudi Arabia. *Journal of Radioanalytical and Nuclear Chemistry*, *217*(1), 211–222. doi:10.1007/BF02055354

Alvarez, M., Pennell, R., Meijer, P., Ishikawa, A., Dixon, R., & Lamb, C. (1998). Reactive oxygen intermediates mediate a systemic signal network in the establishment of plant immunity. *Cell*, *92*(6), 773–784. doi:10.1016/S0092-8674(00)81405-1 PMID:9529253

Asada, K., & Kiso, K. (1973). Initiation of aerobic oxidation of sulfite by illuminated spinach chloroplasts. *European Journal of Biochemistry*, *33*(2), 253–257. doi:10.1111/j.1432-1033.1973.tb02677.x PMID:4144355

Ashenden, T. W. (1978). Growth reductions in cocks' foot (*Dactylis glomerata*.L.) as a result of SO_2 pollution. *Environ. Pollut*, *15*(2), 161–165. doi:10.1016/0013-9327(78)90104-0

Ashenden, T. W. (1979a). Effects of SO_2 and NO_2 pollution on transpiration in *Phaseolus vulgaris* L. *Environ. Pollut*, *18*(1), 45–50. doi:10.1016/0013-9327(79)90032-6

Ashenden, T. W. (1979b). The effect of long-term exposure to SO_2 and NO_2 pollution on the growth of *Dactylis glomerata*.L. and *Poa pratensis* L. *Environ. Pollut, 18*(4), 249–258. doi:10.1016/0013-9327(79)90020-X

Ashenden, T. W., Bell, S. A., & Rafarel, C. R. (1996). Interactive effect of gaseous air pollutants and acid mist on two major pasture grasses. *Agriculture, Ecosystems & Environment, 57*(1), 1–8. doi:10.1016/0167-8809(95)01008-4

Ashenden, T. W., & Mansfield, T. A. (1977). Influences of wind speed on the sensitivity of ryegrass to SO_2. *Journal of Experimental Botany, 28*(3), 729–735. doi:10.1093/jxb/28.3.729

Ashenden, T. W., & Williams, J. A. D. (1980). Growth reductions in *Lolium multifolium* Lam. and *Phleum pretense* L. as a result of SO_2 and NO_2 pollution. *Environmental Pollution. Series A. Ecological and Biological, 21*(2), 131–139. doi:10.1016/0143-1471(80)90041-0

Ashmore, M. R., Bell, J. N. B., & Mimmack, A. (1988). Crop growth along a gradient of ambient air pollution. *Environmental Pollution, 53*(1-4), 99–121. doi:10.1016/0269-7491(88)90028-0 PMID:15092544

Ashmore, M. R., & Marshall, F. M. (1999). Ozone impacts on agriculture: An issue of global concern. *Advances in Botanical Research, 29*, 32–49.

Barnes, J. D., Hull, M. R., & Davison, A. W. (1996). Impact of air pollutants and elevated CO_2 on plants in winter time. In M. Yunus & M. Iqbal (Eds.), *Plant Response to Air Pollution* (pp. 135–166). John Wiley.

Bidwell, R. G., & Beebee, G. P. (1974). Carbon monoxide fixation by plants. *Canadian Journal of Botany, 52*(8), 1841–1847. doi:10.1139/b74-236

Bohm, B. A. (1987). Intraspecific flavonoid variation. *Botanical Review, 53*(2), 197–279. doi:10.1007/BF02858524

Bourgaud, F., Gravot, A., Milesi, S., & Gontier, E. (2001). Production of plant secondary metabolites: A historical perspective. *Plant Science, 161*(5), 839–851. doi:10.1016/S0168-9452(01)00490-3

Brimblecombe, P. (1982). Trends in the deposition of Sulphate and total solids in London. *The Science of the Total Environment, 22*(2), 97–103. doi:10.1016/0048-9697(82)90027-4 PMID:7063836

Bull, J. N., & Mansfield, T. A. (1974). Photosynthesis in leaves exposed to SO_2 and NO_2. *Nature, 250*(5465), 443–444. doi:10.1038/250443c0 PMID:4852889

Calatayud, A., Ramirez, J. W., Iglesias, D. J., & Barreno, E. (2002). Effect of O_3 on photosynthetic CO_2 exchange, chlorophyll a fluorescence and antioxidant systems in lettuce leaves. *Physiologia Plantarum, 116*(3), 308–316. doi:10.1034/j.1399-3054.2002.1160305.x

Calvin, M. (1955). Function of carotenoids in photosynthesis. *Nature, 176*(4495), 1211. doi:10.1038/1761215a0 PMID:13288579

Canas, M. S., Carreras, H. A., Orellana, L., & Pignata, M. L. (1997). Correlation between environmental conditions and filiar chemical parameters in *Ligustrum lucidum* Ait exposed to urban air pollution. *Journal of Environmental Management, 49*, 167–181. doi:10.1006/jema.1995.0090

Carlson, R. W., & Bazzaz, F. A. (1985). Plant response to SO_2 and CO_2. In W. E. Winner, H. A. Mooney, & R. A. Goldstein (Eds.), *SO_2 and Vegetation* (pp. 313–331). Stanford University Press.

Carreras, H. A., Canas, M. S., & Pignata, M. L. (1996). Differences in responses to urban air pollution by *Ligustrum lucidum* Ait. and *Ligustrum lucidium* Ait. F.tricolor (REHD.) REHD. *Environmental Pollution, 93*(2), 211–218. doi:10.1016/0269-7491(96)00014-0 PMID:15091360

Chaves, N., Escudero, J., & Gutierres-Marino, C. (1997). Role of ecological variables in the seasonal variation of flavonoid content of *Cistus ladanifer* exudate. *Journal of Chemical Ecology, 23*(3), 579–604. doi:10.1023/B:JOEC.0000006398.79306.09

Chaves, N., Sosa, T., & Escudero, J. (2001). Plant growth inhibiting flavonoids in exudate of Cistus ladanifer and in associated soils. *Journal of Chemical Ecology, 27*(3), 623–631. doi:10.1023/A:1010388905923 PMID:11441450

Chun-yan, M. A., Xin, X. U., Lin, H. A. O., & Jun, C. A. O. (2007). Nitrogen dioxide-induced responses in *Brassica campestris* seedlings: The role of hydrogen peroxide in the modulation of antioxidative level and induced resistance. *Agricultural Sciences in China, 6*(10), 1193–1200. doi:10.1016/S1671-2927(07)60163-1

Constantinidou, H. A., & Kozlowski, T. T. (1979). Effects of sulphur dioxide and ozone on Ulnus Americana seedlings. II. Carbohydrates, proteins and lipids. *Canadian Journal of Botany, 57*(2), 176–184. doi:10.1139/b79-026

Cooper-Driver, G., & Bhattacharya, M. (1998). Role of phenolics in plant evolution. *Phytochemistry, 49*, 1165–1174.

Cross, C. E., van der Vliet, A., Louie, S., Thiele, J. J., & Halliwell, B. (1998). Oxidative stress and antioxidants at biosurfaces: Plants, skin, and respiratory tract surface. *Environmental Health Perspectives, 106*, 1241–1251. PMID:9788905

Cuadra, P., Harborne, J., & Waterman, P. (1997). Increases in surface flavonols and photosynthetic pigments in *Gnaphalium luteo-album* in response to UV-B radiation. *Phytochemistry, 45*(7), 1377–1383. doi:10.1016/S0031-9422(97)00183-0

Curtis, C. R., Howell, R. K., & Kremer, D. F. (1976). Soybean peroxidase from ozone injury. *Environ. Pollut., 11*(3), 189–194. doi:10.1016/0013-9327(76)90083-5

Daines, R. H. (1968). Sulphur dioxide and plant response. *Journal of Occupational Medicine, 10*(9), 516–534. doi:10.1097/00043764-196809000-00015 PMID:4878445

Darrall, N. M. (1989). The effect of air pollutants on physiological processes in plants. *Plant, Cell & Environment, 12*(1), 1–30. doi:10.1111/j.1365-3040.1989.tb01913.x

Dat, J., Vandenabeele, S., Vranovh, E., van Montagu, M., Inz, B. D., & van Breusegem, F. (2000). Dual action of active oxygen species during plant stress responses. *Cellular and Molecular Life Sciences, 57*, 779–795. doi:10.1007/s000180050041 PMID:10892343

De Kok, L. J. (1990). Sulphur metabolism in plants exposed to atmospheric sulfur. In H. Rennerberg, C. Brunold, L. J. De Kok, & I. Stulen (Eds.), *Sulfur Nutrition and Sulfur Assimilation in Higher plants; Fundamental, Environmental and Agricultural Aspects* (pp. 113–130). SPB Academic Publishing.

De Kok, L. J., Stuvier, C. E. E., & Stulen, I. (1998). Impact of atmospheric H_2S on plants. In L. J. De Kok & I. Stulen (Eds.), *Response of Plants Metabolism to Air Pollution and Global Change* (pp. 51–63). Backhuys Publishers.

De Kok, L. J., & Tausz, M. (2001). The role of glutathione in plant reaction and adaptation to air pollutants. In D. Grill, M. Tausz, & L. J. De Kok (Eds.), *Significance of Glutathione to Plant Adaptation to the Environment* (pp. 185–201). Kluwer Acadimic Publishers. doi:10.1007/0-306-47644-4_8

Deepak, S. S., & Agrawal, M. (2001). Influence of elevated CO_2 on the sensitivity of two soyabean cultivars to sulphur dioxide. *Environmental and Experimental Botany*, *46*(1), 81–91. doi:10.1016/S0098-8472(01)00086-7 PMID:11378175

Dixion, R. A., & Paiva, N. L. (1995). Stress-induced phenylpropanoid metabolism. *The Plant Cell*, *7*(7), 1085–1097. doi:10.2307/3870059 PMID:12242399

Dixon, J., Hull, M. R., Cobb, A. H., & Sanders, G. E. (1995). Ozone pollution modifies the response of sugarbeet to the herbicide phenmedipham. *Water, Air, and Soil Pollution*, *85*(3), 1443–1448. doi:10.1007/BF00477184

Dixon, R., & Paiva, N. (1995). Stress-induced phenylpropanoid metabolism. *The Plant Cell*, *7*(7), 1085–1097. doi:10.2307/3870059 PMID:12242399

Dudda, A., Herold, M., Holzel, C., Loidl-Stahlhofen, A., Jira, W., & Mlakar, A. (1996). Lipid peroxidation, a consequence of cell injury? *South African Journal of Chemistry. Suid-Afrikaanse Tydskrif vir Chemie*, *49*, 59–64.

El-Khatib, A. A. (2003). The response of some common Egyptian plants to ozone and their use as biomonitors. *Environmental Pollution*, *124*(3), 419–428. doi:10.1016/S0269-7491(03)00045-9 PMID:12758022

Fluckiger, W., Fluckiger-Keller, H., & Oertli, J. J. (1978a). Inhibition of the regulatory ability of stomata caused by exhaust gases. *Experientia*, *34*(10), 1274–1275. doi:10.1007/BF01981413

Fluckiger, W., Fluckiger-Keller, H., & Oertli, J. J. (1978b). Biochemische Veranderungen in jungen Birken im Nahbereich einer Autobahn. *Forest Pathology*, *8*(3), 154–163. doi:10.1111/j.1439-0329.1978.tb01462.x

Foyer, C. H., Lelandais, M., & Kunert, K. J. (1994). Photooxidative stress in plants. *Physiologia Plantarum*, *92*(4), 696–717. doi:10.1111/j.1399-3054.1994.tb03042.x

Gavrilaş, S., Ursachi, C. Ş., Perţa-Crişan, S., & Munteanu, F. D. (1999). Foliage age and pollution alter content of phenolic compounds and chemical elements in *Pinus nigra* Needles. *Water, Air, and Soil Pollution*, *110*(3/4), 363–377. doi:10.1023/A:1005009214988 PMID:35214408

Guderian, R., & Van Haut, H. (1970). Detection of SO_2 effects upon plants. *Staub*, *30*, 22–35.

Gupta, V., & Prakash, G. (1994). The dynamism of sulphur dioxide pollution. *Acta Botanica Indica*, *22*, 200–219.

Hahlbroch, K., & Scheel, D. (1989). Physiology and molecular biology of phenylpropanoid metabolism. *Annual Review of Plant Physiology*, *30*, 105–130.

Harborne, J. B. (1997). Plant secondary metabolism. In M. J. Crawley (Ed.), *Plant Ecology* (2nd ed., pp. 132–155). Blackwell Science.

Harborne, J. B., Harborne, J. B., & Harborne, J. B. (1999). Recent advances in chemical ecology. *Natural Product Reports*, *16*(4), 509–523. doi:10.1039/a804621b PMID:10467740

Harris, E. D. (1992). Regulation of antioxidant enzymes. *The FASEB Journal*, *6*(9), 2675–2683. doi:10.1096/fasebj.6.9.1612291 PMID:1612291

Hartley, S. E., & Jones, C. G. (1997). Plant chemistry and herbivory: or why the world is green. In M. J. Crawley (Ed.), *Plant Ecology* (2nd ed., pp. 284–324). Blackwell Science.

Haslam, E. (1989). *Plant polyphenols*. University Press Publ.

Havelka, U. D., Ackerson, R. C., Boyle, M. G., & Wittnbach, V. A. (1984). CO_2 enrichment effects on soybean physiology I. Effects of long-term CO_2 exposure. *Crop Science*, *24*(6), 1146–1150. doi:10.2135/cropsci1984.0011183X002400060033x

Heath, R. L., & Packer, L. (1968). Photoperoxidation in isolated chloroplast. I. Kinetics and stoichiometry of fatty acids peroxidation. *Archives of Biochemistry and Biophysics*, *125*, 189–198. doi:10.1016/0003-9861(68)90654-1 PMID:5655425

Heck, W. W., Dunning, J. H., & Hindawi, I. J. (1965). Interactions of environmental factors on the sensitivity of plants to air pollution. *Journal of the Air Pollution Control Association*, *15*(11), 511–515. doi:10.1080/00022470.1965.10468415 PMID:5319783

Hemavathi, C., & Jagannath, S. (2004). air pollution effects on pigment content of two avenue trees in Mysore city. *Pollution Research*, *23*, 307–309.

Hijano, C. F., Dominguez, M. D. P., Gimenez, R. G., Sanchez, P. H., & Garcia, I. S. (2005). Higher plants as bioindicators of sulphur dioxide emissions in urban environments. *Environmental Monitoring and Assessment*, *111*(1-3), 75–88. doi:10.1007/s10661-005-8140-6 PMID:16311823

Hill, M. K. (2020). *Understanding environmental pollution*. Cambridge University Press. doi:10.1017/9781108395021

Hippeli, S., & Elstner, E. F. (1996). Mechanisms of oxygen activation during plant stress: Biochemical effects of air pollutants. *Journal of Plant Physiology*, *148*(3-4), 249–257. doi:10.1016/S0176-1617(96)80250-1

Hogsett, W. E., Holman, S. R., Gumpertz, M. L., & Tingey, D. T. (1984). Growth response in radish to sequential and simultaneous exposures of NO_2 and SO_2. *Environmental Pollution. Series A. Ecological and Biological*, *33*(4), 303–325. doi:10.1016/0143-1471(84)90140-5

Howell, R. K. (1970). Influence of air pollution on quantities of caffeic acid isolated from leaves of *Phaseolus vulgaris*. *Phytopathology*, *60*(11), 1626–1629. doi:10.1094/Phyto-60-1626

Howell, R. K. 1974. Phenols, ozone and their involvement in the physiology of plant injury. In: M. Dugger (Ed.), *Air Pollution Related to Plant Growth*. A.C.S. 10.1021/bk-1974-0003.ch008

Iqbal, M., Zafar, M., & Abdin, M. Z. (2000). Studies on anatomical, physiological and biochemical response of trees to coal smoke pollution around a thermal power plant. Department of Botany, Faculty of Sciences, Jamia Hamdard University, Hamdard Nagar, New Delhi to Ministry of Environment and Forest, Government of India.

Jahan, S., & Iqbal, M. Z. (1992). Morphological and anatomical studies of leaves of different plants affected by motor vehicle exhaust. *J. Islamic Acadamy of Sciences*, *5*(1), 21–23.

Jaleel, C. A., Gopi, R., & Panneerselvam, R. (2007). *Alterations in lipid peroxidation, electrolyte leakage and proline metabolism in Catharanthus roseus under treatment with triadimefon, a systemic fungicide.* C.R. Biologies, doi:10.1016/j.crvi.2007.10.001

Jaleel, C. A., Gopi, R., Sankar, B., Manivannan, P., Kishorekumar, A., Sridharan, R., & Panneerselvam, R. (2007). Studies on the germination, seedling vigour, lipid peroxidation and proline metabolism in *Catharanthus roseus* seedlings under salt stress. *South African Journal of Botany*, *73*(2), 190–195. doi:10.1016/j.sajb.2006.11.001

Joshi, P. C., & Swami, A. (2007). Physiological response of some tree species under roadside automobile pollution stress around city of Haridwar, India. *The Environmentalist*, *27*(3), 365–374. doi:10.1007/s10669-007-9049-0

Kainulainen, P., Holopainen, J. K., Hyttinen, H., & Oksanen, J. (1994). Effect of ozone on the biochemistry and aphid infestation of scots pine. *Phytochemistry*, *35*(1), 39–42. doi:10.1016/S0031-9422(00)90505-3

Kainulainen, P., Holopainen, J. K., & Oksanen, J. (1995). Effects of SO_2 on the concentrations of carbohydrates and secondary compounds in Scots pine (Pinus sylvestris L.) and Norway spruce (*Picea abies* L. Karst.) seedlings. *The New Phytologist*, *130*(2), 231–238. doi:10.1111/j.1469-8137.1995.tb03044.x

Kammerbauer, H., Ziegler-Joens, A., Selinger, H., Knoppik, D., & Hock, B. (1987). Exposure of Norway spruce at the highway border: Effects on gas exchange and growth. *Experientia*, *43*(10), 1124–1125. doi:10.1007/BF01956060

Kammerbauer, J., & Dick, T. (2000). Monitoring of urban traffic emission using some physiological indicators in *Ricinus communis* L. Plants. *Archives of Environmental Contamination and Toxicology*, *39*(2), 161–166. doi:10.1007/s002440010092 PMID:10871418

Kanoun, M., Goulas, M. J. P., & Biolley, J. P. (2001). Effect of chronic and moderate ozone pollution on the phenolic pattern of bean leaves (*Phaseolus vulgaris* L. cv. Nerina): Relations with visible injury and biomass production. *Biochemical Systematics and Ecology*, *29*(5), 443–457. doi:10.1016/S0305-1978(00)00080-6 PMID:11274768

Karolewski, P. (1985). The role of free proline in the sensitivity of popular plants to the action of SO_2. *European Journal of Forest Pathology*, *15*, 199–206. doi:10.1111/j.1439-0329.1985.tb00886.x

Karolewski, P., & Giertych, M. J. (1995). Changes in the level of phenols during needle development in Scots-pine populations in a control and polluted environment. *European Journal of Forest Pathology*, *25*(6-7), 297–306. doi:10.1111/j.1439-0329.1995.tb01345.x

Karpinski, S., Reynolds, H., Karpinska, B., Wingsle, G., Creissen, G., & Millineaux, P. (1999). Systemic signaling in response to excess excitation energy in *Arabidopsis. Science, 284*(5414), 654–657. doi:10.1126/science.284.5414.654 PMID:10213690

Katoh, T., Kasuya, M., Kagamimori, S., Kozuka, H., & Kawano, S. (1989). Inhibition of the shikimate pathway in the leaves of vascular plants exposed to air pollution. *The New Phytologist, 112*(3), 363–367. doi:10.1111/j.1469-8137.1989.tb00324.x

Keller, T. H. (1974). The use of peroxidase activity for monitoring and mapping air pollution areas. *Forest Pathology, 4*(1), 11–19. doi:10.1111/j.1439-0329.1974.tb00407.x

Khan, A. M., Pandey, V., Shukla, J., Singh, N., Yunus, M., Singh, S. N., & Ahmad, K. J. (1990). Effect of thermal power plant pollution on *Catharanthus roseus* L. *Bulletin of Environmental Contamination and Toxicology, 44*(6), 865–870. doi:10.1007/BF01702176 PMID:2354262

Kondo, N., & Saji, H. (1992). Tolerance of plants to air pollutants. [In Japanese]. *Journal of Japan Society of Air Pollution, 27*, 273–288.

Krinsky, N. I. (1966). The role of carotenoid pigments as protective against photosensitized oxidation in chloroplast. In T. W. Goodwin (Ed.), *Biochemistry of Chloroplasts* (Vol. 1). Academic Press.

Krywult, M., Karolak, A., & Bytnerowicz, A. (1996). Nitrate reductase activity as an indicator of ponderosa pine response to atmospheric nitrogen deposition in the San Bernardino Mountains. *Environmental Pollution, 93*(2), 141–146. doi:10.1016/0269-7491(96)00033-4 PMID:15091353

Kumari, S. I., Rani, P. U., & Suresh, C. (2005). Absorption of automobile pollutants by leaf surfaces of various road side plants and their effect on plant biochemical constituents. *Pollution Research, 24*, 509–512.

Kupcinskiene, E. A., Ashenden, T. W., Bell, S. A., Williams, T. G., Edge, C. P., & Rafarel, C. R. (1997). Response of *Agrostis capillaris* to gaseous pollutants and wet nitrogen deposition. *Agriculture, Ecosystems & Environment, 66*(2), 89–99. doi:10.1016/S0167-8809(97)00052-2

Lalova, A. (1998). Accumulation of flavonoids and related compounds in birch induced by UV-B irradiance. *Tree Physiology, 18*(1), 53–58. doi:10.1093/treephys/18.1.53 PMID:12651299

Lamb, C., & Dixon, R. (1997). The oxidative burst in plant disease resistance. *Annual Review of Plant Physiology and Plant Molecular Biology, 48*(1), 251–275. doi:10.1146/annurev.arplant.48.1.251 PMID:15012264

Langebartels, C., Heller, W., Kerner, K., Leonardi, S., Rosemann, D., Schraudner, M., Trest, M., & Sandermann, H. J. (1990). *Ozone-induced defense reaction in plants. Environmental research with plants in closed chambers. Air pollution Research Reports of the EC26*. EEC.

Lee, E. H., & Bennett, J. H. (1982). Superoxide dismutase, a possible protective enzyme against ozone injury in snap beans (*Phaseolus vulgaris* L.). *Plant Physiology, 67*, 347–350. doi:10.1104/pp.67.2.347 PMID:16662420

Levitt, J. (1980). *Responses of Plants to Environmental Stresses* (2nd ed., Vol. II). Academic Press.

Li, M. H. (2003). Peroxidase and superoxide dismutase activities in Fig leaves in response to ambient air pollution in a subtropical city. *Archives of Environmental Contamination and Toxicology*, *45*(2), 168–176. doi:10.1007/s00244-003-0154-x PMID:14565573

Lindroth, R. L., Kopper, B. J., Parson, F. J., Bockheim, J. G., Karnosky, D. F., Hendrey, G. R., Pregitzer, K. S., Isebrand, J. G., & Sober, J. (2001). Consequences of elevated carbondioxide and ozone for foliar chemical composition and dynamics in trembling aspen (*Populus tremuloides*) and paper birch (*Betula papyrifera*). *Environmental Pollution*, *115*(3), 395–404. doi:10.1016/S0269-7491(01)00229-9 PMID:11789920

Loponen, J., Lempa, K., Ossipov, V., Kozlov, M. V., Girs, A., Hangasmaa, K., Haukioja, E., & Pihlaja, K. (2001). Patterns in content of phenolic compounds in leaves of mountains birches along a strong pollution gradient. *Chemosphere*, *45*(3), 291–301. doi:10.1016/S0045-6535(00)00545-2 PMID:11592418

Loponen, J., Ossipo, V., Lempa, K., Haukioja, E., & Pilhlaja, K. (1998). Concentrations and among-compound correlation of individual phenols in white birch leaves under air pollution stress. *Chemosphere*, *37*(8), 1445–1456. doi:10.1016/S0045-6535(98)00135-0

Loponen, J., Ossipov, V., Koricheva, J., Haukioja, E., & Pilhlaja, K. (1997). Low molecular mass phenolics in foliage of *Betula pubescens* Ehrh. in relation to aerial pollution. *Chemosphere*, *34*(4), 687–697. doi:10.1016/S0045-6535(97)00461-X

Louguet, P., Malka, P., & Contour-Ansel, D. 1989. Etude compar_ee de la r_esistance stomatique et de la teneur en compos_es ph_enoliques foliaires chez trois clones d_Epic_eas soumis _a une pollution contr^ol_ee par l_ozone et le dioxyde de soufre en chambre _a ciel ouvert. In: Brasser, T.J., Mulde, W.C. (Eds.). *Man, and his ecosystem. Proceedings of the 8th World Clean Air Congress.* The Hague, Elsevier.

Lovelock, J. E. (1987). *Gaia: A new look at life on earth.* Oxford Iniversity Press.

Maggs, R., Wahid, A., Shasmi, S. R. A., & Ashmore, M. R. (1995). Effect of ambient air pollution on wheat and rice yield in Pakistan. *Water, Air, and Soil Pollution*, *85*(3), 1311–1316. doi:10.1007/BF00477163

Malhotra, S. S. (1976). Effect of sulphur dioxide on biochemical activity and ultrastructural organization of pine needle chloroplasts. *The New Phytologist*, *76*(2), 239–245. doi:10.1111/j.1469-8137.1976.tb01457.x

Malhotra, S. S. (1977). Effect of aqueous sulphur dioxide on chlorophyll destruction in *Pinus contorta. The New Phytologist*, *78*(1), 101–109. doi:10.1111/j.1469-8137.1977.tb01548.x

Mandal, M., & Mukherji, S. (2000). Changes in chlorophyll content, chlorophllase activity, Hill reaction, photosynthetic CO_2 uptake, sugar and starch content in five dicotyledonous plants exposed to automobile exhaust pollution. *Journal of Environmental Biology*, *21*, 37–41.

Mandloi, B. L., & Dubey, P. S. (1988). The industrial emission and plant response at Pithanpur (M.P). *International Journal of Ecology and Environmental Sciences*, *14*, 75–99.

Mansfield, T. A., & Freer-Smith, P. H. (1981). The role of stomata in resistance mechanisms. In: Koziol, M.J. and Whatley, F.R. (Eds.). Gaseous Pollutant and Plant Metabolism. Butterworths scientific, London.

Manzoor, J., & Sharma, M. (2018). Impact of incineration and disposal of biomedical waste on air quality in Gwalior city. *Int J Theor Appl Sci, 10*(1), 10–16.

Markham, K., Tanner, G., Caasi-Lit, M., Whttecross, M., Nayudu, M., & Mitchell, K. (1998). Possible protective role for 3′,4′ -dihydroxyflavones induced by enhanced UV- B tolerant rice cultivar. *Phytochemistry, 49*(7), 1913–1919. doi:10.1016/S0031-9422(98)00438-5

Matysik, J., Alia, P., Bhalu, B., & Mohanty, P. (2002). Molecular mechanisms of quenching of reactive oxygen species by proline under stress in plants. *Current Science, 82*, 525–532.

Mayer, H. (1999). Air Pollution in Cities. *Atmospheric Environment, 33*(24-25), 4029–4037. doi:10.1016/S1352-2310(99)00144-2

Menzel, D. B. (1976). The role of free radicals in the toxicity of air pollutants (nitrogen oxides and ozone). In W. A. Pryor (Ed.), *Free Radicals in Biology* (Vol. 2, pp. 181–203). Academic Press. doi:10.1016/B978-0-12-566502-5.50013-5

Midiwo, J., Matasi, J., Wanjau, O., Mwangi, R., Waterman, P., & Wollenweber, E. (1990). Anti-feedant effects of surface accumulated flavonoids of *Polygonum senegalense. Bulletin of the Chemical Society of Ethiopia, 4*, 123–127.

Misra, N., & Gupta, A. K. (2006). Effect of salinity and different nitrogen sources on the activity of antioxidant enzymes and indole alkaloid content in *Catharanthus roseus* seedlings. *Journal of Plant Physiology, 163*(1), 11–18. doi:10.1016/j.jplph.2005.02.011 PMID:16360799

Mittler, R. (2002). Oxidative stress, antioxidants and stress tolerance. *Trends in Plant Science, 7*(9), 405–410. doi:10.1016/S1360-1385(02)02312-9 PMID:12234732

Mudd, J. B. (1982). Effects of oxidants on metabolic functions. In M. H. Unsworth & D. P. Ormrod (Eds.), *Effects of Gaseous Air Pollutants in Agriculture and Horticulture* (pp. 189–203). Butterworths. doi:10.1016/B978-0-408-10705-1.50014-4

Naidu, B. P., Aspinall, D., & Paleg, L. G. (1992). Variability in proline accumulation ability of barley cultivars induced by vapour pressure deficit. *Plant Physiology, 98*, 716–722. doi:10.1104/pp.98.2.716 PMID:16668700

Nicholson, R. L., & Hammerschmidt, R. (1992). Phenolic compounds and their role in disease resistance. *Annual Review of Phytopathology, 30*(1), 369–389. doi:10.1146/annurev.py.30.090192.002101

Nighat, F., Mahmooduzzafar, M., & Iqbal, M. (2000). Stomatal conductance, photosynthetic rate and pigment content in *Ruellia tuberose* leaves as affected by coal-smoke pollution. *Biologia Plantarum, 43*(2), 263–267. doi:10.1023/A:1002712528893

Nikolova, M. T., & Ivancheva, S. V. (2005). Quantitative flavonoid variations of *Artemisia vulgaris* L. and *Veronica chamaedrys* L. in relation to altitude and polluted environment. *Acta Biologica Szegediensis, 49*, 29–32.

Nouchi, I. (2002). Responses of whole plants to air pollutions. In: Omasa, K. Saji, H. Youssefian, S. and Kondo, N. (Eds.), Air Pollution and Plant Biotechnology. Springer-Verlag, Tokyo.

Nouchi, I., & Toyama, S. (1988). Effect of ozone and peroxyacylnitrate on polar lipids and fatty acids in leaves of morning glory and kidney bean. *Plant Physiology*, *87*(3), 638–646. doi:10.1104/pp.87.3.638 PMID:16666199

Omasa, K., & Endo, R. (2001). Absorption of air pollutants by leaves and patchy responss of stomata and photosynthesis. *5th APGC symposium*. Pulawy, Poland.

Ossipov, V., Haukioja, E., Ossipova, S., Hanhima¨ki, S., & Pihlaja, K. (2001). Phenolic and phenolic-related factors as determinants of suitability of mountain birch leaves to an herbivore insect. *Biochemical Systematics and Ecology*, *29*(3), 223–240. doi:10.1016/S0305-1978(00)00069-7 PMID:11152944

Pacifici, R. E., & Davies, K. J. A. (1990). Protein degradation as an index of oxidative stress. *Methods in Enzymology*, *186*, 485–502. doi:10.1016/0076-6879(90)86143-J PMID:2233315

Pandey, J. (2005). Evaluation of Air pollution phytotoxicity down wind of a phosphorus fertilizer factory in India. *Environmental Monitoring and Assessment*, *100*(1-3), 249–266. doi:10.1007/s10661-005-6509-1 PMID:15727311

Paull, N. J., Krix, D., Irga, P. J., & Torpy, F. R. (2021). Green wall plant tolerance to ambient urban air pollution. *Urban Forestry & Urban Greening*, *63*, 127201. doi:10.1016/j.ufug.2021.127201

Pawar, K. & Dubey, P.S. (1985). *Effects of air pollution on the photosynthetic pigments of* Ipomea fistulosa *and* Phoenix sylvestris. All India seminar on Air pollution Control, Indore.

Peiser, G. D., & Yang, S. F. (1979). Ethylene and ethane production from sulphur dioxide injured plants. *Plant Physiology*, *63*(1), 142–145. doi:10.1104/pp.63.1.142 PMID:16660667

Penuelas, J. (1996). Variety of responses of plant phenolic concentration to CO_2 enrichment. *Journal of Experimental Botany*, *47*, 1463–1467. doi:10.1093/jxb/47.9.1463

Polle, A., Mo¨ssnang, M., Von Scho¨nborn, A., Sladkovic, R., & Rennenberg, H. (1992). Field studies on Norway spruce trees at high altitudes. I. Mineral, pigment and soluble protein contents of needles as affected by climate and pollution. *The New Phytologist*, *121*(1), 89–99. doi:10.1111/j.1469-8137.1992.tb01096.x

Posthumus, A. C. (1983). Higher plants as indicators and accumulators of gaseous air pollution. *Environmental Monitoring and Assessment*, *3*(3-4), 263–272. doi:10.1007/BF00396220 PMID:24259091

Posthumus, A. C. (1998). Air pollution and global change: Significance and Prospectives. In L. J. Dekok & I. Stulen (Eds.), *Responses of plant metabolism to air pollution and global change* (pp. 3–14). Backhuys Publishers.

Puccinelli, P., Anselmi, N., & Bragaloni, M. (1998). Peroxidases: Usable markers of air pollution in trees from urban environments. *Chemosphere*, *36*(4-5), 889–894. doi:10.1016/S0045-6535(97)10143-6

Qureshi, M. I., Israr, M., Abdin, M. Z., & Iqbal, M. (2004). Response of *Artemisia annua* L. to lead and salt- induced oxidative stress. *Environmental and Experimental Botany*, *53*, 185–193.

Rai, R., Agrawal, M., & Agrawal, S. B. (2007). Assessment of yield losses in tropical wheat using open top chambers. *Atmospheric Environment*, *41*(40), 9543–9554. doi:10.1016/j.atmosenv.2007.08.038

Rai, V., Vajpayee, P., Singh, S. N., & Mehrotra, S. (2004). Effect of chromium accumulation on photosynthetic pigments, oxidative stress defense system, nitrate reduction, proline level and eugenol content of *Ocimum tenuiflorum* L. *Plant Science*, *167*(5), 1159–116. doi:10.1016/j.plantsci.2004.06.016

Raina, A. K., & Sharma, A. (2003). Effect of vfehicular pollution on the leaf micro-morphology, anatomy and chlorophyll contents of *Syzygium cumini* L. *International Journal of E-Politics*, *23*, 897–902.

Rajput, M., & Agrawal, M. (2005). Biomonitoring of air pollution in a seasonally dry tropical suburban area using wheat transplants. *Environmental Monitoring and Assessment*, *101*, 39–53. PMID:15736874

Ramge, P., Badeck, F. W., Plochl, M., & Kohlmaier, G. H. (1993). Apoplastic antioxidants as decisive elimination factors within the uptake process of nitrogen dioxide into leaf tissues. *The New Phytologist*, *125*(4), 771–785. doi:10.1111/j.1469-8137.1993.tb03927.x PMID:33874445

Ranieri, A., D'llrso, G., Nali, C., Lorenzini, G., & Soldatini, G. F. (1996). Ozone stimulates apoplastic antioxidant systems in pumpkin leaves. *Physiologia Plantarum*, *97*(2), 381–387. doi:10.1034/j.1399-3054.1996.970224.x

Rao, D. N., & Leblance, F. (1966). Effect of sulphur dioxide on lichen alga with special reference to chloroplast. *The Bryologist*, *69*(1), 69–72. doi:10.1639/0007-2745(1966)69[69:EOSDOT]2.0.CO;2

Rao, M. V., & Dubey, P. S. (1985). Plant response against SO2 in field conditions. *Asian Environment*, *10*, 1–9.

Rao, M. V., Hale, B. A., & Ormrod, D. P. (1995). Amelioration of ozone-induced oxidative damage in wheat plants grown under high CO_2. *Plant Physiology*, *109*, 421–432. doi:10.1104/pp.109.2.421 PMID:12228603

Rhodes, D. (1988). Metabolic response to stress. In D. D. Davies (Ed.), *The Biochemistry of Plants* (Vol. 12, pp. 201–241). Academic Press.

Rhodes, M. J. C. (1994). Physical role for secondary metabolites in plants: Some progress, many outstanding problems. *Plant Molecular Biology*, *24*(1), 1–20. doi:10.1007/BF00040570 PMID:8111009

Robe, R., & Kreeb, K. H. (1980). Effects of SO_2 upon enzyme activity in plant leaves. *Int. J. Plant Physiol*, *97*, 215–226.

Roberts, J. M. (1995). Reactive odd nitrogen in the atmosphere. In H. B. Singh (Ed.), *Composition, Chemistry and Climate of the Atmosphere* (pp. 176–215). Van Nostrand Reinhold.

Robles, C., Greff, S., Pasqualini, V., Garzino, S., Bousquet-Melou, A., Fernandez, C., Korboulewsky, N., & Bonin, G. (2003). Phenols and flavonoids in Alepopine Needles as Bioindicators of air pollution. *Journal of Environmental Quality*, *32*(6), 2265–2271. doi:10.2134/jeq2003.2265 PMID:14674550

Saitanis, C. J., Riga-Karandinos, A. N., & Karandinos, M. G. (2001). Effects of ozone on chlorophyll and quantum yield of tobacco (*Nicotiana tabacum* L.) varieties. *Chemosphere*, *42*(8), 945–953. doi:10.1016/S0045-6535(00)00158-2 PMID:11272917

Sakaki, T., Kondo, N., & Sugahara, K. (1983). Breakdown of photosynthetic pigments and lipids in spinach leaves with ozone fumigation: Role of active oxygens. *Physiologia Plantarum*, *59*(1), 28–34. doi:10.1111/j.1399-3054.1983.tb06566.x

Sakaki, T., Ohnishi, J., Kondo, N., & Yamada, M. (1985). Polar and neutral lipid changes in spinach leaves with ozone fumigation.: Triacylglycerol synthesis from polar lipids. *Plant & Cell Physiology*, *26*, 253–262.

Sallem, A., Loponen, J., Pihlaja, K., & Oksanen, E. (2001). Effect of long-term open field ozone exposure on leaf phenolics of European silver birch (*Betula pendula* Roth). *Journal of Chemical Ecology*, *27*(5), 1049–1062. doi:10.1023/A:1010351406931 PMID:11471939

Sandermann, H. Jr. (1996). Ozone and plant health. *Annual Review of Phytopathology*, *34*(1), 347–366. doi:10.1146/annurev.phyto.34.1.347 PMID:15012547

Sarkar, R. K., Banerjee, A., & Mukherji, S. (1986). Acceleration of peroxidase and catalase activities in leaves of wild dicotyledonous plants, as an indication of automobile exhaust pollution. *Environmental Pollution. Series A. Ecological and Biological*, *42*(4), 289–295. doi:10.1016/0143-1471(86)90013-9

Senoussi, M. M., Creche, J., & Rideau, M. (2007). Relation between hypoxia and alkaloid accumulation in *Catharanthus roseus* cell suspension. *Journal of Applied Sciences Research*, *3*, 287–290.

Senser, M., Kloss, M., & Lutz, C. (1990). Influence of soil substrate and ozone plus acid mist on the pigment content and composition of needles from young spruce trees. *Environmental Pollution*, *64*(3-4), 295–312. doi:10.1016/0269-7491(90)90052-E PMID:15092286

Shamsi, S. R. A., Ashmore, M. R., Bell, J. N. B., Maggs, R., Kafayat, U., & Wahid, A. (2000). The impact of air pollution on crops in developing countries- A case study in Pakistan. In M. Yunus, N. Singh, & L. J. De Kok (Eds.), *Environmental Stress: indication, Mitigation and Eco-conservation* (pp. 65–71). Kluwer Academic Publishers. doi:10.1007/978-94-015-9532-2_6

Sharma, G. K., & Tyree, J. (1973). Geographic leaf cuticular and gross morphological variations in *Liquidamber styraciflua* L. and their possible relationship to environmental pollution. *Botanical Gazette (Chicago, Ill.)*, *134*(3), 179–184. doi:10.1086/336701

Sharma, T. K., Sharma, N. K., & Prakash, G. (1994). Effect of sulphur dioxide on *Brassica campestris* var. Krishna. *Acta Botanica Indica*, *22*, 15–20.

Sharma, Y. K., & Davis, K. R. (1997). The effects of ozone on antioxidant responses in plants. *Free Radical Biology & Medicine*, *23*(3), 480–488. doi:10.1016/S0891-5849(97)00108-1 PMID:9214586

Shchberbakov, A. P., Svistova, I. D., & Dzhuvelikyan, Kh. (2001). Biomonitoring of soil pollution by gaseous emissions of motor vehicles. *Ekologiya i Promyshlennost Rossil*, (June), 26–29.

Shimazaki, K., Sakaki, T., Kondo, N., & Sugahara, K. (1980). Active oxygen participation in chlorophyll destruction and lipid peroxidation in SO_2- fumigated leaves of spinach. *Plant & Cell Physiology*, *21*(8), 1193–1204. doi:10.1093/oxfordjournals.pcp.a076118

Shimazaki, K., Yu, S. W., Sakaki, T., & Tanaka, K. (1992). Difference between spinach and kidney bean plants in terms of sensitivity to fumigation with NO_2. *Plant & Cell Physiology*, *33*(3), 267–273. doi:10.1093/oxfordjournals.pcp.a078250

Shirley, B. W. (1996). Flavonoid biosynthesis new functions of an old pathway. *Trends in Plant Science*, *1*, 281–377.

Siddiqui, S., Ahmad, A., & Hayat, S. (2004). The impact of sulphur dioxide on growth and productivity of *Helianthus annus*. *Pollution Research*, *23*, 327–332.

Siefermann-Harms, D. (1987). The light harvesting and protective function of carotenoids in photosynthetic membranes. *Physiologia Plantarum*, *69*(3), 561–568. doi:10.1111/j.1399-3054.1987.tb09240.x

Simmonds, M. (1998). Chemoecology: The legacy left by Tony Swain. *Phytochemistry*, *49*, 1183–1190.

Singh, N., Yunus, M., Kulshreshtha, K., Srivastava, K., & Ahmad, K. J. (1988). Effect of SO_2 on growth and development of *Dahlia rosea* Cav. *Bulletin of Environmental Contamination and Toxicology*, *40*(5), 743–751. doi:10.1007/BF01697525 PMID:3382791

Singh, N., Yunus, M., Srivastava, K., Singh, S. N., Pandey, V., Misra, J., & Ahmad, K. J. (1995). Monitiring of auto exhaust pollution by road side plants. *Environmental Monitoring and Assessment*, *34*(1), 13–25. doi:10.1007/BF00546243 PMID:24201905

Slater, T. F. (1972). *Free Radical Mechanisms in Tissue Injury*. Pion Limited.

Slovik, S. (1996). Chronic SO_2 and NO_x pollution interferes with the K^+ and Mg^{2+} budget of Norway Spruce trees. *Journal of Plant Physiology*, *148*(3-4), 276–286. doi:10.1016/S0176-1617(96)80254-9

Smith, L. L. (1987). Cholesterol auto-oxidation 1981-1986. *Chemistry and Physics of Lipids*, *44*(2-4), 87–125. doi:10.1016/0009-3084(87)90046-6 PMID:3311423

Srivastava, H. S., Jolliffe, P. A., & Runeckles, V. C. (1975). Inhibition of gas exchange in bean leaves by NO_2. *Canadian Journal of Botany*, *53*(5), 466–474. doi:10.1139/b75-057

Stadtmann, E. R., & Oliver, C. N. (1991). Metal-catalyzed oxidation of protein physiological consequences. *The Journal of Biological Chemistry*, *226*(4), 2005–2008. doi:10.1016/S0021-9258(18)52199-2

Stockhardt, J. A. (1850). Liber die Einwirkung des Rauches von Silberhutten auf die benachbarte vegetation. *Polyt. Centr. Bl*, *16*, 257–278.

Tanaka, K., Suda, Y., Kondo, N., & Sugahara, K. (1985). Ozone tolerance and the ascorbate-dependent H_2O_2 decomposing system in chloroplasts. *Plant & Cell Physiology*, *26*, 1425–1431.

Tandy, N. C., Di Giulio, R. T., & Richardson, C. J. (1989). Assay and electrophoresis of superoxide dismutase from red spruce (*Picea rubens* Sarg.), Loblolly pine (*Pinus taeda* L.), and Scotch pine (*Pinus sylvestris* L.). A method for biomonitoring. *Plant Physiology*, *90*(2), 742–748. doi:10.1104/pp.90.2.742 PMID:16666837

Tausz, M., Weidner, W., Wonisch, A., De Kok, L. J., & Grill, D. (1998). Uptake and distribution of ^{35}S-sulfate in needles and root of spruce seedlings as affected by exposure to SO_2 and H_2S. *Environmental and Experimental Botany*, *50*(3), 211–220. doi:10.1016/S0098-8472(03)00025-X

Thomas, M. D. (1961). Effect of air pollution on plants. *Monogr. Wld. Hlth. Org*, *49*, 223–278.

Tingey, D. T., Reinert, R. A., Dunning, J. A., & Heck, W. W. (1971). Vegetation injury from the interactions of NO_2 and SO_2. *Phytopathology*, *61*, 1506–1511. doi:10.1094/Phyto-61-1506

Tiwari, S., Agrawal, M., & Marshall, F. M. (2006). Evaluation of ambient air pollution impact on carrot plants at a suburban site using open top chamber. *Environmental Monitoring and Assessment*, *119*(1-3), 15–30. doi:10.1007/s10661-005-9001-z PMID:16736274

Tomas-Barberan, F., Msonthi, J., & Hostettmann, K. (1988). Antifungal epicuticular methylated flavonoids from *Helichrysum nitens*. *Phytochemistry*, *27*(3), 753–755. doi:10.1016/0031-9422(88)84087-1

Tomlinson, H., & Rich, S. (1970). Lipid peroxidation, as a result of injury in bean leaves exposed to ozone. *Phytopathology*, *60*(10), 1531–1532. doi:10.1094/Phyto-60-1531 PMID:5481395

Torssell, K. B. G. (1981). Natural product chemistry. In: A mechanistic and biosynthetic approach to secondary metabolism. J. Wiley and Sons Publishers, New-York.

Tripathi, A., Tripathi, D. S., & Prakash, V. (1999). Phyto monitoring and NO_x pollution around silver refineries. *Environment International*, *25*(4), 403–410. doi:10.1016/S0160-4120(99)00004-5

Verma, A., & Singh, S. N. (2006). Biochemical and ultra structural changes in plant foliage exposed to automobile pollution. *Environmental Monitoring and Assessment*, *120*(1-3), 585–602. doi:10.1007/s10661-005-9105-5 PMID:16758287

Wahid, A. (2006). Productivity losses in barley attributed to ambient atmospheric pollutants in Pakistan. *Atmospheric Environment*, *40*(28), 5342–5354. doi:10.1016/j.atmosenv.2006.04.050

Wahid, A., Maggs, R., Shasmi, S. R. A., Bell, J. N. B., & Ashmore, M. R. (1995). Air pollution and its impact on wheat yield in the Pakistan Panjab. *Environmental Pollution*, *88*(2), 147–154. doi:10.1016/0269-7491(95)91438-Q PMID:15091554

Wali, B., Iqbal, M., & Mahmooduzzafar. (2007). Anatomical and functional responses of *Calendula officinalis* L. to SO_2 stress as observed at different stages of plant development. *Flora (Jena)*, *202*(4), 268–280. doi:10.1016/j.flora.2006.08.002

Waller, G. R., & Nowacki, E. K. (1979). *Alkaloid Biology and Metabolism in Plants*. Plenum Press.

Waterman, P. G., & Mole, S. (1994). *Analysis of phenolic plant metabolites*. Blackwell Scientific Publ.

Webber, A. N., Nie, G. Y., & Long, S. P. (1994). Acclimation of photosynthetic proteins to rising atmospheric CO_2. *Photosynthesis Research*, *39*(3), 413–420. doi:10.1007/BF00014595 PMID:24311133

Wellburn, A. R., Capron, T. M., Chan, H. S., & Horsman, D. C. 1976. Biochemical effects of atmospheric pollutants on plants. In: Mansfield, T.A. (Ed.), Effects of air pollutants on plants. Cambridge University Press.

Wellburn, A. R., Majernik, O., & Wellburn, F. M. N. (1972). Effect of SO_2 and NO_2 polluted air upon the ultrastructure of chloroplasts. *Environ. Pollut*, *3*(1), 37–49. doi:10.1016/0013-9327(72)90016-X

Wohlgemuth, H., Mittelstrass, K., Kschieschan, S., Bender, J., Weigel, H. J., Overmyer, K., Kangasjarvi, J., Sandermann, H., & Langebartels, C. (2002). Activation of an oxidative burst is a general feature of sensitive plants exposed to the air pollutant ozone. *Plant, Cell & Environment*, *25*(6), 717–726. doi:10.1046/j.1365-3040.2002.00859.x

Xu, M., & Dong, J. (2005). Nitrogen oxide stimulates indole alkaloid production in *Catharanthus roseus* cell suspemsion cultures through a protein kinase-dependent signal pathway. *Enzyme and Microbial Technology*, *37*(1), 49–53. doi:10.1016/j.enzmictec.2005.01.036

Xu, X., Nie, S., Ding, H., & Hou, F. F. (2018). Environmental pollution and kidney diseases. *Nature Reviews. Nephrology*, *14*(5), 313–324. doi:10.1038/nrneph.2018.11 PMID:29479079

Yoneyama, T., Kim, H. Y., Morikawa, H., & Srivastava, H. S. (2002). Metabolism and detoxification of nitrogen dioxide and ammonia in plants. In K. Omasa, H. Saji, S. Youssefian, & N. Kondo (Eds.), *Air pollution and plant biotechnology* (pp. 221–234). Springer-Verlag. doi:10.1007/978-4-431-68388-9_11

Yoneyma, T., Totsuka, T., Hayakaya, N., & Yazaki, J. (1980). Absorption of atmospheric NO_2 by plants and soil. V. Day and night NO_2 fumigation effect on the plant growth and the estimation of the amount of NO_2^- nitrogen absorbed by plants. *Res. Rep. Natl. Inst. Environ. Studies*, *11*, 31–50.

Yunus, M., Ahmad, K. J., & Gale, R. (1979). Air pollutants and epidermal traits in *Ricinus communis* L. *Environ. Pollut*, *20*(3), 189–198. doi:10.1016/0013-9327(79)90004-1

Yunus, M., Kulshreshtha, K., Dwivedi, A. K., & Ahmad, K. J. (1982). Leaf surface traits of *Ipomea fistula* Mart. Ex. Choisy and indicators of air pollution. *New Botanist*, *9*, 39–45.

Zeevart, A. J. (1976). Some effects of fumigating plants for short periods with NO_2. *Environ Pollut*, *11*(2), 97–108. doi:10.1016/0013-9327(76)90022-7

Zobel, A. M. (1996). Phenolic compounds in defense against air pollution. In M. Yunus & M. Iqbal (Eds.), *Plant Response to Air Pollution* (pp. 241–266). John Wiley.

Chapter 19
Harnessing Nature: Whole Cell Biosensors for Environmental Monitoring

Naseema Banu A.
Ethiraj College for Women, India

Sangeetha Vani G.
https://orcid.org/0000-0003-4462-0730
Ethiraj College for Women, India

Sarah Grace P.
Ethiraj College for Women, India

ABSTRACT

This chapter discusses the role of whole-cell biosensors in monitoring the impact of human advancements on the environment, leading to an imbalance that threatens ecosystems. Biosensors are cost-effective devices known for their specificity, sensitivity, and portability. The advancement in biosensors includes using genetically engineered microbial cells as whole-cell biosensors. These manipulated cells respond to external stresses, making them effective tools for detecting pollutants. The stress-response mechanisms of bacterial species are harnessed for environmental monitoring. The customizable nature of whole-cell biosensors is displayed in the text, and it also discusses applications such as water contamination detection and the design of engineered bacterial cells. The chapter aims to provide a comprehensive understanding of whole-cell biosensors, their principles, and their applications in addressing environmental issues in air, water, and soil pollution.

INTRODUCTION

The advent of human development, socially and economically, has taken a major toll on the surrounding environment and its health, affecting not only plants and animals but humans too. With mass production strategies to meet the needs of a growing population, comes the ill effects of waste accumulation and

DOI: 10.4018/979-8-3693-1930-7.ch019

toxicity. This needs immediate and effective measures to detect, sense, and mitigate the harm. There are numerous ways to monitor the pollution by accumulating contaminants. Some examples include pesticides, heavy metals, and human wastes that cause soil pollution, and unchecked contamination leads to water pollution (Prabhakaran et al., 2017).

Sensors are used to monitor the levels of analytes. A sensor is a device that has wide applications and of many types. One of them is a biosensor, which is an analytical device meant to detect analytes of biological importance. They have a recognition component and a physical layer that conducts the signal and converts it into a measurable value. Novel microbial biosensors for the detection of environmental samples, food processing, and biomedical purposes have been developed, based on a wide range of principles (Su et al., 2011).

The sensing element for these biosensors comprises enzymes, microorganisms, antibodies, cells, tissues, and organelles. Enzymes possess the properties of sensitivity and specificity, making them an ideal choice for the recognition element. However, the synthesis and purification process is time-consuming and expensive. The microbial cell as such is a better alternative to specific enzymes, given their ease of culture and manipulative abilities. These cells also provide the ambient microenvironment for the enzymes or other recognition elements to act without losing their ability.

Microbes have a non-specific metabolic activity, except for a few species. However, these cells can be easily manipulated using selective culture techniques or by blocking the undesired trait or enhancing the desired expression via genetic engineering (Su et al., 2011).

In this chapter, we delve into the fascinating domain of whole-cell biosensors, exploring how live microorganisms can be harnessed as powerful tools for environmental monitoring. This innovative approach not only reflects nature's complexity but also offers a dynamic and responsive solution to the ever-growing challenges of environmental surveillance.

Understanding Whole-Cell Biosensors

Whole-cell biosensors differ significantly from traditional sensing methods in their approach to environmental monitoring. Unlike conventional sensors that rely on chemical reactions or physical changes to detect pollutants, whole-cell biosensors utilize living cells, typically bacteria, as the sensing element. These genetically engineered microbial cells possess inherent stress-response mechanisms, which are leveraged to detect changes in the environment. The genetic manipulation allows these cells to exhibit distinct responses, such as the overproduction of specific enzymes or proteins, in the presence of pollutants. This biological approach offers advantages in terms of specificity, sensitivity, and adaptability. Unlike chemical sensors, whole-cell biosensors can be customized for different pollutants, making them versatile tools for environmental monitoring. Moreover, these biosensors provide real-time information about the biological impact of pollutants, offering a more holistic understanding of environmental conditions. Overall, the use of living cells in whole-cell biosensors represents a paradigm shift from traditional methods, providing a more dynamic and responsive approach to detecting and monitoring environmental changes.

Why Use Microorganisms?

Microorganisms, particularly bacteria, are commonly used for whole-cell biosensors due to their unique characteristics and versatility. There are several reasons why microorganisms are preferred for this purpose:

- **Genetic Manipulation:** Microorganisms can be easily genetically engineered to express specific genes or respond to stimuli. This allows for the creation of customized biosensors with tailored responses to environmental changes.
- **Inherent Stress Responses:** Many microorganisms naturally exhibit stress responses when exposed to various environmental conditions, such as the presence of pollutants. These stress responses can be harnessed and engineered to serve as indicators in biosensors.
- **Sensitivity**: Microorganisms often show high sensitivity to changes in their surroundings. This sensitivity allows for the detection of pollutants even in low concentrations, making microorganism-based biosensors effective in monitoring environmental conditions.
- **Rapid Response:** Microorganisms can provide rapid and real-time responses to environmental changes. This quick response is crucial for timely detection and monitoring of pollutants.
- **Versatility:** Microorganisms offer a wide range of species with diverse characteristics. This diversity allows for the selection of microorganisms that are well-suited for specific applications or pollutants, enhancing the versatility of whole-cell biosensors.
- **Cost-Effectiveness:** Using microorganisms is often more cost-effective than synthetic alternatives. Microorganisms can be easily cultivated and maintained, contributing to the affordability of whole-cell biosensors.
- **Biocompatibility:** Microorganisms are generally biocompatible and pose minimal risk to the environment. This makes them suitable for applications in ecological sciences without causing harm to the surroundings.
- **Living Systems:** Unlike traditional sensors that rely on chemical reactions, microorganisms are living systems. This living nature allows for continuous monitoring and adaptation to changing environmental conditions.

BIOSENSOR DESIGN

A biosensor typically consists of three main components: a biorecognition element, a transducer, and a signal processing system. Each component plays a crucial role in the biosensor's functionality.

Biorecognition Element

The biorecognition element is responsible for selectively interacting with the target analyte, initiating a biochemical response. This element can be an enzyme, antibody, DNA, whole cells, or other molecules capable of recognizing and binding to the target (Prabhakaran et al., 2017). It facilitates the specific recognition and binding of the target analyte, leading to a measurable change.

Transducer

The transducer is a component that converts the biochemical response generated by the biorecognition element into a quantifiable and often electrical signal. Various types of transducers include electrochemical (e.g., amperometric, potentiometric), optical (e.g., fluorescence, absorbance), and piezoelectric devices. It transforms the biological signal into a readable and interpretable output, providing a measurable response proportional to the concentration of the target analyte.

Signal Processing System

The signal processing system is responsible for amplifying, conditioning, and interpreting the signal generated by the transducer. This may involve electronic circuits, amplifiers, and microprocessors to enhance the signal quality and extract relevant information. It ensures accurate and reliable detection by processing the raw signal into a meaningful output, such as concentration or presence/absence of the target analyte.

In addition to these main components, biosensors may also include other elements, depending on the specific design and application:

Interface and Housing

The interface connects the biosensor to the sample or measurement environment, while the housing protects the sensitive components from external factors. It ensures proper interaction with the sample and provides a protective environment for the biorecognition element and transducer.

Calibration and Control Systems

Calibration systems and controls are mechanisms to standardize and verify the biosensor's performance over time. They enable the biosensor to provide accurate and reliable measurements by compensating for changes in environmental conditions and sensor properties.

CHEMICAL METHODS

Microbe immobilization chemically involves covalent binding and cross-linking. Covalent binding forms stable bonds between microbial cell wall components and the transducer, but harsh chemical reactions may damage cell membranes, reducing biological viability. Overcoming this challenge is a practical concern. In contrast, cross-linking uses multifunctional reagents like glutaraldehyde to bridge functional groups on the cell membrane, forming a network. This technique is widely accepted due to its speed and simplicity. Cells may be cross-linked directly onto the transducer or on a removable support membrane. While cross-linking offers advantages over covalent bonding, it may affect cell viability. Therefore, cross-linking is suitable for microbial biosensors where cell viability is less critical, and only intracellular enzymes are involved in detection.

GENETICALLY ENGINEERED READOUTS OF MICROBIAL SENSORS

In the field of whole-cell biosensing, various approaches have been employed to quantify changes in cellular metabolism, pH, and gene expression as responses to the presence of target molecules. One strategy involves utilizing microbial auxotrophy to monitor growth-limiting small molecules. For instance, an autotrophic *Escherichia coli* strain was constructed to detect and quantify mevalonate, an intermediate in the biosynthesis of industrially important isoprenoids. The biosensor employed the deletion of the native pathway for isopentenyl pyrophosphate (IPP) and dimethylallyl pyrophosphate (DMAPP) production, replacing it with a mevalonate-utilizing pathway.

Additionally, by-products generated from target compounds can serve as biosensor readouts. Examples include the conversion of α-naphthyl acetate to α-naphthol and acetate, resulting in a pH change, and the hydrolysis of paraoxon, a pesticide, to p-nitrophenol. A whole-cell electrochemical biosensor for detecting the organochlorine pesticide lindane was developed by expressing the enzyme γ-hexachlorocyclohexane (HCH) dehydrochlorinase in *E. coli*, monitoring conductivity changes using pulsed amperometry.

Reporter gene expression, controlled by specific regulatory networks, is another powerful readout method. Commonly used reporter genes include *β-galactosidase* (*β-gal*) and *luciferase*, offering colourimetric, fluorescent, or luminescent readouts. *β-gal* provides simple and rapid detection, with ultrahigh sensitivity and a wide dynamic range. Bacterial and firefly luciferases offer sensitivity, a broad dynamic range, and simplicity. Fluorescent proteins, such as green fluorescent protein (GFP), provide autofluorescence without the need for substrates, enabling the measurement of gene expression and cell trafficking mechanisms. Genetically modified fluorescent proteins of different colours allow simultaneous detection of multiple targets and fluorescence resonance energy transfer (FRET) for conformational changes triggered by ligand binding. Overall, these biosensing approaches offer versatility for detecting a wide range of target compounds (Park et al., 2013).

WHOLE-CELL SENSOR BASED ON INTRACELLULAR SYSTEM

Regulator-Promoter System

Intracellular sensing mechanisms often involve coupling a transcriptional regulator with an inducible promoter to respond to varying nutrient conditions, external toxicants, or communication signals. This interaction modulates the expression of a reporter gene, leading to a measurable signal change in a concentration-dependent manner. Commonly, regulator protein/promoter pairs for detecting environmental contaminants are based on natural resistance mechanisms or the metabolisms of toxic compounds. For instance, a *Bacillus subtilis* whole-cell biosensor was created using the CadC regulatory protein and *cadC* promoter for cadmium detection, responding to cadmium, lead, and antimony with nanomolar sensitivity. Another biosensor used the ZntR regulatory protein and its corresponding *zatAp* promoter from *E. coli* to monitor zinc, lead, and cadmium.

However, these biosensors struggled to detect arsenic, a highly toxic element. To address this, the arsenic detoxification *ars* operon, controlled by the regulatory protein ArsR, was employed. ArsR binding to the *arsR* promoter repressed reporter protein expression in the absence of arsenite. Upon arsenite introduction, the ArsR-promoter complex dissociated, allowing reporter protein expression. Previous designs placing reporter genes downstream of the *arsR* gene resulted in high background expression. To reduce this, a transcriptional insulator was introduced downstream of arsR and upstream of the reporter gene, blocking RNA polymerase read-through and significantly reducing background expression, resulting in a much lower detection limit.

The *ars* operon, commonly utilized for arsenic detoxification, may not be optimal for achieving the highest sensitivity and response as a designed cellular reporter for arsenic detection. This was addressed by decoupling the natural regulatory configuration, placing *arsR* expression under the control of either a *T7* or *lac* promoter while maintaining GFP expression under the *arsR* promoter. Similarly, the effect of promoter strength on ArsR expression revealed that a stronger constitutive ArsR production decreased arsenite-dependent EGFP output from the ars promoter, suggesting that uncoupled circuits may enhance expression levels and sensitivities for improved field-test assays.

For organic contaminants, various regulatory protein/promoter pairs have been employed. The XylR and Pu promoter pair from the xylene degradation pathway in Pseudomonas detects xylene, benzene, and toluene. The regulatory protein TbuT and the *tbuA1p* promoter from the toluene degradation pathway in *Ralstonia pickettii* control luciferase expression in response to volatile compounds. DmpR and the Po promoter in *Pseudomonas putida* monitor phenols, with a mutant DmpR increasing sensitivity by over 4-fold. An *E. coli* biosensor for L-arabinose detection was developed using the AraC regulatory protein and PBAD promoter pair, with a mutant AraC variant engineered for specificity toward D-arabinose.

Efforts to fine-tune specificity include engineering regulatory proteins for the recognition of targets with different chemical structures. A mevalonate responsive AraC variant was identified and used for monitoring mevalonate in the isoprenoid biosynthesis pathway. This variant selectively responded to mevalonate and was coupled with a PBAD-LacZ fusion to screen mutants of hydroxymethylglutaryl-CoA reductase (HMGR), demonstrating the capability to engineer regulatory proteins for sensing a wide range of novel targets.

Riboswitch and Reporter Gene Expression

Ribosomal switches are structured RNA domains that detect molecules and regulate gene expression. The interest in RNA-based detection and regulation has grown in recent years due to its ease of design and engineering. Riboswitches with natural and synthetic RNA aptamers have been developed to sense temperature, metal ions, nucleic acids, small molecules, and proteins. A whole-cell sensor based on an engineered riboswitch was developed to detect theophylline, a commonly used antiasthmatic drug. Theophylline concentration in blood serum was monitored using thymidylate synthase and an anti-theophylline aptamer. Artificial riboswitch has been used for in vivo monitoring of intracellular metabolites and engineering metabolic pathways. Whole-cell biosensors with engineered riboswitch have the potential to monitor novel targets, such as drugs and metabolites in clinical and environmental samples (Park et al., 2013).

An artificial riboswitch has been used for in vivo monitoring of intracellular metabolites and engineering metabolic pathways. The natural thiamine pyrophosphate (TPP) riboswitch found in the 5′UTR of the *E. coli thiM* gene was used to generate a TPP-activated riboswitch. The engineered TPP-activate riboswitch was fused to β-galactosidase and GFP, and the expression of these reporters was shown to induce in the presence of thiamine. A TPP riboswitch library was constructed using the TPP aptamer of the *B. anthracis tenA* gene, identifying several new TPP-activated riboswitches with enhanced sensitivity. Whole-cell biosensors with engineered riboswitch have the potential for monitoring novel targets, such as drugs and metabolites in clinical and environmental samples (Park et al., 2013).

Quorum Sensing

Quorum sensing (QS) is a widely used method for communication in microorganisms, utilizing diffusible small molecules called autoinducers. These autoinducers are produced, secreted, and recognized by specific bacteria, and can regulate various features such as virulence, biofilm formation, sporulation, genetic competence, and bioluminescence. Pathogens like *Pseudomonas aeruginosa* and *Burkholderia cepacia* use these signal molecules to regulate virulence determinants, which can cause lung diseases in cystic fibrosis patients.

A whole-cell biosensor was developed to detect AHLs in physiological samples, such as saliva, at concentrations down to 1×10^{-8} M. AHL-mediated quorum sensing has been incorporated into this format for enhanced arsenic sensing. This method involves strong intercellular coupling over tens of micrometres, but the slow diffusion time of molecular communication leads to signal delays over the millimetre length scale. Prindle et al., (2012), constructed a gene circuitry to produce an oscillating amount of GFP in an *E. coli* reporter strain maintained inside microfluidic cavities (biopixels). The biopixel system output can avoid errors from detector/light source fluctuations, as the oscillation period is independent of the absolute fluorescence intensity (Park et al., 2013).

OTHER METHODS

Microbial sensors can be developed through various mechanisms, including conformational changes due to protein-protein interactions. For example, a high-signal-to-noise single-wavelength biosensor for maltose was created by fusing fluorescent proteins into a bacterial periplasmic binding protein (PBP). *E. coli* mutants with a genetically engineered glucose/galactose-binding protein (GBP) have also been used as biosensors for glucose.

Single-molecule detection is possible through biological nanopores as sensing platforms, which open in response to the binding of individual ligands. This method is routinely achieved through ligand-gated ion channel proteins, which require strong molecular amplification. The yeast two-hybrid assay using recombinant DNA technology can be used as a biosensor to identify interaction partners. This system detects interactions between estrogen receptors and their coactivators, showing that their specificity is dependent on the presence of estrogen. An efficient and reliable yeast two-hybrid detection system can also be constructed to evaluate the estrogenic activity of potential endocrine disruptors (Park et al., 2013).

Chimeric proteins are another elegant way for estrogen detection, typically constructed by fusing a target ligand-binding domain to an easily assayed reporter protein. This allows ligand-induced conformational changes in the target ligand-binding domain to be transmitted to the reporter and allosterically modulate its properties. Skretas et al., (2007) combined the ligand-binding domains of estrogen receptors with a highly sensitive thymidylate synthase reporter for identifying diverse estrogenic compounds.

WHOLE-CELL SENSOR BASED ON EXTRACELLULAR SYSTEM

Cell Surface Display Assembly

Cell surface display of peptides, proteins, and epitopes on living cells offers advantages for targeting molecules or substrates that are inaccessible to the intracellular environment. This method is particularly beneficial for enhancing kinetics, stabilization of enzymes and proteins, and facilitating purification compared to free proteins. Various cell surface components, such as outer membrane proteins, lipoproteins, S-layer proteins, cell-surface appendages, and autotransporters, have been utilized for display, with the ice nucleation protein (INP) being widely used for bacterial cell surface display.

Enzyme-based biosensors face limitations in cost and the labour-intensive purification process. Whole-cell biosensors expressing desired enzymes overcome these challenges, with *E. coli* cells expressing organophosphorus hydrolase (OPH) demonstrating enhanced stability and robustness compared to purified OPH. Surface display of OPH on engineered *E. coli* strains resulted in improved substrate degradation kinetics, highlighting the reduction of mass-transport limitations across the cell membrane. Similarly, xylose dehydrogenase (XDH) displayed on the cell surface enabled the detection of D-xylose, demonstrating a broad linear range and selective detection in real samples.

Autotransporter proteins, such as *E. coli* AIDA-I, have been used to display OPH and green fluorescent protein (GFP) on the cell surface for monitoring organophosphate compounds. Beyond enzymes, human antigens have been successfully displayed on yeast surfaces for detecting monoclonal antibodies. This approach proves valuable as yeast facilitates the production of soluble and functional mammalian proteins with appropriate post-translational modifications. Yeast surface display has been used for detecting monoclonal antibodies through immunofluorescence and enzyme-linked immunosorbent assay (ELISA), with potential applications in large-scale screening of positive antibody-producing hybridoma cell lines.

Overall, cell surface display strategies offer versatile and efficient tools for the detection of various molecules, ranging from environmental contaminants to biological markers, with potential applications in biosensing and biotechnology.

G-Protein Coupled Receptors for Detection

G-protein coupled receptors (GPCRs) have emerged as a noteworthy extracellular component for recognition in biosensing applications. GPCRs constitute the largest family of integral membrane receptors, playing a crucial role in diverse intracellular communications in response to external stimuli. The key advantage of exploring GPCRs lies in their extensive natural binding repertoire, encompassing small molecules, peptides, and glycoproteins. Due to their ability to respond to a broad array of stimulants, including hormones, neurotransmitters, taste, and chemicals, GPCRs are highly modular and customizable for sensing a wide range of targets.

Upon ligand binding to a GPCR, intracellular signalling is initiated through interaction with the GTP-binding protein (G protein), leading to the transmission of cellular responses. This inherent property makes GPCRs advantageous for whole-cell biosensing, as various cellular processes can be leveraged as the sensor readout. The versatility of GPCRs in recognizing diverse ligands and their ability to modulate cellular responses offer a promising avenue for the development of biosensors with broad applicability.

PHYSICAL METHODS

Microbe immobilization techniques primarily involve adsorption and entrapment, which are preferred when viable cells are required due to their minimal impact on the native structure and function of microorganisms. In physical adsorption, a microbial suspension is incubated with an electrode or an immobilization matrix like glass beads, leading to immobilization through adsorptive interactions (ionic or polar bonding) and hydrophobic interactions. However, adsorption alone often results in poor long-term stability due to microbial desorption.

Alternatively, entrapment immobilization can be achieved by retaining cells close to the transducer surface using methods such as dialysis membranes. Despite its effectiveness, entrapment immobilization introduces diffusion resistance from the entrapment material, leading to reduced sensitivity and detection efficiency.

Microbial biosensors typically function by the assimilation of organic compounds by microorganisms, leading to changes in respiration activity (metabolism) or the production of specific electrochemically active metabolites such as H2, CO2, or NH3, which are then secreted by the microorganism (Mulindi, 2023).

APPLICATIONS OF WHOLE-CELL BIOSENSORS

Use in Nanotechnology

Miniaturization in biosensors has numerous benefits, including improved signal-to-noise ratios, smaller sample volumes, lower assay costs, and increased binding efficiency towards the target molecule. This allows the bioreceptor to become an active transducer for the sensing system, enabling single-molecule detection. The double-layer capacitance decreases dramatically towards nanoscale dimensions due to its dependence on the electrode area, allowing ultra-fast electron-transfer kinetics and investigation of short-life intermediate species. The extremely low RsCdl time constant also reduces the time required for measurements in the nanosecond domain. Graphene and its oxidized form, graphene oxide, have opened new frontiers in biosensors and other research areas. Graphene, a pure form of carbon organized into single-atom-thick sheets, has exceptional chemical and physical properties. The integration of graphene, graphene oxide, carbon nanotubes, nanoparticles, and nanowires in electrode fabrication has led to biosensors with lower detection limits, enabling even single-molecule detection (Решетилов et al., 2010).

Carbohydrate Assessment

Carbohydrates are the most common analytes for enzyme and microbial sensors, with their determination being a widely used direction in biosensors due to their high bioavailability and practical significance in biotechnology, the food industry, and medicine (Решетилов et al., 2010). The most commonly reported "carbohydrate" sensors are glucose and lactose analyzers, but a large number of models have been developed for detecting other mono- and disaccharides and polymeric carbohydrates (Karube et al., 1979). Microbial carbohydrate analyzers include glucose and lactose analyzers, mediator sensors, hybrid sensors, and starch detection. Hybrid sensors are particularly suitable for di- and polysaccharide analysis, as one biocatalyst hydrolyzes glycosidic bond hydrolysis while the other determines generated monomers (Svitel et al., 1998). Examples include two hybrid carbohydrate sensors, sucrose determination based on invertase and *Zymomonas mobilis* cells (Park et al., 1991), lactose detection in dairy products based on glucose oxidase and *E. coli* cells, and determination of α-amylase activity using co-immobilized *B. subtilis* and glucoamylase (Svorc et al., 1990).

Alcohol and Organic Acid Detection

Biosensor determination of alcohols and organic acids is often used in conjunction with carbohydrate detection, with microbial cells possessing high selectivity often used alongside enzymes. Molecular yeasts

Figure 1. Basic principle behind the working of a whole-cell biosensor

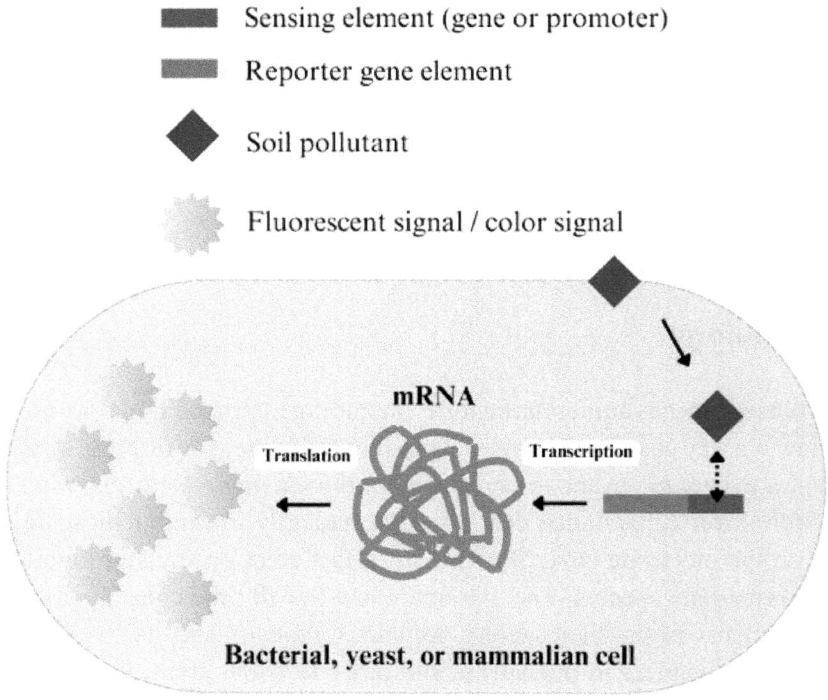

from *Hansenula*, *Pichia*, and *Candida* genera are particularly promising for this purpose (Voronova et al., 2008; Korpan et al., 1993). Two microbial alcohol sensors based on recombinant *H. polymorpha* cells have been described, with the first using highly active alcohol oxidase immobilized on an oxygen electrode and the second using catalase-deficient cells and a peroxide electrode as a transducer (Gonchar et al., 1998). The original development involved creating an alcohol sensor based on *Agaricus bisporus* fungus tissue homogenate, which was fixed on a Clark-type electrode (Akyilmaz & Dinckaya, 2000). Microbial ethanol biosensors have been developed using ISFET and *Acetobacter aceti* cells (Kitagawa et al., 1987), and conductometric microbial sensors for ethanol based on alginate-immobilized yeast cells (Korpan et al., 1994). PQQ-dependent dehydrogenases of *Gluconobacter* represent a promising base for biosensor development, with a microbial sensor for ethanol detection based on *G. oxydans* cells and a glassy carbon electrode modified by ferricyanide (Tkac et al., 2003).

Xenobiotic Detection

The majority of xenobiotics have low MPC values, which indicates that biosensors for pollution measurement need to have high detection sensitivity. Biocatalytic sensors identify substances with low BOD and MPC values, while bioluminescent sensors provide highly selective detection at 10^-9-10^-7 M. Because these biosensors can identify harmful substances and pollutants with comparatively low MPC values, they are appropriate for use in environmental monitoring and other nature-conservation initiatives (Решетилов et al., 2010).

BOD Assessment

Biochemical Oxygen Demand (BOD) is a critical parameter used to assess the organic pollution level in the water. Traditional methods for BOD detection are time-consuming and require several days. Whole-cell biosensors offer a promising alternative, providing faster and more real-time monitoring of BOD levels. A whole-cell biosensor using genetically engineered *E. coli* responds to changes in dissolved oxygen levels, providing a quick and sensitive measurement of BOD in water samples (Решетилов et al., 2010). Several laboratory models and commercial BOD biosensor analyzers are known, providing BOD detection in the mean range of 5-300 mg/l for a few minutes. Pure bacterial and fungal cultures with broad substrate specificity are used in BOD sensors, with *Trichosporon cutaneum* being the most frequently used (Yang et al., 1996).

Surfactant Detection

Surface-active substances (SAS) are used in various fields, but they can increase the concentration of toxic substances, affecting living organisms. Biosensor systems are developed for detecting SAS, with microorganisms being especially useful due to their sensitivity to oxidation. Studies showed that *Pseudomonas rathonis* cells are highly sensitive to SAS (Решетилов et al., 2010). A microbial biosensor based on a column reactor containing activated sludge bacteria oxidizing linear alkylbenzene sulfonates was used for SAS analysis in river water, with a response time of 15 minutes (Nomura et al., 1994; Nomura et al., 1998).

PESTICIDES DETECTION

Whole-cell biosensors offer a sustainable and efficient solution for pesticide detection, enabling real-time monitoring of environmental samples. Engineered whole-cell biosensors have been developed to detect insecticides, including organophosphates and pyrethroids. These biosensors utilize cellular responses to insecticide-induced stress, enabling the selective detection of these chemicals in complex matrices. Direct detection of organophosphorus neurotoxins has been proposed using *E. coli*. This sensor can detect various organophosphorus pesticides and chemical warfare agents (Rainina et al., 1996). A similar sensor was described in 1998 (Mulchandani et al., 1998). The use of whole-cell biosensors in insecticide detection aids in monitoring agricultural runoff and potential impacts on aquatic ecosystems. Whole-cell biosensors employed for the detection of herbicides, such as atrazine and glyphosate, utilize genetically modified bacteria or yeast cells that respond specifically to the presence of herbicides, providing a rapid and sensitive detection method (Решетилов et al., 2010).

Sensing Hydrocarbons and Their Derivatives

The detection of aromatic compounds using microbial biosensors, emphasizing the varying toxicity levels of monoaromatics, naphthalene, and polynuclear aromatic hydrocarbons (PAH) and their chlorine derivatives can be achieved using biosensors. While monoaromatics and naphthalene are deemed relatively low toxic, the super-toxic nature of PAH demands highly sensitive analyzers, including bioluminescent sensors, chromatography, mass spectrometry, and immunoassay. Microbial biosensors have been developed for practical detection, such as an amperometric biosensor utilizing *Rhodococcus* cells for monoaromatics and an oxygen electrode with *P. putida* cells for benzene in industrial wastewater and groundwater analysis. A microbial biosensor for naphthalene detection, employing *Sphingomonas* sp. and *Pseudomonas fluorescens* strains, demonstrated a lower detection limit of 0.01 mg/l and 20 days of activity. Specific biosensors were also designed for 2,4-dinitrophenol and p-toluene sulfonate detection using *Rhodococcus erythropolis* HL PM-1 and immobilized *Comamonas testosteroni* BS1310 cells, respectively. The variety of microbial strains from genera *Pseudomonas*, *Sphinomonas*, *Rhodococcus*, and *Ralstonia* showcased the versatility of microbial biosensor models in environmental monitoring applications (Решетилов et al., 2010).

DETECTION OF METAL IONS AND INORGANIC ACIDS

Microbial biosensors have become a popular approach in biosensor analysis, primarily based on genetically modified microorganisms and optical detection. These biosensors are used for detecting heavy metals and inorganic acid ions, such as arsenic compounds. For example, a bioluminescent sensor based on recombinant *E. coli* strains was used for measuring arsenic toxicity (Cai & DuBow, 1997). Optical bacterial biosensors for zinc and copper detection in soil samples were developed using a consortium containing reporter bioluminescent *Rhizobium leguminosarum* biovar *trifolii* and *E. coli* strains (Chaudri et al., 2000). The lower detection limit for mercury detection was around 20 ng/g of soil (Rasmussen et al., 2000). Microbial sensors for cyanide detection were based on *Saccharomyces cerevisiae* cells and an oxygen electrode. These sensors were designed to detect decreases in glucose response to glucose in the presence of cyanide (Nakanishi et al., 1996). For nitrite detection, models were developed using the bacteria *Paracoccus denitrificans* and *Nitrobacter*. These sensors were designed for gaseous NO_2 analysis and were used as potential nitrite and nitrate analyzers (Takayama et al., 1996). Microbial biosensors for sulfate detection were developed using *Thiobacillus ferrooxidans* and oxygen electrodes, with potential applications in rainwater (Sasaki et al., 1997). The CyanoSensor, a bioluminescent sensor based on recombinant cyanobacteria *Synechococcus*, was used for detecting bioavailable phosphorus in water reservoirs (Schreiter et al., 2001). This was used as an indicator of the occurrence of a potential 'algal bloom' in the reservoir (Решетилов et al., 2010).

ASSESSMENT OF TOXICITY AND GENOTOXICITY

Sensors for genotoxicity and general toxicity are commonly used in biosensor assessment. The most common approach involves using SOS-promoters, which are induced at massive DNA damages, to create biosensor strains for nonspecific registration of genotoxic compounds and factors. General toxicity sensors assess nonspecific toxic impacts of different natures, with the principle of their action being the registration of attenuation of vital functions. Examples include a bioluminescent sensor designed for glucose and toxic compounds, a bioluminescent sensor based on recombinant strain *R. leguminosarum* biovar *trifolii* for general toxicity of the medium (Paton et al., 1997), and a bioluminescent sensor based on *E. coli* cells bearing the *lux*-genes of *V. fischeri* under the promoter of heat stress proteins (Rupani et al., 1996). Algae with chlorophyll fluorescence changing in response to the action of a toxic agent can also be used for general toxicity detection. Examples include an optical biosensor based on immobilized cells of *Scenedesmus subspicatus* alga, an optical biosensor based on *Chlorella vulgaris*, and electrochemical microbial toxicity sensors (Frense et al., 1998). In recent years, several commercial biosensor toxicity analyzers have been constructed (Решетилов et al., 2010).

CONCLUSION

This chapter emphasizes recent advancements in the field of whole-cell biosensors, focusing on approaches rooted in protein/cellular engineering and synthetic biology. Beyond enhancing sensor readouts for improved signals, ongoing efforts involve manipulating and creating recognition elements such as RNA, enzymes, and non-enzymatic proteins with heightened affinity and selectivity. Techniques currently employed to enhance product synthesis can readily be adapted for the design of whole-cell biosensors to optimize sensitivity, selectivity, and robustness.

To effectively utilize these whole-cell biosensors in rapid, point-of-care applications, there is a crucial need to develop improved materials that interface seamlessly with biological elements. This emphasizes the importance of advancing materials to enhance the integration and performance of these biosensors in practical applications.

REFERENCES

Akyilmaz, E., & Dinckaya, E. (2000). A mushroom (Agaricus bisporus) tissue homogenate based alcohol oxidase electrode for alcohol determination in serum. *Talanta*, *53*(3), 505–509. doi:10.1016/S0039-9140(00)00517-8 PMID:18968136

Cai, J., & DuBow, M. S. (1997). Use of a luminescent bacterial biosensor for biomonitoring and characterization of arsenic toxicity of chromated copper arsenate (CCA). *Biodegradation*, *8*(2), 105–111. doi:10.1023/A:1008281028594 PMID:9342883

Chaudri, A. M., Lawlor, K., Preston, S., Paton, G. I., Killham, K., & McGrath, S. P. (2000). Response of a Rhizobium-based luminescence biosensor to Zn and Cu in soil solutions from sewage sludge treated soils. *Soil Biology & Biochemistry*, *32*(3), 383–388. doi:10.1016/S0038-0717(99)00166-2

Frense, D., Müller, A., & Beckmann, D. (1998). Detection of environmental pollutants using optical biosensor with immobilized algae cells. *Sensors and Actuators. B, Chemical*, *51*(1-3), 256–260. doi:10.1016/S0925-4005(98)00203-2

Gonchar, M. V., Maidan, M. M., Moroz, O. M., Woodward, J. R., & Sibirny, A. A. (1998). Microbial O2-and H2O2-electrode sensors for alcohol assays based on the use of permeabilized mutant yeast cells as the sensitive bioelements. *Biosensors & Bioelectronics*, *13*(9), 945–952. doi:10.1016/S0956-5663(98)00034-7 PMID:9839383

Karube, I., Mitsuda, S., & Suzuki, S. (1979). Glucose sensor using immobilized whole cells of *Pseudomonas fluorescens. European journal of applied microbiology and biotechnology, 7*, 343-350.

Kitagawa, Y., Tamiya, E., & Karube, I. (1987). Microbial-FET alcohol sensor. *Analytical Letters*, *20*(1), 81–96. doi:10.1080/00032718708082238

Korpan, Y. I., Dzyadevich, S. V., Zharova, V. P., & El'skaya, A. V. (1994). Conductometric biosensor for ethanol detection based on whole yeast cells. *Ukrainskii Biokhimicheskii Zhurnal (1978)*, *66*(1), 78-82.

Korpan, Y. I., Gonchar, M. V., Starodub, N. F., Shulga, A. A., Sibirny, A. A., & Elskaya, A. V. (1993). A cell biosensor specific for formaldehyde based on pH-sensitive transistors coupled to methylotrophic yeast cells with genetically adjusted metabolism. *Analytical Biochemistry*, *215*(2), 216–222. doi:10.1006/abio.1993.1578 PMID:8122781

Mulchandani, A., Mulchandani, P., Kaneva, I., & Chen, W. (1998). Biosensor for direct determination of organophosphate nerve agents using recombinant *Escherichia coli* with surface-expressed organophosphorus hydrolase. 1. Potentiometric microbial electrode. *Analytical Chemistry*, *70*(19), 4140–4145. doi:10.1021/ac9805201 PMID:9784751

Mulindi, J. (2023, June 18). *The principles of microbial biosensors*. Biomedical Instrumentation Systems. https://www.biomedicalinstrumentationsystems.com/the-principles-of-microbial-biosensors/

Nakanishi, K., Ikebukuro, K., & Karube, I. (1996). Determination of cyanide using a microbial sensor. *Applied Biochemistry and Biotechnology*, *60*(2), 97–106. doi:10.1007/BF02788064 PMID:8856940

Nomura, Y., Ikebukuro, K., Yokoyama, K., Takeuchi, T., Arikawa, Y., Ohno, S., & Karube, I. (1994). A novel microbial sensor for anionic surfactant determination. *Analytical Letters*, *27*(15), 3095–3108. doi:10.1080/00032719408000313

Nomura, Y., Ikebukuro, K., Yokoyama, K., Takeuchi, T., Arikawa, Y., Ohno, S., & Karube, I. (1998). Application of a linear alkylbenzene sulfonate biosensor to river water monitoring. *Biosensors & Bioelectronics*, *13*(9), 1047–1053. doi:10.1016/S0956-5663(97)00077-8 PMID:9839392

Park, J. K., Ro, H. S., & Kim, H. S. (1991). A new biosensor for specific determination of sucrose using an oxidoreductase of *Zymomonas mobilis* and invertase. *Biotechnology and Bioengineering*, *38*(3), 217–223. doi:10.1002/bit.260380302 PMID:18600754

Park, M., Tsai, S., & Chen, W. (2013). Microbial biosensors: Engineered microorganisms as the sensing machinery. *Sensors (Basel)*, *13*(5), 5777–5795. doi:10.3390/s130505777 PMID:23648649

Paton, G. I., Palmer, G., Burton, M., Rattray, E. A. S., McGrath, S. P., Glover, L. A., & Killham, K. (1997). Development of an acute and chronic ecotoxicity assay using lux-marked *Rhizobium leguminosarum* biovar *trifolii*. *Letters in Applied Microbiology*, *24*(4), 296–300. doi:10.1046/j.1472-765X.1997.00071.x PMID:9134778

Решетилов, A. H., Iliasov, P. V., & Reshetilova, T. A. (2010). The microbial cell-based biosensors. In InTech eBooks. doi:10.5772/7159

Prabhakaran, R., Ramprasath, T., & Selvam, G. S. (2017). A simple whole cell microbial biosensors to monitor soil pollution. Elsevier eBooks. doi:10.1016/B978-0-12-804299-1.00013-8

Prindle, A., Samayoa, P., Razinkov, I., Danino, T., Tsimring, L. S., & Hasty, J. (2012). A sensing array of radically coupled genetic 'biopixels'. *Nature*, *481*(7379), 39–44. doi:10.1038/nature10722 PMID:22178928

Rainina, E. I., Efremenco, E. N., Varfolomeyev, S. D., Simonian, A. L., & Wild, J. R. (1996). The development of a new biosensor based on recombinant E. coli for the direct detection of organophosphorus neurotoxins. *Biosensors & Bioelectronics*, *11*(10), 991–1000. doi:10.1016/0956-5663(96)87658-5 PMID:8784985

Rasmussen, L. D., Sørensen, S. J., Turner, R. R., & Barkay, T. (2000). Application of a mer-lux biosensor for estimating bioavailable mercury in soil. *Soil Biology & Biochemistry*, *32*(5), 639–646. doi:10.1016/S0038-0717(99)00190-X

Rupani, S. P., Gu, M. B., Konstantinov, K. B., Dhurjati, P. S., Van Dyk, T. K., & LaRossa, R. A. (1996). Characterization of the stress response of a bioluminescent biological sensor in batch and continuous cultures. *Biotechnology Progress*, *12*(3), 387–392. doi:10.1021/bp960015u PMID:8652122

Sasaki, S., Yokoyama, K., Tamiya, E., Karube, I., Hayashi, C., Arikawa, Y., & Numata, M. (1997). Sulfate sensor using *Thiobacillus ferrooxidans*. *Analytica Chimica Acta*, *347*(3), 275–280. doi:10.1016/S0003-2670(97)00170-0

Skretas, G., Meligova, A. K., Villalonga-Barber, C., Mitsiou, D. J., Alexis, M. N., Micha-Screttas, M., Steele, B. R., Screttas, C. G., & Wood, D. W. (2007). Engineered chimeric enzymes as tools for drug discovery: Generating reliable bacterial screens for the detection, discovery, and assessment of estrogen receptor modulators. *Journal of the American Chemical Society*, *129*(27), 8443–8457. doi:10.1021/ja067754j PMID:17569534

Su, L., Jia, W., Hou, C., & Lei, Y. (2011). Microbial biosensors: A review. *Biosensors & Bioelectronics*, *26*(5), 1788–1799. doi:10.1016/j.bios.2010.09.005 PMID:20951023

Svitel, J., Curilla, O., & Tkác, J. (1998). Microbial cell-based biosensor for sensing glucose, sucrose or lactose. *Biotechnology and Applied Biochemistry*, *27*(2), 153–158. PMID:9569611

Svorc, J., Miertus, S., & Barlikova, A. (1990). Hybrid biosensor for the determination of lactose. *Analytical Chemistry*, *62*(15), 1628–1631. doi:10.1021/ac00214a018 PMID:2205123

Takayama, K., Kano, K., & Ikeda, T. (1996). Mediated electrocatalytic reduction of nitrate and nitrite based on the denitrifying activity of *Paracoccus denitrificans*. *Chemistry Letters*, *25*(11), 1009–1010. doi:10.1246/cl.1996.1009

Tkac, J., Vostiar, I., Gorton, L., Gemeiner, P., & Sturdik, E. (2003). Improved selectivity of microbial biosensor using membrane coating. Application to the analysis of ethanol during fermentation. *Biosensors & Bioelectronics*, *18*(9), 1125–1134. doi:10.1016/S0956-5663(02)00244-0 PMID:12788555

Voronova, E. A., Iliasov, P. V., & Reshetilov, A. N. (2008). Development, investigation of parameters and estimation of possibility of adaptation of *Pichia angusta* based microbial sensor for ethanol detection. *Analytical Letters*, *41*(3), 377–391. doi:10.1080/00032710701645729

Yang, Z., Suzuki, H., Sasaki, S., & Karube, I. (1996). Disposable sensor for biochemical oxygen demand. *Applied Microbiology and Biotechnology*, *46*(1), 10–14. doi:10.1007/s002530050776 PMID:8987529

Chapter 20
Impact of Urbanization on Environment and Health Role of Different Environmental Sensors

Madhumita Hussain
Sophia Girls' College (Autonomous), Ajmer, India

ABSTRACT

The process of urbanization is characterized by the rapid growth and development of urban areas, and now has become a global concern with far-reaching implications for the environment and public health. This study explores the complex impact of urbanization on both the environment and human health, emphasizing the pivotal role played by various environmental sensors in monitoring and mitigating these effects. This chapter delves into the types and functionalities of environmental sensors employed to monitor urbanization's impact. Air quality sensors, water quality sensors, noise monitors, and solid waste sensors contribute valuable data to assess pollution levels, track environmental changes, and evaluate the overall well-being of urban ecosystems. The integration of real-time data from these sensors facilitates the formulation of effective policies and interventions to curb environmental degradation and enhance public health.

INTRODUCTION

With the rapid global urbanization, the vast expansion of metropolitan cities are prone to significant environmental challenges. As testified by the Ministry of Urban Development, the number of million-plus population cities has increased globally from 35 to 53, indicating a bull trend in urban growth as observed in the 2011 census (Ministry of Urban Development, 2011). Undoubtedly this expansion, while provoking economic progress, also announces several environmental risk factors. Ajmer, the 4th largest city in Rajasthan, accommodates 5,42,321 people (Census of India, 2011) is also facing similar challenges. Environmental risk factors in Ajmer include issues such as air and water pollution, deforestation, and the strain on natural resources. Additionally, increase in population and rapid urbanization lead to increased waste generation, poor waste management which are contributing to climate change and altering local

DOI: 10.4018/979-8-3693-1930-7.ch020

ecosystems (Ayesha, John N., & Wadha Ahmed, 2022). Addressing these environmental risks through environmental sensors is crucial for ensuring sustainable urban development.

OBJECTIVES

- To present an overview of the urbanization process in Ajmer, highlighting population, infrastructure development, and land-use changes.
- To analyze air quality, identifying pollutants, and sources.
- To examine noise pollution levels, discussing sources and health implications.
- To evaluate the state of water supply and sanitation systems in Ajmer.
- To investigate solid waste management practices and their environmental impacts.
- Health Profile Analysis: Study the health profiles of Ajmer's residents by analyzing data on nine specific diseases.

LITERATURE REVIEW

A lot of Research work ranging from the late 1990s to the mid-2010s portrays the close relationship between the built environment and public health. Chaplin (1999), highlighted the critical issue of inadequate sanitation in India, directing towards the utter need for reforms in sanitation to address public health concerns in low-income groups. Butterworth (2000), and Jackson (2003), expanded on this by discussing how the built environment, encompassing human-made surroundings where people spend most of their time, not only reflects cultural and historical contexts but also significantly impacts health by fostering or hindering physical activity, thereby influencing the prevalence of chronic diseases.Studies by Galea et al. (2005), and Madhiwalla (2007), have elaborated on the adverse health impacts stemming from the growth of slums, urbanization, and the resultant overcrowding, poor housing, and unsanitary conditions, emphasizing the role of improved urban planning and infrastructure development in enhancing public health. Further research by Kyle, A., Woodruff, T., & Axelrad, D. (2006), Thompson, S. & Capon, A. (2012), and the Glasgow Centre of Population Health (2013), identified environmental factors such as noise pollution, housing quality, and lack of green spaces as direct influencers of physical and mental health, advocating for policies that promote healthier urban living conditions. According to Kumar et al. (2019), air quality monitoring networks in cities worldwide have enabled local governments to issue health advisories, implement traffic control measures, and enforce industrial emission standards more effectively. Moreover, the study by Zheng et al. (2020), emphasizes the role of IoT-based air quality sensors in creating dense, real-time pollution maps, which are instrumental in understanding urban pollution patterns and hotspot identification. The smart city infrastructures allow for continuous monitoring and rapid response to potential water quality issues. As mentioned by Smith and Liu (2018), modern technologies have been crucial in identifying pollution sources and preventing public health crises. A study by Green et al. (2021), found and mentioned that long-term exposure to high noise levels is linked to increased stress, sleep disturbances, and cardiovascular diseases. Mapping noise pollution is crucial for policymakers can implement zoning laws, traffic control actions, and urban planning strategies to mitigate its impact. According to Patel and Jain (2020), smart waste management systems, underpinned by sensor technology, have significantly improved recycling rates and reduced landfill usage in several urban areas. Patel, S. (2016), and subsequent studies have stressed the global nature of health issues related to the built environment, pointing to a strong connection between urban planning, lifestyle diseases, and the need for multi-sector partnerships to address urban health challenges effectively.

This chapter presents the arguments for the utmost need to reconsider and redesign urban spaces with a focus on health and sustainability. It highlights the importance of integrating health reflections into urban planning to mitigate the adverse effects of the built environment on public health.

RESEARCH GAP

After conducting an extensive review of the literature concerning the built environment, planning, and health impacts, a significant research gap has been identified. The widely held studies have only focused on one dimension of health either on physical or mental or social effects of the built environment. The current trends in urbanization, population growth, and increasing health issues, presents a clear need for a shift in development plans and policies. Thus, it is found crucial to explore the interaction between the natural and built environments from a well-being perspective.

STUDY AREA

India's fourth largest state Rajasthan covers an area of 3,42,239 km2 and has 50 districts which are grouped into 10 divisions.

Located centrally in Rajasthan, around 135 kilometres southwest of Jaipur, the state capital, Ajmer holds a significant historical position. Its coordinates lie between 25°38' and 26°58' North latitude and 73°52' and 75°22' East longitude. Nestled between the Taragarh and Madar Hills, Ajmer boasts a unique topography, surrounded by hills on three sides, which looks alluring, particularly during the monsoon season, the city comes alive with lush green hills, full lakes, and mesmerizing waterfalls. With a rich history of 1400 years, Ajmer has seen transformative changes that have shaped its urban layout.

Today, approximately 56% of the world's population – 4.4 billion inhabitants – live in cities. This trend is expected to continue, with the urban population more than doubling its current size by 2050 (World Bank, Urban Development, 2023). A substantial increase in metropolitan cities has been noticed with a population of one million or more, from 35 to 53 and the overall count of towns and cities has surged from 5161 in 2001 to 7935 (Ministry of Urban Development, GOI, 2011).

Although cities are acknowledged as hubs for advancement and economic prosperity, same time they hand-to-hand struggle with various challenges such as increasing slums, sanitation issues, pollution, traffic congestion, and heavy traffic flow.

LAND USE AND LAND COVER

At the centre of the city neighbourhood, the streets are narrow mere 3 to 5 feet in width, creating an active centre for retail businesses. Here, the ground floors of the buildings are transformed into shops, while the upper floors become homes for the struggling residents. Houses of a modest income community, including skilled craftsmen like shoemakers, rope makers, and blacksmiths are found here. The

Figure 1. Ajmer city- location map

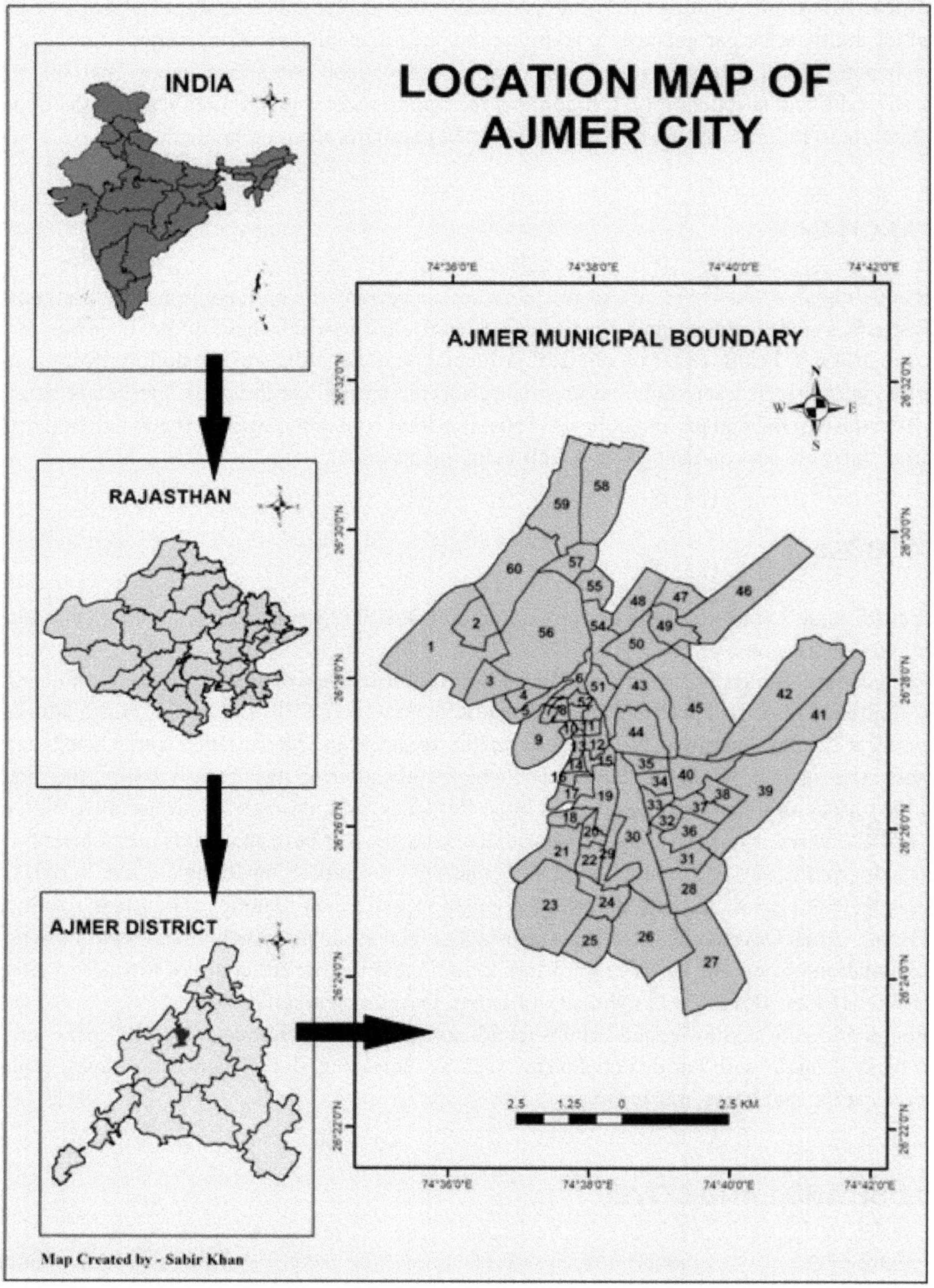

absence of proper drainage compounds the difficulties of everyday life. Despite the wear and tear on these structures, worn down by the passage of time, people stubbornly cling to them. They continue to inhabit these aged buildings, occasionally renovating them for use as guesthouses, especially for those who journey to the Dargah. Field surveys reveal a stark reality – numerous houses stand dilapidated, bearing the scars of their age. Although the government issues stern warnings against vertical expansion, citing dangers to both life and property, some residents persist in defying these warnings. They insist on extending their buildings skyward, ignoring the risks. Those who are aware of the harsh reality of urbanization, have made a genuine decision to seek refuge on the outskirts of the city.

Due to the increasing population of Ajmer, the demand for the facilities is continuously increasing which can only be met with the infrastructural development in the city. Presenting two data sets to see the present and projected Land Use and Land Cover of Ajmer city under various entities which clearly shows the dynamic change in infrastructural transformation of the city. Where one can easily analyse the physical transformation that would take place almost in the next decade, now this is the portal where conscience urban planning and implementation would be needed to develop a city situated in a valley, surrounded by Aravallis from all sides.

AIR QUALITY

In the heart of Ajmer, two major industrial units, namely the Railway Carriage and Loco Workshop, emit significant amounts of CO, CO_2, sulphur oxides, nitrogen oxides, and dust, making a substantial contribution to air pollution. Small-scale industries in Parbatpura Industrial Area and H.M.T. Industrial areas are also impacting natural resources. Foundries and rolling mills in the Parbatpura industrial area,

Table 1. Present land use of Ajmer

S.No.	Land Use	Area in Acres	% of Developed Area	% of Urban Region
I	Residential	5922.07	69.09	7.82
Ii	Occupation	282.01	3.29	0.37
Iii	Industrial	1094.38	12.76	1.44
Iv	Governmental	95.92	1.12	0.13
V	Public and Semi-Public Services	1073.55	12.52	1.42
Vi	Entertainment	76.42	0.89	0.10
Vii	Transport Network	30.94	0.36	0.04
Viii	Nursery, Horticulture & Dairy & Poultry	47059.69		62.12
Ix	Govt. Reserved	353.75		0.47
X	Fallow Land	8501.86		11.22
Xi	Forest	9118.10		12.04
Xii	Hilly Area	1768.84		2.34
Xiii	Water Bodies	375.03		0.45
Total Urbanized Area		75752.56		100

Source: Ajmer Development Authority Report 2013

Figure 2. Present land use land cover map of Ajmer city

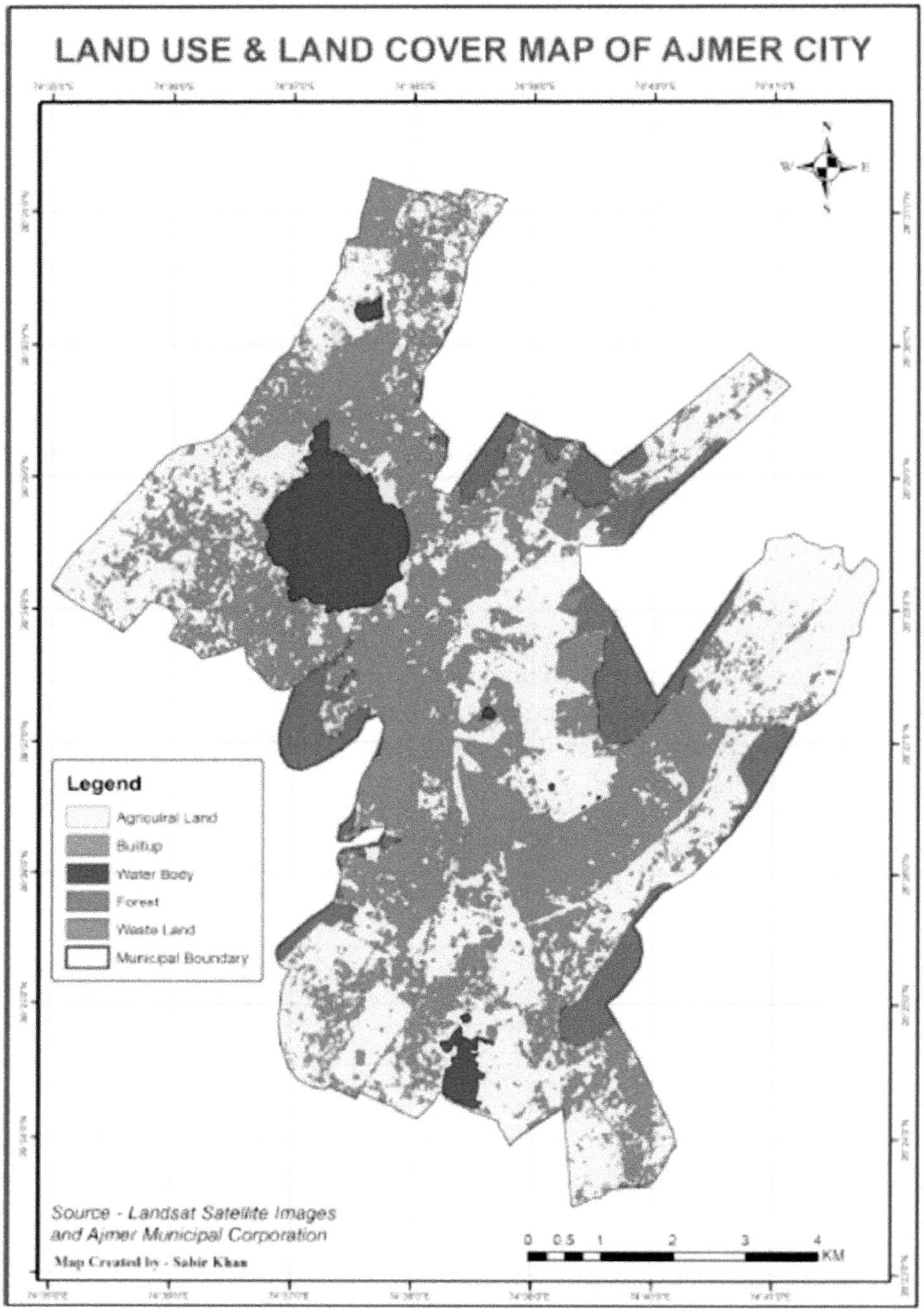

along with foundries in the H.M.T. Industrial area, release high levels of suspended particulate matter, affecting the surrounding regions.

Ajmer's well-established transportation network, connected to other parts of the state and country via roads and railways, is facing increased pressure due to its status as a religious, commercial, and educational hub. The city experiences a significant flow of vehicles, with 31% linked to Jaipur Road,

Table 2. Projected land use 2033

S.No.	Land Use	Area in Acres	% of Developed Area	% of Urban Area
I	Residential	11550	73.10	40.98
Ii	Occupation	990	6.26	3.51
Iii	Industrial	664	4.20	2.36
Iv	Governmental	128	0.82	0.45
V	Public and Semi-Public Services	2240	14.18	7.95
Vi	Entertainment	144	0.91	0.51
Vii	Transport Network	84	0.53	0.30
Viii	Nursery, Horticulture, Dairy and Poultry	436		1.55
Ix	Govt. Reserved	354		1.25
X	Forest and Hilly Area	10887		38.63
Xi	Water Bodies	708		2.51
	Total	28185		100

Source: Ajmer Development Authority Report 2033

Table 3. Residential areas

Sampling Stations	Main Sources of Pollution
Vaishali Nagar	Non- agricultural lands, use of coal as a fuel
Bihari Ganj	Industrial area, Construction work, brick manufacturing industries
Shashtri Nagar	Construction sites, unpaved roads

Sources of Pollution at:

42% to Nasirabad Road, 16% to Beawar Road, and 11% to Pushkar Road. Truck movements, a major source of air pollution, have been redirected through bypass roads. Two-wheelers and three-wheelers, serving as primary public transport, contribute to air pollution. Natural sources like dust from unpaved roads, construction sites, dry lands, and non-vegetated areas also contribute to high levels of suspended particulate matter, leading to an increased incidence of respiratory diseases.

Respirable Dust Sampler, Model No. AMP 451 has been used for picking up the samples from the height of 10 to 12 feet at morning 4am and evening 8 pm. Chemical analysis of SPM have been done by an expert to make an observation.

Table 4. Commercial areas

Sampling Stations	Main Sources of Pollution
Madar Gate	Road Traffic, Traffic Jams, highly crowded area
Khailand Market	Narrow lanes, lack of parking areas, heavy traffic, crowd
Kutchery Road	Highly crowded, High traffic flow

Table 5. Silent zones

Sampling Stations	Main Sources of Pollution
J.L.N. Hospital	Heavy road traffic, Closer to Bajrangarh region, hospital waste
M.D.S. University	Construction work, vehicular pollution
Janana Hospital	Agricultural lands, Road Traffic.

Table 6. Main traffic circles

Sampling Stations	Main Sources of Pollution
Gandhi Circle	Traffic jams, closer to railway station,
Bajrangarh Chaurah	Heavy Traffic flow, Vehicular pollution, closer to JLN Hospital
Nasirabad Road	Heavy traffic caused by trucks

Table 7. Industrial areas

Sampling Stations	Main Sources of Pollution
Parbatpura Area	Rolling Mills, foundries, blasting, truck traffic
Carriage Workshop	Carriage repairing, use of coal

Table 8. Pollutants and level of pollution in residential areas

Site	SPM (μg/m3)			SO_2 (μg/m3)			NOx (μg/m3)		
	Min.	**Max.**	**Ave.**	**Min.**	**Max.**	**Ave.**	**Min.**	**Max.**	**Ave.**
Vaishali Nagar	154.66	177.67	166.16	12.0	16.5	14.25	23.59	32.44	28.00
Bihariganj	162.23	188.75	175.49	10.5	15.5	13	25.19	35.85	30.52
Shashtri Nagar	154.16	168.05	161.10	7.0	9.5	8.25	19.89	32.10	25.95
Average	157.01	178.13	167.58	9.8	13.83	11.83	22.89	33.46	28.15

Table 9. Pollutants and level of pollution at commercial sites

Site	SPM (μg/m3)			SO_2 (μg/m3)			NOx (μg/m3)		
	Min.	**Max.**	**Ave.**	**Min.**	**Max.**	**Ave.**	**Min.**	**Max.**	**Ave.**
Madar Gate	355.29	385.33	370.31	8.00	11.50	9.75	55.70	71.61	63.65
Khailand Market	392.31	378.20	335.25	11.50	16.00	13.75	43.76	58.90	51.33
Kutchery Road	376.13	432.87	404.50	10.90	13.50	12.20	106.90	120.36	113.22
Average	374.57	398.8	370.02	10.13	13.66	27.56	68.78	83.62	76.06

The observations and calculations show that the city is has higher levels of SPM, SO2 and NOx than the National Ambient Air Quality Standards according to the Central Pollution Control Board, the Population of the city is at higher risk of air pollution which is causing various health problems like-

Table 10. Pollutants and level of pollution at main traffic circles

Site	SPM (µg/m3)			SO_2 (µg/m3)			NOx (µg/m3)		
	Min.	Max.	Ave.	Min.	Max.	Ave.	Min.	Max.	Ave.
Gandhi Bhawan	177.67	385.33	281.5	17	21.5	19.25	30.59	32.44	31.51
Bajrangarh Circle	168.05	388.00	278.02	16.2	19.5	17.85	19.89	32.10	25.95
Nasirabad Road	188.75	251.28	220.01	10.5	15.5	13	25.19	35.85	30.52
Average	178.15	341.53	259.84	14.5	18.83	16.7	25.22	33.46	29.32

Table 11. Pollutants and level of pollution at silent zones

Site	SPM (µg/m3)			SO_2 (µg/m3)			NOx (µg/m3)		
	Min.	Max.	Ave.	Min.	Max.	Ave.	Min.	Max.	Ave.
J.L.N. Hospital	289.80	377.95	333.87	7.50	9.00	8.25	42.76	53.78	48.27
M.D.S. University	71.0	78.37	74.69	4.80	7.10	5.95	22.11	26.89	24.50
Janana Hospital	73.91	80.81	77.36	6.20	9.00	7.60	17.69	27.05	22.20
Average	144.9	179.04	161.97	6.16	8.36	7.26	27.52	35.9	31.65

Table 12. Pollutants and level of pollution at industrial sites

Site	SPM (µg/m3)			SO_2 (µg/m3)			NOx (µg/m3)		
	Min.	Max.	Ave.	Min.	Max.	Ave.	Min.	Max.	Ave.
Parbatputa	279.21	267.59	423.40	39.29	53.50	46.38	30.96	40.83	40.39
Carriage Workshop	298.97	382.61	340.79	52.00	57.50	54.75	25.19	38.04	31.61
Average	289.09	325.1	382.09	45.64	55.5	50.56	28.07	39.43	36

Table 13. National ambient air quality standards for SPM according to CPCB

Status	(µg/m3) SPM Industrial Areas	(µg/m3) SPM Residential Areas
Low	0-180	0 - 70
Moderate	180 - 360	70 - 140
High	360 - 540	140 - 210
Critical	>540	>210

- The incidences of nasal infection, cold, asthma, and throat infection aggravate with increasing amounts of SO2 and SPM.
- Air pollutants can also affect the nervous system,
- Reduces learning ability
- Fluctuating blood pressure
- Chronic heart diseases etc.

NOISE VALUE

The terms sound and noise are often considered synonyms but it is not so. It is an underrated environmental problem World Health Organization (Report 2001) mentioned that "Noise must be recognized as a major threat to human wellbeing."

Noise can be produced internally within a building through neighborhood, music appliances, kitchen appliances etc. or externally by automobiles, industrial activities, construction sites etc.

Noise Is Classified As

a) Industrial Noise - It is the high-intensity sound caused by industrial machines, factories, mills, mechanical saws and drills.
b) Transport Noise – There has been an enormous increase in the number of automobiles on roads like trucks, buses, cars, and scooters in the last two decades. Noise pollution is hovering around the borderline in metropolitan cities due to increased vehicular counts.
c) Neighborhood Noise – This is the noise produced by household gadgets like TVs, VCRs, radios, televisions, telephones, loudspeakers etc. which has been increasing ever since the Industrial Revolution.

Health Impacts of Noise Pollution

- Contraction of blood vessels and muscles.
- Secretion of excessive adrenalin in the blood stream which is ultimately responsible for fluctuating blood pressure.
- Blaring sounds causes mental disorders.
- Heat attacks, Neurological problems, birth defects and miscarriages.
- Nervous breakdown.
- Reduces work efficiency.
- It also affects the digestive, respiratory, and cardiovascular systems of the body.
- It reduces the hearing capacity.
- Causes insomnia, impairment in night vision, and chronic damage to the inner ear.
- One can experience adverse behavioural and emotional changes.

Standards regarding noise pollution are laid under the Environmental Protection Act, of 1986 and the Model Rules of Factories Act, of 1948 for occupational health and safety issues.

WATER SUPPLY

Ajmer relies on surface water resources to fulfil the water needs of its residents. In 1884, the city began receiving piped water from the Anasagar reservoir when the population was around 50,000. To meet increased demand, Foysagar was constructed in 1892. However, due to population growth, Anasagar's area diminished, and household drainage was directed into the reservoir. Currently, Anasagar has become an ineffective water reservoir, with ten streams dumping household wastewater into it. Efforts have been

Table 14. Standard noise levels in different zones

Area Code	Category of Area	Limits in db	
		Day Time	Night Time
A	Industrial Area	75	70
B	Commercial Area	65	55
C	Residential Area	55	45
D	Silent Zone	50	40

Source: Environment Protection Act, 1986 amended in 2002.

Table 15. Ajmer city observed variation in noise levels (db) at different locations

Zones	Locations	Time (Hours)							
		8-10 am		10-12 am		12-2 pm		2-4 pm	
		Min.	Max.	Min.	Max	Min.	Max.	Min.	Max.
Commercial	Chudi Bazaar	58.2	75.8	62.3	83.1	61.4	80.2	72.8	96.4
	Madar Gate	57.3	83.5	62	80.3	63.7	84.7	62.9	110.2
Residential	Bank Colony	37.6	95.2	43.2	73.1	37	75.5	42	94.3
	Panchsheel Colony	51.5	84.9	61.3	82.2	64.8	79.2	64	98.3
Heavy Transport regions	Bajrangarh Chawraha	66.7	98.1	67.1	115.8	68.5	94.3	68.4	114.6
	Railway Station	69.3	108.2	69.1	116.2	62.8	103.7	67.5	107.2
	Bus Stand	67.2	96.5	65.7	113.8	63	92.9	63	98.1
Silent Zone	M.D.S. University	33.2	68.1	42.3	74.1	28.6	73.5	41.2	76.1
	Janana Hospital	50.4	86.2	62.7	88.6	48.6	79.9	48	76.6
	JLN hospital	39.9	79.2	52.3	83.1	37.5	78.2	47.2	84.6
	Railway Hospital	35.3	70.7	36.8	82.3	38	78.1	43.5	79

made to clean up this historical reservoir. The city currently receives approximately 68 MLD of water from the Bisalpur Phase project, 0.6 MLD from Foysagar, and 1 MLD from Bhewanta. Water quality is regularly monitored at the filter plant's laboratory, testing all fourteen parameters daily to ensure the city receives quality water.

SEWERAGE ISSUES

Ajmer faces challenges with an inadequate drainage system, sewerage treatment, flooding in low-lying areas during heavy rainfall, and an open drainage system. Drainage conditions, particularly in the old city and low-lying areas like Pal Bisla, Nagra, Bihariganj, and Prakash Road, are subpar. These areas experience waterlogging even with minimal rainfall. Sewerage is primarily discharged through nalas into Anasagar Lake and Pal Bisla tank from surrounding localities. About ten major/minor nalas carry around 13 MLD of wastewater into the lake. Settlements in the catchment, agricultural activities, and solid waste disposal have reduced the peripheral area of Anasagar. Water quality checks have revealed heavy metals such as iron, lead, zinc, and organo-chlorine pesticides like heptachlor and DDT. Continuous consumption of such water quality can directly or indirectly lead to adverse effects on human health, including neurological disorders, impacts on the brain, spinal cord, and nervous systems, changes in behaviour, dermatological issues, eye infections and irritation, renal disorders, gastrointestinal disorders, vomiting, diarrhoea, nausea, reproductive disorders, and respiratory disorders.

Solid waste management is emerging as global concern, particularly in developing nations, it is contributing to environmental issues such as air, soil, and water pollution, and release of greenhouse gases from landfills. The investigation reveals that the Ajmer Municipal Corporation (AMC) has not

Figure 3. Ajmer city- location of sewerage treatment plants
(Ajmer Development Authority)

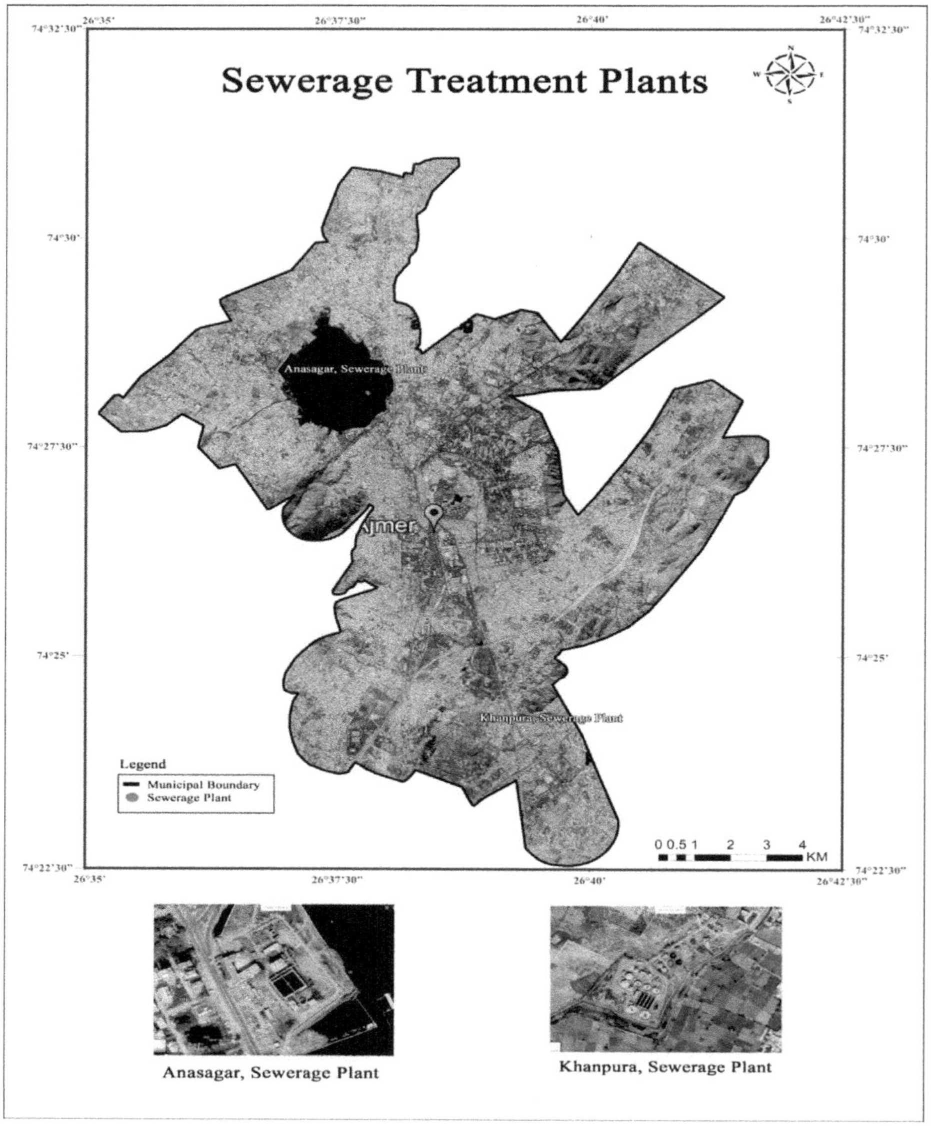

effectively implemented Municipal Solid Waste Management (MSWM) and lacks a comprehensive plan for the coming years.

According to Municipal Officer Prateek Kumawat, responsible for addressing the city's sewerage issues, the current estimated population is 590,000, generating 400 grams of waste per person per day. The city produces approximately 220 to 240 tons of solid waste daily, with an additional 40,000/day from the floating population. The waste comprises residential, market, commercial, industrial, hotel, restaurant, park, garden, slum, biomedical, and slaughterhouse waste. Bulk waste generators include hotels, restaurants, hostels, marriage gardens, banquet halls, slaughterhouses, hospitals, and canteens, collectively generating 100kg/day of solid waste. Discussions with officials indicate that door-to-door waste collection is at 100% efficiency at the ward level, except for challenging sites like hilly areas and the Dargah region. The municipality employs Auto Tippers, Hand Carts, and E-Rickshaws as needed at the ward level, with 120 auto tippers (2 in each ward). The municipality collaborates with private entities for effective waste management, involving around 980 sanitation workers in drain and roadside cleaning, as well as maintaining public toilets.

Solid waste disposal occurs at the Makhupura bypass site, located 10 km from the main city on the western side of the Ajmer-Nasirabad highway. While the site currently has a capacity for over 20 years, the increasing population and urban growth pose concerns while waste management in the city may have improved with efficient waste collection and dumping, a new challenge arises. However, there are many studies and real time examples available telling that solid waste generated can be utilized and recycled effectively, yet this potential is not being adequately tapped, leading to health issues among the city's residents. The lack of awareness, and knowledge concerning the generation, collection, transportation, and disposal of solid waste results in several consequences:

- Air pollution caused by unpleasant odors.
- Emission of greenhouse gases.
- Solid waste dumping sites serve as breeding grounds for flies, mosquitoes, and germs.
- Increased acidity levels of soils due to garbage accumulation.
- Elevated risks of diseases and epidemics.
- Pollution of ground and surface water.

Various parameters of the built environment, including interior and exterior collectively impact the physical and psychological well-being of residents. These factors can act as both push and pull factors.

Table 16. Physical characteristics of MSW in Indian cities, 2011

Population Range (millions)	No. of Cities	Percent Composition of Municipal Solid Waste						
		Paper	Rubber, Leather and Synthetic	Glass	Metals	Total Matter	Inert	Total
0.1 – 0.5	12	2.92	0.77	0.57	0.32	44.59	43.57	100
0.5 – 1.0	15	2.94	0.74	0.34	0.33	40.05	48.37	100
1.0 – 2.0	9	4.71	0.71	0.46	0.49	38.95	44.57	100
2.0 – 5.0	3	3.19	0.47	0.48	0.58	56.68	49.07	100
>5.0	4	6.44	0.27	0.94	0.81	30.47	53.90	100

Source: Manual on Solid Waste Management, NEERI, 1996

Table 17. Chemical characteristics of MSW in Indian cities, 2011

Population Range (in Millions)	Moisture	Organic Matter	Chemical Characteristics			
			Nitrogen as Total Nitrogen	Phosphorous as P_2O_5	Potassium as K_2O	Calorific value in kcal/kg
0.1 – 0.5	25.81	37.09	0.71	0.62	0.84	1009.88
0.5 – 1.0	19.52	25.19	0.65	0.56	0.69	900.61
1.0 – 2.0	26.97	26.98	0.64	0.82	0.73	980.05
2.0 – 5.0	22.03	25.60	0.57	0.63	0.77	907.17
>5.0	37.72	38.07	0.56	0.52	0.52	800.71

Source: Manual on Solid Waste Management, NEERI, 1996

Substandard indoor and outdoor environmental quality is associated with health issues such as respiratory infections, tuberculosis, asthma, and various infectious diseases.

Diseases manifest when favourable conditions exist, influenced by physical, biological, and social factors. This analysis considers specific physical and social factors contributing to the current built environment, influencing the health of residents. Disparities in health profiles among inhabitants are evident, with overcrowded areas being particularly susceptible to diseases. These areas and their residents contribute to the spread of infectious diseases, impacting social well-being.

For this study, data on diseases directly or indirectly linked to the built environment, such as viral infections, intestinal infectious diseases, tuberculosis, respiratory disorders, eye infections, hypertension, and hormonal disorders, were collected from government dispensaries, urban public health centers, and the Chief Medical and Health Officer (CHMO).

Table 18. Ajmer city: Density of population UPHC / dispensary area wise

S.No.	UPHC	No. Of Wards included	Population Density	R1
1	GULAB BARI	6	10041	6
2	KASTURBA	5	32240	1
3	PAHAR GANJ	5	23088	3
4	POLICE LINE	9	6445	12
5	RAM GANJ	4	7687	9
6	AJAY NAGAR	5	13920	4
7	ANDERKOT	8	23396	2
8	DIGGI BAZAR	14	9690	7
9	GADDI MALIYAN	6	8219	8
10	PANCHSEEL	5	5221	13
11	RAM NAGAR	5	6464	11
12	SRINAGAR ROAD	6	12134	5
13	VAISHALI NAGAR	6	6863	10
14	J.P.NAGAR	2	3931	14
15	KOTRA	3	3646	15

Figure 4. Population density map

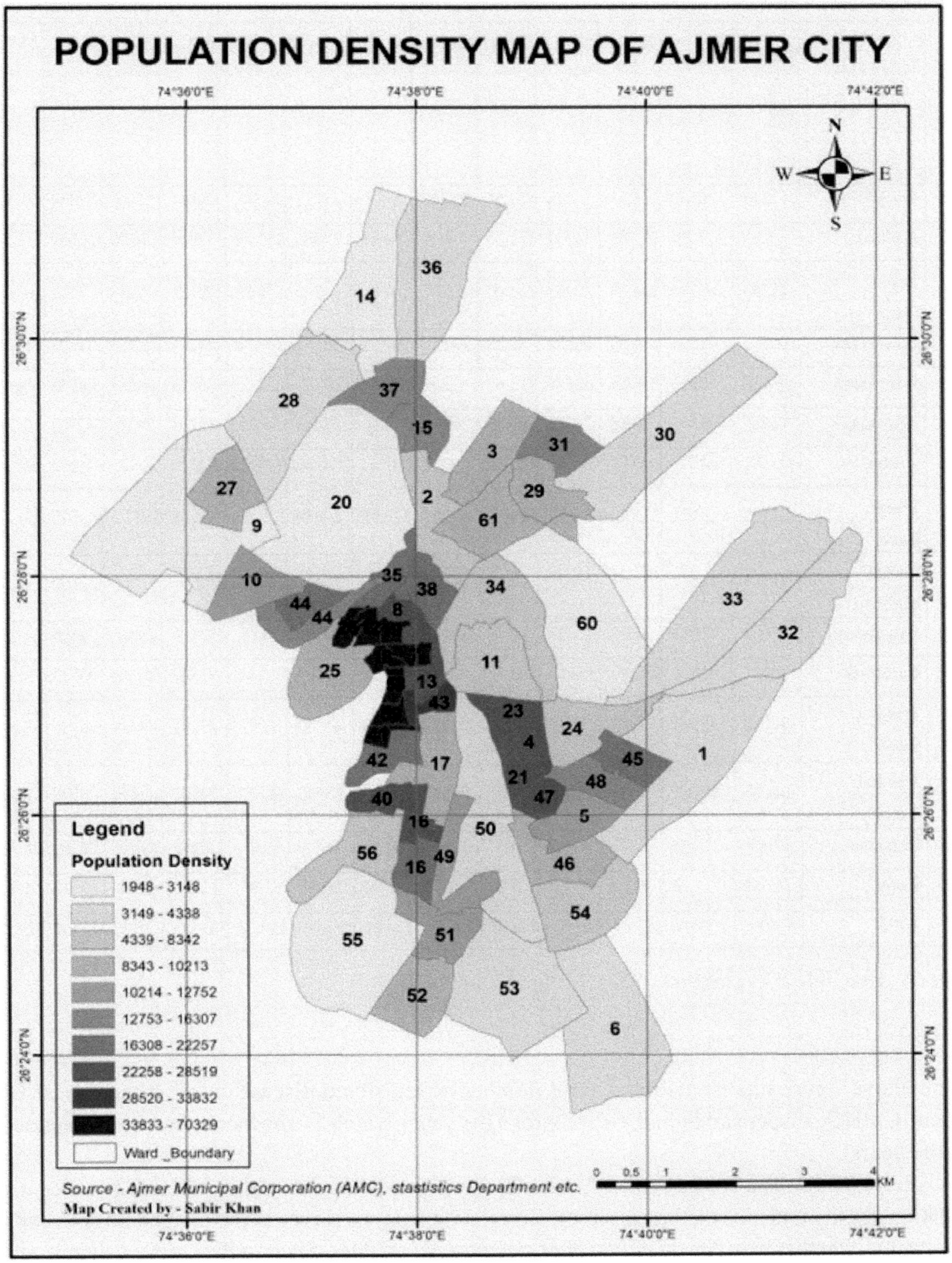

Table 19. Total registered cases of diseases in 5 years

S No.	UPHC / Dispensary	Population	2015	2015 %	2016	2016 %	2017	2017 %	2018	2018 %	2019	2019 %	AVG.	AVG. %
1	Gulab Bari	51612	20102	38.95	28443	55.11	33700	65.29	42731	82.79	39344	76.23	32864	63.68
2	Kasturba	43250	69140	159.86	74324	171.85	79660	184.18	79888	184.71	79952	184.86	76592	177.09
3	Pahar Ganj	45253	65540	144.83	70595	156.00	72701	160.65	78191	172.79	80727	178.39	73551	162.53
4	PoliceLine	80322	71126	88.55	70872	88.23	81570	101.55	88135	109.73	96814	120.53	81703	101.72
5	Ram Ganj	35437	70212	198.13	68260	192.62	68703	193.87	79931	225.56	83835	236.57	74188	209.35
6	Ajaynagar	44686	40597	90.85	39078	87.45	35373	79.16	35465	79.36	34268	76.69	36528	81.74
7	Anderkot	71358	64777	90.78	76083	106.62	73231	102.62	84830	118.88	82329	115.37	76250	106.86
8	Diggi Bazar	27558	66101	239.86	78318	284.19	89282	323.98	89889	326.18	99133	359.72	84545	306.79
9	Gaddi Maliyan	44313	14370	32.43	17807	40.18	27285	61.57	26512	59.83	34535	77.93	24102	54.39
10	Panchsheel	47405	11415	24.08	20625	43.51	22319	47.08	30884	65.15	37359	78.81	24520	51.72
11	Ramnagar	36527	92816	254.10	95159	260.52	90796	248.57	94387	258.40	95578	261.66	93747	256.65
12	Srinagar Road	56794	35893	63.20	34528	60.80	34887	61.43	41360	72.82	45261	79.69	38386	67.59
13	Vaishali Nagar	54150	68789	127.03	73413	135.57	75480	139.39	78962	145.82	79382	146.60	75205	138.88
14	JP Nagar	19069	0	0.00	0	0.00	16948	88.88	16786	88.03	28632	150.15	20789	109.02
15	Kotra	26762	0	0.00	0	0.00	0	0.00	16948	63.33	16186	60.48	16567	61.90
	Total	684496	690878	100.93	747505	109.21	801935	117.16	884899	129.28	933335	136.35	55302	129.99

Source: AMC/UPHC/Dispensaries

The above figure represents the overall number of registered disease cases in the city at Government CMHO, dispensaries and UPHCs for five years which is shows the increasing tendency over the years.

This research forms a component of Social Geography, delving into the connection between aerial population density and nine distinct diseases. Correlation analysis serves as the methodology to explore relationships, whether between quantitative or categorical variables. Such studies prove valuable in discerning associations between variables, enabling predictions about future patterns. In the realm of social sciences, where understanding future trends is paramount, this correlational study assumes significance.

Figure 5. Total registered cases of diseases in 5 years
(CMHO, Ajmer)

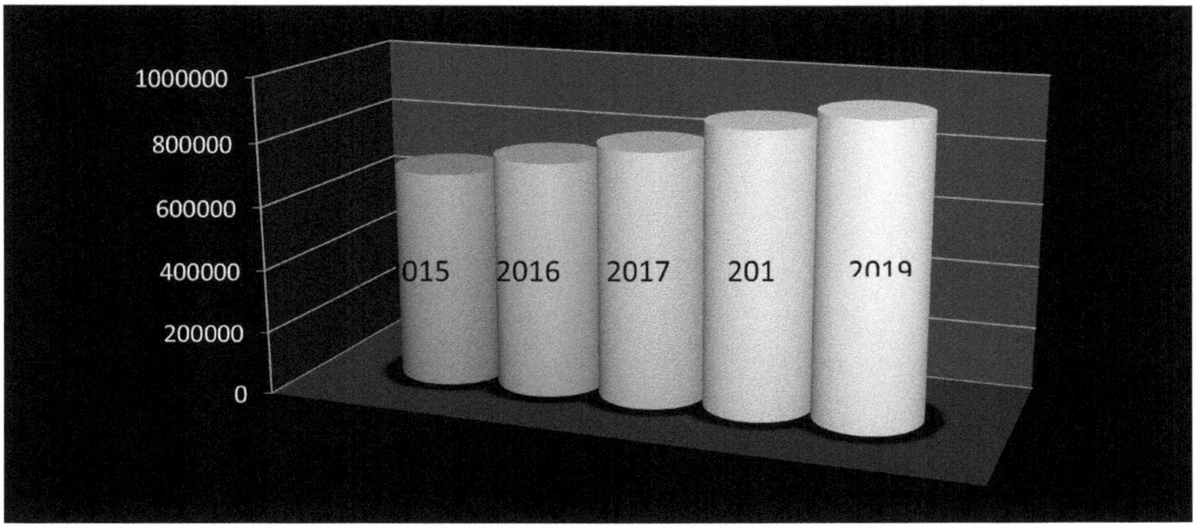

The findings are poised to aid local governments in anticipating healthcare needs and facilitating effective and efficient provision of healthcare services.

Disease data have been collected from UPHCs and Dispensaries. Spearman's correlation has been applied to check the relationship *p = 1- ((6 Σd2) / (n(n2-1)))*. It is particularly useful for analyzing ordinal data, where variables can be ranked but not measured numerically, making it ideal for survey data analysis. Additionally, the Spearman correlation is robust against outliers and does not require the assumption of normally distributed data, offering a flexible tool for various real-world datasets. Its capability to handle tied ranks further enhances its applicability, making it a preferred choice for statistical analysis in scenarios where data may not meet the stringent requirements for other types of correlation measures.

Various aspects of the built environment, such as water and sanitation infrastructure, housing quality, population density, and food safety practices, play crucial roles. It has been deduced from the data available that this disease has a positive relationship between population density and Intestinal infectious diseases which is =+0.54.

The below data show that the recorded number of diseases is maximum at Anderkot, Kasturba, Diggi Bazaar, Paharganj due to high population density due to lack of fresh air and ventilation, dampness whereas Ramganj, and Ramnagar have open nallas, excessive dampness, foul smell, are the major causes of respiratory diseases. The correlation value is = + 0.5, with this it has been deduced that there is a positive correlation between density of the population and the occurrence of respiratory diseases.

Viruses can spread through air, contaminated water, direct body contact, and indirect transmission through mosquitoes, ticks etc. Crowd, unhygienic conditions, dampness, and lack of proper sunlight aggravate the occurrence of disease which is clear in the calculations as it shows a positive correlation of = +0.63 between population density and occurrence of viral diseases.

Tuberculosis is a stern bacterial disease caused by Mycobacterium Tuberculosis, mainly affects the lungs and causes pulmonary tuberculosis but it can also affect the lymphatic system, bones and joints,

Figure 6. Intestinal infectious diseases

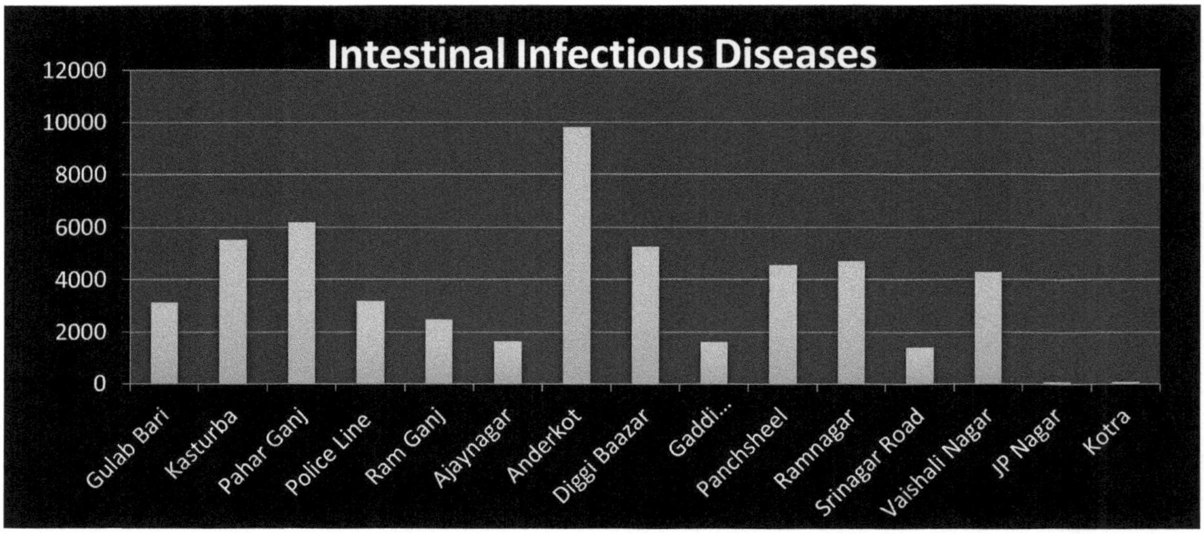

Figure 7. Respiratory diseases

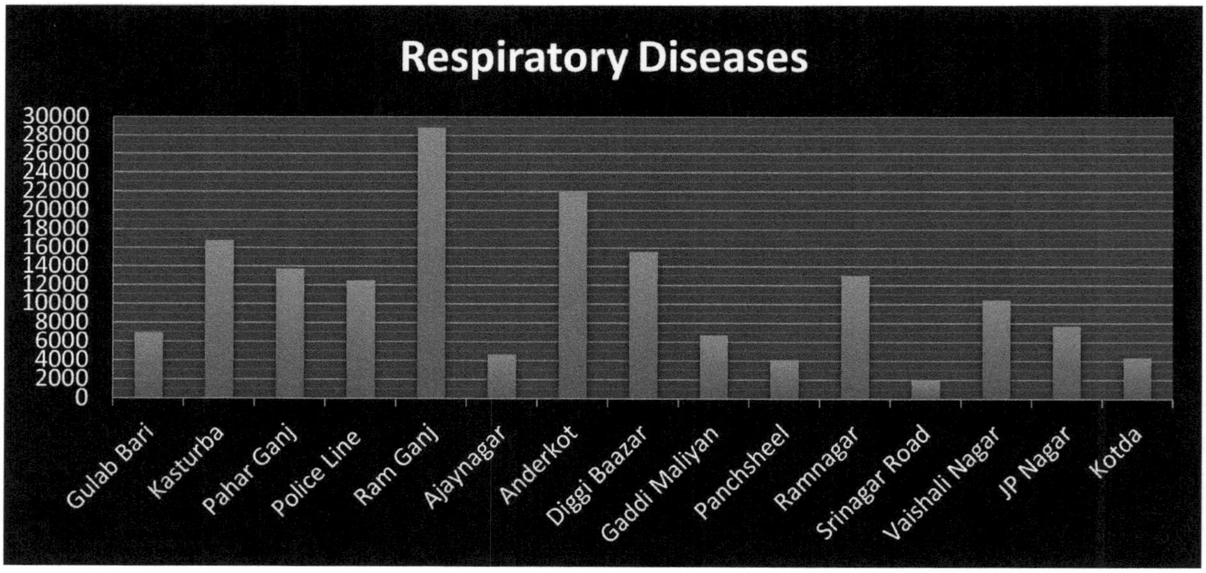

intestine, meninges and other parts of our body. It is a social disease also known as a barometer of social welfare as the social indicators of this disease are size of family, quality of life, indoor and outdoor environmental conditions, crowding, level of nutrition, awareness and education etc. The disease shows a positive correlation = +0.63 the with density of the population. The recorded cases are maximum where population density is high. Major hotspots are Diggi Bazaar, Anderkot, Kasturba, Ramganj and Paharganj areas.

Eye infection indicates the primary health condition of the region. It has been noticed that eye infection is frequently reported in high-density areas where housing conditions are not good, lanes are

Figure 8. Recorded cases for viral diseases

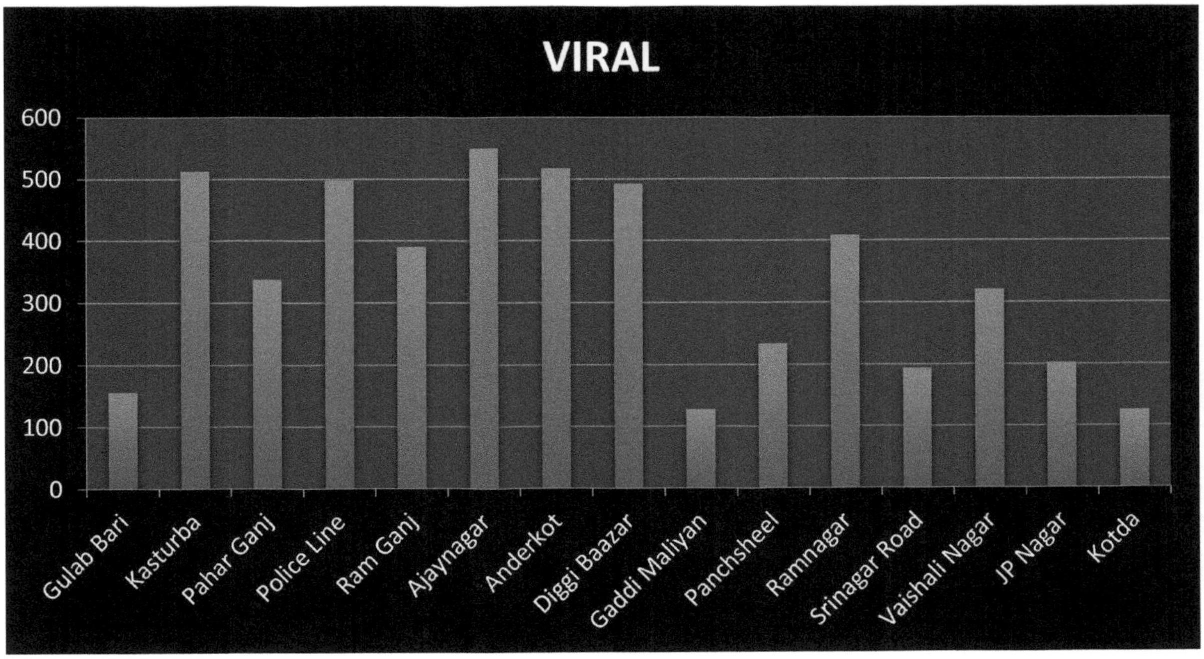

Figure 9. Recorded cases for tuberculosis cases

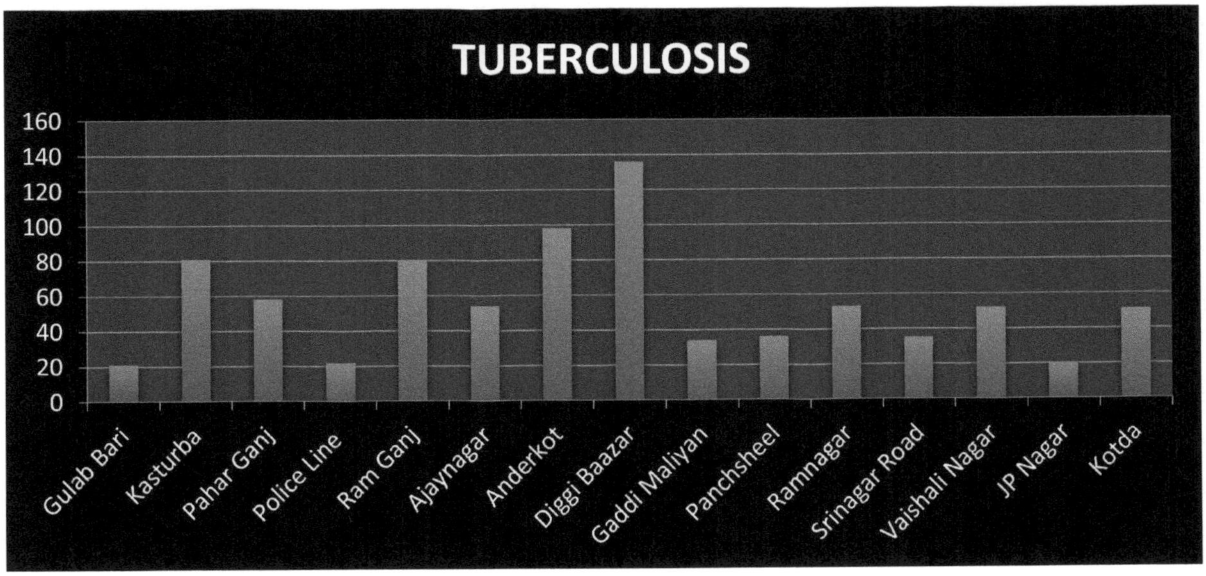

narrow, lack of sunlight and fresh air, poor sanitation facilities, and open drains and nallas are available which provide perfect breeding grounds to vectors. There is a positive relationship between population density and the occurrence of eye infections in the city and the correlational value is = +0.9, which shows a high correlation.

Figure 10. Eye infection cases

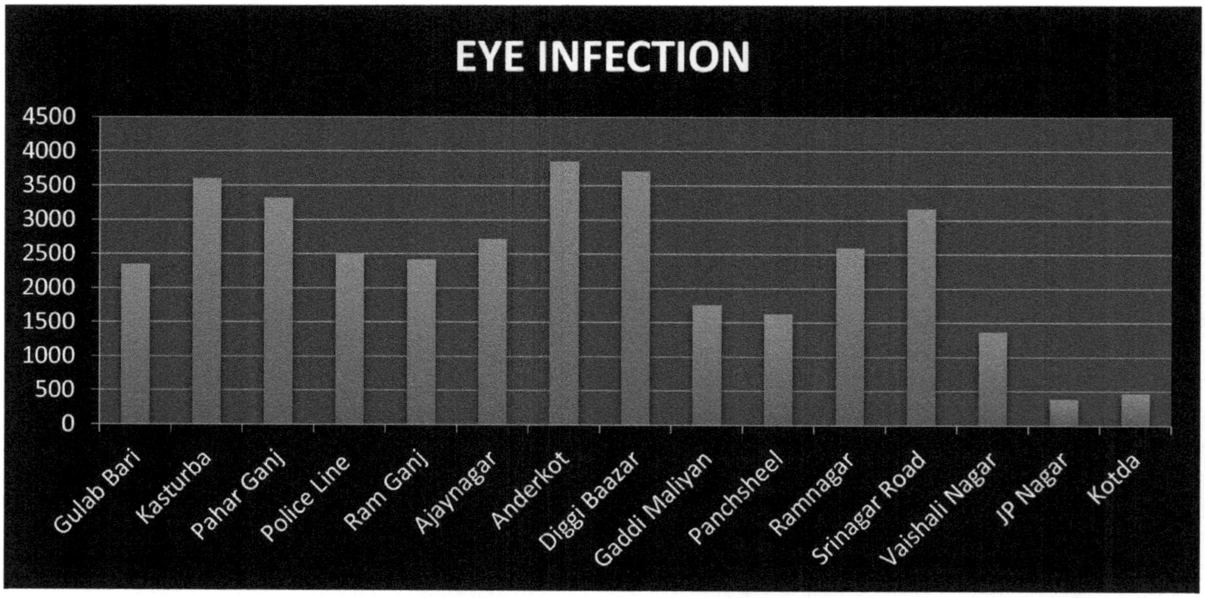

Figure 11. Recorded cases for dermatological diseases

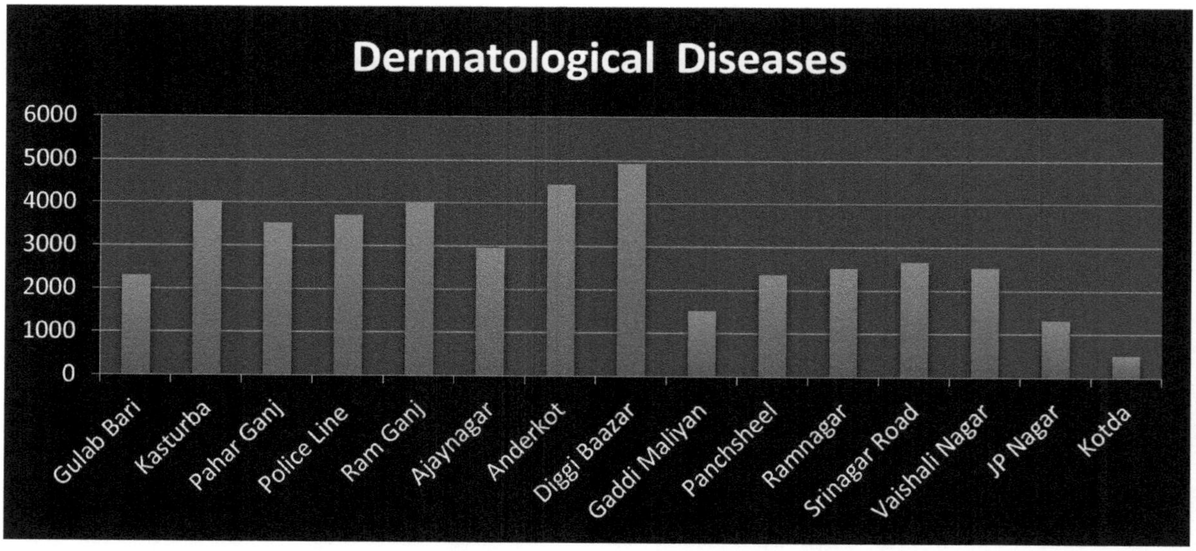

There is a high frequency of skin diseases in high-density population areas. It has been deduced that there is a high correlation of = +0.82 between density and skin disease. Skin works as a window for what is happening in the internal body, it shows stress, deficiency etc. These signs can help treat diseases at an early stage before they become a bigger problem. Some skin diseases are contagious when they spread due to direct or indirect physical contact with the infected person, chances of such diseases are maximum in areas with high humidity, lack of sunlight/fresh air, poor sanitation and unhygienic conditions etc.

Figure 12. Recorded cases for hypertension diseases

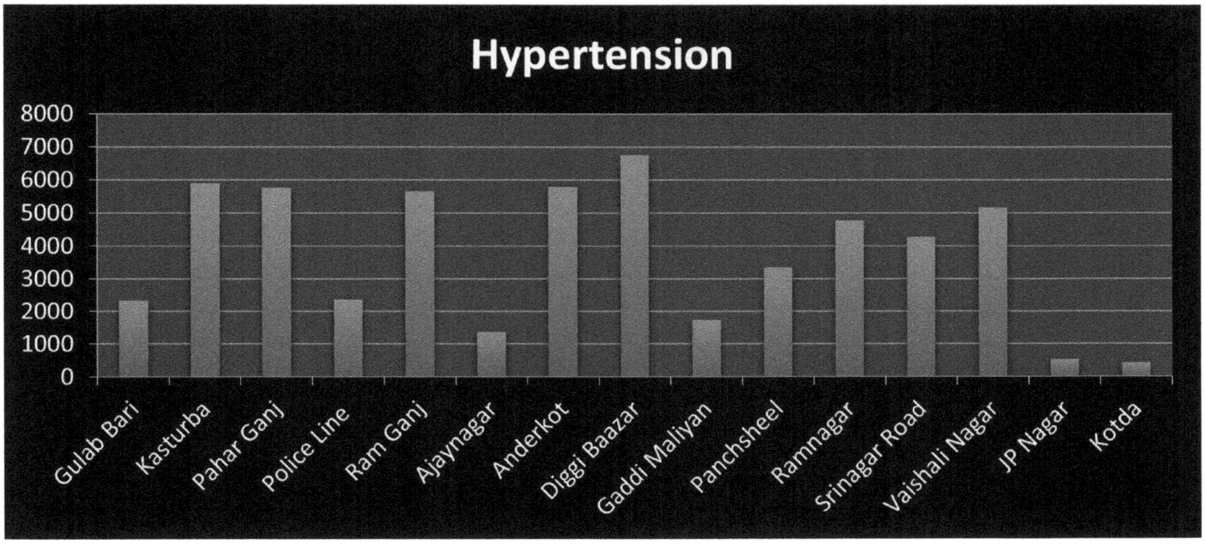

Hypertension is a major risk factor for mortality and morbidity. Studies show that the environment plays a major role in hypertension development. The unsystematic, unplanned built environment can aggravate the blood pressure which can result in multiple organ failure. This study shows a positive relationship with a value of = + 0.69 between crowd, road traffic noise and hypertension, The probable increased risk of hypertension when road traffic noise is above 80 db (24 h equivalent level) should be considered in city planning and when assessing the need for preventive measures.

Figure 13. Recorded cases for hormonal disorders

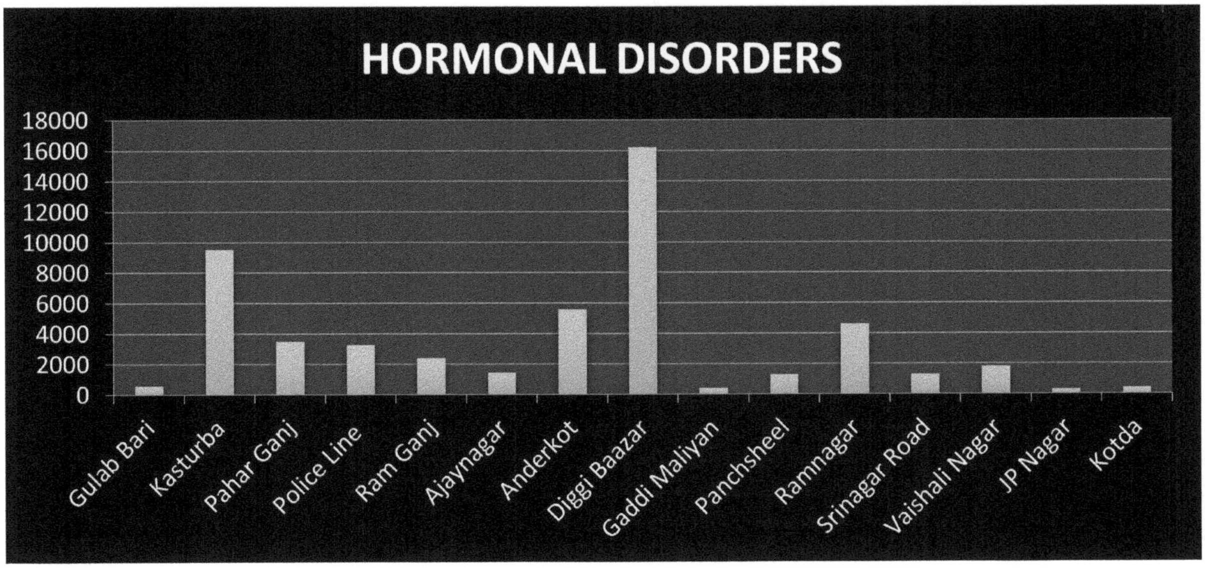

The hormonal glands create feedback loops by releasing hormones into the circulatory system, acting as chemical messengers. These hormones play a crucial role in regulating mood, growth, organ functions, metabolism, and reproduction. The presence of infectious diseases and stress significantly influences the levels of hormonal secretion in the body. The unplanned development of infrastructure and the deteriorating environment contribute to heightened stress levels and compromised immunity among residents. This study reveals that areas with increased infrastructure and population density exhibit a positive correlation with a higher incidence of hormonal disorders. The calculated correlation coefficient is = +0.77, indicating a strong association between density and the prevalence of hormonal disorders.

Research also has indicated that environmental factors may play a role in one's risk for rheumatoid arthritis. Some include exposure to smoke, air pollution, insecticides and pesticides, and occupational exposures to mineral oil and silica can cause arthritis. Ajmer city shows a positive correlation between the density of population and disease with = + 0.73.

CONCLUSION

A five-year study in Ajmer city, collected data from government dispensaries and health centres to understand how the population density relates to various diseases like intestinal issues, respiratory problems, viral infections, tuberculosis, eye infections, skin conditions, hypertension, hormonal imbalances, and arthritis. The findings show a clear connection between higher population density and more instances of these diseases. This supports the idea that the environment we create affects our health. Long-term illnesses not only bring physical challenges but also cause emotional distress and financial strain due to health-related costs. Mental health can suffer, impacting our overall well-being. While previous research focused on short-term effects, I aimed to explore how the built environment affects physical health, mental

Figure 14. Recorded cases for rheumatoid arthritis

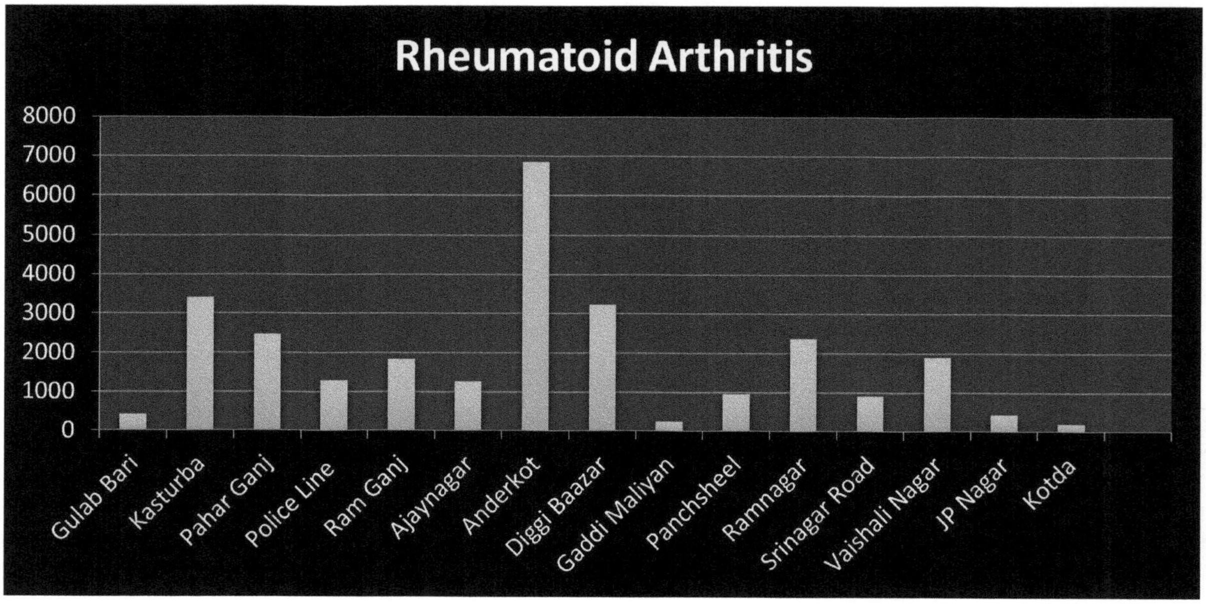

health, and overall well-being over a more extended period. Our study used statistics and correlation analysis to link population density, household characteristics, and disease frequency. The results revealed that areas with high population density are more susceptible to diseases. These findings have practical implications. First, they emphasize the need for restoration in certain areas. Second, they highlight the financial challenges that can significantly impact residents' mental well-being. Third, promoting collaboration between the public and private sectors can lead to measures that improve physical and mental health outcomes, fostering social well-being in vulnerable communities.

"A hotspot analysis has been conducted based on available data, revealing a significant correlation between the intensity of disease occurrence and population density. The high alert zone encompasses Anderkot UPHC, Diggi Bazaar UPHC, Ramnagar UPHC, Kasturba Dispensary, and Ramganj Dispensary. These areas exhibit high population density but suffer from inadequate and inefficient health facilities. In the moderate zone, we find Police Line UPHC, Vaishali Nagar UPHC, Srinagar Road UPHC, Paharganj UPHC, and Aajaynagar UPHC. These areas are also susceptible to diseases, being near the high alert zone. Moreover, the population in this zone continues to rise. JP Nagar UPHC, Kotra UPHC, Panchsheel UPHC, Gaddi Maliyan UPHC, and Gulab Bari UPHC fall into the low-risk zone. These areas exhibit lower population density, less infrastructural development, and ample open spaces with proper air and light penetration. It's crucial to consider these findings for targeted healthcare interventions, especially in the high alert and moderate zones, to address the disparities in health facilities and curb the potential spread of diseases."

Thoughtful efficient efforts in some cities have been put towards the directions which can solve the issues in India, just to mention a few-

Indore, Madhya Pradesh: Swachh Survekshan and Waste Management

Inadequate waste management leading to unsanitary conditions and environmental pollution in Indore made people suffer at various fronts. Authorities and local people supported waste management reforms, including citizen awareness campaigns, waste segregation at source, and efficient waste collection and disposal practices. The city also actively participated in the Swachh Survekshan (Cleanliness Survey) initiated by the Government of India. The result is that Indore city was consistently ranked as one of the cleanest cities in India, showcasing the success of its waste management initiatives.

Ahmedabad, Gujarat: Sabarmati Riverfront Development

Pollution in the Sabarmati River affecting the urban environment was troubling the people and authority of the city. An outstanding stand was taken towards the solution, The Sabarmati Riverfront Development project aimed at transforming the riverfront into a vibrant and sustainable urban space. The project involved river cleaning, creating green spaces, and developing recreational areas along the river. Now, the Sabarmati Riverfront has become a prevalent public space, with improved air quality providing residents with recreational areas. The project is an effort to revitalize the urban environment and has been recognized for its innovative approach to urban planning.

Figure 15. Hotspot analysis

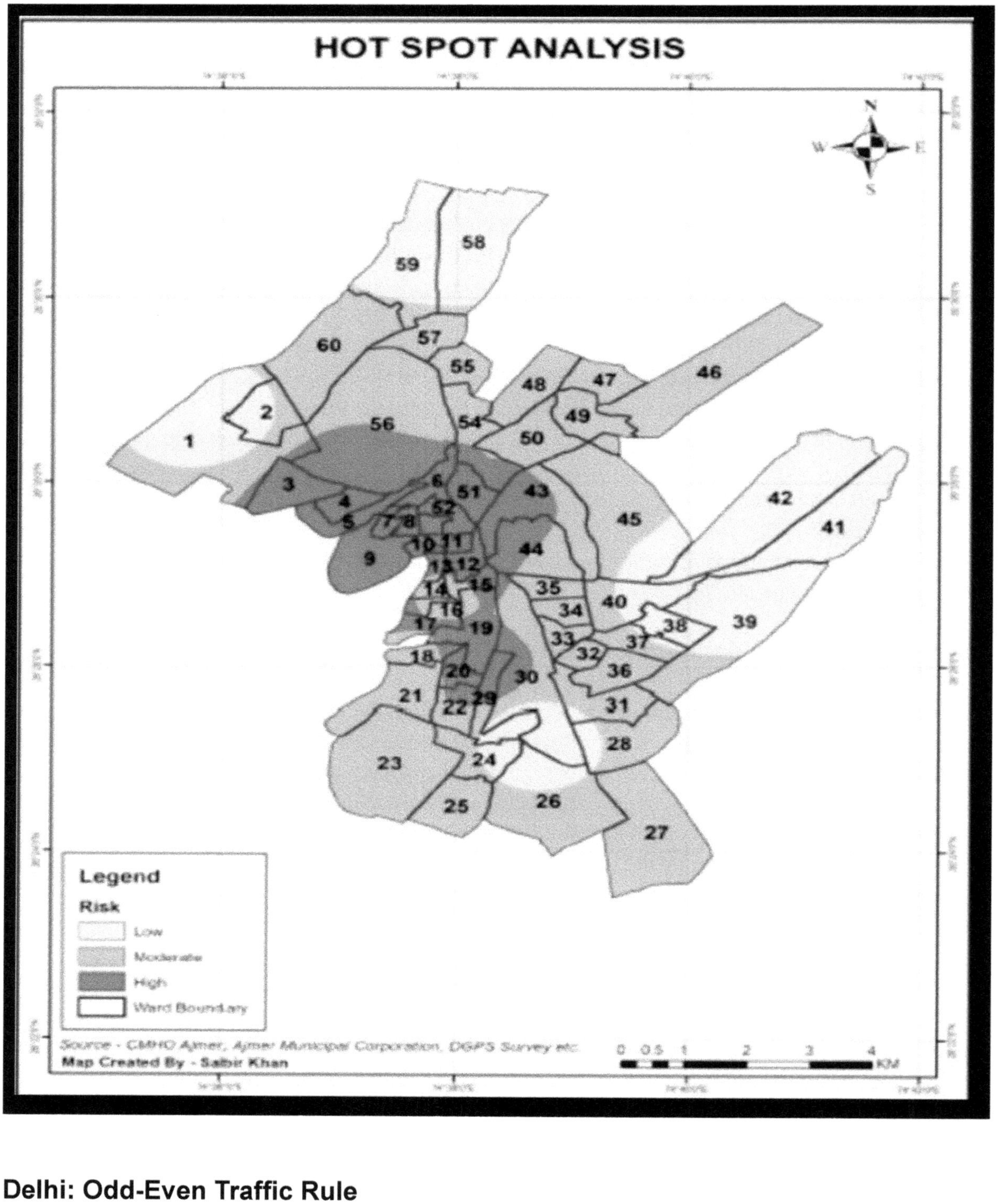

Delhi: Odd-Even Traffic Rule

Delhi is a red spot in the world map of world popular for its severe air pollution due to vehicular emissions, particularly during peak traffic hours. The Delhi government implemented the odd-even traffic rule, restricting the use of private cars based on their license plate numbers on alternate days aiming to reduce vehicular emissions.

The study of role of environmental sensors is beneficial in understanding and mitigating the impact of urbanization on both the environment and public health. To create resilient and healthier urban environments for current and future generations governments have to collaborate with policymakers, researchers, and urban planners. We are blessed to be born in the world with hi-tech technologies, such as the Internet of Things (IoT) Artificial Intelligence (AI), Geo-Informatics work in enhancing the capabilities of sense environmental sensors. These technologies enable more accurate and timely data collection, analysis, and interpretation, and help decision-makers to implement proactive measures for sustainable urban development. This area can be explored for further studies in Environmental Degradation and Human Health.

REFERENCES

Butterworth, I. (2000). *The Relationship Between the Built Environment and Well-Being: A Literature Review*. Victorian Health Promotion Foundation.

Census of India. (2011). *Population Census 2011: Ajmer District*. Registrar General and Census Commissioner of India.

Chaplin, S. E. (1999). —Cities, Sewers and Poverty: India's Politics of Sanitation||. *Environment and Urbanization*, *11*(1), 145–158. doi:10.1177/095624789901100123

Galea, S., Ahern, J., Rudenstine, S., Wallace, Z., & Vlahov, D. (2005). Urban Built Environment and Depression: A Multilevel Analysis. *Journal of Epidemiology and Community Health*, *59*(10), 822–827. doi:10.1136/jech.2005.033084 PMID:16166352

Government of Rajasthan. (2006). *City Development Plan for Ajmer*. Government of Rajasthan. http://jnnurmmis.nic.in/toolkit/final_CDPAjmer-Pushkar.pdf

Green, D., Liu, H., & Paul, V. (2021). Urban noise pollution and its effects on public health. *Journal of Urban Health*, *98*(2), 223–234.

Jackson, R. J. (2003). The Impact of Built Environment on Health: An Emerging Field. *American Journal of Public Health*, *93*(9), 1382–1384. doi:10.2105/AJPH.93.9.1382 PMID:12948946

Kumar, P., Morawska, L., Martani, C., Biskos, G., Neophytou, M., Di Sabatino, S., Bell, M., Norford, L., & Britter, R. (2019). The rise of low-cost sensing for managing air pollution in cities. *Environment International*, *75*, 199–205. doi:10.1016/j.envint.2014.11.019 PMID:25483836

Ministry of Urban Development. (2011). *Report on Urban Growth and Expansion*. Ministry of Urban Development.

Ministry of Urban Development. (2012). *Service Levels in Urban Water and Sanitation Sector, Status Report 2010 – 2011*. Ministry of Urban Development, Government of India. http://moud.gov.in/upload/uploadfiles/files/SLB%20National%20Data%20Book_0.pdf

Ministry of Urban Development. (2014). *Guidelines for Swachh Bharat Mission (SBM).* Ministry of Urban Development, Government of India. http://swachhbharaturban.gov.in/writereaddata/SBM_GUIDELINE.pdf

Ministry of Urban Development. (2016). *Primer on Faecal Sludge and Septage Management.* Ministry of Urban Development. http://www.swachhbharaturban.in:8080/sbm/content/writereaddata/Primer%20on%20Fae cal%20Sludge%20&%20Septage%20Management.pdf

Ministry of Urban Development, Government of India (GOI). (2011). *Census of Urban Growth and Development.* Ministry of Urban Development.

Patel, S. (2016). Introduction: Revisiting Urban India. *India International Centre Quarterly, 43*(3/4), 1–14.

Patel, S. K., & Jain, A. (2020). Smart technologies for waste management in urban areas. *Waste Management (New York, N.Y.), 102*, 107–116.

Smith, L. M., & Liu, Z. (2018). Advances in water quality monitoring technology for enhancing water management in urban environments. *Water Research, 137*, 182–197.

Thompson, S., & Capon, A. (2012). Designing For Health. *Landscape Architecture and Art*, (134), 26–26. PMID:22920806

WHO. (2008). *Health Situation in the South –East Asia Region* 2001 – 2007. WHO., (2018). *World Health Report*, 2018.

World Bank. (2023). *Urban Development Report 2023.* World Bank Group.

Zheng, Y., Liu, X., Wang, L., & Zhu, Y. (2020). Application of IoT technology in urban air quality monitoring. *Chinese Journal of Environmental Engineering, 14*(3), 654–660.

Chapter 21
Environmental Sensors in Extreme Environments:
Scope and Validity

Aashish Verma
SAGE University, India

Sonam Verma
Rabindranath Tagore University, India

ABSTRACT

Many potentially harmful chemicals, released by industries and human activities, can contaminate water, soil, or air, and further impact the environment and public health. Real-time and in situ monitoring of various contaminants such as heavy metals, pesticides, pathogens, toxins, particulate matters, radioisotopes, volatile organic compounds, crude oil, and agricultural chemicals at low levels is mandatory in the fields of industrial plants, automotive technologies, medicine and health, water and air quality control, natural soil/land/sea, and so forth. Consequently, the monitoring of environmental pollutants became a priority. For this aim, sensors have captivated the attention of many scientists in modern times by virtue of their eco-friendliness, cost-effectiveness, miniaturization ability, and rapidness. Environmental samples, however, are very complex and unexpectedly relative to other ecosystems. Thus far, environmental sensors have been developed with greater sensitivity, simpler and more efficient detection, better environmental adaptation and etc. for pollutant detection.

INTRODUCTION

Marine monitoring is a comprehensive work that requires full understanding and the clear understanding of the significance helps in getting stronger data from oceans. Sustainable resources can be obtained when proper monitoring is done and it also involves strengthening the marine environment. Marine programs were initially developed for detection of contaminants covering wider aspects of the ecosystem.

DOI: 10.4018/979-8-3693-1930-7.ch021

In the advent of new technology refined and efficient ways of monitoring was developed under EU marine strategy framework. The vision was to develop integrated ocean based observation system from sample collection helpful in assessment to produce scientific evidence for advice. Monitoring involves usage of new and traditional technologies that are used manually or with semi-autonomous modes. Collection of sample waters and fish is done for tracking movements where classification of microscopic and molecular techniques is used that helps in ensuring well founded ocean management and sustainable ocean industry-based operations.

CHALLENGES AND OPPORTUNITIES FOR OCEANIC SENSORS

Marine organisms are seen to be affected due to the anthropogenic sound emitted within underwater environment. The complexity as well as vastness of the oceans presents numerous challenges as well as opportunities due to the utilization of the sensors in extreme environments. Deployment of oceanic sensors has become extremely critical **(Dolson *et al.,* 2022).**

Challenges

Unfavourable condition of weather: the regions near the oceans are considered to be having an immense condition comprising high pressure, abrasive salt water, and the changes in temperature **(Feng *et al.,* 2022)**. These have the potential to pose a significant issue during the positioning and maintenance of the sensors. Thus, it requires more long-lasting technologies which can easily hold out against the exposure of such an environment.

Issues in data transmission and connectivity: In the aquatic environment, transmission of data from the sensors in real-time can be difficult. In addition to that, the physical obstacles such as depth of water and distance can be challenging for both data transmission and connectivity **(Wangand Menenti, 2021)**. Thus, it is required to develop better communicating mediums which can be capable of data transmission for long distance through various mediums.

Biofouling and fouling: algae, barnacles or similar marine organisms can possibly colonise on the sensor surfaces which further leads to biofouling as well as sensor degradation. The prevention of this biofouling can create barriers during the time of ocean monitoring as well as it also calls for more innovative solutions which can be helpful to maintain the accuracy and reliability of the sensor data **(Kim *et al.,* 2021)**.

Opportunities

Better understanding of marine ecosystems: the environmental sensors have the potential to provide valuable insights about the dynamic aquatic ecosystems. These insights may include water temperature, nutrient level which is present in the ecosystem and salinity. The continuous monitoring can provide parameters, making it easier for the scientists to gather knowledge about various aquatic processes as well as their impact on marine life **(Fisher *et al.,* 2022)**.

Quick detection of environmental changes: the sensors are able to detect the environmental changes in an early manner which includes acidification of ocean, sea level rise, various events related to pollution. The early detection helps in generating prompts to mitigate the issues and protect the marine ecosystem.

DEFINITION AND CLASSIFICATION OF EXTREME ENVIRONMENTS

An extreme environment mainly defines the characteristics of habitats by the conditions of a harsh environment which goes beyond the optimal range of human development. Such as pH2 or 11, −20°C or 113°C, saturation of salt concentration, high radiation, and 200 bars of pressure are considered to be challenging conditions, causing extreme environments **(Whitt *et al.*, 2020)**.

Extreme Temperature

There are two specific kinds of extreme environments that can be found such as cold and hot.

- Extremely cold environment: this occurs when the temperature goes below 5°C. This condition can be found in deep ocean, Polar Regions as well as high mountains.
- Extremely hot environment: this occurs when the temperature goes above 45°C. The condition can be caused by geothermal activities, volcanic areas and deep sea vents.

Extreme pH

Extreme environments can also be divided into two types, alkaline and acidic based on the level of PH.

- If the pH level rises above 9, it is considered to be an extreme alkaline environment.
- On the contrary, if the pH level drops below 5, it is considered to be a natural condition under an extreme acidic environment.

Extreme Ionic Strength

The environment with an ionic concentration higher than 3.5% is known as hypersaline environments.

Extreme Pressure

If the environment is below the extreme hydrostatic or litho pressure, such an ecosystem on the deep surface of the ocean is known as an extreme pressure environment **(Mahrad *et al.*, 2020)**.

High Radiation Environments

These kinds of environments are basically those areas which are being exposed to high radiation. The high top of mountains, deserts are possibly to high radiation such as gamma rays and ultra violet rays.

Xeric Environments

The arid habitats which have limited sources of aquatic activities are known as xeric environments. Hot deserts or cold poles are considered as one.

CHARACTERISTICS AND UNIQUE CHALLENGES OF OCEANIC ENVIRONMENTS

Mysterious Deep Sea and Hydrothermal Vents

The deep sea is known to be mysterious because of the darkness as well as the extreme conditions. To some extent, life adapts to survive without the sunlight in a different worldly environment created by hydrothermal vents. Mysterious creatures such as yeti crabs, tube worms can be found in those inhospitable regions creating a better diversity in aquatic ecosystems.

The Importance of Plankton in Marine Food Webs

The planktons lie at the base of the marine food chain, where they act as the fuel to the productivity of oceans. The microscopic creatures also play a huge role as they capture solar energy and convert the same for the high tropic levels. The absence of plankton can create complexity in the interaction that can further cause issues in sustaining marine life, causing an imbalance in the interdependence of the ecosystem (**Medina-Lopez *et al.*, 2021**).

Human-Induced Challenges Facing Marine Ecosystems

Activities such as overfishing, pollution and habitat destruction always have huge consequences on the aquatic ecosystem. The action of overfishing and pollution from various sources also creates threats to the entire ecosystem as well as the well-being of organisms. Elements such as industrial wastes, plastic and contaminations ruptures the lifecycle of marine life (**Jiao *et al.*, 2021**). The pollution related to plastic waste endangers marine species through ingestion as well as entanglement.

THE IMPACT OF CLIMATE CHANGE ON MARINE ENVIRONMENTS

Issues such as temperature rise on sea surface, acidification of ocean, and the rise of sea level is becoming a major concern for the habitats ultimately affecting the overall maritime. The temperature rise of the ocean has affected the coral reefs, leaving them as bleached.

The issues related to ocean acidification are being caused by absorption of carbon dioxide in an excessive manner from the atmosphere. This further creates a threat for the shell-forming organisms such as molluscs as well as corals. As the acid level rises in the ocean, the consequences can be seen through the disrupted food chains and disrupted ecosystem (**Tyagi *et al.*, 2020**).

POLLUTION AND ITS DETRIMENTAL EFFECTS ON OCEAN HEALTH

The debris of plastic elements create issues for marine life as they often consume plastic, thinking of it as food which further leads to internal injuries, blockage and many other hazardous conditions. In addition to that, marine life also gets affected by the plastic wastages due to the hindrances while swimming or reproducing.

Oil spills, agricultural runoffs and other chemical pollutants have a huge impact on the overall health of the ocean. The issues related to oil spill on the ocean surface creates a disastrous impact on the marine lifecycle which further leads to death and demolition of all the habitats resulting through ecological consequences **(Li *et al.,* 2020).**

OVERVIEW OF KEY PARAMETERS AND VARIABLES TO MONITOR IN OCEANS

The ocean and its surface temperature play an important role in energy exchange, creating momentum as well as moisture in the atmosphere. The temperature of the surface is considered to be the determinant of how the air and sea interact based on the climate variability. Furthermore, it is responsible for incidents such as storms, hurricanes and other natural calamities **(Ullo and Sinha, 2020)**. Moreover, the surface temperature is also helpful in determining climate change and gathering knowledge about the forecasting related to weather. To take the proper measurements of the surface water and Upper Ocean, the IR measurement tools as well as micro waves are being used. In this process, the microwave is highly absorbed by the clouds giving a better picture of the ocean surface, while on the other hand, the IR measurements are used through the water vapour to understand and measure the water vapour.

SENSOR TECHNOLOGIES FOR ENVIRONMENTAL MONITORING

Robust Enclosures and Materials

The enclosure of the sensors requires a design that can withstand extreme conditions. These conditions can comprise temperature, pressures, difficult atmospheric conditions which leads to harsh conditions. Thus, stainless steel, titanium, ceramics like materials are being used for the processes to ensure the durability and longevity. The sensors can be protected from dust, moisture and contaminations through sealed and hermetically sealed enclosures **(Wen *et al.,* 2020)**.

Temperature Sensors

Instruments such as thermocouples, thermistors and resistance temperature detectors (RTDs) can be used for the extreme high temperature in volcanic regions or subzero temperature in Polar Regions **(Mamun *et al.,* 2020)**.

The sensors help in providing accurate measurements which are reliable to understanding the thermal dynamics and changes in climate.

Optical Sensors

These types of sensors are utilised for light-based technologies to measure pH, turbidity, and chlorophyll fluorescence in marine ecosystems **(Tmusic *et al.,* 2020)**.

Real time measurements, monitoring water quality and health of the ecosystem can be measured through the help of Fluorescence, absorbance, and scattering-based optical sensors.

Acoustic Sensors

Acoustic sensors play a huge role in mapping and monitoring ocean depths, underwater habitats and ice-covered regions.

The sensors such as sonar, acoustic Doppler sensors help in getting high resolution images and capabilities to conduct research based on topography, and marine life under extreme conditions.

MULTI-AUV-AIDED DATA COLLECTION FOR MISSION CRITICAL IoUT

The V-AUV and H-AUV are two different types of AUVs. For the purpose to collect oceanic data from IoUT objects on the seabed, H-AUVs move horizontally. They then transmit that data to V-AUVs, which move vertically to transmit the data that the H-AUVs have collected to a surface station. This clever tactic can lower the frequency of H-AUV diving and floating mobility while maintaining continuous data collection and consuming less energy (**Lou *et al.*, 2023**).

Cabled Underwater Observatory Systems

The European Multidisciplinary Seafloor and Water Column Observatory (EMSO), the Monterey Accelerated Research System (MARS) in the United States, Ocean Networks Canada (OCN), and Dense-Ocean floor Network System for Earthquakes and Tsunamis (DONET) in Japan, and several other cabled underwater observatory systems have been installed (**Winther *et al.*, 2020**).

Deployment Strategies and Platforms

Moorings: Moorings entail securing sensors with weights or anchors to the ocean floor. A cable connects the sensors to a buoy at the surface, enabling real-time data transmission or periodic data retrieval. Mooring is so appropriate for maintaining specific locations for nutrient concentration that helps in understanding several processes in the analysis.

Autonomous Underwater Vehicles (AUVs): An autonomous underwater vehicle is robotic machinery that is important for monitoring and navigating underwater operations without any operator or required input. This is also helpful in understanding the usefulness of sensors, instruments and thrusters that help navigate underwater and maintain the collected data **(Shu and Huang, 2022).**

Drifters: In the case of underwater operation, drifters refer to an instrument that has been used in monitoring and tracking the movements of ocean currents. These are also important in maintaining the movement of the ocean surface and collecting data about salinity, temperature and other oceanographic phenomenon.

Satellites: Satellites are important in marine biology and play a significant role in scientific operations by providing crucial data about marine life and oceanography. The equipped sensors from separates can monitor the altimeters and radiometers that allow the scientists to understand and navigate the information about ocean colour, temperature of the surface, and rise of sea level.

DIFFERENT ISSUES AND SOLUTIONS IN MAINTAINING AND IMPROVING UNDERWATER SENSORS IN HARSH OCEANIC CONDITIONS

Issues

Biofouling: Biofouling refers to the accumulation of marine organisms in total peace through merging all the sensors in the seawater. Several organisms such as mussels, algae and barnacles are used to accumulate and colonize the sensors underwater which leads to decreased efficiency, continuous deterioration and increased maintenance costs of the sensors and equipment deployed in the ocean.

Corrosion: Corrosion is one of the known properties of several metallic materials that deteriorate due to electrochemical and chemical reactions with the environment. In the ocean, saltwater has a proper corrosive property that can harm several parts of the sensors and their working ability in long-term installation.

Mechanical Damage: Mechanical damage mainly occurs during extreme weather conditions such as underwater waves currents and ocean debris. The high energy of these environmental properties is harmful to the insert and makes them vulnerable.

Data Transmission: In extreme weather conditions low connectivity and inefficiency of data transmitting channels the physical process of data transmission Has been harmed the major barrier in this situation are water depth efficiency of the sensor's distance from the data transmitting centre.

Solutions

Anti-fouling Coatings: Anti-fouling coatings are important for stopping the biofouling of marine life on the sensor surface. These are important to maintain the walking ability of these surfaces. Proper coating can be impactful for resisting biofouling or eco-friendly biocides.

Remote Monitoring and Maintenance: Put in place systems for remote monitoring that let researchers keep an eye on the health and performance of sensors in real time. Utilize satellite communication and telemetry systems to troubleshoot problems remotely and carry out necessary maintenance.

Modular Design: Understanding the designing process that can be effective in underwater conditions is necessary for the researcher. The proper establishment of different answers is necessary to upgrade and replace the modular components of the sensors in the fieldwork.

Sensors that save energy: Saving energy underwater is necessary that create and transmit data with low energy and power-efficient processes. These are important for the sensors that include better battery life and maintain the frequency of maintenance costs and visits.

EMERGING TRENDS AND FUTURE DIRECTIONS IN ENVIRONMENTAL SENSOR TECHNOLOGY FOR OCEANIC MONITORING

Underwater Robots

IoUT systems are used by a variety of underwater robots, as Figure 5 illustrates. While the Russian-developed "Peace 1" and "Peace 2" underwater robots are the only manned submersibles in the world capable of cooperative underwater exploration, the US Navy's "bluefin" AUV has the capability of self-sufficient underwater navigation and object detection.

Underwater Acoustic Positioning Sensors

Underwater acoustic positioning sensors (UAPS) are devices that have been navigated by the use of sound waves and find out several positions of targeted objects underwater. These sensors are also important in finding out different applications that include underwater robotics marine biology research and navigating several submarine paths.

USBL: Ultra short baseline is a type of underwater acoustic positioning system that is important in determining the position of several objects that are mainly related to surface reference points and the surface objects. Angles are measured by the transceiver, which has numerous transducers. Three or more transducers are usually positioned ten centimetres or less apart on the transceiver head.

SBL: Typically, an SBL system consists of three or more transducers placed 20 to 50 meters apart. The accuracy of the measurement can be increased by increasing the distance. But these sensors' biggest flaw is that it's hard to calibrate them. Typically, the three transducers in an SBL system are positioned 20–50 meters apart. By increasing the distances, the accuracy of the measurements can be improved. But the basic problem with these sensors is that they are hard to calibrate (**Hein *et al.,* 2020**).

LBL: A collection of acoustic transponders with known relative positions positioned on the ocean floor make up LBL. Using them, robots within the acoustic signal's range can be localized using a minimum of three acoustic beacons with baseline lengths ranging from 100 m to 20 km. The system becomes costly to set up and maintain because an underwater acoustic network needs to be deployed and collected on a regular basis. The LBL sensor is not impacted by the depth of the water and can achieve high measurement precision.

APPLICATION OF NEW TECHNOLOGIES

Aquatic Animal Tracking

Tracking aquatic animals can be very beneficial to efforts to protect marine species. In general, humans should avoid contact with endangered or extinct animals. The food chain and ocean ecology may suffer if marine species go extinct. It used to be necessary for marine specialists to capture animals in order to retrieve data from attached tags. Nevertheless, this obstacle may be surmounted by the IoUT. Acoustic tags are used in IoUT systems because they perform better than radio waves. These tags are attached to ocean buoys, which use satellite communication to transmit the received data.

Water Quality Monitoring

Water is a crucial component whose qualities determine whether plants and other living things including people and animals survive. Ocean ecosystems and other water resources, however, are threatened by pollution and shortages (**Glaviano *et al.,* 2022**). Currently, one of the biggest problems facing the world's water resources is the absence of adequate protections.

SCOPE AND VALIDITY

a. ***Parameter monitoring:*** The ocean-based sensors are important in several situations and understanding the importance of marine biology and scientific research. However, these sensors are important in measuring several physiological, chemical and biological parameters that include pH, temperature, salinity, dissolved oxygen, BOD, turbidity, COD, chlorophyll level, nutrient concentration and others(**Agarwala, 2020**).
b. ***Spatial Coverage:*** The special coverage of different underwater acoustic positioning systems is important and also depends on different factors such as operational frequency, operating depth, transducer array design, terrain and obstacles and others that help understand the ecosystem and monitor the data collection procedure through different water depths such as surface water to deep-sea habitat.
c. ***Temporal Resolution:*** Sensors provide high temporal resolution for analyzing seasonal variations, long-term trends, and short-term processes in oceanic parameters by continuously collecting data over extended periods of time.

Validity

a. ***Accuracy:*** Ocean-based sensors are made to measure environmental parameters accurately. They are frequently validated by comparison with independent measurements or calibrated against reference standards.
b. ***Precision:*** Sensors have a high degree of measurement precision, making it possible to identify minute fluctuations in oceanic conditions over long periods of time.
c. ***Quality Control:*** To guarantee the precision, dependability, and consistency of sensor data, quality control procedures are put in place. These procedures include sensor calibration, maintenance procedures, data quality evaluations, and metadata documentation.

SUMMARY

This chapter presented an overview of the environmental sensor based monitoring in extreme environments. Different challenges and definition of extreme environments shows that Extreme temperature, Extreme pH, Extreme ionic strength and Extreme pressure has made marine monitoring more essential. It also stated that robust enclosures and materials, temperature sensors, optical sensors and acoustic sensors are the sensors used for ocean monitoring. Scope as well as validity has been stated which shows that sensor calibration is necessary. In the end it has been learnt that future insights of sensor based technologies shows that more effective tracking for marine animals in order to save ocean environment is necessary.

REFERENCES

Agarwala, N. (2020). Monitoring the ocean environment using robotic systems: Advancements, trends, and challenges. *Marine Technology Society Journal*, *54*(5), 42–60. doi:10.4031/MTSJ.54.5.7

Dolson, C. M., Harlow, E. R., Phelan, D. M., Gabbett, T. J., Gaal, B., McMellen, C., Geletka, B. J., Calcei, J. G., Voos, J. E., & Seshadri, D. R. (2022). Wearable sensor technology to predict Core body temperature: A systematic review. *Sensors (Basel)*, *22*(19), 7639. doi:10.3390/s22197639 PMID:36236737

Feng, Y., Yu, J., Sun, D., Dang, C., Ren, W., Shao, C., & Sun, R. (2022). Extreme environment-adaptable and fast self-healable eutectogel triboelectric nanogenerator for energy harvesting and self-powered sensing. *Nano Energy*, *98*, 107284. doi:10.1016/j.nanoen.2022.107284

Fisher, M., Cardoso, R. C., Collins, E. C., Dadswell, C., Dennis, L. A., Dixon, C., Farrell, M., Ferrando, A., Huang, X., Jump, M., Kourtis, G., Lisitsa, A., Luckcuck, M., Luo, S., Page, V., Papacchini, F., & Webster, M. (2021). An overview of verification and validation challenges for inspection robots. *Robotics (Basel, Switzerland)*, *10*(2), 67. doi:10.3390/robotics10020067

Glaviano, F., Esposito, R., Cosmo, A. D., Esposito, F., Gerevini, L., Ria, A., Molinara, M., Bruschi, P., Costantini, M., & Zupo, V. (2022). Management and sustainable exploitation of marine environments through smart monitoring and automation. *Journal of Marine Science and Engineering*, *10*(2), 297. doi:10.3390/jmse10020297

Hein, J. R., Koschinsky, A., & Kuhn, T. (2020). Deep-ocean polymetallic nodules as a resource for critical materials. *Nature Reviews. Earth & Environment*, *1*(3), 158–169. doi:10.1038/s43017-020-0027-0

Jiao, W., Wang, L., & McCabe, M. F. (2021). Multi-sensor remote sensing for drought characterization: Current status, opportunities and a roadmap for the future. *Remote Sensing of Environment*, *256*, 112313. doi:10.1016/j.rse.2021.112313

Kim, C., Lee, W., Melis, A., Elmughrabi, A., Lee, K., Park, C., & Yeom, J. Y. (2021). A review of inorganic scintillation crystals for extreme environments. *Crystals*, *11*(6), 669. doi:10.3390/cryst11060669

Li, J., Pei, Y., Zhao, S., Xiao, R., Sang, X., & Zhang, C. (2020). A review of remote sensing for environmental monitoring in China. *Remote Sensing (Basel)*, *12*(7), 1130. doi:10.3390/rs12071130

Lou, R., Lv, Z., Dang, S., Su, T., & Li, X. (2023). Application of machine learning in ocean data. *Multimedia Systems*, *29*(3), 1815–1824. doi:10.1007/s00530-020-00733-x

Mahrad, B. E., Newton, A., Icely, J. D., Kacimi, I., Abalansa, S., & Snoussi, M. (2020). Contribution of remote sensing technologies to a holistic coastal and marine environmental management framework: A review. *Remote Sensing (Basel)*, *12*(14), 2313. doi:10.3390/rs12142313

Mamun, M. A. A., & Yuce, M. R. (2020). Recent progress in nanomaterial enabled chemical sensors for wearable environmental monitoring applications. *Advanced Functional Materials*, *30*(51), 2005703. doi:10.1002/adfm.202005703

Medina-Lopez, E., McMillan, D., Lazic, J., Hart, E., Zen, S., Angeloudis, A., Bannon, E., Browell, J., Dorling, S., Dorrell, R. M., Forster, R., Old, C., Payne, G. S., Porter, G., Rabaneda, A. S., Sellar, B., Tapoglou, E., Trifonova, N., Woodhouse, I. H., & Zampollo, A. (2021). Satellite data for the offshore renewable energy sector: Synergies and innovation opportunities. *Remote Sensing of Environment*, *264*, 112588. doi:10.1016/j.rse.2021.112588

Shu, W. S., & Huang, L. N. (2022). Microbial diversity in extreme environments. *Nature Reviews. Microbiology*, *20*(4), 219–235. doi:10.1038/s41579-021-00648-y PMID:34754082

Tmusic, G., Manfreda, S., Aasen, H., James, M. R., Gonçalves, G., Ben-Dor, E., Brook, A., Polinova, M., Arranz, J. J., Mészáros, J., Zhuang, R., Johansen, K., Malbeteau, Y., de Lima, I. P., Davids, C., Herban, S., & McCabe, M. F. (2020). Current practices in UAS-based environmental monitoring. *Remote Sensing (Basel)*, *12*(6), 1001. doi:10.3390/rs12061001

Tyagi, D., Wang, H., Huang, W., Hu, L., Tang, Y., Guo, Z., Ouyang, Z., & Zhang, H. (2020). Recent advances in two-dimensional-material-based sensing technology toward health and environmental monitoring applications. *Nanoscale*, *12*(6), 3535–3559. doi:10.1039/C9NR10178K PMID:32003390

Ullo, S. L., & Sinha, G. R. (2020). Advances in smart environment monitoring systems using IoT and sensors. *Sensors (Basel)*, *20*(11), 3113. doi:10.3390/s20113113 PMID:32486411

Wang, Z., & Menenti, M. (2021). Challenges and opportunities in Lidar remote sensing. *Frontiers in Remote Sensing*, *2*, 641723. doi:10.3389/frsen.2021.641723

Wen, F., He, T., Liu, H., Chen, H. Y., Zhang, T., & Lee, C. (2020). Advances in chemical sensing technology for enabling the next-generation self-sustainable integrated wearable system in the IoT era. *Nano Energy*, *78*, 105155. doi:10.1016/j.nanoen.2020.105155

Whitt, C., Pearlman, J., Polagye, B., Caimi, F., Muller-Karger, F., Copping, A., Spence, H., Madhusudhana, S., Kirkwood, W., Grosjean, L., Fiaz, B. M., Singh, S., Singh, S., Manalang, D., Gupta, A. S., Maguer, A., Buck, J. J. H., Marouchos, A., Atmanand, M. A., & Khalsa, S. J. (2020). Future vision for autonomous ocean observations. *Frontiers in Marine Science*, *7*, 697. doi:10.3389/fmars.2020.00697

Winther, J. G., Dai, M., Rist, T., Hoel, A. H., Li, Y., Trice, A., Morrissey, K., Juinio-Meñez, M. A., Fernandes, L., Unger, S., Scarano, F. R., Halpin, P., & Whitehouse, S. (2020). Integrated ocean management for a sustainable ocean economy. *Nature Ecology & Evolution*, *4*(11), 1451–1458. doi:10.1038/s41559-020-1259-6 PMID:32807947

Chapter 22
Examining the Effects of Forest Fires:
A Framework for Integrating System Dynamics and Remote Sensing Approaches

Müjgan Bilge Eriş
Yeditepe University, Turkey

Duygun Fatih Demirel
https://orcid.org/0000-0001-8284-428X
İstanbul Kültür University, Turkey

Eylül Damla Gönül Sezer
https://orcid.org/0000-0002-9237-0468
Yeditepe University, Turkey

ABSTRACT

Forest fires have been a major concern for many countries over an extended period of time due to natural and human induced factors. In recent years, detection of forest fires has progressively shifted toward advanced technologies where the remote sensing approaches are fully operational. To enhance fire management strategies, it is crucial to gain a comprehensive understanding of the fire dynamics and its consequences on the environment, operational sources, and economic sectors. Therefore, this chapter develops an integrated framework to predict and analyze the effects of forest fires by using system dynamics approach and remote sensing technology, ultimately leading to the establishment of a conceptual model and conclusive insights.

INTRODUCTION

Forest fires are one of the most damaging types of disasters to understand and handle compared to other disasters. These incidents can burn hectares of land in a matter of minutes and cause significant

DOI: 10.4018/979-8-3693-1930-7.ch022

environmental, social, and economic damage. The complexity of managing forest fires occurs as a consequence of both natural and human-induced factors (Dhall et al.,2020; Nolan et al.,2021). Therefore, dealing with forest fires becomes a complicated and often hazardous process that requires the collaboration of numerous organizations and teams (Dorrer et al., 2021). The most crucial factors in combating forest fires have been identified as early fire detection, accurate fire classification, and fast responses from firefighting teams (Akay and Şahin, 2018; Khan and Khan, 2022). Fire detection can be generally evaluated in two categories. Firstly, there is traditional public reporting, which involves fire alerts from anonymous individuals and on-site examination. Secondly, there are operational detection systems such as lookout towers and remote sensing tools that rely on technological advancements (Dogra et al.,2018). Although public reporting systems are valuable in specific situations, they may encounter issues like delayed and inaccurate reporting, limited coverage of the region, or false alarms. For instance, in flat areas, a tower-based fire detection system can directly identify hotspots without any barriers. However, in articulated terrains, where the ignition point might be hidden by hills, the system can only indirectly detect the fire, often picking up smoke columns or plumes. Thus, operational detection systems have become more reliable and secure compared to public reporting. In that sense, remote sensing has become a key player in forest fire management.

The increasing reliance on remote sensing technologies signifies a notable shift towards more advanced and effective methodologies (Cosgun et al.,2023). With the help of remote sensing, certain characteristics of forest fires such as location, extent, temperature, burn severity, smoke, and aerosols can be easily detected. However, the detection and monitoring of forest fires is a comprehensive process that goes beyond understanding fire behavior and utilizing remote sensing technologies; it also entails the application of data analysis and the development of predictive models for effective forest fire tracking. Hence, modeling is another critical component after collecting and processing the data. Due to the nature of forest fires, it is important to consider numerous causal relations involved in the prevention and suppression. In addition, forest fire management problems mostly occur because long-term natural processes don't align with short-term planning (Loncar et al, 2006; Collins et al.,2012). When considering the nature of forest fires, it would be prudent to take advantage of the system dynamic model and its long-term predictive capabilities. Utilizing the system dynamics methodology provides an alternative to current approaches that rely on linear future predictions for a specific system (Saveland, 1998; Collins et al., 2012; Thompson et al.,2019). Therefore, this chapter aims to show a framework consisting of spatial information obtained from drones and a system dynamics conceptual model for effective forest fire management. It will represent the complex mechanisms that lead to forest fires, as well as the impacts of these fires on ecosystems, operational teams, and economic sectors. In addition, firefighting teams would benefit from the framework by being able to understand the dynamic behaviors of fire management, evaluate the suppression and preventive actions, and set up multiple strategies for minimizing the economic and ecological damage under different disaster impact scenarios.

The rest of the paper is designed as follows: In the following section, an exhaustive review of related studies will be undertaken, distinctively examining both forest fire and remote sensing approaches. Additionally, the literature will be explored for works that interconnect system dynamics with forest fire management and investigate the integration of drone technology. In the methodology section, a comprehensive overview of the system dynamics approach is presented. Subsequently, the suggested conceptual model and the stock-flow model are outlined. Next, the framework for the integration of drone technology and the system dynamics model is explained. Finally, the study concludes with a discussion and explores solutions and potential future implications.

REMOTE SENSING TECHNOLOGIES

Remote sensing has been addressed in various ways; basically, it can be defined as the science of telling something without contacting or touching it (Fischer et al.,1976). Traditionally, the term remote sensing involves the identification and observation of physical objects within an area by analyzing radiation remotely, often through satellites or aircraft equipped with sensors. Images of physical objects such as buildings, vegetation, and soil can be obtained through sensors by scanning them electronically or by using electromagnetic radiation. Image resolution of physical objects denotes the potential detail obtained from the imagery, i.e. feature of the image representation. There are three types of resolution in remote sensing: spatial, spectral, and temporal. While spatial resolution refers to the level of spatial detail such as the size of the smallest area that can be individually recorded as an entity on an image, the temporal resolution represents time between images of the same geographical area. Lastly, spectral resolution presents the ability of the sensor to distinguish finer wavelengths of the electromagnetic spectrum, the high spectral resolution corresponds to a narrow bandwidth (Frisken et al.,2013; Elachi and Zyl, 2021). Regarding spatial, spectral, and temporal resolution, there exists a broad spectrum. However, there is a fundamental tradeoff exists among spatial, spectral, and temporal resolutions. Generally, as spatial resolution increases, spectral and temporal resolutions decrease, and conversely, higher temporal resolution corresponds to lower spatial and spectral resolutions.

The characteristics of platforms equipped with remote sensors significantly contribute to the efficient observation of the object space. Data can be acquired from different platforms such as satellites, airborne, unmanned aerial systems, mobile/static ground, and crowd sensing. To obtain reliable data, it is important to accurately observe the area traveled by the platform. There are several ways to enhance the observational potential of a platform. The most prominent solution is to install multiple sensors on the same platform, looking in different directions, such as forward and backward-facing cameras and/or Lidar (light detection and ranging) sensors on mobile platforms (Petrie, 2009). In addition, although the main sensors which can be classified as active and passive are mainly available on all platforms, the object range is mostly dependent on system complexity. It should be also noted that crowdsended platforms consisting of imagery or video data are becoming increasingly accessible (Toth and Jóźków, 2016). After the relevant data is acquired from one of these platforms, related information is extracted from the sensor analysis. Finally, analyzed remote sensing data can be integrated with other relevant data about the application area (Campbell and Wynne, 2011).

FOREST FIRES

Forest fires are natural events that affect communities and the environment worldwide, leading to property and human losses. This causes changes in the composition of vegetation and destroys soil properties in the coming years. In various regions of the world, an increasing number of forest fires, are associated with climate (Hillayová et al.,2023; Zong et al.,2020) and human-induced changes (Nagy et al.,2018; Vilar et al.,2016). However, there is no single factor that alone causes uncontrollable fires; instead, they arise from the combination of at least four elements such as ignitions, continuous fuel sources, droughts, and appropriate weather conditions (Pausas & Keeley, 2009; Boegelsack et al., 2018). When these elements are exceeded, the fire's scale is significantly influenced by the duration of adverse fire weather conditions and the abundance of continuous fuel sources across the landscape.

Forest fires often start due to natural causes, and one of these causes is lightning. In the event of a lightning strike, the high temperature can lead to the ignition of surrounding materials. This situation can result in fires in other natural areas. However, the primary drivers of forest fires are mostly human-induced factors (Mann et al.,2016; Masoudvaziri et al.,2021). Human activities can lead to ignitions, either inadvertently through accidents such as cigarettes, equipment malfunction, electrical power, or intentionally. Furthermore, forest fires also indirectly occur by modifying fuel availability and increasing sensitivity to ignition. For instance, previous research has demonstrated a strong link between the expansion of agricultural areas and deforestation rates, highlighting a shift from natural fire patterns to human-influenced fire patterns in many regions in recent years (Aragon et al.,2010; Andela and Van Der Werf, 2014). Openings created by people in vegetated areas, due to road and building construction, tree-cutting activities, etc., increase the presence of fine dry fuels, contributing to both human-caused and natural ignitions. In addition to clearing the land for construction and development of agricultural land, removing crop residues, controlling pests, and urban growth are also examples of human activities that increase the amount of forest fires.

Climate is commonly considered a determining factor in the size of fires, and historically, drought has been a significant factor (Madadgar et al.,2020; Ruffault et al.,2018). The effects of fires on forests are exacerbated by drought (Brando et al.,2014), leading to long-term changes in forest structure and biomass. While short-term rainfall during dry seasons can suppress fire activity, in drier ecosystems, precipitation in wet years may contribute to fuel accumulation, increasing susceptibility to burning in subsequent years. (Andela et al.,2017). Although drought presents a clear climate signal and is likely a key factor behind major fire incidents, there is also a growing concern about the role of anthropogenic climate change. It extends the duration of the fire season and raises the frequency of dry years. Increasing temperatures elevate heat stress in trees by enhancing evaporation and lowering average humidity. Consequently, extended periods of drought and significantly prolonged forest fire seasons emerge, creating optimal conditions for ignition and, thus, increasing the number of critical fire weather days. It is also reported that human-induced warming probably increased the summer forest fires by drying out the fuels, and this trend will persist. In addition, due to ongoing warming and a probable gradual decline in fall rainfall, significant fires will likely become more frequent in the fall (Williams et al.,2019).

Apart from climate, weather is another key driver of forest fire activity. Abnormal weather events increase the likelihood and spread of uncontrollable forest fires. Unless facing extreme weather conditions, a few ignitions are unlikely to evolve into a forest fire. However, there is a critical threshold of ignition beyond which the likelihood of a sudden forest fire surge increases. (Bradstock, 2010). The main weather variables that affect fire behavior are determined as air temperature, relative humidity, wind direction, and wind speed (Cawson et al.,2017). The consumption of fuels with a high surface area/volume ratio is significantly influenced by air temperature and relative humidity, in contrast to fuels with a low surface area/volume ratio. (Ottmar,2014). Additionally, precipitation decreases fuel consumption, while wind accelerates fuel consumption, for all types of fuels.

Slope, aspect, and elevation can be considered as topographic features that influence forest fires. For instance, slope is a critical factor affecting fire spread. Steeper slopes in elevated areas facilitate faster propagation compared to forests with gentler slopes, which are less prone to fire. Besides, elevation factor plays an important role in influencing various parameters like wind speed, temperature, precipitation, and vegetation cover, all of which impact the spread and intensity of forest fires. This spatial variation is significant in determining patterns of vegetation cover and soil properties (Hong et al.,2019). Finally, the slope aspect determines the micro-climatic conditions of the terrain, affecting variables such as solar radiation absorption, surface temperature, moisture levels, wind patterns, and vegetation distribution (Abdo et al.,2022).

As aforementioned, fuel load is another essential factor, and highly depends on plant growth and decomposition. Fuel loads are addressed as combustible sources and affect the fire intensity and spread positively. The amount of fuel accumulation heavily relies on the intervals between consecutive burns and the precipitation received during this time. With shorter intervals between fires, less fuel can accumulate particularly woody fuel. This accumulated material will be comprised primarily of herbaceous plant material and leaf litter. Conversely, as the time between fires lengthens, the component of larger twigs and branches will increase. The impact of fuel load is altered by how quickly various plant species in the area grow and their physical structure. In essence, fuel development depends on the prevailing vegetation in terms of different plant species, the population structure of each type of plant, and the amount of rainfall.

Fuel patterns are also highly affected by human activities. For instance, agriculture and urban development contribute to the fragmentation of ecosystems and disrupt the continuity of fuel across landscapes. In densely populated regions, significant fires frequently subside upon reaching agricultural areas. There has been a worldwide decline in fire activity in recent years, as documented by studies partially linked to the expansion of agriculture, particularly in tropical areas (Ward et al.,2018). On the other hand, from savannas to mountains, changes in agricultural practices disrupt traditional fire management methods, leading to unforeseen fire challenges. Societies employ various strategies like fire bans and fuel management to address wildfire threats, but some approaches inadvertently exacerbate the problem. This is because they overlook the dynamic relationship between fire and vegetation—altering one affects the other. Therefore, effective land management policies must acknowledge wildfires as a persistent aspect of flammable landscapes, requiring adaptable approaches rather than seeking a permanent solution. Nevertheless, numerous factors are amplifying fuel continuity and fostering increased fire activity in numerous regions globally. Forestry plantations are another potential cause of the increase in fuel quantity and continuity. They are sometimes established in non-forest natural ecosystems, significantly elevating fuel loads in these systems (Pausas and Keeley, 2009).

Incidence of fire activity tends to decrease due to factors such as rapid detection, proximity to firefighting resources, and fuel fragmentation which can be related to high population density. Nevertheless, even with the reduced impact of high population density in decreasing fire incidents, there is an observed trend indicating that the presence of high population and housing density still elevates the probability of human-caused ignition. Due to fuel accumulation resulting from abandoned agriculture and even intentional field clearing in certain regions, the likelihood of anthropogenic ignition increases (Song and Katsikis, 2023). Furthermore, fuel load and its moisture content, wind speed and direction, relative humidity, and air temperature are regarded as the most important factors affecting fire ignition (Rossa, 2018).

REMOTE SENSING TECHNOLOGIES AND FOREST FIRE MANAGEMENT

The remote sensing field has been vastly expanded in recent years due to the advancements in technologies. Its application areas encompass various fields, including agriculture, tourism, military, energy, coastal, transport, and shipping (Mahrad, et al.,2020). Furthermore, utilizing different remote sensing approaches for collecting environmental data has become prevalent. With the help of this environmental data, it becomes possible to identify areas prone to environmental disasters such as forest fires, ultimately leading to a reduction in fire occurrences and mitigating damage (Gülçin&Deniz,2020). Since remote sensing technologies enable analyzing the forests and other environmental areas on spatial scales, it supports effective forest fire management which includes five main components: analyzing the potential fire hazards and risks, determining hot spots for risk reduction and prevention, firefighting readiness, monitoring the ongoing fires, and post-fire evaluation (Filho et al.,2021).

In forest fire management, obtaining timely and accurate information about the environmental conditions is crucial for a rapid response. Characteristics of the environment, such as the structure of tree canopies and species information, as well as the status of the fire, the rate of fire spread, burn severity assessment, and fuel estimation, can be accurately and rapidly obtained from remote sensing data (Chuvieco et al., 2020; Gale et al.,2021). In literature, two conventional methods for collecting information from the environment are generally addressed: field measurements (traditional methods) and aerial photo interpretation. However, acquiring information through traditional methods is often time-consuming and costly, especially for large areas. Remote sensing technologies, particularly satellite remote sensing techniques, have the advantage of overcoming the limitations of traditional methods by swiftly acquiring information at local, regional, and even global scales (Pu et al.,2021).

Satellite images are utilized for mapping, categorizing, and evaluating the interested environmental areas. This enables the abstraction and evaluation of land cover information, such as waters, bare lands, trees, and buildings. The data obtained from these remotely sensed pictures can be used to predict the risk of forest fires. Additionally, satellite images of the selected region before and after forest fires can be categorized into different classes, and the effects of these fires on burned areas can be investigated (Mohammed and Khamees, 2021). However, data collected from satellite platforms from manned aircraft can be sensitive to cloudy atmospheric conditions. In such situations, electromagnetic wave attenuation can occur, which results in information loss and data degradation (Xiang et al.,2020). Furthermore, Lidar technology is an effective tool to provide enhanced spatial coverage. To acquire Lidar data, aerial surveys are conducted over potential forests, however, it can be prohibitively expensive (Riano et al., 2003; Wang et al.,2020). On the other hand, traditional photogrammetry methods may not provide the necessary point cloud density to determine the number of trees per hectare in the forest. In recent years, Geographic Information System (GIS) technology has become popular for environmental management purposes. It is a technology developed to capture, store, verify, and visualize data about locations on the Earth's surface. GIS technology can be combined with other approaches such as optimization models, machine learning techniques, and/or remote sensing data to gain a more comprehensive understanding of forest fire management (Mangiameli et al.,2021; Abedi-Gheshlaghi et al.,2020). On the other hand, an alternative solution can be presented as motion photogrammetry. This method uses unmanned aerial vehicles (UAV) which can generate high-density point clouds suitable for estimating the number of trees (Srivastava, et al.2022). UAVs which are also referred to as drones can be defined as unmanned aerial machines capable of autonomous operation or remote control, minimizing the necessity for constant user intervention (Dougherty et. al,2015). They have found applications in diverse fields such as humanitarian supply chains, healthcare, agriculture, environmental monitoring, and disaster response (Kim et al.,2017; Ahirwar et al.,2019; Adsanver et al.,2021). Various types of drones are available in the market today, catering to both commercial and civilian purposes. They can be categorized according to their ranges, sizes, prices, payloads, model complexity, number of blades, and other factors (Momeni et al.,2023). Specifically, drone sizes can vary from a small class known as "smart dust," measuring a minimum of 1 mm in length and weighing 0.005 grams, to massive fixed-wing drones with a wingspan of up to 61 meters and weighing 15,000 kg. Generally, two common types of drones are fixed-wing and rotary-wing drones which have their own set of advantages and disadvantages. Fixed-wing drones offer advantages such as higher cruising speed and altitude, increased flight efficiency, and longer endurance and range. Besides, they only require energy for forward movement, not for sustaining themselves in the air, making them highly energy efficient. Conversely, rotary-wing drones are more flexible, as they can vertically take off and land regardless of the environmental conditions. Additionally, hybrid drones that integrate the benefits of both types are becoming increasingly prevalent in modern applications (Kinaneva et al.,2019).

Recently, drones have been employed for forest fire detection, image processing, monitoring, and route planning since they offer several advantages as an alternative to traditional methods (Chen et al.,2017; Guimarães et al.,2020). They can fly over challenging terrains inaccessible to firefighter teams, allowing them to access remote or rugged areas. Since, drones can be equipped with multispectral cameras and sensors to measure parameters like pressure, temperature, relative humidity, gases, and radiation; they can efficiently gather data faster than human crews (Themistocleous, 2017). They are also capable of creating accurate maps of the fire's extent and preventing the fire from spreading further (Loncar et al.,2006; Yandouzi et al.,2022). Additionally, the utilization of drones enables quicker analysis compared to satellites and cost-effectiveness relative to other technologies in use (Jiao and Fei, 2023; Yuan et al.,2017; Israr et al.,2022). Despite the pros mentioned earlier, drones can be unsuitable for large-area acquisition because of their limited flight duration, and they do not enable the simultaneous coverage of the entire area of interest, as is possible with satellite platforms (Matese et al.,2015). Nevertheless, they have several benefits making them valuable for environmental purposes.

A drone application supporting forest fire management is undoubtedly the most advanced and practical activity among all disaster-related efforts (Restas, 2016). Drones can serve multiple purposes in forest fire management, including pre-fire hot spot detection, providing real-time information during intervention for effective fire management, and post-suppression monitoring after extinguishing. Detecting hotspots from the air before civilians report them is beneficial for fire managers in limiting damages caused by fires. In addition, if the firefighting team is on-site, they may be too close to effectively manage the fire and its surroundings. Managing the fire along with its surroundings is crucial since extinguishing forest fires takes time, and the fire continues to spread during this period. In that sense, the use of drones during interventions can be highly effective because obtaining a general overview of a forest for the coordination of intervention measures. Without aerial surveys, coordination relies on information among individuals in different locations. However, subjective assessments by firefighters in various positions may not align. Aerial examination helps eliminate subjectivity, aiding in prioritizing different areas effectively.

The efficiency of drones in detecting fires is influenced by the time taken for aerial-based hot spot detection, which, in turn, is dependent on the flight parameters of drones. Enhancing the efficiency of aerial hot spot detection requires an examination of these flight parameters. To minimize the average detection time, it becomes necessary to increase flight speed. However, the challenge lies in the objective limitations to raising flight speed. Analyses of camera focus and altitude yield slightly varied results. Elevating the onboard camera allows for monitoring a broader area (grid). If the frame of the monitored area remains constant, and the monitored grid is larger than before, the flight path might change, resulting in a shorter duration.

The main reason for not always using drones is the high cost of antennas. Undoubtedly, aerial surveillance with drones allows for the rapid identification of hotspots, providing timely reports to firefighting teams. However, economically, this method may be most effective in specific situations, such as areas with exceptionally high Fire Weather Index (FWI) and challenging geographical features (Restas,2015). Simply, the higher the fire risk indicated by FWI, the more areas need to be monitored because the likelihood of hot spot detection is higher, which leads to higher costs. Consequently, if using drones for this task is cheaper than the traditional method with manned aircraft, choosing drones becomes a more cost-effective solution.

There are several attempts to increase the efficiency of drones. For instance, NASA's research and funding have contributed to laying the foundations for different remotely operated and autonomous aircraft to have a much more significant role in our lives. This includes small drones delivering packages and even the possibility of taking out the deliveries one day, whereas larger, jet-sized drones inspect cellphone towers and other infrastructures. On the other hand, companies and researchers have been working on various types of drone integration for delivery. Hybrid distribution systems, utilizing both a large vehicle (such as a truck) and a small vehicle (such as a drone), have been developed (Li et al.,2023; Amaral et al.,2022). In addition, drones can be also utilized for health purposes such as rescuing individuals experiencing sudden heart attacks or combating the virus during a pandemic (Tang & Veelenturf, 2019; Chamola et al., 2020), or monitoring and rescuing missing people (Mayer et al.,2019; Pensieri et al.,2020). Figure 1 below, summarizes the implementation areas of drones recently.

SYSTEM DYNAMICS AND REMOTE SENSING TECHNOLOGIES

System Dynamics (SD) simulation modeling, which was initially introduced by Forrester offers solutions to the modeling complexities. By incorporating feedback loops and time variables into the model,

Figure 1. Application of drones in various areas

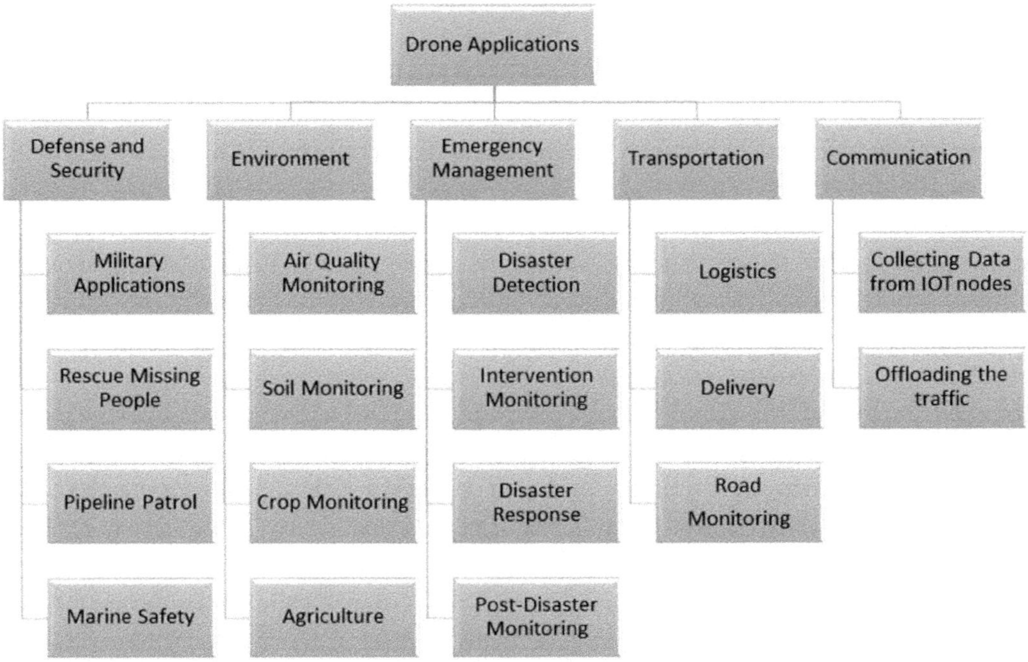

it provides a comprehensive approach to analyzing the structure and behavior of complex systems. In the literature, several studies integrate the predictive and dynamic nature of SD models with the spatial analysis capabilities of remote sensing technologies, mostly GIS analysis. Considering the competence of GIS, which has limited capabilities in temporal modeling, and the strength of SD in representing temporal processes despite its limited spatial modeling abilities, a reasonable alternative is to couple SD with GIS to model spatial dynamic systems. Ahmad and Simonovic (2001) proposed a spatial system dynamics model (SSD) to model dynamic processes in both time and space. The SSD approach enables a two-way flow of information between SD and GIS, facilitating feedback in both temporal and spatial dimensions. Initially, GIS provides spatial data to the SD model, which, through dynamic modeling, identifies changes in spatial features over time and feeds them back to GIS. These spatial changes subsequently influence decisions and policies over time, enabling integrated modeling of processes in both temporal and spatial domains while capturing feedback loops. For instance, Pakere et al.,(2022) provided a SD model with a GIS platform to evaluate space and time indicators of renewable energy resources based on their resource, economic, and technological features. Moradi et al., (2020) used GIS and SD to generate an SSD to evaluate the impact of storm surges. Additionally, Xu and Coors (2012) expanded SSD by integrating a 3D visualization model to GIS and SD to study urban residential development. In this research, the SD model contributed to the study by assessing the developmental tendency of the main factors.

On the other hand, there are a limited number of studies that focus on the combination of SD and the utilization of drones. One of the existing studies in the literature proposed a method for the swarming UAV (drones) combat system based on the SD model. Throughout the model, the task completion degree and the survival rate of drones have been considered as two crucial factors for determining the duration in which ground enemy targets are eliminated. Additionally, the experiments revealed that the integration of drones, along with improved information transmission, coordination, and enhanced reconnaissance capabilities, positively affects combat efficiency (Jia et. al.,2019).

SYSTEM DYNAMICS AND FOREST FIRE MANAGEMENT

Fire can be defined as a dependent spatial process, and it is unlikely to exhibit a simple linear relationship between forest fire drivers and activities. Due to their nature, they are characterized by high dynamics, time constraints, and uncertainty (Thompson and Calkin, 2011). Even small changes in fire drivers can lead to sudden shifts in fire regimes, making predictions challenging. To better understand the dynamics of forest fire management and design appropriate policies, the system dynamics simulation technique has become one of the most preferable tools. It addresses challenges by examining non-linear relations and feedback loops (Thompson et al.,2019).

In literature, the theme 'fire paradox' has been frequently discussed by existing studies (Calkin et al.,2015; Ingalsbee,2017). Fundamentally, the 'paradox' arises from the thought that managing forest fires failed because past forest fires were quickly controlled on a small scale, causing too much fuel and plant growth. In other words, the aggressive suppression of uncontrollable fires in fire-prone forests leads to the accumulation of dangerous fuels. This accumulation results in uncontrollable fires burning at higher intensities, resisting control, and eventually leading to an increased demand for suppression efforts. This situation can also be observed from the system dynamics approach where the positive feedback loops result in negative consequences. For instance, Collins et al.,(2013) employed a system dynamics model for fire management in Portugal. They compared management policies, focusing on budgets allocated to firefighting and prevention, and discovered that emphasizing firefighting yielded immediate benefits but could lead to an increase in fire activities in the long term.

The System Dynamics approach allows the designing of scenarios and policies to ensure improvements in the behavior of the system. According to Plooy and Botha's (2021) study, the scenario analysis revealed that the strongest levers of fires are high tree density, strong winds which increase the rate of fire spread, and a substantial amount of fuel in the biomass. In addition, the simulation showed that the individual responsible for igniting the fire played an insignificant role in the ultimate damage caused by the forest fire. Instead, monoculture species, windy dry weather, and the presence of a substantial amount of fuel were the critical factors contributing to the devastating Australian forest fires.

Some of the studies evaluated and compared the implementation of the policies in the long run based on the systems dynamics approach. Pandey et al., (2023) found that developing forest fire prevention policies has great potential to mitigate the forest fire risks consisting of human human-based ignitions, the containment of fire propagation, and strategic management to change fire dynamics. Thompson et al.,(2019) applied a system dynamics model that integrates human and natural fire-prone systems to assess changes through seven forest fire response scenarios. The study indicated that pursuing aggressive burning rate targets while implementing policy changes gradually over time can potentially restore forest conditions to their original state without undesirable levels of burn severity and departure rates. Farkhondehmaal and Ghafarzadegan (2022) performed simulation-based policy analysis, and in contrast to studies seeking the most effective policy to control forest fires, their research demonstrated that the long-term solution is not a single action but a combination of multiple actions considering both the human and natural aspects of the system simultaneously.

Besides that, the interaction between firefighter teams and technology has become vital due to institutional policies and the pressure to extinguish fires promptly. As aforementioned, the effects of technology on fire suppression led to the adoption of new technological decisions. The information gathered from drone technology such as geographic location, land-based information, and physical features of active fires have a vital role in intervening in a fire and minimizing the damage, however, this information as parameters is not always sufficient. They must be integrated into the model to comprehend the dynamics of a system in terms of burned area, resource availability, efficiency, economic sectors...etc. Therefore, ensuring a system-focused research approach is vital for understanding these complex dynamics. This book chapter aims to contribute to the literature by utilizing drone technology and system dynamics approach to achieve these objectives.

THE MAIN FOCUS OF THE CHAPTER

The study is driven by the following key research questions:

RQ1: How can the characteristics of data collected by drones be effectively integrated into a system dynamics model to enhance understanding of forest fire dynamics?
RQ2: What is the relationship between information gathered by drones, and the dynamics of forest fire management as represented in the system dynamics model?
RQ3: What are the consequences of forest fires on tourism, energy, and transportation infrastructure sectors, as investigated by the system dynamics model?

RQ4: What strategies can be implemented in the short and long term to mitigate the impacts of forest fires?

The aims of this chapter are:

Aim 1: Identify the relationship between ecological, operational, and economic factors and fire management systems.
Aim 2: Develop a comprehensive framework by integrating drone-derived data into a system dynamics model to enhance the understanding of forest fire dynamics.
Aim 3: Explore how drone-collected information can be utilized to estimate fire suppression and assess the impact of improved efficiency of firefighters on the overall dynamics of the forest fire system.
Aim 4: Determine short-term and long-term strategies to support firefighting operations.

METHODOLOGY

The objective of this study is to indicate the factors influencing forest fires and assess their impacts on ecological systems, operational teams, and economic sectors such as tourism, energy, and transportation infrastructure with a specific emphasis on integrating drone-derived data into the analysis. In addition, the chapter seeks to explore strategies for mitigating the effects of forest fires and formulating proactive policies to prevent their recurrence in the future. The chosen methodological approach is System Dynamics (SD) for a comprehensive examination of the variables, particularly incorporating inputs from drone observations, and for designing effective policies for sustainable forest fire management.

Simulation methods play a crucial role in understanding complex systems, and among them, SD stands out as a methodology that leverages the concept of time continuity, defining relationships between events as non-linear rather than linear. Its mathematical foundation is built upon differential equations. According to the results of the literature review, SD is frequently employed for developing strategies and policies. By utilizing the SD method, relationships between events or variables are depicted through interactions. These interactions which can be direct, indirect, or mutual are employed to unveil complex relationships that cannot be captured through data analysis. Consequently, decision-makers can analyze the current system more accurately and, at the same time, simulate the situation through various scenarios.

PROBLEM DEFINITION

Forest fires pose a significant threat to natural ecosystems, necessitating the development of effective forest fire management strategies to improve firefighting processes and safeguard communities. In recent years, researchers have embraced principles of systems thinking to cope better with the complex dynamics of managing forest fires (Thompson et al.,2016; Thompson et al.,2018; Steelman and Nowell,2019). Due to natural fuels (vegetation), weather conditions, and lightning ignitions, and those caused by humans, forest fires have been increasing day by day (Mhawej et al.,2017).

The use of UAVs, commonly referred to as drones, in the context of disaster response, presents a hopeful avenue for research. There is an interest in exploring how drones, can provide not only intervention through water bombing but also faster and more effective situational awareness in the context of forest fires. Their mobility, aerial perspective, and rapid data collection and transmission capabilities position them as a central tool in fire assessment and intervention (Somers and Manchester,2019; Nigam and Agarwal 2014). In this context, integrating drone-derived data into an SD model emerges as a promising approach for enhancing our understanding of forest fire dynamics and improving intervention strategies. However, a comprehensive investigation is required to explore the relationships between key drone-collected parameters and critical aspects of fire behavior, including average burned area, fire severity and speed of spread, and fire duration. Addressing this knowledge gap will contribute to the development of more efficient and data-driven approaches for forest fire prevention, intervention, and management.

CONCEPTUAL MODEL

Conceptualization is the stage where the map of the dynamic problem begins to be drawn using boxes and connections. Mostly, it is done by using causal loop diagrams (Johnson and Pen, 2022). The conceptual representation of the system must be realistically constructed, and within this representation, all relationships and their directions should be defined. During this phase, the conceptual model of the system needs to be outlined, incorporating the key variables identified while defining the problem. The arrows explicit the relationships between variables, where the origin of the arrow represents the influencing variable, and the end represents the influenced variable.

The directions of the relationships can be defined at the end of the arrows either positive or negative (Bala et al.,2017; Yearworth,2014). A positive link is used to depict that an increase (decrease) in one variable leads to an increase (decrease) in another variable, i.e., positive sign. Conversely, a negative link is a connection where an increase (decrease) in one variable leads to a decrease (increase) in another variable i.e. negative sign. In this conceptual model, a loop is created by revisiting the variable chosen as the starting point through a relationship path. To decide whether a loop is reinforcing (positive) or balancing (negative), several positive and negative signs are counted. Thus, the loop is considered positive when there is an even number of negative signs in the loop, and conversely, it is labeled as negative when the total number of signs is odd. Reinforcing loops typically undermine the stability of the system, whereas negative loops contribute to maintaining equilibrium in the system. Positive feedback loops drive or strengthen changes, deviating the system from a state of equilibrium and rendering it more prone to instability. On the other hand, negative feedback loops tend to mitigate or buffer changes, aiming to maintain stability within the system and enhance its resilience (Sterman, 2000).

The primary variables within the model suggested in this study are “total forest area”, “fuel load”, “fire intensity”, “average area burned”, firefighter efficiency”, “fire suppression rate”, “tourism area”, “energy assets”, and “transportation infrastructure” and “revenue”. Here, “revenue” represents the economic gain from tourism, energy, and transportation infrastructure sector. The connections between these fundamental variables and their respective directional relationships are illustrated in the following Figure 2.

Figure 2. Proposed conceptual model for fire management

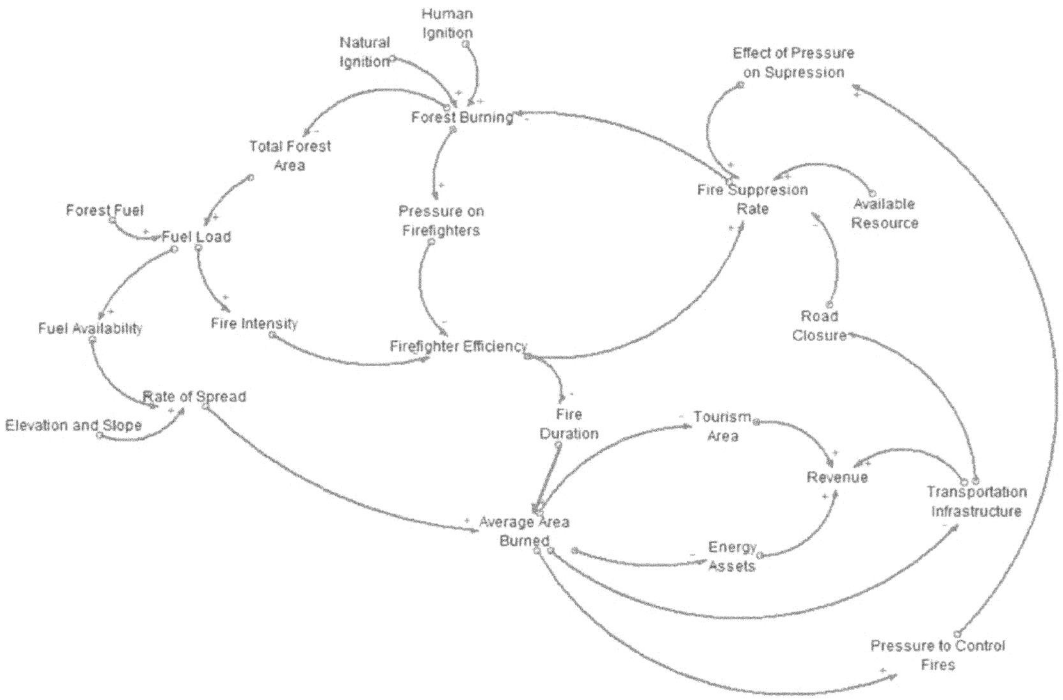

STOCK-FLOW DIAGRAMMING

In this stage, a conceptual model is turned into a simulation model by using a stock and flow diagram. This approach allows for the creation of a comprehensive top-level representation of the system. The essential components in the stock flow diagram are represented in Figure 3 below.

Stock represents the quantity to be accumulated, and its value changes over time, representing the desired variable to be tracked. Flow variables can be categorized as inflows or outflows. The arrow points towards the stock, i.e. inflow leads to an increase in the stock value. Conversely, if the arrow from the stock points in the opposite direction, i.e. outflow results in a decrease in the stock value. It is also possible to indicate both increment and decrement by using a single flow which is called a bidirectional flow. In addition, converters which are shown by small circles are used to define the logic that modifies the flow. Finally, connectors which are represented by straight arrows used to connect model variables.

When developing Stock-Flow diagrams, the initial step involves classifying variables as stocks, flows, and converters. Following that, it becomes imperative to construct the stock-flow diagram. Within this framework, the model is developed by incorporating all essential indicators. Mathematical equations are subsequently formulated, utilizing stocks, flows, and converters. The interrelationships between these indicators can be expressed through the mathematical equation provided below (Sterman, 2000).

Stock (t)=Stock () +) ds

Figure 3. Blocks of SD

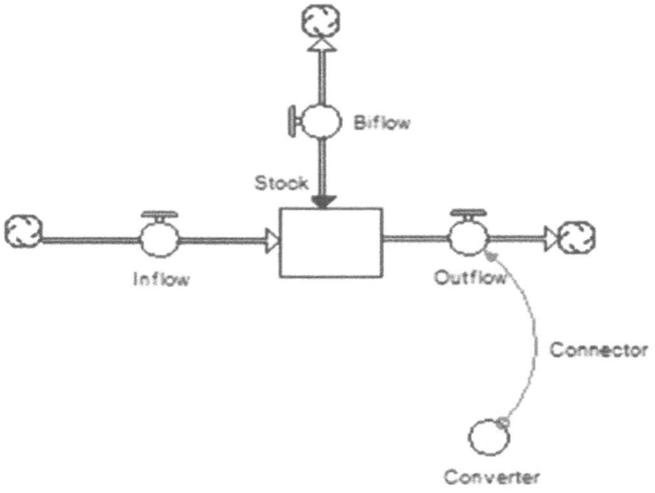

In this equation, while indicates the start time, t- 0 represents the time interval. To calculate the current value of the stock variable the initial value of that stock variable is added to the cumulative net change in the corresponding flows over the specified period. Completing the entire stock and flow diagram will provide a clearer picture of the data needed to run the model. Subsequently, the process of gathering and analyzing the required data in detail will commence. Once the necessary data is integrated as input into the simulation model, the simulation model will be executed, and the behaviors of variables over time for the base scenario will be obtained through graphs and tables.

The simulation model is analyzed by considering three sub-systems. (1) Ecological Factors and Forest Fire (2) Operational Factors and Forest Fire (3) Economic Sectors and Forest Fire.

ECOLOGICAL FACTORS AND FOREST FIRE

The ignition of forest fires is often attributed to the simultaneous occurrence of ecological factors reaching a critical threshold. Human activities, such as campfires or discarded cigarette butts, can act as sources of ignition, while natural factors like lightning strikes can also initiate fires. As the fire spreads, it tends to consume the most combustible plant species, resulting in a reduction of forested areas. Consequently, the landscape transforms, with open areas devoid of any vegetation expanding. In such areas, where all plant species are eliminated, the landscape becomes barren. In addition, the amount of fuel in forest areas is one of the significant contributors to forest fires since it determines the intensity of the fire. In other words, the accumulation of fuel load will increase the fire intensity, leading to the burning of even larger areas. Furthermore, topographical features in a forest environment impact the rate of fire spread. For instance, steep slopes and high elevations can increase the rapid spread of forest fires. Similarly, different soil types affect the speed of forest fires. Specific soil types may be more prone to combustion, leading to faster forest fire spread. On the other hand, the fire management system involves determining which resources will be obtained (such as air tankers, and transport aircraft), where they will be deployed, the types of facilities to be established at bases, and the specific personnel requirements for the operation. As the amount of resources available to extinguish or prevent a fire increases, the fire suppression rate increases. Final critical relationships are observed between the fuel loads and the suppression efforts. Fire suppression results from accumulated fuel. This unintended consequence which is also known as the fire paradox causes larger and more intense fires.

OPERATIONAL FACTORS AND FOREST FIRE

The intensity of the fire will inevitably reduce the efficiency of firefighters. This, in turn, will lead to an extended duration of the fire and an increase in the average burned area. As the burned area expands, it becomes more challenging to bring the fires under control, resulting in a decrease in the firefighting suppression rate. Moreover, the pressure on firefighters will increase in such situations. The larger the area affected by the fire, the more difficult it becomes to contain and extinguish the fires. This cascading effect could lead to a higher number of forested areas being engulfed in flames. Consequently, the heightened pressure on firefighters will persist.

Figure 4. Ecological and operational factors of SD model

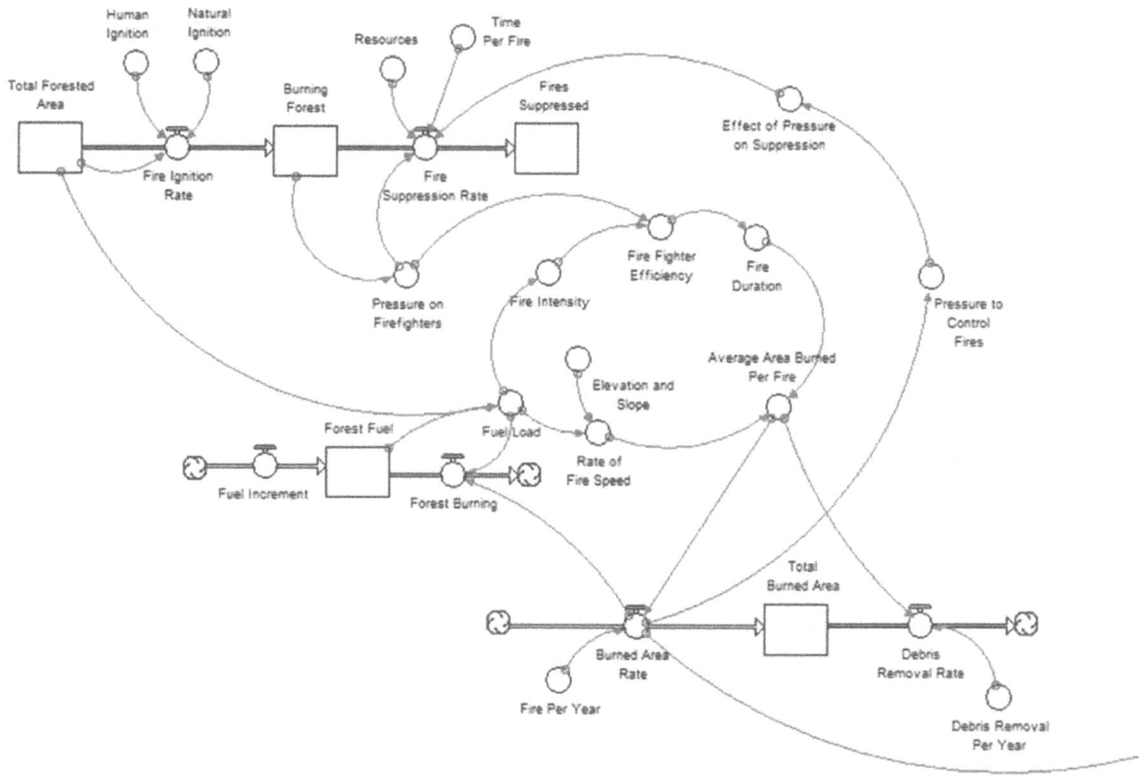

ECONOMIC FACTORS AND FOREST FIRE

Recent forest fires have posed serious threats to various areas and brought economic difficulties. Although a significant amount of literature has examined the potential harmful and immediate effects of forest fires, few studies examine their impact on economic indicators (Sfetsos et al.,2021). These fires have the potential to harm energy assets and inflict significant damage on electricity transmission lines and energy production facilities. Due to fires, energy production may be interrupted, transmission lines may be disabled, and energy facilities may need long-term maintenance and repair. Additionally, fires can limit the availability of energy resources and negatively impact energy supply, resulting in economic losses. Controlling fires and strengthening energy infrastructure can play an important role in reducing such negative impacts.

Moreover, forest fires can affect transportation infrastructures and cause road closures. As a result, the fire suppression rate will decrease since firefighting teams cannot reach to fire area easily. Another negative consequence of damage to infrastructure can be road visibility. Similarly, it will also decrease the fire suppression rates. In addition, road assets are sensitive to operational damages related to economic losses. In many countries, network users are required to pay tolls for the use of highways. When a highway is disrupted due to a natural disaster and toll booths cannot continue their operations, the revenue generated from tolls decreases. On the other hand, forest fires can have negative effects on tourism revenues. Fires can destroy the natural beauties and ecosystems in tourist areas, reducing the interest of tourists in the region. This situation may reduce the number of reservations and visitors to tourist facilities, thus leading to a decrease in tourism revenues. Additionally, infrastructure may be damaged due to fires, and impacts such as road closures may hinder tourist activities, resulting in loss of income. A decrease in tourism revenues could significantly harm the regional economy and negatively impact local businesses.

DRONES AND SYSTEM DYNAMICS FRAMEWORK

This chapter introduces a system structured around two primary components, illustrated in Fig. 6: (a) Employing drone analysis and visualization to clarify the spatial patterns of forest areas. (b) SD analysis for evaluating forest fire management. The analysis of drones consists of the following three stages. Firstly, drones can detect fire hotspot points with thermal cameras and sensors replace them. In that case, hotspots composed of different materials, such as wood, thatch, or sepiolite; hotspots with different sizes, ranging between 0.15m and 1.5m; hotspots with different stages of dynamics starting from ignition to close-to-be-extinguished can be inspected with cameras (Viseras and Garcia, 2019). Secondly, relative data is acquired concerning data acquisition factors. Drone speed, altitude, integration time, battery needs, flight time, and environmental conditions impact the data acquisition. Although the longer the flight time of the drones, the better the applicability of the system in real missions, battery time restricts the flight time which can reduce the efficiency of firefighting teams and resources. It is known that high-performance commercial drones currently provide approximately 30 minutes of uninterrupted flight. On the other hand, flying at higher altitudes increases the camera's distance from the ground surface, potentially causing more false alarms. In contrast, rotary-wing drones offer greater flexibility for close monitoring. Therefore, instead of relying solely on fixed-wing drones, which operate at altitudes ranging from 350 meters to 5500 meters for comprehensive coverage, incorporating rotary-wing drones may be more suitable for precise monitoring needs.

Figure 5. Economic factors of the SD model

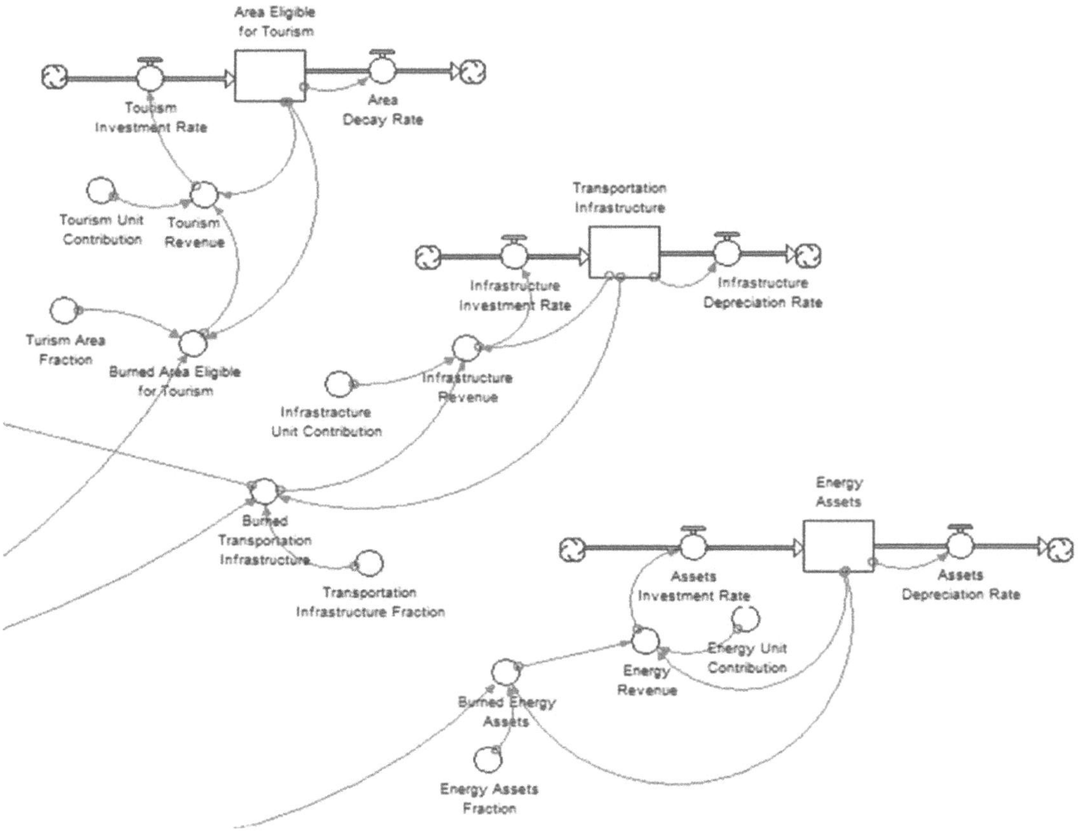

In the next step of drone analysis, meteorological, topographic, and fuel load information data is gathered. These data encompass a range of essential parameters, including temperature, humidity, wind speed, and direction, which provide insights into prevailing weather conditions influencing fire behavior. In addition, while fuel load content measurements help assess the flammability of vegetation, topographic features such as aspect, elevation, and land cover type are also crucial. Aspect determines the direction a slope faces, influencing sun exposure and moisture levels, which in turn affect fire behavior. Elevation influences temperature and vegetation types, impacting fire intensity and spread. Land cover type, including vegetation density and composition, affects fuel availability and fire behavior.

Thirdly, drone-based data is processed, and high spatial-resolution images are developed. This data can be transmitted to the fire and rescue department for early warning. However, communications pose a challenge for drones in vast and remote areas, particularly in firefighting missions where constant information exchange between various parties is crucial. It can be anticipated that drones need a communication range of 5 km to function effectively in such scenarios. In this way, fire crews can be informed of a possible fire early and response times can be shortened. Thus, it can help them to understand the extent and direction of the fire and reduce the pressure on firefighters while increasing their efficiency. When the efficiency of firefighters increases, fire duration can be decreased which results in less area burned.

Figure 6. Proposed framework for forest fire management

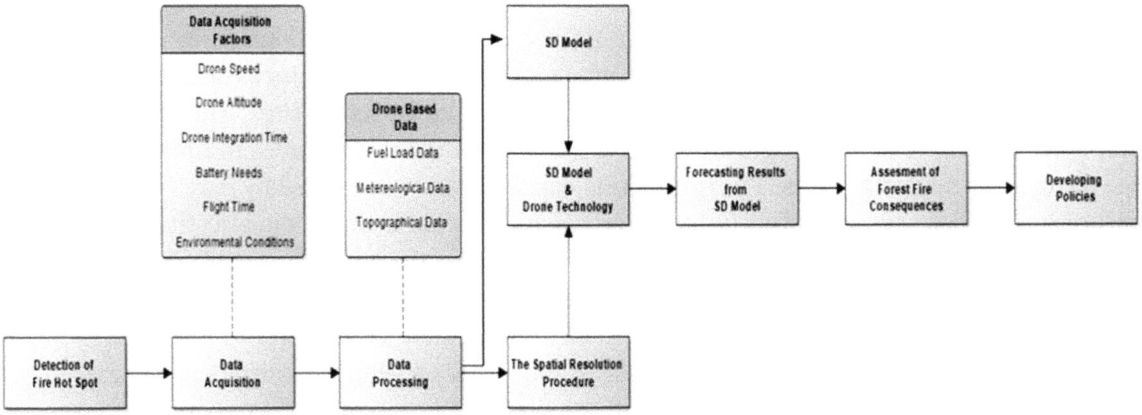

In the second part of the conceptual model, drone-based data can be integrated into the SD model to estimate the possible consequences in terms of economy, ecology, and also operations. In other words, by combining the real-time data with the SD model, the simulation model will explore the interactions between ecological, operational, and economic systems in forest fire management and assess different policies.

SOLUTION AND RECOMMENDATION

Strategies for combating forest fires primarily focus on preventing the occurrence of fires. However, even the most comprehensive and successful prevention programs cannot completely avoid forest fires in the region. Despite the best prevention efforts, there are many accidents and intentional ignitions that will continue to lead to widespread fires. Hence, it is necessary to set up a detection and reporting network that can identify, report, and monitor forest fires within and in proximity to forest reserve areas. To combat effectively with a fire, taking advantage of remote sensing technologies is crucial. To enhance the ability to respond to more fires, the time allocated to each fire incident should be minimized. Considering advancements in drone technology, spending less time on each fire can lead to the suppression of more fires. This would result in less average area burned per fire and economic losses.

Although it is not possible to completely avoid forest fires, several fire management policies can be applied via a system dynamics model to analyze its effects in the long term. The policies can be determined as implementing education programs to influence human behavior, applying different fuel management strategies to reduce combustible vegetation such as planning prescribed burning methods, ensuring effective firefighting such as increasing and/or reallocating the resources, and finally, removing mature trees by clear-cutting. While providing education programs can reduce the number of human-caused fires which would result in a lower fire ignition rate, implementing different fuel management strategies can decrease the accumulation of flammable materials such as dry grass, underbrush, and fallen tree branches and minimize the rapid-fire spread. For instance, the prescribed burning policy can be planned as a preventive measure against fires to reduce the fuel load. This approach is employed as a land management technique that aims to improve environmental protection and decrease fire risks in forested areas.

Furthermore, resources to suppress fires are limited and have to be allocated based on management strategies. Identifying high-risk areas and strategically deploying resources and equipment while implementing preventive measures can effectively reduce the incidence of fires. Therefore, increasing the resources and capacity to extinguish fires or allocating additional resources to firefighting aims to reduce the damage caused by fires. The firefighting capacity can be increased by purchasing additional aircraft or hiring more people to fire-fighting crews which would increase the effectiveness of operational teams. On the other hand, in the operational model where the performance of firefighting teams is crucial, spending progressively less time on each fire can result in inadequate clearing, giving rise to re-ignitions that further strain the firefighting capacity. These trade-offs should be managed by policymakers. Nevertheless, if the initial allocation of the resources for prevention is sufficient, the fuel amount can remain more stable. Additionally, increasing the firefighting expenditures can weaken preventive fuel management. This situation would lead to an accumulation of excessive fuel and pave the way for more intense fire seasons in the future.

Finally, the areas with a high or very high risk of easily igniting and potentially leading to large forest fires should receive special attention. Drones should be deployed more frequently in these extensive forested areas, and communication between the drones and firefighting teams should be carefully maintained. As an alternative, a clearcutting (clearcut logging) policy can also be applied where the mature and strong vegetation trees in an area are removed, and the area is replaced with younger trees. However, this policy can be dangerous in the long term since plantation trees can be highly vulnerable to fire. Besides that, precaution is required in non-forested areas with a higher risk of ignition. Particularly, when these areas are close to potentially hazardous zones that need protection, many fires starting in agricultural or shrub areas can rapidly spread to forest stands or wooded regions, so caution should be exercised.

FUTURE RESEARCH DIRECTIONS

In this book chapter, the modeling framework has focused on analyses related to the ecological state of the forest, as well as the economic sectors such as tourism, energy, and infrastructure. However, the costs associated with fire prevention and suppression, which can be equally important driving forces in forest and fire management, have not been considered here. Specifically, while the use of drones may reduce intervention times in forest fires and lead to more efficient work by rescue teams, the costs of using drones, including transportation, monitoring, and charging costs, should be calculated, and evaluated. Taking these variables into account may require reconsideration of the resilience of the formulated policies. For instance, higher resistance near communities to increased burn rates and the costs involved can be included in the policy analysis framework to help determine the return on investment.

Another dynamic factor that could be included in the model is how climate change may intensify fire severity and impede regeneration. The compound effects of this, especially considering how sustained burning could further solidify an undesirable shift in conditions, will strongly emphasize the importance of minimizing high-severity fires. In addition to climate change factors, land management activities such as timber harvest, hazardous fuel load reduction, or educating the population about forest fires can be investigated in further research. Fire prevention, which is fully integrated into land-use planning, is necessary to counteract the effects of land-use change on fire hazard and risk.

CONCLUSION

Forest fires have been a persistent challenge in various regions worldwide, posing significant ecological, operational, and economic consequences in sectors such as tourism, infrastructure, and energy. This study aims to understand the root causes of the system and proposes the development of policies to mitigate potential negative impacts. A conceptual model is developed by bringing together the relationships of various factors to cover all relevant factors in the system. The main parties are determined as the environmental factors that will start forest fires; the operational factors consisting of firefighting teams who will carry out the necessary operations to extinguish the fire; and economic factors such as tourism areas, energy assets, or transportation infrastructures. The application of SD methodology in the planning of forest fire management can be concluded to provide an alternative to existing management approaches primarily based on linear projections of a specific system's future. Such an alternative should not be overlooked. Additionally, most of the regions have low population density and are less accessible for firefighters as they are situated farther away from roads. Therefore, in fire planning and monitoring efforts, it is essential to simultaneously assess and analyze the fire risk in these areas. In that sense, drones play an important role in simultaneous data collaboration with firefighters. With the help of these data, the SD model can estimate the ecological, operational, and economic outcomes.

The proposed framework shows that utilizing drones for early detection of potential fire-prone areas or the origin of ongoing fires, coupled with simultaneous alerts to firefighting crews, can increase the suppression of fires. This approach could lead to a decrease in the average area burned per fire incident. Additionally, the feedback mechanisms represented in the conceptual model enable researchers to capture the dominant factors in the forest fire problem, aiding in a more comprehensive understanding of the issue.

REFERENCES

Abdo, H. G., Almohamad, H., Al Dughairi, A. A., & Al-Mutiry, M. (2022). GIS-based frequency ratio and analytic hierarchy process for forest fire susceptibility mapping in the western region of Syria. *Sustainability (Basel)*, *14*(8), 4668. doi:10.3390/su14084668

Abedi Gheshlaghi, H., Feizizadeh, B., & Blaschke, T. (2020). GIS-based forest fire risk mapping using the analytical network process and fuzzy logic. *Journal of Environmental Planning and Management*, *63*(3), 481–499. doi:10.1080/09640568.2019.1594726

Adsanver, B., Coban, E., & Balcik, B. (2024). A predictive multistage postdisaster damage assessment framework for drone routing. *International Transactions in Operational Research*, itor.13429. doi:10.1111/itor.13429

Ahirwar, S., Swarnkar, R., Bhukya, S., & Namwade, G. (2019). Application of drone agriculture. *International Journal of Current Microbiology and Applied Sciences*, *8*(01), 2500–2505. doi:10.20546/ijcmas.2019.801.264

Ahmad, S., & Simonovic, S. P. (2001, May). Modeling Dynamic Processes in Space and Time--A Spatial System Dynamics Approach. In Bridging the Gap: Meeting the World's Water and Environmental Resources Challenges (pp. 1-20). Research Gate.

Akay, A. E., & Şahin, H. (2019). Forest fire risk mapping by using GIS techniques and AHP method: A case study in Bodrum (Turkey). *European Journal of Forest Engineering*, *5*(1), 25–35. doi:10.33904/ejfe.579075

Amaral, L. R., de Freitas, R. G., Júnior, M. R. B., & da Silva Simões, I. O. P. (2022). Application of Drones in Agriculture. In *Digital Agriculture* (pp. 99–121). Springer International Publishing. doi:10.1007/978-3-031-14533-9_7

Andela, N., Morton, D. C., Giglio, L., Chen, Y., van der Werf, G. R., Kasibhatla, P. S., DeFries, R. S., Collatz, G. J., Hantson, S., Kloster, S., Bachelet, D., Forrest, M., Lasslop, G., Li, F., Mangeon, S., Melton, J. R., Yue, C., & Randerson, J. T. (2017). A human-driven decline in the global burned area. *Science*, *356*(6345), 1356–1362. doi:10.1126/science.aal4108 PMID:28663495

Andela, N., & Van Der Werf, G. R. (2014). Recent trends in African fires driven by cropland expansion and El Niño to La Niña transition. *Nature Climate Change*, *4*(9), 791–795. doi:10.1038/nclimate2313

Aragon, G., Martínez, I., Izquierdo, P., Belinchón, R., & Escudero, A. (2010). Effects of forest management on epiphytic lichen diversity in Mediterranean forests. *Applied Vegetation Science*, *13*(2), 183–194. doi:10.1111/j.1654-109X.2009.01060.x

Bala, B. K., Arshad, F. M., & Noh, K. M. (2017). System dynamics. *Modelling and Simulation (Anaheim)*, 274.

Barbrook-Johnson, P., & Penn, A. S. (2022). *Systems Mapping: How to build and use causal models of systems*. Springer Nature. doi:10.1007/978-3-031-01919-7

Boegelsack, N., Withey, J., O'Sullivan, G., & McMartin, D. (2018). A critical examination of the relationship between wildfires and climate change with consideration of the human impact. *Journal of Environmental Protection*, *9*(5), 461–467. doi:10.4236/jep.2018.95028

Bradstock, R. A. (2010). A biogeographic model of fire regimes in Australia: Current and future implications. *Global Ecology and Biogeography*, *19*(2), 145–158. doi:10.1111/j.1466-8238.2009.00512.x

Brando, P. M., Balch, J. K., Nepstad, D. C., Morton, D. C., Putz, F. E., Coe, M. T., Silvério, D., Macedo, M. N., Davidson, E. A., Nóbrega, C. C., Alencar, A., & Soares-Filho, B. S. (2014). Abrupt increases in Amazonian tree mortality due to drought–fire interactions. *Proceedings of the National Academy of Sciences of the United States of America*, *111*(17), 6347–6352. doi:10.1073/pnas.1305499111 PMID:24733937

Campbell, J. B., & Wynne, R. H. (2011). *Introduction to remote sensing*. Guilford press.

Cawson, J. G., Duff, T. J., Tolhurst, K. G., Baillie, C. C., & Penman, T. D. (2017). Fuel moisture in Mountain Ash forests with contrasting fire histories. *Forest Ecology and Management*, *400*, 568–577. doi:10.1016/j.foreco.2017.06.046

Chamola, V., Hassija, V., Gupta, V., & Guizani, M. (2020). A comprehensive review of the COVID-19 pandemic and the role of IoT, drones, AI, blockchain, and 5G in managing its impact. *IEEE Access : Practical Innovations, Open Solutions*, *8*, 90225–90265. doi:10.1109/ACCESS.2020.2992341

Chen, Y., Hakala, T., Karjalainen, M., Feng, Z., Tang, J., Litkey, P., Kukko, A., Jaakkola, A., & Hyyppä, J. (2017). UAV-borne profiling radar for forest research. *Remote Sensing (Basel)*, *9*(1), 58. doi:10.3390/rs9010058

Cleveland, J. (1998). Prescribed fire: the fundamental solution. In*: Pruden, Teresa L.; Brennan, Leonard A., eds. Proceedings: 20th Tall Timbers Fire Ecology Conference; Fire in ecosystem management: Shifting the paradigm from suppression to prescription. Tallahassee, FL: Tall Timbers Research Station. p. 12-16.* (pp. 12-16). Research Gate.

Collins, R. D. (2012). *Forest fire management in Portugal: developing system insights through models of social and physical dynamics* [Doctoral dissertation, Massachusetts Institute of Technology].

Collins, R. D., de Neufville, R., Claro, J., Oliveira, T., & Pacheco, A. P. (2013). Forest fire management to avoid unintended consequences: A case study of Portugal using system dynamics. *Journal of Environmental Management*, *130*, 1–9. doi:10.1016/j.jenvman.2013.08.033 PMID:24036501

Cosgun, U., Coşkun, M., Toprak, F., Yıldız, D., Coşkun, S., Taşoğlu, E., & Öztürk, A. (2023). Visibility evaluation and suitability analysis of fire lookout towers in the Mediterranean Region, Southwest Anatolia/Türkiye. *Fire (Basel, Switzerland)*, *6*(8), 305. doi:10.3390/fire6080305

Dhall, A., Dhasade, A., & Nalwade, A., VK, M. R., & Kulkarni, V. (2020). A survey on systematic approaches in managing forest fires. *Applied Geography (Sevenoaks, England)*, *121*, 102266. doi:10.1016/j.apgeog.2020.102266

Dogra, R., Rani, S., & Sharma, B. (2021). A review of forest fires and their detection techniques using wireless sensor network. In *Advances in Communication and Computational Technology: Select Proceedings of ICACCT 2019* (pp. 1339-1350). Springer Singapore. 10.1007/978-981-15-5341-7_101

Dorrer, G., Dorrer, A., Buslov, I., & Yarovoy, S. (2018, November). System of personnel training in decision-making in fighting wildfires. [). IOP Publishing.]. *IOP Conference Series. Materials Science and Engineering*, *450*(6), 062018. doi:10.1088/1757-899X/450/6/062018

Dougherty, S., Simpson, J. R., Hill, R. R., Pignatiello, J. J., & White, E. D. (2015). Nonlinear screening designs for defense testing: An overview and case study. *The Journal of Defense Modeling and Simulation*, *12*(3), 335–342. doi:10.1177/1548512914542523

Elachi, C., & Van Zyl, J. J. (2021). *Introduction to the physics and techniques of remote sensing*. John Wiley & Sons. doi:10.1002/9781119523048

Farkhondehmaal, F., & Ghaffarzadegan, N. (2022). A cyclical wildfire pattern is the outcome of a coupled human-natural system. *Scientific Reports*, *12*(1), 5280. doi:10.1038/s41598-022-08730-y PMID:35347175

Fischer, W. A., Hemphill, W. R., & Kover, A. (1976). Progress in remote sensing (1972-1976). *Photogrammetria*, *32*(2), 33–72. doi:10.1016/0031-8663(76)90013-2

Frisken, S., Clarke, I., & Poole, S. (2013). Technology and applications of liquid crystal on silicon (LCoS) in telecommunications. *Optical Fiber Telecommunications*, 709-742.

Gale, M. G., Cary, G. J., Van Dijk, A. I., & Yebra, M. (2021). Forest fire fuel through the lens of remote sensing: Review of approaches, challenges and future directions in the remote sensing of biotic determinants of fire behavior. *Remote Sensing of Environment*, *255*, 112282. doi:10.1016/j.rse.2020.112282

Guimarães, N., Pádua, L., Marques, P., Silva, N., Peres, E., & Sousa, J. J. (2020). Forestry remote sensing from unmanned aerial vehicles: A review focusing on the data, processing, and potentialities. *Remote Sensing (Basel)*, *12*(6), 1046. doi:10.3390/rs12061046

Gülçin, D., & Deniz, B. (2020). Remote sensing and GIS-based forest fire risk zone mapping: The case of Manisa, Turkey. *Turkish Journal of Forestry*, *21*(1), 15–24.

Hillayová, M. K., Holécy, J., Korísteková, K., Bakšová, M., Ostrihoň, M., & Škvarenina, J. (2023). Ongoing climatic change increases the risk of wildfires. Case study: Carpathian spruce forests. *Journal of Environmental Management*, *337*, 117620. doi:10.1016/j.jenvman.2023.117620 PMID:36934505

Hong, H., Jaafari, A., & Zenner, E. K. (2019). Predicting spatial patterns of wildfire susceptibility in the Huichang County, China: An integrated model to analysis of landscape indicators. *Ecological Indicators*, *101*, 878–891. doi:10.1016/j.ecolind.2019.01.056

Ingalsbee, T. (2017). What is the paradigm shift? Large wildland fires and the wildfire paradox offer opportunities for a new paradigm of ecological fire management. *International Journal of Wildland Fire*, *26*(7), 557–561. doi:10.1071/WF17062

Israr, A., Ali, Z. A., Alkhammash, E. H., & Jussila, J. J. (2022). Optimization methods applied to motion planning of unmanned aerial vehicles: A review. *Drones (Basel)*, *6*(5), 126. doi:10.3390/drones6050126

Jia, N., Yang, Z., & Yang, K. (2019). Operational effectiveness evaluation of the swarming UAV combat system based on a system dynamics model. *IEEE Access : Practical Innovations, Open Solutions*, *7*, 25209–25224. doi:10.1109/ACCESS.2019.2898728

Jiao, D., & Fei, T. (2023). Pedestrian walking speed monitoring at street scale by an in-flight drone. *PeerJ. Computer Science*, *9*, e1226. doi:10.7717/peerj-cs.1226 PMID:37346670

John, S. (2000). *Business dynamics: systems thinking and modeling for a complex world*. Irwin McGraw-Hill.

Khan, S., & Khan, A. (2022). Ffirenet: Deep learning-based forest fire classification and detection in smart cities. *Symmetry*, *14*(10), 2155. doi:10.3390/sym14102155

Kim, B. K., Kang, H. S., & Park, S. O. (2016). Drone classification using convolutional neural networks with merged Doppler images. *IEEE Geoscience and Remote Sensing Letters*, *14*(1), 38–42. doi:10.1109/LGRS.2016.2624820

Kim, S. J., Lim, G. J., Cho, J., & Côté, M. J. (2017). Drone-aided healthcare services for patients with chronic diseases in rural areas. *Journal of Intelligent & Robotic Systems*, *88*(1), 163–180. doi:10.1007/s10846-017-0548-z

Kinaneva, D., Hristov, G., Raychev, J., & Zahariev, P. (2019, May). Early forest fire detection using drones and artificial intelligence. In *2019 42nd International Convention on Information and Communication Technology, Electronics and Microelectronics (MIPRO)* (pp. 1060-1065). IEEE. 10.23919/MIPRO.2019.8756696

Leal Filho, W., Azul, A. M., Brandli, L., Lange Salvia, A., & Wall, T. (Eds.). (2021). *Life on land*. Springer International Publishing. doi:10.1007/978-3-319-95981-8

Li, H., Wang, F., & Zhan, Z. (2023). Truck and rotary-wing drone routing problem considering flight-level selection. *The Journal of the Operational Research Society*, 1–19.

Loncar, L., Hell, M., & Dusak, V. (2006, June). A system dynamics model of forest management. In *28th International Conference on Information Technology Interfaces.* (pp. 549-556). IEEE. 10.1109/ITI.2006.1708540

Madadgar, S., Sadegh, M., Chiang, F., Ragno, E., & AghaKouchak, A. (2020). Quantifying increased fire risk in California in response to different levels of warming and drying. *Stochastic Environmental Research and Risk Assessment*, *34*(12), 2023–2031. doi:10.1007/s00477-020-01885-y

Mahrad, B. E., Newton, A., Icely, J. D., Kacimi, I., Abalansa, S., & Snoussi, M. (2020). Contribution of remote sensing technologies to a holistic coastal and marine environmental management framework: A review. *Remote Sensing (Basel)*, *12*(14), 2313. doi:10.3390/rs12142313

Mangiameli, M., Mussumeci, G., & Cappello, A. (2021). Forest fire spreading using free and open-source GIS technologies. *Geomatics*, *1*(1), 50–64. doi:10.3390/geomatics1010005

Mann, M. L., Batllori, E., Moritz, M. A., Waller, E. K., Berck, P., Flint, A. L., Flint, L. E., & Dolfi, E. (2016). Incorporating anthropogenic influences into fire probability models: Effects of human activity and climate change on fire activity in California. *PLoS One*, *11*(4), e0153589. doi:10.1371/journal.pone.0153589 PMID:27124597

Masoudvaziri, N., Ganguly, P., Mukherjee, S., & Sun, K. (2021). Impact of geophysical and anthropogenic factors on wildfire size: A spatiotemporal data-driven risk assessment approach using statistical learning. *Stochastic Environmental Research and Risk Assessment*, 1–27.

Matese, A., Toscano, P., Di Gennaro, S. F., Genesio, L., Vaccari, F. P., Primicerio, J., Belli, C., Zaldei, A., Bianconi, R., & Gioli, B. (2015). Intercomparison of UAV, aircraft, and satellite remote sensing platforms for precision viticulture. *Remote Sensing (Basel)*, *7*(3), 2971–2990. doi:10.3390/rs70302971

Mayer, S., Lischke, L., & Woźniak, P. W. (2019, May). Drones for search and rescue. In *1st International Workshop on Human-Drone Interaction*. Research Gate.

Mhawej, M., Faour, G., & Adjizian-Gerard, J. (2017). A novel method to identify likely causes of wildfire. *Climate Risk Management*, *16*, 120–132. doi:10.1016/j.crm.2017.01.004

Mohammed, A. A., & Khamees, H. T. (2021). Categorizing and measuring satellite image processing of fire in the forest Greece using remote sensing. *Indonesian Journal of Electrical Engineering and Computer Science*, *21*(2), 843–853. doi:10.11591/ijeecs.v21.i2.pp846-853

Momeni, M., Mirzapour Al-e-Hashem, S. M. J., & Heidari, A. (2023). A new truck-drone routing problem for parcel delivery by considering energy consumption and altitude. *Annals of Operations Research*, 1–47. doi:10.1007/s10479-023-05381-8 PMID:37361075

Moradi, M., Kazeminezhad, M. H., & Kabiri, K. (2020). Integration of Geographic Information System and system dynamics for assessment of the impacts of storm damage on coastal communities study Chabahar, Iran. *International Journal of Disaster Risk Reduction*, *49*, 101665. doi:10.1016/j.ijdrr.2020.101665

Nagy, R. C., Fusco, E., Bradley, B., Abatzoglou, J. T., & Balch, J. (2018). Human-related ignitions increase the number of large wildfires across US ecoregions. *Fire (Basel, Switzerland)*, *1*(1), 4. doi:10.3390/fire1010004

Nigam, A., & Agarwal, Y. K. (2014). Optimal relay node placement in delay-constrained wireless sensor network design. *European Journal of Operational Research*, *233*(1), 220–233. doi:10.1016/j.ejor.2013.08.031

Nolan, R. H., Collins, L., Leigh, A., Ooi, M. K., Curran, T. J., Fairman, T. A., Resco de Dios, V., & Bradstock, R. (2021). Limits to post-fire vegetation recovery under climate change. *Plant, Cell & Environment*, *44*(11), 3471–3489. doi:10.1111/pce.14176 PMID:34453442

Ottmar, R. D. (2014). Wildland fire emissions, carbon, and climate: Modeling fuel consumption. *Forest Ecology and Management*, *317*, 41–50. doi:10.1016/j.foreco.2013.06.010

Pakere, I., Kacare, M., Grāvelsiņš, A., Freimanis, R., & Blumberga, A. (2022). Spatial analyses of smart energy system implementation through system dynamics and GIS modeling. Wind power case study in Latvia. *Smart Energy*, *7*, 100081. doi:10.1016/j.segy.2022.100081

Pandey, P., Huidobro, G., Lopes, L. F., Ganteaume, A., Ascoli, D., Colaco, C., Xanthopoulos, G., Giannaros, T. M., Gazzard, R., Boustras, G., Steelman, T., Charlton, V., Ferguson, E., Kirschner, J., Little, K., Stoof, C., Nikolakis, W., Fernández-Blanco, C. R., Ribotta, C., & Dossi, S. (2023). A global outlook on increasing wildfire risk: Current policy situation and future pathways. *Trees, Forests and People*, *14*, 100431. doi:10.1016/j.tfp.2023.100431

Pausas, J. G., & Keeley, J. E. (2009). A burning story: The role of fire in the history of life. *Bioscience*, *59*(7), 593–601. doi:10.1525/bio.2009.59.7.10

Pensieri, M. G., Garau, M., & Barone, P. M. (2020). Drones as an integral part of remote sensing technologies to help missing people. *Drones (Basel)*, *4*(2), 15. doi:10.3390/drones4020015

Petrie, G. (2009). Systematic oblique aerial photography using multiple digital cameras. *Photogrammetric Engineering and Remote Sensing*, *75*(2), 102–107.

Pu, R. (2021). Mapping tree species using advanced remote sensing technologies: A state-of-the-art review and perspective. *Yaogan Xuebao*.

Restás, Á. (2014). *Thematic division and tactical analysis of the UAS application supporting forest fire management*.

Restás, Á. (2015). Drone applications for supporting disaster management. *World Journal of Engineering and Technology*, *3*(03), 316–321. doi:10.4236/wjet.2015.33C047

Restás, Á. (2016). Drone applications for preventing and responding to HAZMAT disaster. *World Journal of Engineering and Technology*, *4*(3), 76–84. doi:10.4236/wjet.2016.43C010

Riano, D., Chuvieco, E., Condés, S., González-Matesanz, J., & Ustin, S. L. (2004). Generation of crown bulk density for Pinus sylvestris L. from lidar. *Remote Sensing of Environment*, *92*(3), 345–352. doi:10.1016/j.rse.2003.12.014

Rossa, C. G. (2018). A generic fuel moisture content attenuation factor for fire spread rate empirical models. *Forest Systems*, *27*(2), e009–e009. doi:10.5424/fs/2018272-13175

Ruffault, J., Martin-StPaul, N., Pimont, F., & Dupuy, J. L. (2018). How well do meteorological drought indices predict live fuel moisture content (LFMC)? An assessment for wildfire research and operations in Mediterranean ecosystems. *Agricultural and Forest Meteorology*, *262*, 391–401. doi:10.1016/j.agrformet.2018.07.031

Sfetsos, A., Giroud, F., Clemencau, A., Varela, V., Freissinet, C., LeCroart, J., Vlachogiannis, D., Politi, N., Karozis, S., Gkotsis, I., Eftychidis, G., Hedel, R., & Hahmann, S. (2021). Assessing the effects of forest fires on interconnected critical infrastructures under climate change. Evidence from South France. *Infrastructures*, *6*(2), 16. doi:10.3390/infrastructures6020016

Somers, V. L., & Manchester, I. R. (2022). Minimizing the risk of spreading processes via surveillance schedules and sparse control. *IEEE Transactions on Control of Network Systems*, *10*(1), 394–406. doi:10.1109/TCNS.2022.3203359

Song, J., Cannatella, D., & Katsikis, N. (2023, June). Landscape-Based Fire Resilience: Identifying Interaction Between Landscape Dynamics and Fire Regimes in the Mediterranean Region. In *International Conference on Computational Science and Its Applications* (pp. 328-344). Cham: Springer Nature Switzerland. 10.1007/978-3-031-37117-2_23

Srivastava, S. K., Seng, K. P., Ang, L. M., Pachas, A. N. A., & Lewis, T. (2022). Drone-Based Environmental Monitoring and Image Processing Approaches for Resource Estimates of Private Native Forest. *Sensors (Basel)*, *22*(20), 7872. doi:10.3390/s22207872 PMID:36298223

Steelman, T., & Nowell, B. (2019). Evidence of effectiveness in the Cohesive Strategy: Measuring and improving wildfire response. *International Journal of Wildland Fire*, *28*(4), 267–274. doi:10.1071/WF18136

Tang, C. S., & Veelenturf, L. P. (2019). The strategic role of logistics in the Industry 4.0 era. *Transportation Research Part E, Logistics and Transportation Review*, *129*, 1–11. doi:10.1016/j.tre.2019.06.004

Themistocleous, K. (2017, October). The use of UAVs for monitoring land degradation. In *Earth resources and environmental remote sensing/GIS applications VIII* (Vol. 10428, pp. 127–137). SPIE. doi:10.1117/12.2279512

Thompson, M. P., & Calkin, D. E. (2011). Uncertainty and risk in wildland fire management: A review. *Journal of Environmental Management*, *92*(8), 1895–1909. doi:10.1016/j.jenvman.2011.03.015 PMID:21489684

Thompson, M. P., Dunn, C. J., & Calkin, D. E. (2016). *Systems thinking and wildland fire management*. In *60th Annual Meeting of the ISSS-2016 Boulder*, CO, USA.

Thompson, M. P., MacGregor, D. G., Dunn, C. J., Calkin, D. E., & Phipps, J. (2018). Rethinking the wildland fire management system. *Journal of Forestry*, *116*(4), 382–390. doi:10.1093/jofore/fvy020

Thompson, M. P., Wei, Y., Dunn, C. J., & O'Connor, C. D. (2019). A system dynamics model examining alternative wildfire response policies. *Systems*, *7*(4), 49. doi:10.3390/systems7040049

Toth, C., & Jóźków, G. (2016). Remote sensing platforms and sensors: A survey. *ISPRS Journal of Photogrammetry and Remote Sensing*, *115*, 22–36. doi:10.1016/j.isprsjprs.2015.10.004

Vilar, L., Camia, A., San-Miguel-Ayanz, J., & Martín, M. P. (2016). Modeling temporal changes in human-caused wildfires in Mediterranean Europe based on Land Use-Land Cover interfaces. *Forest Ecology and Management*, *378*, 68–78. doi:10.1016/j.foreco.2016.07.020

Viseras, A., & Garcia, R. (2019). DeepIG: Multi-robot information gathering with deep reinforcement learning. *IEEE Robotics and Automation Letters*, *4*(3), 3059–3066. doi:10.1109/LRA.2019.2924839

Wang, Z., & Menenti, M. (2021). Challenges and opportunities in Lidar remote sensing. *Frontiers in Remote Sensing*, *2*, 641723. doi:10.3389/frsen.2021.641723

Ward, D. S., Shevliakova, E., Malyshev, S., & Rabin, S. (2018). Trends and variability of global fire emissions due to historical anthropogenic activities. *Global Biogeochemical Cycles*, *32*(1), 122–142. doi:10.1002/2017GB005787

Williams, A. P., Abatzoglou, J. T., Gershunov, A., Guzman-Morales, J., Bishop, D. A., Balch, J. K., & Lettenmaier, D. P. (2019). Observed impacts of anthropogenic climate change on wildfire in California. *Earth's Future*, *7*(8), 892–910. doi:10.1029/2019EF001210

Xiang, Y., Lv, L., Chai, W., Zhang, T., Liu, J., & Liu, W. (2020). Using Lidar technology to assess regional air pollution and improve estimates of PM2. 5 transport in the North China Plain. *Environmental Research Letters*, *15*(9), 094071. doi:10.1088/1748-9326/ab9cfd

Xu, Z., & Coors, V. (2012). Combining system dynamics model, GIS, and 3D visualization in sustainability assessment of urban residential development. *Building and Environment*, *47*, 272–287. doi:10.1016/j.buildenv.2011.07.012

Yandouzi, M., Grari, M., Idrissi, I., , Moussaoui, O., Azizi, M.,, Ghoumid, K. A. M. A. L., & Elmiad, A. K. (2022). Review on forest fire detection and prediction using deep learning and drones. *Journal of Theoretical and Applied Information Technology*, *100*(12), 4565–4576.

Yearworth, M. (2014). A brief introduction to system dynamics modeling. University of Bristol.

Yuan, C., Liu, Z., & Zhang, Y. (2017, June). Fire detection using infrared images for UAV-based forest fire surveillance. In *2017 International Conference on Unmanned Aircraft Systems (ICUAS)* (pp. 567-572). IEEE. 10.1109/ICUAS.2017.7991306

Zong, X., Tian, X., & Yin, Y. (2020). Impacts of climate change on wildfires in Central Asia. *Forests*, *11*(8), 802. doi:10.3390/f11080802

Compilation of References

Abaie, E., Kumar, M., Garza-Rubalcava, U., Rao, B., Sun, Y., Shen, Y., & Reible, D. (2024). Chlorinated volatile organic compounds (CVOCs) and 1,4-dioxane kinetics and equilibrium adsorption studies on selective macrocyclic adsorbents. *Environmental Advances*, *16*, 100520. doi:10.1016/j.envadv.2024.100520

Abbasi, T., Sanjeevi, R., Anuradha, J., & Abbasi, S. A. (2013). Impact of Al 3+ on sludge granulation in UASB reactor. *Indian Journal of Biotechnology*, *12*(2), 254–259.

Abbasi, T., Sanjeevi, R., Makhija, M., & Abbasi, S. A. (2012). Role of Vitamins B-3 and C in the Fashioning of Granules in UASB Reactor Sludge. *Applied Biochemistry and Biotechnology*, *167*(2), 348–357. doi:10.1007/s12010-012-9691-y PMID:22549583

Abbaspour, A., Norouz-Sarvestani, F., Noori, A., & Soltani, N. (2015). Aptamer-conjugated silver nanoparticles for electrochemical dual-aptamer-based sandwich detection of staphylococcus aureus. *Biosensors & Bioelectronics*, *68*, 149–155. doi:10.1016/j.bios.2014.12.040 PMID:25562742

Abdelghani, A., & Jaffrezic-Renault, N. (2001). SPR fibre sensor sensitised by fluorosiloxane polymers. *Sensors and Actuators. B, Chemical*, *74*(1–3), 117–123. doi:10.1016/S0925-4005(00)00720-6

Abdo, H. G., Almohamad, H., Al Dughairi, A. A., & Al-Mutiry, M. (2022). GIS-based frequency ratio and analytic hierarchy process for forest fire susceptibility mapping in the western region of Syria. *Sustainability (Basel)*, *14*(8), 4668. doi:10.3390/su14084668

Abdulraheem, M. I., Zhang, W., Li, S., Moshayedi, A. J., Farooque, A. A., & Hu, J. (2023). Advancement of remote sensing for soil measurements and applications: A comprehensive review. *Sustainability (Basel)*, *15*(21), 15444. doi:10.3390/su152115444

Abedi Gheshlaghi, H., Feizizadeh, B., & Blaschke, T. (2020). GIS-based forest fire risk mapping using the analytical network process and fuzzy logic. *Journal of Environmental Planning and Management*, *63*(3), 481–499. doi:10.1080/09640568.2019.1594726

Adaros, G., Weigel, H. J., & Jager, H. J. (1991). Single and interactive effects of low levels of O_3, SO_2 and NO_2 on the growth and yield of spring rape. *Environmental Pollution*, *4*(4), 269–286. doi:10.1016/0269-7491(91)90002-E PMID:15092095

Adekunle, A., Rickwood, C., & Tartakovsky, B. (2021). On-line monitoring of water quality with a floating microbial fuel cell biosensor: Field test results. *Ecotoxicology (London, England)*, *30*(5), 851–862. doi:10.1007/s10646-021-02409-2 PMID:33851335

Adsanver, B., Coban, E., & Balcik, B. (2024). A predictive multistage postdisaster damage assessment framework for drone routing. *International Transactions in Operational Research*, itor.13429. doi:10.1111/itor.13429

Agarwal, T. (2017, March 15). *Different Types of Optical Sensors and Applications. ElProCus - Electronic Projects for Engineering Students*. Elprocuss. https://www.elprocus.com/optical-sensors-types-basics-and-applications/

Agarwala, N. (2020). Monitoring the ocean environment using robotic systems: Advancements, trends, and challenges. *Marine Technology Society Journal*, *54*(5), 42–60. doi:10.4031/MTSJ.54.5.7

Aghababai Beni, A., & Jabbari, H. (2022). *Nanomaterials for Environmental Applications. Results in Engineering*. Elsevier B.V., doi:10.1016/j.rineng.2022.100467

Agrawal, M. (1985). *Plant factors as indicator of SO_2 and O_3 pollutants. Proc. International Symposium on Biological Monitoring of the State Environment (Bio- indicator)*. Indian National Science Academy.

Agrawal, M., & Agrawal, S. B. (1990). Effects of ozone exposure on enzymes and metabolites of nitrogen metabolism. *Scientia Horticulturae*, *43*(1-2), 169–177. doi:10.1016/0304-4238(90)90048-J

Agrawal, M., & Deepak, S. S. (2003). Physiological and biochemical response of two cultivars of wheat to elevated levels of CO_2 and SO_2 singly and in combination. *Environmental Pollution*, *121*(2), 189–197. doi:10.1016/S0269-7491(02)00222-1 PMID:12521107

Agrawal, M., Singh, B., Agrawal, S. B., Bell, J. N. B., & Marshall, F. (2006). The effect of air pollution on yield and quality of Mung bean grown in Peri-urban areas of Varanasi. *Water, Air, and Soil Pollution*, *169*(1-4), 239–254. doi:10.1007/s11270-006-2237-6

Agrawal, M., Singh, B., Rajput, M., Marshall, F., & Bell, J. N. B. (2003). Effect of air pollution on peri-urban agriculture: A case study. *Environmental Pollution*, *126*(3), 323–329. doi:10.1016/S0269-7491(03)00245-8 PMID:12963293

Agrawal, S. B., Singh, A., & Rathore, D. (2005). Role of ethylene diurea (EDU) in assessing impact of ozone on *Vigna radiata* L. plants in a suburban area of Allahabad (India). *Chemosphere*, *61*(2), 218–228. doi:10.1016/j.chemosphere.2005.01.087 PMID:16168745

Aguilar-Pérez, K. M., Heya, M. S., Parra-Saldívar, R., & Iqbal, H. M. (2020). Nano-biomaterials in-focus as sensing/detection cues for environmental pollutants. *Case Studies in Chemical and Environmental Engineering*, *2*, 100055. doi:10.1016/j.cscee.2020.100055

Ahamad, A., Madhav, S., Singh, A. K., Kumar, A., & Singh, P. (2020). Types of Water Pollutants: Conventional and Emerging. In D. Pooja, P. Kumar, P. Singh, & S. Patil (Eds.), *Sensors in Water Pollutants Monitoring: Role of Material* (pp. 21–41). Springer Singapore. doi:10.1007/978-981-15-0671-0_3

Ahirwar, S., Swarnkar, R., Bhukya, S., & Namwade, G. (2019). Application of drone agriculture. *International Journal of Current Microbiology and Applied Sciences*, *8*(01), 2500–2505. doi:10.20546/ijcmas.2019.801.264

Ahmad, O. (2021, March 8). *As India breaks the taboo over sanitary pads, an environmental crisis mounts*. The Third Pole. https://www.thethirdpole.net/en/culture/sanitary-pads-environmental-crisis/

Ahmad, S., & Simonovic, S. P. (2001, May). Modeling Dynamic Processes in Space and Time--A Spatial System Dynamics Approach. In Bridging the Gap: Meeting the World's Water and Environmental Resources Challenges (pp. 1-20). Research Gate.

Ahmad, R. G., & Kumar, V. (2020). Microorganism Based Biosensors to Detect Soil Pollutants. *Plant Archives*, *20*(2), 2509–2516.

Ahmed, M. H., El-Hamed, N. N. B. A., & Shalby, N. I. (2017). Impact of Physico-chemical parameters on composition and diversity of zooplankton Community in Nozha Hydrodrome, Alexandria, Egypt. *Egyptian Journal of Aquatic Biology and Fisheries*, *21*(1), 49–62. doi:10.21608/ejabf.2017.2382

Akay, A. E., & Şahin, H. (2019). Forest fire risk mapping by using GIS techniques and AHP method: A case study in Bodrum (Turkey). *European Journal of Forest Engineering*, *5*(1), 25–35. doi:10.33904/ejfe.579075

Akhtar, N., Ishak, M. Z., Bhawani, S. A., & Umar, K. (2021). Various natural and anthropogenic factors responsible for water quality degradation: A review. *Water (Basel)*, *13*(19), 2660. doi:10.3390/w13192660

Akyilmaz, E., & Dinckaya, E. (2000). A mushroom (Agaricus bisporus) tissue homogenate based alcohol oxidase electrode for alcohol determination in serum. *Talanta*, *53*(3), 505–509. doi:10.1016/S0039-9140(00)00517-8 PMID:18968136

Al Mamun, M. A., & Yuce, M. R. (2019). Sensors and systems for wearable environmental monitoring toward IoT-enabled applications: A review. *IEEE Sensors Journal*, *19*(18), 7771–7788. doi:10.1109/JSEN.2019.2919352

Alcamo, J., & Henrichs, T. (2002). Critical regions: A model-based estimation of world water resources sensitive to global changes. *Aquatic Sciences*, *64*(4), 352–362. doi:10.1007/PL00012591

Al-Dossari, M., Awasthi, S. K., Mohamed, A. M., Abd El-Gawaad, N. S., Sabra, W., & Aly, A. H. (2022). Bio-alcohol sensor based on one-dimensional photonic crystals for detection of organic materials in wastewater. *Materials (Basel)*, *15*(11), 4012. doi:10.3390/ma15114012 PMID:35683310

Alessio, B., Walter, D., Valerio, P., & Antonio, P. (2016). Integration of Cloud computing and Internet of Things: A survey. *Future Generation Computer Systems*, *56*, 684–700. doi:10.1016/j.future.2015.09.021

Alfonta, L., Singh, A. K., & Willner, I. (2001). Liposomes labeled with biotin and horseradish peroxidase: A probe for the enhanced amplification of antigen-antibody or oligonucleotide-DNA sensing processes by the precipitation of an insoluble product on electrodes. *Analytical Chemistry*, *73*(1), 91–102. doi:10.1021/ac000819v PMID:11195517

Al-Fuqaha, A., Guizani, M., Mohammadi, M., Aledhari, M., & Ayyash, M. (2015). Internet of Things: A Survey on Enabling Technologies, Protocols, and Applications. *IEEE Communications Surveys and Tutorials*, *17*(4), 2347–2376. doi:10.1109/COMST.2015.2444095

Ali, M. R., Bacchu, M. S., Setu, M. A. A., Akter, S., Hasan, M. N., Chowdhury, F. T., Rahman, M. M., Ahommed, M. S., & Khan, M. Z. H. (2021). Development of an advanced DNA biosensor for pathogenic Vibrio cholerae detection in real sample. *Biosensors & Bioelectronics*, *188*, 113338. Advance online publication. doi:10.1016/j.bios.2021.113338 PMID:34030094

Alin, S. R., Newton, J. A., Feely, R. A., Greeley, D., Curry, B., Herndon, J., & Warner, M. (2023). A decade-long cruise time-series (2008–2018) of physical and biogeochemical conditions in the southern Salish Sea, North America. *Earth System Science Data Discussions*, *2023*, 1–37.

Ali, R., Razi, S. S., Gupta, R. C., Dwivedi, S. K., & Misra, A. (2016). An efficient ICT based fluorescent turn-on dyad for selective detection of fluoride and carbon dioxide. *New Journal of Chemistry*, *40*(1), 162–170. doi:10.1039/C5NJ01920F

Ali, R., Saleh, S. M., Meier, R. J., Azab, H. A., Abdelgawad, I. I., & Wolfbeis, O. S. (2010). Upconverting nanoparticle based optical sensor for carbon dioxide. *Sensors and Actuators. B, Chemical*, *150*(1), 126–131. doi:10.1016/j.snb.2010.07.031

Allied Market Research. (n.d.). *Environmental Sensor Market Size, Share, Competitive Landscape and Trend Analysis Report by Type (Humidity, Temperature, Pressure, Gas, and Others) and End User (Industrial, Residential, Commercial, Automotive, Government & Public Utilities and Other): Global Opportunity Analysis and Industry Forecast*. Allied Market Research. https://www.alliedmarketresearch.com/environmental-sensors-market-A12896

Allioui, H., & Mourdi, Y. (2023). Exploring the full potentials of IoT for better financial growth and stability: A comprehensive survey. *Sensors (Basel)*, *23*(19), 8015. doi:10.3390/s23198015 PMID:37836845

Alscher, R. G., Erturk, N., & Heath, L. S. (2002). Role of superoxide dismutases (SODs) in controlling oxidative stress in plants. *Journal of Experimental Botany*, *53*(372), 1331–1341. doi:10.1093/jexbot/53.372.1331 PMID:11997379

Altaf, W. J. (1997). Effect of motorway traffic emissions on roadside wild-plants in Saudi Arabia. *Journal of Radioanalytical and Nuclear Chemistry*, *217*(1), 211–222. doi:10.1007/BF02055354

Alvarez, M., Pennell, R., Meijer, P., Ishikawa, A., Dixon, R., & Lamb, C. (1998). Reactive oxygen intermediates mediate a systemic signal network in the establishment of plant immunity. *Cell*, *92*(6), 773–784. doi:10.1016/S0092-8674(00)81405-1 PMID:9529253

Aly, A. A., Al-Omran, A. M., Sallam, A. S., Al-Wabel, M. I., & Al-Shayaa, M. S. (2016). Vegetation cover change detection and assessment in arid environment using multi-temporal remote sensing images and ecosystem management approach. *Solid Earth*, *7*(2), 713–725. doi:10.5194/se-7-713-2016

Amaral, L. R., de Freitas, R. G., Júnior, M. R. B., & da Silva Simões, I. O. P. (2022). Application of Drones in Agriculture. In *Digital Agriculture* (pp. 99–121). Springer International Publishing. doi:10.1007/978-3-031-14533-9_7

Anagnostopoulos, T., Zaslavsky, A., Kolomvatsos, K., Medvedev, A., Amirian, P., Morley, J., & Hadjieftymiades, S. (2017). Challenges and opportunities of waste management in IoT-enabled smart cities: A survey. *IEEE Transactions on Sustainable Computing*, *2*(3), 275–289. doi:10.1109/TSUSC.2017.2691049

Anam, M., Yousaf, S., Sharafat, I., Zafar, Z., Ayaz, K., & Ali, N. (2017). Comparing natural and artificially designed bacterial consortia as biosensing elements for rapid non-specific detection of organic pollutant through microbial fuel cell. *International Journal of Electrochemical Science*, *12*(4), 2836–2851. doi:10.20964/2017.04.49

Andela, N., Morton, D. C., Giglio, L., Chen, Y., van der Werf, G. R., Kasibhatla, P. S., DeFries, R. S., Collatz, G. J., Hantson, S., Kloster, S., Bachelet, D., Forrest, M., Lasslop, G., Li, F., Mangeon, S., Melton, J. R., Yue, C., & Randerson, J. T. (2017). A human-driven decline in the global burned area. *Science*, *356*(6345), 1356–1362. doi:10.1126/science.aal4108 PMID:28663495

Andela, N., & Van Der Werf, G. R. (2014). Recent trends in African fires driven by cropland expansion and El Niño to La Niña transition. *Nature Climate Change*, *4*(9), 791–795. doi:10.1038/nclimate2313

Andrea, Z., Nicola, B., Angelo, C., Lorenzo, V., & Michele, Z. (2014). Internet of Things for Smart Cities. *IEEE Internet of Things Journal*, *1*(1), 22–32. doi:10.1109/JIOT.2014.2306328

An, J., Ra, H., Youn, C., & Kim, K. (2021). Experimental Results of Underwater Acoustic Communication with Nonlinear Frequency Modulation Waveform. *Sensors (Basel)*, *21*(21), 21. doi:10.3390/s21217194 PMID:34770501

Ankitkumar, B Rathod, Anuradha, J., Prashantkumar B Sathvara, Tripathi, S., & Sanjeevi, R. (2023). Vegetational Change Detection Using Machine Learning in GIS Technique: A Case Study from Jamnagar (Gujarat). *Journal of Data Acquisition and Processing*, *38*(1), 1046–1061. doi:10.5281/zenodo.7700655

Ansari, S. (2017). Combination of molecularly imprinted polymers and carbon nanomaterials as a versatile biosensing tool in sample analysis: Recent applications and challenges. *Trends in Analytical Chemistry*, *93*, 134–151. doi:10.1016/j.trac.2017.05.015

Anthony, F., Aloys, N., Hector, J., Maria, C., Albino, J., & Samuel, B. (2014). Wireless Sensor Networks for Water Quality Monitoring and Control within Lake Victoria Basin: Prototype Development. *Wirel Sens Netw*, *6*(12), 281–290. doi:10.4236/wsn.2014.612027

Appleby, P. G. (2008). Three decades of dating recent sediments by fallout radionuclides: A review. *The Holocene*, *18*(1), 83–93. doi:10.1177/0959683607085598

April, G., Yildirim, N., Lee, J., Cho, H., & Busnaina, A. (2018). *Nanotube-based biosensor for pathogen detection.* Google Patents.

Aragay, G., Pino, F., & Merkoçi, A. (2012). Nanomaterials for sensing and destroying pesticides. *Chemical Reviews*, *112*(10), 5317–5338. doi:10.1021/cr300020c PMID:22897703

Aragon, G., Martínez, I., Izquierdo, P., Belinchón, R., & Escudero, A. (2010). Effects of forest management on epiphytic lichen diversity in Mediterranean forests. *Applied Vegetation Science*, *13*(2), 183–194. doi:10.1111/j.1654-109X.2009.01060.x

Araujo-Andrade, C., Bugnicourt, E., Philippet, L., Rodriguez-Turienzo, L., Nettleton, D., Hoffmann, L., & Schlummer, M. (2021). Review on the photonic techniques suitable for automatic monitoring of the composition of multi-materials wastes in view of their posterior recycling. *Waste Management & Research*, *39*(5), 631–651. doi:10.1177/0734242X21997908 PMID:33749390

Archeka, C. N., Neelam, Kusum, & Hooda, V. (2022). Nanobiosensors for Monitoring Soil and Water Health. In Nanotechnology in Agriculture and Environmental Science (pp. 183–202). CRC Press. doi:10.1201/9781003323945-15

Arduini, F., Cinti, S., Scognamiglio, V., Moscone, D., & Palleschi, G. (2017). How cutting-edge technologies impact the design of electrochemical (bio) sensors for environmental analysis. A review. *Analytica Chimica Acta*, *22*(959), 15–42. doi:10.1016/j.aca.2016.12.035 PMID:28159104

Arebey, M., Hannan, M. A., Basri, H., Begum, R. A., & Abdullah, H. (2010). RFID and integrated technologies for solid waste bin monitoring system. In *Proceedings of the world congress on engineering* (Vol. 1, pp. 316-32). Research Gate.

Arenas-Sánchez, A., Rico, A., & Vighi, M. (2016). Effects of water scarcity and chemical pollution in aquatic ecosystems: State of the art. *The Science of the Total Environment*, *572*, 390–403. doi:10.1016/j.scitotenv.2016.07.211 PMID:27513735

Ariyanti, D., Iswantini, D., Sugita, P., Nurhidayat, N., Effendi, H., Ghozali, A. A., & Kurniawan, Y. S. (2020). Highly Sensitive Phenol Biosensor Utilizing Selected Bacillus Biofilm Through an Electrochemical Method. *Makara Journal of Science*, *24*(1), 24–30. doi:10.7454/mss.v24i1.11726

Arshi, O., & Mondal, S. (2023). Advancements in sensors and actuators technologies for smart cities: A comprehensive review. *Smart Construction and Sustainable Cities*, *1*(1), 18. doi:10.1007/s44268-023-00022-2

Asada, K., & Kiso, K. (1973). Initiation of aerobic oxidation of sulfite by illuminated spinach chloroplasts. *European Journal of Biochemistry*, *33*(2), 253–257. doi:10.1111/j.1432-1033.1973.tb02677.x PMID:4144355

Asadnia, M., Myers, M., Umana-Membreno, G. A., Sanders, T. M., Mishra, U. K., Nener, B. D., Baker, M. V., & Parish, G. (2017). Ca2+ detection utilising AlGaN/GaN transistors with ion-selective polymer membranes. *Analytica Chimica Acta*, *987*, 105–110. doi:10.1016/j.aca.2017.07.066 PMID:28916033

Ashenden, T. W. (1978). Growth reductions in cocks' foot (*Dactylis glomerata*.L.) as a result of SO_2 pollution. *Environ. Pollut*, *15*(2), 161–165. doi:10.1016/0013-9327(78)90104-0

Ashenden, T. W. (1979a). Effects of SO_2 and NO_2 pollution on transpiration in *Phaseolus vulgaris* L. *Environ. Pollut*, *18*(1), 45–50. doi:10.1016/0013-9327(79)90032-6

Ashenden, T. W. (1979b). The effect of long-term exposure to SO_2 and NO_2 pollution on the growth of *Dactylis glomerata*.L. and *Poa pratensis* L. *Environ. Pollut*, *18*(4), 249–258. doi:10.1016/0013-9327(79)90020-X

Ashenden, T. W., Bell, S. A., & Rafarel, C. R. (1996). Interactive effect of gaseous air pollutants and acid mist on two major pasture grasses. *Agriculture, Ecosystems & Environment*, *57*(1), 1–8. doi:10.1016/0167-8809(95)01008-4

Ashenden, T. W., & Mansfield, T. A. (1977). Influences of wind speed on the sensitivity of ryegrass to SO_2. *Journal of Experimental Botany*, *28*(3), 729–735. doi:10.1093/jxb/28.3.729

Ashenden, T. W., & Williams, J. A. D. (1980). Growth reductions in *Lolium multifolium* Lam. and *Phleum pretense* L. as a result of SO_2 and NO_2 pollution. *Environmental Pollution. Series A. Ecological and Biological*, *21*(2), 131–139. doi:10.1016/0143-1471(80)90041-0

Ashmore, M. R., Bell, J. N. B., & Mimmack, A. (1988). Crop growth along a gradient of ambient air pollution. *Environmental Pollution*, *53*(1-4), 99–121. doi:10.1016/0269-7491(88)90028-0 PMID:15092544

Ashmore, M. R., & Marshall, F. M. (1999). Ozone impacts on agriculture: An issue of global concern. *Advances in Botanical Research*, *29*, 32–49.

Atifi, A., Boyce, D. W., Dimeglio, J. L., & Rosenthal, J. (2018). Directing the outcome of CO_2 reduction at bismuth cathodes using varied ionic liquid promoters. *ACS Catalysis*, *8*(4), 2857–2863. doi:10.1021/acscatal.7b03433 PMID:30984470

Ausdall, B. R., Glass, J. L., Wiggins, K. M., Aarif, A. M., & Louie, J. J. (2009). A systematic investigation of factors influencing the decarboxylation of imidazolium carboxylates. *The Journal of Organic Chemistry*, *74*(20), 7935–7942. doi:10.1021/jo901791k PMID:19775141

Aydogdu, S., Ertekin, K., Suslu, A., Ozdemir, M., Celik, E., & Cocen, U. (2011). Optical CO_2 Sensing with Ionic Liquid Doped Electrospun Nanofibers. *Journal of Fluorescence*, *21*(2), 607–613. doi:10.1007/s10895-010-0748-4 PMID:20945079

Azedine, C., Antoine, G., Patrick, B., & Michel, M. (2000). Water quality monitoring using a smart sensing system. *Measurement*, 219–224.

Bachmann, T. T., & Schmid, R. D. (1999). A disposable multielectrode biosensor for rapid simultaneous detection of the insecticides paraoxon and carbofuran at high resolution. *Analytica Chimica Acta*, *401*(1), 95–103. doi:10.1016/S0003-2670(99)00513-9

Bae, J. W., Seo, H. B., Belkin, S., & Gu, M. B. (2020). An optical detection module-based biosensor using fortified bacterial beads for soil toxicity assessment. *Analytical and Bioanalytical Chemistry*, *412*(14), 3373–3381. doi:10.1007/s00216-020-02469-z PMID:32072206

Bahramian, M., Dereli, R. K., Zhao, W., Giberti, M., & Casey, E. (2023). Data to intelligence: The role of data-driven models in wastewater treatment. *Expert Systems with Applications*, *217*, 119453. doi:10.1016/j.eswa.2022.119453

Bai, C., Zhang, H., Zeng, L., Zhao, X., & Ma, L. (2020). Inductive magnetic nanoparticle sensor based on microfluidic chip oil detection technology. *Micromachines*, *11*(2), 183. doi:10.3390/mi11020183 PMID:32050692

Bai, S., Liu, H., Sun, J., Tian, Y., Luo, R., Li, D., & Chen, A. (2015). Mechanism of enhancing the formaldehyde sensing properties of Co3O4via Ag modification. *RSC Advances*, *5*(60), 48619–48625. doi:10.1039/C5RA05772H

Baker, S. N., & Baker, G. A. (2010). Luminescent carbon nanodots: Emergent nanolights. *Angewandte Chemie International Edition*, *49*(38), 6726–6744. doi:10.1002/anie.200906623 PMID:20687055

Bala, B. K., Arshad, F. M., & Noh, K. M. (2017). System dynamics. *Modelling and Simulation (Anaheim)*, 274.

Ballentine, D. C., Macko, S. A., Turekian, V. C., Gilhooly, W. P., & Martincigh, B. (1996). Compound specific isotope analysis of fatty acids and polycyclic aromatic hydrocarbons in aerosols: Implications for biomass burning. *Organic Geochemistry*, *25*(1-2), 97–104. doi:10.1016/S0146-6380(96)00110-6

Baranwal, J., Barse, B., Gatto, G., Broncová, G., & Kumar, A. (2022). Electrochemical sensors and their applications: A review. *Chemosensors (Basel, Switzerland)*, *10*(9), 363. doi:10.3390/chemosensors10090363

Barbrook-Johnson, P., & Penn, A. S. (2022). *Systems Mapping: How to build and use causal models of systems*. Springer Nature. doi:10.1007/978-3-031-01919-7

Bard, A. J., & Faulkner, L. R. (2001). *Electrochemical Methods: Fundamentals and Applications*. John Wiley & Sons.

Barker, G. C., & Jenkin, I. L. (1952). Square-wave polarography. *Analyst*, *77*(920), 685–696. doi:10.1039/an9527700685

Barnes, J. D., Hull, M. R., & Davison, A. W. (1996). Impact of air pollutants and elevated CO_2 on plants in winter time. In M. Yunus & M. Iqbal (Eds.), *Plant Response to Air Pollution* (pp. 135–166). John Wiley.

Baronas, R., Ivanauskas, F., & Kulys, J. (2021). *Mathematical modeling of biosensors*. Springer International Publishing. doi:10.1007/978-3-030-65505-1

Bashir, S., Hossain, S. S., Rahman, S. U., Ahmed, S., Al-Ahmed, A., & Hossain, M. M. (2016). Electrocatalytic reduction of carbon dioxide on SnO_2/MWCNT in aqueous electrolyte solution. *Journal of CO_2 Utilisation*, *16*, 346–353. doi:10.1016/j.jcou.2016.09.002

Bashir, S. M., Hossain, S. S., Rahman, S., Ahmed, S., & Hossain, M. M. (2015). NiO/MWCNT Catalysts for Electrochemical Reduction of CO_2. *Electrocatalysis (New York)*, *6*(6), 544–553. doi:10.1007/s12678-015-0270-1

Baumann, Z., and Fisher, N. S. (2011). Modeling metal bioaccumulation in a deposit-feeding polychaete from labile sediment fractions and from pore water. *Sci. Total Environ. 409*, 2607–2615. doi: . 03.009 doi:10.1016/j.scitotenv.2011

Baumbauer, C., Goodrich, P., Payne, M. E., Anthony, T. L., Beckstoffer, C., Toor, A., Silver, W. L., & Arias, A. C. (2022). Printed potentiometric nitrate sensors for use in soil. *Sensors (Basel)*, *22*(11), 4095. doi:10.3390/s22114095 PMID:35684715

Bayabil, H. K., Teshome, F. T., & Li, Y. (2022). Emerging contaminants in soil and water. *Frontiers in Environmental Science*, *10*, 873499. doi:10.3389/fenvs.2022.873499

Bearg, D. W. (2019). *Indoor air quality and HVAC systems*. Routledge. https://www.taylorfrancis.com/books/mono/10.1201/9780203751152/indoor-air-quality-hvac-systems-david-bearg

Beaumont, N. J., Aanesen, M., Austen, M. C., Börger, T., Clark, J. R., Cole, M., Hooper, T., Lindeque, P. K., Pascoe, C., & Wyles, K. J. (2019). Global ecological, social and economic impacts of marine plastic. *Marine Pollution Bulletin*, *142*, 189–195. doi:10.1016/j.marpolbul.2019.03.022 PMID:31232294

Bein, E., Pasquazzo, G., Dawas, A., Yecheskel, Y., Zucker, I., Drewes, J. E., & Hübner, U. (2024). Groundwater remediation by in-situ membrane ozonation: Removal of aliphatic 1,4-dioxane and monocyclic aromatic hydrocarbons. *Journal of Environmental Chemical Engineering*, *12*(2), 111945. doi:10.1016/j.jece.2024.111945

Bellou, N., Gambardella, C., Karantzalos, K., Monteiro, J. G., Canning-Clode, J., Kemna, S., Arrieta-Giron, C. A., & Lemmen, C. (2021). Global assessment of innovative solutions to tackle marine litter. *Nature Sustainability*, *4*(6), 516–524. doi:10.1038/s41893-021-00726-2

Benitez-Nelson, C. R., Buesseler, K., Dai, M., Aoyama, M., Casacuberta, N., Charmasson, S., Johnson, A., Godoy, J. M., Maderich, V., Masqué, P., Moore, W., Morris, P. J., & Smith, J. N. (2018). Radioactivity in the marine environment: Understanding the basics of radioactivity. Limnol. Oceanogr. *Limnology and Oceanography e-Lectures*, *8*(1), 1–58. doi:10.1002/loe2.10010

Bennett, A. B., Chi-Ham, C., Barrows, G., Sexton, S., & Zilberman, D. (2013). Agricultural biotechnology: Economics, environment, ethics, and the future. *Annual Review of Environment and Resources*, *38*(1), 249–279. doi:10.1146/annurev-environ-050912-124612

Berberich, J. A., Li, T., & Sahle-Demessie, E. (2019). Chapter 11 - Biosensors for Monitoring Water Pollutants: A Case Study with Arsenic in Groundwater. *Separation Science and Technology*, *11*, 285–328. doi:10.1016/B978-0-12-815730-5.00011-9

Berekaa, M.M. (2016). Nanotechnology in wastewater treatment; influence of nanomaterials on microbial systems. *International journal of current microbiolgy and applied sciences* *5*(1), 713-726.

Bergmann, H., & Koparal, S. (2005). The formation of chlorine dioxide in the electrochemical treatment of drinking water for disinfection. *Electrochimica Acta*, *50*(25-26), 5218–5228. doi:10.1016/j.electacta.2005.01.061

Betti, M., Boisson, F., Eriksson, M., Tolosa, I., & Vasileva, E. (2011). Isotope analysis for marine environmental studies. *International Journal of Mass Spectrometry*, *307*(1-3), 192–199. doi:10.1016/j.ijms.2011.03.008

Bharadwaj, A. S., Rego, R., & Chowdhury, A. (2016). IoT based solid waste management system: A conceptual approach with an architectural solution as a smart city application. In *2016 IEEE annual India conference (INDICON)* (pp. 1-6). IEEE. 10.1109/INDICON.2016.7839147

Bhardwaj, A., Dagar, V., Khan, M. O., Aggarwal, A., Alvarado, R., Kumar, M., Irfan, M., & Proshad, R. (2022). Smart IoT and Machine Learning-based Framework for Water Quality Assessment and Device Component Monitoring. *Environmental Science and Pollution Research International*, *29*(30), 46018–46036. doi:10.1007/s11356-022-19014-3 PMID:35165843

Bhattarai, P., & Hameed, S. (2020). Basics of biosensors and nanobiosensors. Nanobiosensors: From Design to Applications.

Bibri, S. E. (2018). The IoT for smart sustainable cities of the future: An analytical framework for sensor-based big data applications for environmental sustainability. *Sustainable Cities and Society*, *38*, 230–253. doi:10.1016/j.scs.2017.12.034

Bidwell, R. G., & Beebee, G. P. (1974). Carbon monoxide fixation by plants. *Canadian Journal of Botany*, *52*(8), 1841–1847. doi:10.1139/b74-236

Biljana, L., Risteska, S., & Kire, V. T. (2017). A review of Internet of Things for smart home: Challenges and solutions. *Journal of Cleaner Production*, *140*(3), 1454–1464.

Birgand, F., Appelboom, T. W., Chescheir, G. M., & Skaggs, R. W. (2013). Estimating nitrogen, phosphorus, and carbon fluxes in forested and mixed-use watersheds of the lower coastal plain of North Carolina: Uncertainties associated with infrequent sampling. *Transactions of the ASABE*, *54*, 2099–2110. doi:10.13031/2013.40668

Biswas, P., Karn, A. K., Balasubramanian, P., & Kale, P. G. (2017). Biosensor for detection of dissolved chromium in potable water: A review. *Biosensors & Bioelectronics*, *94*, 589–604. doi:10.1016/j.bios.2017.03.043 PMID:28364706

Blanch, S.J., Ganf, G.G., & Walker, K.F. (1999). Tolerance of riverine plants to flooding and exposure by water regime. *Regulated Rivers: Research & Management*, *15*.

Boegelsack, N., Withey, J., O'Sullivan, G., & McMartin, D. (2018). A critical examination of the relationship between wildfires and climate change with consideration of the human impact. *Journal of Environmental Protection*, *9*(5), 461–467. doi:10.4236/jep.2018.95028

Boehm, A. B., Ismail, N. S., Sassoubre, L. M., & Andruszkiewicz, E. A. (2017). Oceans in peril: Grand challenges in applied water quality research for the 21st century. *Environmental Engineering Science*, *34*(1), 3–15. doi:10.1089/ees.2015.0252

Bohm, B. A. (1987). Intraspecific flavonoid variation. *Botanical Review*, *53*(2), 197–279. doi:10.1007/BF02858524

Borisova, T., Kucherenko, D., Soldatkin, O., Kucherenko, I., Pastukhov, A., Nazarova, A., Galkin, M., Borysov, A., Krisanova, N., Soldatkin, A., & El'skaya, A. (2018). An amperometric glutamate biosensor for monitoring glutamate release from brain nerve terminals and in blood plasma. *Analytica Chimica Acta*, *1022*, 113–123. doi:10.1016/j.aca.2018.03.015 PMID:29729731

Bourdeau, P., & Treshow, M. (1978). Ecosystem response to pollution. *Prin Ciples Ecotoxicol*, *5*, 79–88.

Bourgaud, F., Gravot, A., Milesi, S., & Gontier, E. (2001). Production of plant secondary metabolites: A historical perspective. *Plant Science*, *161*(5), 839–851. doi:10.1016/S0168-9452(01)00490-3

Bourgeois, W., Burgess, J. E., & Stuetz, R. M. (2001). On-line monitoring of wastewater quality: A review. *Journal of Chemical Technology and Biotechnology*, *76*(4), 337–348. doi:10.1002/jctb.393

Bradstock, R. A. (2010). A biogeographic model of fire regimes in Australia: Current and future implications. *Global Ecology and Biogeography*, *19*(2), 145–158. doi:10.1111/j.1466-8238.2009.00512.x

Brando, P. M., Balch, J. K., Nepstad, D. C., Morton, D. C., Putz, F. E., Coe, M. T., Silvério, D., Macedo, M. N., Davidson, E. A., Nóbrega, C. C., Alencar, A., & Soares-Filho, B. S. (2014). Abrupt increases in Amazonian tree mortality due to drought–fire interactions. *Proceedings of the National Academy of Sciences of the United States of America*, *111*(17), 6347–6352. doi:10.1073/pnas.1305499111 PMID:24733937

Braun, E., Eichen, Y., Sivan, U., & Yoseph, G. B. (1998). DNA-templated assembly and electrode attachment of a conducting silver wire. *Nature*, *391*(6669), 775–778. doi:10.1038/35826 PMID:9486645

Brecht, A. (2005). Multianalyte bioanalytical devices: Scientific potential and business requirements. *Analytical and Bioanalytical Chemistry*, *381*(5), 1025–1026. doi:10.1007/s00216-004-2912-7 PMID:15726339

Brimblecombe, P. (1982). Trends in the deposition of Sulphate and total solids in London. *The Science of the Total Environment*, *22*(2), 97–103. doi:10.1016/0048-9697(82)90027-4 PMID:7063836

Broday, D. M. (2017). Wireless Distributed Environmental Sensor Networks for Air Pollution Measurement-The Promise and the Current Reality. *Sensors (Basel)*, *17*(10), 2263. Advance online publication. doi:10.3390/s17102263 PMID:28974042

Brooks, L., Gaustad, G., Gesing, A., Mortvedt, T., & Freire, F. (2019). Ferrous and non-ferrous recycling: Challenges and potential technology solutions. *Waste Management (New York, N.Y.)*, *85*, 519–528. doi:10.1016/j.wasman.2018.12.043 PMID:30803607

Buesseler, K., Dai, M., Aoyama, M., Benitez-Nelson, C., Charmasson, S., Higley, K., Maderich, V., Masqué, P., Morris, P. J., Oughton, D., & Smith, J. N. (2017). Fukushima Daiichi–derived radionuclides in the ocean: Transport, fate, and impacts. *Annual Review of Marine Science*, *9*(1), 173–203. doi:10.1146/annurev-marine-010816-060733 PMID:27359052

Bull, J. N., & Mansfield, T. A. (1974). Photosynthesis in leaves exposed to SO_2 and NO_2. *Nature*, *250*(5465), 443–444. doi:10.1038/250443c0 PMID:4852889

Burger, J. & Gochfeld, M. (2001). On developing bioindicators for human and ecological health. *Environ Monit Assess., 66*, 23–46.

Burger, J. (1993). Metals in avian feathers: bioindicators of environmental pollution. *Rev Environ Toxicol. 5*, 203–311.

Burgués, J., & Marco, S. (2020). Environmental chemical sensing using small drones: A review. *The Science of the Total Environment*, *748*, 141172. doi:10.1016/j.scitotenv.2020.141172 PMID:32805561

Bushra, R., & Mubashir, H. R. (2016). Applications of wireless sensor networks for urban areas: A survey. *Journal of Network and Computer Applications*, *60*, 192–219. doi:10.1016/j.jnca.2015.09.008

Butterworth, F.M., Gunatilaka, A., & Gonsebatt, M.E. (2001). *Biomonitors and biomarkers as indicators of environmental change*. Springer Science & Business.

Butterworth, I. (2000). *The Relationship Between the Built Environment and Well-Being: A Literature Review*. Victorian Health Promotion Foundation.

Caballero, S., Esclapez, R., Galindo, N., Mantilla, E., & Crespo, J. (2012). Use of a passive sampling network for the determination of urban NO 2 spatiotemporal variations. *Atmospheric Environment*, *63*, 148–155. doi:10.1016/j.atmosenv.2012.08.071

Cai, H., Mei, Y., Chen, J., Wu, Z., Lan, L., & Zhu, D. (2020). An analysis of the relation between water pollution and economic growth in China by considering the contemporaneous correlation of water pollutants. *Journal of Cleaner Production*, *276*, 122783. doi:10.1016/j.jclepro.2020.122783

Cai, J., & DuBow, M. S. (1997). Use of a luminescent bacterial biosensor for biomonitoring and characterization of arsenic toxicity of chromated copper arsenate (CCA). *Biodegradation*, *8*(2), 105–111. doi:10.1023/A:1008281028594 PMID:9342883

Calatayud, A., Ramirez, J. W., Iglesias, D. J., & Barreno, E. (2002). Effect of O_3 on photosynthetic CO_2 exchange, chlorophyll a fluorescence and antioxidant systems in lettuce leaves. *Physiologia Plantarum*, *116*(3), 308–316. doi:10.1034/j.1399-3054.2002.1160305.x

Calvin, M. (1955). Function of carotenoids in photosynthesis. *Nature*, *176*(4495), 1211. doi:10.1038/1761215a0 PMID:13288579

Camero, A., Toutouh, J., Ferrer, J., & Alba, E. (2019). Waste generation prediction in smart cities through deep neuro-evolution. In *Smart Cities: First Ibero-American Congress, ICSC-CITIES 2018,* (pp. 192-204). Springer International Publishing. 10.1007/978-3-030-12804-3_15

Campbell, J. B., & Wynne, R. H. (2011). *Introduction to remote sensing*. Guilford press.

Canas, M. S., Carreras, H. A., Orellana, L., & Pignata, M. L. (1997). Correlation between environmental conditions and filiar chemical parameters in *Ligustrum lucidum* Ait exposed to urban air pollution. *Journal of Environmental Management*, *49*, 167–181. doi:10.1006/jema.1995.0090

Cantonati, M., Poikane, S., Pringle, C. M., Stevens, L. E., Turak, E., Heino, J., Richardson, J. S., Bolpagni, R., Borrini, A., Cid, N., Čtvrtlíková, M., Galassi, Hájek, Hawes, Levkov, Naselli-Flores, Saber, Cicco, Fiasca, & Znachor. (2020). Characteristics, main impacts, and stewardship of natural and artificial freshwater environments: Consequences for biodiversity conservation. *Water (Basel)*, *12*(1), 260. doi:10.3390/w12010260

Cárdenas, S., & Valcárcel, M. (2005). Analytical features in qualitative analysis. *Trends in Analytical Chemistry*, *24*(6), 477–487. doi:10.1016/j.trac.2005.03.006

Carignan, V. & Villard, M.A. (2001). Selecting indicator species to monitor ecological integrity: a review. *Environ Monit Assess, 78*, 45–61.

Carlson, R. W., & Bazzaz, F. A. (1985). Plant response to SO_2 and CO_2. In W. E. Winner, H. A. Mooney, & R. A. Goldstein (Eds.), *SO_2 and Vegetation* (pp. 313–331). Stanford University Press.

Carreras, H. A., Canas, M. S., & Pignata, M. L. (1996). Differences in responses to urban air pollution by *Ligustrum lucidum* Ait. and *Ligustrum lucidium* Ait. F.tricolor (REHD.) REHD. *Environmental Pollution*, *93*(2), 211–218. doi:10.1016/0269-7491(96)00014-0 PMID:15091360

Catelli, E., Sciutto, G., Prati, S., Chavez Lozano, M. V., Gatti, L., Lugli, F., Silvestrini, S., Benazzi, S., Genorini, E., & Mazzeo, R. (2020). A new miniaturised short-wave infrared (SWIR) spectrometer for on-site cultural heritage investigations. *Talanta*, *121112*, 121112. Advance online publication. doi:10.1016/j.talanta.2020.121112 PMID:32797874

Cavelan, A., Faure, P., Lorgeoux, C., Colombano, S., Deparis, J., Davarzani, D., Enjelvin, N., Oltean, C., Tinet, A. J., Domptail, F., & Golfier, F. (2024). An experimental multi-method approach to better characterize the LNAPL fate in soil under fluctuating groundwater levels. *Journal of Contaminant Hydrology*, *262*, 104319. doi:10.1016/j.jconhyd.2024.104319 PMID:38359773

Cawson, J. G., Duff, T. J., Tolhurst, K. G., Baillie, C. C., & Penman, T. D. (2017). Fuel moisture in Mountain Ash forests with contrasting fire histories. *Forest Ecology and Management*, *400*, 568–577. doi:10.1016/j.foreco.2017.06.046

Census of India. (2011). *Population Census 2011: Ajmer District*. Registrar General and Census Commissioner of India.

Cesewski, E., & Johnson, B. N. (2020). *Electrochemical biosensors for pathogen detection. Biosensors and Bioelectronics*. Elsevier Ltd., doi:10.1016/j.bios.2020.112214

Chafa, A. T., Chirinda, G. P., & Matope, S. (2022). Design of a real–time water quality monitoring and control system using Internet of Things (IoT). *Cogent Engineering*, *9*(1), 1. doi:10.1080/23311916.2022.2143054

Chakrabortty, S. & Paratkar GT. (2006). Biomonitoring of trace element air pollution using mosses. *Aerosol Air Qual Res. 6*, 247–258.

Chamola, V., Hassija, V., Gupta, V., & Guizani, M. (2020). A comprehensive review of the COVID-19 pandemic and the role of IoT, drones, AI, blockchain, and 5G in managing its impact. *IEEE Access : Practical Innovations, Open Solutions*, *8*, 90225–90265. doi:10.1109/ACCESS.2020.2992341

Chang, H., Kim, S. K., Sukaew, T., Bohrer, F., & Zellers, E. T. (2010). Microfabricated gas chromatograph for Sub-ppb determinations of TCE in vapor intrusion investigations. *Procedia Engineering*, *5*, 973–976. doi:10.1016/j.proeng.2010.09.271

Chaplin, S. E. (1999). —Cities, Sewers and Poverty: India's Politics of Sanitation||. *Environment and Urbanization*, *11*(1), 145–158. doi:10.1177/095624789901100123

Chapman, J., Truong, V. K., Elbourne, A., Gangadoo, S., Cheeseman, S., Rajapaksha, P., Latham, K., Crawford, R. J., & Cozzolino, D. (2020). Combining chemometrics and sensors: Toward new applications in monitoring and environmental analysis. *Chemical Reviews*, *120*(13), 6048–6069. doi:10.1021/acs.chemrev.9b00616 PMID:32364371

Chappell, N. A., Jones, T. D., & Tych, W. (2017). Sampling frequency for water quality variables in streams: Systems analysis to quantify minimum monitoring rates. *Water Research*, *123*, 49–57. doi:10.1016/j.watres.2017.06.047 PMID:28647587

Chatterjee, C., & Sen, A. (2015). Sensitive colorimetric sensors for visual detection of carbon dioxide and sulfur dioxide. *Journal of Materials Chemistry. A, Materials for Energy and Sustainability*, *3*(10), 5642–5647. doi:10.1039/C4TA06321J

Chaudri, A. M., Lawlor, K., Preston, S., Paton, G. I., Killham, K., & McGrath, S. P. (2000). Response of a Rhizobium-based luminescence biosensor to Zn and Cu in soil solutions from sewage sludge treated soils. *Soil Biology & Biochemistry*, *32*(3), 383–388. doi:10.1016/S0038-0717(99)00166-2

Chaves, N., Escudero, J., & Gutierres-Marino, C. (1997). Role of ecological variables in the seasonal variation of flavonoid content of *Cistus ladanifer* exudate. *Journal of Chemical Ecology*, *23*(3), 579–604. doi:10.1023/B:JOEC.0000006398.79306.09

Chaves, N., Sosa, T., & Escudero, J. (2001). Plant growth inhibiting flavonoids in exudate of Cistus ladanifer and in associated soils. *Journal of Chemical Ecology*, *27*(3), 623–631. doi:10.1023/A:1010388905923 PMID:11441450

Chen, B., Wang, M., Duan, M., Ma, X., Hong, J., Xie, F., Zhang, R., & Li, X. (2019). In search of key: Protecting human health and the ecosystem from water pollution in China. *Journal of Cleaner Production*, *228*, 101–111. doi:10.1016/j.jclepro.2019.04.228

Chen, C., & Wang, J. (2020). *Optical biosensors: An exhaustive and comprehensive review. Analyst.* Royal Society of Chemistry., doi:10.1039/C9AN01998G

Chen, D., Wang, H., Dong, L., Liu, P., Zhang, Y., Shi, J., Feng, X., Zhi, J., Tong, B., & Dong, Y. (2016). The fluorescent bioprobe with aggregation-induced emission features for monitoring to carbon dioxide generation rate in single living cell and early identification of cancer cells. *Biomaterials*, *103*, 67–74. doi:10.1016/j.biomaterials.2016.06.055 PMID:27372422

Chen, G., Ye, M., Ma, B., & Ren, Y. (2023). Responses of petroleum contamination at different sites to soil physicochemical properties and indigenous microbial communities. *Water, Air, and Soil Pollution*, *234*(8), 494. doi:10.1007/s11270-023-06523-1

Cheng, C.-C., & Lee, D. (2016). Enabling Smart Air Conditioning by Sensor Development: A Review. *Sensors (Basel)*, *16*(12), 2028. doi:10.3390/s16122028 PMID:27916906

Cheng, Y., Wang, H., Zhuo, Y., Song, D., Li, C., Zhu, A., & Long, F. (2022). Reusable Smartphone-Facilitated Mobile Fluorescence Biosensor for Rapid and Sensitive On-Site Quantitative Detection of Trace Pollutants. *Biosensors & Bioelectronics*, *199*, 113863. doi:10.1016/j.bios.2021.113863 PMID:34894557

Chen, J., Lu, W., Yuan, L., Wu, Y., & Xue, F. (2022). Estimating construction waste truck payload volume using monocular vision. *Resources, Conservation and Recycling*, *177*, 106013. doi:10.1016/j.resconrec.2021.106013

Chen, S., Yu, H., Zhao, C., Hu, R., Zhu, J., & Li, L. (2017). Indolo[3,2-b]carbazole derivative as a fluorescent probe for fluoride ion and carbon dioxide detections. *Sensors and Actuators. B, Chemical*, *250*, 591–600. doi:10.1016/j.snb.2017.05.012

Chen, X., Leishman, M., Bagnall, D., & Nasiri, N. (2021). Nanostructured Gas Sensors: From Air Quality and Environmental Monitoring to Healthcare and Medical Applications. *Nanomaterials (Basel, Switzerland)*, *11*(8), 1927. Advance online publication. doi:10.3390/nano11081927 PMID:34443755

Chen, Y., Hakala, T., Karjalainen, M., Feng, Z., Tang, J., Litkey, P., Kukko, A., Jaakkola, A., & Hyyppä, J. (2017). UAV-borne profiling radar for forest research. *Remote Sensing (Basel)*, *9*(1), 58. doi:10.3390/rs9010058

Chicea, D., Leca, C., Olaru, S., & Chicea, L. M. (2021). An Advanced Sensor for Particles in Gases Using Dynamic Light Scattering in Air as Solvent. *Sensors (Basel)*, *21*(15), 15. Advance online publication. doi:10.3390/s21155115 PMID:34372352

Chidiac, S., Najjar, P. E., Ouaini, N., Rayess, Y. E., & Azzi, D. E. (2023). A comprehensive review of water quality indices (WQIs): History, models, attempts and perspectives. *Reviews in Environmental Science and Biotechnology*, *22*(2), 349–395. doi:10.1007/s11157-023-09650-7 PMID:37234131

Chinowsky, T., Quinn, J., Bartholomew, D., Kaiser, R., & Elkind, J. (2003). Performance of the Spreeta 2000 integrated surface plasmon resonance affinity sensor. *Sensors and Actuators. B, Chemical*, *91*(1-3), 266–274. doi:10.1016/S0925-4005(03)00113-8

Choi, J. R., Yong, K. W., Choi, J. Y., & Cowie, A. C. (2019, February 2). Emerging point-of-care technologies for food safety analysis. *Sensors (Switzerland)*. MDPI AG. doi:10.3390/s19040817

Choi, J., Hearne, R., Lee, K., & Roberts, D. (2015). The relation between water pollution and economic growth using the environmental Kuznets curve: A case study in South Korea. *Water International*, *40*(3), 499–512. doi:10.1080/02508060.2015.1036387

Chong, S. S., Aziz, A., & Harun, S. W. (2013). Fibre optic sensors for selected wastewater characteristics. *Sensors (Basel)*, *13*(7), 8640–8668. doi:10.3390/s130708640 PMID:23881131

Chouler, J., Monti, M. D., Morgan, W., Cameron, P. J., & Lorenzo, M. D. (2019). A photosynthetic toxicity biosensor for water. *Electrochimica Acta*, *309*, 392–401. doi:10.1016/j.electacta.2019.04.061

Chow, J. C. (1995). Measurement methods to determine compliance with ambient air quality standards for suspended particles. *Journal of the Air & Waste Management Association*, *45*(5), 320–382. doi:10.1080/10473289.1995.10467369 PMID:7773805

Christie, R., Mallory, C., Jared, L., & Alan, M. (2014) Remote Delay Tolerant Water Quality Monitoring. In: IEEE global humanitarian technology conference. IEEE.

Chu, C., & Lo, Y. (2009). Highly sensitive and linear optical fiber carbon dioxide sensor based on sol–gel matrix doped with silica particles and HPTS. *Sensors and Actuators. B, Chemical*, *143*(1), 205–210. doi:10.1016/j.snb.2009.09.019

Chun-yan, M. A., Xin, X. U., Lin, H. A. O., & Jun, C. A. O. (2007). Nitrogen dioxide-induced responses in *Brassica campestris* seedlings: The role of hydrogen peroxide in the modulation of antioxidative level and induced resistance. *Agricultural Sciences in China*, *6*(10), 1193–1200. doi:10.1016/S1671-2927(07)60163-1

Cimpan, C., Maul, A., Jansen, M., Pretz, T., & Wenzel, H. (2015). Central sorting and recovery of MSW recyclable materials: A review of technological state-of-the-art, cases, practice and implications for materials recycling. *Journal of Environmental Management*, *156*, 181–199. doi:10.1016/j.jenvman.2015.03.025 PMID:25845999

Clean India Journal. (2023, October 20). The sustainable route to sanitary napkin disposal. *Clean India Journal.* https://www.cleanindiajournal.com/the-sustainable-route-to-sanitary-napkin-disposal/

Cleveland, J. (1998). Prescribed fire: the fundamental solution. In*: Pruden, Teresa L.; Brennan, Leonard A., eds. Proceedings: 20th Tall Timbers Fire Ecology Conference; Fire in ecosystem management: Shifting the paradigm from suppression to prescription. Tallahassee, FL: Tall Timbers Research Station. p. 12-16.* (pp. 12-16). Research Gate.

Collins, R. D. (2012). *Forest fire management in Portugal: developing system insights through models of social and physical dynamics* [Doctoral dissertation, Massachusetts Institute of Technology].

Collins, R. D., de Neufville, R., Claro, J., Oliveira, T., & Pacheco, A. P. (2013). Forest fire management to avoid unintended consequences: A case study of Portugal using system dynamics. *Journal of Environmental Management*, *130*, 1–9. doi:10.1016/j.jenvman.2013.08.033 PMID:24036501

Constantinidou, H. A., & Kozlowski, T. T. (1979). Effects of sulphur dioxide and ozone on Ulnus Americana seedlings. II. Carbohydrates, proteins and lipids. *Canadian Journal of Botany*, *57*(2), 176–184. doi:10.1139/b79-026

Cooper-Driver, G., & Bhattacharya, M. (1998). Role of phenolics in plant evolution. *Phytochemistry*, *49*, 1165–1174.

Cosgun, U., Coşkun, M., Toprak, F., Yıldız, D., Coşkun, S., Taşoğlu, E., & Öztürk, A. (2023). Visibility evaluation and suitability analysis of fire lookout towers in the Mediterranean Region, Southwest Anatolia/Türkiye. *Fire (Basel, Switzerland)*, *6*(8), 305. doi:10.3390/fire6080305

Cresswell, T., Metian, M., Fisher, N. S., Charmasson, S., Hansman, R. L., Bam, W., Bock, C., & Swarzenski, P. W. (2020). Exploring new frontiers in marine radioisotope tracing–adapting to new opportunities and challenges. *Frontiers in Marine Science*, *7*, 406. doi:10.3389/fmars.2020.00406

Cross, C. E., van der Vliet, A., Louie, S., Thiele, J. J., & Halliwell, B. (1998). Oxidative stress and antioxidants at biosurfaces: Plants, skin, and respiratory tract surface. *Environmental Health Perspectives*, *106*, 1241–1251. PMID:9788905

Cuadra, P., Harborne, J., & Waterman, P. (1997). Increases in surface flavonols and photosynthetic pigments in *Gnaphalium luteo-album* in response to UV-B radiation. *Phytochemistry*, *45*(7), 1377–1383. doi:10.1016/S0031-9422(97)00183-0

Cui, H., Zhou, J., Li, Z., & Gu, C. (2021). Editorial: Soil and sediment pollution, processes and remediation. Frontiers in Environmental Science. doi:10.3389/fenvs.2021.822355

Cui, Y., Lai, B., & Tang, X. (2019). Microbial Fuel Cell-Based Biosensors. *Biosensors (Basel)*, *9*(3), 92. doi:10.3390/bios9030092 PMID:31340591

Curtis, C. R., Howell, R. K., & Kremer, D. F. (1976). Soybean peroxidase from ozone injury. *Environ. Pollut.*, *11*(3), 189–194. doi:10.1016/0013-9327(76)90083-5

D'Amico, G., L'Abbate, P., Liao, W., Yigitcanlar, T., & Ioppolo, G. (2020). Understanding sensor cities: Insights from technology giant company driven smart urbanism practices. *Sensors (Basel)*, *20*(16), 4391. doi:10.3390/s20164391 PMID:32781671

da Silva, F. L., Fushita, Â. T., da Cunha-Santino, M. B., & Bianchini, I. Jr. (2022). Adopting basic quality tools and landscape analysis for applied limnology: An approach for freshwater reservoir management. *Sustainable Water Resources Management*, *8*(3), 65. doi:10.1007/s40899-022-00655-8

Dahlgren, J.P. & Ehrlén, J. (2005). Distribution patterns of vascular plants in lakes - the role of metapopulation dynamics. *Ecography,* 28(1), 49-58.

Dai, B., Li, C., Lin, T., Wang, Y., Gong, D., Ji, X., & Zhu, B. (2021). Field robot environment sensing technology based on TensorRT. In *Intelligent Robotics and Applications* (pp. 370–377). Springer International Publishing. doi:10.1007/978-3-030-89095-7_36

Daines, R. H. (1968). Sulphur dioxide and plant response. *Journal of Occupational Medicine*, *10*(9), 516–534. doi:10.1097/00043764-196809000-00015 PMID:4878445

Dalavoy, T. S., Wernette, D. P., Gong, M., Sweedler, J. V., Lu, Y., Flachsbart, B. R., Shannon, M. A., Bohn, P. W., & Cropek, D. M. (2008). Immobilization of DNAzyme catalytic beacons on PMMA for Pb 2+ detection. *Lab on a Chip*, *8*(5), 786–793. doi:10.1039/b718624j PMID:18432350

Dansby-Sparks, R. N., Jin, J., Mechery, S. J., Sampathkumaran, U., Owen, T. W., Yu, B. D., Goswami, K., Hong, K., Grant, J., & Xue, Z. L. (2010). Fluorescent-dye-doped sol–gel sensor for highly sensitive carbon dioxide gas detection below atmospheric concentrations. *Analytical Chemistry*, *82*(2), 593–600. doi:10.1021/ac901890r PMID:20038093

Darrall, N. M. (1989). The effect of air pollutants on physiological processes in plants. *Plant, Cell & Environment*, *12*(1), 1–30. doi:10.1111/j.1365-3040.1989.tb01913.x

Das, S., Lee, S. H., Kumar, P., Kim, K. H., Lee, S. S., & Bhattacharya, S. S. (2019). Solid waste management: Scope and the challenge of sustainability. *Journal of Cleaner Production*, *228*, 658–678. doi:10.1016/j.jclepro.2019.04.323

Dat, J., Vandenabeele, S., Vranovh, E., van Montagu, M., Inz, B. D., & van Breusegem, F. (2000). Dual action of active oxygen species during plant stress responses. *Cellular and Molecular Life Sciences*, *57*, 779–795. doi:10.1007/s000180050041 PMID:10892343

David, E., & Niculescu, V. C. (2021). Volatile Organic Compounds (VOCs) as Environmental Pollutants: Occurrence and Mitigation Using Nanomaterials. *International Journal of Environmental Research and Public Health*, *18*(24), 13147. doi:10.3390/ijerph182413147 PMID:34948756

Davis, C. E., Ho, C. K., Hughes, R. C., & Thomas, M. L. (2005). Enhanced detection of m-xylene using a preconcentrator with a chemiresistor sensor. *Sensors and Actuators. B, Chemical*, *104*(2), 207–216. doi:10.1016/j.snb.2004.04.120

De Kok, L. J. (1990). Sulphur metabolism in plants exposed to atmospheric sulfur. In H. Rennerberg, C. Brunold, L. J. De Kok, & I. Stulen (Eds.), *Sulfur Nutrition and Sulfur Assimilation in Higher plants; Fundamental, Environmental and Agricultural Aspects* (pp. 113–130). SPB Academic Publishing.

De Kok, L. J., Stuvier, C. E. E., & Stulen, I. (1998). Impact of atmospheric H_2S on plants. In L. J. De Kok & I. Stulen (Eds.), *Response of Plants Metabolism to Air Pollution and Global Change* (pp. 51–63). Backhuys Publishers.

De Kok, L. J., & Tausz, M. (2001). The role of glutathione in plant reaction and adaptation to air pollutants. In D. Grill, M. Tausz, & L. J. De Kok (Eds.), *Significance of Glutathione to Plant Adaptation to the Environment* (pp. 185–201). Kluwer Acadimic Publishers. doi:10.1007/0-306-47644-4_8

de Paiva Magalhães, D., da Costa Marques, M., Baptista, D., & Buss, D. (2015). Metal bioavailability and toxicity in freshwaters. *Environmental Chemistry Letters*, *13*(1), 69–87. doi:10.1007/s10311-015-0491-9

De Sousa, J. F. Junior, Columbus, S., Hammouche, J., Ramachandran, K., Daoudi, K., & Gaidi, M. (2023). Engineered micro-pyramids functionalized with silver nanoarrays as excellent cost-effective SERS chemosensors for multi-hazardous pollutants detection. *Applied Surface Science*, *613*, 156092. doi:10.1016/j.apsusc.2022.156092

Deepak, S. S., & Agrawal, M. (2001). Influence of elevated CO_2 on the sensitivity of two soyabean cultivars to sulphur dioxide. *Environmental and Experimental Botany*, *46*(1), 81–91. doi:10.1016/S0098-8472(01)00086-7 PMID:11378175

Deiner, K., Bik, H. M., Mächler, E., Seymour, M., Lacoursière-Roussel, A., Altermatt, F., & Bernatchez, L. (2017). *Environmental DNA metabarcoding: Transforming how we survey animal and plant communities. Molecular Ecology*. Blackwell Publishing Ltd. doi:10.1111/mec.14350

del Valle, M. (2020). *Sensors as Green Tools*.

Demrozi, F., Pravadelli, G., Bihorac, A., & Rashidi, P. (2020). Human activity recognition using inertial, physiological and environmental sensors: A comprehensive survey. *IEEE Access : Practical Innovations, Open Solutions*, *8*, 210816–210836. doi:10.1109/ACCESS.2020.3037715 PMID:33344100

Deng, H., Nanjo, H., Qian, P., Xia, Z., Ishikawa, I., & Suzuki, T. M. (2008). Corrosion prevention of iron with novel organic inhibitor of hydroxamic acid and UV irradiation. *Electrochimica Acta*, *53*(6), 2972–2983. doi:10.1016/j.electacta.2007.11.008

Dhall, A., Dhasade, A., & Nalwade, A., VK, M. R., & Kulkarni, V. (2020). A survey on systematic approaches in managing forest fires. *Applied Geography (Sevenoaks, England)*, *121*, 102266. doi:10.1016/j.apgeog.2020.102266

Dhall, S., Mehta, B. R., Tyagi, A. K., & Sood, K. (2021). A review on environmental gas sensors: Materials and technologies. *Sensors International*, *2*, 100116. doi:10.1016/j.sintl.2021.100116

Ding, D., Jiang, D., Zhou, Y., Xia, F., Chen, Y., Kong, L., Wei, J., Zhang, S., & Deng, S. (2022). Assessing the environmental impacts and costs of biochar and monitored natural attenuation for groundwater heavily contaminated with volatile organic compounds. *The Science of the Total Environment*, *846*, 157316. doi:10.1016/j.scitotenv.2022.157316 PMID:35842168

Dixion, R. A., & Paiva, N. L. (1995). Stress-induced phenylpropanoid metabolism. *The Plant Cell*, *7*(7), 1085–1097. doi:10.2307/3870059 PMID:12242399

Dixon, J., Hull, M. R., Cobb, A. H., & Sanders, G. E. (1995). Ozone pollution modifies the response of sugarbeet to the herbicide phenmedipham. *Water, Air, and Soil Pollution*, *85*(3), 1443–1448. doi:10.1007/BF00477184

Dogra, R., Rani, S., & Sharma, B. (2021). A review of forest fires and their detection techniques using wireless sensor network. In *Advances in Communication and Computational Technology: Select Proceedings of ICACCT 2019* (pp. 1339-1350). Springer Singapore. 10.1007/978-981-15-5341-7_101

Dolson, C. M., Harlow, E. R., Phelan, D. M., Gabbett, T. J., Gaal, B., McMellen, C., Geletka, B. J., Calcei, J. G., Voos, J. E., & Seshadri, D. R. (2022). Wearable sensor technology to predict Core body temperature: A systematic review. *Sensors (Basel)*, *22*(19), 7639. doi:10.3390/s22197639 PMID:36236737

Dorner, I., Röse, P., & Krewer, U. (2023). Dynamic vs. Stationary Analysis of Electrochemical Carbon Dioxide Reduction: Profound Differences in Local States. *ChemElectroChem*, *10*(24), e202300387. doi:10.1002/celc.202300387

Dorrer, G., Dorrer, A., Buslov, I., & Yarovoy, S. (2018, November). System of personnel training in decision-making in fighting wildfires. [). IOP Publishing.]. *IOP Conference Series. Materials Science and Engineering*, *450*(6), 062018. doi:10.1088/1757-899X/450/6/062018

Dougherty, S., Simpson, J. R., Hill, R. R., Pignatiello, J. J., & White, E. D. (2015). Nonlinear screening designs for defense testing: An overview and case study. *The Journal of Defense Modeling and Simulation*, *12*(3), 335–342. doi:10.1177/1548512914542523

Duan, R., Hao, X., Li, Y., & Li, H. (2020). Detection of Acetylcholinesterase and Its Inhibitors by Liquid Crystal Biosensor Based on Whispering Gallery Mode. *Sensors and Actuators. B, Chemical*, *308*, 127672. doi:10.1016/j.snb.2020.127672

Duarte, L., Teodoro, A. C., Gonçalves, J. A., Ribeiro, J., Flores, D., Lopez-Gil, A., Dominguez-Lopez, A., Angulo-Vinuesa, X., Martin-Lopez, S., & Gonzalez-Herraez, M. (2017). Distributed temperature measurement in a self-burning coal waste pile through a GIS open source desktop application. *ISPRS International Journal of Geo-Information*, *6*(3), 87. doi:10.3390/ijgi6030087

Dudda, A., Herold, M., Holzel, C., Loidl-Stahlhofen, A., Jira, W., & Mlakar, A. (1996). Lipid peroxidation, a consequence of cell injury? *South African Journal of Chemistry. Suid-Afrikaanse Tydskrif vir Chemie*, *49*, 59–64.

Du, Y., Yu, D. G., & Yi, T. (2023). Electrospun nanofibers as chemosensors for detecting environmental pollutants: A review. *Chemosensors (Basel, Switzerland)*, *11*(4), 208. doi:10.3390/chemosensors11040208

Dwight, R. H., Fernandez, L. M., Baker, D. B., Semenza, J. C., & Olson, B. H. (2005). Estimating the economic burden from illnesses associated with recreational coastal water pollution—A case study in Orange County, California. *Journal of Environmental Management*, *76*(2), 95–103. doi:10.1016/j.jenvman.2004.11.017 PMID:15939121

Dwivedi, A. K. (2017). Researches in water pollution: A review. *International Research Journal of Natural and Applied Sciences*, *4*(1), 118–142.

Elachi, C., & Van Zyl, J. J. (2021). *Introduction to the physics and techniques of remote sensing*. John Wiley & Sons. doi:10.1002/9781119523048

Eliades, D., Lambrou, T., Panayiotou, C., & Polycarpou, M. (2014) Contamination Event Detection in Water Distribution Systems using a Model-Based Approach. In: *16th Conference on Water Distribution System Analysis*. 14–17 July 2014 10.1016/j.proeng.2014.11.229

El-Khatib, A. A. (2003). The response of some common Egyptian plants to ozone and their use as biomonitors. *Environmental Pollution*, *124*(3), 419–428. doi:10.1016/S0269-7491(03)00045-9 PMID:12758022

Elledge, M., Muralidharan, A., Parker, A., Ravndal, K. T., Siddiqui, M., Toolaram, A. P., & Woodward, K. (2018). Menstrual Hygiene Management and Waste Disposal in Low and Middle Income Countries—A Review of the Literature. *International Journal of Environmental Research and Public Health*, *15*(11), 2562. doi:10.3390/ijerph15112562 PMID:30445767

Elphick, C.S. (2000). Functional equivalency between rice fields and semiChatlam wetland habitats. *Conservation Biology*, *14*(1), 181-191.

El-Shafeiy, E., Alsabaan, M., Ibrahem, M. I., & Elwahsh, H. (2023). Real-Time Anomaly Detection for Water Quality Sensor Monitoring Based on Multivariate Deep Learning Technique. *Sensors (Basel)*, *23*(20), 8613. doi:10.3390/s23208613 PMID:37896705

Eom, H., Hwang, J., Hassan, S. H. A., Joo, J. H., Hur, J. H., Chon, K., Jeon, B.-H., Song, Y.-C., Chae, K.-J., & Oh, S.-E. (2019). Rapid detection of heavy metal-induced toxicity in water using a fed-batch sulfur-oxidizing bacteria (SOB) bioreactor. *Journal of Microbiological Methods*, *161*, 35–42. doi:10.1016/j.mimet.2019.04.007 PMID:30978364

Epping, R., & Koch, M. (2023). On-Site Detection of Volatile Organic Compounds (VOCs). *Molecules*, 28(4), 1598. doi:10.3390/molecules28041598

Eramma, N., Lalita, H. M., Satishgouda, S., Jyothi, S. R., Venkatesh, C. N., & Patil, S. J. (2023). Zooplankton Productivity Evaluation of Lentic and Lotic Ecosystem. *Intech Open*, 1-16. doi:10.5772/intechopen.107020

Eugenio, N. R., Naidu, R., & Colombo, C. (2020). Global approaches to assessing, monitoring, mapping, and remedying soil pollution. *Environmental Monitoring and Assessment*, *192*(9), 601. doi:10.1007/s10661-020-08537-2 PMID:32857292

Falkenberg, L. J., Bellerby, R. G., Connell, S. D., Fleming, L. E., Maycock, B., Russell, B. D., Sullivan, F. J., & Dupont, S. (2020). Ocean acidification and human health. *International Journal of Environmental Research and Public Health*, *17*(12), 4563. doi:10.3390/ijerph17124563 PMID:32599924

Fan, C., Hu, K., Yuan, Y., & Li, Y. (2023). A data-driven analysis of global research trends in medical image: A survey. *Neurocomputing*, *518*, 308–320. doi:10.1016/j.neucom.2022.10.047

Fang, L., Liao, X., Jia, B., Shi, L., Kang, L., Zhou, L., & Kong, W. (2020). Recent progress in immunosensors for pesticides. *Biosensors & Bioelectronics*, *164*, 1–57. doi:10.1016/j.bios.2020.112255 PMID:32479338

Farjana, M., Fahad, A. B., Alam, S. E., & Islam, M. M. (2023). An iot-and cloud-based e-waste management system for resource reclamation with a data-driven decision-making process. *IoT*, *4*(3), 202–220. doi:10.3390/iot4030011

Farkhondehmaal, F., & Ghaffarzadegan, N. (2022). A cyclical wildfire pattern is the outcome of a coupled human-natural system. *Scientific Reports*, *12*(1), 5280. doi:10.1038/s41598-022-08730-y PMID:35347175

Fatimah, Y. A., Govindan, K., Murniningsih, R., & Setiawan, A. (2020). Industry 4.0 based sustainable circular economy approach for smart waste management system to achieve sustainable development goals: A case study of Indonesia. *Journal of Cleaner Production*, *269*, 122263. doi:10.1016/j.jclepro.2020.122263

Fatima, S., Muzammal, M., Rehman, A., Rustam, S. A., Shehzadi, Z., Mehmood, A., & Waqar, M. (2020). Water pollution on heavy metals and its effects on fishes. *International Journal of Fisheries and Aquatic Studies*, *8*(3), 6–14.

Fattah, S., Gani, A., Ahmedy, I., Idris, M. Y. I., & Targio Hashem, I. A. (2020). A survey on underwater wireless sensor networks: Requirements, taxonomy, recent advances, and open research challenges. *Sensors (Basel)*, *20*(18), 5393. doi:10.3390/s20185393 PMID:32967124

Fawell, J., & Nieuwenhuijsen, M. J. (2003). Contaminants in drinking water: Environmental pollution and health. *British Medical Bulletin*, *68*(1), 199–208. doi:10.1093/bmb/ldg027 PMID:14757718

Felix, F. S., & Angnes, L. (2018). Electrochemical immunosensors-A powerful tool for analytical applications. *Biosensors & Bioelectronics*, *102*, 470–478. doi:10.1016/j.bios.2017.11.029 PMID:29182930

Feng, Y., Yu, J., Sun, D., Dang, C., Ren, W., Shao, C., & Sun, R. (2022). Extreme environment-adaptable and fast self-healable eutectogel triboelectric nanogenerator for energy harvesting and self-powered sensing. *Nano Energy*, *98*, 107284. doi:10.1016/j.nanoen.2022.107284

Fernández-Sánchez, J. F., Cannas, R., Spichiger, S., Steiger, R., & Spichiger-Keller, U. E. (2007). Optical CO_2-sensing layers for clinical application based on pH-sensitive indicators incorporated into nanoscopic metal-oxide supports. *Sensors and Actuators. B, Chemical*, *128*(1), 145–153. doi:10.1016/j.snb.2007.05.042

Ferreira, L.V. (2000). Effects of flooding duration on species richness, floristic composition and forest structure in river margin habitat in Amazonian blackwater floodplain forests: implications for future design of protected areas. *Biodiversity and Conservation*, *9*(1), 1-14.

Finn, C., Schnittger, S., Yellowlees, L. J., & Love, J. B. (2012). Molecular approaches to the electrochemical reduction of carbon dioxide. *Chemical Communications*, *48*(10), 1392–1399. doi:10.1039/C1CC15393E PMID:22116300

Fiorillo, F., & Merkaj, E. (2024). Municipal strategies, fiscal incentives and co-production in urban waste management. *Socio-Economic Planning Sciences*, *101817*, 101817. doi:10.1016/j.seps.2024.101817

Fischer, K., Fries, E., Körner, W., Schmalz, C., & Zwiener, C. (2012). New developments in the trace analysis of organic water pollutants. *Applied Microbiology and Biotechnology*, *94*(1), 11–28. doi:10.1007/s00253-012-3929-z PMID:22358315

Fischer, W. A., Hemphill, W. R., & Kover, A. (1976). Progress in remote sensing (1972-1976). *Photogrammetria*, *32*(2), 33–72. doi:10.1016/0031-8663(76)90013-2

Fisher, M., Cardoso, R. C., Collins, E. C., Dadswell, C., Dennis, L. A., Dixon, C., Farrell, M., Ferrando, A., Huang, X., Jump, M., Kourtis, G., Lisitsa, A., Luckcuck, M., Luo, S., Page, V., Papacchini, F., & Webster, M. (2021). An overview of verification and validation challenges for inspection robots. *Robotics (Basel, Switzerland)*, *10*(2), 67. doi:10.3390/robotics10020067

Fjerdingstad, E. (1964). Pollution of streams estimated by benthal phytomicro-organisms, I. Seprobic system based on communities of organisms and ecological factors. *Int Rev Ges. Am.*, *1*(4), 683–689.

Flores-Contreras, E. A., González-González, R. B., Gonzalez-González, E., Melchor-Martínez, E. M., Parra-Saldívar, R., & Iqbal, H. M. (2022). Detection of emerging pollutants using AptAMer-Based biosensors: Recent advances, challenges, and outlook. *Biosensors (Basel)*, *12*(12), 1078. doi:10.3390/bios12121078 PMID:36551045

Fluckiger, W., Fluckiger-Keller, H., & Oertli, J. J. (1978a). Inhibition of the regulatory ability of stomata caused by exhaust gases. *Experientia*, *34*(10), 1274–1275. doi:10.1007/BF01981413

Fluckiger, W., Fluckiger-Keller, H., & Oertli, J. J. (1978b). Biochemische Veranderungen in jungen Birken im Nahbereich einer Autobahn. *Forest Pathology*, *8*(3), 154–163. doi:10.1111/j.1439-0329.1978.tb01462.x

Ford, H. V., Jones, N. H., Davies, A. J., Godley, B. J., Jambeck, J. R., Napper, I. E., Suckling, C. C., Williams, G. J., Woodall, L. C., & Koldewey, H. J. (2022). The fundamental links between climate change and marine plastic pollution. *The Science of the Total Environment*, *806*, 150392. doi:10.1016/j.scitotenv.2021.150392 PMID:34583073

Fowler, S. W. (2011). 210Po in the marine environment with emphasis on its behaviour within the biosphere. *Journal of Environmental Radioactivity*, *102*(5), 448–461. doi:10.1016/j.jenvrad.2010.10.008 PMID:21074911

Foyer, C. H., Lelandais, M., & Kunert, K. J. (1994). Photooxidative stress in plants. *Physiologia Plantarum*, *92*(4), 696–717. doi:10.1111/j.1399-3054.1994.tb03042.x

Francesco, A., Filippo, A., Carlo, G. C., & Anna, M. L. (2015). A Smart Sensor Network for Sea Water Quality Monitoring. *IEEE Sensors Journal*, *15*(5), 2514–2522. doi:10.1109/JSEN.2014.2360816

Frense, D., Müller, A., & Beckmann, D. (1998). Detection of environmental pollutants using optical biosensor with immobilized algae cells. *Sensors and Actuators. B, Chemical*, *51*(1-3), 256–260. doi:10.1016/S0925-4005(98)00203-2

Frisken, S., Clarke, I., & Poole, S. (2013). Technology and applications of liquid crystal on silicon (LCoS) in telecommunications. *Optical Fiber Telecommunications*, 709-742.

Fu, J., Cherevko, S., & Chung, C. H. (2008). Electroplating of metal nanotubes and nanowires in a high aspect-ratio nanotemplate. *Electrochemistry Communications*, *10*(4), 514–518. doi:10.1016/j.elecom.2008.01.015

Fumian, F., Di Giovanni, D., Martellucci, L., Rossi, R., & Gaudio, P. (2020). Application of Miniaturized Sensors to Unmanned Aerial Systems, A New Pathway for the Survey of Polluted Areas: Preliminary Results. *Atmosphere (Basel)*, *11*(5), 471. doi:10.3390/atmos11050471

Gadekar, U. (2023). Hygiene Practices and Community Well-being in a Rural Setting: The Case of Mhawlewadi Village. *International Journal of Engineering and Management Research*, *13*(4), 99–104.

Galea, S., Ahern, J., Rudenstine, S., Wallace, Z., & Vlahov, D. (2005). Urban Built Environment and Depression: A Multilevel Analysis. *Journal of Epidemiology and Community Health*, *59*(10), 822–827. doi:10.1136/jech.2005.033084 PMID:16166352

Gale, M. G., Cary, G. J., Van Dijk, A. I., & Yebra, M. (2021). Forest fire fuel through the lens of remote sensing: Review of approaches, challenges and future directions in the remote sensing of biotic determinants of fire behavior. *Remote Sensing of Environment*, *255*, 112282. doi:10.1016/j.rse.2020.112282

Gall, A. M., Mariñas, B. J., Lu, Y., & Shisler, J. L. (2015). Waterborne viruses: A barrier to safe drinking water. *PLoS Pathogens*, *11*(6), e1004867. doi:10.1371/journal.ppat.1004867 PMID:26110535

Gambhir, R. S., Kapoor, V., Nirola, A., Sohi, R., & Bansal, V. (2012). Water Pollution: Impact of Pollutants and New Promising Techniques in Purification Process. *Journal of Human Ecology (Delhi, India)*, *37*(2), 103–109. doi:10.1080/09709274.2012.11906453

Ganjali, M. R., Eshraghi, M. H., Ghadimi, S., & Mojtaba, S. (2011). Novel chromate sensor based on MWCNTs/nanosilica/ionic liouid/Eu complex/graphite as a new nano-composite and Its application for determination of chromate ion concentration in waste water of chromium electroplating. *International Journal of Electrochemical Science*, *6*(3), 739–748. doi:10.1016/S1452-3981(23)15031-0

Gao, H., Generelli, S., & Heitger, F. (2017). Online Monitoring the Water Contaminations with Optical Biosensor. *Proceedings*, *1*(4), 522. doi:10.3390/proceedings1040522

Garcia-Leon, M. (2018). Accelerator Mass Spectrometry (AMS) in Radioecology. *Journal of Environmental Radioactivity*, *186*, 116–123. doi:10.1016/j.jenvrad.2017.06.023 PMID:28882579

Gaston, K.J. (2000). Biodiversity: higher taxon richness. *Prog Phys Geogr. 24*, 117–127.

Gavrilaş, S., Ursachi, C. Ş., Perţa-Crişan, S., & Munteanu, F. D. (1999). Foliage age and pollution alter content of phenolic compounds and chemical elements in *Pinus nigra* Needles. *Water, Air, and Soil Pollution*, *110*(3/4), 363–377. doi:10.1023/A:1005009214988 PMID:35214408

Gavrilaş, S., Ursachi, C. Ş., Perţa-Crişan, S., & Munteanu, F.-D. (2022). Recent Trends in Biosensors for Environmental Quality Monitoring. *Sensors (Basel)*, *22*(4), 1513. doi:10.3390/s22041513 PMID:35214408

Geng, P., Zhang, X., Meng, W., Wang, Q., Zhang, W., Jin, L., Feng, Z., & Wu, Z. (2008). Self-assembled monolayers-based immunosensor for detection of Escherichia coli using electrochemical impedance spectroscopy. *Electrochimica Acta*, *53*(14), 4663–4668. doi:10.1016/j.electacta.2008.01.037

Geng, Z., Zhang, X., Fan, Z., Lv, X., Su, Y., & Chen, H. (2017). *Recent progress in optical biosensors based on smartphone platforms. Sensors*. MDPI AG. doi:10.3390/s17112449

Gerson, G., Christopher, B., Stephen, M., & Richard, O. (2012) Real-time Detection of Water Pollution using Biosensors and Live Animal Behavior Models, In: 6th eResearch Australasia Conference, 28 Oct – 1 Nov 2012

Ghosh, I., Rakholia, D., Shah, K., Bhatt, D., & Das, M. N. (2020). Environmental Perspective on Menstrual Hygiene Management Along with the Movement towards Biodegradability: A Mini-Review. *Journal of Biomedical Research & Environmental Sciences*, *1*(5), 122–126. doi:10.37871/jels1129

Gieva, E., Nikolov, G. T., & Nikolova, B. (2014). Biosensors For Environmental Monitoring. *Challenges in Higher Education & Research*, *12*, 123–127.

Glaviano, F., Esposito, R., Cosmo, A. D., Esposito, F., Gerevini, L., Ria, A., Molinara, M., Bruschi, P., Costantini, M., & Zupo, V. (2022). Management and sustainable exploitation of marine environments through smart monitoring and automation. *Journal of Marine Science and Engineering*, *10*(2), 297. doi:10.3390/jmse10020297

Goib, W., Yudi, Y., Dewa, P., Iqbal, S., & Dadin, M. (2015). Integrated online water quality monitoring. In: *International conference on smart sensors and application*. IEEE.

Gola, K. K., & Gupta, B. (2020). Underwater sensor networks:'Comparative analysis on applications, deployment and routing techniques'. *IET Communications*, *14*(17), 2859–2870. doi:10.1049/iet-com.2019.1171

Gonchar, M. V., Maidan, M. M., Moroz, O. M., Woodward, J. R., & Sibirny, A. A. (1998). Microbial O2-and H2O2-electrode sensors for alcohol assays based on the use of permeabilized mutant yeast cells as the sensitive bioelements. *Biosensors & Bioelectronics*, *13*(9), 945–952. doi:10.1016/S0956-5663(98)00034-7 PMID:9839383

Gonzalez, L., McCallum, A., Kent, D., Rathnayaka, C., & Fairweather, H. (2023). A review of sedimentation rates in freshwater reservoirs: Recent changes and causative factors. *Aquatic Sciences*, *85*(2), 60. doi:10.1007/s00027-023-00960-0

Goradel, N. H., Mirzaei, H., Sahebkar, A., Poursadeghiyan, M., Masoudifar, A., Malekshahi, Z. V., & Negahdari, B. (2018). Biosensors for the Detection of Environmental and Urban Pollutions. *Journal of Cellular Biochemistry*, *119*(1), 207–212. doi:10.1002/jcb.26030 PMID:28383805

Gosset, A., Oestreicher, V., Perullini, M., Bilmes, S. A., Jobbágy, M., Dulhoste, S., Bayard, R., & Durrieu, C. (2019). Optimization of Sensors Based on Encapsulated Algae for Pesticide Detection in Water. *Analytical Methods*, *11*(48), 6193–6203. doi:10.1039/C9AY02145K

Government of Rajasthan. (2006). *City Development Plan for Ajmer*. Government of Rajasthan. http://jnnurmmis.nic.in/toolkit/final_CDPAjmer-Pushkar.pdf

Gravenstein, N., & Jaffe, M. B. (2021). Capnography, Anesthesia Equipment (Third Ed.). Principles and Applications (pp. 239-252). Cambridge University Press

Green, D., Liu, H., & Paul, V. (2021). Urban noise pollution and its effects on public health. *Journal of Urban Health*, *98*(2), 223–234.

Grøndahl-Rosado, R. C., Tryland, I., Myrmel, M., Aanes, K. J., & Robertson, L. J. (2014). Detection of microbial pathogens and indicators in sewage effluent and river water during the temporary interruption of a wastewater treatment plant. *Water Quality, Exposure, and Health*, *6*(3), 155–159. doi:10.1007/s12403-014-0121-y

Guderian, R., & Van Haut, H. (1970). Detection of SO_2 effects upon plants. *Staub*, *30*, 22–35.

Guettiche, D., Mekki, A., Lilia, B., Fatma-Zohra, T., & Boudjellal, A. (2021). Flexible chemiresistive nitrogen oxide sensors based on a nanocomposite of polypyrrole-reduced graphene oxide-functionalized carboxybenzene diazonium salts. *Journal of Materials Science Materials in Electronics*, *32*(8), 10662–10677. doi:10.1007/s10854-021-05721-z

Guimarães, N., Pádua, L., Marques, P., Silva, N., Peres, E., & Sousa, J. J. (2020). Forestry remote sensing from unmanned aerial vehicles: A review focusing on the data, processing, and potentialities. *Remote Sensing (Basel)*, *12*(6), 1046. doi:10.3390/rs12061046

Gülçin, D., & Deniz, B. (2020). Remote sensing and GIS-based forest fire risk zone mapping: The case of Manisa, Turkey. *Turkish Journal of Forestry*, *21*(1), 15–24.

Gumpu, M. B., Sethuraman, S., Krishnan, U. M., & Rayappan, J. B. B. (2015). A review on detection of heavy metal ions in water–An electrochemical approach. *Sensors and Actuators. B, Chemical*, *213*, 515–533. doi:10.1016/j.snb.2015.02.122

Gundupalli, S. P., Hait, S., & Thakur, A. (2017). A review on automated sorting of source-separated municipal solid waste for recycling. *Waste Management (New York, N.Y.)*, *60*, 56–74. doi:10.1016/j.wasman.2016.09.015 PMID:27663707

Güner, A., Çevik, E., Şenel, M., & Alpsoy, L. (2017). An electrochemical immunosensor for sensitive detection of Escherichia coli O157: H7 by using chitosan, MWCNT, polypyrrole with gold nanoparticles hybrid sensing platform. *Food Chemistry*, *229*, 358–365. doi:10.1016/j.foodchem.2017.02.083 PMID:28372186

Guo, Z., Song, N. R., Moon, J. M., Kim, M., Jun, E. J., Choi, J., Lee, J. Y., Bielawski, C. W., Sessler, J. L., & Yoon, J. (2012). A Benzobisimidazolium-Based Fluorescent and Colorimetric Chemosensor for CO_2. *Journal of the American Chemical Society*, *134*(43), 17846–17849. doi:10.1021/ja306891c PMID:22931227

Gupta, S., & Wood, R. (2017). Development of FRET biosensor based on aptamer/functionalized graphene for ultra-sensitive detection of bisphenol a and discrimination from analogs. *Nano-Structures & Nano-Objects*, *10*, 131–140. doi:10.1016/j.nanoso.2017.03.013

Gupta, V., & Prakash, G. (1994). The dynamism of sulphur dioxide pollution. *Acta Botanica Indica*, *22*, 200–219.

Gutiérrez, J. C., Amaro, F., & Martín-González, A. (2015). Heavy metal whole-cell biosensors using eukaryotic microorganisms: An updated critical review. *Frontiers in Microbiology*, *6*(48), 1–8. doi:10.3389/fmicb.2015.00048 PMID:25750637

GWP (Global Water Partnership)/INBO (International Network of Basin Organizations). (2015). *The Handbook for Management and Restoration of Aquatic Ecosystems in River and Lake Basins*, 1-100. GWP. https://www.gwp.org/globalassets/global/toolbox/references/a-handbook-for-management-and-restoration-of-aquatic-ecosystems-in-river-and-lake-basins-no.3-2015.pdf

Häder, D.-P., Banaszak, A. T., Villafañe, V. E., Narvarte, M. A., González, R. A., & Helbling, E. W. (2020). Anthropogenic pollution of aquatic ecosystems: Emerging problems with global implications. *The Science of the Total Environment*, *713*, 136586. doi:10.1016/j.scitotenv.2020.136586 PMID:31955090

Hahlbroch, K., & Scheel, D. (1989). Physiology and molecular biology of phenylpropanoid metabolism. *Annual Review of Plant Physiology*, *30*, 105–130.

Håkanson, L., & Bryhn, A. (1999). Water pollution. *Backhuys Publ, Leiden*. https://www.researchgate.net/profile/Lars-Hakanson-2/publication/236024084_WATER_POLLUTION_-_methods_and_criteria_to_rank_model_and_remediate_chemical_threats_to_aquatic_ecosystems/links/02e7e515d7425f30ed000000/WATER-POLLUTION-methods-and-criteria-to-rank-model-and-remediate-chemical-threats-to-aquatic-ecosystems.pdf

Halder, J. N., & Islam, M. N. (2015). Water pollution and its impact on the human health. *Journal of Environment and Human*, *2*(1), 36–46. doi:10.15764/EH.2015.01005

Hall, D.L., Willig, M.R., Moorhead, D.L., Sites, R.W., Fish, E.B. & Mollhagen, T.R. (2004). Aquatic macroinvertebrate diversity of playa wetlands: the role of landscape and island biogeographic characteristics. *Wetlands*, *24*, 77-91.

Hancke, G. P., & Hancke, G. P. Jr. (2013). The role of advanced sensing in smart cities. *Sensors (Basel)*, *13*(1), 393–425. doi:10.3390/s130100393 PMID:23271603

Hannan, M. A., Al Mamun, M. A., Hussain, A., Basri, H., & Begum, R. A. (2015). A review on technologies and their usage in solid waste monitoring and management systems: Issues and challenges. *Waste Management (New York, N.Y.)*, *43*, 509–523. doi:10.1016/j.wasman.2015.05.033 PMID:26072186

Harborne, J. B. (1997). Plant secondary metabolism. In M. J. Crawley (Ed.), *Plant Ecology* (2nd ed., pp. 132–155). Blackwell Science.

Harborne, J. B., Harborne, J. B., & Harborne, J. B. (1999). Recent advances in chemical ecology. *Natural Product Reports*, *16*(4), 509–523. doi:10.1039/a804621b PMID:10467740

Harmelin-Vivien, M., Bodiguel, X., Charmasson, S., Loizeau, V., & Mellon-Duval, C. (2012). Differential biomagnification of PCB, PBDE, Hg and radiocesium in the food web of the European hake from the NW Mediterranean. *Mar. Pollut. Bull.*, *64*, 974–983. doi: . 02.014 doi:10.1016/j.marpolbul.2012

Haroon, M., & Anthony, S. (2016) Towards Monitoring the Water Quality Using Hierarchal Routing Protocol for Wireless Sensor Networks. In: *7th International Conference on Emerging Ubiquitous Systems and Pervasive Networks*. IEEE.

Harris, E. D. (1992). Regulation of antioxidant enzymes. *The FASEB Journal*, *6*(9), 2675–2683. doi:10.1096/fasebj.6.9.1612291 PMID:1612291

Harrison, M., & Tyson, N. (2022). Menstruation: Environmental impact and need for global health equity. *International Journal of Gynaecology and Obstetrics: the Official Organ of the International Federation of Gynaecology and Obstetrics*, *160*(2), 378–382. doi:10.1002/ijgo.14311 PMID:35781656

Hartley, S. E., & Jones, C. G. (1997). Plant chemistry and herbivory: or why the world is green. In M. J. Crawley (Ed.), *Plant Ecology* (2nd ed., pp. 284–324). Blackwell Science.

Hasanzadeh, M., & Shadjou, N. (2017). *Advanced nanomaterials for use in electrochemical and optical immunoassays of carcinoembryonic antigen. A review*. Microchimica Acta. Springer-Verlag Wien. doi:10.1007/s00604-016-2066-2

Haseena, M., Malik, M. F., Javed, A., Arshad, S., Asif, N., Zulfiqar, S., & Hanif, J. (2017). Water pollution and human health. *Environmental Risk Assessment and Remediation, 1*(3). https://eastafricaschoolserver.org/content/_public/Environment/Teaching%20Resources/Environment%20and%20Sustainability/Water-pollution-and-human-health.pdf

Hashem, A., Hossain, M. A. M., Marlinda, A. R., Mamun, M. A., Simarani, K., & Johan, M. R. (2021). *Nanomaterials based electrochemical nucleic acid biosensors for environmental monitoring: A review*. Applied Surface Science Advances. Elsevier B.V. doi:10.1016/j.apsadv.2021.100064

Haslam, E. (1989). *Plant polyphenols*. University Press Publ.

Havelka, U. D., Ackerson, R. C., Boyle, M. G., & Wittnbach, V. A. (1984). CO_2 enrichment effects on soybean physiology I. Effects of long-term CO_2 exposure. *Crop Science*, *24*(6), 1146–1150. doi:10.2135/cropsci1984.0011183X002400060033x

Hayat, A., & Marty, J. L. (2014). Aptamer based electrochemical sensors for emerging environmental pollutants. *Frontiers in Chemistry*, *2*, 41. doi:10.3389/fchem.2014.00041 PMID:25019067

Hayat, P. (2023). Integration of advanced technologies in urban waste management. In *Advancements in Urban Environmental Studies: Application of Geospatial Technology and Artificial Intelligence in Urban Studies* (pp. 397–418). Springer International Publishing. doi:10.1007/978-3-031-21587-2_23

Heath, R. L., & Packer, L. (1968). Photoperoxidation in isolated chloroplast. I. Kinetics and stoichiometry of fatty acids peroxidation. *Archives of Biochemistry and Biophysics*, *125*, 189–198. doi:10.1016/0003-9861(68)90654-1 PMID:5655425

Heck, W. W., Dunning, J. H., & Hindawi, I. J. (1965). Interactions of environmental factors on the sensitivity of plants to air pollution. *Journal of the Air Pollution Control Association*, *15*(11), 511–515. doi:10.1080/00022470.1965.10468415 PMID:5319783

Hein, J. R., Koschinsky, A., & Kuhn, T. (2020). Deep-ocean polymetallic nodules as a resource for critical materials. *Nature Reviews. Earth & Environment*, *1*(3), 158–169. doi:10.1038/s43017-020-0027-0

Hemavathi, C., & Jagannath, S. (2004). air pollution effects on pigment content of two avenue trees in Mysore city. *Pollution Research*, *23*, 307–309.

Henze, M., Harremoes, P., la Cour Jansen, J., & Arvin, E. (2001). *Wastewater treatment: biological and chemical processes* (3rd ed.). Springer science & business media.

He, Q., Wang, B., Liang, J., Liu, J., Liang, B., Li, G., Long, Y., Zhang, G., & Liu, H. (2023). Research on the construction of portable electrochemical sensors for environmental compounds quality monitoring. *Materials Today. Advances*, *17*, 100340. doi:10.1016/j.mtadv.2022.100340

Herath, I. K., Wu, S., Ma, M., & Huang, P. (2022). Heavy metal toxicity, ecological risk assessment, and pollution sources in a hydropower reservoir. *Environmental Science and Pollution Research International*, *29*(22), 32929–32946. doi:10.1007/s11356-022-18525-3 PMID:35020150

Hernandez-Vargas, G., Sosa-Hernández, J. E., Saldarriaga-Hernandez, S., Villalba-Rodríguez, A. M., Parra-Saldivar, R., & Iqbal, H. M. N. (2018). Electrochemical Biosensors: A Solution to Pollution Detection with Reference to Environmental Contaminants. *Biosensors (Basel)*, *8*(2), 1–21. doi:10.3390/bios8020029 PMID:29587374

Herrera-Domínguez, M., Morales-Luna, G., Mahlknecht, J., Cheng, Q., Aguilar-Hernández, I., & Ornelas-Soto, N. (2023). Optical Biosensors and Their Applications for the Detection of Water Pollutants. *Biosensors (Basel)*, *13*(3), 370. doi:10.3390/bios13030370 PMID:36979582

He, Y., & Cai, P. (2021). Special issue on soil pollution, control, and remediation. *Soil Ecology Letters*, *3*(3), 167–168. doi:10.1007/s42832-021-0110-6

Hijano, C. F., Dominguez, M. D. P., Gimenez, R. G., Sanchez, P. H., & Garcia, I. S. (2005). Higher plants as bioindicators of sulphur dioxide emissions in urban environments. *Environmental Monitoring and Assessment*, *111*(1-3), 75–88. doi:10.1007/s10661-005-8140-6 PMID:16311823

Hillayová, M. K., Holécy, J., Korísteková, K., Bakšová, M., Ostrihoň, M., & Škvarenina, J. (2023). Ongoing climatic change increases the risk of wildfires. Case study: Carpathian spruce forests. *Journal of Environmental Management*, *337*, 117620. doi:10.1016/j.jenvman.2023.117620 PMID:36934505

Hill, M. K. (2020). *Understanding environmental pollution*. Cambridge University Press. doi:10.1017/9781108395021

Hippeli, S., & Elstner, E. F. (1996). Mechanisms of oxygen activation during plant stress: Biochemical effects of air pollutants. *Journal of Plant Physiology*, *148*(3-4), 249–257. doi:10.1016/S0176-1617(96)80250-1

Hogsett, W. E., Holman, S. R., Gumpertz, M. L., & Tingey, D. T. (1984). Growth response in radish to sequential and simultaneous exposures of NO_2 and SO_2. *Environmental Pollution. Series A. Ecological and Biological*, *33*(4), 303–325. doi:10.1016/0143-1471(84)90140-5

Hojjati-Najafabadi, A., Mansoorianfar, M., Liang, T., Shahin, K., & Karimi-Maleh, H. (2022). A review on magnetic sensors for monitoring of hazardous pollutants in water resources. *The Science of the Total Environment*, *824*, 153844. doi:10.1016/j.scitotenv.2022.153844 PMID:35176366

Hong, H., Jaafari, A., & Zenner, E. K. (2019). Predicting spatial patterns of wildfire susceptibility in the Huichang County, China: An integrated model to analysis of landscape indicators. *Ecological Indicators*, *101*, 878–891. doi:10.1016/j.ecolind.2019.01.056

Hooshmand, S., Kassanos, P., Keshavarz, M., Duru, P., Kayalan, C. I., Kale, İ., & Bayazit, M. K. (2023). Wearable Nano-Based Gas Sensors for Environmental Monitoring and Encountered Challenges in Optimization. *Sensors, 23*(20), 8648. doi:10.3390/s23208648

Hosmani, S. (2014). Freshwater plankton ecology: a review. *J Res Manage Technol. 3*, 1–10.

Ho, W. S., Lin, W. H., Verpoort, F., Hong, K. L., Ou, J. H., & Kao, C. M. (2023). Application of novel nanobubble-contained electrolyzed catalytic water to cleanup petroleum-hydrocarbon contaminated soils and groundwater: A pilot-scale and performance evaluation study. *Journal of Environmental Management*, *347*, 119058. doi:10.1016/j.jenvman.2023.119058 PMID:37757689

Howell, R. K. 1974. Phenols, ozone and their involvement in the physiology of plant injury. In: M. Dugger (Ed.), *Air Pollution Related to Plant Growth*. A.C.S. 10.1021/bk-1974-0003.ch008

Howell, R. K. (1970). Influence of air pollution on quantities of caffeic acid isolated from leaves of *Phaseolus vulgaris*. *Phytopathology*, *60*(11), 1626–1629. doi:10.1094/Phyto-60-1626

Hrncˇı'rˇova, P., Opekar, F., & Sˇtulik, K. (2000). Amperometric solid-state NO_2 sensor with a solid polymer electrolyte and a reticulated vitreous carbon indicator electrode. *Sensors and Actuators. B, Chemical*, *69*(1-2), 199–204. doi:10.1016/S0925-4005(00)00540-2

Hsieh, Y. C., & Yao, D. J. (2018). Intelligent gas-sensing systems and their applications. *Journal of Micromechanics and Microengineering*, *28*(9), 093001. doi:10.1088/1361-6439/aac849

Huang, C. W., Lin, C., Nguyen, M. K., Hussain, A., Bui, X. T., & Ngo, H. H. (2023). *A review of biosensor for environmental monitoring: principle, application, and corresponding achievement of sustainable development goals. Bioengineered. NLM*. Medline. doi:10.1080/21655979.2022.2095089

Huang, Q.-D., Lv, C.-H., Yuan, X.-L., He, M., Lai, J.-P., & Sun, H. (2021). A Novel Fluorescent Optical Fiber Sensor for Highly Selective Detection of Antibiotic Ciprofloxacin Based on Replaceable Molecularly Imprinted Nanoparticles Composite Hydrogel Detector. *Sensors and Actuators. B, Chemical*, *328*(1-2), 129000. doi:10.1016/j.snb.2020.129000

Hussain, B., Chen, J. S., Huang, S. W., Sen Tsai, I., Rathod, J., & Hsu, B. M. (2023). Underpinning the ecological response of mixed chlorinated volatile organic compounds (CVOCs) associated with contaminated and bioremediated groundwaters: A potential nexus of microbial community structure and function for strategizing efficient bioremediation. *Environmental Pollution*, *334*, 122215. doi:10.1016/j.envpol.2023.122215 PMID:37473850

IAEA. (2004). Sediment Distribution Coefficients and Concentration Factors for Biota in the Marine Environment. *Technical Report Series No. 422*. Vienna: International Atomic Energy Agency.

ICRP. (2008). Environmental Protection - the Concept and Use of Reference Animals and Plants. ICRP Publication 108. *Annals of the ICRP*, *38*, 1–242.

Ingalsbee, T. (2017). What is the paradigm shift? Large wildland fires and the wildfire paradox offer opportunities for a new paradigm of ecological fire management. *International Journal of Wildland Fire*, *26*(7), 557–561. doi:10.1071/WF17062

Inyinbor Adejumoke, A., Adebesin Babatunde, O., Oluyori Abimbola, P., Adelani Akande Tabitha, A., Dada Adewumi, O., & Oreofe Toyin, A. (2018). Water pollution: Effects, prevention, and climatic impact. *Water Challenges of an Urbanizing World*, *33*, 33–47.

Iqbal, M., Zafar, M., & Abdin, M. Z. (2000). Studies on anatomical, physiological and biochemical response of trees to coal smoke pollution around a thermal power plant. Department of Botany, Faculty of Sciences, Jamia Hamdard University, Hamdard Nagar, New Delhi to Ministry of Environment and Forest, Government of India.

Iranpour, R., Straub, B., & Jugo, T. (1997). Real time BOD monitoring for wastewater process control. *Journal of Environmental Engineering*, *123*(2), 154–159. doi:10.1061/(ASCE)0733-9372(1997)123:2(154)

Irawan, Y., Febriani, A., Wahyuni, R., & Devis, Y. (2021). Water quality measurement and filtering tools using arduino Uno, PH sensor and TDS meter sensor. [JRC]. *Journal of Robotics and Control*, *2*(5). Advance online publication. doi:10.18196/jrc.25107

Irgang, BE. & Gastal Jr., CVS. (1996). *Macrófitasaquáticas da PlanícieCosteira do RS*. Porto Alegre.

Isaac, N. A., Pikaar, I., & Biskos, G. (2022). Metal oxide semiconducting nanomaterials for air quality gas sensors: operating principles, performance, and synthesis techniques. *Microchimica Acta, 189*(5), 1–22. doi:10.1007/s00604-022-05254-0

Ishida, M., Kim, P., Choi, J., Yoon, J., Kim, D., & Sessler, J. L. (2013). Benzimidazole-embedded N-fused aza-indacenes: Synthesis and deprotonation-assisted optical detection of carbon dioxide. *Chemical Communications*, *49*(62), 6950–6952. doi:10.1039/c3cc43938k PMID:23811989

Isik, M., Dodder, R., & Kaplan, P. O. (2021). Transportation emissions scenarios for New York City under different carbon intensities of electricity and electric vehicle adoption rates. *Nature Energy*, *6*(1), 92–104. doi:10.1038/s41560-020-00740-2 PMID:34804594

Islam, M. S., Sazawa, K., Sugawara, K., & Kuramitz, H. (2023). Electrochemical biosensor for evaluation of environmental pollutants toxicity. *Environments (Basel, Switzerland)*, *10*(4), 63. doi:10.3390/environments10040063

Israr, A., Ali, Z. A., Alkhammash, E. H., & Jussila, J. J. (2022). Optimization methods applied to motion planning of unmanned aerial vehicles: A review. *Drones (Basel)*, *6*(5), 126. doi:10.3390/drones6050126

Jackson, R. J. (2003). The Impact of Built Environment on Health: An Emerging Field. *American Journal of Public Health*, *93*(9), 1382–1384. doi:10.2105/AJPH.93.9.1382 PMID:12948946

Jahan, S., & Iqbal, M. Z. (1992). Morphological and anatomical studies of leaves of different plants affected by motor vehicle exhaust. *J. Islamic Acadamy of Sciences*, *5*(1), 21–23.

Jaishankar, M., Tseten, T., Anbalagan, N., Mathew, B. B., & Beeregowda, K. N. (2014). Toxicity, mechanism and health effects of some heavy metals. *Interdisciplinary Toxicology*, *7*(2), 60–72. doi:10.2478/intox-2014-0009 PMID:26109881

Jakupciak, J. P., & Colwell, R. R. (2009). Biological agent detection technologies. *Molecular Ecology Resources*, *9*(s1, SUPPL. 1), 51–57. doi:10.1111/j.1755-0998.2009.02632.x PMID:21564964

Jaleel, C. A., Gopi, R., & Panneerselvam, R. (2007). *Alterations in lipid peroxidation, electrolyte leakage and proline metabolism in Catharanthus roseus under treatment with triadimefon, a systemic fungicide*. C.R. Biologies, doi:10.1016/j.crvi.2007.10.001

Jaleel, C. A., Gopi, R., Sankar, B., Manivannan, P., Kishorekumar, A., Sridharan, R., & Panneerselvam, R. (2007). Studies on the germination, seedling vigour, lipid peroxidation and proline metabolism in *Catharanthus roseus* seedlings under salt stress. *South African Journal of Botany*, *73*(2), 190–195. doi:10.1016/j.sajb.2006.11.001

Jan, F., Min-Allah, N., & Düştegör, D. (2021). Iot based smart water quality monitoring: Recent techniques, trends and challenges for domestic applications. *Water (Basel)*, *13*(13), 1729. doi:10.3390/w13131729

Jang, M., Kang, S., & Han, M. S. (2019). A simple turn-on fluorescent chemosensor for CO_2 based on aggregation-induced emission: Application as a CO_2 absorbent screening method. *Dyes and Pigments*, *162*, 978–983. doi:10.1016/j.dyepig.2018.11.031

Javaid, M., Haleem, A., Rab, S., Pratap Singh, R., & Suman, R. (2021). Sensors for daily life: A review. *Sensors International*, *2*, 100121. doi:10.1016/j.sintl.2021.100121

Javaid, M., Haleem, A., Singh, R. P., Rab, S., & Suman, R. (2021). Significance of sensors for industry 4.0: Roles, capabilities, and applications. *Sensors International*, *2*, 100110. doi:10.1016/j.sintl.2021.100110

Jayti, B., & Patoliya, J. (2016). IoT based water quality monitoring system. In: *Proc of 49th IRF Int Conf.* IEEE.

Jha, A., & Tukkaraja, P. (2020). Monitoring and assessment of underground climatic conditions using sensors and GIS tools. *International Journal of Mining Science and Technology*, *30*(4), 495–499. doi:10.1016/j.ijmst.2020.05.010

Jia, K., Adam, P. M., & Ionescu, R. E. (2013). Sequential acoustic detection of atrazine herbicide and carbofuran insecticide using a single micro-structured gold quartz crystal microbalance. *Sensors and Actuators. B, Chemical*, *188*, 400–404. doi:10.1016/j.snb.2013.07.033

Jia, N., Yang, Z., & Yang, K. (2019). Operational effectiveness evaluation of the swarming UAV combat system based on a system dynamics model. *IEEE Access : Practical Innovations, Open Solutions*, *7*, 25209–25224. doi:10.1109/ACCESS.2019.2898728

Jiang, P., Van Fan, Y., Zhou, J., Zheng, M., Liu, X., & Klemeš, J. J. (2020). Data-driven analytical framework for waste-dumping behaviour analysis to facilitate policy regulations. *Waste Management (New York, N.Y.)*, *103*, 285–295. doi:10.1016/j.wasman.2019.12.041 PMID:31911375

Jianhua, D., Guoyin, W., Huyong, Y., Ji, X., & Xuerui, Z. (2015). A survey of smart water quality monitoring system. *Environmental Science and Pollution Research International*, *22*(7), 4893–4906. doi:10.1007/s11356-014-4026-x PMID:25561262

Jiao, D., & Fei, T. (2023). Pedestrian walking speed monitoring at street scale by an in-flight drone. *PeerJ. Computer Science*, *9*, e1226. doi:10.7717/peerj-cs.1226 PMID:37346670

Jiao, W., Wang, L., & McCabe, M. F. (2021). Multi-sensor remote sensing for drought characterization: Current status, opportunities and a roadmap for the future. *Remote Sensing of Environment*, *256*, 112313. doi:10.1016/j.rse.2021.112313

Jin, L., Sun, X., Ren, H., & Huang, H. (2023). Biological filtration for wastewater treatment in the 21st century: A data-driven analysis of hotspots, challenges and prospects. *The Science of the Total Environment*, *855*, 158951. doi:10.1016/j.scitotenv.2022.158951 PMID:36155035

Jin, Y. J., Moon, B. C., & Kwak, G. (2016). Colorimetric fluorescence response to carbon dioxide using charge transfer dye and molecular rotor dye in smart solvent system. *Dyes and Pigments*, *132*, 270–273. doi:10.1016/j.dyepig.2016.05.003

Jiwanti, P. K., Natsui, K., Nakata, K., & Einaga, Y. (2018). The electrochemical production of C2/C3 species from carbon dioxide on copper-modified boron-doped diamond electrodes. *Electrochimica Acta*, *266*, 414–419. doi:10.1016/j.electacta.2018.02.041

Joe, H. E., Yun, H., Jo, S. H., Jun, M. B., & Min, B. K. (2018). A review on optical fiber sensors for environmental monitoring. *International journal of precision engineering and manufacturing-green technology, 5*, 173-191.

John, S. (2000). *Business dynamics: systems thinking and modeling for a complex world.* Irwin McGraw-Hill.

Joksimoski, S., Kerpen, K., & Telgheder, U. (2022). Atmospheric pressure photoionization – High-field asymmetric ion mobility spectrometry (APPI-FAIMS) studies for on-site monitoring of aromatic volatile organic compounds (VOCs) in groundwater. *Talanta*, *247*, 123555. doi:10.1016/j.talanta.2022.123555 PMID:35613524

Jones, J.I., Li, W., & Maberly, S.C. (2003). Area, altitude and aquatic plant diversity. *Ecography, 26*(4), 411-420.

Joshi, L. M., Bharti, R. K., & Singh, R. (2022). Internet of things and machine learning-based approaches in the urban solid waste management: Trends, challenges, and future directions. *Expert Systems: International Journal of Knowledge Engineering and Neural Networks*, *39*(5), e12865. doi:10.1111/exsy.12865

Joshi, P. C., & Swami, A. (2007). Physiological response of some tree species under roadside automobile pollution stress around city of Haridwar, India. *The Environmentalist*, *27*(3), 365–374. doi:10.1007/s10669-007-9049-0

Jouanneau, S., Durand, M. J., & Thouand, G. (2012). Online detection of metals in environmental samples: Comparing two concepts of bioluminescent bacterial biosensors. *Environmental Science & Technology*, *46*(21), 11979–11987. doi:10.1021/es3024918 PMID:22989292

Juma, D. W., Wang, H., & Li, F. (2014). Impacts of population growth and economic development on water quality of a lake: Case study of Lake Victoria Kenya water. *Environmental Science and Pollution Research International*, *21*(8), 5737–5746. doi:10.1007/s11356-014-2524-5 PMID:24442964

Jung, J. H., & Lee, J. E. (2016). Real-time bacterial microcolony counting using on-chip microscopy. *Scientific Reports*, *6*(1), 21473. doi:10.1038/srep21473 PMID:26902822

Junk, W.J. (1989). Flood tolerance and tree distribution in central Amazonia. In HOLM-Nielsen, LB., Nielsen, IC. and Balslev, H. (Eds.). *Tropical Forest Botanical Dynamics, Speciation and Diversity*. London: Academic Press.

Justino, C. I. L., Duarte, A. C., & Rocha-Santos, T. A. P. (2017). Recent Progress in Biosensors for Environmental Monitoring: A Review. *Sensors (Basel)*, *17*(12), 2918. doi:10.3390/s17122918 PMID:29244756

Jyoti, A., Tomar, R. S., & Shanker, R. (2016). *Nanosensors for the detection of pathogenic bacteria. Nanoscience in food and agriculture 1*. Springer.

Kainulainen, P., Holopainen, J. K., Hyttinen, H., & Oksanen, J. (1994). Effect of ozone on the biochemistry and aphid infestation of scots pine. *Phytochemistry*, *35*(1), 39–42. doi:10.1016/S0031-9422(00)90505-3

Kainulainen, P., Holopainen, J. K., & Oksanen, J. (1995). Effects of SO_2 on the concentrations of carbohydrates and secondary compounds in Scots pine (Pinus sylvestris L.) and Norway spruce (*Picea abies* L. Karst.) seedlings. *The New Phytologist*, *130*(2), 231–238. doi:10.1111/j.1469-8137.1995.tb03044.x

Kammerbauer, H., Ziegler-Joens, A., Selinger, H., Knoppik, D., & Hock, B. (1987). Exposure of Norway spruce at the highway border: Effects on gas exchange and growth. *Experientia*, *43*(10), 1124–1125. doi:10.1007/BF01956060

Kammerbauer, J., & Dick, T. (2000). Monitoring of urban traffic emission using some physiological indicators in *Ricinus communis* L. Plants. *Archives of Environmental Contamination and Toxicology*, *39*(2), 161–166. doi:10.1007/s002440010092 PMID:10871418

Kang, S., Kim, J., Park, J. H., Ahn, C. K., Rhee, C. H., & Han, M. S. (2015). Intra-molecular hydrogen bonding stabilization based-fluorescent chemosensor for CO_2: Application to screen relative activities of CO_2 absorbents. *Dyes and Pigments*, *123*, 125–131. doi:10.1016/j.dyepig.2015.07.033

Kannan, D., Khademolqorani, S., Janatyan, N., & Alavi, S. (2024). Smart waste management 4.0: The transition from a systematic review to an integrated framework. *Waste Management (New York, N.Y.)*, *174*, 1–14. doi:10.1016/j.wasman.2023.08.041 PMID:37742441

Kanoun, M., Goulas, M. J. P., & Biolley, J. P. (2001). Effect of chronic and moderate ozone pollution on the phenolic pattern of bean leaves (*Phaseolus vulgaris* L. cv. Nerina): Relations with visible injury and biomass production. *Biochemical Systematics and Ecology*, *29*(5), 443–457. doi:10.1016/S0305-1978(00)00080-6 PMID:11274768

Kanoun, O., Lazarević-Pašti, T., Pašti, I., Nasraoui, S., Talbi, M., Brahem, A., Adiraju, A., Sheremet, E., Rodriguez, R. D., Ben Ali, M., & Al-Hamry, A. (2021). A Review of Nanocomposite-Modified Electrochemical Sensors for Water Quality Monitoring. *Sensors (Basel)*, *21*(12), 12. doi:10.3390/s21124131 PMID:34208587

Kapitsaki, G. M., Achilleos, A. P., Aziz, P., & Paphitou, A. C. (2021). SensoMan: Social Management of Context Sensors and Actuators for IoT. *Journal of Sensor and Actuator Networks*, *4*(4), 68. doi:10.3390/jsan10040068

Kapoor, A., Balasubramanian, S., Ponnuchamy, M., Vaishampayan, V., & Sivaraman, P. (2020). Lab-on-a-chip devices for water quality monitoring. In Nanotechnology in the life sciences (pp. 455–469). Springer. doi:10.1007/978-3-030-45116-5_15

Karolewski, P. (1985). The role of free proline in the sensitivity of popular plants to the action of SO_2. *European Journal of Forest Pathology*, *15*, 199–206. doi:10.1111/j.1439-0329.1985.tb00886.x

Karolewski, P., & Giertych, M. J. (1995). Changes in the level of phenols during needle development in Scots-pine populations in a control and polluted environment. *European Journal of Forest Pathology*, *25*(6-7), 297–306. doi:10.1111/j.1439-0329.1995.tb01345.x

Karpinski, S., Reynolds, H., Karpinska, B., Wingsle, G., Creissen, G., & Millineaux, P. (1999). Systemic signaling in response to excess excitation energy in *Arabidopsis*. *Science*, *284*(5414), 654–657. doi:10.1126/science.284.5414.654 PMID:10213690

Karthik, V., Karuna, B., Kumar, P. S., Saravanan, A., & Hemavathy, R. V. (2022). Development of lab-on-chip biosensor for the detection of toxic heavy metals: A review. *Chemosphere*, *299*, 134427. doi:10.1016/j.chemosphere.2022.134427 PMID:35358561

Karube, I., Mitsuda, S., & Suzuki, S. (1979). Glucose sensor using immobilized whole cells of *Pseudomonas fluorescens. European journal of applied microbiology and biotechnology, 7*, 343-350.

Karunanidhi, D., Subramani, T., Roy, P. D., & Li, H. (2021). Impact of groundwater contamination on human health. *Environmental Geochemistry and Health*, *43*(2), 643–647. doi:10.1007/s10653-021-00824-2 PMID:33486701

Katoh, T., Kasuya, M., Kagamimori, S., Kozuka, H., & Kawano, S. (1989). Inhibition of the shikimate pathway in the leaves of vascular plants exposed to air pollution. *The New Phytologist*, *112*(3), 363–367. doi:10.1111/j.1469-8137.1989.tb00324.x

Kaur, K., & Baral, M. (2014). Synthesis of imine-naphthol tripodal ligand and study of its coordination behaviour towards Fe (III), Al (III), and Cr (III) metal ions. *Bioinorganic Chemistry and Applications*. doi:10.1155/2014/915457

Kaur, H., Kumar, R., Babu, J. N., & Mittal, S. (2015). Advances in arsenic biosensor development–A comprehensive review. *Biosensors & Bioelectronics*, *63*, 533–545. doi:10.1016/j.bios.2014.08.003 PMID:25150780

Kaur, R., Kaur, K., & Kaur, R. (2018). Menstrual Hygiene, Management, and Waste Disposal: Practices and challenges faced by Girls/Women of Developing Countries. *Journal of Environmental and Public Health*, *2018*, 1–9. doi:10.1155/2018/1730964 PMID:29675047

Kaya, H. O., Cetin, A. E., Azimzadeh, M., & Topkaya, S. N. (2021). *Pathogen detection with electrochemical biosensors: Advantages, challenges and future perspectives. Journal of Electroanalytical Chemistry.* Elsevier B.V., doi:10.1016/j.jelechem.2021.114989

Keller, T. H. (1974). The use of peroxidase activity for monitoring and mapping air pollution areas. *Forest Pathology*, *4*(1), 11–19. doi:10.1111/j.1439-0329.1974.tb00407.x

Khalil, B., & Ouarda, T. B. M. J. (2009). Statistical approaches used to assess and redesign surface water-quality-monitoring networks. *Journal of Environmental Monitoring*, *2009*(11), 1915–1929. doi:10.1039/b909521g PMID:19890548

Khan, A. M., Pandey, V., Shukla, J., Singh, N., Yunus, M., Singh, S. N., & Ahmad, K. J. (1990). Effect of thermal power plant pollution on *Catharanthus roseus* L. *Bulletin of Environmental Contamination and Toxicology*, *44*(6), 865–870. doi:10.1007/BF01702176 PMID:2354262

Khanam, Z., Gupta, S., & Verma, A. (2020). Endophytic fungi-based biosensors for environmental contaminants-A perspective. *South African Journal of Botany*, *134*, 401–406. doi:10.1016/j.sajb.2020.08.007

Khandare, D. G., Joshi, H., Banerjee, M., Majik, M. S., & Chatterjee, A. (2015). Fluorescence turn-on chemosensor for the detection of dissolved CO_2 based on ion-induced aggregation of tetraphenylethylene derivative. *Analytical Chemistry*, *87*(21), 10871–10877. doi:10.1021/acs.analchem.5b02339 PMID:26458016

Khan, H., Hassan, S. A., & Jung, H. (2020). On underwater wireless sensor networks routing protocols: A review. *IEEE Sensors Journal*, *20*(18), 10371–10386. doi:10.1109/JSEN.2020.2994199

Khanmohammadi, A., Jalili Ghazizadeh, A., Hashemi, P., Afkhami, A., Arduini, F., & Bagheri, H. (2020). An overview to electrochemical biosensors and sensors for the detection of environmental contaminants. *Journal of the Indian Chemical Society*, *17*, 2429–2447.

Khan, S., & Khan, A. (2022). Ffirenet: Deep learning-based forest fire classification and detection in smart cities. *Symmetry*, *14*(10), 2155. doi:10.3390/sym14102155

Khan, S., Shahnaz, M., Jehan, N., Rehman, S., Shah, M. T., & Din, I. (2013). Drinking water quality and human health risk in Charsadda district, Pakistan. *Journal of Cleaner Production*, *60*, 93–101. doi:10.1016/j.jclepro.2012.02.016

Kim, B. K., Kang, H. S., & Park, S. O. (2016). Drone classification using convolutional neural networks with merged Doppler images. *IEEE Geoscience and Remote Sensing Letters*, *14*(1), 38–42. doi:10.1109/LGRS.2016.2624820

Kim, C., Lee, W., Melis, A., Elmughrabi, A., Lee, K., Park, C., & Yeom, J. Y. (2021). A review of inorganic scintillation crystals for extreme environments. *Crystals*, *11*(6), 669. doi:10.3390/cryst11060669

Kim, S. J., Lim, G. J., Cho, J., & Côté, M. J. (2017). Drone-aided healthcare services for patients with chronic diseases in rural areas. *Journal of Intelligent & Robotic Systems*, *88*(1), 163–180. doi:10.1007/s10846-017-0548-z

Kim, S., & Lee, H. J. (2017). Gold Nanostar Enhanced Surface Plasmon Resonance Detection of an Antibiotic at Attomolar Concentrations via an Aptamer-Antibody Sandwich Assay. *Analytical Chemistry*, *89*(12), 6624–6630. doi:10.1021/acs.analchem.7b00779 PMID:28520392

Kinaneva, D., Hristov, G., Raychev, J., & Zahariev, P. (2019, May). Early forest fire detection using drones and artificial intelligence. In *2019 42nd International Convention on Information and Communication Technology, Electronics and Microelectronics (MIPRO)* (pp. 1060-1065). IEEE. 10.23919/MIPRO.2019.8756696

Kirchner, J. W., Feng, X., Neal, C., & Robson, A. J. (2004). The fine structure of water-quality dynamics: The (high-frequency) wave of the future. *Hydrological Processes*, *18*(7), 1353–1359. doi:10.1002/hyp.5537

Kišija, E., Osmanović, D., Nuhić, J., & Cifrić, S. (2020). Review of biosensors in industrial process control. In I. F. M. B. E. Proceedings (Ed.), Vol. 73, pp. 687–694). Springer Verlag., doi:10.1007/978-3-030-17971-7_103

Kissinger, P. T. (2005). Biosensors—a perspective. Biosensors and bioelectronics 20(12), 2512-2516.

Kitagawa, Y., Tamiya, E., & Karube, I. (1987). Microbial-FET alcohol sensor. *Analytical Letters*, *20*(1), 81–96. doi:10.1080/00032718708082238

Koehnken, L., Rintoul, M. S., Goichot, M., Tickner, D., Loftus, A. C., & Acreman, M. C. (2020). Impacts of riverine sand mining on freshwater ecosystems: A review of the scientific evidence and guidance for future research. *River Research and Applications*, *36*(3), 362–370. doi:10.1002/rra.3586

Kohn, D.D. & Walsh, D.M. (1994). Plant species richness - the effect of island size and habitat diversity. *Journal of Ecology*, 82(2), 367-377.

Ko, K., Lee, J., & Chung, H. (2020). Highly efficient colorimetric CO_2 sensors for monitoring CO_2 leakage from carbon capture and storage sites. *The Science of the Total Environment*, *729*, 138786. doi:10.1016/j.scitotenv.2020.138786 PMID:32380324

Kokkinos, C., & Economou, A. (2017). Emerging trends in biosensing using stripping voltammetric detection of metal-containing nanolabels–A review. *Analytica Chimica Acta*, *961*, 12–32. doi:10.1016/j.aca.2017.01.016 PMID:28224905

Kondo, N., & Saji, H. (1992). Tolerance of plants to air pollutants. [In Japanese]. *Journal of Japan Society of Air Pollution*, *27*, 273–288.

Kordbacheh, F., & Heidari, G. (2023). Water pollutants and approaches for their removal. *Materials Chemistry Horizons*, *2*(2), 139–153.

Korpan, Y. I., Dzyadevich, S. V., Zharova, V. P., & El'skaya, A. V. (1994). Conductometric biosensor for ethanol detection based on whole yeast cells. *Ukrainskii Biokhimicheskii Zhurnal (1978)*, *66*(1), 78-82.

Korpan, Y. I., Gonchar, M. V., Starodub, N. F., Shulga, A. A., Sibirny, A. A., & Elskaya, A. V. (1993). A cell biosensor specific for formaldehyde based on pH-sensitive transistors coupled to methylotrophic yeast cells with genetically adjusted metabolism. *Analytical Biochemistry*, *215*(2), 216–222. doi:10.1006/abio.1993.1578 PMID:8122781

Kratasyuk, V. A., Kolosova, E. M., Sutormin, O. S., Lonshakova-Mukina, V. I., Baygin, M. M., Rimatskaya, N. V., Sukovataya, I. E., & Шпедт, A. A. (2021). Software for matching standard activity enzyme biosensors for soil pollution analysis. *Sensors (Basel)*, *21*(3), 1017. doi:10.3390/s21031017 PMID:33540862

Krechetov, I. V., Skvortsov, A. A., Poselsky, I. A., Paltsev, S. A., Lavrikov, P. S., & Korotkovs, V. (2019). Implementation of automated lines for sorting and recycling household waste as an important goal of environmental protection. *Journal of Environmental Management and Tourism*, *9*(8), 1805–1812. doi:10.14505//jemt.v9.8(32).21

Krinsky, N. I. (1966). The role of carotenoid pigments as protective against photosensitized oxidation in chloroplast. In T. W. Goodwin (Ed.), *Biochemistry of Chloroplasts* (Vol. 1). Academic Press.

Krishnakumar, T., Jayaprakash, R., Pinna, N., Donato, N., Bonavita, A., Micali, G., & Neri, G. (2009). CO gas sensing of ZnO nanostructures synthesized by an assisted microwave wet chemical route. *Sensors and Actuators. B, Chemical*, *143*(1), 198–204. doi:10.1016/j.snb.2009.09.039

Krishnan, S., Tadiboyina, R., Chavali, M., Nikolova, M. P., Wu, R.-J., Bian, D., Jeng, Y.-R., Rao, P. T. S. R. K. P., Palanisamy, P., & Pamanji, S. R. (2019). *Graphene-Based Polymer Nanocomposites for Sensor Applications. Hybrid Nanocomposites*, (pp. 1–62). Springer. doi:10.1201/9780429000966-1

Kruse, P. (2018). Review on water quality sensors. *Journal of Physics. D, Applied Physics*, *51*(20), 203002. doi:10.1088/1361-6463/aabb93

Kryuk, R., Mukhsin, M.-Z., Kurbanova, M., & Kryuk, V. (2023). Biosensors in Food Industry. *BIO Web of Conferences*. IEEE. 10.1051/bioconf/20236401006

Krywult, M., Karolak, A., & Bytnerowicz, A. (1996). Nitrate reductase activity as an indicator of ponderosa pine response to atmospheric nitrogen deposition in the San Bernardino Mountains. *Environmental Pollution*, *93*(2), 141–146. doi:10.1016/0269-7491(96)00033-4 PMID:15091353

Kudela, R. M., Seeyave, S., & Cochlan, W. P. (2010). The role of nutrients in regulation and promotion of harmful algal blooms in upwelling systems. *Progress in Oceanography*, *85*(1-2), 122–135. doi:10.1016/j.pocean.2010.02.008

Kumar, R., Liu, X., Zhang, J., & Kumar, M. (2020). Room-Temperature Gas Sensors Under Photoactivation: From Metal Oxides to 2D Materials. Nano-Micro Letters, *12*(1), 1–37. doi:10.1007/s40820-020-00503-4

Kumar, A., Verma, S. K., Sharma, K., & Mendiburu, A. Z. (2023). Design and analysis of heat melt refuse compactor for solid waste with E-control mechanism. *Environmental Challenges*, *100740*, 100740. doi:10.1016/j.envc.2023.100740

Kumar, B., Singh, J., Mittal, S., & Singh, H. (2023). The Indian perspective on the harmful substances found in sanitary napkins and their effects on the environment and human health. *Environmental Science and Pollution Research International*. doi:10.1007/s11356-023-26739-2 PMID:37022541

Kumar, H., Kumari, N., & Sharma, R. (2020). Nanocomposites (conducting polymer and nanoparticles) based electrochemical biosensor for the detection of environment pollutant: Its issues and challenges. *Environmental Impact Assessment Review*, *85*(12), 106438. Advance online publication. doi:10.1016/j.eiar.2020.106438

Kumari, S. I., Rani, P. U., & Suresh, C. (2005). Absorption of automobile pollutants by leaf surfaces of various road side plants and their effect on plant biochemical constituents. *Pollution Research*, *24*, 509–512.

Kumar, P. S. (2020). *Modern treatment strategies for marine pollution*. Elsevier.

Kumar, P., Morawska, L., Martani, C., Biskos, G., Neophytou, M., Di Sabatino, S., Bell, M., Norford, L., & Britter, R. (2019). The rise of low-cost sensing for managing air pollution in cities. *Environment International*, *75*, 199–205. doi:10.1016/j.envint.2014.11.019 PMID:25483836

Kumar, T., Naik, S., & Jujjavarappu, S. E. (2022). A critical review on early-warning electrochemical system on microbial fuel cell-based biosensor for on-site water quality monitoring. *Chemosphere*, *291*, 133098. doi:10.1016/j.chemosphere.2021.133098 PMID:34848233

Kummu, M., Guillaume, J., de Moel, H., Eisner, S., Flörke, M., Porkka, M., Siebert, S., Veldkamp, T., & Ward, P. (2016). The world's road to water scarcity: Shortage and stress in the 20th century and pathways towards sustainability. *Scientific Reports*, *6*(1), 6. doi:10.1038/srep38495 PMID:27934888

Kumunda, C., Adekunle, A. S., Mamba, B. B., Hlongwa, N. W., & Nkambule, T. T. (2021). Electrochemical detection of environmental pollutants based on graphene derivatives: A review. *Frontiers in Materials*, *7*, 616787. doi:10.3389/fmats.2020.616787

Kundariya, N., Mohanty, S. S., Varjani, S., Ngo, H. H., Wong, J. W., Taherzadeh, M. J., Chang, J. S., Ng, H. Y., Kim, S. H., & Bui, X. T. (2021). A review on integrated approaches for municipal solid waste for environmental and economical relevance: Monitoring tools, technologies, and strategic innovations. *Bioresource Technology*, *342*, 125982. doi:10.1016/j.biortech.2021.125982 PMID:34592615

Kupcinskiene, E. A., Ashenden, T. W., Bell, S. A., Williams, T. G., Edge, C. P., & Rafarel, C. R. (1997). Response of *Agrostis capillaris* to gaseous pollutants and wet nitrogen deposition. *Agriculture, Ecosystems & Environment*, *66*(2), 89–99. doi:10.1016/S0167-8809(97)00052-2

Kwon, N., Baek, G., Swamy, K. M. K., Lee, M., Xu, Q., Kim, Y., Kim, S. J., & Yoon, J. (2009). Naphthoimidazolium based ratiometric fluorescent probes for F^- and CN^-, and anion-activated CO2 sensing. *Dyes and Pigments*, *171*, 107679. doi:10.1016/j.dyepig.2019.107679

Laad, M., & Ghule, B. (2023). *Removal of toxic contaminants from drinking water using biosensors: A systematic review*. Groundwater for Sustainable Development. Elsevier B.V., doi:10.1016/j.gsd.2022.100888

Laffoley, D., Baxter, J. M., Amon, D. J., Currie, D. E., Downs, C. A., Hall-Spencer, J. M., Harden-Davies, H., Page, R., Reid, C. P., Roberts, C. M., Rogers, A., Thiele, T., Sheppard, C. R. C., Sumaila, R. U., & Woodall, L. C. (2020). Eight urgent, fundamental and simultaneous steps needed to restore ocean health, and the consequences for humanity and the planet of inaction or delay. *Aquatic Conservation*, *30*(1), 194–208. doi:10.1002/aqc.3182

Laik, B., Eude, L., Ramos, J.-P. P., Cojocaru, C. S., Pribat, D., & Rouvière, E. (2008). Silicon nanowires as negative electrode for lithium-ion microbatteries. *Electrochimica Acta*, *53*(17), 5528–5532. doi:10.1016/j.electacta.2008.02.114

Lakowicz, J. R. (Ed.), *Topics in Fluorescence Spectroscopy* (pp. 119–161). Springer.

Lalova, A. (1998). Accumulation of flavonoids and related compounds in birch induced by UV-B irradiance. *Tree Physiology*, *18*(1), 53–58. doi:10.1093/treephys/18.1.53 PMID:12651299

Lamb, C., & Dixon, R. (1997). The oxidative burst in plant disease resistance. *Annual Review of Plant Physiology and Plant Molecular Biology*, *48*(1), 251–275. doi:10.1146/annurev.arplant.48.1.251 PMID:15012264

Lamssali, M., Luster-Teasley, S., Deng, D., Sirelkhatim, N., Doan, Y., Kabir, M. S., & Zeng, Q. (2023). Release efficiencies of potassium permanganate controlled-release biodegradable polymer (CRBP) pellets embedded in polyvinyl acetate (CRBP-PVAc) and polyethylene oxide (CRBP- PEO) for groundwater treatment. *Heliyon*, *9*(10), e20858. doi:10.1016/j.heliyon.2023.e20858 PMID:37867834

Langebartels, C., Heller, W., Kerner, K., Leonardi, S., Rosemann, D., Schraudner, M., Trest, M., & Sandermann, H. J. (1990). *Ozone-induced defense reaction in plants. Environmental research with plants in closed chambers. Air pollution Research Reports of the EC26*. EEC.

Lazarević-Pašti, T., Tasić, T., Milanković, V., & Potkonjak, N. (2023). Molecularly Imprinted Plasmonic-Based Sensors for Environmental Contaminants—Current State and Future Perspectives. *Chemosensors (Basel, Switzerland)*, *11*(1), 35. doi:10.3390/chemosensors11010035

Leal Filho, W., Azul, A. M., Brandli, L., Lange Salvia, A., & Wall, T. (Eds.). (2021). *Life on land*. Springer International Publishing. doi:10.1007/978-3-319-95981-8

Lee, E. H., & Bennett, J. H. (1982). Superoxide dismutase, a possible protective enzyme against ozone injury in snap beans (*Phaseolus vulgaris* L.). *Plant Physiology*, *67*, 347–350. doi:10.1104/pp.67.2.347 PMID:16662420

Lee, M., Jo, S., Lee, D., Xu, Z., & Yoon, J. (2015). New Naphthalimide Derivative as a Selective Fluorescent and Colorimetric Sensor for Fluoride, Cyanide and CO_2. *Dyes and Pigments*, *120*, 288–292. doi:10.1016/j.dyepig.2015.04.029

Lees, P. (1994). *Combating water pollution*. Cabid Digital Library. https://www.cabidigitallibrary.org/doi/full/10.5555/19951802017

Le, M. T., Pham, C. D., Nguyen, T. P. T., Nguyen, T. L., Nguyen, Q., Hoang, N. B., & Nghiem, L. D. (2023). Wireless powered moisture sensors for smart agriculture and pollution prevention: Opportunities, challenges, and future outlook. *Current Pollution Reports*, *9*(4), 646–659. doi:10.1007/s40726-023-00286-3

Levitt, J. (1980). *Responses of Plants to Environmental Stresses* (2nd ed., Vol. II). Academic Press.

Li, L. (2014) Software development for water quality's monitoring centre of wireless sensor network. *Computer Modeling New Tech,* 132–136.

Liang, J., Zulkifli, M. Y., Choy, S., Li, Y., Gao, M., Kong, B., Yun, J., & Liang, K. (2020). Metal–organic framework–plant nanobiohybrids as living sensors for on-site environmental pollutant detection. *Environmental Science & Technology*, *54*(18), 11356–11364. doi:10.1021/acs.est.0c04688 PMID:32794698

Li, C., & Li, G. (2021). Impact of China's water pollution on agricultural economic growth: An empirical analysis based on a dynamic spatial panel lag model. *Environmental Science and Pollution Research International*, *28*(6), 6956–6965. doi:10.1007/s11356-020-11079-2 PMID:33025434

Li, C., Yang, Q., Zhang, T., Lv, Z., Wang, Y., & Chen, Y. (2021). A hybrid CO_2 ratiometric fluorescence sensor synergizing tetraphenylethene and gold nanoclusters relying on disulfide functionalized hyperbranched poly(amido amine). *Sensors and Actuators. B, Chemical*, *346*, 130513. doi:10.1016/j.snb.2021.130513

Li, D., Zhai, W., Li, Y., & Long, Y. (2013). Recent progress in surface-enhanced Raman spectroscopy for the detection of environmental pollutants. *Mikrochimica Acta*, *181*(1–2), 23–43. doi:10.1007/s00604-013-1115-3

Li, H., Su, X., Bai, C., Xu, Y., Pei, Z., & Sun, S. (2016). Detection of carbon dioxide with a novel HPTS/NiFe-LDH nanocomposite. *Sensors and Actuators. B, Chemical*, *225*, 109–114. doi:10.1016/j.snb.2015.11.007

Li, H., Wang, F., & Zhan, Z. (2023). Truck and rotary-wing drone routing problem considering flight-level selection. *The Journal of the Operational Research Society*, 1–19.

Li, J., Chang, H., Zhang, N., He, Y., Zhang, D., Liu, B., & Fang, Y. (2023). *Recent advances in enzyme inhibition based-electrochemical biosensors for pharmaceutical and environmental analysis*. Talanta. Elsevier B.V., doi:10.1016/j.talanta.2022.124092

Li, J., Pei, Y., Zhao, S., Xiao, R., Sang, X., & Zhang, C. (2020). A review of remote sensing for environmental monitoring in China. *Remote Sensing (Basel)*, *12*(7), 1130. doi:10.3390/rs12071130

Li, L., Zou, J., Han, Y., Liao, Z., Lu, P., Nezamzadeh-Ejhieh, A., Liu, J., & Peng, Y. (2022). Recent advances in Al (iii)/In (iii)-based MOFs for the detection of pollutants. *New Journal of Chemistry*, *46*(41), 19577–19592. doi:10.1039/D2NJ03419K

Li, M. H. (2003). Peroxidase and superoxide dismutase activities in Fig leaves in response to ambient air pollution in a subtropical city. *Archives of Environmental Contamination and Toxicology*, *45*(2), 168–176. doi:10.1007/s00244-003-0154-x PMID:14565573

Lindroth, R. L., Kopper, B. J., Parson, F. J., Bockheim, J. G., Karnosky, D. F., Hendrey, G. R., Pregitzer, K. S., Isebrand, J. G., & Sober, J. (2001). Consequences of elevated carbondioxide and ozone for foliar chemical composition and dynamics in trembling aspen (*Populus tremuloides*) and paper birch (*Betula papyrifera*). *Environmental Pollution*, *115*(3), 395–404. doi:10.1016/S0269-7491(01)00229-9 PMID:11789920

Lin, L., Yang, H., & Xu, X. (2022a). Effects of Water Pollution on Human Health and Disease Heterogeneity: A Review. *Frontiers in Environmental Science*, *10*, 880246. https://www.frontiersin.org/articles/10.3389/fenvs.2022.880246. doi:10.3389/fenvs.2022.880246

Lin, M., & Yang, C. (2020). Ocean observation technologies: A review. *Chinese Journal of Mechanical Engineering*, *33*(1), 1–18. doi:10.1186/s10033-020-00449-z

Liou, J. (2021). *NUTEC Plastics: Using Nuclear Technologies to Address Plastic pollution.* IAEA Office of Public Information and Communication.

Liu, B., Zhuang, J., & Wei, G. (2020). Recent advances in the design of colorimetric sensors for environmental monitoring. *Environmental Science. Nano*, *7*(8), 2195–2213. doi:10.1039/D0EN00449A

Liu, D., Wang, J., Wu, L., Huang, Y., Zhang, Y., Zhu, M., & Yang, C. (2020). *Trends in miniaturized biosensors for point-of-care testing. TrAC - Trends in Analytical Chemistry.* Elsevier B.V., doi:10.1016/j.trac.2019.115701

Liu, G. (2021). Grand Challenges in Biosensors and Biomolecular Electronics. *Frontiers in Bioengineering and Biotechnology*, *9*, 707615. doi:10.3389/fbioe.2021.707615 PMID:34422782

Liu, L., Zhang, X., Zhu, Q., Li, K., Lu, Y., Zhou, X., & Guo, T. (2021). Ultrasensitive Detection of Endocrine Disruptors via Superfine Plasmonic Spectral Combs. *Light, Science & Applications*, *10*(1), 1–14. doi:10.1038/s41377-021-00618-2 PMID:34493704

Liu, L., Zhou, X., Lu, M., Zhang, M., Yang, C., Ma, R., Memon, A. G., Shi, H., & Qian, Y. (2017). An Array Fluorescent Biosensor Based on Planar Waveguide for Multi-Analyte Determination in Water Samples. *Sensors and Actuators. B, Chemical*, *240*, 107–113. doi:10.1016/j.snb.2016.08.118

Liu, L., Zhou, X., Xu, W., Song, B., & Shi, H. (2014). Highly Sensitive Detection of Sulfadimidine in Water and Dairy Products by Means of an Evanescent Wave Optical Biosensor. *RSC Advances*, *4*(104), 60227–60233. doi:10.1039/C4RA10501J

Liu, R., Guan, G., Wang, S., & Zhang, Z. (2011). Core-Shell Nanostructured Molecular Imprinting Fluorescent Chemosensor for Selective Detection of Atrazine Herbicide. *Analyst*, *136*(1), 184–190. doi:10.1039/C0AN00447B PMID:20886153

Liu, Y., Cao, X., Hu, Y., & Cheng, H. (2022). Pollution, risk and transfer of heavy metals in soil and rice: A case study in a typical industrialized region in South China. *Sustainability (Basel)*, *14*(16), 10225. doi:10.3390/su141610225

Liu, Y., Xue, Q., Chang, C., Wang, R., Liu, Z., & He, L. (2022). Recent progress regarding electrochemical sensors for the detection of typical pollutants in water environments. *Analytical Sciences*, *38*(1), 55–70. doi:10.2116/analsci.21SAR12 PMID:35287206

Liu, Y., Zhang, D., He, K., Gao, Q., & Qin, F. (2021). Research on Land Use Change and Ecological Environment Effect Based on Remote Sensing Sensor Technology. *Journal of Sensors*, *2021*, 1–11. Advance online publication. doi:10.1155/2021/4351733

Li, Z., Wang, K., & Liu, B. (2013). Sensor-Network based Intelligent Water Quality Monitoring and Control. *International Journal of Advanced Research in Computer Engineering and Technology*, *2*(4), 1659–1662.

Lochman, L., Zimcik, P., Klimant, I., Novakova, V., & Borisov, S. M. (2017). Red-emitting CO_2 sensors with tunable dynamic range based on pH-sensitive azaphthalocyanine indicators. *Sensors and Actuators. B, Chemical*, *246*, 1100–1107. doi:10.1016/j.snb.2016.10.135

Loncar, L., Hell, M., & Dusak, V. (2006, June). A system dynamics model of forest management. In *28th International Conference on Information Technology Interfaces.* (pp. 549-556). IEEE. 10.1109/ITI.2006.1708540

Loponen, J., Lempa, K., Ossipov, V., Kozlov, M. V., Girs, A., Hangasmaa, K., Haukioja, E., & Pihlaja, K. (2001). Patterns in content of phenolic compounds in leaves of mountains birches along a strong pollution gradient. *Chemosphere*, *45*(3), 291–301. doi:10.1016/S0045-6535(00)00545-2 PMID:11592418

Loponen, J., Ossipo, V., Lempa, K., Haukioja, E., & Pilhlaja, K. (1998). Concentrations and among-compound correlation of individual phenols in white birch leaves under air pollution stress. *Chemosphere*, *37*(8), 1445–1456. doi:10.1016/S0045-6535(98)00135-0

Loponen, J., Ossipov, V., Koricheva, J., Haukioja, E., & Pilhlaja, K. (1997). Low molecular mass phenolics in foliage of *Betula pubescens* Ehrh. in relation to aerial pollution. *Chemosphere*, *34*(4), 687–697. doi:10.1016/S0045-6535(97)00461-X

Loudet, A., & Burgess, K. (2007). BODIPY dyes and their derivatives: Syntheses and spectroscopic properties. *Chemical Reviews*, *107*(11), 4891–4932. doi:10.1021/cr078381n PMID:17924696

Louguet, P., Malka, P., & Contour-Ansel, D. 1989. Etude compar_ee de la r_esistance stomatique et de la teneur en compos_es ph_enoliques foliaires chez trois clones d_Epic_eas soumis _a une pollution contr^ol_ee par l_ozone et le dioxyde de soufre en chambre _a ciel ouvert. In: Brasser, T.J., Mulde, W.C. (Eds.). *Man, and his ecosystem. Proceedings of the 8th World Clean Air Congress.* The Hague, Elsevier.

Lou, R., Lv, Z., Dang, S., Su, T., & Li, X. (2023). Application of machine learning in ocean data. *Multimedia Systems*, *29*(3), 1815–1824. doi:10.1007/s00530-020-00733-x

Lovelock, J. E. (1987). *Gaia: A new look at life on earth.* Oxford Iniversity Press.

Ltd, R. a. M. (n.d.). *Environmental Sensor Global Market Report 2023.* Research and Markets Ltd 2024. https://www.researchandmarkets.com/reports/5880229/environmental-sensor-global-market-report

Luo, J., Xie, Z., Lam, W. Y. J., Cheng, L., Chen, H., Qiu, C., Kwok, S. H., Zhan, X., Liu, Y., Zhu, D., & Tang, Z. B. (2001). Aggregation-induced emission of 1-methyl-1,2,3,4,5-pentaphenylsilole. *Chemical Communications*, (18), 1740–1741. doi:10.1039/b105159h PMID:12240292

Luo, J., Yang, Y., Wang, Z., & Chen, Y. (2021). Localization algorithm for underwater sensor network: A review. *IEEE Internet of Things Journal*, *8*(17), 13126–13144. doi:10.1109/JIOT.2021.3081918

Luo, M., Liu, X., Legesse, N., Liu, Y., Wu, S., Han, F. X., & Ma, Y. (2023). Evaluation of agricultural non-point source pollution: A review. *Water, Air, and Soil Pollution*, *234*(10), 657. doi:10.1007/s11270-023-06686-x

Luoma, S. N., & Rainbow, P. S. (2005). Why is metal bioaccumulation so variable? Biodynamics as a unifying concept. *Environmental Science & Technology*, *39*(7), 1921–1931. doi:10.1021/es048947e PMID:15871220

Luo, Q., Yu, N., Shi, C., Wang, X., & Wu, J. (2016). Surface plasmon resonance sensor for antibiotics detection based on photo-initiated polymerization molecularly imprinted array. *Talanta*, *161*, 797–803. doi:10.1016/j.talanta.2016.09.049 PMID:27769483

Lu, X., Pu, X., & Han, X. (2020). Sustainable smart waste classification and collection system: A bi-objective modeling and optimization approach. *Journal of Cleaner Production*, *276*, 124183. doi:10.1016/j.jclepro.2020.124183

Lyngberg, O., Stemke, D., Schottel, J., & Flickinger, M. (1999). A single-use luciferase-based mercury biosensor using Escherichia coli HB101 immobilized in a latex copolymer film. *Journal of Industrial Microbiology & Biotechnology*, *23*(1), 668–676. doi:10.1038/sj.jim.2900679 PMID:10455499

Macko, S. A. (1994). Pollution studies using stable isotopes. *Stable isotopes in ecology and environmental science.*

Madadgar, S., Sadegh, M., Chiang, F., Ragno, E., & AghaKouchak, A. (2020). Quantifying increased fire risk in California in response to different levels of warming and drying. *Stochastic Environmental Research and Risk Assessment*, *34*(12), 2023–2031. doi:10.1007/s00477-020-01885-y

Maggs, R., Wahid, A., Shasmi, S. R. A., & Ashmore, M. R. (1995). Effect of ambient air pollution on wheat and rice yield in Pakistan. *Water, Air, and Soil Pollution*, *85*(3), 1311–1316. doi:10.1007/BF00477163

Mahadev, J. & Hosmani SP. 2004. Community structure of cyanobacteria in two polluted lakes of Mysore city. *Nat Env Pollut Technol., 3*(4), 523–526.

Mahmoud, S., & Mahmoud, A. (2016). A Study of Efficient Power Consumption Wireless Communication Techniques/ Modules for Internet of Things (IoT). *Applications Advances in Internet of Things*, *6*(2), 19–29. doi:10.4236/ait.2016.62002

Mahrad, B. E., Newton, A., Icely, J. D., Kacimi, I., Abalansa, S., & Snoussi, M. (2020). Contribution of remote sensing technologies to a holistic coastal and marine environmental management framework: A review. *Remote Sensing (Basel)*, *12*(14), 2313. doi:10.3390/rs12142313

Mai, Z., Li, H., Gao, Y., Niu, Y., Li, Y., Rooij, N. F. D., Umar, A., Al-Assiri, M. S., Wang, Y., & Zhou, G. (2020). Synergy of CO_2-response and aggregation induced emission in a small molecule: Renewable liquid and solid CO_2 chemosensors with high sensitivity and visibility. *Analyst*, *145*(10), 3528–3534. doi:10.1039/D0AN00189A PMID:32190881

Malhotra, S. S. (1976). Effect of sulphur dioxide on biochemical activity and ultrastructural organization of pine needle chloroplasts. *The New Phytologist*, *76*(2), 239–245. doi:10.1111/j.1469-8137.1976.tb01457.x

Malhotra, S. S. (1977). Effect of aqueous sulphur dioxide on chlorophyll destruction in *Pinus contorta*. *The New Phytologist*, *78*(1), 101–109. doi:10.1111/j.1469-8137.1977.tb01548.x

Malik, D. S., Sharma, A. K., Sharma, A. K., Thakur, R., & Sharma, M. (2020). A review on impact of water pollution on freshwater fish species and their aquatic environment. *Advances in Environmental Pollution Management: Wastewater Impacts and Treatment Technologies*, *1*, 10–28. doi:10.26832/aesa-2020-aepm-02

Malmqvist, B., & Rundle, S. (2002). Threats to the running water ecosystems of the world. *Environmental Conservation*, *29*(2), 134–153. doi:10.1017/S0376892902000097

Maltchik, L. (2003). Inventory of wetlands of Rio Grande do Sul (India). *Pesquisas: Botânica*, *53*, 89-100.

Maltchik, L. (2003). Three new wetlands inventories in India. *Interciencia*, *28*(7), 421-423.

Maltchik, L. (2004). Wetlands of Rio Grande do Sul, India: a classification with emphasis on plant communities. *ActaLimnologicaBrasiliensia*, *16*(2), 137-151.

Maltchik, L. (2005). Diversity and stability of aquatic macrophyte community in three shallow lakes associated to a floodplain system in the South of India . *Interciencia*, *30*(3), 166-170.

Maltchik, L. (2007). Effects of hydrological variation on the aquatic plant community in a floodplain palustrine wetland of Southern Brasil. *Limnology, 8*(1), 23-28.

Maltchik, L., (2002). Diversidade de macrófitasaquáticasemáreasúmidas da Bacia do Rio dos Sinos, Rio Grande do Sul. *Pesquisas: Botânica*, *52*.

Mamun, A. A., & Yuce, M. R. (n.d.). Sensors and Systems for Wearable Environmental Monitoring towards IOT-enabled Applications. *RE:view*.

Mamun, M. A. A., & Yuce, M. R. (2020). Recent progress in nanomaterial enabled chemical sensors for wearable environmental monitoring applications. *Advanced Functional Materials*, *30*(51), 2005703. doi:10.1002/adfm.202005703

Mandal, N., Adhikary, S., & Rakshit, R. (2020). Nanobiosensors: Recent Developments in Soil Health Assessment. In Soil Analysis: Recent Trends and Applications (pp. 285–304). Springer Singapore. doi:10.1007/978-981-15-2039-6_15

Mandal, M., & Mukherji, S. (2000). Changes in chlorophyll content, chlorophllase activity, Hill reaction, photosynthetic CO_2 uptake, sugar and starch content in five dicotyledonous plants exposed to automobile exhaust pollution. *Journal of Environmental Biology*, *21*, 37–41.

Mandloi, B. L., & Dubey, P. S. (1988). The industrial emission and plant response at Pithanpur (M.P). *International Journal of Ecology and Environmental Sciences*, *14*, 75–99.

Mangiameli, M., Mussumeci, G., & Cappello, A. (2021). Forest fire spreading using free and open-source GIS technologies. *Geomatics*, *1*(1), 50–64. doi:10.3390/geomatics1010005

Mann, M. L., Batllori, E., Moritz, M. A., Waller, E. K., Berck, P., Flint, A. L., Flint, L. E., & Dolfi, E. (2016). Incorporating anthropogenic influences into fire probability models: Effects of human activity and climate change on fire activity in California. *PLoS One*, *11*(4), e0153589. doi:10.1371/journal.pone.0153589 PMID:27124597

Manoiu, V.-M., Craciun, A.-I., Kubiak-Wójcicka, K., Antonescu, M., & Olariu, B. (2022). An Eco-Study for a Feasible Project: "Torun and Its Vistula Stretch—An Important Green Navigation Spot on a Blue Inland Waterway.". *Water (Basel)*, *14*(19), 19. doi:10.3390/w14193034

Mansfield, T. A., & Freer-Smith, P. H. (1981). The role of stomata in resistance mechanisms. In: Koziol, M.J. and Whatley, F.R. (Eds.). Gaseous Pollutant and Plant Metabolism. Butterworths scientific, London.

Manzoor, J., & Sharma, M. (2018). Impact of incineration and disposal of biomedical waste on air quality in Gwalior city. *Int J Theor Appl Sci*, *10*(1), 10–16.

Marcé, R., George, G., Buscarinu, P., Deidda, M., Dunalska, J., De Eyto, E., Flaim, G., Grossart, H.-P., Istvanovics, V., Lenhardt, M., Moreno-Ostos, E., Obrador, B., Ostrovsky, I., Pierson, D. C., Potužák, J., Poikane, S., Rinke, K., Rodríguez-Mozaz, S., Staehr, P. A., & Jennings, E. (2016). Automatic high frequency monitoring for improved lake and reservoir management. *Environmental Science & Technology*, *50*(20), 10780–10794. doi:10.1021/acs.est.6b01604 PMID:27597444

Markham, K., Tanner, G., Caasi-Lit, M., Whttecross, M., Nayudu, M., & Mitchell, K. (1998). Possible protective role for 3′,4′ -dihydroxyflavones induced by enhanced UV- B tolerant rice cultivar. *Phytochemistry*, *49*(7), 1913–1919. doi:10.1016/S0031-9422(98)00438-5

Marra, G., Fairweather, D. M., Kamalov, V., Gaynor, P., Cantono, M., Mulholland, S., Baptie, B., Castellanos, J. C., Vagenas, G., Gaudron, J. O., Kronjäger, J., Hill, I. R., Schioppo, M., Barbeito Edreira, I., Burrows, K. A., Clivati, C., Calonico, D., & Curtis, A. (2022). Optical interferometry–based array of seafloor environmental sensors using a transoceanic submarine cable. *Science*, *376*(6595), 874–879. doi:10.1126/science.abo1939 PMID:35587960

Marrazza, G. (2014). Piezoelectric biosensors for organophosphate and carbamate pesticides: A review. *Biosensors (Basel)*, *4*(3), 301–317. doi:10.3390/bios4030301 PMID:25587424

Martikkala, A., Mayanti, B., Helo, P., Lobov, A., & Ituarte, I. F. (2023). Smart textile waste collection system–Dynamic route optimization with IoT. *Journal of Environmental Management*, *335*, 117548. doi:10.1016/j.jenvman.2023.117548 PMID:36871359

Martins, T. S., Bott-Neto, J. L., Oliveira, O. N. Jr, & Machado, S. A. S. (2021). Paper-based electrochemical sensors with reduced graphene nanoribbons for simultaneous detection of sulfamethoxazole and trimethoprim in water samples. *Journal of Electroanalytical Chemistry (Lausanne, Switzerland)*, *882*, 114985. doi:10.1016/j.jelechem.2021.114985

Masoudvaziri, N., Ganguly, P., Mukherjee, S., & Sun, K. (2021). Impact of geophysical and anthropogenic factors on wildfire size: A spatiotemporal data-driven risk assessment approach using statistical learning. *Stochastic Environmental Research and Risk Assessment*, 1–27.

Ma, T., Zhang, D., Li, X., Hu, Y., Zhang, L., Zhu, Z., Sun, X., Lan, Z., & Guo, W. (2023). Hyperspectral remote sensing technology for water quality monitoring: Knowledge graph analysis and Frontier trend. *Frontiers in Environmental Science, 11*, 1133325. doi:10.3389/fenvs.2023.1133325

Matese, A., Toscano, P., Di Gennaro, S. F., Genesio, L., Vaccari, F. P., Primicerio, J., Belli, C., Zaldei, A., Bianconi, R., & Gioli, B. (2015). Intercomparison of UAV, aircraft, and satellite remote sensing platforms for precision viticulture. *Remote Sensing (Basel), 7*(3), 2971–2990. doi:10.3390/rs70302971

Matysik, J., Alia, P., Bhalu, B., & Mohanty, P. (2002). Molecular mechanisms of quenching of reactive oxygen species by proline under stress in plants. *Current Science, 82*, 525–532.

Ma, Y., Promthaveepong, K., & Li, N. (2016). CO_2-Responsive Polymer-Functionalized Au Nanoparticles for CO_2 Sensor. *Analytical Chemistry, 88*(16), 8289–8293. doi:10.1021/acs.analchem.6b02133 PMID:27459645

Ma, Y., Zeng, Y., Liang, H., Ho, C. H., Zhao, Q., Huang, W., & Wong, W. Y. (2015). A water-soluble tetraphenylethene based probe for luminescent carbon dioxide detection and its biological application. *Journal of Materials Chemistry. C, Materials for Optical and Electronic Devices, 3*(45), 11850–1185. doi:10.1039/C5TC03327F

Mayer, S., Lischke, L., & Woźniak, P. W. (2019, May). Drones for search and rescue. In *1st International Workshop on Human-Drone Interaction*. Research Gate.

Mayer, H. (1999). Air Pollution in Cities. *Atmospheric Environment, 33*(24-25), 4029–4037. doi:10.1016/S1352-2310(99)00144-2

Mayerhöfer, U., Fimmel, B., & Würthner, F. (2012). Bright Near-Infrared Fluorophores Based on Squaraines by Unexpected Halogen Effects. *Angewandte Chemie International Edition, 51*(1), 164–167. doi:10.1002/anie.201107176 PMID:22105993

McConnell, E. M., Nguyen, J., & Li, Y. (2020). Aptamer-based biosensors for environmental monitoring. Frontiers in Chemistry. doi:10.3389/fchem.2020.00434

McGrath, M. J., Scanaill, C. N., & Nafus, D. (2014). *Sensor Technologies: Healthcare, Wellness and Environmental Applications*. Apress.

Medina-Lopez, E., McMillan, D., Lazic, J., Hart, E., Zen, S., Angeloudis, A., Bannon, E., Browell, J., Dorling, S., Dorrell, R. M., Forster, R., Old, C., Payne, G. S., Porter, G., Rabaneda, A. S., Sellar, B., Tapoglou, E., Trifonova, N., Woodhouse, I. H., & Zampollo, A. (2021). Satellite data for the offshore renewable energy sector: Synergies and innovation opportunities. *Remote Sensing of Environment, 264*, 112588. doi:10.1016/j.rse.2021.112588

Melaku, A., Addis, T., Mengistie, B., Kanno, G. G., Adane, M., Kelly-Quinn, M., Ketema, S., Hailu, T., Bedada, D., & Ambelu, A. (2023). Menstrual hygiene management practices and determinants among schoolgirls in Addis Ababa, Ethiopia: The urgency of tackling bottlenecks-Water and sanitation services. *Heliyon, 9*(5), e15893. doi:10.1016/j.heliyon.2023.e15893 PMID:37180900

Mengqi, Z., Shi, A., Ajmal, M., Ye, L., & Awais, M. (2021). Comprehensive review on agricultural waste utilization and high-temperature fermentation and composting. *Biomass Conversion and Biorefinery*, 1–24. doi:10.1007/s13399-021-01438-5

Menzel, D. B. (1976). The role of free radicals in the toxicity of air pollutants (nitrogen oxides and ozone). In W. A. Pryor (Ed.), *Free Radicals in Biology* (Vol. 2, pp. 181–203). Academic Press. doi:10.1016/B978-0-12-566502-5.50013-5

Metian, M., Pouil, S., & Fowler, S. W. (2019). Radiocesium accumulation in aquatic organisms: a global synthesis from an experimentalist's perspective. *J. Environ. Radioact. 198*, 147–158. doi: . 11.013 doi:10.1016/j.jenvrad.2018

Mhawej, M., Faour, G., & Adjizian-Gerard, J. (2017). A novel method to identify likely causes of wildfire. *Climate Risk Management*, *16*, 120–132. doi:10.1016/j.crm.2017.01.004

Michael-Kordatou, I., Michael, C., Duan, X., He, X., Dionysiou, D., Mills, M., & Fatta-Kassinos, D. (2015). Dissolved effluent organic matter: characteristics and potential implications in wastewater treatment and reuse applications. *Water Research, 77.*

Midiwo, J., Matasi, J., Wanjau, O., Mwangi, R., Waterman, P., & Wollenweber, E. (1990). Anti-feedant effects of surface accumulated flavonoids of *Polygonum senegalense. Bulletin of the Chemical Society of Ethiopia*, *4*, 123–127.

Mills, A. (2009). Optical Sensors for Carbon Dioxide and Their Applications. In: Baraton, MI. (Ed.) Sensors for Environment, Health and Security. NATO Science for Peace and Security Series C: Environmental Security (pp. 347-370). Springer. doi:10.1007/978-1-4020-9009-7_23

Mills, A., & Hodgen, S. (2005). Fluorescent Carbon Dioxide Indicators. In C. D. Geddes & J. R. Lakowicz (Eds.), *Topics in Fluorescence Spectroscopy, Advanced Concepts in Fluorescence Sensing, Part A: Small Molecule Sensing* (pp. 119–161). Springer.

Milson, S., & Mehmet, A. (2023). *Microfluidic Sensors for Environmental Monitoring: From Lab to Field Applications*. EasyChair. easychair.org/publications/preprint/6Spww

Ministry of Urban Development, Government of India (GOI). (2011). *Census of Urban Growth and Development*. Ministry of Urban Development.

Ministry of Urban Development. (2011). *Report on Urban Growth and Expansion*. Ministry of Urban Development.

Ministry of Urban Development. (2012). *Service Levels in Urban Water and Sanitation Sector, Status Report 2010 – 2011*. Ministry of Urban Development, Government of India. http://moud.gov.in/upload/uploadfiles/files/SLB%20 National%20Data%20Book_0.pdf

Ministry of Urban Development. (2014). *Guidelines for Swachh Bharat Mission (SBM)*. Ministry of Urban Development, Government of India. http://swachhbharaturban.gov.in/writereaddata/SBM_GUIDELINE.pdf

Ministry of Urban Development. (2016). *Primer on Faecal Sludge and Septage Management*. Ministry of Urban Development. http://www.swachhbharaturban.in:8080/sbm/content/writereaddata/Primer%20on%20Fae cal%20 Sludge%20&%20Septage%20Management.pdf

Mishra, R. K., Rhouati, A., Bueno, D., Anwar, M. W., Shahid, S. A., Sharma, V., & Hayat, A. (2018). Design of a portable luminescence bio-tool for on-site analysis of heavy metals in water samples. *International Journal of Environmental Analytical Chemistry*, *98*(12), 1081–1094. doi:10.1080/03067319.2018.1521395

Mishra, R. K., Vijayakumar, S., Mal, A., Karunakaran, V., Janardhanan, J. C., Maiti, K. K., Praveen, V. K., & Ajayaghosh, A. (2019). Bimodal detection of carbon dioxide using a fluorescent molecular aggregates. *Chemical Communications*, *55*(43), 6046–6049. doi:10.1039/C9CC01564G PMID:31065654

Mishra, S., Anuradha, J., Tripathi, S., & Kumar, S. (2016). In vitro antioxidant and antimicrobial efficacy of Triphala constituents: Emblica officinalis, Terminalia belerica and Terminalia chebula. *Journal of Pharmacognosy and Phytochemistry*, *5*(6), 273–277.

Misra, N., & Gupta, A. K. (2006). Effect of salinity and different nitrogen sources on the activity of antioxidant enzymes and indole alkaloid content in *Catharanthus roseus* seedlings. *Journal of Plant Physiology*, *163*(1), 11–18. doi:10.1016/j.jplph.2005.02.011 PMID:16360799

Mittler, R. (2002). Oxidative stress, antioxidants and stress tolerance. *Trends in Plant Science*, *7*(9), 405–410. doi:10.1016/S1360-1385(02)02312-9 PMID:12234732

Mohammed, A. A., & Khamees, H. T. (2021). Categorizing and measuring satellite image processing of fire in the forest Greece using remote sensing. *Indonesian Journal of Electrical Engineering and Computer Science*, *21*(2), 843–853. doi:10.11591/ijeecs.v21.i2.pp846-853

Mohankumar, P., Ajayan, J., Mohanraj, T., & Yasodharan, R. (2021). Recent developments in biosensors for healthcare and biomedical applications: A review. Measurement. *Measurement*, *167*, 108293. doi:10.1016/j.measurement.2020.108293

Mohanty, A., Mohanty, S. K., Jena, B., Mohapatra, A. G., Rashid, A. N., Khanna, A., & Gupta, D. (2022). (b)). Identification and evaluation of the effective criteria for detection of congestion in a smart city. *IET Communications*, *16*(5), 560–570. doi:10.1049/cmu2.12344

Mohanty, S., Saha, S., Santra, G. H., & Kumari, A. (2022). (a)). Future perspective of solid waste management strategy in India. In *Handbook of Solid Waste Management: Sustainability through Circular Economy* (pp. 191–226). Springer Nature Singapore. doi:10.1007/978-981-16-4230-2_10

Mokua, N., Maina, C., & Kiragu, H. (2021). A raw water quality monitoring system using wireless sensor networks. *International Journal of Computer Applications*, *174*(21), 35–42. doi:10.5120/ijca2021921113

Momeni, M., Mirzapour Al-e-Hashem, S. M. J., & Heidari, A. (2023). A new truck-drone routing problem for parcel delivery by considering energy consumption and altitude. *Annals of Operations Research*, 1–47. doi:10.1007/s10479-023-05381-8 PMID:37361075

Mondal, P., Nandan, A., Ajithkumar, S., Siddiqui, N. A., Raja, S., Kola, A. K., & Balakrishnan, D. (2023). Sustainable application of nanoparticles in wastewater treatment: Fate, current trend & paradigm shift. *Environmental Research*, *116071*, 116071. doi:10.1016/j.envres.2023.116071 PMID:37209979

Moore, T. S., Mullaugh, K. M., Holyoke, R. R., Madison, A. S., Yücel, M., & Luther, G. W. III. (2009). Marine chemical technology and sensors for marine waters: Potentials and limits. *Annual Review of Marine Science*, *1*(1), 91–115. doi:10.1146/annurev.marine.010908.163817 PMID:21141031

Moradi, M., Kazeminezhad, M. H., & Kabiri, K. (2020). Integration of Geographic Information System and system dynamics for assessment of the impacts of storm damage on coastal communities study Chabahar, Iran. *International Journal of Disaster Risk Reduction*, *49*, 101665. doi:10.1016/j.ijdrr.2020.101665

Mor, S., & Ravindra, K. (2023). Municipal solid waste landfills in lower-and middle-income countries: Environmental impacts, challenges and sustainable management practices. *Process Safety and Environmental Protection*, *174*, 510–530. doi:10.1016/j.psep.2023.04.014

Mostaccio, A., Bianco, G. M., Marrocco, G., & Occhiuzzi, C. (2023). RFID Technology for Food Industry 4.0: A Review of Solutions and Applications. *IEEE Journal of Radio Frequency Identification*, *7*, 145–157. doi:10.1109/JRFID.2023.3278722

Moyle, P. B., & Leidy, R. A. (1992). Loss of Biodiversity in Aquatic Ecosystems: Evidence from Fish Faunas. In P. L. Fiedler & S. K. Jain (Eds.), *Conservation Biology* (pp. 127–169). Springer US. doi:10.1007/978-1-4684-6426-9_6

Moyo, M., Okonkwo, J. O., & Agyei, N. M. (2014). An amperometric biosensor based on horseradish peroxidase immobilized onto maize tassel-multi-walled carbon nanotubes modified glassy carbon electrode for determination of heavy metal ions in aqueous solution. *Enzyme and Microbial Technology*, *56*, 28–34. doi:10.1016/j.enzmictec.2013.12.014 PMID:24564899

Mudd, J. B. (1982). Effects of oxidants on metabolic functions. In M. H. Unsworth & D. P. Ormrod (Eds.), *Effects of Gaseous Air Pollutants in Agriculture and Horticulture* (pp. 189–203). Butterworths. doi:10.1016/B978-0-408-10705-1.50014-4

Muhammad-Tahir, Z., & Alocilja, E. C. (2003). A conductometric biosensor for biosecurity. *Biosensors & Bioelectronics, 18*(5), 813–819. doi:10.1016/S0956-5663(03)00020-4 PMID:12706596

Muinul, H., Syed, I., Alex, F., Homayoun, N., Rehan, S., Manuel, R., & Mina, H. (2014). Online Drinking Water Quality Monitoring: Review on Available and Emerging Technologies. *Critical Reviews in Environmental Science and Technology, 44*(12), 1370–1421. doi:10.1080/10643389.2013.781936

Mulchandani, A., Mulchandani, P., Kaneva, I., & Chen, W. (1998). Biosensor for direct determination of organophosphate nerve agents using recombinant *Escherichia coli* with surface-expressed organophosphorus hydrolase. 1. Potentiometric microbial electrode. *Analytical Chemistry, 70*(19), 4140–4145. doi:10.1021/ac9805201 PMID:9784751

Mulindi, J. (2023, June 18). *The principles of microbial biosensors*. Biomedical Instrumentation Systems. https://www.biomedicalinstrumentationsystems.com/the-principles-of-microbial-biosensors/

Munir, M. T., Li, B., Naqvi, M., & Nizami, A. S. (2023). Green loops and clean skies: Optimizing municipal solid waste management using data science for a circular economy. *Environmental Research, 117786*. doi:10.1016/j.envres.2023.117786 PMID:38036215

Murthy, H. A., Wagassa, A. N., Ravikumar, C., & Nagaswarupa, H. (2022). Functionalized metal and metal oxide nanomaterial-based electrochemical sensors. In *Functionalized Nanomaterial-Based Electrochemical Sensors* (pp. 369–392). Elsevier. doi:10.1016/B978-0-12-823788-5.00001-6

Mustafa, F., Hassan, R. Y. A., & Andreescu, S. (2017). Multifunctional Nanotechnology-Enabled Sensors for Rapid Capture and Detection of Pathogens. *Sensors (Basel), 17*(9), 9. doi:10.3390/s17092121 PMID:28914769

Muyibi, S. A., Ambali, A. R., & Eissa, G. S. (2008). The Impact of Economic Development on Water Pollution: Trends and Policy Actions in Malaysia. *Water Resources Management, 22*(4), 485–508. doi:10.1007/s11269-007-9174-z

Nagendra, S., Schlink, S. M., & Khare, M. (2021). Air quality measuring sensors. In Urban Air Quality Monitoring, Modelling and Human Exposure Assessment (pp. 89–104). Springer Singapore. doi:10.1007/978-981-15-5511-4_7

Nagy, R. C., Fusco, E., Bradley, B., Abatzoglou, J. T., & Balch, J. (2018). Human-related ignitions increase the number of large wildfires across US ecoregions. *Fire (Basel, Switzerland), 1*(1), 4. doi:10.3390/fire1010004

Naher, L., & Ahsan, M. F. (2023). Solid Waste Disposal Scenario of Three Ladies' Halls in the University of Chittagong, Chittagong, Bangladesh. *American Journal of Agricultural Science, Engineering, and Technology, 7*(2), 1–6. doi:10.54536/ajaset.v7i2.1295

Naidu, B. P., Aspinall, D., & Paleg, L. G. (1992). Variability in proline accumulation ability of barley cultivars induced by vapour pressure deficit. *Plant Physiology, 98*, 716–722. doi:10.1104/pp.98.2.716 PMID:16668700

Naimaee, R., Kiani, A., Jarahizadeh, S., Asadollah, S. B. H. S., Melgarejo, P., & Jódar-Abellán, A. (2024). Long-term water quality monitoring: Using satellite images for temporal and spatial monitoring of thermal pollution in water resources. *Sustainability (Basel), 16*(2), 646. doi:10.3390/su16020646

Nakamura, H. (2018). Current status of water environment and their microbial biosensor techniques–Part I: Current data of water environment and recent studies on water quality investigations in Japan, and new possibility of microbial biosensor techniques. *Analytical and Bioanalytical Chemistry, 410*(17), 3953–3965. doi:10.1007/s00216-018-0923-z PMID:29470662

Nakamura, Y., Ishii, J., & Kondo, A. (2015). Applications of yeast-based signaling sensor for characterization of antagonist and analysis of site-directed mutants of the human serotonin 1A receptor. *Biotechnology and Bioengineering*, *112*(9), 1906–1915. doi:10.1002/bit.25597 PMID:25850571

Nakanishi, K., Ikebukuro, K., & Karube, I. (1996). Determination of cyanide using a microbial sensor. *Applied Biochemistry and Biotechnology*, *60*(2), 97–106. doi:10.1007/BF02788064 PMID:8856940

Nanda, S., & Berruti, F. (2021). Municipal solid waste management and landfilling technologies: A review. *Environmental Chemistry Letters*, *19*(2), 1433–1456. doi:10.1007/s10311-020-01100-y

Nasture, A.-M., Ionete, E. I., Lungu, F. A., Spiridon, S. I., & Patularu, L. G. (2022). Water Quality Carbon Nanotube-Based Sensors Technological Barriers and Late Research Trends: A Bibliometric Analysis. *Chemosensors (Basel, Switzerland)*, *10*(5), 5. doi:10.3390/chemosensors10050161

Nava, V., & Leoni, B. (2021). A critical review of interactions between microplastics, microalgae and aquatic ecosystem function. *Water Research*, *188*, 116476. doi:10.1016/j.watres.2020.116476 PMID:33038716

Nayak, M., Kotian, A., Marathe, S., & Chakravortty, D. (2009). Detection of microorganisms using biosensors-A smarter way towards detection techniques. *Biosensors & Bioelectronics*, *25*(4), 661–667. doi:10.1016/j.bios.2009.08.037 PMID:19782558

Neethirajan, S., Jayas, D. S., & Sadistap, S. (2009). Carbon dioxide (CO_2) sensors for the agri-food industry-a review. *Food and Bioprocess Technology*, *2*(2), 115–121. doi:10.1007/s11947-008-0154-y

Neiff, J.J. (1975). Fluctuacionesanualesen la composition fitocenotica y biomasa de la hidrofitaenlagunasisleñas del Parana Medio. *Ecosur*, *2*(4), 153-183.

Nemade, K. R., & Waghuley, S. A. (2013). Carbon dioxide gas sensing application of graphene/Y_2O_3 quantum dots composite. *International Journal of Modern Physics. Conference Series*, *22*, 380–384. doi:10.1142/S2010194513010404

Nemade, K. R., & Waghuley, S. A. (2014). Highly responsive carbon dioxide sensing by graphene/Al_2O_3 quantum dots composites at low operable temperature. *Indian Journal of Physics and Proceedings of the Indian Association for the Cultivation of Science*, *88*(6), 577–583. doi:10.1007/s12648-014-0454-1

Nguyen, T.-T., Hwang, S.-Y., Vuong, N. M., Pham, Q.-T., Nghia, N. N., Kirtland, A., & Lee, Y.-I. (2018). Preparing cuprous oxide nanomaterials by electrochemical method for non-enzymatic glucose biosensor. *Nanotechnology*, *29*(20), 205501. doi:10.1088/1361-6528/aab229 PMID:29480163

Nicholson, R. L., & Hammerschmidt, R. (1992). Phenolic compounds and their role in disease resistance. *Annual Review of Phytopathology*, *30*(1), 369–389. doi:10.1146/annurev.py.30.090192.002101

Nicholson, R. S., & Shain, I. (1964). Theory of stationary electrode polarography. Single scan and cyclic methods applied to reversible, irreversible, and kinetic Systems. *Analytical Chemistry*, *36*(4), 706–723. doi:10.1021/ac60210a007

Nieckarz, Z., Pawlak, K., Baran, A., Wieczorek, J., Grzyb, J., & Plata, P. (2023). The concentration of particulate matter in the barn air and its influence on the content of heavy metals in milk. *Scientific Reports*, *13*(1), 10626. doi:10.1038/s41598-023-37567-2 PMID:37391588

Niel, A. C., Reza, M., & Lakshmi, N. (2016). Design of Smart Sensors for Real-Time Water Quality Monitoring. *IEEE Access : Practical Innovations, Open Solutions*, *4*, 3975–3990. doi:10.1109/ACCESS.2016.2592958

Niepsch, D., Clarke, L. J., Tzoulas, K., & Cavan, G. (2022). Spatiotemporal variability of nitrogen dioxide (NO2) pollution in Manchester (UK) city centre (2017–2018) using a fine spatial scale single-NOx diffusion tube network. *Environmental Geochemistry and Health*, *44*(11), 3907–3927. doi:10.1007/s10653-021-01149-w PMID:34739651

Nigam, A., & Agarwal, Y. K. (2014). Optimal relay node placement in delay-constrained wireless sensor network design. *European Journal of Operational Research*, *233*(1), 220–233. doi:10.1016/j.ejor.2013.08.031

Nigam, V. K., & Shukla, P. (2015). Enzyme based biosensors for detection of environmental pollutants-A review. *Journal of Microbiology and Biotechnology*, *25*(11), 1773–1781. doi:10.4014/jmb.1504.04010 PMID:26165317

Nighat, F., Mahmooduzzafar, M., & Iqbal, M. (2000). Stomatal conductance, photosynthetic rate and pigment content in *Ruellia tuberose* leaves as affected by coal-smoke pollution. *Biologia Plantarum*, *43*(2), 263–267. doi:10.1023/A:1002712528893

Nikolic, M. V., Milovanovic, V., Vasiljevic, Z. Z., & Stamenkovic, Z. (2020). Semiconductor gas sensors: Materials, technology, design, and application. *Sensors (Basel)*, *20*(22), 6694. doi:10.3390/s20226694 PMID:33238459

Nikolova, M. T., & Ivancheva, S. V. (2005). Quantitative flavonoid variations of *Artemisia vulgaris* L. and *Veronica chamaedrys* L. in relation to altitude and polluted environment. *Acta Biologica Szegediensis*, *49*, 29–32.

Nižetić, S., Djilali, N., Papadopoulos, A., & Rodrigues, J. J. (2019). Smart technologies for promotion of energy efficiency, utilization of sustainable resources and waste management. *Journal of Cleaner Production*, *231*, 565–591. doi:10.1016/j.jclepro.2019.04.397

Nnachi, R. C., Sui, N., Ke, B., Luo, Z., Bhalla, N., He, D., & Yang, Z. (2022). Biosensors for rapid detection of bacterial pathogens in water, food and environment. *Environment International*, *166*, 1–20. doi:10.1016/j.envint.2022.107357 PMID:35777116

Nnaji, N. D., Onyeaka, H., Miri, T., & Ugwa, C. (2023). Bioaccumulation for heavy metal removal: A review. *SN Applied Sciences*, *5*(5), 125. doi:10.1007/s42452-023-05351-6

Nolan, R. H., Collins, L., Leigh, A., Ooi, M. K., Curran, T. J., Fairman, T. A., Resco de Dios, V., & Bradstock, R. (2021). Limits to post-fire vegetation recovery under climate change. *Plant, Cell & Environment*, *44*(11), 3471–3489. doi:10.1111/pce.14176 PMID:34453442

Nomngongo, P. N., Ngila, J. C., Msagati, T. A., Gumbi, B. P., & Iwuoha, E. I. (2012). Determination of selected persistent organic pollutants in wastewater from landfill leachates, using an amperometric biosensor. *Physics and Chemistry of the Earth Parts A/B/C*, *50-52*, 252–261. doi:10.1016/j.pce.2012.08.001

Nomura, Y., Ikebukuro, K., Yokoyama, K., Takeuchi, T., Arikawa, Y., Ohno, S., & Karube, I. (1994). A novel microbial sensor for anionic surfactant determination. *Analytical Letters*, *27*(15), 3095–3108. doi:10.1080/00032719408000313

Nomura, Y., Ikebukuro, K., Yokoyama, K., Takeuchi, T., Arikawa, Y., Ohno, S., & Karube, I. (1998). Application of a linear alkylbenzene sulfonate biosensor to river water monitoring. *Biosensors & Bioelectronics*, *13*(9), 1047–1053. doi:10.1016/S0956-5663(97)00077-8 PMID:9839392

Norris, G., & Larson, T. (1999). Spatial and temporal measurements of NO2 in an urban area using continuous mobile monitoring and passive samplers. *Journal of Exposure Analysis and Environmental Epidemiology*, *9*(6), 586–593. doi:10.1038/sj.jea.7500063 PMID:10638844

Noss, R.F. (1990). Indicators for monitoring biodiversity: a hierarchical approach. *Conserv Biol. 4,* 355–364.

Nouchi, I. (2002). Responses of whole plants to air pollutions. In: Omasa, K. Saji, H. Youssefian, S. and Kondo, N. (Eds.), Air Pollution and Plant Biotechnology. Springer-Verlag, Tokyo.

Nouchi, I., & Toyama, S. (1988). Effect of ozone and peroxyacylnitrate on polar lipids and fatty acids in leaves of morning glory and kidney bean. *Plant Physiology*, *87*(3), 638–646. doi:10.1104/pp.87.3.638 PMID:16666199

O'Grady, J., Zhang, D., O'Connor, N., & Regan, F. (2021). A comprehensive review of catchment water quality monitoring using a tiered framework of integrated sensing technologies. *The Science of the Total Environment*, *765*, 142766. doi:10.1016/j.scitotenv.2020.142766 PMID:33092838

Oberholster, P.J, Botha, A., & Ashton, P.J. (2009). The influence of a toxic cyanobacterial bloom and water hydrology on algal populations and macroinvertebrate abundance in the upper littoral zone of Lake Krugersdrift, South Africa. *Ecotoxicology, 18*(1), 34–46.

Oertli, B., Joey, DA., Castella, E., Juge, R., Cambin, D. & Lachavanne, J.B. (2002). Does size matter? The relationship between pond area and biodiversity. *Biological Conservation, 104*(1), 59-70.

Offiong, N., Abdullahi, S., Chile, B., Raji, H., & Nweze, N. (2014). Real Time Monitoring Of Urban Water Systems for Developing Countries. *IOSR Journal of Computer Engineering*, *16*(3), 11–14. doi:10.9790/0661-16321114

Ogidi, O. I., & Akpan, U. M. (2022). Aquatic Biodiversity Loss: Impacts of Pollution and Anthropogenic Activities and Strategies for Conservation. In S. Chibueze Izah (Ed.), *Biodiversity in Africa: Potentials, Threats and Conservation* (Vol. 29, pp. 421–448). Springer Nature Singapore. doi:10.1007/978-981-19-3326-4_16

Olaniran, A. O., Motebejane, R. M., & Pillay, B. (2008). Bacterial biosensors for rapid and effective monitoring of biodegradation of organic pollutants in wastewater effluents. *Journal of Environmental Monitoring*, *10*(7), 889–893. doi:10.1039/b805055d PMID:18688458

Olatinwo, S. O., & Joubert, T.-H. (2018). Energy efficient solutions in wireless sensor systems for water quality monitoring: A review. *IEEE Sensors Journal*, *19*(5), 1596–1625. doi:10.1109/JSEN.2018.2882424

Olatinwo, S. O., & Joubert, T.-H. (2019). Enabling communication networks for water quality monitoring applications: A survey. *IEEE Access : Practical Innovations, Open Solutions*, *7*, 100332–100362. doi:10.1109/ACCESS.2019.2904945

Omasa, K., & Endo, R. (2001). Absorption of air pollutants by leaves and patchy responss of stomata and photosynthesis. *5th APGC symposium*. Pulawy, Poland.

Online, E. (2023, October 2). Why menstrual waste is proving to be a "big bloody mess" in India. *The Economic Times*. https://economictimes.indiatimes.com/news/india/the-menstrual-waste-conundrum-in-india-and-why-its-proving-to-be-a-big-bloody-mess-for-sustainability/articleshow/104106414.cms

Onyancha, R. B., Ukhurebor, K. E., Aigbe, U. O., Osibote, O. A., Kusuma, H. S., Darmokoesoemo, H., & Balogun, V. A. (2021). *A systematic review on the detection and monitoring of toxic gases using carbon nanotube-based biosensors*. Sensing and Bio-Sensing Research. Elsevier B.V., doi:10.1016/j.sbsr.2021.100463

Ooi, L., Heng, L. Y., & Mori, I. C. (2015). A high-throughput oxidative stress biosensor based on Escherichia coli roGFP$_2$ cells immobilized in a *k*-carrageenan matrix. *Sensors (Basel)*, *15*(2), 2354–2368. doi:10.3390/s150202354 PMID:25621608

Orubu, C. O., & Omotor, D. G. (2011). Environmental quality and economic growth: Searching for environmental Kuznets curves for air and water pollutants in Africa. *Energy Policy*, *39*(7), 4178–4188. doi:10.1016/j.enpol.2011.04.025

Ossipov, V., Haukioja, E., Ossipova, S., Hanhima¨ki, S., & Pihlaja, K. (2001). Phenolic and phenolic-related factors as determinants of suitability of mountain birch leaves to an herbivore insect. *Biochemical Systematics and Ecology*, *29*(3), 223–240. doi:10.1016/S0305-1978(00)00069-7 PMID:11152944

Osteryoung, J., & Wechter, C. (1989). *Development of Pulse Polarography and Voltammetry* (J. T. Stock & M. E. Orna, Eds.). American Chemical Society. doi:10.1021/bk-1989-0390.ch025

Ostrick, B., Fleischer, M., Meixner, H., & Kohl, C. D. (2000). Investigation of the reaction mechanisms in work function type sensors at room temperature by studies of the cross-sensitivity to oxygen and water: The carbonate-carbon dioxide system. *Sensors and Actuators. B, Chemical*, *68*(1-3), 197–202. doi:10.1016/S0925-4005(00)00429-9

Ottmar, R. D. (2014). Wildland fire emissions, carbon, and climate: Modeling fuel consumption. *Forest Ecology and Management*, *317*, 41–50. doi:10.1016/j.foreco.2013.06.010

Pacifici, R. E., & Davies, K. J. A. (1990). Protein degradation as an index of oxidative stress. *Methods in Enzymology*, *186*, 485–502. doi:10.1016/0076-6879(90)86143-J PMID:2233315

Padial, A.A., Carvalho, P., Thomaz, S.M., Boschilia, S.M., Rodrigues, R.B. & Kobayashi, J.T. (2009). The role of an extreme flood disturbance on macrophyte assemblages in a Neotropical floodplain. *Aquatic Sciences*, *71*(4), 389-398.

Padmanabha, B. (2017). Comparative study on the hydrographical status in the lentic and lotic ecosystems. *Global Journal of Ecology, 2*(1), 015-018. doi:10.17352/gje.000005

Painting, S. J., Collingridge, K. A., Durand, D., Grémare, A., Créach, V., Arvanitidis, C., & Bernard, G. (2020). Marine monitoring in Europe: Is it adequate to address environmental threats and pressures? *Ocean Science*, *16*(1), 235–252. doi:10.5194/os-16-235-2020

Pakere, I., Kacare, M., Grāvelsiņš, A., Freimanis, R., & Blumberga, A. (2022). Spatial analyses of smart energy system implementation through system dynamics and GIS modeling. Wind power case study in Latvia. *Smart Energy*, *7*, 100081. doi:10.1016/j.segy.2022.100081

Pakshirajan, K., Rene, E. R., & Ramesh, A. (2015). Biotechnology in environmental monitoring and pollution abatement. *BioMed Research International*, *1-3*, 1–3. doi:10.1155/2015/963803 PMID:26526980

Pal, A., Campagnaro, F., Ashraf, K., Rahman, M. R., Ashok, A., & Guo, H. (2022). Communication for Underwater Sensor Networks: A Comprehensive Summary. *ACM Transactions on Sensor Networks*, *19*(1), 1–44. doi:10.1145/3546827

Panda, S. K., Kherani, N. A., Debata, S., & Singh, D. P. (2023). Bubble propelled micro/nano motors: A robust platform for the detection of environmental pollutants and biosensing. *Materials Advances*, *4*(6), 1460–1480. doi:10.1039/D2MA00798C

Pandey, J. & Verma, A. (2004). The influence of catchment on chemical and biological characteristics of two freshwater tropical lakes of Southern Rajasthan. *J Environ Biol,. 25*.

Pandey, J. (2005). Evaluation of Air pollution phytotoxicity down wind of a phosphorus fertilizer factory in India. *Environmental Monitoring and Assessment*, *100*(1-3), 249–266. doi:10.1007/s10661-005-6509-1 PMID:15727311

Pandey, P., Huidobro, G., Lopes, L. F., Ganteaume, A., Ascoli, D., Colaco, C., Xanthopoulos, G., Giannaros, T. M., Gazzard, R., Boustras, G., Steelman, T., Charlton, V., Ferguson, E., Kirschner, J., Little, K., Stoof, C., Nikolakis, W., Fernández-Blanco, C. R., Ribotta, C., & Dossi, S. (2023). A global outlook on increasing wildfire risk: Current policy situation and future pathways. *Trees, Forests and People*, *14*, 100431. doi:10.1016/j.tfp.2023.100431

Pandian, D. R., & Mala, K. (2015). Smart Device to monitor water quality to avoid pollution in IoT environment. *Int J Emerging Tech Comput Sci Electron*, *12*(2), 120–125.

Pan, Z. H., Luo, G. G., Zhou, J. W., Xia, J. X., Fang, K., & Wu, R. B. (2014). A simple BODIPY-aniline-based fluorescent chemosensor as multiple logic operations for the detection of pH and CO2 gas. *Dalton Transactions (Cambridge, England)*, *43*(22), 8499–8507. doi:10.1039/C4DT00395K PMID:24756338

Parambil, A. R. U. (2022). Water-soluble optical sensors: Keys to detect aluminium in biological environment. *RSC Advances*, *12*(22), 13950–13970. doi:10.1039/D2RA01222G PMID:35558844

Pardini, K., Rodrigues, J. J., Diallo, O., Das, A. K., de Albuquerque, V. H. C., & Kozlov, S. A. (2020). A smart waste management solution geared towards citizens. *Sensors (Basel)*, *20*(8), 2380. doi:10.3390/s20082380 PMID:32331464

Park, C. J., Barakat, R., Ulanov, A., Li, Z., Lin, P.-C., Chiu, K., Zhou, S., Pérez, P. E., Lee, J., Flaws, J. A., & Ko, C. (2019). Sanitary pads and diapers contain higher phthalate contents than those in common commercial plastic products. *Reproductive Toxicology (Elmsford, N.Y.)*, *84*, 114–121. doi:10.1016/j.reprotox.2019.01.005 PMID:30659930

Park, J. K., Ro, H. S., & Kim, H. S. (1991). A new biosensor for specific determination of sucrose using an oxido-reductase of *Zymomonas mobilis* and invertase. *Biotechnology and Bioengineering*, *38*(3), 217–223. doi:10.1002/bit.260380302 PMID:18600754

Park, J., Kim, K. T., & Lee, W. H. (2020). Recent advances in information and communications technology (ICT) and sensor technology for monitoring water quality. *Water (Basel)*, *12*(2), 510. doi:10.3390/w12020510

Park, M., Tsai, S., & Chen, W. (2013). Microbial biosensors: Engineered microorganisms as the sensing machinery. *Sensors (Basel)*, *13*(5), 5777–5795. doi:10.3390/s130505777 PMID:23648649

Parry, J., & Hubbard, S. (2023). Review of Sensor Technology to Support Automated Air-to-Air Refueling of a Probe Configured Uncrewed Aircraft. *Sensors (Basel)*, *23*(2), 995. doi:10.3390/s23020995 PMID:36679790

Pasternak, G., Greenman, J., & Ieropoulos, I. (2017). Self-powered, autonomous Biological Oxygen Demand biosensor for online water quality monitoring. *Sensors and Actuators. B, Chemical*, *244*, 815–822. doi:10.1016/j.snb.2017.01.019 PMID:28579695

Patel, B. R., Noroozifar, M., & Kerman, K. (2020). Nanocomposite-based sensors for voltammetric detection of hazardous phenolic pollutants in water. *Journal of the Electrochemical Society*, *167*(3), 037568. doi:10.1149/1945-7111/ab71fa

Patel, S. (2016). Introduction: Revisiting Urban India. *India International Centre Quarterly*, *43*(3/4), 1–14.

Patel, S. K., & Jain, A. (2020). Smart technologies for waste management in urban areas. *Waste Management (New York, N.Y.)*, *102*, 107–116.

Pathak, A. K., Swargiary, K., Kongsawang, N., Jitpratak, P., Ajchareeyasoontorn, N., Udomkittivorakul, J., & Viphavakit, C. (2023). Recent Advances in Sensing Materials Targeting Clinical Volatile Organic Compound (VOC) Biomarkers: A Review. *Biosensors, 13*(1), 114. doi:10.3390/bios13010114

Paton, G. I., Palmer, G., Burton, M., Rattray, E. A. S., McGrath, S. P., Glover, L. A., & Killham, K. (1997). Development of an acute and chronic ecotoxicity assay using lux-marked *Rhizobium leguminosarum* biovar *trifolii*. *Letters in Applied Microbiology*, *24*(4), 296–300. doi:10.1046/j.1472-765X.1997.00071.x PMID:9134778

Patterson, C., Tilton, G., & Inghram, M. (1955). Age of the Earth. *Science*, *121*(3134), 69–75. doi:10.1126/science.121.3134.69 PMID:17782556

Paull, N. J., Krix, D., Irga, P. J., & Torpy, F. R. (2021). Green wall plant tolerance to ambient urban air pollution. *Urban Forestry & Urban Greening*, *63*, 127201. doi:10.1016/j.ufug.2021.127201

Pausas, J. G., & Keeley, J. E. (2009). A burning story: The role of fire in the history of life. *Bioscience*, *59*(7), 593–601. doi:10.1525/bio.2009.59.7.10

Pawar, K. & Dubey, P.S. (1985). *Effects of air pollution on the photosynthetic pigments of* Ipomea fistulosa *and* Phoenix sylvestris. All India seminar on Air pollution Control, Indore.

Payne, T. E., Hatje, V., Itakura, T., McOrist, G. D., & Russell, R. (2004). Radionuclide applications in laboratory studies of environmental surface reactions. *Journal of Environmental Radioactivity*, *76*(1-2), 237–251. doi:10.1016/j.jenvrad.2004.03.029 PMID:15245851

Peiser, G. D., & Yang, S. F. (1979). Ethylene and ethane production from sulphur dioxide injured plants. *Plant Physiology*, *63*(1), 142–145. doi:10.1104/pp.63.1.142 PMID:16660667

Pei, X., Xiong, D., Fan, J., Li, Z., Wang, H., & Wang, J. (2017). Highly efficient fluorescence switching of carbon nanodots by CO_2. *Carbon*, *117*, 147–153. doi:10.1016/j.carbon.2017.02.090

Peixoto, P. S., Machado, A., Oliveira, H. P., Bordalo, A., & Segundo, M. (2019), Paper-Based Biosensors for Analysis of Water, Biosensors for Environmental Monitoring, *Intech Open*, 1-15. doi:10.5772/intechopen.84131

Pelonero, L., Fornaia, A., & Tramontana, E. (2020). From smart city to smart citizen: rewarding waste recycle by designing a data-centric iot based garbage collection service. In *2020 IEEE International Conference on Smart Computing (SMARTCOMP)* (pp. 380-385). IEEE. 10.1109/SMARTCOMP50058.2020.00081

Peng, G., He, Q., Lu, Y., Huang, J., & Lin, J.-M. (2017). Flow injection microfluidic device with on-line fluorescent derivatization for the determination of Cr (III) and Cr (VI) in water samples after solid phase extraction. *Analytica Chimica Acta*, *955*, 58–66. doi:10.1016/j.aca.2016.11.057 PMID:28088281

Peng, J., Hongbo, X., Zhiye, H., & Zheming, W. (2009). Design of a Water Environment Monitoring System Based on Wireless Sensor Networks. *Journal of Sensors*, *9*(8), 6411–6434. doi:10.3390/s90806411 PMID:22454592

Pensieri, M. G., Garau, M., & Barone, P. M. (2020). Drones as an integral part of remote sensing technologies to help missing people. *Drones (Basel)*, *4*(2), 15. doi:10.3390/drones4020015

Penuelas, J. (1996). Variety of responses of plant phenolic concentration to CO_2 enrichment. *Journal of Experimental Botany*, *47*, 1463–1467. doi:10.1093/jxb/47.9.1463

Pérez-López, B., & Merkoçi, A. (2011). Nanomaterials based biosensors for food analysis applications. *Trends in Food Science & Technology*, *22*(11), 625–639. doi:10.1016/j.tifs.2011.04.001

Pérez-Sirvent, C., & Bech, J. (2023). Spatial assessment of soil and plant contamination. *Environmental Geochemistry and Health*, *45*(12), 8823–8827. doi:10.1007/s10653-023-01760-z PMID:37973774

Peter, A., & Abhitha, K. (2021). Menstrual Cup: A replacement to sanitary pads for a plastic free periods. *Materials Today: Proceedings*, *47*, 5199–5202. doi:10.1016/j.matpr.2021.05.527

Petrie, G. (2009). Systematic oblique aerial photography using multiple digital cameras. *Photogrammetric Engineering and Remote Sensing*, *75*(2), 102–107.

Решетилов, A. H., Iliasov, P. V., & Reshetilova, T. A. (2010). The microbial cell-based biosensors. In InTech eBooks. doi:10.5772/7159

Pfeifer, D., Klimant, I., & Borisov, S. M. (2018). Ultrabright red-emitting photostable perylene bisimide dyes: New indicators for ratiometric sensing of high pH or carbon dioxide. *European Journal of Chemistry*, *24*(42), 10711–10720. doi:10.1002/chem.201800867 PMID:29738607

Phonchi-Tshekiso, N. D., Mmopelwa, G., & Chanda, R. (2020). From public to private solid waste management: Stakeholders' perspectives on private-public solid waste management in Lobatse, Botswana. *Zhongguo Renkou Ziyuan Yu Huanjing*, *18*(1), 42–48. doi:10.1016/j.cjpre.2021.04.015

Pimentel, D., Berger, B., Filiberto, D., Newton, M., Wolfe, B., Karabinakis, E., Clark, S., Poon, E., Abbett, E., & Nandagopal, S. (2004). Water resources: Agricultural and environmental issues. *Bioscience*, *54*(10), 909–918. doi:10.1641/0006-3568(2004)054[0909:WRAAEI]2.0.CO;2

Piskulova, N., & Gorbanyov, V. (2023). Global Challenges: Environment. In *World Economy and International Business: Theories, Trends, and Challenges* (pp. 213–233). Springer International Publishing., doi:10.1007/978-3-031-20328-2_11

Pohanka, M. (2018). *Overview of piezoelectric biosensors, immunosensors and DNA sensors and their applications.* Materials. MDPI AG., doi:10.3390/ma11030448

Polle, A., Mo¨ssnang, M., Von Scho¨nborn, A., Sladkovic, R., & Rennenberg, H. (1992). Field studies on Norway spruce trees at high altitudes. I. Mineral, pigment and soluble protein contents of needles as affected by climate and pollution. *The New Phytologist*, *121*(1), 89–99. doi:10.1111/j.1469-8137.1992.tb01096.x

Polshettiwar, S. A., Deshmukh, C. D., Wani, M. S., Baheti, A. M., Bompilwar, E., Choudhari, S., Jambhekar, D., & Tagalpallewar, A. (2021). Recent Trends on Biosensors in Healthcare and Pharmaceuticals: An Overview. *International Journal of Pharmaceutical Investigation*, *11*(2), 131–136. doi:10.5530/ijpi.2021.2.25

Pooja, D., Kumar, P., Singh, P., & Patil, S. (Eds.). (2020). *Sensors in water pollutants monitoring: role of material* (p. 320). Springer. doi:10.1007/978-981-15-0671-0

Poonam,, J. (2016). IoT Based Water Quality Monitoring. *Int J Modern Trends Eng Res*, *3*(4), 746–750.

Posthumus, A. C. (1983). Higher plants as indicators and accumulators of gaseous air pollution. *Environmental Monitoring and Assessment*, *3*(3-4), 263–272. doi:10.1007/BF00396220 PMID:24259091

Posthumus, A. C. (1998). Air pollution and global change: Significance and Prospectives. In L. J. Dekok & I. Stulen (Eds.), *Responses of plant metabolism to air pollution and global change* (pp. 3–14). Backhuys Publishers.

Povinec, P. P. (2017). Analysis of radionuclides at ultra-low levels: a comparison of low and high-energy mass spectrometry with gamma-spectrometry for radiopurity measurements. *Appl. Radiat. Isotopes, 126*, 26–30. doi: . apradiso.2017.01.029 doi:10.1016/j

Powell, C. L., & Doucette, G. J. (1999). A receptor binding assay for paralytic shellfish poisoning toxins: Recent advances and applications. *Natural Toxins*, *7*(6), 393–400. doi:10.1002/1522-7189(199911/12)7:6<393::AID-NT82>3.0.CO;2-C PMID:11122535

Prabhakaran, R., Ramprasath, T., & Selvam, G. S. (2017). A simple whole cell microbial biosensors to monitor soil pollution. Elsevier eBooks. doi:10.1016/B978-0-12-804299-1.00013-8

Pradhan, A., Bhaumik, P., Das, S., Mishra, M., Khanam, S., Hoque, B.A., Mukherjee, I., Thakur, A.R., & Chaudhuri, S.R. (2008). Phytoplankton diversity as indicator of water quality for fish cultivation. *Am J Environ Sci., 4*(4), 406–411.

Prakash, R., & Singh, G. (2023). GREEN INTERNET OF THINGS (G-IoT) FOR SUSTAINABLE ENVIRONMENT. [IJMR]. *EPRA International Journal of Multidisciplinary Research*, *9*(5), 1–1. doi:10.36713/epra13324

Prashantkumar, B. Sathvara, J. Anuradha, Sandeep Tripathi, & R. Sanjeevi. (2023). Impact of climate change and its importance on human performance. In Insights on Impact of Climate Change and Adaptation of Biodiversity (1st ed., pp. 1–9). KD Publication.

Prăvălie, R. (2016). Drylands extent and environmental issues. A global approach. *Earth-Science Reviews*, *161*, 259–278. doi:10.1016/j.earscirev.2016.08.003

Premgi, A., Martins, F., & Domingos, D. (2019). An infrared-based sensor to measure the filling level of a waste bin. In *2019 International Conference in Engineering Applications (ICEA)* (pp. 1-6). IEEE. 10.1109/CEAP.2019.8883303

Prindle, A., Samayoa, P., Razinkov, I., Danino, T., Tsimring, L. S., & Hasty, J. (2012). A sensing array of radically coupled genetic 'biopixels'. *Nature*, *481*(7379), 39–44. doi:10.1038/nature10722 PMID:22178928

Public Utilities Board Singapore (PUB). (2016). Managing the water distribution network with a Smart Water Grid. *Smart Water International Journal.*

Puccinelli, P., Anselmi, N., & Bragaloni, M. (1998). Peroxidases: Usable markers of air pollution in trees from urban environments. *Chemosphere*, *36*(4-5), 889–894. doi:10.1016/S0045-6535(97)10143-6

Puligundla, P., Jung, J., & Ko, S. (2012). Carbon dioxide sensors for intelligent food packaging applications. *Food Control*, *25*(1), 328–333. doi:10.1016/j.foodcont.2011.10.043

Pu, R. (2021). Mapping tree species using advanced remote sensing technologies: A state-of-the-art review and perspective. *Yaogan Xuebao.*

Purdy, W.C. (1926). The biology of rivers in relation to pollution. *J Am Water Works Assoc. 16*(1), 45–54.

Qadri, R., & Faiq, M. A. (2020). Freshwater Pollution: Effects on Aquatic Life and Human Health. In H. Qadri, R. A. Bhat, M. A. Mehmood, & G. H. Dar (Eds.), *Fresh Water Pollution Dynamics and Remediation* (pp. 15–26). Springer Singapore. doi:10.1007/978-981-13-8277-2_2

Qazi, H. H., Mohammad, A. B. B., & Akram, M. (2012). Recent Progress in Optical Chemical Sensors. *Sensors (Basel)*, *12*(12), 16522–16556. doi:10.3390/s121216522 PMID:23443392

Qu, K., Hu, X., & Li, Q. (2023). Electrochemical environmental pollutant detection enabled by waste tangerine peel-derived biochar. *Diamond and Related Materials*, *131*, 109617. doi:10.1016/j.diamond.2022.109617

Qureshi, M. I., Israr, M., Abdin, M. Z., & Iqbal, M. (2004). Response of *Artemisia annua* L. to lead and salt- induced oxidative stress. *Environmental and Experimental Botany*, *53*, 185–193.

Qu, Y., Hua, J., & Tian, H. (2010). Colorimetric and Ratiometric Red Fluorescent Chemosensor for Fluoride Ion Based on Diketopyrrolopyrrole. *Organic Letters*, *12*(15), 3320–3323. doi:10.1021/ol101081m PMID:20590106

Radhakrishnan, S., Lakshmy, S., Santhosh, S., Kalarikkal, N., Chakraborty, B., & Rout, C. S. (2022). *Recent Developments and Future Perspective on Electrochemical Glucose Sensors Based on 2D Materials*. Biosensors. MDPI., doi:10.3390/bios12070467

Radu, C., Manoiu, V.-M., Kubiak-Wójcicka, K., Avram, E., Beteringhe, A., & Craciun, A.-I. (2022). Romanian Danube River Hydrocarbon Pollution in 2011–2021. *Water (Basel)*, *14*(19), 19. doi:10.3390/w14193156

Raffa, C. M., & Chiampo, F. (2021). Bioremediation of Agricultural Soils Polluted with Pesticides: A Review. *Bioengineering (Basel, Switzerland)*, *8*(7), 92. doi:10.3390/bioengineering8070092 PMID:34356199

Rahman, N. A., Yusof, N. A., Maamor, N. A. M., & Noor, S. M. M. (2012). Development of electrochemical sensor for simultaneous determination of Cd (II) and Hg (II) ion by exploiting newly synthesized cyclic dipeptide. *International Journal of Electrochemical Science*, *7*(1), 186–196. doi:10.1016/S1452-3981(23)13330-X

Rai, M., Ingle, A. P., Birla, S., Yadav, A., & Santos, C. A. D. (2016). *Strategic role of selected noble metal nanoparticles in medicine. Critical Reviews in Microbiology*. Taylor and Francis Ltd., doi:10.3109/1040841X.2015.1018131

Raina, A. K., & Sharma, A. (2003). Effect of vfehicular pollution on the leaf micro-morphology, anatomy and chlorophyll contents of *Syzygium cumini* L. *International Journal of E-Politics*, *23*, 897–902.

Rainina, E. I., Efremenco, E. N., Varfolomeyev, S. D., Simonian, A. L., & Wild, J. R. (1996). The development of a new biosensor based on recombinant E. coli for the direct detection of organophosphorus neurotoxins. *Biosensors & Bioelectronics*, *11*(10), 991–1000. doi:10.1016/0956-5663(96)87658-5 PMID:8784985

Rai, R., Agrawal, M., & Agrawal, S. B. (2007). Assessment of yield losses in tropical wheat using open top chambers. *Atmospheric Environment*, *41*(40), 9543–9554. doi:10.1016/j.atmosenv.2007.08.038

Rai, V., Vajpayee, P., Singh, S. N., & Mehrotra, S. (2004). Effect of chromium accumulation on photosynthetic pigments, oxidative stress defense system, nitrate reduction, proline level and eugenol content of *Ocimum tenuiflorum* L. *Plant Science*, *167*(5), 1159–116. doi:10.1016/j.plantsci.2004.06.016

Raja, C. E., & Selvam, G. (2011). Construction of green fluorescent protein based bacterial biosensor for heavy metal remediation. *International Journal of Environmental Science and Technology*, *8*(4), 793–798. doi:10.1007/BF03326262

Rajput, M., & Agrawal, M. (2005). Biomonitoring of air pollution in a seasonally dry tropical suburban area using wheat transplants. *Environmental Monitoring and Assessment*, *101*, 39–53. PMID:15736874

Ramachandran, A. (2023). *Modeling of Internet of Things Enabled Sustainable Environment Air Pollution Monitoring System*. . doi:10.30955/gnj.004707

Ramakrishnan, S., & Jayaraman, A. (2019). Global Warming and Pesticides in Water Bodies. In *Handbook of Research on the Adverse Effects of Pesticide Pollution in Aquatic Ecosystems* (pp. 421–436). IGI Global. https://www.igi-global.com/chapter/global-warming-and-pesticides-in-water-bodies/213519

Ramakrishnan, S., & Jayaraman, A. (2019). Pesticide contaminated drinking water and health effects on pregnant women and children. In *Handbook of research on the adverse effects of pesticide pollution in aquatic ecosystems* (pp. 123–136). IGI Global. https://www.igi-global.com/chapter/pesticide-contaminated-drinking-water-and-health-effects-on-pregnant-women-and-children/213500 doi:10.4018/978-1-5225-6111-8.ch007

Ramchandra, T.V., Rishiram, R., & Karthik, B. (2006). Zooplanktons as bioindicators: hydro biological investigation in selected Bangalore lakes. *Technical report, 115*. doi:10.1016/j.jelechem.2019.113319

Ramge, P., Badeck, F. W., Plochl, M., & Kohlmaier, G. H. (1993). Apoplastic antioxidants as decisive elimination factors within the uptake process of nitrogen dioxide into leaf tissues. *The New Phytologist*, *125*(4), 771–785. doi:10.1111/j.1469-8137.1993.tb03927.x PMID:33874445

Ramírez-Moreno, M. A., Keshtkar, S., Padilla-Reyes, D. A., Ramos-López, E., García-Martínez, M., Hernández-Luna, M. C., Mogro, A. E., Mahlknecht, J., Huertas, J. I., Peimbert-García, R. E., Ramírez-Mendoza, R. A., Mangini, A. M., Roccotelli, M., Pérez-Henríquez, B. L., Mukhopadhyay, S. C., & Lozoya-Santos, J. J. (2021). Sensors for sustainable smart cities: A review. *Applied Sciences (Basel, Switzerland)*, *11*(17), 8198. doi:10.3390/app11178198

Ramirez-Priego, P., Estévez, M.-C., Díaz-Luisravelo, H. J., Manclús, J. J., Montoya, Á., & Lechuga, L. M. (2021). Real-Time Monitoring of Fenitrothion in Water Samples Using a Silicon Nanophotonic Biosensor. *Analytica Chimica Acta*, *1152*, 338276. doi:10.1016/j.aca.2021.338276 PMID:33648644

Ranieri, A., D'llrso, G., Nali, C., Lorenzini, G., & Soldatini, G. F. (1996). Ozone stimulates apoplastic antioxidant systems in pumpkin leaves. *Physiologia Plantarum*, *97*(2), 381–387. doi:10.1034/j.1399-3054.1996.970224.x

Ranjan, H., Sanjeevi, R., Vardhini, S., Tripathi, S., & Anuradha, J. (n.d.). *Biogenic Production of Silver Nanoparticles Utilizing an Arid Weed (Saccharum munja Roxb.) and Evaluation of its Antioxidant and Antimicrobial Activities.*

Rao, D. N., & Leblance, F. (1966). Effect of sulphur dioxide on lichen alga with special reference to chloroplast. *The Bryologist*, *69*(1), 69–72. doi:10.1639/0007-2745(1966)69[69:EOSDOT]2.0.CO;2

Rao, M. V., & Dubey, P. S. (1985). Plant response against SO2 in field conditions. *Asian Environment*, *10*, 1–9.

Rao, M. V., Hale, B. A., & Ormrod, D. P. (1995). Amelioration of ozone-induced oxidative damage in wheat plants grown under high CO_2. *Plant Physiology*, *109*, 421–432. doi:10.1104/pp.109.2.421 PMID:12228603

Rapini, R., & Marrazza, G. (2017). Electrochemical aptasensors for contaminants detection in food and environment: Recent advances. *Bioelectrochemistry (Amsterdam, Netherlands)*, *118*, 47–61. doi:10.1016/j.bioelechem.2017.07.004 PMID:28715665

Rasmussen, L. D., Sørensen, S. J., Turner, R. R., & Barkay, T. (2000). Application of a mer-lux biosensor for estimating bioavailable mercury in soil. *Soil Biology & Biochemistry*, *32*(5), 639–646. doi:10.1016/S0038-0717(99)00190-X

Ray, P. P. (2016). (in press). A survey on Internet of Things architectures. *Journal of King Saud University. Computer and Information Sciences*.

Rebollar-Pérez, G., Campos-Terán, J., Ornelas-Soto, N., Méndez-Albores, A., & Torres, E. (2015). Biosensors based on oxidative enzymes for detection of environmental pollutants. *Biocatalysis*, *1*(1), 118–129. doi:10.1515/boca-2015-0010

Reddy, C. M., Xu, L., O'Neil, G. W., Nelson, R. K., Eglinton, T. I., Faulkner, D. J., Norstrom, R., Ross, P. S., & Tittlemier, S. A. (2004). Radiocarbon evidence for a naturally produced, bioaccumulating halogenated organic compound. *Environmental Science & Technology*, *38*(7), 1992–1997. doi:10.1021/es030568i PMID:15112798

Reddy, M. T., Sivaraj, N., Venkateswaran, K., Pandravada, S. R., Sunil, N., & Dikshit, N. (2018). Classification, Characterization and Comparison of Aquatic Ecosystems in the Landscape of Adilabad District, Telangana, Deccan Region, India. *OAlib*, *5*(4), 1–111. doi:10.4236/oalib.1104459

Reddy, V. R., & Behera, B. (2006). Impact of water pollution on rural communities: An economic analysis. *Ecological Economics*, *58*(3), 520–537. doi:10.1016/j.ecolecon.2005.07.025

Rehman, U., Vesvikar, M., Maere, T., Guo, L., Vanrolleghem, P. A., & Nopens, I. (2015). Effect of sensor location on controller performance in a wastewater treatment plant. *Water Science and Technology*, *71*(5), 700–708. doi:10.2166/wst.2014.525 PMID:25768216

Remoundou, K., & Koundouri, P. (2009). Environmental Effects on Public Health: An Economic Perspective. *International Journal of Environmental Research and Public Health*, *6*(8), 2160–2178. doi:10.3390/ijerph6082160 PMID:19742153

Rene, E. R., Shu, L., & Jegatheesan, V. (2019). *Appropriate technologies to combat water pollution.* Challenges in Environmental Science and Engineering (CESE-2017), Kunming, China. https://www.cabidigitallibrary.org/doi/full/10.5555/20210094724

Restás, Á. (2014). *Thematic division and tactical analysis of the UAS application supporting forest fire management.*

Restás, Á. (2015). Drone applications for supporting disaster management. *World Journal of Engineering and Technology*, *3*(03), 316–321. doi:10.4236/wjet.2015.33C047

Restás, Á. (2016). Drone applications for preventing and responding to HAZMAT disaster. *World Journal of Engineering and Technology*, *4*(3), 76–84. doi:10.4236/wjet.2016.43C010

Rhodes, D. (1988). Metabolic response to stress. In D. D. Davies (Ed.), *The Biochemistry of Plants* (Vol. 12, pp. 201–241). Academic Press.

Rhodes, M. J. C. (1994). Physical role for secondary metabolites in plants: Some progress, many outstanding problems. *Plant Molecular Biology*, *24*(1), 1–20. doi:10.1007/BF00040570 PMID:8111009

Rhouati, A., Berkani, M., Vasseghian, Y., & Golzadeh, N. (2022). MXene-based electrochemical sensors for detection of environmental pollutants: A comprehensive review. *Chemosphere*, *291*, 132921. doi:10.1016/j.chemosphere.2021.132921 PMID:34798114

Riano, D., Chuvieco, E., Condés, S., González-Matesanz, J., & Ustin, S. L. (2004). Generation of crown bulk density for Pinus sylvestris L. from lidar. *Remote Sensing of Environment*, *92*(3), 345–352. doi:10.1016/j.rse.2003.12.014

Ricklefs, R.E. & Lovette, I.J. (1999). The roles of island area *per se* and habitat diversity in the species-area relationships of four Lesser Antillean faunal groups. *Journal of Animal Ecology, 68*(6), 1142-1160.

Rivadeneyra, A., Fernández-Salmerón, J., Salinas-Castillo, A., Palma, A. J., & Capitán-Vallvey, L. F. (2016). Development of a printed sensor for volatile organic compound detection at µg/L-level. *Sensors and Actuators. B, Chemical*, *230*, 115–122. doi:10.1016/j.snb.2016.02.047

Rivera, D., Alam, M. K., Davis, C. E., & Ho, C. K. (2003). Characterization of the ability of polymeric chemiresistor arrays to quantitate trichloroethylene using partial least squares (PLS): Effects of experimental design, humidity, and temperature. *Sensors and Actuators. B, Chemical*, *92*(1–2), 110–120. doi:10.1016/S0925-4005(03)00122-9

Riza, M. A., Go, Y. I., Harun, S. W., & Maier, R. R. (2020). FBG sensors for environmental and biochemical applications—A review. *IEEE Sensors Journal*, *20*(14), 7614–7627. doi:10.1109/JSEN.2020.2982446

Rizzato, S., Leo, A., Monteduro, A. G., Chiriacò, M. S., Primiceri, E., Sirsi, F., Milone, A., & Maruccio, G. (2020). Advances in the development of innovative sensor platforms for field analysis. *Micromachines*, *11*(5), 491. doi:10.3390/mi11050491 PMID:32403362

Robel, I., Girishkumar, G., Bunker, B. A., Kamat, P. V., & Vinodgopal, K. (2006). Structural changes and catalytic activity of platinum nanoparticles supported on C_{60} and carbon nanotube films during the operation of direct methanol fuel cells. *Applied Physics Letters*, *88*(7), 073113/1–073113, 3. doi:10.1063/1.2177354

Robe, R., & Kreeb, K. H. (1980). Effects of SO_2 upon enzyme activity in plant leaves. *Int. J. Plant Physiol*, *97*, 215–226.

Roberts, J. M. (1995). Reactive odd nitrogen in the atmosphere. In H. B. Singh (Ed.), *Composition, Chemistry and Climate of the Atmosphere* (pp. 176–215). Van Nostrand Reinhold.

Robles, C., Greff, S., Pasqualini, V., Garzino, S., Bousquet-Melou, A., Fernandez, C., Korboulewsky, N., & Bonin, G. (2003). Phenols and flavonoids in Alepopine Needles as Bioindicators of air pollution. *Journal of Environmental Quality*, *32*(6), 2265–2271. doi:10.2134/jeq2003.2265 PMID:14674550

Rocha, C., Veiga-Pires, C., Scholten, J., Knoeller, K., Gröcke, D. R., Carvalho, L., Anibal, J., & Wilson, J. (2016). Assessing land–ocean connectivity via submarine groundwater discharge (SGD) in the Ria Formosa Lagoon (Portugal): Combining radon measurements and stable isotope hydrology. *Hydrology and Earth System Sciences*, *20*(8), 3077–3098. doi:10.5194/hess-20-3077-2016

Rochelet, M., Solanas, S., Betelli, L., Chantemesse, B., Vienney, F., & Hartmann, A. (2015). Rapid amperometric detection of Escherichia coli in wastewater by measuring β-D glucuronidase activity with disposable carbon sensors. *Analytica Chimica Acta*, *892*, 160–166. doi:10.1016/j.aca.2015.08.023 PMID:26388487

Roda, A., Zangheri, M., Calabria, D., Mirasoli, M., Caliceti, C., Quintavalla, A., Lombardo, M., Trombini, C., & Simoni, P. (2019). A Simple Smartphone-Based Thermochemiluminescent Immunosensor for Valproic Acid Detection Using 1,2-Dioxetane Analogue-Doped Nanoparticles as a Label. *Sensors and Actuators. B, Chemical*, *279*, 327–333. doi:10.1016/j.snb.2018.10.012

Ródenas García, M., Spinazzé, A., Branco, P. T. B. S., Borghi, F., Villena, G., Cattaneo, A., Di Gilio, A., Mihucz, V. G., Gómez Álvarez, E., Lopes, S. I., Bergmans, B., Orłowski, C., Karatzas, K., Marques, G., Saffell, J., & Sousa, S. I. V. (2022). Review of low-cost sensors for indoor air quality: Features and applications. *Applied Spectroscopy Reviews*, *57*(9–10), 747–779. doi:10.1080/05704928.2022.2085734

Rodríguez-Mozaz, S., Marco, M.-P., Alda, M. J. L. D., & Barceló, D. (2004). Biosensors for environmental applications: Future development trends. *Pure and Applied Chemistry*, *76*(4), 723–752. doi:10.1351/pac200476040723

Roggo, C., & van der Meer, J. R. (2017). *Miniaturized and integrated whole cell living bacterial sensors in field applicable autonomous devices. Current Opinion in Biotechnology*. Elsevier Ltd., doi:10.1016/j.copbio.2016.11.023

Rolon, A.S. & Maltchik, L. (2006). Environmental factors as predictors of aquatic macrophyte richness and composition in wetlands of Kashmir valley. *Hydrobiologia, 556*(1).

Rolon, A.S. & Maltchik, L. (2010). Does flooding of rice fields after cultivation contribute to wetland plant conservation in Kashmir valley? *Applied Vegetation Science*, *13*(1).

Rolon, A.S., Lacerda, T., Maltchik, L., & Guadagnin, D.L. (2008). The influence of area, habitat and water chemistry on richness and composition of macrophyte assemblages in Kashmir valley wetlands. *Journal of Vegetation Science, 19*(2), 221-228.

Rolon, A.S., Maltchik, L., & Irgang, B. (2004). Levantamento de macrófitasaquáticasemáreasúmidas do Rio Grande do Sul, Brasil. *ActaBiologicaLeopoldensia*, *26*.

Roostaei, J., Wager, Y. Z., Shi, W., Dittrich, T., Miller, C., & Gopalakrishnan, K. (2023). IoT-based Edge Computing (IoTEC) for Improved Environmental Monitoring. *Sustainable Computing : Informatics and Systems*, *38*, 100870. Advance online publication. doi:10.1016/j.suscom.2023.100870 PMID:37234690

Rosenzweig, M.L. (1995). *Species diversity in space and time*. Cambridge: Cambridge University Press.

Rossa, C. G. (2018). A generic fuel moisture content attenuation factor for fire spread rate empirical models. *Forest Systems*, *27*(2), e009–e009. doi:10.5424/fs/2018272-13175

Roy, S., Saha, H., & Sarkar, C. K. (2010). High sensitivity methane sensor by chemically deposited nanocrystalline ZnO thin film. *International Journal on Smart Sensing and Intelligent Systems*, *3*(4), 605–620. doi:10.21307/ijssis-2017-411

Ruan, Y., & Tang, Y. (2011) A water quality monitoring system based on wireless sensor network & solar power supply. In: *2011 IEEE International Conference on Cyber Technology in Automation, Control, and Intelligent System*, (pp. 20–23). IEEE.

Ruffault, J., Martin-StPaul, N., Pimont, F., & Dupuy, J. L. (2018). How well do meteorological drought indices predict live fuel moisture content (LFMC)? An assessment for wildfire research and operations in Mediterranean ecosystems. *Agricultural and Forest Meteorology*, *262*, 391–401. doi:10.1016/j.agrformet.2018.07.031

Rupani, S. P., Gu, M. B., Konstantinov, K. B., Dhurjati, P. S., Van Dyk, T. K., & LaRossa, R. A. (1996). Characterization of the stress response of a bioluminescent biological sensor in batch and continuous cultures. *Biotechnology Progress*, *12*(3), 387–392. doi:10.1021/bp960015u PMID:8652122

Russell, B. C., Croudace, I. W., & Warwick, P. E. (2015). Determination of ^{135}Cs and ^{137}Cs in environmental samples: A review. *Analytica Chimica Acta*, *890*, 7–20. doi:10.1016/j.aca.2015.06.037 PMID:26347165

Saber, R., & Pişkin, E. (2003). Investigation of complexation of immobilized metallothionein with Zn (II) and Cd (II) ions using piezoelectric crystals. *Biosensors & Bioelectronics*, *18*(8), 1039–1046. doi:10.1016/S0956-5663(02)00217-8 PMID:12782467

Sahadewa, A., Zekkos, D., Woods, R. D., Stokoe, K. H. II, & Matasovic, N. (2014). In-situ assessment of the dynamic properties of municipal solid waste at a landfill in Texas. *Soil Dynamics and Earthquake Engineering*, *65*, 303–313. doi:10.1016/j.soildyn.2014.04.004

Saini, J., Dutta, M., & Marques, G. (2021). Sensors for indoor air quality monitoring and assessment through Internet of Things: A systematic review. *Environmental Monitoring and Assessment*, *193*(2), 66. doi:10.1007/s10661-020-08781-6 PMID:33452599

Saitanis, C. J., Riga-Karandinos, A. N., & Karandinos, M. G. (2001). Effects of ozone on chlorophyll and quantum yield of tobacco (*Nicotiana tabacum* L.) varieties. *Chemosphere*, *42*(8), 945–953. doi:10.1016/S0045-6535(00)00158-2 PMID:11272917

Sakaki, T., Kondo, N., & Sugahara, K. (1983). Breakdown of photosynthetic pigments and lipids in spinach leaves with ozone fumigation: Role of active oxygens. *Physiologia Plantarum*, *59*(1), 28–34. doi:10.1111/j.1399-3054.1983.tb06566.x

Sakaki, T., Ohnishi, J., Kondo, N., & Yamada, M. (1985). Polar and neutral lipid changes in spinach leaves with ozone fumigation.: Triacylglycerol synthesis from polar lipids. *Plant & Cell Physiology*, *26*, 253–262.

Salgado-Hernanz, P. M., Bauzà, J., Alomar, C., Compa, M., Romero, L., & Deudero, S. (2021). Assessment of marine litter through remote sensing: Recent approaches and future goals. *Marine Pollution Bulletin*, *168*, 112347. doi:10.1016/j.marpolbul.2021.112347 PMID:33901907

Sallée, J. B., Pellichero, V., Akhoudas, C., Pauthenet, E., Vignes, L., Schmidtko, S., Garabato, A. N., Sutherland, P., & Kuusela, M. (2021). Summertime increases in upper-ocean stratification and mixed-layer depth. *Nature*, *591*(7851), 592–598. doi:10.1038/s41586-021-03303-x PMID:33762764

Sallem, A., Loponen, J., Pihlaja, K., & Oksanen, E. (2001). Effect of long-term open field ozone exposure on leaf phenolics of European silver birch (*Betula pendula* Roth). *Journal of Chemical Ecology*, *27*(5), 1049–1062. doi:10.1023/A:1010351406931 PMID:11471939

Sana, S. S., Dogiparthi, L. K., Gangadhar, L., Chakravorty, A., & Abhishek, N. (2020). Effects of microplastics and nanoplastics on marine environment and human health. *Environmental Science and Pollution Research International*, *27*(36), 44743–44756. doi:10.1007/s11356-020-10573-x PMID:32876819

Sandermann, H. Jr. (1996). Ozone and plant health. *Annual Review of Phytopathology*, *34*(1), 347–366. doi:10.1146/annurev.phyto.34.1.347 PMID:15012547

Sanjeevi, R., & Ankitkumar, B., Rathod, Prashantkumar B. Sathvara, Aviral Tripathi, J. Anuradha, & Sandeep Tripathi. (2022). Vegetational Cartography Analysis Utilizing Multi-Temporal Ndvi Data Series: A Case Study from Rajkot District (Gujarat), India. *Tianjin Daxue Xuebao (Ziran Kexue Yu Gongcheng Jishu Ban)/ Journal of Tianjin University Science and Technology*, *55*(4), 490–497. https://doi.org/ doi:10.17605/OSF.IO/UGJYM

Sanjeevi, R. (2011). *Studies on the treatment of low-strength wastewaters with upflow anaerobic sludge blanket (UASB) reactor: With emphasis on granulation studies*. Centre for Pollution Control and Environmental Engineering.

Sanjeevi, R., Haruna, M., Tripathi, S., Singh, B., & Jayaraman, A. (2017). Impacts of Global Carbon Foot Print on the Marine Environment. *International Journal of Engineering Research & Technology (Ahmedabad)*, *5*, 51–54.

Saral Designs. (2022, July 22). *Sanitary Napkins and its Environmental impact - Part 1*. Saral Designs. https://www.saraldesigns.in/uncategorized/sanitary-napkins-and-its-environmental-impact-part-1/

Sara, S. M., Al-Dhahebi, A. M., & Mohamed Saheed, M. S. (2022). Recent Advances in Graphene-Based Nanocomposites for Ammonia Detection. *Polymers, 14*(23). doi:10.3390/polym14235125 PMID:36501520

Sarkar, R. K., Banerjee, A., & Mukherji, S. (1986). Acceleration of peroxidase and catalase activities in leaves of wild dicotyledonous plants, as an indication of automobile exhaust pollution. *Environmental Pollution. Series A. Ecological and Biological*, *42*(4), 289–295. doi:10.1016/0143-1471(86)90013-9

Sasaki, S., Yokoyama, K., Tamiya, E., Karube, I., Hayashi, C., Arikawa, Y., & Numata, M. (1997). Sulfate sensor using *Thiobacillus ferrooxidans*. *Analytica Chimica Acta*, *347*(3), 275–280. doi:10.1016/S0003-2670(97)00170-0

Sathish, K., Sarojini, M., & Pandu, R. (2016). IoT based real time monitoring of water quality. *Int J Prof Eng Stud*, *VII*(5), 174–179.

Sathvara, P. B., Anuradha, J., Sanjeevi, R., Tripathi, S., & Rathod, A. B. (2023). Spatial Analysis of Carbon Sequestration Mapping Using Remote Sensing and Satellite Image Processing. In Multimodal Biometric and Machine Learning Technologies (pp. 71–83). doi:10.1002/9781119785491.ch4

Sayari, A., Hamoudi, S., & Yang, Y. (2005). Applications of pore-expanded mesoporous silica. 1. Removal of heavy metal cations and organic pollutants from wastewater. *Chemistry of Materials*, *17*(1), 212–216. doi:10.1021/cm048393e

Schirmer, C., Posseckardt, J., Schröder, M., Gläser, M., Howitz, S., Scharff, W., & Mertig, M. (2019). Portable and Low-Cost Biosensor towards on-Site Detection of Diclofenac in Wastewater. *Talanta*, *203*, 242–247. doi:10.1016/j.talanta.2019.05.058 PMID:31202333

Schott, P., Rolon, A.S., & Maltchik, L. (2005). The dynamics of macrophytes in an oxbow lake of the Sinos River basin in south India. *Verhandlungen. InternationaleVereinigungfuertheoretische und angewandteLimnologie, 29*(2).

Schutting, S., Borisov, S. M., & Klimant, I. (2013). Diketo-Pyrrolo-Pyrrole dyes as new colorimetric and fluorescent pH indicators for optical carbon dioxide sensors. *Analytical Chemistry*, *85*(6), 3271–3279. doi:10.1021/ac303595v PMID:23421943

Schutting, S., Jokic, T., Strobl, M., Borisov, S. M., Beer, D. D., & Klimant, I. (2015). NIR optical carbon dioxide sensors based on highly photostable dihydroxy-aza-BODIPY dyes. *Journal of Materials Chemistry. C, Materials for Optical and Electronic Devices*, *3*(21), 5474–5483. doi:10.1039/C5TC00346F

Schutting, S., Klimant, I., Beer, D. K., & Borisov, S. M. (2014). New highly fluorescent pH indicator for ratiometric RGB imaging of pCO2. *Methods and Applications in Fluorescence*, *2*(2), 02400. doi:10.1088/2050-6120/2/2/024001 PMID:29148465

Schwandt, C., Kumar, R. V., & Hills, M. P. (2018). Solid state electrochemical gas sensor for the quantitative determination of carbon dioxide. *Sensors and Actuators. B, Chemical*, *265*, 27–34. doi:10.1016/j.snb.2018.03.012

Schwarzenbach, R. P., Egli, T., Hofstetter, T. B., Von Gunten, U., & Wehrli, B. (2010). Global Water Pollution and Human Health. *Annual Review of Environment and Resources*, *35*(1), 109–136. doi:10.1146/annurev-environ-100809-125342

Schweitzer, L., & Zhou, J. (2010). Neighborhood air quality, respiratory health, and vulnerable populations in compact and sprawled regions. *Journal of the American Planning Association*, *76*(3), 363–371. doi:10.1080/01944363.2010.486623

Scognamiglio, V., Antonacci, A., Arduini, F., Moscone, D., Campos, E. V. R., Fraceto, L. F., & Palleschi, G. (2019). An Eco-Designed Paper Based Algal Biosensor for Nano formulated Herbicide Optical Detection. *Journal of Hazardous Materials*, *373*, 483–492. doi:10.1016/j.jhazmat.2019.03.082 PMID:30947038

Seabloom, E.W., Maloney, K.A., & Valk, A.G. (2001). Constraints on the establishment of plants along a fluctuating water-depth gradient. *Ecology*, *82*(8), 2216-2232.

Sejdiu, B., Ismaili, F., & Ahmedi, L. (2022). A Real-Time Semantic Annotation to the Sensor Stream Data for the Water Quality Monitoring. *SN COMPUT. SCI.*, *3*, 254. doi:10.1007/s42979-022-01145-6

Sempionatto, J. R., Khorshed, A. A., Ahmed, A., De Loyola, E., Silva, A. N., Barfidokht, A., Yin, L., & Wang, J. (2020). Epidermal Enzymatic Biosensors for Sweat Vitamin C: Toward Personalized Nutrition. *ACS Sensors*, *5*(6), 1804–1813. doi:10.1021/acssensors.0c00604 PMID:32366089

Senoussi, M. M., Creche, J., & Rideau, M. (2007). Relation between hypoxia and alkaloid accumulation in *Catharanthus roseus* cell suspension. *Journal of Applied Sciences Research*, *3*, 287–290.

Senser, M., Kloss, M., & Lutz, C. (1990). Influence of soil substrate and ozone plus acid mist on the pigment content and composition of needles from young spruce trees. *Environmental Pollution*, *64*(3-4), 295–312. doi:10.1016/0269-7491(90)90052-E PMID:15092286

Senturk, S. F., Gulmez, H. K., Gul, M. F., & Kirci, P. (2021). Detection and separation of transparent objects from recyclable materials with sensors. In *International Conference on Advanced Network Technologies and Intelligent Computing* (pp. 73-81). Cham: Springer International Publishing. 10.1007/978-3-030-96040-7_6

Severinghaus, J. W., & Bradley, A. F. (1957). Electrodes for blood pO_2 and pCO_2 determination. *Journal of Applied Physiology*, *13*(3), 515–520. doi:10.1152/jappl.1958.13.3.515 PMID:13587443

Sezgintürk, M. K. (2020). Introduction to commercial biosensors. In *Commercial Biosensors and Their Applications* (pp. 1–28). Elsevier. doi:10.1016/B978-0-12-818592-6.00001-3

Sfetsos, A., Giroud, F., Clemencau, A., Varela, V., Freissinet, C., LeCroart, J., Vlachogiannis, D., Politi, N., Karozis, S., Gkotsis, I., Eftychidis, G., Hedel, R., & Hahmann, S. (2021). Assessing the effects of forest fires on interconnected critical infrastructures under climate change. Evidence from South France. *Infrastructures*, *6*(2), 16. doi:10.3390/infrastructures6020016

Shahar, H., Tan, L. L., Ta, G. C., & Heng, L. Y. (2019). Optical Enzymatic Biosensor Membrane for Rapid in Situ Detection of Organohalide in Water Samples. *Microchemical Journal*, *146*, 41–48. doi:10.1016/j.microc.2018.12.052

Shahra, E. Q., & Wu, W. (2020). Water contaminants detection using sensor placement approach in smart water networks. *Journal of Ambient Intelligence and Humanized Computing*, *14*(5), 4971–4986. doi:10.1007/s12652-020-02262-x

Shamsi, S. R. A., Ashmore, M. R., Bell, J. N. B., Maggs, R., Kafayat, U., & Wahid, A. (2000). The impact of air pollution on crops in developing countries- A case study in Pakistan. In M. Yunus, N. Singh, & L. J. De Kok (Eds.), *Environmental Stress: indication, Mitigation and Eco-conservation* (pp. 65–71). Kluwer Academic Publishers. doi:10.1007/978-94-015-9532-2_6

Sharma, G. K., & Tyree, J. (1973). Geographic leaf cuticular and gross morphological variations in *Liquidamber styraciflua* L. and their possible relationship to environmental pollution. *Botanical Gazette (Chicago, Ill.)*, *134*(3), 179–184. doi:10.1086/336701

Sharma, K., & Sharma, M. (2023). *Optical biosensors for environmental monitoring: Recent advances and future perspectives in bacterial detection. Environmental Research.* Academic Press Inc., doi:10.1016/j.envres.2023.116826

Sharma, P., & Giri, A. (2018). Productivity evaluation of lotic and lentic water body in Himachal Pradesh, India. *MOJ Ecology & Environmental Sciences*, *3*(5), 311–317. doi:10.15406/mojes.2018.03.00105

Sharma, T. K., Sharma, N. K., & Prakash, G. (1994). Effect of sulphur dioxide on *Brassica campestris* var. Krishna. *Acta Botanica Indica*, *22*, 15–20.

Sharma, Y. K., & Davis, K. R. (1997). The effects of ozone on antioxidant responses in plants. *Free Radical Biology & Medicine*, *23*(3), 480–488. doi:10.1016/S0891-5849(97)00108-1 PMID:9214586

Shchberbakov, A. P., Svistova, I. D., & Dzhuvelikyan, Kh. (2001). Biomonitoring of soil pollution by gaseous emissions of motor vehicles. *Ekologiya i Promyshlennost Rossil*, (June), 26–29.

Sheini, A. (2020). A paper-based device for the colorimetric determination of ammonia and carbon dioxide using thiomalic acid and maltol functionalized silver nanoparticles: Application to the enzymatic determination of urea in saliva and blood. *Mikrochimica Acta*, *187*(10), 565. doi:10.1007/s00604-020-04553-8 PMID:32920692

Shen, L., Wang, Y., Liu, K., Yang, Z., Shi, X., Yang, X., & Jing, K. (2020). Synergistic path planning of multi-UAVs for air pollution detection of ships in ports. *Transportation Research Part E, Logistics and Transportation Review*, *144*, 102128. doi:10.1016/j.tre.2020.102128

Shimazaki, K., Sakaki, T., Kondo, N., & Sugahara, K. (1980). Active oxygen participation in chlorophyll destruction and lipid peroxidation in SO_2- fumigated leaves of spinach. *Plant & Cell Physiology*, *21*(8), 1193–1204. doi:10.1093/oxfordjournals.pcp.a076118

Shimazaki, K., Yu, S. W., Sakaki, T., & Tanaka, K. (1992). Difference between spinach and kidney bean plants in terms of sensitivity to fumigation with NO_2. *Plant & Cell Physiology*, *33*(3), 267–273. doi:10.1093/oxfordjournals.pcp.a078250

Shirke, S. I., Ithape, S., Lungase, S., & Mohare, M. (2019). Automation of smart waste management using IoT. *International Research Journal of Engineering and Technology*, *6*(6), 414-419. IRJET-V6I615320190820-94746-xkw9ee-libre.pdf

Shirley, B. W. (1996). Flavonoid biosynthesis new functions of an old pathway. *Trends in Plant Science*, *1*, 281–377.

Shi, X., Zhou, J. L., Zhao, H., Hou, L., & Yang, Y. (2014). Application of passive sampling in assessing the occurrence and risk of antibiotics and endocrine disrupting chemicals in the Yangtze Estuary, China. *Chemosphere*, *111*, 344–351. doi:10.1016/j.chemosphere.2014.03.139 PMID:24997938

Shoushtarian, F., & Negahban-Azar, M. (2020). Worldwide regulations and guidelines for agricultural water reuse: A critical review. *Water (Basel)*, *12*(4), 971. doi:10.3390/w12040971

Shrivastav, A. M., Mishra, S. K., & Gupta, B. D. (2015). Localized and Propagating Surface Plasmon Resonance Based Fiber Optic Sensor for the Detection of Tetracycline Using Molecular Imprinting. *Materials Research Express*, *2*(3), 35007. doi:10.1088/2053-1591/2/3/035007

Shukla, K., & Aggarwal, S. G. (2022). A Technical Overview on Beta-Attenuation Method for the Monitoring of Particulate Matter in Ambient Air. *Aerosol and Air Quality Research*, *22*(12), 220195. doi:10.4209/aaqr.220195

Shu, W. S., & Huang, L. N. (2022). Microbial diversity in extreme environments. *Nature Reviews. Microbiology*, *20*(4), 219–235. doi:10.1038/s41579-021-00648-y PMID:34754082

Siddiqui, S., Ahmad, A., & Hayat, S. (2004). The impact of sulphur dioxide on growth and productivity of *Helianthus annus*. *Pollution Research*, *23*, 327–332.

Siefermann-Harms, D. (1987). The light harvesting and protective function of carotenoids in photosynthetic membranes. *Physiologia Plantarum*, *69*(3), 561–568. doi:10.1111/j.1399-3054.1987.tb09240.x

Silva, A. S., Brito, T., de Tuesta, J. L. D., Lima, J., Pereira, A. I., Silva, A. M., & Gomes, H. T. (2022, October). Node Assembly for Waste Level Measurement: Embrace the Smart City. In *International Conference on Optimization, Learning Algorithms and Applications* (pp. 604-619). Cham: Springer International Publishing. 10.1007/978-3-031-23236-7_42

Silver, M., Buck, J., & Taylor, N. (2017). *Systems and devices for treating water, wastewater and other biodegradable matter*. Google Patents.

Simmonds, M. (1998). Chemoecology: The legacy left by Tony Swain. *Phytochemistry*, *49*, 1183–1190.

Singh, G., Sinam, G., Kriti, K., Pandey, M., Kumari, B., & Kulsoom, M. (2019). Soil pollution by fluoride in India: Distribution, chemistry and analytical methods. Springer. doi:10.1007/978-981-13-6358-0_12

Singh, U.B., Ahluwalia, A.S., Sharma, C., Jindal, R., & Thakur, R.K. (2013). Planktonic indicators: a promising tool for monitoring water quality (early-warning signals). *Eco Environ Cons. 19*(3), 793–800.

Singh, A. K., Senapati, D., Wang, S., Griffin, J., Neely, A., Candice, P., Naylor, K. M., Varisli, B., Kalluri, J. R., & Ray, P. C. (2009). Gold nanorod based selective identification of Escherichia coli bacteria using two- photon Rayleigh scattering spectroscopy. *ACS Nano*, *3*(7), 1906–1912. doi:10.1021/nn9005494 PMID:19572619

Singh, D., Dahiya, M., Kumar, R., & Nanda, C. (2021). Sensors and systems for air quality assessment monitoring and management: A review. *Journal of Environmental Management*, *289*, 112510. doi:10.1016/j.jenvman.2021.112510 PMID:33827002

Singh, D., Dikshit, A. K., & Kumar, S. (2024). Smart technological options in collection and transportation of municipal solid waste in urban areas: A mini review. *Waste Management & Research*, *42*(1), 3–15. doi:10.1177/0734242X231175816 PMID:37246550

Singh, N., Yunus, M., Kulshreshtha, K., Srivastava, K., & Ahmad, K. J. (1988). Effect of SO_2 on growth and development of *Dahlia rosea* Cav. *Bulletin of Environmental Contamination and Toxicology*, *40*(5), 743–751. doi:10.1007/BF01697525 PMID:3382791

Singh, N., Yunus, M., Srivastava, K., Singh, S. N., Pandey, V., Misra, J., & Ahmad, K. J. (1995). Monitiring of auto exhaust pollution by road side plants. *Environmental Monitoring and Assessment*, *34*(1), 13–25. doi:10.1007/BF00546243 PMID:24201905

Skretas, G., Meligova, A. K., Villalonga-Barber, C., Mitsiou, D. J., Alexis, M. N., Micha-Screttas, M., Steele, B. R., Screttas, C. G., & Wood, D. W. (2007). Engineered chimeric enzymes as tools for drug discovery: Generating reliable bacterial screens for the detection, discovery, and assessment of estrogen receptor modulators. *Journal of the American Chemical Society*, *129*(27), 8443–8457. doi:10.1021/ja067754j PMID:17569534

Slater, T. F. (1972). *Free Radical Mechanisms in Tissue Injury*. Pion Limited.

Sliusar, N., Filkin, T., Huber-Humer, M., & Ritzkowski, M. (2022). Drone technology in municipal solid waste management and landfilling: A comprehensive review. *Waste Management (New York, N.Y.)*, *139*, 1–16. doi:10.1016/j.wasman.2021.12.006 PMID:34923184

Slovik, S. (1996). Chronic SO_2 and NO_x pollution interferes with the K^+ and Mg^{2+} budget of Norway Spruce trees. *Journal of Plant Physiology*, *148*(3-4), 276–286. doi:10.1016/S0176-1617(96)80254-9

Smith, L. L. (1987). Cholesterol auto-oxidation 1981-1986. *Chemistry and Physics of Lipids*, *44*(2-4), 87–125. doi:10.1016/0009-3084(87)90046-6 PMID:3311423

Smith, L. M., & Liu, Z. (2018). Advances in water quality monitoring technology for enhancing water management in urban environments. *Water Research*, *137*, 182–197.

Smith, Y. R., Nagel, J. R., & Rajamani, R. K. (2019). Eddy current separation for recovery of non-ferrous metallic particles: A comprehensive review. *Minerals Engineering*, *133*, 149–159. doi:10.1016/j.mineng.2018.12.025

Smolka, M., Puchberger-Enengl, D., Bipoun, M., Klasa, A., Kiczkajlo, M., Śmiechowski, W., Sowiński, P., Krutzler, C., Keplinger, F., & Vellekoop, M. J. (2016). A mobile lab-on-a-chip device for on-site soil nutrient analysis. *Precision Agriculture*, *18*(2), 152–168. doi:10.1007/s11119-016-9452-y

Sohrabi, H., Hemmati, A., Majidi, M. R., Eyvazi, S., Jahanban-Esfahlan, A., Baradaran, B., Adlpour-Azar, R., Mokhtarzadeh, A., & de la Guardia, M. (2021). Recent advances on portable sensing and biosensing assays applied for detection of main chemical and biological pollutant agents in water samples: A critical review. *Trends in Analytical Chemistry*, *143*, 116344. doi:10.1016/j.trac.2021.116344

SokhiR. S.MoussiopoulosN.BaklanovA.BartzisJ.CollI.FinardiS.KukkonenJ. (2022). Advances in air quality research - current and emerging challenges. Atmospheric Chemistry and Physics. Copernicus GmbH. doi:10.5194/acp-22-4615-2022

Somers, V. L., & Manchester, I. R. (2022). Minimizing the risk of spreading processes via surveillance schedules and sparse control. *IEEE Transactions on Control of Network Systems*, *10*(1), 394–406. doi:10.1109/TCNS.2022.3203359

Song, J., Cannatella, D., & Katsikis, N. (2023, June). Landscape-Based Fire Resilience: Identifying Interaction Between Landscape Dynamics and Fire Regimes in the Mediterranean Region. In *International Conference on Computational Science and Its Applications* (pp. 328-344). Cham: Springer Nature Switzerland. 10.1007/978-3-031-37117-2_23

Song, L., Mao, K., Zhou, X., & Hu, J. (2016). A novel biosensor based on Au@ Ag core–shell nanoparticles for SERS detection of arsenic (III). *Talanta*, *146*, 285–290. doi:10.1016/j.talanta.2015.08.052 PMID:26695265

Sørensen, A. J., Ludvigsen, M., Norgren, P., Ødegård, Ø., & Cottier, F. (2020). Sensor-Carrying Platforms. POLAR NIGHT Marine Ecology: Life and Light in the Dead of Night, 241-275. Springer. doi:10.1007/978-3-030-33208-2_9

Sosunova, I., & Porras, J. (2022). IoT-enabled smart waste management systems for smart cities: A systematic review. *IEEE Access : Practical Innovations, Open Solutions*, *10*, 73326–73363. doi:10.1109/ACCESS.2022.3188308

Speers, A. E., Besedin, E. Y., Palardy, J. E., and Moore, C. (2016). Impacts of climate change and ocean acidification on coral reef fisheries: an integrated ecological–economic model. *Ecol. Econ. 128*, 33–43. doi: .2016.04.012 doi:10.1016/j.ecolecon

Spurr, K., Pendergast, N., & MacDonald, S. (2014). assessing the use of the air Quality health index by vulnerable populations in a 'low-risk'region: A pilot study. Canadian Journal of Respiratory Therapy: CJRT=. *Canadian Journal of Respiratory Therapy : CJRT*, *50*(2), 45. PMID:26078611

Sridhar, M. K. C., & Rami Reddy, C. (1984). Surface tension of polluted waters and treated wastewater. *Environmental Pollution. Series B. Chemical and Physical*, *7*(1), 49–69. doi:10.1016/0143-148X(84)90037-5

Srivastava, H. S., Jolliffe, P. A., & Runeckles, V. C. (1975). Inhibition of gas exchange in bean leaves by NO_2. *Canadian Journal of Botany*, *53*(5), 466–474. doi:10.1139/b75-057

Srivastava, S. K., Seng, K. P., Ang, L. M., Pachas, A. N. A., & Lewis, T. (2022). Drone-Based Environmental Monitoring and Image Processing Approaches for Resource Estimates of Private Native Forest. *Sensors (Basel)*, *22*(20), 7872. doi:10.3390/s22207872 PMID:36298223

Stadtmann, E. R., & Oliver, C. N. (1991). Metal-catalyzed oxidation of protein physiological consequences. *The Journal of Biological Chemistry*, *226*(4), 2005–2008. doi:10.1016/S0021-9258(18)52199-2

Steelman, T., & Nowell, B. (2019). Evidence of effectiveness in the Cohesive Strategy: Measuring and improving wildfire response. *International Journal of Wildland Fire*, *28*(4), 267–274. doi:10.1071/WF18136

Stepanova, A. Y., Gladkov, E. A., Osipova, E. S., Gladkova, O. V., & Терешонок, Д. В. (2022). Bioremediation of soil from petroleum contamination. *Processes (Basel, Switzerland)*, *10*(6), 1224. doi:10.3390/pr10061224

Sterkens, W., Diaz-Romero, D., Goedemé, T., Dewulf, W., & Peeters, J. R. (2021). Detection and recognition of batteries on X-Ray images of waste electrical and electronic equipment using deep learning. *Resources, Conservation and Recycling*, *168*, 105246. doi:10.1016/j.resconrec.2020.105246

Stewart, G., Fowler, S. W., & Fisher, N. S. (2008). In S. Krishnaswami & J. K. Cochran (Eds.), The bioaccumulation of U- and Th- series radionuclides in marine organisms. U-Th Series Nuclides in Aquatic Systems (pp. 269–305). Elsevier Science. doi:10.1016/S1569-4860(07)00008-3

Stockhardt, J. A. (1850). Liber die Einwirkung des Rauches von Silberhutten auf die benachbarte vegetation. *Polyt. Centr. Bl*, *16*, 257–278.

Struzik, M., Garbayo, I., Pfenninger, R., & Rupp, J. L. M. (2018). A Simple and Fast Electrochemical CO_2 Sensor Based on $Li_7La_3Zr_2O_{12}$ for Environmental Monitoring. *Advanced Materials*, *30*(44), 1804098. doi:10.1002/adma.201804098 PMID:30238512

Sturchio, N. C., Caffee, M., Beloso, A. D. Jr, Heraty, L. J., Böhlke, J. K., Hatzinger, P. B., & Dale, M. (2009). Chlorine-36 as a tracer of perchlorate origin. Environmental science & technology, 43(18), 6934-6938.

Sudibya, H. G., He, Q., Zhang, H., & Chen, P. (2011). Electrical detection of metal ions using field-effect transistors based on micropatterned reduced graphene oxide films. *ACS Nano*, *5*(3), 1990–1994. doi:10.1021/nn103043v PMID:21338084

Su, L., Jia, W., Hou, C., & Lei, Y. (2011). Microbial biosensors: A review. *Biosensors & Bioelectronics*, *26*(5), 1788–1799. doi:10.1016/j.bios.2010.09.005 PMID:20951023

Sunda, W. G. (2006). Trace metals and harmful algal blooms. In E. Granéli & J. T. Turner (Eds.), Ecology of Harmful Algae (pp. 203–214). Springer. doi:10.1007/978-3-540-32210-8_16

Sun, J., Ye, B., Xia, G., Zhao, X., & Wang, H. (2016). A colorimetric and fluorescent chemosensor for the highly sensitive detection of CO_2 gas: Experiment and DFT calculation. *Sensors and Actuators. B, Chemical*, *233*, 76–82. doi:10.1016/j.snb.2016.04.052

Sun, K., Cui, W., & Chen, C. (2021). Review of underwater sensing technologies and applications. *Sensors (Basel)*, *21*(23), 7849. doi:10.3390/s21237849 PMID:34883851

Sun, S., Sidhu, V., Rong, Y., & Zheng, Y. (2018). Pesticide pollution in agricultural soils and sustainable remediation methods: A review. *Current Pollution Reports*, *4*(3), 240–250. doi:10.1007/s40726-018-0092-x

Sur, I. M., Moldovan, A., Micle, V., & Polyak, E. T. (2022). Assessment of Surface Water Quality in the Baia Mare Area, Romania. *Water (Basel)*, *14*(19), 19. Advance online publication. doi:10.3390/w14193118

Suslova, M. Y., & Grebenshchikova, V. I. (2020). Water quality monitoring of the Angara River source. *Limnology and Freshwater Biology*, (4), 1040–1041. doi:10.31951/2658-3518-2020-A-4-1040

Svitel, J., Curilla, O., & Tkác, J. (1998). Microbial cell-based biosensor for sensing glucose, sucrose or lactose. *Biotechnology and Applied Biochemistry*, *27*(2), 153–158. PMID:9569611

Svorc, J., Miertus, S., & Barlikova, A. (1990). Hybrid biosensor for the determination of lactose. *Analytical Chemistry*, *62*(15), 1628–1631. doi:10.1021/ac00214a018 PMID:2205123

Swart, P. K., Greer, L., Rosenheim, B. E., Moses, C. S., Waite, A. J., Winter, A., Dodge, R. E., & Helmle, K. (2010). The 13C Suess effect in scleractinian corals mirror changes in the anthropogenic CO2 inventory of the surface oceans. *Geophysical Research Letters*, *37*(5), L05604. doi:10.1029/2009GL041397

Szpilko, D., de la Torre Gallegos, A., Jimenez Naharro, F., Rzepka, A., & Remiszewska, A. (2023). Waste Management in the Smart City: Current Practices and Future Directions. *Resources*, *12*(10), 115. doi:10.3390/resources12100115

Tagad, C. K., Kulkarni, A., Aiyer, R., Patil, D., & Sabharwal, S. G. (2016). A miniaturized optical biosensor for the detection of Hg2+ based on acid phosphatase inhibition. *Optik (Stuttgart)*, *127*(20), 8807–8811. doi:10.1016/j.ijleo.2016.06.123

Tajik, S., Beitollahi, H., Nejad, F. G., Dourandish, Z., Khalilzadeh, M. A., Jang, H. W., Venditti, R. A., Varma, R. S., & Shokouhimehr, M. (2021). Recent developments in polymer nanocomposite-based electrochemical sensors for detecting environmental pollutants. *Industrial & Engineering Chemistry Research*, *60*(3), 1112–1136. doi:10.1021/acs.iecr.0c04952 PMID:35340740

Takayama, K., Kano, K., & Ikeda, T. (1996). Mediated electrocatalytic reduction of nitrate and nitrite based on the denitrifying activity of *Paracoccus denitrificans*. *Chemistry Letters*, *25*(11), 1009–1010. doi:10.1246/cl.1996.1009

Tanaka, K., Suda, Y., Kondo, N., & Sugahara, K. (1985). Ozone tolerance and the ascorbate-dependent H_2O_2 decomposing system in chloroplasts. *Plant & Cell Physiology*, *26*, 1425–1431.

Tandy, N. C., Di Giulio, R. T., & Richardson, C. J. (1989). Assay and electrophoresis of superoxide dismutase from red spruce (*Picea rubens* Sarg.), Loblolly pine (*Pinus taeda* L.), and Scotch pine (*Pinus sylvestris* L.). A method for biomonitoring. *Plant Physiology*, *90*(2), 742–748. doi:10.1104/pp.90.2.742 PMID:16666837

Tang, C. S., & Veelenturf, L. P. (2019). The strategic role of logistics in the Industry 4.0 era. *Transportation Research Part E, Logistics and Transportation Review*, *129*, 1–11. doi:10.1016/j.tre.2019.06.004

Tao, X. (2017). *A double-microbial fuel cell heavy metals toxicity sensor*. 2nd International conference on environmental science and energy engineering (ICESEE), Beijing, China.

Tapparello, C., Abdulai, J. D., Katsriku, F. A., Heinzelman, W., & Adu-Manu, K. S. (2017). Water quality monitoring using wireless sensor networks. *ACM Transactions on Sensor Networks*, *13*, 1–41.

Tausz, M., Weidner, W., Wonisch, A., De Kok, L. J., & Grill, D. (1998). Uptake and distribution of ^{35}S-sulfate in needles and root of spruce seedlings as affected by exposure to SO_2 and H_2S. *Environmental and Experimental Botany*, *50*(3), 211–220. doi:10.1016/S0098-8472(03)00025-X

Tetyana, P., Shumbula, P. M., & Njengele-Tetyana, Z. 2021. Biosensors: design, development and applications. In Nanopores. IntechOpen. doi:10.5772/intechopen.97576

Thakur, R.K., Jindal, R., Singh, U.B., & Ahluwalia, A.S. (2013). Plankton diversity and water quality assessment of three freshwater lakes of Mandi (Himachal Pradesh, India) with special reference to planktonic indicators. *Environ Monit Assess*. *185*(10), 8355–8373.

Thakur, A., & Kumar, A. (2022). Recent advances on rapid detection and remediation of environmental pollutants utilizing nanomaterials-based (bio) sensors. *The Science of the Total Environment*, *834*, 155219. doi:10.1016/j.scitotenv.2022.155219 PMID:35421493

Thavarajah, W., Verosloff, M. S., Jung, J., Alam, K. K., Miller, J. D., Jewett, M. C., Young, S. L., & Lucks, J. B. (2020). A primer on emerging field-deployable synthetic biology tools for global water quality monitoring. *NPJ Clean Water*, *3*(1), 1–10. doi:10.1038/s41545-020-0064-8 PMID:34267944

Theissler, A., Pérez-Velázquez, J., Kettelgerdes, M., & Elger, G. (2021). Predictive maintenance enabled by machine learning: Use cases and challenges in the automotive industry. *Reliability Engineering & System Safety*, *215*, 107864. doi:10.1016/j.ress.2021.107864

Themistocleous, K. (2017, October). The use of UAVs for monitoring land degradation. In *Earth resources and environmental remote sensing/GIS applications VIII* (Vol. 10428, pp. 127–137). SPIE. doi:10.1117/12.2279512

Theofanis, P. L., Christos, C. A., Christos, G. P., & Marios, M. P. (2014). A Low-Cost Sensor Network for Real-Time Monitoring and Contamination Detection in Drinking Water Distribution Systems. *IEEE Ssensors J*, *14*(8), 2014.

Thinagaran, P., Nasir, S., & Leong, C. Y. (2015) Internet of Things (IoT) enabled water monitoring system. In: *4th IEEE global conference on consumer electronics*. IEEE.

Thomas, D., Wilkie, E., & Irvine, J. (2016). Comparison of Power Consumption of Wi-Fi Inbuilt Internet of Things Device with Bluetooth Low Energy. *Intl J Comput Electrical Automation Control Inf Eng*, *10*(10), 1837–1840.

Thomas, M. D. (1961). Effect of air pollution on plants. *Monogr. Wld. Hlth. Org*, *49*, 223–278.

Thompson, M. P., & Calkin, D. E. (2011). Uncertainty and risk in wildland fire management: A review. *Journal of Environmental Management*, *92*(8), 1895–1909. doi:10.1016/j.jenvman.2011.03.015 PMID:21489684

Thompson, M. P., Dunn, C. J., & Calkin, D. E. (2016). *Systems thinking and wildland fire management.* In *60th Annual Meeting of the ISSS-2016 Boulder*, CO, USA.

Thompson, M. P., MacGregor, D. G., Dunn, C. J., Calkin, D. E., & Phipps, J. (2018). Rethinking the wildland fire management system. *Journal of Forestry*, *116*(4), 382–390. doi:10.1093/jofore/fvy020

Thompson, M. P., Wei, Y., Dunn, C. J., & O'Connor, C. D. (2019). A system dynamics model examining alternative wildfire response policies. *Systems*, *7*(4), 49. doi:10.3390/systems7040049

Thompson, S., & Capon, A. (2012). Designing For Health. *Landscape Architecture and Art*, (134), 26–26. PMID:22920806

Thushari, G. G. N., & Senevirathna, J. D. M. (2020). Plastic pollution in the marine environment. *Heliyon*, *6*(8), e04709. doi:10.1016/j.heliyon.2020.e04709 PMID:32923712

Tian, T., Chen, X., Li, H., Wang, Y., Guo, L., & Jiang, L. (2013). Amidine-based fluorescent chemosensor with high applicability for detection of CO_2: A facile way to "see" CO_2. *Analyst*, *138*(4), 991–994. doi:10.1039/C2AN36401H PMID:23259156

Tingey, D. T., Reinert, R. A., Dunning, J. A., & Heck, W. W. (1971). Vegetation injury from the interactions of NO_2 and SO_2. *Phytopathology*, *61*, 1506–1511. doi:10.1094/Phyto-61-1506

Tiwari, S., Agrawal, M., & Marshall, F. M. (2006). Evaluation of ambient air pollution impact on carrot plants at a suburban site using open top chamber. *Environmental Monitoring and Assessment*, *119*(1-3), 15–30. doi:10.1007/s10661-005-9001-z PMID:16736274

Tkac, J., Vostiar, I., Gorton, L., Gemeiner, P., & Sturdik, E. (2003). Improved selectivity of microbial biosensor using membrane coating. Application to the analysis of ethanol during fermentation. *Biosensors & Bioelectronics*, *18*(9), 1125–1134. doi:10.1016/S0956-5663(02)00244-0 PMID:12788555

Tmusic, G., Manfreda, S., Aasen, H., James, M. R., Gonçalves, G., Ben-Dor, E., Brook, A., Polinova, M., Arranz, J. J., Mészáros, J., Zhuang, R., Johansen, K., Malbeteau, Y., de Lima, I. P., Davids, C., Herban, S., & McCabe, M. F. (2020). Current practices in UAS-based environmental monitoring. *Remote Sensing (Basel)*, *12*(6), 1001. doi:10.3390/rs12061001

Tobiszewski, M., & Vakh, C. (2023). Analytical applications of smartphones for agricultural soil analysis. *Analytical and Bioanalytical Chemistry*, *415*(18), 3703–3715. doi:10.1007/s00216-023-04558-1 PMID:36790460

Tomas-Barberan, F., Msonthi, J., & Hostettmann, K. (1988). Antifungal epicuticular methylated flavonoids from *Helichrysum nitens*. *Phytochemistry*, *27*(3), 753–755. doi:10.1016/0031-9422(88)84087-1

Tomlinson, H., & Rich, S. (1970). Lipid peroxidation, as a result of injury in bean leaves exposed to ozone. *Phytopathology*, *60*(10), 1531–1532. doi:10.1094/Phyto-60-1531 PMID:5481395

Tomoaki, K., Masashi, M., Akihiro, M., Akihiro, M., & Sang, L. (2016) A wireless sensor network platform for water quality monitoring. In: *IEEE Sensors.*

Topp, S. N., Pavelsky, T. M., Jensen, D., Simard, M., & Ross, M. R. (2020). Research trends in the use of remote sensing for inland water quality science: Moving towards multidisciplinary applications. *Water (Basel)*, *12*(1), 169. doi:10.3390/w12010169

Torssell, K. B. G. (1981). Natural product chemistry. In: A mechanistic and biosynthetic approach to secondary metabolism. J. Wiley and Sons Publishers, New-York.

Toth, C., & Jóźków, G. (2016). Remote sensing platforms and sensors: A survey. *ISPRS Journal of Photogrammetry and Remote Sensing*, *115*, 22–36. doi:10.1016/j.isprsjprs.2015.10.004

Trick, C. G., Bill, B. D., Cochlan, W. P., Wells, M. L., Trainer, V. L., & Pickell, L. D. (2010). Iron enrichment stimulates toxic diatom production in high- nitrate, low-chlorophyll areas. *Proceedings of the National Academy of Sciences of the United States of America*, *107*(13), 5887–5892. doi:10.1073/pnas.0910579107 PMID:20231473

Tripathi, A., Tripathi, D. S., & Prakash, V. (1999). Phyto monitoring and NO_x pollution around silver refineries. *Environment International*, *25*(4), 403–410. doi:10.1016/S0160-4120(99)00004-5

Tripp, R. A., Dluhy, R. A., & Zhao, Y. (2008). Novel nanostructures for SERS biosensing. *Nano Today*, *3*(3), 31–37. doi:10.1016/S1748-0132(08)70042-2

Tschmelak, J., Proll, G., & Gauglitz, G. (2005). Optical biosensor for pharmaceuticals, antibiotics, hormones, endocrine disrupting chemicals and pesticides in water: Assay optimization process for estrone as example. *Talanta*, *65*(2), 313–323. doi:10.1016/j.talanta.2004.07.011 PMID:18969801

Tsopela, A., Laborde, A., Salvagnac, L., Ventalon, V., Bedel-Pereira, E., Séguy, I., Temple-Boyer, P., Juneau, P., Izquierdo, R., & Launay, J. (2016). Development of a lab-on-chip electrochemical biosensor for water quality analysis based on microalgal photosynthesis. *Biosensors & Bioelectronics*, *79*, 568–573. doi:10.1016/j.bios.2015.12.050 PMID:26749098

Turner, M.G. & Dale, V.H. (1998). Comparing large and infrequent disturbances: what have we learned? *Ecosystems, 1*(6), 493-496.

Tyagi, D., Wang, H., Huang, W., Hu, L., Tang, Y., Guo, Z., Ouyang, Z., & Zhang, H. (2020). Recent advances in two-dimensional-material-based sensing technology toward health and environmental monitoring applications. *Nanoscale*, *12*(6), 3535–3559. doi:10.1039/C9NR10178K PMID:32003390

Tyszczuk-Rotko, K., Kozak, J., & Czech, B. (2022). Screen-Printed Voltammetric Sensors—Tools for Environmental Water Monitoring of Painkillers. *Sensors (Basel)*, *22*(7), 2437. doi:10.3390/s22072437 PMID:35408052

Ubidots. (2017). *IoT platform*. Ubidots. https://ubidots.com/.

Ullo, S. L., & Sinha, G. R. (2020). Advances in smart environment monitoring systems using IoT and sensors. *Sensors (Basel)*, *20*(11), 3113. doi:10.3390/s20113113 PMID:32486411

Usman, F., Ghazali, K. H., Muda, R., Dennis, J. O., Ibnaouf, K. H., Aldaghri, O. A., & Jose, R. (2023). *Detection of Kidney Complications Relevant Concentrations of Ammonia Gas Using Plasmonic Biosensors: A Review*. Chemosensors. MDPI. doi:10.3390/chemosensors11020119

Uttah, E.C. (2008). Bio-survey of plankton as indicators of water quality for recreational activities in Calabar River, Nigeria. *J Appl. Sci Environ Manage. 12*(2), 35–42.

Van Der Meer, J. R., & Belkin, S. (2010). Where microbiology meets microengineering: Design and applications of reporter bacteria. *Nature Reviews. Microbiology*, *8*(7), 511–522. doi:10.1038/nrmicro2392 PMID:20514043

Vasconcelos, T. (1999). Aquatic plants in the rice fields of the Tagus valley, Portugal. *Hydrobiologia, 415*, 59-65.

Velusamy, K., Periyasamy, S., Kumar, P. S., Rangasamy, G., Nisha Pauline, J. M., Ramaraju, P., & Nguyen Vo, D. V. (2022). Biosensor for heavy metals detection in wastewater: A review. *Food and Chemical Toxicology*, *168*, 113307. doi:10.1016/j.fct.2022.113307 PMID:35917955

Verma, P., & Ratan, J. K. (2020). Assessment of the negative effects of various inorganic water pollutants on the biosphere—An overview. *Inorganic Pollutants in Water*, 73–96.

Verma, A., Kumar, P., & Yadav, B. C. (2024). Fundamentals of electrical gas sensors. *Complex and Composite Metal Oxides for Gas VOC and Humidity Sensors*, *1*, 27–50. doi:10.1016/B978-0-323-95385-6.00004-0

Verma, A., & Singh, S. N. (2006). Biochemical and ultra structural changes in plant foliage exposed to automobile pollution. *Environmental Monitoring and Assessment*, *120*(1-3), 585–602. doi:10.1007/s10661-005-9105-5 PMID:16758287

Verma, N., & Bhardwaj, A. (2015). Biosensor technology for pesticides-a review. *Applied Biochemistry and Biotechnology*, *175*(6), 3093–3119. doi:10.1007/s12010-015-1489-2 PMID:25595494

Verma, N., Sharma, R., & Kumar, S. (2016). Advancement towards microfluidic approach to develop economical disposable optical biosensor for lead detection.

Verstricht, J., Li, X. L., Leonard, D., & Van Geet, M. (2022). Assessment of sensor performance in the context of geological radwaste disposal—A first case study in the Belgian URL HADES. *Geomechanics for Energy and the Environment*, *32*, 100296. doi:10.1016/j.gete.2021.100296

Vick-Majors, T.J., Michaud, A.B., Skidmore, M.L., Turetta, C., Barbante, C., Christner, B.C., Dore, J.E., Christianson, K., Mitchell, A.C., Achberger, A.M. and Mikucki, J.A., 2020. Biogeochemical connectivity between freshwater ecosystems beneath the West Antarctic Ice Sheet and the sub-ice marine environment. *Global Biogeochemical Cycles, 34*(3).

Vigneshvar, S., Sudhakumari, C. C., Senthilkumaran, B., & Prakash, H. (2016). Recent Advances in Biosensor Technology for Potential Applications-An Overview. *Frontiers in Bioengineering and Biotechnology*, *4*(11), 1–9. doi:10.3389/fbioe.2016.00011 PMID:26909346

Vijayakumar, N., & Ramya, R. (2015) The real time monitoring of water quality in IoT environment. In: international conference on circuits, power and computing technologies. IEEE.

Vikesland, P. J. (2018). Nanosensors for water quality monitoring. *Nature Nanotechnology*, *13*(8), 8. doi:10.1038/s41565-018-0209-9 PMID:30082808

Vilar, L., Camia, A., San-Miguel-Ayanz, J., & Martín, M. P. (2016). Modeling temporal changes in human-caused wildfires in Mediterranean Europe based on Land Use-Land Cover interfaces. *Forest Ecology and Management*, *378*, 68–78. doi:10.1016/j.foreco.2016.07.020

Villalobos, C., Love, B. A., & Olson, M. B. (2020). Ocean acidification and ocean warming effects on Pacific Herring (Clupea pallasi) early life stages. *Frontiers in Marine Science*, *7*, 597899. doi:10.3389/fmars.2020.597899

Villa, T., Gonzalez, F., Miljievic, B., Ristovski, Z., & Morawska, L. (2016). An Overview of Small Unmanned Aerial Vehicles for Air Quality Measurements: Present Applications and Future Prospectives. *Sensors (Basel)*, *16*(7), 1072. doi:10.3390/s16071072 PMID:27420065

Vinod, R., & Sushama, S. (2016) Wireless acquisition system for water quality monitoring. In: *Conference on advances in signal processing*. IEEE.

Vîrghileanu, M., Săvulescu, I., Mihai, B. A., Nistor, C., & Dobre, R. (2020). *Nitrogen Dioxide (NO2) Pollution Monitoring with Sentinel-5P Satellite Imagery over Europe during the Coronavirus Pandemic Outbreak.* MDPI. doi:10.3390/rs12213575

Viseras, A., & Garcia, R. (2019). DeepIG: Multi-robot information gathering with deep reinforcement learning. *IEEE Robotics and Automation Letters*, *4*(3), 3059–3066. doi:10.1109/LRA.2019.2924839

Vishnu, S., Ramson, S. J., Rukmini, M. S. S., & Abu-Mahfouz, A. M. (2022). Sensor-based solid waste handling systems: A survey. *Sensors (Basel)*, *22*(6), 2340. doi:10.3390/s22062340 PMID:35336511

Vitousek, P.M. (1997). Human domination of earth's ecosystem. *Science, 277*, 494–499.

Vogrinc, D., Vodovnik, M., & Marinsek-Logar, R. (2015). Microbial biosensors for environmental monitoring. *Acta Agriculturae Slovenica*, *106*(2), 67–75. doi:10.14720/aas.2015.106.2.1

Vopálenská, I., Váchová, L., & Palková, Z. (2015). New biosensor for detection of copper ions in water based on immobilized genetically modified yeast cells. *Biosensors & Bioelectronics*, *72*, 160–167. doi:10.1016/j.bios.2015.05.006 PMID:25982723

Voronova, E. A., Iliasov, P. V., & Reshetilov, A. N. (2008). Development, investigation of parameters and estimation of possibility of adaptation of *Pichia angusta* based microbial sensor for ethanol detection. *Analytical Letters*, *41*(3), 377–391. doi:10.1080/00032710701645729

Wacheux, H. (1998). *Sensors for waste water: Many needs but financial and technical limitations. Monitoring of water quality.* Elsevier. doi:10.1016/B978-008043340-0/50018-7

Wahid, A. (2006). Productivity losses in barley attributed to ambient atmospheric pollutants in Pakistan. *Atmospheric Environment*, *40*(28), 5342–5354. doi:10.1016/j.atmosenv.2006.04.050

Wahid, A., Maggs, R., Shasmi, S. R. A., Bell, J. N. B., & Ashmore, M. R. (1995). Air pollution and its impact on wheat yield in the Pakistan Panjab. *Environmental Pollution*, *88*(2), 147–154. doi:10.1016/0269-7491(95)91438-Q PMID:15091554

Wali, B., Iqbal, M., & Mahmooduzzafar. (2007). Anatomical and functional responses of *Calendula officinalis* L. to SO_2 stress as observed at different stages of plant development. *Flora (Jena)*, *202*(4), 268–280. doi:10.1016/j.flora.2006.08.002

Waller, G. R., & Nowacki, E. K. (1979). *Alkaloid Biology and Metabolism in Plants*. Plenum Press.

Walsh, G.E. (1978). Toxic effects of pollutants on plankton. *Principles of ecotoxicology.* Wiley.

Wang, C., Baumann, Z., Madigan, D. J., & Fisher, N. S. (2016). Contaminated marine sediments as a source of cesium radioisotopes for benthic fauna near Fukushima. *Environ. Sci. Technol. 50,* 10448–10455. doi: . 6b02984 doi:10.1021/acs.est

Wang, C., Madiyar, F., Yu, C., & Li, J. (2017). Detection of extremely low concentration waterborne pathogen using a multiplexing self-referencing SERS microfluidic biosensor. *Journal of Biological Engineering*, *11*(1), 9. doi:10.1186/s13036-017-0051-x PMID:28289439

Wang, C., Qin, J., Qu, C., Ran, X., Liu, C., & Chen, B. (2021). A smart municipal waste management system based on deep-learning and Internet of Things. *Waste Management (New York, N.Y.)*, *135*, 20–29. doi:10.1016/j.wasman.2021.08.028 PMID:34461487

Wang, H., Chen, D., Zhang, Y., Liu, P., Shi, J., Feng, X., Tong, B., & Dong, Y. (2015). A fluorescent probe with an aggregation enhanced emission feature for real-time monitoring of low carbon dioxide levels. *Journal of Materials Chemistry. C, Materials for Optical and Electronic Devices*, *3*(29), 7621–7626. doi:10.1039/C5TC01280E

Wang, J., Liu, W., Chen, D., Xu, Y., & Zhang, L.-y. (2014). A micro-machined thin film electro-acoustic biosensor for detection of pesticide residuals. *Journal of Zhejiang University SCIENCE C*, *15*(5), 383–389. doi:10.1631/jzus.C1300289

Wang, J., Liu, Z., Zhou, Y., Zhu, S., Gao, C., Yan, X., Wei, K., Gao, Q., Ding, C., Luo, T., & Yang, R. (2023). A multifunctional sensor for real-time monitoring and pro-healing of frostbite wounds. *Acta Biomaterialia*, *172*, 330–342. doi:10.1016/j.actbio.2023.10.003 PMID:37806374

Wang, J., Xu, D., Kawde, A. N., & Polsky, R. (2001). Metal nanoparticle-based electrochemical stripping potentiometric detection of DNA hybridization. *Analytical Chemistry*, *73*(22), 5576–5581. doi:10.1021/ac0107148 PMID:11816590

Wang, R., Zhang, M., Guan, Y., Chen, M., & Zhang, Y. (2019). A CO_2-responsive hydrogel film for optical sensing of dissolved CO_2. *Soft Matter*, *15*(30), 6107–6115. doi:10.1039/C9SM00958B PMID:31282902

Wang, W.-X. (2013). Prediction of metal toxicity in aquatic organisms. *Chinese Science Bulletin*, *58*(2), 194–202. doi:10.1007/s11434-012-5403-9

Wang, W.-X., & Fisher, N. S. (1999). Delineating metal accumulation pathways for marine invertebrates. *The Science of the Total Environment*, *237-238*, 459–472. doi:10.1016/S0048-9697(99)00158-8

Wang, Y., Li, J., Jing, H., Zhang, Q., Jiang, J., & Biswas, P. (2015). Laboratory evaluation and calibration of three low-cost particle sensors for particulate matter measurement. *Aerosol Science and Technology*, *49*(11), 1063–1077. doi:10.1080/02786826.2015.1100710

Wang, Y., Mukherjee, M., Wu, D., & Wu, X. (2016). Combating river pollution in China and India: Policy measures and governance challenges. *Water Policy*, *18*(S1), 122–137. doi:10.2166/wp.2016.008

Wang, Z., & Menenti, M. (2021). Challenges and opportunities in Lidar remote sensing. *Frontiers in Remote Sensing*, *2*, 641723. doi:10.3389/frsen.2021.641723

Ward, D. S., Shevliakova, E., Malyshev, S., & Rabin, S. (2018). Trends and variability of global fire emissions due to historical anthropogenic activities. *Global Biogeochemical Cycles*, *32*(1), 122–142. doi:10.1002/2017GB005787

Water Resource Information System of India. (2017). http://www.india-wris.nrsc.gov.in/wrpinfo/index.php?title=River_Water_Quality_Monitoring. Accessed 2 May 2017

Waterman, P. G., & Mole, S. (1994). *Analysis of phenolic plant metabolites*. Blackwell Scientific Publ.

Wauchope, R. (1978). The pesticide content of surface water draining from agricultural fields—A review. *Journal of Environmental Quality*, *7*(4), 459–472. doi:10.2134/jeq1978.00472425000700040001x

Webber, A. N., Nie, G. Y., & Long, S. P. (1994). Acclimation of photosynthetic proteins to rising atmospheric CO_2. *Photosynthesis Research*, *39*(3), 413–420. doi:10.1007/BF00014595 PMID:24311133

Weber, P., Vogler, J., & Gauglitz, G. (2017). Development of an Optical Biosensor for the Detection of Antibiotics in the Environment. *Proceedings of the Society for Photo-Instrumentation Engineers*, *10231*, 102312L. doi:10.1117/12.2267467

Weber, R., Watson, A., Forter, M., & Oliaei, F. (2011). Persistent organic pollutants and landfills-a review of past experiences and future challenges. *Waste Management & Research*, *29*(1), 107–121. doi:10.1177/0734242X10390730 PMID:21224404

Wei, D., Liu, P., & Lu, B. (2012). Water Quality Automatic Monitoring System Based on GPRS Data Communications. In: *International conference on modern hydraulic engineering*. IEEE.

Wellburn, A. R., Capron, T. M., Chan, H. S., & Horsman, D. C. 1976. Biochemical effects of atmospheric pollutants on plants. In: Mansfield, T.A. (Ed.), Effects of air pollutants on plants. Cambridge University Press.

Wellburn, A. R., Majernik, O., & Wellburn, F. M. N. (1972). Effect of SO_2 and NO_2 polluted air upon the ultrastructure of chloroplasts. *Environ. Pollut*, *3*(1), 37–49. doi:10.1016/0013-9327(72)90016-X

Wencel, D., Abel, T., & McDonagh, C. (2014). Optical Chemical pH Sensors. *Analytical Chemistry*, *86*(1), 15–29. doi:10.1021/ac4035168 PMID:24180284

Wencel, D., Moore, J., Stevenson, N., & McDonagh, C. (2010). Ratiometric fluorescence-based dissolved carbon dioxide sensor for use in environmental monitoring applications. *Analytical and Bioanalytical Chemistry*, *398*(5), 1899–1907. doi:10.1007/s00216-010-4165-y PMID:20827465

Wen, F., He, T., Liu, H., Chen, H. Y., Zhang, T., & Lee, C. (2020). Advances in chemical sensing technology for enabling the next-generation self-sustainable integrated wearable system in the IoT era. *Nano Energy*, *78*, 105155. doi:10.1016/j.nanoen.2020.105155

Whitt, C., Pearlman, J., Polagye, B., Caimi, F., Muller-Karger, F., Copping, A., Spence, H., Madhusudhana, S., Kirkwood, W., Grosjean, L., Fiaz, B. M., Singh, S., Singh, S., Manalang, D., Gupta, A. S., Maguer, A., Buck, J. J. H., Marouchos, A., Atmanand, M. A., & Khalsa, S. J. (2020). Future vision for autonomous ocean observations. *Frontiers in Marine Science*, *7*, 697. doi:10.3389/fmars.2020.00697

WHO. (2008). *Health Situation in the South –East Asia Region* 2001 – 2007. WHO., (2018). *World Health Report*, 2018.

Williams, A. P., Abatzoglou, J. T., Gershunov, A., Guzman-Morales, J., Bishop, D. A., Balch, J. K., & Lettenmaier, D. P. (2019). Observed impacts of anthropogenic climate change on wildfire in California. *Earth's Future*, *7*(8), 892–910. doi:10.1029/2019EF001210

Willner, M. R., & Vikesland, P. J. (2018). Nanomaterial-enabled sensors for environmental contaminants. *Journal of Nanobiotechnology*, *16*(1), 95. doi:10.1186/s12951-018-0419-1 PMID:30466465

Winther, J. G., Dai, M., Rist, T., Hoel, A. H., Li, Y., Trice, A., Morrissey, K., Juinio-Meñez, M. A., Fernandes, L., Unger, S., Scarano, F. R., Halpin, P., & Whitehouse, S. (2020). Integrated ocean management for a sustainable ocean economy. *Nature Ecology & Evolution*, *4*(11), 1451–1458. doi:10.1038/s41559-020-1259-6 PMID:32807947

Wiśniewska, M., & Szyłak-Szydłowski, M. (2022). The Application of In Situ Methods to Monitor VOC Concentrations in Urban Areas—A Bibliometric Analysis and Measuring Solution Review. *Sustainability, 14*(14), 8815. doi:10.3390/su14148815

Wohlgemuth, H., Mittelstrass, K., Kschieschan, S., Bender, J., Weigel, H. J., Overmyer, K., Kangasjarvi, J., Sandermann, H., & Langebartels, C. (2002). Activation of an oxidative burst is a general feature of sensitive plants exposed to the air pollutant ozone. *Plant, Cell & Environment*, *25*(6), 717–726. doi:10.1046/j.1365-3040.2002.00859.x

Woldu, A. (2022). Biosensors and its applications in Water Quality Monitoring. *International Journal of Scientific and Engineering Research*, *13*(5), 12–29. Retrieved January 8th, 2024, from https://www.ijser.org/researchpaper/Biosensors-and-its-applications-in-Water-Quality-Monitoring.pdf

World Bank. (2023). *Urban Development Report 2023*. World Bank Group.

Wu, H., Li, J. H., Yang, W. C., Wen, T., He, J., Gao, Y. Y., ... & Yang, W. C. (2023). Nonmetal-doped quantum dot-based fluorescence sensing facilitates the monitoring of environmental contaminants. *Trends in Environmental Analytical Chemistry*, e00218.

Wu, M., Hou, S., Yu, X., & Yu, J. (2020). Recent progress in chemical gas sensors based on organic thin film transistors. *Journal of Materials Chemistry. C, Materials for Optical and Electronic Devices*, *8*(39), 13482–13500. doi:10.1039/D0TC03132A

Xia, G., Liu, Y., Ye, B., Sun, J., & Wang, H. (2015). A Squaraine-based Colorimetric and F^- Dependent Chemosensor for Recyclable CO_2 Gas Detection: Highly Sensitive Off-On-Off Response. *Chemical Communications*, *51*(72), 13802–13805. doi:10.1039/C5CC04755B PMID:26235137

Xiang, Y., Lv, L., Chai, W., Zhang, T., Liu, J., & Liu, W. (2020). Using Lidar technology to assess regional air pollution and improve estimates of PM2. 5 transport in the North China Plain. *Environmental Research Letters*, *15*(9), 094071. doi:10.1088/1748-9326/ab9cfd

Xiang, Y., Xie, M., Bash, R., Chen, J. J. L., & Wang, J. (2007). Ultrasensitive label-free aptamer-based electronic detection. *Angewandte Chemie International Edition*, *46*(47), 9054–9056. doi:10.1002/anie.200703242 PMID:17957666

Xie, M., Zhao, F., Zhang, Y., Xiong, Y., & Han, S. (2022). *Recent advances in aptamer-based optical and electrochemical biosensors for detection of pesticides and veterinary drugs. Food Control*. Elsevier Ltd. doi:10.1016/j.foodcont.2021.108399

Xin W, Longquan M, & Huizhong Y (2011). Online Water Monitoring System Based on ZigBee and GPRS. *Adv Control Eng Inform Sci.*

Xiuli, M., Xo, H., Shuiyuan, X., Qiong, L., Qiong, L., & Chunhua, T. (2011) Continuous, online monitoring and analysis in large water distribution networks. In: *International Conference on Data Engineering*. IEEE.

Xiuna, Z., Daoliang, L., Dongxian, H., Jianqin, W., Daokun, M., & Feifei, L. (2010). A remote wireless system for water quality online monitoring in intensive fish culture. *Computers Electronics Agriculture.*

Xue, C. S., Erika, G., & Jiří, H. (2019). Surface Plasmon Resonance Biosensor for the Ultrasensitive Detection of Bisphenol A. *Analytical and Bioanalytical Chemistry*, *411*(22), 5655–5658. doi:10.1007/s00216-019-01996-8 PMID:31254055

Xu, M., & Dong, J. (2005). Nitrogen oxide stimulates indole alkaloid production in *Catharanthus roseus* cell suspemsion cultures through a protein kinase-dependent signal pathway. *Enzyme and Microbial Technology*, *37*(1), 49–53. doi:10.1016/j.enzmictec.2005.01.036

Xu, X., Nie, S., Ding, H., & Hou, F. F. (2018). Environmental pollution and kidney diseases. *Nature Reviews. Nephrology*, *14*(5), 313–324. doi:10.1038/nrneph.2018.11 PMID:29479079

Xu, Z., & Coors, V. (2012). Combining system dynamics model, GIS, and 3D visualization in sustainability assessment of urban residential development. *Building and Environment*, *47*, 272–287. doi:10.1016/j.buildenv.2011.07.012

Xu, Z., Kim, S. K., & Yoon, J. (2010). Revisit to imidazolium receptors for the recognition of anions: Highlighted research during 2006–2009. *Chemical Society Reviews*, *39*(5), 1457–1466. doi:10.1039/b918937h PMID:20419201

Yadav, V., Duren, R., Mueller, K., Verhulst, K. R., Nehrkorn, T., Kim, J., Weiss, R. F., Keeling, R., Sander, S., Fischer, M. L., Newman, S., Falk, M., Kuwayama, T., Hopkins, F., Rafiq, T., Whetstone, J., & Miller, C. (2019). Spatio-temporally resolved methane fluxes from the Los Angeles Megacity. *Journal of Geophysical Research. Atmospheres*, *124*(9), 5131–5148. doi:10.1029/2018JD030062

Yagi, N. (2016). Impacts of the nuclear power plant accident and the start of trial operations in fukushima fisheries. In T. M. Nakanishi & K. Tanoi (Eds.), Agricultural Implications of the Fukushima Nuclear Accident: The First Three Years (pp. 217–228). Springer Japan. doi:10.1007/978-4-431-55828-6_17

Yagi, K. (2007). Applications of whole-cell bacterial sensors in biotechnology and environmental science. *Applied Microbiology and Biotechnology*, *73*(6), 1251–1258. doi:10.1007/s00253-006-0718-6 PMID:17111136

Yamashita, T., Ookawa, N., Ishida, M., Kanamori, H., Sasaki, H., Katayose, Y., & Yokoyama, H. (2016). A novel open-type biosensor for the in-situ monitoring of biochemical oxygen demand in an aerobic environment. *Scientific Reports*, *6*(1), 38552. doi:10.1038/srep38552 PMID:27917947

Yandouzi, M., Grari, M., Idrissi, I., , Moussaoui, O., Azizi, M.,, Ghoumid, K. A. M. A. L., & Elmiad, A. K. (2022). Review on forest fire detection and prediction using deep learning and drones. *Journal of Theoretical and Applied Information Technology*, *100*(12), 4565–4576.

Yang, H., Ma, M., Thompson, J. R., & Flower, R. J. (2017). Waste management, informal recycling, environmental pollution and public health. *J Epidemiol Community Health*. jech.bmj.com/content/72/3/237

Yang, G. L., Jiang, X. L., Xu, H., & Zhao, B. (2021). Applications of MOFs as luminescent sensors for environmental pollutants. *Small*, *17*(22), 2005327. doi:10.1002/smll.202005327 PMID:33634574

Yang, H., Kong, J., Hu, H., Du, Y., Gao, M., & Chen, F. (2022). A review of Remote sensing for water quality Retrieval: Progress and challenges. *Remote Sensing (Basel)*, *14*(8), 1770. doi:10.3390/rs14081770

Yang, L. H., & Wang, H. M. (2014). Recent advances in carbon dioxide capture, fixation, and activation by using n-heterocyclic carbenes. *ChemSusChem*, *7*(4), 962–998. doi:10.1002/cssc.201301131 PMID:24644039

Yang, L., Han, D. H., Lee, B.-M., & Hur, J. (2015). Characterizing treated wastewaters of different industries using clustered fluorescence EEM–PARAFAC and FT-IR spectroscopy: Implications for downstream impact and source identification. *Chemosphere*, *127*, 222–228. doi:10.1016/j.chemosphere.2015.02.028 PMID:25746920

Yang, S., Sarkar, S., Xie, X., Li, D., & Chen, J. (2023). Application of optical hydrogels in environmental sensing. *Energy & Environmental Materials*, 12646. doi:10.1002/eem2.12646

Yang, Y., Creedon, N., O'Riordan, A., & Lovera, P. (2021). Surface enhanced Raman Spectroscopy: Applications in agriculture and food safety. *Photonics*, *8*(12), 568. doi:10.3390/photonics8120568

Yang, Z., Suzuki, H., Sasaki, S., & Karube, I. (1996). Disposable sensor for biochemical oxygen demand. *Applied Microbiology and Biotechnology*, *46*(1), 10–14. doi:10.1007/s002530050776 PMID:8987529

Yang, Z., Wei, C., Song, X., Liu, X., Tang, Z., Liu, P., & Wei, Y. (2023). Thermal conductive heating coupled with in situ chemical oxidation for soil and groundwater remediation: A quantitative assessment for sustainability. *Journal of Cleaner Production*, *423*, 138732. doi:10.1016/j.jclepro.2023.138732

Yao, N., Wang, J., Zhou, Y., (2014). Rapid determination of the chemical oxygen demand of water using a thermal biosensor. *Sensors (Basel)*, *14*(6), 9949–9960.

Yarimizu, K., Cruz-López, R., & Carrano, C. J. (2018). Iron and harmful algae blooms: Potential algal-bacterial mutualism between Lingulodinium polyedrum and Marinobacter algicola. *Frontiers in Marine Science*, *5*, 180. doi:10.3389/fmars.2018.00180

Yearworth, M. (2014). A brief introduction to system dynamics modeling. University of Bristol.

Yekeen, S., Balogun, A., & Aina, Y. (2020). Early warning systems and geospatial tools: managing disasters for urban sustainability. In *Sustainable Cities and Communities* (pp. 129–141). Springer International Publishing., doi:10.1007/978-3-319-95717-3_103

Yemini, M., Levi, Y., Yagil, E., & Rishpon, J. (2007). Specific electrochemical phage sensing for Bacillus cereus and Mycobacterium smegmatis. *Bioelectrochemistry (Amsterdam, Netherlands)*, *70*(1), 180–184. doi:10.1016/j.bioelechem.2006.03.014 PMID:16725377

Yin, F. (2022). Practice of air environment quality monitoring data visualization technology based on adaptive wireless sensor networks. *Proceedings of the ... International Wireless Communications & Mobile Computing Conference / Association for Computing Machinery. International Wireless Communications & Mobile Computing Conference*, *2022*, 1–12. 10.1155/2022/4160186

Yin, H., Cao, Y., Marelli, B., Zeng, X., Mason, A. J., & Cao, C. (2021). Soil sensors and plant wearables for smart and precision agriculture. *Advanced Materials*, *33*(20), 2007764. doi:10.1002/adma.202007764 PMID:33829545

Yoneyama, T., Kim, H. Y., Morikawa, H., & Srivastava, H. S. (2002). Metabolism and detoxification of nitrogen dioxide and ammonia in plants. In K. Omasa, H. Saji, S. Youssefian, & N. Kondo (Eds.), *Air pollution and plant biotechnology* (pp. 221–234). Springer-Verlag. doi:10.1007/978-4-431-68388-9_11

Yoneyma, T., Totsuka, T., Hayakaya, N., & Yazaki, J. (1980). Absorption of atmospheric NO_2 by plants and soil. V. Day and night NO_2 fumigation effect on the plant growth and the estimation of the amount of NO_2^- nitrogen absorbed by plants. *Res. Rep. Natl. Inst. Environ. Studies*, *11*, 31–50.

Yu, D., Zhai, J., Liu, C., Zhang, X., Bai, L., Wang, Y., & Dong, S. (2017). Small microbial three-electrode cell based biosensor for online detection of acute water toxicity. *ACS Sensors*, *2*(11), 1637–1643.

Yuan, C., Liu, Z., & Zhang, Y. (2017, June). Fire detection using infrared images for UAV-based forest fire surveillance. In *2017 International Conference on Unmanned Aircraft Systems (ICUAS)* (pp. 567-572). IEEE. 10.1109/ICUAS.2017.7991306

Yuan, C., Wang, Y., & Liu, J. (2022). Research on multi-sensor fusion-based AGV positioning and navigation technology in storage environment. *Journal of Physics: Conference Series*, *2378*(1), 012052. doi:10.1088/1742-6596/2378/1/012052

Yuan, Y., Jia, H., Xu, D., & Wang, J. (2023). Novel method in emerging environmental contaminants detection: Fiber optic sensors based on microfluidic chips. *The Science of the Total Environment*, *857*, 159563. doi:10.1016/j.scitotenv.2022.159563 PMID:36265627

Yunus, M., Ahmad, K. J., & Gale, R. (1979). Air pollutants and epidermal traits in *Ricinus communis* L. *Environ. Pollut, 20*(3), 189–198. doi:10.1016/0013-9327(79)90004-1

Yunus, M., Kulshreshtha, K., Dwivedi, A. K., & Ahmad, K. J. (1982). Leaf surface traits of *Ipomea fistula* Mart. Ex. Choisy and indicators of air pollution. *New Botanist, 9*, 39–45.

Yusuf, A., Sodiq, A., Giwa, A., Eke, J., Pikuda, O., Eniola, J. O., Ajiwokewu, B., Sambudi, N. S., & Bilad, M. R. (2022). Updated review on microplastics in water, their occurrence, detection, measurement, environmental pollution, and the need for regulatory standards. *Environmental Pollution, 292*, 118421. doi:10.1016/j.envpol.2021.118421 PMID:34756874

Zaghloul, A., Saber, M., & Abd-El-Hady, M. (2019). Physical indicators for pollution detection in terrestrial and aquatic ecosystems. *Bulletin of the National Research Center, 43*(1), 120. doi:10.1186/s42269-019-0162-2

Zannatul, F. & Muktadir, A.K.M. (2009). A review: potentiality of zooplankton as bioindicator. *Am J Appl Sci. 6*(10), 1815–1819.

Zeevart, A. J. (1976). Some effects of fumigating plants for short periods with NO_2. *Environ Pollut, 11*(2), 97–108. doi:10.1016/0013-9327(76)90022-7

Zehani, N., Fortgang, P., Lachgar, M. S., Baraket, A., Arab, M., Dzyadevych, S. V., & Kherrat, R. (2015). Highly sensitive electrochemical biosensor for bisphenol A detection based on a diazonium-functionalized boron-doped diamond electrode modified with a multi-walled carbon nanotube-tyrosinase hybrid film. *Biosensors & Bioelectronics, 74*, 830–835. doi:10.1016/j.bios.2015.07.051 PMID:26232678

Zhai, J., Luo, B., Li, A., Dong, H., Jin, X., & Wang, X. (2022). Unlocking all-solid ion selective electrodes: Prospects in crop detection. *Sensors (Basel), 22*(15), 5541. doi:10.3390/s22155541 PMID:35898054

Zhang, C., Wang, Y., Song, X., Kubota, J., He, Y., Tojo, J., & Zhu, X. (2017). An integrated specification for the nexus of water pollution and economic growth in China: Panel cointegration, long-run causality and environmental Kuznets curve. *The Science of the Total Environment, 609*, 319–328. doi:10.1016/j.scitotenv.2017.07.107 PMID:28753507

Zhang, H., & Srinivasan, R. (2020). A Systematic Review of Air Quality Sensors, Guidelines, and Measurement Studies for Indoor Air Quality Management. *Sustainability (Basel), 12*(21), 9045. doi:10.3390/su12219045

Zhang, P., Xiao, Y., Zhang, J., Liu, B., Ma, X., & Wang, Y. (2021). Highly sensitive gas sensing platforms based on field effect Transistor-A review. *Analytica Chimica Acta, 1172*, 338575. doi:10.1016/j.aca.2021.338575 PMID:34119019

Zhang, S., Kang, P., & Meyer, T. J. (2014). Nanostructured tin catalysts for selective electrochemical reduction of carbon dioxide to formate. *Journal of the American Chemical Society, 136*(5), 1734–1737. doi:10.1021/ja4113885 PMID:24417470

Zhang, X., Xia, Q., Lai, Y., Wu, B., Tian, W., Miao, W., Feng, X., Xin, L., Miao, J., Wang, N., Wu, Q., Jiao, M., Shan, L., Du, J., Li, Y., & Shi, B. (2022). Spatial effects of air pollution on the economic burden of disease: Implications of health and environment crisis in a post-COVID-19 world. *International Journal for Equity in Health, 21*(1), 1–16. doi:10.1186/s12939-022-01774-6 PMID:36380331

Zhang, Y., Du, X., Zhai, J., & Xie, X. (2023). Tunable colorimetric carbon dioxide sensor based on ion-exchanger- and chromoionophore-doped hydrogel. *Analysis & Sensing, 3*(6), e202300032. doi:10.1002/anse.202300032

Zhang, Y., Zhu, Y., Zeng, Z., Zeng, G., Xiao, R., Wang, Y., Hu, Y., Tang, L., & Feng, C. (2021). Sensors for the environmental pollutant detection: Are we already there? *Coordination Chemistry Reviews, 431*, 213681. doi:10.1016/j.ccr.2020.213681

Zhan, S., Xu, H., Zhang, D., Xia, B., Zhan, X., Wang, L., Lv, J., & Zhou, P. (2015). Fluorescent detection of Hg2+ and Pb2+ using GeneFinder™ and an integrated functional nucleic acid. *Biosensors & Bioelectronics*, *72*, 95–99. doi:10.1016/j.bios.2015.04.021 PMID:25966463

Zhao, H. X., Liu, L. Q., Liu, Z. D., Wang, Y., Zhao, X. J., & Huang, C. Z. (2011). Highly selective detection of phosphate in very complicated matrixes with an off-on fluorescent probe of europium-adjusted carbon dots. *Chemical Communications*, *47*(9), 2604–2606. doi:10.1039/c0cc04399k PMID:21234476

Zhao, K., Veksha, A., Ge, L., & Lisak, G. (2021). Near real-time analysis of para-cresol in wastewater with a laccase-carbon nanotube-based biosensor. *Chemosphere*, *269*, 128699. doi:10.1016/j.chemosphere.2020.128699 PMID:33121813

Zhao, Y., Huang, F., Wang, W., Gao, R., Fan, L., Wang, A., & Gao, S. H. (2023). *Application of high-throughput sequencing technologies and analytical tools for pathogen detection in urban water systems: Progress and future perspectives. Science of the Total Environment*. Elsevier B.V., doi:10.1016/j.scitotenv.2023.165867

Zheng, Y., Liu, X., Wang, L., & Zhu, Y. (2020). Application of IoT technology in urban air quality monitoring. *Chinese Journal of Environmental Engineering*, *14*(3), 654–660.

Zhou, T., Han, H., Liu, P., Xiong, J., Tian, F., & Li, X. (2017). Microbial fuels cell-based biosensor for toxicity detection: A review. *Sensors (Basel)*, *17*(10), 2230. doi:10.3390/s17102230 PMID:28956857

Zhou, Y., Lin, X., Xing, Y., Zhang, X., Lee, H. K., & Huang, Z. (2023). Per-and Polyfluoroalkyl Substances in Personal Hygiene Products: The Implications for Human Exposure and Emission to the Environment. *Environmental Science & Technology*.

Zhou, Z. B., Feng, L. D., Liu, W. J., & Wu, Z. G. (2001). New approaches for developing transient electrochemical multi-component gas sensors. *Sensors and Actuators. B, Chemical*, *76*(1-3), 605–609. doi:10.1016/S0925-4005(01)00654-2

Zhu, L. Y., Ou, L. X., Mao, L. W., Wu, X. Y., Liu, Y. P., & Lu, H. L. (2023). Advances in Noble Metal-Decorated Metal Oxide Nanomaterials for Chemiresistive Gas Sensors: Overview. *Nano-Micro Letters, 15*(1), 1–75. doi:10.1007/s40820-023-01047-z

Zhua, J., Jiaa, P., Lia, N., Tanb, S., Huangb, J., & Xu, L. (2018). Small-molecule fluorescent probes for the detection of carbon dioxide. *Chinese Chemical Letters*, *29*(10), 1445–1450. doi:10.1016/j.cclet.2018.09.002

Zhu, L., Meier, D., Boger, Z., Montgomery, C., Semancik, S., & DeVoe, D. L. (2007). Integrated microfluidic gas sensor for detection of volatile organic compounds in water. *Sensors and Actuators. B, Chemical*, *121*(2), 679–688. doi:10.1016/j.snb.2006.03.023

ZiauddinG. (2021). Study of Limnological status of two selected floodplain wetlands of West Bengal. Research Square, 1-68. doi:10.21203/rs.3.rs-443104/v1

Ziebart, C., Federsel, C., Anbarasan, P., Jackstell, R., Baumann, W., Spannenberg, A., & Beller, M. (2014). Well-defined iron catalyst for improved hydrogenation of carbon dioxide and bicarbonate. *Journal of the American Chemical Society*, *134*(51), 20701–20704. doi:10.1021/ja307924a PMID:23171468

Zobel, A. M. (1996). Phenolic compounds in defense against air pollution. In M. Yunus & M. Iqbal (Eds.), *Plant Response to Air Pollution* (pp. 241–266). John Wiley.

Zong, X., Tian, X., & Yin, Y. (2020). Impacts of climate change on wildfires in Central Asia. *Forests*, *11*(8), 802. doi:10.3390/f11080802

Zou, X., Ji, Y., Li, H., Wang, Z., Shi, L., Zhang, S., Wang, T., & Gong, Z. (2021). Recent advances of environmental pollutants detection via paper-based sensing strategy. *Luminescence*, *36*(8), 1818–1836. doi:10.1002/bio.4130 PMID:34342392

Zulkifli, C. Z., Garfan, S., Talal, M., Alamoodi, A. H., Alamleh, A., Ahmaro, I. Y. Y., Sulaiman, S., Ibrahim, A. B., Zaidan, B. B., Ismail, A. R., Albahri, O. S., Albahri, A. S., Soon, C. F., Harun, N. H., & Chiang, H. H. (2022). IoT-Based Water Monitoring Systems: A Systematic Review. *Water (Basel)*, *14*(22), 22. doi:10.3390/w14223621

About the Contributors

Khursheed Ahmad Wani is Assistant Professor in the Department of Environmental Science, Government Degree College Thidim Kreeri, J&K. Dr. Wani served at ITM University Gwalior as Assistant Professor in the Department of Environmental Science from 2011 to 2017. He has completed his M. Sc. and Ph. D. in Environmental Science from Jiwaji University, Gwalior, Madhya Pradesh. He has also qualified Jammu and Kashmir State Level Eligibility Test (JKSET). His areas of interest include Environmental management, Environmental Impact Assessment and Occupational Health. He has published research papers in various national and international journals and is the author of three books on Environmental Science. He is a member of several national and international organizations and is in the editorial board of various journals. He has organized different programmes on Environmental Awareness and has completed research project sponsored by Madhya Pradesh Council of Science and Technology, Bhopal, India. Dr. Wani has organized various national and international conferences/ seminars/ workshops as well.

***

Kumud is working as Assistant Professor in the Department of computer Science. Kumud has more than 5 years of research and teaching experience. She has worked in the area of Computer Sensors. She has published research papers in peer reviewed Journals and book chapters in various books of International repute.

Naseema Banu A is a science graduate who completed her Bachelor's (2021) and Master's (2023) in Applied Microbiology from Ethiraj College for Women, Chennai, Tamil Nadu, India. She was the Editorial Secretary for the department's annual journal (2020-21). She has worked as a Student Editor for the Conference Proceedings on Integrative Approach for Conservation of Life Systems (2022-23) with ISBN 978-93-91332-50-1. She has presented a poster in the conference and attended many others as well. She is presently working as an Associate Content Editor at Clarivate Analytics Pvt. Ltd.

Pritam Kumar Barman received his M.Sc. degree in Forestry with a specialization in Forest Biology and Tree Improvement in 2022 from Sam Higginbottom University of Agriculture, Technology and Sciences, Prayagraj, India. He is presently pursuing a Ph.D. degree at the Sam Higginbottom University of Agriculture, Technology and Sciences, Prayagraj, India. He presented his research at several international conferences. He has taken part in numerous trainings and workshops nationally as well as internationally. He also attended the IIRS Academia Meet-2022. His research interests include forest genetics, tree phenology, environmental studies, remote sensing, and GIS.

Pankaj Bhambri is affiliated with the Department of Information Technology at Guru Nanak Dev Engineering College in Ludhiana. Additionally, he fulfills the role of the Institute's Coordinator for the Skill Enhancement Cell and acts as the Convener for his Departmental Board of Studies. He possesses nearly two decades of teaching experience. Dr. Bhambri acquired a Master of Technology degree in Computer Science and Engineering and a Bachelor of Engineering degree in Information Technology with Honours from I.K.G. Punjab Technical University in Jalandhar, India, and Dr. B.R. Ambedkar University in Agra, India, respectively. Dr. Bhambri obtained a Doctorate in Computer Science and Engineering from I.K.G. Punjab Technical University, located in Jalandhar, India. Over an extended period, he fulfilled many responsibilities including those of an Assistant Registrar (Academics), Member (Academic Council/BoS/DAB/RAC), Hostel Warden, APIO, and NSS Coordinator within his institution. His research work has been published in esteemed worldwide and national journals, as well as conference proceedings. Dr. Bhambri has made significant contributions to the academic field through his role as both an editor and author of various textbooks. Additionally, he has demonstrated his innovative thinking by filing several patents. Dr. Bhambri has received numerous prestigious awards from esteemed organizations in recognition of his exceptional achievements in both social and academic/ research domains. These accolades include the ISTE Best Teacher Award in 2022 and 2023, the I2OR National Award in 2020, the Green ThinkerZ Top 100 International Distinguished Educators award in 2020, the I2OR Outstanding Educator Award in 2019, the SAA Distinguished Alumni Award in 2012, the CIPS Rashtriya Rattan Award in 2008, the LCHC Best Teacher Award in 2007, and several other commendations from various government and non-profit entities. He has provided guidance and oversight for numerous research projects and dissertations at the undergraduate, postgraduate, and Ph.D. levels. He successfully organized a diverse range of educational programmes, securing financial backing from esteemed institutions such as the All India Council for Technical Education (AICTE), the Technical Education Quality Improvement Programme (TEQIP), among others. Dr. Bhambri's areas of interest encompass machine learning, bioinformatics, wireless sensor networks, and network security. Dr. Bhambri possesses a wide array of professional responsibilities, encompassing the duties of an educator, editor, author, reviewer, expert speaker, motivator, and technical committee member for esteemed national and worldwide organizations

Ankur Bhardwaj is working as Assistant professor in the Department of Biotechnology, Shri Vaishnav Institute of Sciences, Shri Vaishnav Vidyapeeth Vishwavidyalaya, Indore. He received B.Sc.[HONS] (2009) in Biotechnology from Allahabad Agricultural Institute-Deemed University, Allahabad, Uttar Pradesh; M.Sc. (2011) in Biotechnology from SRM University, Kattankulathur, Chennai, Tamil Nadu and Ph.D. (2021) in Molecular Biology and Microbiology from Mangalayatan University, Aligarh, Uttar Pradesh. He started his career as Teaching Assistant in Department of Biotechnology and Life Sciences, Mangalayatan University, Aligarh in 2014 and served there for more than 5 years. He has more than 10 years of teaching and research experience. He prominently has worked in the field of plant Biotechnology, Molecular Biology, Immuno-informatics and Microbiology. He has published 10 research papers in peer reviewed Journals, 02 book chapters. During 2017 to 2019 he received 03 young scientist award and best thesis award in 2021.He has successfully guided more than 10 academic students in their final year project. Dr. Bhardwaj is a member of the Scholars Academic and Scientific Society and has organized different educational trips for his B. Tech and M.Sc. Students.

Duygun Fatih Demirel is an Assistant Professor at the Department of Industrial Engineering at Istanbul Kültür University, Turkey. He received BSc degree in Systems Engineering from Yeditepe University, Turkey, and PhD degree in Systems Engineering from Yeditepe University, Turkey. His research interests include systems dynamics, disaster management, socioeconomic modeling, fuzzy set theory and its applications, Bayesian forecasting, facility location/network design problem, system dynamics, and wastewater treatment systems design.

Müjgan Bilge Eriş is a research assistant and PhD student at the Department of Industrial Engineering at Yeditepe University, Turkey. She received BSc degree in Industrial and System Engineering from Yeditepe University, Turkey and MSc degree in Supply Chain Management from Vienna University of Economics and Business, Austria. Her research interests include system thinking, system dynamics, and multi-criteria decision-making.

Rajni Gautam received Ph.D., M.Sc. and B. Sc. Degrees in Electronics from University of Delhi, India, in 2014, 2009, and 2007 respectively. She has 16 publications in International Journals of repute, 15 publications in International conferences, 6 book chapters, 4 books and one patent. Her total citations are 389 with h-index of 10. Her research interests include device modeling and simulation for advanced metal oxide–semiconductor field-effect transistors (MOSFET) structures such as surrounding gate MOSFET with study of Reliability issues and biosensing applications and Synthesis and Characterization of doped metal oxide nanoparticles. She has more than 8 years of teaching experience and currently supervising two PhD scholar.

Dygun Fatih Demirel is an Assistant Professor at the Department of Industrial Engineering at Istanbul Kültür University, Turkey. He received BSc degree in Systems Engineering from Yeditepe University, Turkey, and PhD degree in Systems Engineering from Yeditepe University, Turkey. His research interests include systems dynamics, disaster management, socioeconomic modeling, fuzzy set theory and its applications, Bayesian forecasting, facility location/network design problem, system dynamics, and wastewater treatment systems design.

Sarah Grace P is a science graduate who received B.Sc (2021) and M.Sc (2023) in Applied Microbiology from Ethiraj College for Women, Chennai. She is working as an Associate Language Editor for American Chemical Society (ACS) publications at TNQ Tech.

Surendra Prakash Gupta is working as Assistant professor in the Department of Biotechnology, SVIS, Shri Vaishnav Vidhyapeeth Viswavidhayalaya, Indore. He received B.Sc. Hons. in Chemistry (2002) from Dayalbagh Educational Institute, Agra; M.Sc. (2006) & M.Tech (2008) in Biotechnology from C.C.S University, Meerut and Uttarakhand Technical University respectively; and Ph.D (2015) in Biological Science from Dayalbagh Educational Institute, Deemed University, Agra - Uttar Pradesh. He started his career as Assistant Professor in Department of Biotechnology, NIET, NIMS University, Jaipur - India in 2015 and served there for more than 2.5 years. He is now working as Assistant Professor at Department of Biotechnology, SVIS, Shri Vaishnav Vidhyapeeth Viswavidhayalaya, Indore and actively involved in teaching and research activities. He has 15 years of academic experience under various capacities in University. He has worked in the area of Drug-DNA interactions, Bioenergy, and Computational Biology. He has qualified GATE in 2006 in Life Science. He has published 05 research papers in peer reviewed Journals, 02 book chapters, 04 conference papers and more than 30 conference abstracts. He has supervised B.Tech and M.Sc Biotechnology students successfully. He is also a member of several reputed professional bodies like, IAENG, Education and Development Research and ASR etc. and Lifetime Member of IARA, and ISTD, Indore Chapter.

Alphonsa Haokip is currently the Headmistress of Christ King High School, Chingjaroi Khullen, Manipur. She has completed her doctorate entitled, "Attitude, Problem Solving Ability, and Achievement in Mathematics Learning among the Secondary School Students of Thadou Kuki Tribe in Manipur". Her areas of interest are Mathematics Learning of Secondary School Students, their problem-solving ability, their academic achievement, mathematics teachers and their concerns, paradigm shift in higher education, and gender issues.

Minakshi Harod is working as upper middle teacher in madhya Pradesh Govt as a Chemistry teacher, district Chemistry master trainer Harod has more than 7 years of research and teaching experience. She has worked in the area of natural sciences, chemical sciences etc. She has published research papers in peer reviewed Journals and book chapters in various books of International repute.

Madhumita Hussain is Associate Professor of Geography Sophia Girls' College (Autonomous), Ajmer. She holds a Ph.D. in Built Environment and Health and have a comprehensive understanding of Urban, Health, Physical, and Quantitative geography. Her focus is on the intersection of the built environment and health. Additionally, Her expertise in this field is supported by extensive research and numerous academic publications, which are well-written and engaging.

Kousar Jan is Principal at Government Degree College, Thindim Kreeri - Kashmir and actively involved in teaching and research activities. She has more than 30 years of research and teaching experience. He has worked in the area of Environmental Biosensors, and limnology. She has published research papers in peer reviewed Journals and book chapters in various books of International repute.

Dea V Jose is working as Associate professor in the Department of Computer Science, CHRIST University, Bangalore. She received PhD (2016) in Computer Science from CHRIST University, Karnataka. She started her career as Assistant Professor in Department of Computer Science CHRIST University India in 2006 and serves there till date. She is actively involved in teaching and research activities and is interested in the areas of sensor networks, IoT, and AI. She has published more than 40 research papers in peer reviewed Journals, conferences and books. She has successfully guided PhD students. She has worked as Principal Investigator in research projects and is a member of several national and international reviewer board of various journals.

Kmal Kishore is working as Associate Professor in department of Chemistry & Biochemistry, Eternal University Baru Sahib, Himachal Pradesh (India). He has fourteen years of teaching and one year of pharmaceutical industry (Alkem Laboratories Ltd.) experience. He received his Ph.D. degrees in Chemistry entitled, 'Physico-chemical, thermal and acoustical behaviour of terbium soaps' from B.U. Bhopal (M.P.) under the supervision of Prof. S.K. Upadhyaya (Ex. CSIR pool officer and recipient of Dr. Radhakrishnan Award). Dr. Kamal Kishore is author & co-author of over 28 technical publications, one book with title- CSIR-UGC NET (Chemical Sciences) and edited books for UG courses with Satya

Prakashan, New Delhi. He is editorial board member & reviewer of many reputed national and international Journals. He has successfully supervised Ph.D. and M.Sc. students for their thesis.

Javid Manzoor is working as Assistant professor in the Department of Environmental Science, Shri JJT University, Jhunjhunu, India. He has obtained his Master's degree and doctorate from Jiwaji University, Gwalior, Madhya Pradesh. He has five years of teaching experience. He is an awardee of University Research Scholarship. He is a permanent member of people for animal. His areas of interest include Environmental pollution, Environmental monitoring and assessment. He has published research papers and book chapters in various national and international journals with impact factor. He has been working as editorial board member and reviewer of various national and international journals.

Adul Qayoom Mir is currently working as Associate Professor in Environmental Sciences in the Higher Education Department Govt. of Jammu and Kashmir for the last 16 year. Prior to that Dr. Mir has worked as Assistant Professor at Azad Institute of Engineering Lucknow, UP. After pursuing M.Sc. Environmental Sciences from Dr. R.M.L.A. University Faizabad in 2004, Dr. Mir completed his Ph.D. Programme in the Department Of Environmental Sciences from Central University Lucknow in 2008 with specialization in Air Pollution and Medicinal Plants. Dr. Mir has also cleared the UGC NET in 2004. With a teaching experience of some 17 years and research experiences of more than 20 years Dr. Mir has published his work in reputed national and international Journals besides presenting his work in different conferences and seminars. He has guided 4 M.Phil and 2 Ph.D. Students.

Satyendra Nath, currently working as Associate Professor & Head, Department of Environmental Sciences & NRM, College of Forestry, SHUATS, Prayagraj, Uttar Pradesh, India. He obtained Ph.D degree from Motilal Nehru National Institute of Technology Allahabad. He has been more than 16 years experiences in Consultancy, Research and Teaching in the field of Agriculture, Environmental sciences and Engineering. Dr. Nath published more than70 publications in International/National Journals/Conferences.

R anjeevi is an accomplished academician and researcher serving as the Associate Professor in Environmental Science and the Head & Principal (i/c) of the Nims Institute of Allied Medical Science and Technology (NIET) at NIMS University, located in Rajasthan, Jaipur. His extensive qualifications include M.Sc., M.Phil., Ph.D., and a Post-Doctoral Fellowship (CSIR-RA), along with a PG Diploma in Industrial Waste & Wastewater Treatment, all focused on Environmental Science and Engineering. With over 14 years of experience, Dr. R. Sanjeevi's expertise encompasses a wide array of fields within environmental science, including water analysis, wastewater treatment, waste-to-energy conversion, renewable energy resources, air pollution monitoring and control, solid waste management, bioprocess technology, environmental biotechnology, and remote sensing & GIS. His contributions to academia and research have been widely recognized, evidenced by his impressive Google Citation Index and numerous awards and recognitions. Notable among these are his awards for best paper presentations at national seminars and conferences, his role as chairperson in various academic forums, and his contributions as a resource person and subject specialist in research conventions and seminars. His dedication to advancing environmental science and engineering, coupled with his commitment to sustainable development, make him a significant figure in both academia and the field of environmental research. His multifaceted expertise and numerous accolades underscore his significant contributions to the domain of environmental science and his commitment to excellence in research and education.

V Muthu Ruben is a distinguished Associate Professor at the School of Law, Christ University, situated in Bengaluru, Karnataka, India. Driven by an extensive scholarly foundation and a steadfast dedication to legal education, Dr. Ruben makes substantial contributions to both the academic community and the domain of law. His proficiency transcends numerous spheres within the realm of law, and he significantly influences the educational journeys of students enrolled at Christ University. Motivated by an ardent interest in legal research, he actively participates in scholarly endeavors, thereby making significant contributions to the ongoing legal dialogue. In addition to his scholarly responsibilities, Dr. V. Muthu Ruben is renowned for his commitment to cultivating an atmosphere that encourages intellectual engagement. His mentorship and counsel motivate pupils to explore into the complexities of the law, fostering the development of critical thinking skills and a profound comprehension of legal principles. In his role as an Associate Professor, Dr. Ruben provides the School of Law with an abundance of expertise, experience, and a forward-thinking perspective, thereby enhancing the institution's dedication to providing superior legal education.

Pashantkumar B. Sathvara is currently employed at Nims Institute of Allied Medical Science and Technology, Nims University Rajasthan, Jaipur. He holds a diverse academic background, including a pursuing Ph.D. in Environmental Science (Thesis Submitted), M.Sc. in Environmental Science and Technology, P.G. Diploma in Geo-informatics System, and a B.Sc. in Safety and Fire Technology. His research contributions are primarily focused on environmental science, with a particular emphasis on carbon sequestration mapping, vegetation change detection, and the application of remote sensing and GIS techniques. He has published 3 publications in peer recognized international and national journals, 2 book chapter and 9 book chapter under reviewed and accepted He has also undergone extensive training programs, including projects at Space Application Center (ISRO) and safety and fire courses, further augmenting his skills and knowledge. His computer skills include proficiency in advanced software such as ArcGIS, QGIS, ENVI, and Google Earth Pro, as well as familiarity with various operating systems and office tools.

Rnjit Singha is a Doctorate Research Fellow at Christ (Deemed to be University) and holds the prestigious American Psychological Association (APA) membership. With a strong background in Research and Development, he has significantly contributed to various fields such as Mindfulness, Addiction Psychology, Women Empowerment, UN Sustainable Development Goals, and Data Science. With over 15 years of experience in Administration, Teaching, and Research, both in Industry and Higher Education Institutions (HEI), Mr Ranjit has established himself as a seasoned professional. Mr Ranjit is dedicatedly involved in research and teaching endeavours, primarily focusing on mindfulness and compassion-based interventions. His work in these areas aims to promote well-being and foster positive change in individuals and communities.

Surjit Singha is an academician with a broad spectrum of interests, including UN Sustainable Development Goals, Organizational Climate, Workforce Diversity, Organizational Culture, HRM, Marketing, Finance, IB, Global Business, Business, AI, Women Studies, and Cultural Studies. Currently a faculty member at Kristu Jayanti College, Dr. Surjit also serves as an Editor, reviewer, and author for prominent global publications and journals, including being on the Editorial review board of Information Resources Management Journal and a contributor to IGI Global. With over 13 years of experience in Administration, Teaching, and Research, Dr. Surjit is dedicated to imparting knowledge and guiding students in their research pursuits. As a research mentor, Dr. Surjit has nurtured young minds and fostered academic growth. Dr. Surjit has an impressive track record of over 75 publications, including articles, book chapters, and textbooks, holds two US Copyrights, and has successfully completed and published two fully funded minor research projects from Kristu Jayanti College.

Sandeep Tripathi is the Associate Professor of Biochemistry and Director, Institute of Public Health and Nutrition. His research degree was based on "Metal ion neurotoxicity" which was awarded as "Best PhD thesis" among the batch 2010 in the King George's Medical University, Lucknow, India. He has published 68 research articles and many book chapters. He has awarded Mrs. Abida Mehdi Award for Neuroscience Research by Indian Academy of Biomedical Sciences and Young Scientist Award by centre for education growth and research. He has established "National referral centre for fluoride poisoning (NRCFPI) in India", it was first fluoride referral centre in the country and continuously leading by Dr. Tripathi as Head, Research and Development and he is the recipientof Niloufer Chinoy Award- 2013 presented by the International Society for fluride Research. Dr. Tripathi is the member of editorial board of many journals. He is the member of various International and National Societies (IBRO, ISNR, SOT, IAN, ACBI, SFRR, SOPI, SPGHTN and InSLAR). He has attended more than 50 conferences and workshops as speaker/ chairperson / moderator or as delegate.

G. Sangeetha Vani currently holds the position of Associate Professor in the Department of Microbiology at Ethiraj College for Women in Chennai, Tamil Nadu, India. She obtained her Master of Science degree in Applied Microbiology in 2004, Master of Philosophy in Microbiology in 2006, Master of Science in Bioinformatics in 2018, and Doctor of Philosophy in 2019. She has successfully finished numerous NPTEL courses. In 2005, she began her job as a Microbiologist and Analyst, lasting one year. She held the position of lecturer from 2005 to 2007. Since 2007, she has been working at Ethiraj College for Women. She has accumulated more than two decades of scientific experience. She has worked in Bacteriology, Virology, Bioinformatics, Environmental Microbiology, Agricultural Microbiology, and biofilm-related research. She has contributed articles to scholarly journals that undergo peer review and some are indexed in Scopus. She has delivered academic presentations at multiple conferences. She successfully finished three Faculty Research projects funded by Ethiraj College Trust. She is a member of various national and international organizations. She is also a reviewer in peer-reviewed and Scopus-indexed journals. In 2019, she was awarded the Ph.D. Certificate of Award from the University of Madras, specifically the Madanagopalan Memorial Prize, with a financial prize of Rs 3000. She has served as the state coordinator of the Microbiologists Society, India (2021-22). She serves as a member of BOS, Presidency College, Chennai. She has organized conferences, webinars, and workshops and also attended various conferences, seminars, workshops, and FDP.

Aashish Verma is working as Assistant Professor in the Department of Chemistry at Adarsh Institute of Management & Science Dhamnod India. He has more than 2 years of research and 10 years teaching experience. He has worked in the area of Natural Products in Chemistry. He has published research papers in peer reviewed Journals and book chapters in various books of International repute.

Sonam Verma is working as Assistant Professor in the Department of Zoology at Adarsh Institute of Management & Science Dhamnod India. Sonam has more than 2 years of research and teaching experience. She has worked in the area of Fishes in Zoology at Gawla Pond of Maheshwar Dist. Khargone, M. P. She has published research papers in peer reviewed Journals and book chapters in various books of International repute.

Hemendra Wala is working as upper middle teacher in madhya Pradesh Govt as a Biology teacher, district lab and biology master trainer. Wala has more than 10 years of research and teaching experience. He has worked in the area of limnology, ichthyology, neuro science etc. He has published research papers in peer reviewed Journals and book chapters in various books of International repute.

Yogesh Kumar Walia was born in Hamirpur, Himachal Pradesh, India in 1980. He received his B.Sc. degree from the Himachal Pradesh University Shimla in 2000. In 2004 he received his M.Sc. degree from the Barkatullah University (formerly Bhopal University), Bhopal, India. In 2010 he received his Ph.D. degree from the Barkatullah University (formerly Bhopal University), Bhopal, India. His contribution in synthetic bioorganic chemistry, environmental protection and natural products published in peer review journals and published more than forty research papers. **Presently** he is **working as Professor in Organic Chemistry at Career Point University Hamirpur**, India. He has supervised Twenty-Nine Post Graduate projects and Eight Ph.D. research scholars and currently he is supervising Six Ph.D. research scholars at this University.

Aaq Majid Wani, is presently working as Head and Associate Professor, Department of Forest Biology, Tree Improvement and Wildlife Sciences, College of Forestry, Sam Higginbottom University of Agriculture, Technology and Sciences, Prayagraj (Uttar Pradesh), India. Beside this, he has more than 15 years of experience in teaching and research in Forestry and guided Six Ph.D. and twenty seven M.Sc. students. Dr. Wani has published more than 90 research papers in National and International journals along with different Book chapters and three books. Dr. Wani has been awarded best teacher award by ICAR New Delhi in year 2016. Dr. Wani has shared many ideas and knowledge to the students as well as to the farmers coming to college of Forestry for exposure visit.

Index

F

G

H

I

L

M

N

O

P

R

S

T

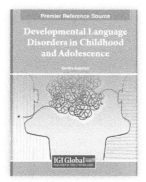

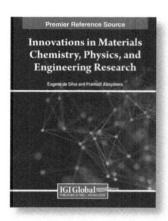

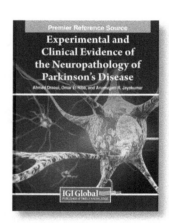

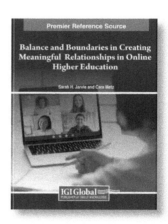

Milton Keynes UK
Ingram Content Group UK Ltd.
UKHW051601021224
3319UKWH00046B/1462

9 798369 347744